Virtual Fundamental Cycles in Symplectic Topology

Mathematical
Surveys
and
Monographs

Volume 237

Virtual Fundamental Cycles in Symplectic Topology

John W. Morgan, editor
Dusa McDuff
Mohammad Tehrani
Kenji Fukaya
Dominic Joyce

AMS AMERICAN MATHEMATICAL SOCIETY
Providence, Rhode Island

SIMONS CENTER FOR GEOMETRY AND PHYSICS

2010 *Mathematics Subject Classification.* Primary 53D45, 53D37, 58J10, 57R17, 57R18.

For additional information and updates on this book, visit
www.ams.org/bookpages/surv-237

Library of Congress Cataloging-in-Publication Data

Names: Morgan, John, 1946– editor. | McDuff, Dusa, 1945– | Simons Center for Geometry and Physics (Stony Brook University)

Title: Virtual fundamental cycles in symplectic topology / John Morgan, editor ; Dusa McDuff [and three others].

Description: Providence, Rhode Island : American Mathematical Society, [2019] | Series: Mathematical surveys and monographs ; volume 237 | "Simons Center for Geometry and Physics, Stony Brook, New York." | Includes bibliographical references and index.

Identifiers: LCCN 2018057233 | ISBN 9781470450144 (alk. paper)

Subjects: LCSH: Symplectic geometry. | Geometry, Differential. | AMS: Differential geometry – Symplectic geometry, contact geometry – Gromov-Witten invariants, quantum cohomology, Frobenius manifolds. msc | Differential geometry – Symplectic geometry, contact geometry – Mirror symmetry, symplectic aspects; homological mirror symmetry; Fukaya category. msc | Global analysis, analysis on manifolds – Partial differential equations on manifolds; differential operators – Differential complexes. msc | Manifolds and cell complexes – Differential topology – Symplectic and contact topology. msc | Manifolds and cell complexes – Differential topology – Topology and geometry of orbifolds. msc

Classification: LCC QA665 .V57 2019 | DDC 516.3/6–dc23

LC record available at https://lccn.loc.gov/2018057233

∞ The paper used in this book is acid-free and falls within the guidelines
established to ensure permanence and durability.
Visit the AMS home page at https://www.ams.org/

10 9 8 7 6 5 4 3 2 1 24 23 22 21 20 19

Contents

Introduction

In 2013–2014, the Simons Center for Geometry and Physics hosted a one-year program to discuss various problems related to the foundations of that part of symplectic geometry which concerns the theory of moduli spaces of pseudo-holomorphic curves. An important part of this program were lecture courses on the 'Virtual fundamental chain and cycle' determined by these moduli spaces, a technique of central importance in symplectic geometry. This volume consists of written and expanded versions of these lecture courses.

Our goal in bringing together experts in these techniques to give the lecture courses and in producing this follow-up volume is to facilitate a wider understanding of the various approaches to the basics of the virtual fundamental chain and cycle theory in symplectic geometry and the assumptions that each approach makes. We believe that future major advances in symplectic geometry will require use of these techniques without restrictive topological or geometric assumptions on the symplectic manifold. It is our hope that this volume will contribute to the wider application of these methods.

The Foundational Work of Gromov and Floer. The method of using pseudo-holomorphic curves in the study of global symplectic geometry was introduced in a ground-breaking paper by M. Gromov, [**3**]. From the appearance of this paper until today, this approach has been the most important tool in global symplectic geometry. The method is based on the following facts: (i) Symplectic manifolds have compatible almost complex structures (though the manifold often does not have any complex structure). (ii) The equation for a map from an almost complex structure on a smooth 2-manifold to an almost complex structure on a higher dimensional manifold is an elliptic equation so the usual elliptic theory applies. (iii) Since almost complex structures on smooth 2-manifolds are automatically integrable, pseudo-holomorphic maps from smooth 2-manifolds to almost complex manifolds are in fact holomorphic maps from complex curves[1]. As a result, the rich and highly developed holomorphic curve theory applies in the context of pseudo-holomorphic maps from a complex one-dimensional manifold to a symplectic manifold.

Even though there are many almost complex structures compatible with a given symplectic structure, the space of all such is contractible. Together with Gromov's compactness theorem for these pseudo-holomorphic maps, which relies essentially on the fact that we are mapping to a symplectic manifold not just an almost complex manifold, this implies that the cobordism class of the moduli space of pseudo-holomorphic maps from a complex one-dimensional manifolds to a given symplectic manifold is independent of the choice of the compatible almost complex structure. Such a cobordism class is actually an invariant of the symplectic structure, not just of the almost complex structure or its homotopy class.

Combined with physicist's view of the topological sigma model, see [**7**], this discovery by Gromov produces celebrated invariants of symplectic manifolds, which are now called *Gromov-Witten invariants*. There is also the quantum cup product on cohomology defined using these invariants. Some of the basic properties,

[1]Typically there is no holomorphic map from a complex manifold of dimension > 1 to an almost complex manifold.

especially the associativity of the quantum cup product, were first observed by physicists working on String Theory.

Another important application of Gromov's theory of pseudo-holomorphic curves is Floer homology. A. Floer observed that the gradient line of the classical action functional of periodic Hamiltonian system can be identified with a pseudo-holomorphic curve (with an order zero perturbation term). This makes it possible to apply the theory of pseudo-holomorphic curves to the problems of studying periodic solutions of periodic Hamiltonian systems (which is one of the most important problems in symplectic geometry), see [**1**]. Using similar techniques he also studied the problem of giving a lower bound for the number of intersection points of two Lagrangian submanifolds.

In symplectic geometry nowadays there are various 'Floer-type' homology theories defined using moduli spaces of pseudo-holomorphic curves and their variants.

Early applications of the theory. In early days of the development of these theories, the moduli spaces of pseudo-holomorphic curves were first studied under various additional simplifying assumptions. A typical assumption is monotonicity of the symplectic manifold (X, ω), meaning the existence of a positive number c such that the equality

$$\langle [\omega], \alpha \rangle = c \langle c_1(TX), \alpha \rangle \tag{1}$$

holds for any $\alpha \in \pi_2(X)$, or semi-positivity, meaning that

$$\langle c_1(TX), \alpha \rangle \geq 0$$

for any $\alpha \in \pi_2(X)$ with $\langle [\omega], \alpha \rangle > 0$. In the case of Floer homology associated with Lagrangian submanifolds $L_i \subset X$, $i = 1, 2$ typical assumptions are monotonicity (a similar equality to (1) above) or exactness, meaning the existence of one form θ on X such that $d\theta = \omega$ and $[\theta] = 0$ in the de-Rham cohomology of the L_i.)

These assumptions are used to establish the transversality of the moduli space of pseudo-holomorphic curves. Here, transversality means that the linearization of the Cauchy-Riemann equation is a surjective linear operator. Under this condition the relevant moduli space of pseudo-holomorphic curves is a smooth manifold and its dimension is calculated by the topological data using the Atiyah-Singer index theorem or the Riemann-Roch theorem.

In the case of monotone or semi-positive symplectic manifolds (or monotone or exact Lagrangian submanifolds), a standard perturbation argument shows that transversality holds for generic compatible almost complex structure.

Beyond these cases, an obstruction to being able to achieve the transversality arises from ramified multiple covers of transversal solutions. The problem is that the formal dimension of these covers can be negative even though the formal dimension of the underlying simple curve is non-negative. The main motivation for introducing the virtual fundamental chain and cycle technique (the main theme of this book) is to resolve this problem.

An Example. Let us give an explicit example of this phenomenon. We consider the moduli space of pseudo-holomorphic maps $u \colon S^2 \to X$ from a two-sphere with three marked points, taken to be $0, 1, \infty$, to a symplectic manifold X of real dimension $2n$, and denote by α the homology class $u_*[S^2]$. We assume that $\langle c_1(X), \alpha \rangle > 0$.

The dimension of this moduli space is

$$2\langle c_1(TX), \alpha\rangle + 2n.$$

We denote this moduli space by $\mathcal{M}_{0,3}(\alpha)$. (Here the first subscript, 0, is the genus of the curve and the second subscript, 3, is the number of marked points on the curve.) The map $u \mapsto (u(0), u(1), u(\infty)) \in X \times X \times X$ is called the *evaluation map*. We take cycles $Q_1, Q_2, Q_3 \subset X$ such that

$$\sum_{i=1}^{3}(2n - \dim Q_i) = 2\langle c_1(TX), \alpha\rangle + 2n. \tag{2}$$

Then, the formal dimension of the moduli space

$$\mathcal{M}_{0,3}(\alpha; Q_1, Q_2, Q_3) = \{u \in \mathcal{M}_{0,3}(\alpha) \mid (u(0), u(1), u(\infty)) \in Q_1 \times Q_2 \times Q_3\}$$

is 0. If this moduli space and its compactification are cut out transversally, then the sum over the (finite set of) points of the moduli space of signs coming from orientations is a typical example of the Gromov-Witten invariant. Indeed, because of the assumption $\langle c_1(TX), \alpha\rangle > 0$, for a generic compatible almost complex structure the moduli space $\mathcal{M}_{0,3}(\alpha; Q_1, Q_2, Q_3)$ is cut out transversally, so that it is a smooth manifold of dimension 0. However, it is not in general true that this moduli space is compact for such a generic choice of almost complex structure.

Gromov's compactness theorem holds and implies that if we consider a sequence u_i of elements of $\mathcal{M}_{0,3}(\alpha; Q_1, Q_2, Q_3)$ we may choose a subsequence so that its limit becomes a stable map of a holomorphic curve to X.

An example of stable curve which appears as a limit is drawn in Figure 1. It

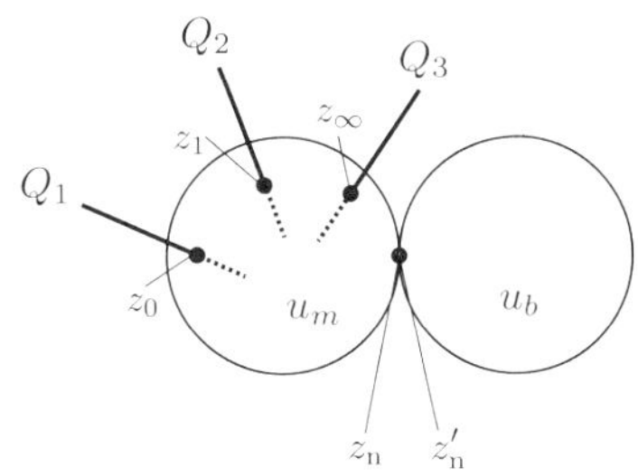

FIGURE 1. Limiting Configuration

is a union of curves $u_m : S^2 \to X$, $u_b : S^2 \to X$. The domain S^2 of u_m comes with four special points $(z_0, z_1, z_\infty, z_{\mathrm{n}})$ where z_0, z_1, z_∞ are limits of $0, 1, \infty$ in the domain of u_i, respectively. The domain of u_b comes with one special point z'_{n}, and we require $u_m(z_{\mathrm{n}}) = u_b(z'_{\mathrm{n}})$. So we may regard the source of our map u_∞ as a union of two spheres jointed at $z_{\mathrm{n}} = z'_{\mathrm{n}}$. We put $\alpha_m = (u_m)_*([S^2])$ and $\alpha_b = (u_b)_*([S^2])$, giving $\alpha_m + \alpha_b = \alpha$.

Then $(u_m; z_0, z_1, z_\infty, z_{\mathrm{n}})$ and $(u_b; z'_{\mathrm{n}})$ represent elements of the moduli spaces $\mathcal{M}_{0,4}(\alpha_m; Q_1, Q_2, Q_3)$ and $\mathcal{M}_{0,1}(\alpha_b)$, respectively.

The 'virtual' dimensions of the moduli spaces are given by

$$\begin{aligned} \dim \mathcal{M}_{0,4}(\alpha_m; Q_1, Q_2, Q_3) &= 2\langle c_1(TX), \alpha_m\rangle + 2n + 2 - \sum_{i=1}^{3}(2n - \dim Q_i), \\ \dim \mathcal{M}_{0,1}(\alpha_b) &= 2\langle c_1(TX), \alpha_b\rangle + 2n - 4. \end{aligned} \tag{3}$$

The numbers $+2$ and -4 in the above dimension formulae are explained as follows. The source curve $(S^2; z_0, z_1, z_\infty, z_{\rm n})$ of u_m has a 2-dimensional moduli space: The source curve $(S^2; z'_{\rm n})$ of u_b has an automorphism group of dimension 4. The condition $u_m(z_{\rm n}) = u_b(z'_{\rm n})$ adds $2n$ constraints. On the other hand, by Equations (2) and (3)

$$\dim \mathcal{M}_{0,4}(\alpha_m; Q_1, Q_2, Q_3) + \dim \mathcal{M}_{0,1}(\alpha_b) - 2n = -2.$$

This implies that if all the moduli spaces involved are transversal then this configuration does not appear. In order to prove the compactness of $\mathcal{M}_{0,3}(\alpha; Q_1, Q_2, Q_3)$ we need these limiting moduli spaces to be cut out transversally (so that they are empty).

However it is not in general true that $\mathcal{M}_{0,1}(\alpha_b)$ is transversal for generic almost complex structure. We consider the case when $\alpha_b = k\alpha'_b$ with

$$2\langle c_1(TX), \alpha'_b\rangle < 0.$$

Then, it can happen that there is a pseudo-holomorphic curve with

$$\begin{aligned}&2\langle c_1(TX), \alpha'_b\rangle + 2n - 4 \geq 0 \quad \text{and}\\ &2\langle c_1(TX), \alpha_b\rangle + 2n - 4 = 2\langle c_1(TX), k\alpha'_b\rangle + 2n - 4 < 0.\end{aligned}$$

One explicit numerical example is $n = 3$, $\langle c_1(TX), \alpha'_b\rangle = -1$, and $k = 2$.

Because of the first inequality, we can not show $\mathcal{M}_{0,1}(\alpha'_b) = \emptyset$ by dimension counting, and indeed it may be non-empty for every almost complex structure. But if $(u', 0)$, $u' \colon S^2 \to X$ represents an element of $\mathcal{M}_{0,1}(\alpha'_b)$ then $(u'', 0)$ with $u''(z) = u'(z^k)$ represents an element of $\mathcal{M}_{0,1}(\alpha_b)$. In particular $\mathcal{M}_{0,1}(\alpha_b)$ is non-empty. Thus, the space $\mathcal{M}_{0,1}(\alpha_b)$ with negative 'virtual' dimension is non-empty for every almost complex structure and so is not cut out transversally for any such structure.

This is a typical example of the problem of multiple covers of negative formal dimension. Avoiding this problem is the reason one often makes certain (sometimes restrictive) assumptions in the applications of pseudo-holomorphic curves in symplectic geometry.

The Virtual Fundamental Chain and Cycle Approach. Because it is not possible to use perturbation of the almost complex structure to produce the fundamental class of the moduli space when there are multiple covers of formally negative dimension, one is led to introduce the virtual fundamental chain or cycle technique. This approach resolves this problem in practically all the situations that appear in the study of pseudo-holomorphic curves in symplectic geometry. However, the method comes with the drawback that one is forced to use the rational numbers as a coefficient ring. Namely, Gromov-Witten invariants obtained in this way are necessarily rational numbers rather than integers and coefficients for the Floer homology is the field of rational numbers. (The reason why this approach does not work over integers is similar to the reason that Euler number of an orbifold is, in general, a rational number rather than an integer.)

Let us elaborate this point a bit more below. An orbifold is a space that looks like U/Γ locally, where U is an open ball centered at the origin of $\mathbb{R}^n$ and Γ is a finite group acting orthogonally on $\mathbb{R}^n$. We consider an orbifold X which is a union of $\mathbb{C}/\mathbb{Z}_2$ and $\mathbb{C}/\mathbb{Z}_3$. We identify $[z^2] \in \mathbb{C}/\mathbb{Z}_2$ with $[w^3] = [1/z^2] \in \mathbb{C}/\mathbb{Z}_3$. This orbifold X has two singular points which correspond to $z = 0 \in \mathbb{C}/\mathbb{Z}_2$ and

$w = 0 \in \mathbb{C}/\mathbb{Z}_3$, respectively. There exists a vector field on X which vanishes only at these two singular points. Because of the presence of non-trivial isotropy group we need to count these zeros with multiplicity $1/2$ and $1/3$, respectively. Therefore the orbifold Euler number of X is

$$\frac{1}{2} + \frac{1}{3} = \frac{5}{6}.$$

There are various versions of virtual fundamental chain or cycle technique. Some versions work with infinite dimensional spaces directly, other versions are based on local, finite dimensional reductions. All three articles in this book work in the context of finite dimensional reductions. In these versions the 'virtual fundamental chain or cycle technique' consists of the following three steps.

Step 1: Represent a moduli space locally as a zero set of certain section s of an orbibundle E over an orbifold U.

Step 2: Regard such a local description of the moduli space as a 'coordinate chart' of certain geometric object and introduce an appropriate notion of 'coordinate change'.

Step 3: Either (a) 'perturb' the section s in a way compatible with the coordinate changes so that it becomes transversal to 0 and then glue the zero sets of perturbed sections in 'various coordinate charts' by the coordinate change (this is the method used in the first two articles) or (b) develop and use general techniques from 'derived topology' to produce a fundamental chain or cycle from the geometric object in Step 2. An approach along these 'derived' lines is described in the third article. The application of the material in that article to produce the fundamental cycle and chain will appear in an article in preparation by Joyce.

It is worth pointing out that the arguments for Step 3 use only the formal properties of the geometric objects constructed in Step 2.

Brief descriptions of the articles in this volume. As we indicated, the goal of all the articles in this volume is to produce a fundamental cycle and chain associated to the moduli space of pseudo-holomorphic maps of curves to a symplectic manifold X equipped with a compatible almost complex structure. The starting point is a local description of the moduli space as the zeros of a smooth section of a smooth finite dimensional vector bundle over a smooth finite dimensional base modulo a finite group action – the *Kuranishi chart*. If the sections were transverse, the moduli space would inherit the structure of a finite dimensional smooth orbifold, which is oriented and thus has a (rational) fundamental class. But one must deal with non-transverse sections, especially in the case when the finite groups are non-trivial.

The article by Dusa McDuff describes her joint work with Katrin Wehrheim, [**6**]. Their goal is to produce the virtual fundamental cycle for the moduli space of pseudo-holomorphic maps. As they say in the introduction, "However, in practice, [the moduli space] usually has a more complicated structure since the operator is not transverse to zero. Intuitively the fundamental class is therefore the zero set of a suitable perturbation of the Fredholm operator: all the difficulty in constructing it lies in finding a suitable framework in which to build this perturbation." The framework described in this article is for the transitions between Kurannishi charts, formulated in what they call a *weak SS Kuranishi atlas*. In this article they show that any moduli space of genus zero pseudo-holomorphic curves in a compact symplectic manifold with a compatible almost complex structure has such an atlas and any two are compatible. From such an atlas, they produce a weighted branched topological manifold and a virtual fundamental class in the Cech homology of this branched manifold, unique up to cobordism. This then gives the virtual fundamental class for such moduli spaces. The case of moduli spaces of higher genus curves is covered by papers referenced in this article.

The second article is written by Mohammad Farajzadeh Tehrani partially using materials taken from a series of talks by Kenji Fukaya at the Simons Center for Geometry and Physics and also using material from papers written by Fukaya-Y.-G. Oh, H. Ohta and K. Ono [**2**]. The first half of the article is on the abstract theory of Kuranishi structures. After the definition of a Kuranishi structure is explained, the author provides a detailed account of the construction of the virtual fundamental class on a space with a Kuranishi structure. The first part of this process is a construction of a system with a finite number of charts (called a dimensionally graded system, DGS). (The initial Kuranishi structure can, and usually does, consist of an infinite number of charts.) DGS is a special case of a so called 'good coordinate system' appearing in [**2**]. The construction of a compatible system of multi-sections (multi-valued perturbations) on the those finitely many charts (transversal to zero) then are described. The triangulation of the zero set of such multi-section is given and as a result one is able to define a singular homology class, that is the virtual fundamental class. In the second half of this article, the construction of a Kuranishi structure on the moduli space of pseudo holomorphic curves is discussed. The author starts by explaining such moduli spaces and their topologies. Examples are given to show how the construction works. This article provides a comprehensive introduction to the construction of the Gromov-Witten invariant for general symplectic manifold.

The third article is a survey by Joyce of his paper [**5**] and his in-progress multivolume book "Kuranishi spaces and Symplectic Geometry." Preliminary versions of volumes I, II of this book are available at:

http://people.maths.ox.ac.uk/ joyce/Kuranishi.html.

This article proposes a new definition of Kuranishi space that has the nice property that the resulting objects form a 2-category, denoted **Kur**.

Any Fukaya–Oh–Ohta–Ono Kuranishi space $\mathbf{X}$, as in the second article, can be made into a Kuranishi space $\mathbf{X}'$ in this sense, uniquely up to equivalence in **Kur**. The same holds for McDuff and Wehrheim's 'Kuranishi atlases' described in the first article in this volume and also holds for the Hofer, Wysocki and Zehnder's 'Polyfold Fredholm structures', [**4**].

The Kuranishi spaces as defined in this article, i.e., the objects of **Kur**, are objects in derived differential geometry, a concept introduced by the author in earlier work. Derived differential geometry is the study of classes of derived manifolds and derived orbifolds that the author calls 'd-manifolds' and 'd-orbifolds', respectively. There is an equivalence of 2-categories $\mathbf{Kur} \sim \mathbf{dOrb}$, where **dOrb** is the 2-category of d-orbifolds. So Kuranishi spaces are really a form of derived orbifolds. As such they have their own differential geometry, with notions of orientation, immersions, submersions, transverse fibre products, and the other standard properties of good geometric categories.

References

[1] A. Floer, *Symplectic fixed points and holomorphic spheres*, Comm. Math. Phys. **120** (1989), no. 4, 575–611. MR987770

[2] K. Fukaya, Y-G. Oh, H. Ohta, K. Ono, *Technical details on Kuranishi structure and virtual fundamental chain, preprint* arXiv:1209.4410

[3] M. Gromov, *Pseudo holomorphic curves in symplectic manifolds*, Invent. Math. **82** (1985), no. 2, 307–347, DOI 10.1007/BF01388806. MR809718

[4] H. H. W. Hofer, *Polyfolds and Fredholm theory*, Lectures on geometry, Clay Lect. Notes, Oxford Univ. Press, Oxford, 2017, pp. 87–158. MR3676594

[5] D. Joyce, *A new definition of Kuranishi space, preprint* arXiv:1409.6908.

[6] D. McDuff and K. Wehrheim, *The fundamental class of smooth Kuranishi atlases with trivial isotropy*, J. Topol. Anal. **10** (2018), no. 1, 71–243, DOI 10.1142/S1793525318500048. MR3737511

[7] E. Witten, *Topological sigma models*, Comm. Math. Phys. **118** (1988), no. 3, 411–449. MR958805

John W. Morgan
October 7, 2017

Notes on Kuranishi Atlases

Dusa McDuff

1. Introduction

These notes aim to explain a joint project with Katrin Wehrheim [**MW1**, **MW2**, **MW3**, **MW4**, **MW5**] that uses finite dimensional reductions to construct a virtual fundamental class (VFC) for Gromov–Witten moduli spaces X of closed curves. Here X is a compactified moduli space of J-holomorphic stable maps, and so can locally be described as the zero set of a Fredholm operator on the space of sections of a bundle over a nodal Riemann surface, modulo the action of a Lie group that acts with finite stabilizers. Thus in the best case scenario X would be a compact finite dimensional orbifold, which carries a natural orientation and so has a fundamental class. However, in practice, X usually has a more complicated structure since the operator is not transverse to zero. Intuitively the fundamental class is therefore the zero set of a suitable perturbation of the Fredholm operator: all the difficulty in constructing it lies in finding a suitable framework in which to build this perturbation.

Many of the possible approaches to this question are discussed in [**MW2**]. For example, in the polyfold approach Hofer–Wysocki–Zehnder develop a radically new analytic framework in order to build a suitable ambient space that contains all the relevant objects. One then perturbs the section to get a transverse zero set. In contrast, we use finite dimensional reduction which constructs a local model (or chart) for small open subsets $F \subset X$ via traditional analysis, and then builds a suitable ambient space from these local charts using some relatively nontrivial point set topology.[1] Our method is based on early work by Fukaya–Ono [**FO**] and Fukaya–Oh–Ohta–Ono [**FOOO**]; see also [**FOOO1**]. However we have reformulated their ideas in order both to build the virtual neighbourhood of X and to clarify the formal structures underlying the construction. We make explicit all important choices (of tamings, shrinkings and reductions), thus creating tools with which to give an explicit proof that the virtual class $[X]^{vir}_{\mathcal{K}}$ is independent of these choices.

We start from the idea that there is a good local model for a sufficiently small open subset F of X, that we call a **basic chart K** with **footprint** $F \subset X$. Since typically there is no direct map from one basic chart to another we relate them via **transition data** given by "transition (or sum) charts" and coordinate changes. A **Kuranishi atlas** $\mathcal{K}$ is made from a finite covering family of basic charts together

The author was partially supported by NSF grant DMS 1308669.

[1] There seems to be a law of conservation of difficulty: in each approach the intrinsic difficulties in the problem appear in different guise, and are solved either using more analysis as in [**HWZ1**, **HWZ2**, **HWZ3**, **HWZ4**], or more topology as here, or more sheaf theory as in [**P**].

with suitable transition data. For simplicity, in these notes we will work throughout in the smooth category (though, as we explain below, one can get by with less smoothness).

The papers [**MW1**, **MW2**, **MW3**, **MW4**, **MW5**] prove the following theorems.

Theorem A. *Let (M^{2n}, ω, J) be a $2n$-dimensional symplectic manifold with tame almost complex structure J, let $\overline{\mathcal{M}}_{g,k}(A, J)$ be the compact space of nodal J-holomorphic genus g stable maps in class A with k marked points modulo reparametrization, and let $d = 2n(1-g) + 2c_1(A) + 2k + 6(g-1)$. Then $X := \overline{\mathcal{M}}_{g,k}(A, J)$ has an oriented, d-dimensional, weak Kuranishi atlas $\mathcal{K}$ that is well defined modulo oriented concordance (i.e. cobordism over $[0, 1] \times X$).*

Theorem B. *Let $\mathcal{K}$ be an oriented, d-dimensional, weak Kuranishi atlas on a compact metrizable space X. Then $\mathcal{K}$ determines a cobordism class of oriented, compact weighted branched manifolds, and an element $[X]^{vir}_{\mathcal{K}}$ in the Čech homology group $\check{H}_d(X; \mathbb{Q})$. Both depend only on the oriented cobordism class of $\mathcal{K}$.*

In these notes, we will concentrate on explaining what goes into the proof of Theorem B, i.e. on the development of the theory of smooth Kuranishi atlases. The proof of Theorem A is sketched in §4.1.

- If the curves in X have trivial isotropy and have smooth (i.e. non nodal) domains, we construct the invariant as an oriented cobordism class of compact smooth manifolds, and then take an appropriate inverse limit to get the Čech homology class.[2] If there is nontrivial isotropy we construct it as an oriented cobordism class of weighted branched (smooth) manifolds $(Z_{\mathcal{K}}, \Lambda_{\mathcal{K}})$, that is the realization of an explicit weighted nonsingular branched (wnb) groupoid; see Definition 3.4.5 ff. This abstract scheme applies equally well whether or not the isotropy acts effectively. It also applies in the nodal case, though in this situation one needs a gluing theorem in order to build the charts.
- Cobordism classes of oriented weighted branched manifolds contain more topological information than the fundamental class. For example, Pontriagin numbers are cobordism invariants; see [**M6**, Remark 4.7].
- One aim of our project is to construct the virtual moduli cycle using traditional tools as far as possible, and in particular to prove Theorem A using the approach to gluing in [**MS**]. This provides continuity of the gluing map as the gluing parameter $\mathbf{a}$ converges to zero, but gives no control over derivatives with respect to $\mathbf{a}$ near $\mathbf{a} = 0$. With this approach, the charts are only stratified smooth (abbreviated SS), i.e. they are topological spaces that are unions of even dimensional, smooth strata. As we explain in [**MW5**], this is sufficient to prove Theorem B for atlases of dimension $d = 0$ using perturbation sections as described below, and so is sufficient if one simply wants to calculate the number of curves going through some homological constraint. However, if one wants to establish the Kontsevich–Manin axioms one must work with atlases of arbitrary dimension. In order to obtain control over the first derivative (which is

[2] We could get an integral class if we used Steenrod homology as developed in Milnor [**Mi**]. However, as noted in [**MW2**, Remark 8.2.4], integral Čech homology does not even satisfy the basic axioms of a homology theory.

all that is needed in this context), one must improve the gluing theorem in [**MS**]. This is accomplished by Castellano, who proves a $\mathcal{C}^1$ version of the gluing theorem in [**C1**], and uses it to establish the Kontsevich–Manin axioms for genus zero invariants in [**C2**]. One virtue of this low tech approach is that one can use a logarithmic gluing profile (instead of the exponential one used in polyfold theory), which helps when proving the axioms, since all the natural forgetful maps are then smooth.

- A major difficulty with the proof of Theorem A as stated is that the construction of a smooth atlas needs a smooth gluing theorem, and for that it is most efficient to use the theory of polyfolds. As explained in [**MW4**], one can always extract an atlas from a polyfold description of X. But it is legitimate to ask: why do that? Why not simply work with polyfolds? One possible answer is that finite dimensional models are easier to work with than polyfolds. Another point is that they provide a bridge between traditional and polyfold constructions, allowing one to prove rigorously that the two approaches give the same VFC. For further discussion of these issues, see Remark 1.1.2 (ii),(iii) below.
- The construction in Theorem A of the atlas $\mathcal{K}$ on $\overline{\mathcal{M}}_{g,k}(A,J)$ extends in a natural way to a construction of a cobordism atlas on the moduli space $\bigcup_{t\in[0,1]}\overline{\mathcal{M}}_{g,k}(A,J_t)$ for paths $(J_t)_{t\in[0,1]}$ of ω-tame almost complex structures. Hence the weighted branched manifolds that represent $[\mathcal{M}_{g,k}(A,J_\alpha)]^{vir}_{\mathcal{K}}$ for $\alpha=0,1$ are oriented cobordant. It follows that all invariants constructed using the class $[\mathcal{M}_{g,k}(A,J)]^{vir}_{\mathcal{K}}$ are independent of the choice of ω-tame J.

These notes begin by developing the abstract theory of Kuranishi atlases, that is on proving theorem B for smooth atlases. The first two sections give precise statements of the main definitions and results from [**MW1, MW2, MW3**], and sketches of the most important proofs. For simplicity we first discuss the smooth case with trivial isotropy. Since the branching is caused by the presence of isotropy, in this case the zero set is a manifold: see Proposition 2.4.6. We then discuss the case of nontrivial isotropy, where we allow arbitrary, even noneffective, actions of finite groups. Throughout, we illustrate the general theorems by examples. In particular, Examples 3.4.6 and Example 3.4.7 give simple examples of weighted branched manifolds, while Example 3.4.8 explains our construction in a case when the isotropy group acts noneffectively.

Section 4 outlines the proof of Theorem A; more details of the construction in the genus zero case will be given in [**MW5**]. Section 5 discusses further examples. We begin with the case of atlases with trivial obstruction spaces, whose underlying space X is therefore an orbifold, explaining some results in [**M6**]. We show that in this case the theory is equivalent to the standard way of thinking of an orbifold as the realization of an ep groupoid; see Moerdijk [**Mo**].

Proposition C. *Each compact orbifold has a Kuranishi atlas with trivial obstruction spaces. Moreover, there is a bijective correspondence between commensurability classes of such Kuranishi atlases and Morita equivalence classes of ep groupoids.*

The proof constructs from each groupoid representative $\mathbf{G}$ of X an atlas that maps to an explicitly describable subcategory (i.e. submonoid) of $\mathbf{G}$ with realization X; see Proposition 5.1.6. Thus, this atlas is a simpler model for the groupoid that captures all essential information. We then consider the case when X is a compact manifold with obstruction bundle E. We show how to construct an atlas in this case, proving in Proposition 5.2.3 that the resulting VFC agrees with the usual notion of the Euler class of a bundle. We then explain how to extend this result to a general orbibundle $E \to X$ over a compact orbifold.

Finally, in Section 6 we discuss some modifications of the basic definitions that are useful when considering products. The point here is that the product of two Kuranishi atlases is not an atlas in the sense of our original definition. On the other hand, there are many geometric situations (such as in Floer theory) in which one wants to use induction to build a coherent family of atlases on spaces where the boundary of one space X is the product of two or more spaces that, via an inductive process, are already provided with atlases. We explain how to do this in our current framework, by generalizing the order structure underlying our notion of atlas. Some details of the adjustments needed to deal with Hamiltonian Floer theory may be found in Remark 6.3.9. These ideas will be developed in future work.

1.1. Outline of the main ideas. As explained above, a **Kuranishi atlas** $\mathcal{K}$ is made from a finite covering family of basic charts, that are related to each other via transition charts and coordinate changes. Our first aim is to unite all these charts into a **topological category $\mathbf{B}_{\mathcal{K}}$**, akin to the étale proper groupoids[3] often used to model orbifolds. If we ignore questions of smoothness, the space of objects $\mathrm{Obj}_{\mathbf{B}_{\mathcal{K}}}$ of such a topological category is the disjoint union $\bigsqcup_I U_I$ of smooth manifolds of different dimensions. There are at most a finite number of morphisms between any two points. Therefore the space $|\mathcal{K}|$ obtained by quotienting $\mathrm{Obj}_{\mathbf{B}_{\mathcal{K}}}$ by the equivalence relation generated by the morphisms looks something like an orbifold. In fact, in good cases this space, called the **virtual neighbourhood of X**, is a finite union of (non disjoint) orbifolds; see Remark 2.3.16. This ambient space $|\mathcal{K}|$ supports a "bundle" $|\mathrm{pr}| : |\mathbf{E}_{\mathcal{K}}| \to |\mathcal{K}|$ with canonical section $|\mathfrak{s}| : |\mathcal{K}| \to |\mathbf{E}_{\mathcal{K}}|$. The latter is the finite dimensional remnant of the original Fredholm operator, and its zero set can be canonically identified with a copy $\iota_{\mathcal{K}}(X)$ of X. Hence the idea is that the virtual moduli cycle $[X]^{vir}_{\mathcal{K}}$ should be represented by the zero set of a perturbed (multi)section $\mathfrak{s} + \nu$ that is chosen to be transverse to zero.

EXAMPLE 1.1.1. If X were a manifold, we could take each basic chart to be an open subset $F_i \subset X$, while the "transition chart" relating F_i to F_j would simply be the intersection $F_{ij} := F_i \cap F_j$. In this case the atlas $\mathcal{K}$ would consist of a finite open covering $(F_i)_{i=1,\dots,N}$ of X together with the collection of nonempty intersections $\big(F_I := \bigcap_{i\in I} F_i\big)$ related to each other by the obvious inclusions. As we will see, one advantage of the categorical framework is that it gives a succinct way of expressing the compatibility conditions between all spaces and maps of interest. $\Diamond$

The needed abstract structure is easiest to understand if we assume that there are no nodal curves and that all isotropy groups are trivial. Therefore we begin in §2 by considering smooth atlases with trivial isotropy. We consider nontrivial isotropy in §3.

We now outline the main steps in the construction of $[X]^{vir}_{\mathcal{K}}$.

[3]The definition of a topological category is given at the beginning of §2.2, while ep groupoids are defined at the beginning of §5.1 with examples in Example 3.1.17. Besides fitting in well with the notion of an orbifold as an ep groupoid, we will see that this categorical language provides a succinct way to describe the appropriate equivalences between the different charts in the atlas $\mathcal{K}$.

- The first difficulty in realizing this idea is that in practice one cannot actually construct atlases; instead one constructs a weak atlas, which is like an atlas except that one has less control of the domains of the charts and coordinate changes. But a weak atlas does not even define a category, let alone one whose realization $|\mathbf{B}_{\mathcal{K}}| = \mathrm{Obj}_{\mathcal{B}_{\mathcal{K}}}/{\sim} =: |\mathcal{K}|$ has good topological properties. For example, we would like $|\mathcal{K}|$ to be Hausdorff and (in order to make local constructions possible) for the projection $\pi_{\mathcal{K}} : U_I \to |\mathcal{K}|$ to be a homeomorphism to its image.

  In §2.2 we formulate the **taming conditions** for a weak atlas. Our main results are:
  - Proposition 2.3.13, which shows that the realization of a tame atlas has these good topological properties, and
  - Proposition 2.3.17, which shows that every weak smooth atlas can be tamed.

  We prove these results in §2.3 for **filtered topological atlases**, in order that they also apply to the case of smooth atlases with nontrivial isotropy. Here, filtration is a generalization of the notion of additivity that is already built into the definition of an atlas. As explained in Remark 2.3.19, some version of this condition is crucial here. (See §6 where this notion of additivity is weakened to a notion that is compatible with products.)
- The taming procedure gives us two categories $\mathbf{B}_{\mathcal{K}}$ and $\mathbf{E}_{\mathcal{K}}$ with a projection functor $\mathrm{pr} : \mathbf{E}_{\mathcal{K}} \to \mathbf{B}_{\mathcal{K}}$ (the "obstruction bundle") and section functor $\mathfrak{s} : \mathbf{B}_{\mathcal{K}} \to \mathbf{E}_{\mathcal{K}}$ (defined by the Cauchy–Riemann operator). However the category has too many morphisms (i.e. the chart domains overlap too much) for us to be able to construct a perturbation functor $\nu : \mathbf{B}_{\mathcal{K}} \to \mathbf{E}_{\mathcal{K}}$ such that $\mathfrak{s}+\nu$ is transverse to zero (written $\mathfrak{s}+\nu \pitchfork 0$). We therefore pass to a full subcategory $\mathbf{B}_{\mathcal{K}}|_{\mathcal{V}}$ of $\mathbf{B}_{\mathcal{K}}$ with objects $\mathcal{V} := \bigsqcup V_I$ that does support suitable functors $\nu : \mathbf{B}_{\mathcal{K}}|_{\mathcal{V}} \to \mathbf{E}_{\mathcal{K}}|_{\mathcal{V}}$; see Definition 2.4.1. This subcategory $\mathbf{B}_{\mathcal{K}}|_{\mathcal{V}}$ is called a **reduction** of $\mathcal{K}$. Its realization $|\mathbf{B}_{\mathcal{K}}|_{\mathcal{V}}|$ injects continuously onto the subspace $\pi_{\mathcal{K}}(\mathcal{V}) \subset |\mathcal{K}|$. Constructing it is similar to passing from the covering of a triangulated space by the stars of its vertices to the covering by the stars of its first barycentric subdivision. It is the analog of a "good coordinate system" (now called a "dimensionally graded system" in [**FF**]) in the theory of Kuranishi structures.
- We next define the notion of a reduced perturbation $\nu : \mathbf{B}_{\mathcal{K}}|_{\mathcal{V}} \to \mathbf{E}_{\mathcal{K}}|_{\mathcal{V}}$ of $\mathfrak{s}$ (see Definition 2.4.5), and show that, if ν is precompact in a suitable sense, the realization $\left|(\mathfrak{s}|_{\mathcal{V}} + \nu)^{-1}(0)\right|$ of the zero set is compact. The intricate construction of ν is one of the most difficult parts of the general theory. To explain the ideas, we give a fairly detailed description in Proposition 2.4.10, though still do not do quite enough for a complete proof. In the trivial isotropy case the zero set is a closed submanifold of $|\mathcal{K}|$ lying in the precompact "neighbourhood"[4] $\pi_{\mathcal{K}}(\mathcal{V}) \subset |\mathcal{K}|$ of $\iota_{\mathcal{K}}(X)$. The final step is to construct the fundamental class $[X]^{vir}_{\mathcal{K}}$ from this zero set. We define this class to lie in rational Čech homology rather than the

[4]In fact, $\iota_{\mathcal{K}}(X)$ does *not* have a compact neighbourhood in $|\mathcal{K}|$; as explained in Remark 2.4.4 we should think of $\pi_{\mathcal{K}}(\mathcal{V})$ as the closest we can come to a compact neighbourhood of $\iota_{\mathcal{K}}(X)$.

more familiar singular theory, because the former theory has the needed continuity properties under inverse limits.[5]

- As we will see in §3 the above ideas adapt readily to the case of nontrivial isotropy via the notion of the **intermediate category**, which is a filtered topological category and so can be tamed by the results in §2.3. Although the needed perturbation ν is multivalued when considered as a section of $|\mathbf{E}_{\mathcal{K}}| \to |\mathcal{K}|$, it is the realization of a *single valued* map $\nu : \mathcal{V} \to \mathbf{E}_{\mathcal{K}}|_{\mathcal{V}}$, which satisfies some compatibility conditions between charts but is *not* a functor.[6] This allows us to give a very explicit description of the zero set $(\mathfrak{s}|_{\mathcal{V}}+\nu)^{-1}(0)$, which forms an étale (but nonproper) weighted nonsingular groupoid in a natural and functorial way; see Proposition 3.4.4.
- Of course, to obtain a fundamental class one also needs to discuss orientations, and in order to prove uniqueness of this class one also needs to set up an adequate cobordism theory. Cobordisms are discussed briefly in §2.2 (see Definition 2.2.11 ff.) and orientations in §3.2 (also see Definition 2.2.15).

REMARK 1.1.2. (i) Note that although the cobordism relation is all one needs when proving the uniqueness of $[X]^{vir}_{\mathcal{K}}$ since this is just a homology class, it does not seem to be the "correct" relation, in the sense that rather different moduli problems might well give rise to cobordant atlases. The construction in §4.1 for an atlas on a fixed GW moduli space X builds an atlas whose **commensurability class** (see Definition 4.1.4) is independent of all choices. However, the construction takes place in a context where one can define a transition chart between any two charts with overlapping footprints, a property that has no very natural abstract description.[7] Therefore commensurability is probably not the optimal relation either, though (as shown by Proposition C) it is optimal if all obstruction spaces vanish. It may be that Joyce's notion of a Kuranishi space [**J14, J15**] best captures the Fredholm index condition on X, at least in the smooth context; see also Yang [**Y14**]. The aim of our work is not to tackle such an abstract problem, but to develop a complete and explicit theory that can be used in practice to construct and calculate GW invariants.

(ii) Pardon's very interesting approach to the construction of the GW virtual fundamental class in [**P**] uses atlases that have many of the features of the theory presented here. In particular, his notion of implicit atlas (developed on the basis of [**MW0**]) includes transition charts and coordinate changes that are essentially the same as ours, though he has no need for a smooth structure. However he avoids making choices by considering *all* charts, and he avoids the taming problems we encounter, firstly by using properties of the ambient space of stable maps, considering *all* solutions to the given equation with specified stable domain, and secondly by

[5] One could obtain an integral fundamental class in the Steenrod homology developed in [**Mi**]. However, rational Čech homology is simply the homology theory dual to rational Čech cohomology, and so is easier to understand; see [**MW2**, Remark 8.2.4]. Also, one must use rational coefficients if there is nontrivial isotropy.

[6] However ν is a functor from the "pruned" category $\mathbf{B}_{\mathcal{K}}|_{\mathcal{V}}^{\setminus\Gamma}$ to $\mathbf{E}_{\mathcal{K}}|_{\mathcal{V}}$, where $\mathbf{B}_{\mathcal{K}}|_{\mathcal{V}}^{\setminus\Gamma}$ is a (non-full!) subcategory of $\mathbf{B}_{\mathcal{K}}|_{\mathcal{V}}$ obtained by discarding appropriate morphisms.

[7] This property is a consequence of the fact that in this case all charts can be considered as subsets of a single polyfold, which as we will see in [**MW4**] gives an overall enveloping structure.

using a different more algebraic way to define the VFC (via a version of sheaf theory and some homological algebra) that does not involve considering the quotient space $|\mathcal{K}|$. Further, instead of using a reduction in order to thin out the footprint covering $(F_I)_{I\in\mathcal{I}_{\mathcal{K}}}$ sufficiently to construct a perturbation section ν, he relates the charts corresponding to different indices $I \subset J$ via a construction he calls "deformation to the normal cone". One virtue of this construction is that there is now no need to rescale the gluing parameters, so that Teichmüller space, for example, can keep its usual smooth structure.

(iii) **Further developments:** Our categorical approach to the basic definitions fits in very well with polyfold theory as developed by Hofer–Wysocki–Zehnder [**HWZ4**, **HWZ7**]. In particular, the fact that the ambient space of stable maps is incorporated into the polyfold framework means that one can use it to give a direct construction of a tame atlas. Because polyfolds admit partitions of unity, it turns out that one can go yet further, constructing a (generalized) Kuranishi atlas for the solution set of a sc-Fredholm operator that has just *one* chart that still has a trivialized bundle $U \times E$ but where the domain U is a *weighted branched manifold* as in Definition 3.4.5 below: see [**MW4**]. A similar circle of ideas is developed in [**M7**] where it is shown how to construct such a global chart starting from a nonsmooth atlas as in Pardon, using a rather intricate construction that is based on the techniques explained below in §2, 3. As in [**P**], one can then give a purely homological definition of the VFC without constructing a transverse perturbation section, but also without using the sophisticated sheaf theory in [**P**].

Such a global finite dimensional model for the infinite dimensional Fredholm problem has a very simple structure that allows one to avoid the difficulties in the theory of atlases caused by the fact that the chart domains have varying dimensions. However, both approaches to constructing this global chart rely in their different ways on the existence of a partition of unity, a construct that does not exist in the algebraic world, for example. As will be explained in later work, it is possible to construct a chain level version of the VFC using a nonsmooth atlas together with Pardon's construction of the "deformation to the normal cone". This construction of the VFC is perhaps the most natural and elementary, using the least extra structure. ◊

Remark 1.1.3. (i) As already said, our approach grew out of the work of Fukaya–Ono in [**FO**] and Fukaya–Oh–Ohta–Ono in [**FOOO**]. In particular, we use essentially the same Fredholm analysis that is implicit in these papers (though we make it much more explicit). Similarly, our construction of charts is closely related to theirs, but more explicit and more global. The main technical differences are the following:

- We use "Fredholm stabilization" (see Remark 4.1.3) in order to avoid some genericity issues with the obstruction bundles.
- We make a more explicit choice of the labeling of the added marked points, giving more control over the isotropy groups and permitting the construction of transition charts $\mathbf{K}_I$ with footprint equal to the whole intersection F_I.
- The development in [**FO**, **FOOO**, **FF**] goes straight from a Kuranishi structure to a "good coordinate system" or DGS (in our language, accomplishing taming and reduction in one step), thus apparently bypassing

some of the topological questions involving in constructing the category $\mathbf{B}_{\mathcal{K}}$ and virtual neighbourhood $|\mathcal{K}|$.

Our approach does involve developing more basic topology. One advantage is that we can exhibit the virtual fundamental class $[X]^{vir}_{\mathcal{K}}$ as an element of the homology of X as stated in Theorem A, rather than simply as a cobordism class of "regularized moduli spaces" $\overline{\mathcal{M}}^{\nu}$ with a well defined image under an appropriate extension of the natural map $X := \overline{\mathcal{M}}_{0,k}(M, A, J) \stackrel{\text{stab}\times\text{eval}}{\longrightarrow} \overline{\mathcal{M}}_{0,k} \times M^k$. See Remark 3.1.19 for further discussion of the relation between our work and that of Fukaya et al. Note finally that the new more detailed version of the Kuranishi structure approach explained in Farajzadeh Tehrani–Fukaya [**FF**] (which is, of course, influenced by our work) also permits the definition of $[X]^{vir}_{\mathcal{K}}$ as an element in the homology of X via a notion of "thickening".

(ii) Our ideas were first explained in the 2012 preprint [**MW0**] which constructed the fundamental cycle for smooth atlases with trivial isotropy. This paper has now been rewritten and separated into the two papers [**MW1**, **MW2**], the first of which deals the topological questions in more generality than [**MW0**], and the second of which constructs the fundamental cycle in the trivial isotropy case (including a discussion of orientations). A third paper [**MW3**] extends these results to the case of smooth atlases with nontrivial isotropy. The lectures [**M4**] (from 2013) and [**M5**] (from 2014) give an overview of the whole construction. ◊

2. Kuranishi atlases with trivial isotropy

Throughout these notes X is assumed to be a compact and metrizable space. Further in this chapter we assume (usually without explicit mention) that the isotropy is trivial. The proof of Theorem B in this case is completed at the end of §2.4. For the general case see §3.

2.1. Smooth Kuranishi charts, coordinate changes and atlases.

This section gives basic definitions.

Definition 2.1.1. *Let $F \subset X$ be a nonempty open subset. A* **(smooth) Kuranishi chart** *for X with* **footprint** *F (and trivial isotropy) is a tuple $\mathbf{K} = (U, E, s, \psi)$ consisting of*

- *the* **domain** *U, which is a smooth k-dimensional manifold;*[8]
- *the* **obstruction space** *E, which is a finite dimensional real vector space;*
- *the* **section** *$\mathfrak{s} : U \to U \times E, x \mapsto (x, s(x))$ which is given by a smooth map $s : U \to E$;*
- *the* **footprint map** *$\psi : s^{-1}(0) \to X$, which is a homeomorphism to the footprint $\psi(s^{-1}(0)) = F$, which is an open subset of X.*

The **dimension** *of $\mathbf{K}$ is $\dim \mathbf{K} := \dim U - \dim E$.*

Definition 2.1.2. *A* **map** *$\widehat{\Phi} : \mathbf{K} \to \mathbf{K}'$ between Kuranishi charts is a pair $(\phi, \widehat{\phi})$ consisting of an embedding $\phi : U \to U'$ and a linear injection $\widehat{\phi} : E \to E'$ such that*

[8]We assume throughout that manifolds are second countable and without boundary, unless explicit mention is made to the contrary.

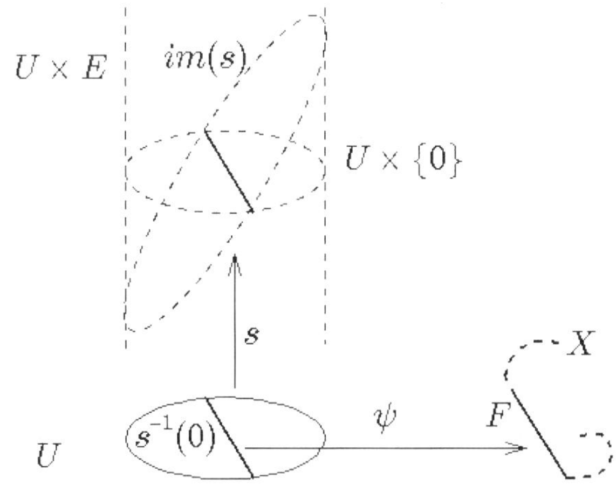

FIGURE 2.1.1. A Kuranishi chart

(1) *the embedding restricts to* $\phi|_{s^{-1}(0)} = \psi'^{-1} \circ \psi : s^{-1}(0) \to s'^{-1}(0)$, *the transition map induced from the footprints in* X;
(2) *the embedding intertwines the sections,* $s' \circ \phi = \widehat{\phi} \circ s$, *on the entire domain* U.

That is, the following diagrams commute:

$$
(2.1.1) \qquad \begin{array}{ccc} U \times E & \stackrel{\phi \times \widehat{\phi}}{\longrightarrow} & U' \times E' \\ \uparrow \mathfrak{s} & & \uparrow \mathfrak{s}' \\ U & \stackrel{\phi}{\longrightarrow} & U' \end{array} \qquad\qquad \begin{array}{ccc} s^{-1}(0) & \stackrel{\phi}{\longrightarrow} & s'^{-1}(0) \\ \downarrow \psi & & \downarrow \psi' \\ X & \stackrel{\mathrm{id}}{\longrightarrow} & X. \end{array}
$$

The dimension of the obstruction space E typically varies as the footprint $F \subset X$ changes. Indeed, the maps $\phi, \widehat{\phi}$ need not be surjective. However, as we will see in Definition 2.1.5, the maps allowed as coordinate changes are carefully controlled in the normal direction.

DEFINITION 2.1.3. *Let* $\mathbf{K}$ *be a Kuranishi chart and* $F' \subset F$ *an open subset of the footprint. A* **restriction of K to** F' *is a Kuranishi chart of the form*

$$
\mathbf{K}' = \mathbf{K}|_{U'} := \big(U', E' = E, s' = s|_{U'}, \psi' = \psi|_{s'^{-1}(0)} \big)
$$

given by a choice of open subset $U' \subset U$ *of the domain such that* $U' \cap s^{-1}(0) = \psi^{-1}(F')$. *In particular,* $\mathbf{K}'$ *has footprint* $\psi'(s'^{-1}(0)) = F'$.

By [**MW2**, Lemma 5.1.6], we may restrict to any open subset of the footprint. If moreover $F' \sqsubset F$ is precompact, then U' can be chosen to be precompact in U, written $U' \sqsubset U$.

The next step is to construct a coordinate change $\widehat{\Phi}_{IJ} : \mathbf{K}_I \to \mathbf{K}_J$ between two charts with nested footprints $F_I \supset F_J$. For simplicity we will formulate the definition in the situation that is relevant to Kuranishi atlases. That is, we suppose that a finite set of Kuranishi charts $(\mathbf{K}_i)_{i\in\{1,\dots,N\}}$ is given such that for each $I \subset \{1,\dots,N\}$ with $F_I := \bigcap_{i\in I} F_i \neq \emptyset$ we have another Kuranishi chart $\mathbf{K}_I$ (called a **transition (or sum) chart**) with

(2.1.2) obstruction space $E_I = \prod_{i\in I} E_i$, and footprint $F_I := \bigcap_{i\in I} F_i$.

REMARK 2.1.4. Since we assume in an atlas that

$$\dim U_I - \dim E_I =: \dim \mathbf{K}_I = \dim \mathbf{K}_i = \dim U_i - \dim E_i, \quad \forall i \in I,$$

in general the domain of the transition chart U_I has dimension strictly larger than $\dim U_i$ for $i \in I$. Further, U_I usually cannot be built in some topological way from the U_i (e.g. by taking products). Indeed in the Gromov–Witten situation U_I consists (very roughly speaking) of pairs $((e_i)_{i\in I}, u)$ where u is a solution to an equation of the form $\overline{\partial}_J u = \sum_{i\in I} \lambda(e_i)$, and so cannot be made directly from the U_i, which for each i consists of pairs (e_i, u) of solutions to the individual equation $\overline{\partial}_J u = \lambda(e_i)$. Note also that we choose the obstruction spaces E_i to cover the cokernel of the linearization of $\overline{\partial}_J$ at the points in U_i. Thus each domain U_i is a manifold that is cut out transversally by the equation. Since the function $s_I : U_I \to E_I$ is the finite dimensional reduction of $\overline{\partial}_J$, for each $I \subset K$ the derivative $\mathrm{d}_v s_K$ at a point $v \in \operatorname{im} \phi_{IK}$ has kernel contained in $\mathrm{T}_v(\operatorname{im} \phi_{IK})$ and cokernel that is covered by $\widehat{\phi}_{IK}(E_I)$. This explains the index condition in Definition 2.1.5 and Remark 2.1.6 below. $\Diamond$

When $I \subset J$ we write $\widehat{\phi} := \widehat{\phi}_{IJ} : E_I \to E_J$ for the natural inclusion, omitting it where no confusion is possible.[9]

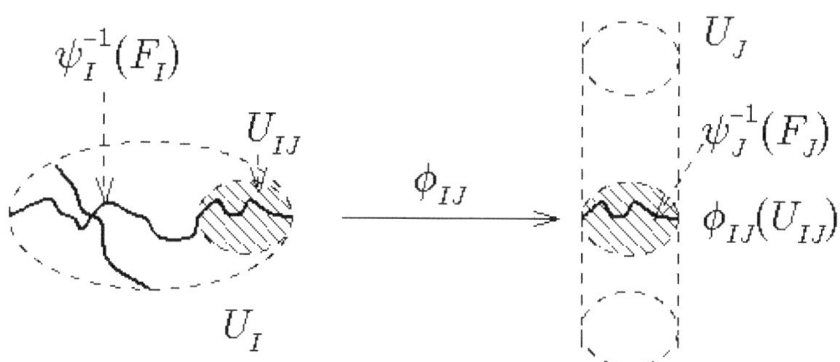

FIGURE 2.1.2. A coordinate change in which $\dim U_J = \dim U_I + 1$. Both U_{IJ} and its image $\phi_{IJ}(U_{IJ})$ are shaded.

DEFINITION 2.1.5. *For $I \subset J$, let $\mathbf{K}_I$ and $\mathbf{K}_J$ be Kuranishi charts as above, with domains U_I, U_J and footprints $F_I \supset F_J$. A* **coordinate change** *from $\mathbf{K}_I$ to $\mathbf{K}_J$ with* **domain** *U_{IJ} is a map $\widehat{\Phi} : \mathbf{K}_I|_{U_{IJ}} \to \mathbf{K}_J$, which satisfies the* **index condition** *in (i),(ii) below, and whose domain is an open subset $U_{IJ} \subset U_I$ such that $\psi_I(s_I^{-1}(0) \cap U_{IJ}) = F_J$.*

(1) *The embedding $\phi : U_{IJ} \to U_J$ underlying the map $\widehat{\Phi}$ identifies the kernels,*
$$\mathrm{d}_u \phi\big(\ker \mathrm{d}_u s_I\big) = \ker \mathrm{d}_{\phi(u)} s_J \qquad \forall u \in U_{IJ};$$
(2) *the linear embedding $\widehat{\phi} : E_I \to E_J$ given by the map $\widehat{\Phi}$ identifies the cokernels,*
$$\forall u \in U_{IJ}: \qquad E_I = \operatorname{im} \mathrm{d}_u s_I \oplus C_{u,I} \quad \Longrightarrow \quad E_J = \operatorname{im} \mathrm{d}_{\phi(u)} s_J \oplus \widehat{\phi}(C_{u,I}).$$

[9]Note that the assumption $E_I = \prod_{i\in I} E_i$ means that the family is **additive** in the sense of [**MW2**, Definition 6.1.5]. Therefore all the atlases that we now consider are additive, and for simplicity we no longer mention this condition explicitly. We discuss a weakened version in §6.

REMARK 2.1.6. We show in [**MW2**, Lemma 5.2.5] that the index condition is equivalent to the **tangent bundle condition**, which requires isomorphisms for all $v = \phi(u) \in \phi(U_{IJ})$,

$$\text{(2.1.3)} \qquad \mathrm{d}_v\, s_J : {}^{\mathrm{T}_v U_J}\Big/_{\mathrm{d}_u \phi(\mathrm{T}_u U_I)} \xrightarrow{\cong} {}^{E_J}\Big/_{\widehat{\phi}(E_I)},$$

or equivalently at all (suppressed) base points as above

$$\text{(2.1.4)} \qquad E_J = \operatorname{im} \mathrm{d}\, s_J + \operatorname{im} \widehat{\phi}_{IJ} \qquad \text{and} \qquad \operatorname{im} \mathrm{d}\, s_J \cap \operatorname{im} \widehat{\phi}_{IJ} = \widehat{\phi}_{IJ}(\operatorname{im} \mathrm{d}\, s_I).$$

Moreover, the index condition implies that $\phi(U_{IJ})$ is an open subset of $s_J^{-1}(\widehat{\phi}(E_I))$, and that the charts $\mathbf{K}_I, \mathbf{K}_J$ have the same dimension. ◊

DEFINITION 2.1.7. *Let X be a compact metrizable space.*

- *A* **covering family of basic charts** *for X is a finite collection $(\mathbf{K}_i)_{i=1,\ldots,N}$ of Kuranishi charts for X whose footprints cover $X = \bigcup_{i=1}^N F_i$.*
- **Transition data** *for a covering family $(\mathbf{K}_i)_{i=1,\ldots,N}$ is a collection of Kuranishi charts $(\mathbf{K}_J)_{J\in\mathcal{I}_\mathcal{K},|J|\geq 2}$ and coordinate changes $(\widehat{\Phi}_{IJ})_{I,J\in\mathcal{I}_\mathcal{K},I\subsetneq J}$ as follows:*
  1. *$\mathcal{I}_\mathcal{K}$ denotes the set of nonempty subsets $I \subset \{1,\ldots,N\}$ for which the intersection of footprints is nonempty,*
  $$\mathcal{I}_\mathcal{K} := \left\{\emptyset \neq I \subset \{1,\ldots,N\} \;:\; F_I := \textstyle\bigcap_{i\in I} F_i \;\neq\; \emptyset\right\};$$
  2. *$\mathbf{K}_J$ is a Kuranishi chart for X with footprint $F_J = \bigcap_{i\in J} F_i$ for each $J \in \mathcal{I}_\mathcal{K}$ with $|J| \geq 2$, and for one element sets $J = \{i\}$ we denote $\mathbf{K}_{\{i\}} := \mathbf{K}_i$;*
  3. *$\widehat{\Phi}_{IJ}$ is a coordinate change $\mathbf{K}_I \to \mathbf{K}_J$ for every $I, J \in \mathcal{I}_\mathcal{K}$ with $I \subsetneq J$.*

According to Definition 2.1.5 the domain U_{IJ} of $\widehat{\Phi}_{IJ}$ is part of the transition data for a covering family. Further, this data automatically satisfies a cocycle condition on the zero sets since, due to the footprint maps to X, we have for $I \subset J \subset K$:

$$\phi_{JK} \circ \phi_{IJ} = \psi_K^{-1} \circ \psi_J \circ \psi_J^{-1} \circ \psi_I = \psi_K^{-1} \circ \psi_I = \phi_{IK} \qquad \text{on } s_I^{-1}(0) \cap U_{IK}.$$

Further, the composite maps $\phi_{JK} \circ \phi_{IJ}, \widehat{\phi}_{JK} \circ \widehat{\phi}_{IJ} = \widehat{\phi}_{IK}$ automatically satisfy the intertwining relations in Definition 2.1.2. Hence one can always define a composite coordinate change $\widehat{\Phi}_{JK} \circ \widehat{\Phi}_{IJ}$ from $\mathbf{K}_I$ to $\mathbf{K}_K$ with domain $U_{IJ} \cap \phi_{IJ}^{-1}(U_{JK})$. (For details, see [**MW2**, Lemma 5.2.7] and [**MW1**, Lemma 2.2.5].) But in general this domain may have little relation to the domain U_{IK} of ϕ_{IK}, apart from the fact that these two sets have the same intersection with the zero set $s_I^{-1}(0)$. Since there is no natural ambient topological space into which the entire domains of the Kuranishi charts map, the cocycle condition on the complement of the zero sets has to be added as an axiom. There are three natural notions of cocycle condition with varying requirements on the domains of the coordinate changes.

DEFINITION 2.1.8. *Let $\mathcal{K} = (\mathbf{K}_I, \widehat{\Phi}_{IJ})_{I,J\in\mathcal{I}_\mathcal{K},I\subsetneq J}$ be a tuple of basic charts and transition data. Then for any $I, J, K \in \mathcal{I}_\mathcal{K}$ with $I \subsetneq J \subsetneq K$ we define the composed coordinate change $\widehat{\Phi}_{JK} \circ \widehat{\Phi}_{IJ} : \mathbf{K}_I \to \mathbf{K}_K$ as above with domain $\phi_{IJ}^{-1}(U_{JK}) \subset U_{IJ}$. We say that the triple of coordinate changes $\widehat{\Phi}_{IJ}, \widehat{\Phi}_{JK}, \widehat{\Phi}_{IK}$ satisfies the*

- **weak cocycle condition** *if* $\widehat{\Phi}_{JK} \circ \widehat{\Phi}_{IJ} \approx \widehat{\Phi}_{IK}$, *i.e. the coordinate changes are equal on the overlap; in particular if*

$$\phi_{JK} \circ \phi_{IJ} = \phi_{IK} \qquad on\ \phi_{IJ}^{-1}(U_{JK}) \cap U_{IK};$$

- **cocycle condition** *if* $\widehat{\Phi}_{JK} \circ \widehat{\Phi}_{IJ} \subset \widehat{\Phi}_{IK}$, *i.e.* $\widehat{\Phi}_{IK}$ *extends the composed coordinate change; in particular if*

$$\phi_{JK} \circ \phi_{IJ} = \phi_{IK} \qquad on\ \phi_{IJ}^{-1}(U_{JK}) \subset U_{IK}; \tag{2.1.5}$$

- **strong cocycle condition** *if* $\widehat{\Phi}_{JK} \circ \widehat{\Phi}_{IJ} = \widehat{\Phi}_{IK}$ *are equal as coordinate changes; in particular if*

$$\phi_{JK} \circ \phi_{IJ} = \phi_{IK} \qquad on\ \phi_{IJ}^{-1}(U_{JK}) = U_{IK}. \tag{2.1.6}$$

The following diagram of sets and maps between them might be useful in decoding the cocycle conditions.

$$\begin{array}{ccccc} \phi_{IJ}^{-1}(U_{JK}) \cap U_{IJ} & \xrightarrow{\ \phi_{IJ}\ } & U_{JK} & \xrightarrow{\ \phi_{JK}\ } & U_K \\ \big\downarrow & & & \nearrow{\scriptstyle \phi_{IK}} & \\ U_I & \hookleftarrow & U_{IK}. & & \end{array}$$

The relevant distinction between these versions of the cocycle condition is that the weak condition can be achieved in practice by constructions of finite dimensional reductions for holomorphic curve moduli spaces, whereas the strong condition is needed for our construction of a virtual moduli cycle from perturbations of the sections in the Kuranishi charts. The cocycle condition is an intermediate notion which is too strong to be constructed in practice and too weak to induce a VFC, but it does allow us to formulate Kuranishi atlases categorically. This in turn gives rise, via a topological realization of a category, to a virtual neighbourhood of X into which all Kuranishi domains map.

DEFINITION 2.1.9. *A* **weak Kuranishi atlas of dimension d** *on a compact metrizable space* X *is a tuple*

$$\mathcal{K} = \big(\mathbf{K}_I, \widehat{\Phi}_{IJ}\big)_{I,J\in\mathcal{I}_{\mathcal{K}}, I\subsetneq J}$$

consisting of a covering family of basic charts $(\mathbf{K}_i)_{i=1,\dots,N}$ *of dimension* d *and transition data* $(\mathbf{K}_J)_{|J|\geq 2}$, $(\widehat{\Phi}_{IJ})_{I\subsetneq J}$ *for* $(\mathbf{K}_i)$ *as in Definition* 2.1.7, *that satisfy the weak cocycle condition* $\widehat{\Phi}_{JK} \circ \widehat{\Phi}_{IJ} \approx \widehat{\Phi}_{IK}$ *for every triple* $I, J, K \in \mathcal{I}_K$ *with* $I \subsetneq J \subsetneq K$. *A weak Kuranishi atlas* $\mathcal{K}$ *is called a* **Kuranishi atlas** *if it satisfies the cocycle condition of* (2.1.5).

REMARK 2.1.10. Very similar definitions apply if the isotropy groups are nontrivial, or if X is stratified (for example, it consists of nodal J-holomorphic curves). In the former case we must modify the coordinate changes (see Definition 3.1.10), while in the latter case the domains of the charts are stratified smooth (SS) spaces, which means that we must develop an adequate theory of SS maps. In §2.3 we introduce a notion of **topological atlas** that will provide a common context for the topological constructions. Therefore, for clarity we will sometimes call the atlases of Definition 2.1.9 **smooth** and **with trivial isotropy**. ◊

2.2. The Kuranishi category and virtual neighbourhood $|\mathcal{K}|$. After defining the Kuranishi category $\mathbf{B}_{\mathcal{K}}$ of a Kuranishi atlas $\mathcal{K}$ and the associated realization $|\mathcal{K}|$, we state the main results about the topological space $|\mathcal{K}|$, giving all the relevant definitions. Most proofs are deferred to §2.3 where they are carried out in the broader context of topological atlases.

It is useful to think of the domains and obstruction spaces of a Kuranishi atlas as forming the following categories. Recall that a **topological category $\mathbf{C}$** is a small category (i.e. the collections $\mathrm{Obj}_{\mathbf{C}}, \mathrm{Mor}_{\mathbf{C}}$ of objects and morphisms are sets) in which $\mathrm{Obj}_{\mathbf{C}}$ and $\mathrm{Mor}_{\mathbf{C}}$ are provided with topologies in such a way that all structural maps, such as source and target maps $s, t : \mathrm{Mor}_{\mathcal{K}} \to \mathrm{Obj}_{\mathcal{K}}$ as well as composition (and inverse in the case of a groupoid), are continuous. If $\mathcal{K}$ is a smooth atlas as defined in Definition 2.1.9, then the spaces $\mathrm{Obj}_{\mathbf{C}}$ and $\mathrm{Mor}_{\mathbf{C}}$ are disjoint unions of smooth finite dimensional manifolds of varying dimensions, and the structural maps are smooth embeddings, but later we also consider topological atlases. As we will see, the language of categories is a good way to describe how a space (such as $|\mathcal{K}|$) is built from simpler pieces.

DEFINITION 2.2.1. *Given a Kuranishi atlas $\mathcal{K}$ we define its* **domain category** $\mathbf{B}_{\mathcal{K}}$ *to consist of the space of objects*[10]

$$\mathrm{Obj}_{\mathbf{B}_{\mathcal{K}}} := \bigsqcup_{I \in \mathcal{I}_{\mathcal{K}}} U_I \;=\; \big\{(I, x) \,\big|\, I \in \mathcal{I}_{\mathcal{K}}, x \in U_I\big\}$$

and the space of morphisms

$$\mathrm{Mor}_{\mathbf{B}_{\mathcal{K}}} := \bigsqcup_{I, J \in \mathcal{I}_{\mathcal{K}}, I \subset J} U_{IJ} \;=\; \big\{(I, J, x) \,\big|\, I, J \in \mathcal{I}_{\mathcal{K}}, I \subset J, x \in U_{IJ}\big\}.$$

Here we denote $U_{II} := U_I$ for $I = J$, and for $I \subsetneq J$ use the domain $U_{IJ} \subset U_I$ of the restriction $\mathbf{K}_I|_{U_{IJ}}$ to F_J that is part of the coordinate change $\widehat{\Phi}_{IJ} : \mathbf{K}_I|_{U_{IJ}} \to \mathbf{K}_J$.

Source and target of these morphisms are given by

$$(I, J, x) \in \mathrm{Mor}_{\mathbf{B}_{\mathcal{K}}}\big((I, x), (J, \phi_{IJ}(x))\big),$$

where $\phi_{IJ} : U_{IJ} \to U_J$ is the embedding given by $\widehat{\Phi}_{IJ}$, and we denote $\phi_{II} := \mathrm{id}_{U_I}$. Composition[11] *is defined by*

$$\big(I, J, x\big) \circ \big(J, K, y\big) := \big(I, K, x\big)$$

for any $I \subset J \subset K$ and $x \in U_{IJ}, y \in U_{JK}$ such that $\phi_{IJ}(x) = y$.

The **obstruction category** $\mathbf{E}_{\mathcal{K}}$ *is defined in complete analogy to $\mathbf{B}_{\mathcal{K}}$ to consist of the spaces of objects $\mathrm{Obj}_{\mathbf{E}_{\mathcal{K}}} := \bigsqcup_{I \in \mathcal{I}_{\mathcal{K}}} U_I \times E_I$ and morphisms*

$$\mathrm{Mor}_{\mathbf{E}_{\mathcal{K}}} := \big\{(I, J, x, e) \,\big|\, I, J \in \mathcal{I}_{\mathcal{K}}, I \subset J, x \in U_{IJ}, e \in E_I\big\}.$$

We may also express the further parts of a Kuranishi atlas in categorical terms:

- The obstruction category $\mathbf{E}_{\mathcal{K}}$ is a bundle over $\mathbf{B}_{\mathcal{K}}$ in the sense that there is a functor $\mathrm{pr}_{\mathcal{K}} : \mathbf{E}_{\mathcal{K}} \to \mathbf{B}_{\mathcal{K}}$ that is given on objects and morphisms by projection $(I, x, e) \mapsto (I, x)$ and $(I, J, x, e) \mapsto (I, J, x)$ with locally trivial fiber E_I.

[10] When forming categories such as $\mathbf{B}_{\mathcal{K}}$, we take always the space of objects to be the disjoint union of the domains U_I, even if we happen to have defined the sets U_I as subsets of some larger space such as $\mathbb{R}^2$ or a space of maps as in the Gromov–Witten case. Similarly, the morphism space is a disjoint union of the U_{IJ} even though $U_{IJ} \subset U_I$ for all $J \supset I$.

[11] Note that this is written in the categorical ordering.

- The sections s_I induce a smooth section of this bundle, i.e. a functor $\mathfrak{s}_{\mathcal{K}} : \mathbf{B}_{\mathcal{K}} \to \mathbf{E}_{\mathcal{K}}$ which acts smoothly on the spaces of objects and morphisms, and whose composite with the projection $\mathrm{pr}_{\mathcal{K}} : \mathbf{E}_{\mathcal{K}} \to \mathbf{B}_{\mathcal{K}}$ is the identity. More precisely, it is given by $(I,x) \mapsto (I,x,s_I(x))$ on objects and by $(I,J,x) \mapsto (I,J,x,s_I(x))$ on morphisms.
- The zero sets of the sections $\bigsqcup_{I\in\mathcal{I}_{\mathcal{K}}} \{I\} \times s_I^{-1}(0) \subset \mathrm{Obj}_{\mathbf{B}_{\mathcal{K}}}$ form a very special strictly full subcategory $\mathfrak{s}_{\mathcal{K}}^{-1}(0)$ of $\mathbf{B}_{\mathcal{K}}$. Namely, $\mathbf{B}_{\mathcal{K}}$ splits into the subcategory $\mathfrak{s}_{\mathcal{K}}^{-1}(0)$ and its complement (given by the full subcategory with objects $\{(I,x) \,|\, s_I(x) \neq 0\}$) in the sense that there are no morphisms of $\mathbf{B}_{\mathcal{K}}$ between the two underlying sets of objects. (This holds by (2.2.1) below.)
- The footprint maps ψ_I give rise to a surjective functor $\psi_{\mathcal{K}} : \mathfrak{s}_{\mathcal{K}}^{-1}(0) \to \mathbf{X}$ to the category $\mathbf{X}$ with object space X and trivial morphism space, i.e. consisting only of identity maps. It is given by $(I,x) \mapsto \psi_I(x)$ on objects and by $(I,J,x) \mapsto \mathrm{id}_{\psi_I(x)}$ on morphisms.

We denote the **topological realization of the category $\mathbf{B}_{\mathcal{K}}$** by $|\mathbf{B}_{\mathcal{K}}|$, often abbreviated to $|\mathcal{K}|$. This is the space formed as the quotient of $\mathrm{Obj}_{\mathbf{B}_{\mathcal{K}}} = \bigsqcup_I U_I$ by the equivalence relation generated by the morphisms, and is given the quotient topology. Thus, for example, the realization of the category $\mathbf{X}$ is the space X itself. Since $\mathfrak{s}_{\mathcal{K}}$ is a functor, the equivalence relation on $\mathrm{Obj}_{\mathbf{B}_{\mathcal{K}}}$ preserves the zero sets. More precisely, the fact that the morphisms in $\mathbf{B}_{\mathcal{K}}$ intertwine the zero sets $s_I^{-1}(0)$ and the footprint maps implies that

$$(I,x) \sim (J,y),\, s_I(x) = 0 \implies s_J(y) = 0, \psi_I(x) = \psi_J(y). \tag{2.2.1}$$

Hence $\mathfrak{s}_{\mathcal{K}}^{-1}(0)$ can be considered as a full subcategory of $\mathbf{B}_{\mathcal{K}}$, and this inclusion induces a natural continuous bijection from the realization $|\mathfrak{s}_{\mathcal{K}}^{-1}(0)|$ of the subcategory $\mathfrak{s}_{\mathcal{K}}^{-1}(0)$ (with its quotient topology) to the zero set $|\mathfrak{s}_{\mathcal{K}}|^{-1}(0) \subset |\mathcal{K}|$ of the function $|\mathfrak{s}_{\mathcal{K}}| : |\mathcal{K}| \to |\mathbf{E}_{\mathcal{K}}|$ (with the subspace topology). As in [**MW1**, Lemma 2.4.2], one can prove directly from the definitions that the inverse is continuous.

LEMMA 2.2.2. *The inverse of the footprint maps $\psi_I^{-1} : F_I \to U_I$ fit together to give an injective map*

$$\iota_{\mathcal{K}} : X \to |\mathfrak{s}_{\mathcal{K}}|^{-1}(0) \subset |\mathcal{K}| \tag{2.2.2}$$

that is a homeomorphism to its image $|\mathfrak{s}_{\mathcal{K}}|^{-1}(0)$.

Thus the zero set $|\mathfrak{s}_{\mathcal{K}}|^{-1}(0) \subset |\mathcal{K}|$ has the expected topology. However, as is shown by Example 2.2.16 below, the topology on $|\mathcal{K}|$ itself can be very wild; it is not in general Hausdorff and the natural maps $\pi_{\mathcal{K}} : U_I \to |\mathcal{K}|$ need not be injective, let alone homeomorphisms to their images. Further, even though the isotropy is trivial, the fibers of the projection $|\,\mathrm{pr}\,| : |\mathbf{E}_{\mathcal{K}}| \to |\mathcal{K}|$ need not be vector spaces. In order to remedy these problems we introduce the notions of **tameness, shrinking, metrizability, and cobordism** for weak atlases and prove the following result.

THEOREM 2.2.3 (see Theorem 6.3.9 in [**MW2**]). *Let $\mathcal{K}$ be a weak Kuranishi atlas (with trivial isotropy) on a compact metrizable space X. Then an appropriate shrinking of $\mathcal{K}$ provides a metrizable tame Kuranishi atlas $\mathcal{K}'$ with domains $(U_I' \subset U_I)_{I\in\mathcal{I}_{\mathcal{K}'}}$ such that the realizations $|\mathcal{K}'|$ and $|\mathbf{E}_{\mathcal{K}'}|$ are Hausdorff in the quotient topology. In addition, for each $I \in \mathcal{I}_{\mathcal{K}'} = \mathcal{I}_{\mathcal{K}}$ the projection maps $\pi_{\mathcal{K}'} : U_I' \to |\mathcal{K}'|$*

and $\pi_{\mathcal{K}'} : U'_I \times E_I \to |\mathbf{E}_{\mathcal{K}'}|$ *are homeomorphisms onto their images and fit into a commutative diagram*

$$\begin{array}{ccc} U'_I \times E_I & \xrightarrow{\pi_{\mathcal{K}'}} & |\mathbf{E}_{\mathcal{K}'}| \\ \downarrow & & \downarrow {\scriptstyle |\operatorname{pr}_{\mathcal{K}'}|} \\ U'_I & \xrightarrow{\pi_{\mathcal{K}'}} & |\mathcal{K}'|. \end{array}$$

where the horizontal maps intertwine the vector space structure on E_I *with a vector space structure on the fibers of* $|\operatorname{pr}_{\mathcal{K}'}|$.

Moreover, any two such shrinkings are cobordant by a metrizable tame Kuranishi cobordism whose realization also has the above Hausdorff, homeomorphism, and linearity properties.

We give most of the details of the proof in the next section §2.3. In fact, we will prove a more general version that will be relevant when we come to consider smooth atlases with nontrivial isotropy. Thus, we will introduce a notion of **topological atlas** and discuss what taming means in that context. In the rest of this section, we explain the new notions in the smooth context, stating more precise versions of the above theorem for this case.

Definition 2.2.4. *A weak Kuranishi atlas is said to be* **tame** *if for all* $I, J, K \in \mathcal{I}_{\mathcal{K}}$ *we have*[12]

$$U_{IJ} \cap U_{IK} = U_{I(J\cup K)} \qquad \forall I \subset J, K; \tag{2.2.3}$$

$$\phi_{IJ}(U_{IK}) = U_{JK} \cap s_J^{-1}\big(\widehat{\phi}_{IJ}(E_I)\big) \qquad \forall I \subset J \subset K. \tag{2.2.4}$$

Here we allow equalities, using the notation $U_{II} := U_I$ *and* $\phi_{II} := \mathrm{Id}_{U_I}$. *Further, to allow for the possibility that* $J \cup K \notin \mathcal{I}_{\mathcal{K}}$, *we define* $U_{IL} := \emptyset$ *for* $L \subset \{1, \dots, N\}$ *with* $L \notin \mathcal{I}_{\mathcal{K}}$. *Therefore* (2.2.3) *includes the condition*

$$U_{IJ} \cap U_{IK} \neq \emptyset \implies F_J \cap F_K \neq \emptyset \qquad (\iff J \cup K \in \mathcal{I}_{\mathcal{K}}).$$

The notion of tameness generalizes the identities $F_J \cap F_K = F_{J\cup K}$ and $\psi_J^{-1}(F_K) = U_{JK} \cap s_J^{-1}(0_J)$ between the footprints and zero sets, which we can include into (2.2.3) and (2.2.4) as the case $I = \emptyset$, by using the notation

$$U_{\emptyset J} := F_J, \qquad \phi_{\emptyset J} := \psi_J^{-1}. \tag{2.2.5}$$

Indeed, the first tameness condition (2.2.3) extends the identity for intersections of footprints – which is equivalent to $\psi_I^{-1}(F_J) \cap \psi_I^{-1}(F_K) = \psi_I^{-1}(F_{J\cup K})$ for all $I \subset J, K$ – to the domains of the transition maps in U_I. In particular, with $J \subset K$ it implies nesting of the domains of the transition maps,

$$U_{IK} \subset U_{IJ} \qquad \forall I \subset J \subset K. \tag{2.2.6}$$

(This in turn generalizes the $I = \emptyset$ case $F_K \subset F_J$ for $J \subset K$.) The second tameness condition (2.2.4) extends the relation between footprints and zero sets – equivalent to $\phi_{IJ}(\psi_I^{-1}(F_K)) = U_{JK} \cap s_J^{-1}(0)$ for all $I \subset J$ – to a relation between domains of transition maps and preimages of corresponding subbundles by the appropriate

[12] In (2.2.4) below we write $s_J^{-1}\big(\widehat{\phi}_{IJ}(E_I)\big)$ instead of $s_J^{-1}(E_I)$ for clarity; but we usually use the shorter notation, identifying the subspace $\{(e_j)_{j\in J} \mid e_j = 0 \ \forall j \in J \smallsetminus I\} \subset E_J$ with E_I when $I \subset J$.

section $s_\bullet$. In particular, with $J = K$ it controls the image of the transition maps, generalizing the $I = \emptyset$ case $\psi_J^{-1}(F_J) = s_J^{-1}(0)$ to

$$\text{(2.2.7)} \qquad \operatorname{im} \phi_{IJ} := \phi_{IJ}(U_{IJ}) = s_J^{-1}(E_I) \qquad \forall I \subset J.$$

It follows that that the image of each transition map ϕ_{IJ} is a closed subset of the Kuranishi domain U_J. Further, if $I \subset J \subset K$ we may combine conditions (2.2.4) and (2.2.7) to obtain the identity $\phi_{IJ}(U_{IK}) = U_{JK} \cap s_J^{-1}(E_I) = U_{JK} \cap (\operatorname{im} \phi_{IJ})$, which pulls back via ϕ_{IJ}^{-1} to $U_{IK} = U_{IJ} \cap \phi_{IJ}^{-1}(U_{JK})$. This proves the first part of the following proposition.

PROPOSITION 2.2.5. *Let $\mathcal{K}$ be a tame weak Kuranishi atlas (with trivial isotropy). Then the following holds.*

(1) *$\mathcal{K}$ satisfies the strong cocycle condition; in particular it is a Kuranishi atlas.*
(2) *Both $|\mathcal{K}|$ and $|\mathbf{E}_\mathcal{K}|$ are Hausdorff, and for each $I \in \mathcal{I}_\mathcal{K}$ the quotient maps $\pi_\mathcal{K}|_{U_I} : U_I \to |\mathcal{K}|$ and $\pi_\mathcal{K}|_{U_I \times E_I} : U_I \times E_I \to |\mathbf{E}_\mathcal{K}|$ are homeomorphisms onto their image.*
(3) *There is a unique linear structure on the fibers of $|\operatorname{pr}_\mathcal{K}| : |\mathbf{E}_\mathcal{K}| \to |\mathcal{K}|$ such that for every $I \in \mathcal{I}_\mathcal{K}$ the embedding $\pi_\mathcal{K} : U_I \times E_I \to |\mathbf{E}_\mathcal{K}|$ is linear on the fibers.*

PROOF. See Proposition 2.3.15. □

Thus the quotient topology on the realization $|\mathcal{K}|$ of a tame atlas is reasonably well behaved. Nevertheless it is almost never metrizable: indeed if there is a coordinate change $\widehat{\Phi}_{IJ} : \mathbf{K}_I \to \mathbf{K}_J$ with $\dim U_I < \dim U_J$ and such that the subset F_J is not closed in F_I then for each $x_I \in \psi_I^{-1}\big(F_I \cap (\overline{F}_J \smallsetminus F_J)\big) \subset U_I$ the point $\pi_\mathcal{K}(x_I) \in |\mathcal{K}|$ does *not* have a countable neighbourhood base in the quotient topology; cf. Example 2.2.16 below.

DEFINITION 2.2.6 (Definition 6.1.14 in [**MW2**]). *A Kuranishi atlas $\mathcal{K}$ is called* **metrizable** *if there is a bounded metric d on the set $|\mathcal{K}|$ such that for each $I \in \mathcal{I}_\mathcal{K}$ the pullback metric $d_I := (\pi_\mathcal{K}|_{U_I})^* d$ on U_I induces the given topology on U_I. In this situation we call d an* **admissible metric** *on $|\mathcal{K}|$. A* **metric Kuranishi atlas** *is a pair $(\mathcal{K}, d)$ consisting of a metrizable Kuranishi atlas and a choice of admissible metric d.*

REMARK 2.2.7. We will use this metric on $|\mathcal{K}|$ when constructing the perturbation section ν in order to control its domain and size and hence ensure that the perturbed zero set has compact realization. ◊

Before stating the existence result, it is convenient to introduce the further notion of a **shrinking.** We write $V' \sqsubset V$ to denote that V' is precompact in V, i.e. the closure (written $\overline{V'}$ or $cl_V(V')$) of V' in V is compact.

DEFINITION 2.2.8. *Let $(F_i)_{i=1,\dots,N}$ be an open cover of a compact space X. We say that $(F_i')_{i=1,\dots,N}$ is a* **shrinking** *of (F_i) if $F_i' \sqsubset F_i$ are precompact open subsets, which cover $X = \bigcup_{i=1,\dots,N} F_i'$, and are such that for all subsets $I \subset \{1, \dots, N\}$ we have*

$$\text{(2.2.8)} \qquad F_I := \textstyle\bigcap_{i \in I} F_i \neq \emptyset \qquad \Longrightarrow \qquad F_I' := \textstyle\bigcap_{i \in I} F_i' \neq \emptyset.$$

DEFINITION 2.2.9. *Let $\mathcal{K} = \big(\mathbf{K}_I, \widehat{\Phi}_{IJ}\big)_{I,J\in\mathcal{I}_\mathcal{K}, I\subsetneq J}$ be a weak Kuranishi atlas. We say that a weak Kuranishi atlas $\mathcal{K}' = (\mathbf{K}'_I, \widehat{\Phi}'_{IJ})_{I,J\in\mathcal{I}_{\mathcal{K}'}, I\subsetneq J}$ is a* **shrinking** *of $\mathcal{K}$, and write $\mathcal{K}' \sqsubset \mathcal{K}$, if*

(1) *the footprint cover $(F'_i)_{i=1,\dots,N'}$ is a shrinking of the cover $(F_i)_{i=1,\dots,N}$, in particular the numbers $N = N'$ of basic charts agree, and so do the index sets $\mathcal{I}_{\mathcal{K}'} = \mathcal{I}_\mathcal{K}$;*
(2) *for each $I \in \mathcal{I}_\mathcal{K}$ the chart $\mathbf{K}'_I$ is the restriction of $\mathbf{K}_I$ to a precompact domain $U'_I \subset U_I$ as in Definition 2.1.3;*
(3) *for each $I, J \in \mathcal{I}_\mathcal{K}$ with $I \subsetneq J$ the coordinate change $\widehat{\Phi}'_{IJ}$ is the restriction of $\widehat{\Phi}_{IJ}$ to the open subset $U'_{IJ} := \phi_{IJ}^{-1}(U'_J)\cap U'_I$ (see* [**MW2**, Lemma 5.2.6]*).*

In order to construct metric tame Kuranishi atlases, we will find it useful to consider tame shrinkings $\mathcal{K}_{sh}$ of a weak Kuranishi atlas $\mathcal{K}$ that are obtained as shrinkings of an intermediate tame shrinking $\mathcal{K}'$ of $\mathcal{K}$. For short we will call such $\mathcal{K}_{sh}$ a **preshrunk tame shrinking** of $\mathcal{K}$ and write $\mathcal{K}_{sh} \sqsubset \mathcal{K}' \sqsubset \mathcal{K}$.

PROPOSITION 2.2.10. *Every weak Kuranishi atlas $\mathcal{K}$ has a shrinking $\mathcal{K}_{sh}$ that is a tame Kuranishi atlas – for short called a* **tame shrinking**. *Moreover if $\mathcal{K}_{sh}$ is preshrunk with $\mathcal{K}_{sh} \sqsubset \mathcal{K}' \sqsubset \mathcal{K}$ where $\mathcal{K}'$ is also tame, then the following holds:*

- *the induced map $|\mathcal{K}_{sh}| \to |\mathcal{K}'|$ is a continuous injection;*
- *the atlas $\mathcal{K}_{sh}$ is metrizable;*
- *we may choose the metric d so that the metric topology on $(|\mathcal{K}_{sh}|, d)$ equals its topology as a subspace of $|\mathcal{K}'|$ with the quotient topology.*

Propositions 2.2.5 and 2.2.10 contain all the results in the first part of Theorem 2.2.3, and are proved in §2.3. The next important concept used in Theorem 2.2.3 is that of cobordism. We develop an appropriate theory of **topological cobordism Kuranishi atlases** in [**MW1**, §4] over spaces Y with collared boundary; for the smooth theory see [**MW2**, §6.2]. To give its flavor, we now quote a few key definitions for smooth atlases, where for simplicity we restrict to the case of **concordances**, i.e. cobordisms over $[0, 1] \times X$.

DEFINITION 2.2.11.
- *Let $\mathbf{K}^\alpha = (U^\alpha, E^\alpha, s^\alpha, \psi^\alpha)$ be a Kuranishi chart on X, and let $A \subset [0, 1]$ be a relatively open interval. Then we define the* **product chart** *for $[0, 1] \times X$ with footprint $A \times F_I^\alpha$ as*

$$A \times \mathbf{K}^\alpha := \big(A \times U^\alpha, E^\alpha, s^\alpha \circ \mathrm{pr}_{U^\alpha}, \mathrm{id}_A \times \psi^\alpha\big).$$

- *A* **Kuranishi chart with collared boundary** *on $[0, 1] \times X$ is a tuple $\mathbf{K} = (U, E, s, \psi)$ as in Definition 2.1.1, with the following variations:*
  (1) *The footprint $F = \psi(s^{-1}(0)) \subset [0, 1] \times X$ intersects the boundary $\{0, 1\} \times X$.*
  (2) *The domain U is a smooth manifold whose boundary splits into two parts $\partial U = \partial^0 U \sqcup \partial^1 U$ such that $\partial^\alpha U$ is nonempty iff F intersects $\{\alpha\} \times X$.*
  (3) *If $\partial^\alpha U \neq \emptyset$ then there is a relatively open neighbourhood $A^\alpha \subset [0, 1]$ of α and an embedding $\iota^\alpha : A^\alpha \times \partial^\alpha U \hookrightarrow U$ onto a neighbourhood of $\partial^\alpha U \subset U$ such that*

$$\big(\{\alpha\} \times \partial^\alpha U\,,\, E\,,\, s \circ \iota^\alpha\,,\, \psi \circ \iota^\alpha\big) \;=\; A^\alpha \times \partial^\alpha \mathbf{K}$$

is the product of A^α with a Kuranishi chart $\partial^\alpha \mathbf{K}$ for X (called the **restriction of K to the boundary***) with footprint $F^\alpha \subset X$ such that $(A^\alpha \times X) \cap F = A^\alpha \times F$.*

DEFINITION 2.2.12. • *Let $\mathbf{K}_I, \mathbf{K}_J$ be Kuranishi charts on $[0,1] \times X$ such that only $\mathbf{K}_I$ or both $\mathbf{K}_I, \mathbf{K}_J$ have collared boundary. Then a* **coordinate change with collared boundary** *$\widehat{\Phi}_{IJ} : \mathbf{K}_I \to \mathbf{K}_J$ is a tuple $\widehat{\Phi}_{IJ} = (U_{IJ}, \phi_{IJ}, \widehat{\phi}_{IJ})$ of domain and embeddings as in Definition 2.1.5, with the following boundary variations and collar form requirement:*

(1) *The domain is a relatively open subset $U_{IJ} \subset U_I$ with boundary components $\partial^\alpha U_{IJ} := U_{IJ} \cap \partial^\alpha U_I$;*

(2) *If $F_J \cap \{\alpha\} \times X \neq \emptyset$ for $\alpha = 0$ or 1 (so that $F_I \cap \{\alpha\} \times X \neq \emptyset$), there is a relatively open neighbourhood $B^\alpha \subset [0,1]$ of α such that*

$$\begin{aligned} (\iota_I^\alpha)^{-1}(U_{IJ}) \cap \big(B^\alpha \times \partial^\alpha U_I\big) &= B^\alpha \times \partial^\alpha U_{IJ}, \\ (\iota_J^\alpha)^{-1}(\operatorname{im} \phi_{IJ}) \cap \big(B^\alpha \times \partial^\alpha U_J\big) &= \phi_{IJ}(B^\alpha \times \partial^\alpha U_{IJ}), \end{aligned}$$

and

$$\big(B^\alpha \times \partial^\alpha U_{IJ} \,,\, (\iota_J^\alpha)^{-1} \circ \phi_{IJ} \circ \iota_I^\alpha \,,\, \widehat{\phi}_{IJ} \big) = \mathrm{id}_{B^\alpha} \times \partial^\alpha \widehat{\Phi}_{IJ},$$

where $\partial^\alpha \widehat{\Phi}_{IJ} : \partial^\alpha \mathbf{K}_I \to \partial^\alpha \mathbf{K}_J$ is a coordinate change.

(3) *If $\partial^\alpha F_J = \emptyset$ but $\partial^\alpha F_I \neq \emptyset$ for $\alpha = 0$ or 1 there is a neighbourhood $B^\alpha \subset [0,1]$ of α such that*

$$U_{IJ} \cap \iota_I^\alpha\big(B^\alpha \times \partial^\alpha U_I\big) = \emptyset.$$

• *For any coordinate change with collared boundary $\widehat{\Phi}_{IJ}$ on $[0,1] \times X$ we call the uniquely determined coordinate changes $\partial^\alpha \widehat{\Phi}_{IJ}$ for X the* **restrictions of $\widehat{\Phi}_{IJ}$ to the boundary** *for $\alpha = 0, 1$.*

DEFINITION 2.2.13. *A* **(weak) Kuranishi cobordism** *on $[0,1] \times X$ is a tuple*

$$\mathcal{K} = \big(\mathbf{K}_I, \widehat{\Phi}_{IJ}\big)_{I,J \in \mathcal{I}_\mathcal{K}}$$

of basic charts and transition data as in Definition 2.1.9 with the following boundary variations and collar form requirements:

- *The charts of $\mathcal{K}$ are either Kuranishi charts with collared boundary or standard Kuranishi charts whose footprints are precompactly contained in $(0,1) \times X$.*
- *The coordinate changes $\widehat{\Phi}_{IJ} : \mathbf{K}_I \to \mathbf{K}_J$ are either standard coordinate changes on $(0,1) \times X$ between pairs of standard charts, or coordinate changes with collared boundary between pairs of charts, of which at least the first has collared boundary.*

Moreover, we call $\mathcal{K}$ **tame** *if it satisfies the tameness conditions of Definition 2.2.4.*

The essential feature of our definitions is that the charts are now manifolds with collared boundary, and that we require compatibility of this collar structure with coordinate changes and all other structures, such as metrics. In particular, a **metric Kuranishi concordance**[13] $(\mathcal{K}, d)$ from $\mathcal{K}^0$ to $\mathcal{K}^1$ is a metric atlas $(\mathcal{K}, d)$ over $[0,1] \times X$ that for $\alpha = 0, 1$ restricts to the atlas $\mathcal{K}^\alpha =: \partial^\alpha \mathcal{K}$ on X, and near

[13]Concordance is called deformation equivalence in [**FF**].

each boundary has an *isometric* identification with the product $A^\alpha_\varepsilon \times \mathcal{K}^\alpha$ where $A^0_\varepsilon = [0,\varepsilon), A^1_\varepsilon = (1-\varepsilon, 1]$. Here is the main existence result.

PROPOSITION 2.2.14. *Let $\mathcal{K}$ be a weak Kuranishi concordance on $[0,1]\times X$, and let $\mathcal{K}^0_{sh}, \mathcal{K}^1_{sh}$ be preshrunk tame shrinkings of $\partial^0\mathcal{K}$ and $\partial^1\mathcal{K}$ with admissible metrics d^α. Then there is a preshrunk tame shrinking $\mathcal{K}_{sh}$ of $\mathcal{K}$ that provides a metric tame Kuranishi concordance from $\mathcal{K}^0_{sh}$ to $\mathcal{K}^1_{sh}$. Further, we may choose the metric on $\mathcal{K}_{sh}$ so that for $\alpha = 0,1$ it restricts to d^α on $\partial^\alpha\mathcal{K}_{sh}$.*

The proof is a fairly routine generalization of that of Proposition 2.2.10, although it turns out to be surprisingly hard to interpolate between two given metrics in this way. (Note that we are *not* considering Riemannian metrics.) The necessary details are given in [**MW1**, Proposition 4.2.4]. An analogous statement is also true for Kuranishi cobordisms.

A final basic ingredient of the proof of Theorem B is that of **orientation**. When the atlas $\mathcal{K}$ is smooth with trivial isotropy we show in §3.3 below that there is a real line bundle $|\Lambda| \to |\mathcal{K}|$ whose pull back $\pi^*_{\mathcal{K}}(|\Lambda|)|_{U_I}$ restricts on each domain U_I to the orientation bundle $\Lambda^{\max} \mathrm{T}U_I \otimes (\Lambda^{\max} E_I)^*$ of $U_I \times E_I$. (Here, $\Lambda^{\max} V$ denotes the maximal exterior power of the vector space V.)

DEFINITION 2.2.15. *We define an orientation of $\mathcal{K}$ to be a nonvanishing section σ of $|\Lambda| \to |\mathcal{K}|$.*

The pullback of σ to $\pi^*_{\mathcal{K}}(|\Lambda|)|_{U_I}$ then determines an orientation of $U_I \times E_I$, that is preserved by the coordinate changes because the bundle $|\Lambda|$ is defined over $|\mathcal{K}|$, and as we show in Lemma 3.3.9 induces compatible orientations on the zero sets of a perturbed transverse section.

We next give a simple example of a tame smooth atlas whose realization is neither metrizable nor locally compact.

EXAMPLE 2.2.16 (Failure of metrizability and local compactness). For simplicity we will give an example with noncompact $X = \mathbb{R}$. (A similar example can be constructed with $X = S^1$.) We construct a Kuranishi atlas $\mathcal{K}$ on X with two basic charts, $\mathbf{K}_1 = (U_1 = \mathbb{R}, E_1 = \{0\}, s = 0, \psi_1 = \mathrm{id})$ and

$$\mathbf{K}_2 = \big(U_2 = (0,\infty)\times\mathbb{R},\ E_2 = \mathbb{R},\ s_2(x,y) = y,\ \psi_2(x,y) = x\big),$$

one transition chart $\mathbf{K}_{12} = \mathbf{K}_2|_{U_{12}}$ with domain $U_{12} := U_2$, and the coordinate changes $\widehat{\Phi}_{i,12}$ induced by the natural embeddings of the domains $U_{1,12} := (0,\infty) \hookrightarrow (0,\infty)\times\{0\}$ and $U_{2,12} := U_2 \hookrightarrow U_2$. Then as a set $|\mathcal{K}| = \big(U_1 \sqcup U_2 \sqcup U_{12}\big)/{\sim}$ can be identified with $\big(\mathbb{R}\times\{0\}\big) \cup \big((0,\infty)\times\mathbb{R}\big) \subset \mathbb{R}^2$. However, the quotient topology at $(0,0) \in |\mathcal{K}|$ is strictly stronger than the subspace topology. That is, for any open set $O \subset \mathbb{R}^2$ the induced subset $O \cap |\mathcal{K}| \subset |\mathcal{K}|$ is open, but some open subsets of $|\mathcal{K}|$ cannot be represented in this way. In fact, for any $\varepsilon > 0$ and continuous function $f : (0,\varepsilon) \to (0,\infty)$, the set

$$U_{f,\varepsilon} := \big\{[x] \,\big|\, x \in U_1, |x| < \varepsilon\big\} \cup \big\{[(x,y)] \,\big|\, (x,y) \in U_2, |x| < \varepsilon, |y| < f(x)\big\} \subset |\mathcal{K}|$$

is open in the quotient topology. It is shown in [**MW1**, Example 2.4.5] that these sets form an (uncountable) basis for the neighbourhoods of $[(0,0)]$ in the quotient topology.

Notice that this atlas $\mathcal{K}$ is tame. Therefore taming by itself does not give a quotient with manageable topology. On the other hand, the only bad point $|\mathcal{K}|$ is $(0,0)$. Indeed, according to Proposition 2.2.10 the realization of any shrinking

$\mathcal{K}'$ of $\mathcal{K}$ injects into $|\mathcal{K}|$ and is metrizable when given the corresponding subspace topology. For example, we could take $U_1' := (-\infty, 2) \sqsubset U_1$, $U_2' := (1, \infty) \times \mathbb{R} \sqsubset U_2$ and $U_{12}' := U_2'$. $\Diamond$

REMARK 2.2.17. (i) The categories $\mathbf{B}_{\mathcal{K}}, \mathbf{E}_{\mathcal{K}}$ have many similarities with the étale[14] categories used to model orbifolds i.e. the spaces of objects and morphisms are smooth manifolds and all structural maps (such as the source map, composition and so on) are smooth. Moreover, all sets of morphisms in $\mathbf{B}_{\mathcal{K}}$ or $\mathbf{E}_{\mathcal{K}}$ between fixed objects are finite. However, because there could be coordinate changes $\phi_{IJ} : U_{IJ} \to U_J$ with $\dim U_{IJ} < \dim U_J$, the target map $t : \mathrm{Mor}_{\mathbf{B}_{\mathcal{K}}} \to \mathrm{Obj}_{\mathbf{B}_{\mathcal{K}}}$ is not in general a local diffeomorphism, although it is locally injective. Moreover, one cannot in general complete $\mathbf{B}_{\mathcal{K}}$ to a groupoid by adding inverses and compositions, while keeping the property that the morphism space is a union of smooth manifolds. The problem here is that the inclusion of inverses of the coordinate changes, and their compositions, may yield singular spaces of morphisms. Indeed, coordinate changes $\mathbf{K}_I \to \mathbf{K}_K$ and $\mathbf{K}_J \to \mathbf{K}_K$ with the same target chart are given by embeddings $\phi_{IK} : U_{IK} \to U_K$ and $\phi_{JK} : U_{JK} \to U_K$, whose images may not intersect transversely (for example, often their intersection is contained only in the zero set $s_K^{-1}(0)$); yet this intersection would be a component of the space of morphisms from U_I to U_J. We show in Proposition 5.1.5 below that in the special case when all obstruction spaces E_I are trivial, i.e. $E_I = \{0\}$, one can adjoin these compositions and inverses to $\mathbf{B}_{\mathcal{K}}$, obtaining an étale proper groupoid whose realization is an orbifold.

(ii) When, as here, the isotropy groups are trivial, one can think of the category $\mathbf{B}_{\mathcal{K}}$ as a topological poset; in other words the relation $\le$ on $\mathrm{Obj}_{\mathcal{B}_{\mathcal{K}}}$ defined by setting

$$(I, x) \le (J, y) \Longleftrightarrow \mathrm{Mor}_{\mathbf{B}_{\mathcal{K}}}\big((I, x), (J, y)\big) \ne \emptyset$$

is a partial order. As we will see in Lemma 2.3.12, the taming conditions impose further regularity on this order relation. Indeed part (a) shows that if two elements $(J, x), (K, y)$ have a lower bound (i.e. an element (I, w) that is less than or equal to them both), then they have both a unique greatest lower bound and a unique least upper bound. The notion of reduction (see Definition 2.4.1) further simplifies this partial order; indeed at this point one could reduce to a family of subsets with partial ordering given by the integer lengths $|I|$; cf. the notion of dimensionally graded system (DGS) in [**FF**]. $\Diamond$

2.3. Tame topological atlases. This section explains the proof of a generalization of Theorem 2.2.3. The main results are Proposition 2.3.13, that establishes the main properties of tame atlases and Proposition 2.3.17, that constructs them. Here we explain the key points of the proof. Further details and analogous results for cobordisms may be found in [**MW1**].

We begin by defining topological Kuranishi charts and atlases.

DEFINITION 2.3.1. *A* **topological Kuranishi chart** *for X with open footprint $F \subset X$ is a tuple $\mathbf{K} = (U, \mathbb{E}, 0, \mathfrak{s}, \psi)$ consisting of*

- *the* **domain** *U, which is a separable, locally compact metric space;*
- *the* **obstruction "bundle"** $\mathrm{pr} : \mathbb{E} \to U$, *which is a continuous map between separable, locally compact metric spaces;*

[14] For relevant definitions see §5.1.

- *the* **zero section** $0: U \to \mathbb{E}$, *which is a continuous map with* $\mathrm{pr} \circ 0 = \mathrm{id}_U$*;*
- *the* **section** $\mathfrak{s} : U \to \mathbb{E}$, *which is a continuous map with* $\mathrm{pr} \circ \mathfrak{s} = \mathrm{id}_U$*;*
- *the* **footprint map** $\psi : \mathfrak{s}^{-1}(0) \to X$*, which is a homeomorphism between the* **zero set** $\mathfrak{s}^{-1}(0) := \mathfrak{s}^{-1}(\mathrm{im}\, 0) = \{x \in U \,|\, \mathfrak{s}(x) = 0(x)\}$ *and the* **footprint** $\psi(\mathfrak{s}^{-1}(0)) = F$.

REMARK 2.3.2. Because finite dimensional manifolds are separable, locally compact, and metrizable, each smooth Kuranishi chart with trivial isotropy induces a topological Kuranishi chart with $\mathbb{E} := U \times E$, $0(x) := (x, 0)$, and $\mathfrak{s}(x) = (x, s(x))$. As we will see in §3, a smooth local chart near a point with nontrivial isotropy is a tuple (U, E, Γ, s, ψ) where U, E are as in Definition 2.1.1, Γ is a finite group that acts on U and E, the function $s : U \to E$ is equivariant, and the footprint map $\psi : s^{-1}(0) \to X$ induces a homeomorphism from ${}^{s^{-1}(0)}/_{\Gamma}$ onto an open subset of X. Such a tuple gives rise to a topological chart with domain ${}^{U}/_{\Gamma}$ and bundle $\mathrm{pr} : \mathbb{E} := {}^{U\times E}/_{\Gamma} \to {}^{U}/_{\Gamma}$, where Γ acts diagonally on $U \times E$. Note that the fibers of this bundle at points $x \in {}^{U}/_{\Gamma}$ with nontrivial stabilizer Stab_x are quotients ${}^{E}/_{\mathrm{Stab}_x}$, and hence in general do not have a linear structure. Nevertheless there is a well defined zero section as well as a section $\mathfrak{s} : U \to \mathbb{E}$ induced by $s : U \to E$. ◊

Although one can define restrictions as before, one must adapt the notion of coordinate change as follows.

DEFINITION 2.3.3. *Let* $\mathbf{K}_I$ *and* $\mathbf{K}_J$ *be topological Kuranishi charts such that* $F_J \subset F_I$*. A* **topological coordinate change** *from* $\mathbf{K}_I$ *to* $\mathbf{K}_J$ *is a map* $\widehat{\Phi} : \mathbf{K}_I|_{U_{IJ}} \to \mathbf{K}_J$ *defined on a restriction of* $\mathbf{K}_I$ *to* F_J*. More precisely:*

- *The* **domain** *of the coordinate change is an open subset* $U_{IJ} \subset U_I$ *such that* $\mathfrak{s}_I^{-1}(0_I) \cap U_{IJ} = \psi_I^{-1}(F_J)$.
- *The* **map** *of the coordinate change is a topological embedding (i.e. homeomorphism to its image)* $\widehat{\Phi} : \mathbb{E}_I|_{U_{IJ}} := \mathrm{pr}_I^{-1}(U_{IJ}) \to \mathbb{E}_J$ *with the following properties.*
  1. *It is a bundle map, i.e. we have* $\mathrm{pr}_J \circ \widehat{\Phi} = \phi \circ \mathrm{pr}_I|_{\mathrm{pr}_I^{-1}(U_{IJ})}$ *for a topological embedding* $\phi : U_{IJ} \to U_J$*, and it is linear in the sense that* $0_J \circ \phi = \widehat{\Phi} \circ 0_I|_{U_{IJ}}$.
  2. *It intertwines the sections, i.e.* $\mathfrak{s}_J \circ \phi = \widehat{\Phi} \circ \mathfrak{s}_I|_{U_{IJ}}$.
  3. *It restricts to the transition map induced from the footprints in* X*, i.e.*
  $$\phi|_{\psi_I^{-1}(F_I \cap F_J)} = \psi_J^{-1} \circ \psi_I : U_{IJ} \cap \mathfrak{s}_I^{-1}(0_I) \to \mathfrak{s}_J^{-1}(0_J).$$

In particular, the following diagrams commute:

$$\begin{array}{ccc} \mathbb{E}_I|_{U_{IJ}} & \xrightarrow{\widehat{\Phi}} & \mathbb{E}_J \\ \downarrow \mathrm{pr}_I & & \downarrow \mathrm{pr}_J \\ U_{IJ} & \xrightarrow{\phi} & U_J \end{array} \qquad \begin{array}{ccc} \mathbb{E}_I|_{U_{IJ}} & \xrightarrow{\widehat{\Phi}} & \mathbb{E}_J \\ \uparrow 0_I & & \uparrow 0_J \\ U_{IJ} & \xrightarrow{\phi} & U_J \end{array}$$

$$\begin{array}{ccc} \mathbb{E}_I|_{U_{IJ}} & \xrightarrow{\widehat{\Phi}} & \mathbb{E}_J \\ \uparrow \mathfrak{s}_I & & \uparrow \mathfrak{s}_J \\ U_{IJ} & \xrightarrow{\phi} & U_J \end{array} \qquad \begin{array}{ccc} U_{IJ} \cap \mathfrak{s}_I^{-1}(0_I) & \xrightarrow{\phi} & \mathfrak{s}_J^{-1}(0_J) \\ \downarrow \psi_I & & \downarrow \psi_J \\ X & \xrightarrow{\mathrm{Id}} & X. \end{array}$$

REMARK 2.3.4. The above definition does not provide charts with enough structure to be able to formulate any equivalent to the index condition. Indeed, because

$\mathrm{pr}: \mathbb{E}_I \to U_I$ is not assumed to be a vector bundle, there is as yet no notion to replace the quotient bundle $U_J \times \left({}^{E_J}/_{\widehat{\phi}_{IJ}(E_I)} \right)$ that appears in (2.1.3). We will formulate a weaker but adequate replacement for this condition when we introduce the notion of filtration; see Definition 2.3.6 and Lemma 2.3.8 below. $\Diamond$

It is immediate that smooth coordinate changes between smooth charts satisfy the above conditions. Moreover, one can define restriction and composition of coordinate changes as before, as well as the various versions of the cocycle condition.

DEFINITION 2.3.5. *A* **(weak) topological Kuranishi atlas** *on a compact metrizable space X is a tuple*

$$\mathcal{K} = \left(\mathbf{K}_I, \widehat{\Phi}_{IJ}\right)_{I,J\in\mathcal{I}_{\mathcal{K}}, I\subsetneq J}$$

of a covering family of basic charts $(\mathbf{K}_i)_{i=1,\dots,N}$ and transition data $(\mathbf{K}_J)_{|J|\geq 2}$, $(\widehat{\Phi}_{IJ})_{I\subsetneq J}$ for $(\mathbf{K}_i)$ as in Definition 2.1.7, that consists of topological Kuranishi charts and topological coordinate changes satisfying the (weak) cocycle condition.

Because $\widehat{\Phi}$ is a bundle map, a weak atlas is an atlas if and only if $U_{IJ} \cap \phi_{IJ}^{-1}(U_{JK}) \subset U_{IK}$ for all $I \subsetneq J \subsetneq K$. Further, if $\mathcal{K}$ is a topological atlas we can form the topological categories $\mathbf{B}_{\mathcal{K}}, \mathbf{E}_{\mathcal{K}}$ as before, and then assemble the maps $\mathrm{pr}_I, 0_I, \mathfrak{s}_I$ into functors denoted $\mathrm{pr}_{\mathcal{K}}, 0_{\mathcal{K}}, \mathfrak{s}_{\mathcal{K}}$. In particular, the functor $\mathrm{pr}_{\mathcal{K}}: \mathbf{E}_{\mathcal{K}} \to \mathbf{B}_{\mathcal{K}}$ induces a continuous map

$$|\mathrm{pr}_{\mathcal{K}}| : |\mathbf{E}_{\mathcal{K}}| \to |\mathcal{K}|,$$

which we call the **obstruction bundle** of $\mathcal{K}$, although its fibers generally do not have the structure of a vector space. Similarly, the functors $0_{\mathcal{K}}$ and $\mathfrak{s}_{\mathcal{K}}$ induce continuous maps

$$|0_{\mathcal{K}}| : |\mathcal{K}| \to |\mathbf{E}_{\mathcal{K}}|, \qquad |\mathfrak{s}_{\mathcal{K}}| : |\mathcal{K}| \to |\mathbf{E}_{\mathcal{K}}|,$$

that are sections in the sense that $|\mathrm{pr}_{\mathcal{K}}| \circ |\mathfrak{s}_{\mathcal{K}}| = |\mathrm{pr}_{\mathcal{K}}| \circ |0_{\mathcal{K}}| = \mathrm{id}_{|\mathcal{K}|}$. Moreover, the realization of the full subcategory $\mathfrak{s}_{\mathcal{K}}^{-1}(0_{\mathcal{K}})$ is homeomorphic to the zero section $|\mathfrak{s}_{\mathcal{K}}|^{-1}(|0_{\mathcal{K}}|) \subset |\mathcal{K}|$ and the analog of Lemma 2.2.2 holds. In other words, the footprint functor $\psi_{\mathcal{K}} : \mathfrak{s}_{\mathcal{K}}^{-1}(0_{\mathcal{K}}) \to \mathbf{X}$ descends to a homeomorphism $|\psi_{\mathcal{K}}| : |\mathfrak{s}_{\mathcal{K}}|^{-1}(|0_{\mathcal{K}}|) \to X$ with inverse given by

$$\iota_{\mathcal{K}} := |\psi_{\mathcal{K}}|^{-1} : X \longrightarrow |\mathfrak{s}_{\mathcal{K}}|^{-1}(|0_{\mathcal{K}}|) \subset |\mathcal{K}|, \qquad p \mapsto [(I, \psi_I^{-1}(p))],$$

where $[(I, \psi_I^{-1}(p))]$ is independent of the choice of $I \in \mathcal{I}_{\mathcal{K}}$ with $p \in F_I$.

Notice that the above definitions contain no analog of the additivity condition that we imposed in the smooth case, namely that for each $I \in \mathcal{I}_{\mathcal{K}}$ the obstruction space E_I for $\mathbf{K}_I$ is the product $\prod_{i\in I} E_i$. As will become clearer below, this condition is needed in order to construct shrinkings that satisfy the second tameness equation $\phi_{IJ}(U_{IK}) = U_{JK} \cap s_J^{-1}\big(\widehat{\phi}_{IJ}(E_I)\big)$, which in turn is a key ingredient of the proof of Proposition 2.2.5. The topological analog of additivity is the following filtration condition.

DEFINITION 2.3.6. *Let $\mathcal{K}$ be a weak topological Kuranishi atlas. We say that $\mathcal{K}$ is* **filtered** *if it is equipped with a* **filtration**, *that is a tuple of closed subsets $\mathbb{E}_{IJ} \subset \mathbb{E}_J$ for each $I, J \in \mathcal{I}_{\mathcal{K}}$ with $I \subset J$, that satisfy the following conditions:*

(1) $\mathbb{E}_{JJ} = \mathbb{E}_J$ *and* $\mathbb{E}_{\emptyset J} = \mathrm{im}\, 0_J$ *for all* $J \in \mathcal{I}_{\mathcal{K}}$;

(2) $\widehat{\Phi}_{JK}(\mathbb{E}_{IJ}) = \mathbb{E}_{IK} \cap \mathrm{pr}_K^{-1}(\mathrm{im}\, \phi_{JK})$ *for all* $I, J, K \in \mathcal{I}_{\mathcal{K}}$ *with* $I \subset J \subsetneq K$;

(3) $\mathbb{E}_{IJ} \cap \mathbb{E}_{HJ} = \mathbb{E}_{(I\cap H)J}$ *for all* $I, H, J \in \mathcal{I}_{\mathcal{K}}$ *with* $I, H \subset J$;

(4) *im* ϕ_{IJ} *is an open subset of* $\mathfrak{s}_J^{-1}(\mathbb{E}_{IJ})$ *for all* $I, J \in \mathcal{I}_{\mathcal{K}}$ *with* $I \subsetneq J$.

REMARK 2.3.7. Applying condition (ii) above to any triple $I = I \subsetneq J$ for $J \in \mathcal{I}_{\mathcal{K}}$ gives

$$\widehat{\Phi}_{IJ}(\mathbb{E}_I) = \mathbb{E}_{IJ} \cap \mathrm{pr}_J^{-1}(\mathrm{im}\ \phi_{IJ}). \tag{2.3.1}$$

In particular, by the compatibility of coordinate changes $\mathfrak{s}_J \circ \phi_{IJ} = \widehat{\Phi}_{IJ} \circ \mathfrak{s}_I|_{U_{IJ}}$ we obtain

$$\mathfrak{s}_J\big(\phi_{IJ}(U_{IJ})\big) = \widehat{\Phi}_{IJ}(\mathfrak{s}_I(U_{IJ})) \subset \widehat{\Phi}_{IJ}(\mathbb{E}_I) \subset \mathbb{E}_{IJ}.$$

In other words, the first three conditions above imply the inclusion $\phi_{IJ}(U_{IJ}) \subset \mathfrak{s}_J^{-1}(\mathbb{E}_{IJ})$. Condition (iv) strengthens this by saying the image is open. The proof of the next lemma shows that this can be viewed as topological version of the index condition. $\diamondsuit$

LEMMA 2.3.8. (a) *Any weak smooth Kuranishi atlas is filtered with filtration* $\mathbb{E}_{IJ} := U_J \times \widehat{\phi}_{IJ}(E_I)$, *using the conventions* $E_\emptyset := \{0\}$ *and* $\widehat{\phi}_{JJ} := \mathrm{id}_{E_J}$.

(b) *For any filtration* $(\mathbb{E}_{IJ})_{I \subset J}$ *on a weak topological Kuranishi atlas* $\mathcal{K}$ *we have for any* $H, I, J \in \mathcal{I}_{\mathcal{K}}$

$$H, I \subset J \quad \Longrightarrow \quad \mathfrak{s}_J^{-1}(\mathbb{E}_{IJ}) \cap \mathfrak{s}_J^{-1}(\mathbb{E}_{HJ}) = \mathfrak{s}_J^{-1}(\mathbb{E}_{(I\cap H)J}), \tag{2.3.2}$$

in particular

$$H \cap I = \emptyset \quad \Longrightarrow \quad \mathfrak{s}_J^{-1}(\mathbb{E}_{IJ}) \cap \mathfrak{s}_J^{-1}(\mathbb{E}_{HJ}) = \mathfrak{s}_J^{-1}(0_J). \tag{2.3.3}$$

PROOF. To check (a), first note that $U_J \times \widehat{\phi}_{IJ}(E_I) \subset U_J \times E_J$ is closed since $U_J \subset U_J$ and $\widehat{\phi}_{IJ}(E_I) \subset E_J$ are closed. Property (i) in Definition 2.3.6 holds by definition, while property (iii) holds because $\widehat{\phi}_{IJ}(E_I) \cap \widehat{\phi}_{HJ}(E_H) = \widehat{\phi}_{(I\cap H)J}(E_{I\cap H})$. Since $\widehat{\Phi}_{JK} = \phi_{JK} \times \widehat{\phi}_{JK}$, we can use the weak cocycle condition to derive property (ii) as follows:

$$\begin{aligned}
&\widehat{\Phi}_{JK}\big(U_J \times \widehat{\phi}_{IJ}(E_I)\big) \\
&\quad = \mathrm{im}\ \phi_{JK} \times \widehat{\phi}_{JK}\big(\widehat{\phi}_{IJ}(E_I)\big) \ = \ \mathrm{im}\ \phi_{JK} \times \widehat{\phi}_{IK}(E_I) \\
&\quad = \big(U_K \times \widehat{\phi}_{IK}(E_I)\big) \cap \big(\mathrm{im}\ \phi_{JK} \times E_K\big) \ = \ \mathbb{E}_{IK} \cap (\mathrm{pr}_K^{-1}(\mathrm{im}\ \phi_{JK})).
\end{aligned}$$

Finally, property (iv) follows from the index condition for smooth coordinate changes; see Remark 2.1.6. The displayed statements in (b) all follow from applying $\mathfrak{s}_J$ to the defining property (ii), and making use of (i) in case $H \cap I = \emptyset$. $\square$

The definition of tameness given in the smooth case in Definition 2.2.4 readily extends to topological atlases; in fact the only change is that the set $s_J^{-1}(\widehat{\phi}_{IJ}(E_I))$ in (2.2.4) is replaced by $\mathfrak{s}_J^{-1}(\mathbb{E}_{IJ})$.

EXAMPLE 2.3.9. The prototypical example of a smooth atlas that is not filtered is one that has two basic charts $\mathbf{K}_1, \mathbf{K}_2$ with overlapping (but distinct) footprints, sections $\mathfrak{s}_1, \mathfrak{s}_2$ whose zero sets $\mathfrak{s}_1^{-1}(0), \mathfrak{s}_2^{-1}(0)$ have empty interiors, and the *same* obstruction space $E_i := E \neq \{0\}$, which we understand to mean that their transition chart also has obstruction space $E_{12} = E$, so that the linear maps $\widehat{\phi}_{i(12)}$ are isomorphisms. Then the index condition for smooth atlases implies that $\phi_{i(12)}(U_{i(12)})$ is an open submanifold of $U_{12} = \mathfrak{s}_{12}^{-1}(E) = \mathfrak{s}_{12}^{-1}\big(\mathrm{im}\,(\widehat{\phi}_{i(12)})\big)$ that contains $\mathfrak{s}_{12}^{-1}(0)$.

Further filtration condition (i) implies that $\mathbb{E}_i = U_i \times E$ for $i = 1, 2$, which implies by (ii) that $\mathbb{E}_{i(12)}$ contains im $\phi_{i(12)} \times E$. But then (iii) implies that

$$\big(\text{im } \phi_{1(12)} \times E\big) \cap \big(\text{im } \phi_{2(12)} \times E\big) \subset \mathbb{E}_{1(12)} \cap \mathbb{E}_{2(12)} = \mathbb{E}_{\emptyset(12)} = \mathfrak{s}_{12}^{-1}(0),$$

which means that $\mathfrak{s}_{12}^{-1}(0)$ is the intersection of two open subsets of U_{12} and hence is open. Therefore $\phi_{1(12)}^{-1}\big(\mathfrak{s}_{12}^{-1}(0)\big) = \psi_i^{-1}(F_{12})$ is also open in U_i which contradicts the choice of section $\mathfrak{s}_i$.

We discuss in §6 how to deal with situations (for example that of products) where the filtration assumption does not hold. For example, in the situation above it is possible to redefine the transition chart so that its obstruction space is the sum of two copies of E; see method 2 in §5.2 and the proof of Proposition 6.3.1. $\Diamond$

DEFINITION 2.3.10. *A weak topological Kuranishi atlas is* **tame** *if* (2.2.3) *holds, i.e.* $U_{IJ} \cap U_{IK} = U_{I(J \cup K)}$ *for* $I \subset J \subset K$*, and it is equipped with a filtration* $(\mathbb{E}_{IJ})_{I \subset J}$ *such that for all* $I, J, K \in \mathcal{I}_{\mathcal{K}}$

$$\phi_{IJ}(U_{IK}) = U_{JK} \cap \mathfrak{s}_J^{-1}(\mathbb{E}_{IJ}) \qquad \forall I \subset J \subset K. \tag{2.3.4}$$

REMARK 2.3.11. The filtration conditions (i) – (iv) work together with the two tameness identities (2.2.3) and (2.3.4) in a rather subtle way. If conditions (i)–(iii) hold as well as the tameness identities, then we can use the identity

$$\phi_{IK}(U_{IK}) \cap \phi_{JK}(U_{JK}) = \phi_{(I\cap J)K}(U_{(I\cap J)K})$$

of (2.3.7) to simplify the equivalence relation. This is the key tool in the proof that tame atlases have well behaved realizations. On the other hand, as we see in the discussion following (2.3.11), condition (iv) for filtrations is just strong enough to allow the inductive construction of a tame shrinking of an atlas that also satisfies filtration conditions (i)–(iii). $\Diamond$

We now turn to the proof of Proposition 2.2.5 which lists the good properties of tame atlases. We first prove part (ii) for topological atlases in Proposition 2.3.13, and part (iii) in Proposition 2.3.15. The key fact is that in the tame case the equivalence relation on $\text{Obj}_{\mathbf{B}_{\mathcal{K}}}$ simplifies significantly. Let $\preceq$ denote the partial order on $\text{Obj}_{\mathbf{B}_{\mathcal{K}}}$ given by

$$(I, x) \preceq (J, y) \quad \Longleftrightarrow \quad \text{Mor}_{\mathbf{B}_{\mathcal{K}}}((I, x), (J, y)) \neq \emptyset.$$

That is, we have $(I, x) \preceq (J, y)$ iff $x \in U_{IJ}$ and $y = \phi_{IJ}(x)$. Since morphisms in $\mathbf{B}_{\mathcal{K}}$ are closed under composition, $(I, x) \sim (J, y)$ iff there are elements (I_j, x_j) such that

$$(I, x) = (I_0, x_0) \preceq (I_1, x_1) \succeq (I_2, x_2) \preceq \ldots \succeq (I_k, x_k) = (J, y). \tag{2.3.5}$$

For any $I, J \in \mathcal{I}_{\mathcal{K}}$ and subset $S_I \subset U_I$ we denote the subset of points in U_J that are equivalent to a point in S_I by

(2.3.6)

$$\varepsilon_J(S_I) := \pi_{\mathcal{K}}^{-1}(\pi_{\mathcal{K}}(S_I)) \cap U_J = \big\{y \in U_J \,\big|\, \exists x \in S_I : (I, x) \sim (J, y)\big\} \subset U_J.$$

The next result is adapted from Lemma 3.2.3 in [**MW1**].

LEMMA 2.3.12. *Let* $\mathcal{K}$ *be a tame topological Kuranishi atlas.*

(a) *For* $(I, x), (J, y) \in \text{Obj}_{\mathbf{B}_{\mathcal{K}}}$ *the following are equivalent.*
   (i) $(I, x) \sim (J, y)$*;*

(ii) $(I,x) \preceq (I\cup J,z) \succeq (J,y)$ *for some* $z\in U_{I\cup J}$ *(in particular* $I\cup J\in \mathcal{I}_{\mathcal{K}}$*);*

(iii) *either* $(I,x)\succeq (I\cap J,w)\preceq (J,y)$ *for some* $w\in U_{I\cap J}$ *(in particular* $\emptyset\neq I\cap J\in\mathcal{I}_{\mathcal{K}}$*), or* $I\cap J=\emptyset$*,* $\mathfrak{s}_I(x)=0_I(x), \mathfrak{s}_J(y)=0_J(y)$*, and* $\psi_I(x)=\psi_J(y)$*.*

(b) $\pi_{\mathcal{K}}: U_I\to|\mathcal{K}|$ *is injective for each* $I\in\mathcal{I}_{\mathcal{K}}$*, that is* $(I,x)\sim(I,y)$ *implies* $x=y$ *In particular, the elements* z *and* w *in (a) are automatically unique.*

(c) *For any* $I,J\in\mathcal{I}_{\mathcal{K}}$ *and* $S_I\subset U_I$ *we have*

$$\begin{aligned}\varepsilon_J(S_I) \;:=\; U_J\cap\pi_{\mathcal{K}}^{-1}\big(\pi_{\mathcal{K}}(S_I)\big) &= \phi_{J(I\cup J)}^{-1}\big(\phi_{I(I\cup J)}(U_{I(I\cup J)}\cap S_I)\big)\\ &= \phi_{(I\cap J)J}\big(U_{(I\cap J)I}\cap\phi_{(I\cap J)I}^{-1}(S_I)\big),\end{aligned}$$

where in case $I\cap J=\emptyset$ *the right hand side is* $\varepsilon_J(S_I)=\psi_J^{-1}\big(\psi_I(\mathfrak{s}_I^{-1}(0_I)\cap S_I)\big)$*, consistently with* (2.2.5)*.*

PROOF. The key step in the proof of (a) is to show that the taming conditions imply the equivalence of (ii) and (iii). In fact we need slightly more than this: the existence of a chain $(I,x)\preceq(K,z)\succeq(J,y)$ implies (iii), while the existence of a chain $(I,x)\succeq(H,w)\preceq(J,y)$ implies (ii).

To prove these claims, suppose first that there is a chain $(I,x)\succeq(H,w)\preceq(J,y)$. Then $w\in U_{HI}\cap U_{HJ}=U_{H(I\cup J)}$ by (2.2.3). But then

$$x=\phi_{HI}(w)\in\phi_{HI}(U_{H(I\cup J)})=U_{I(I\cup J)}\cap\mathfrak{s}_I^{-1}(\mathbb{E}_{IH})$$

by (2.3.4), so that $\phi_{I(I\cup J)}(x)$ is defined. Moreover,

$$z:=\phi_{I(I\cup J)}(x)=\phi_{I(I\cup J)}\circ\phi_{HI}(w)=\phi_{H(I\cup J)}(w)\in U_{I\cup J},$$

by the cocycle condition. A similar argument shows that $z=\phi_{H(I\cup J)}(w)=\phi_{J(I\cup J)}(y)$. Hence (ii) holds. Secondly, if there is a chain $(I,x)\preceq(K,z)\succeq(J,y)$ then

$$z\in\phi_{IK}(U_{IK})\cap\phi_{JK}(U_{JK})=\mathfrak{s}_K^{-1}(\mathbb{E}_{IK})\cap\mathfrak{s}_K^{-1}(\mathbb{E}_{JK})=\mathfrak{s}_K^{-1}(\mathbb{E}_{(I\cap J)K})$$

by tameness (2.3.4) and the filtration condition (2.3.2). Since $\mathbb{E}_{\emptyset K}=\mathfrak{s}_K^{-1}(0_K)$ by definition, if $\mathfrak{s}_K(z)\neq 0_K(z)$ we must have $I\cap J\neq\emptyset$. Further, whenever $I\cap J\neq\emptyset$ we have

$$z\in\phi_{IK}(U_{IK})\cap\phi_{JK}(U_{JK})=\mathfrak{s}_K^{-1}(\mathrm{im}\,(\widehat{\phi}_{(I\cap J)K}))=\phi_{(I\cap J)K}(U_{(I\cap J)K}), \tag{2.3.7}$$

which implies the existence of suitable $w\in U_{(I\cap J)K}$. On the other hand, if $I\cap J=\emptyset$ then $\mathfrak{s}_K(z)=0_K(z)$ and the equivalence of (ii) and (iii) follows because as in (2.2.1)

$$(I,x)\sim(J,y),\mathfrak{s}_I(x)=0_I(x)\implies\mathfrak{s}_J(y)=0_J(y),\psi_I(x)=\psi_J(y).$$

Using this together with the fact that by composing morphisms one can get rid of consecutive occurrences of $\preceq$ and of $\succeq$, one shows that if $\mathfrak{s}_I(x)\neq 0_I(x)$ any chain as in (2.3.5) can be simplified either to $(I,x)\preceq(K,z)\succeq(J,y)$ or to $(I,x)\succeq(H,w)\preceq(J,y)$. Applying the results once more, one reduces to the case when $K=I\cup J$ and $H=I\cap J$. When $\mathfrak{s}_I(x)=0_I(x)$, a similar argument shows that (i) is equivalent to (ii), and hence also to (iii). This proves (a). Statement (b) then holds by applying (i) with $I=J$. Finally, the formulas for $\varepsilon_J(S_I)$ follow from the equivalent definitions of $\sim$ in (a). □

The above lemma is the basis for the proof of the following version of Proposition 2.2.5 (ii), taken from [**MW1**, Proposition 3.1.13].

PROPOSITION 2.3.13. *Suppose that the topological Kuranishi atlas $\mathcal{K}$ is tame. Then $|\mathcal{K}|$ and $|\mathbf{E}_{\mathcal{K}}|$ are Hausdorff, and for each $I \in \mathcal{I}_{\mathcal{K}}$ the quotient maps $\pi_{\mathcal{K}}|_{U_I} : U_I \to |\mathcal{K}|$ and $\pi_{\mathcal{K}}|_{U_I \times E_I} : U_I \times E_I \to |\mathbf{E}_{\mathcal{K}}|$ are homeomorphisms onto their image.*

PROOF. We prove the claims about $|\mathcal{K}|$. The arguments for $|\mathbf{E}_{\mathcal{K}}|$ are similar, based on the analog of Lemma 2.3.12 for the category $\mathbf{E}_{\mathcal{K}}$.

To see that $|\mathcal{K}|$ is Hausdorff, note first that the equivalence relation on $O := \mathrm{Obj}_{\mathbf{B}_{\mathcal{K}}} = \bigsqcup_{I \in \mathcal{I}_{\mathcal{K}}} U_I$ is closed, i.e. the subset

$$R := \big\{ \big((I,x),(J,y)\big) \mid (I,x) \sim (J,y) \big\} \subset O \times O$$

is closed. Since $\mathcal{I}_{\mathcal{K}}$ is finite and $O \times O$ is the disjoint union of the metrizable sets $U_I \times U_J$, this will follow if we show that for all pairs I, J and all convergence sequences $x^\nu \to x^\infty$ in U_I, $y^\nu \to y^\infty$ in U_J with $(I,x^\nu) \sim (J,y^\nu)$ for all ν, we have $(I,x^\infty) \sim (J,y^\infty)$. For that purpose denote $H := I \cap J$. If $H = \emptyset$, then all the points x^ν, y^ν lie in the appropriate zero set and the result is immediate. Hence suppose that $H \neq \emptyset$. Then by Lemma 2.3.12(a) there is a sequence $w^\nu \in U_H$ such that $x^\nu = \phi_{HI}(w^\nu)$ and $y^\nu = \phi_{HJ}(w^\nu)$. Now it follows from the tameness condition (2.2.7) that x^∞ lies in the relatively closed subset $\phi_{HI}(U_{HI}) = \mathfrak{s}_I^{-1}(\mathbb{E}_{HI}) \subset U_I$, and since ϕ_{HI} is a homeomorphism to its image we deduce convergence $w^\nu \to w^\infty \in U_{HI}$ to a preimage of $x^\infty = \phi_{HI}(w^\infty)$. Then by continuity of the transition map we obtain $\phi_{HJ}(w^\infty) = y^\infty$, so that $(I,x^\infty) \sim (J,y^\infty)$ as claimed. Thus R is closed. Hence $|\mathcal{K}|$ is Hausdorff by [**Bbk**] (see Exercise 19, §10, Chapter 1), which states that if a Hausdorff space O can be written as the union $\bigcup_{k\ge 1} O_k$ of precompact open subsets O_k with $\overline{O_k} \subset O_{k+1}$ for all k, then its quotient by a closed relation is Hausdorff.

To show that $\pi_{\mathcal{K}}|_{U_I}$ is a homeomorphism onto its image, first recall that it is injective by Lemma 2.3.12 (c). It is moreover continuous since $|\mathcal{K}|$ is equipped with the quotient topology. Hence it remains to show that $\pi_{\mathcal{K}}|_{U_I}$ is an open map to its image, i.e. for a given open subset $S_I \subset U_I$ we must find an open subset $\mathcal{W} \subset |\mathcal{K}|$ such that $\mathcal{W} \cap \pi_{\mathcal{K}}(U_I) = \pi_{\mathcal{K}}(S_I)$. Equivalently, we set $\mathcal{W} := |\mathcal{K}| \smallsetminus \mathcal{Q}$ and construct the complement

$$\mathcal{Q} := \big(\iota_{\mathcal{K}}(X) \cup \textstyle\bigcup_{H \subset I} \pi_{\mathcal{K}}(U_H)\big) \smallsetminus \pi_{\mathcal{K}}(S_I) \;\subset\; |\mathcal{K}|.$$

With that the intersection identity follows from $\mathcal{Q} \cap \pi_{\mathcal{K}}(U_I) = \pi_{\mathcal{K}}(U_I) \smallsetminus \pi_{\mathcal{K}}(S_I)$, so it remains to show that $U_J \cap \pi_{\mathcal{K}}^{-1}(\mathcal{Q})$ is closed for each J. In case $J \subset I$ we have $U_J \cap \pi_{\mathcal{K}}^{-1}(\mathcal{Q}) = U_J \smallsetminus \varepsilon_J(S_I)$, which is closed iff $\varepsilon_J(S_I)$ is open. Indeed, Lemma 2.3.12 (d) gives $\varepsilon_J(S_I) = \phi_{JI}^{-1}(S_I) \subset U_J$, and this is open since $S_I \subset U_I$ and hence $S_I \cap U_{IJ} \subset U_{IJ}$ is open and ϕ_{JI} is continuous.

In case $J \not\subset I$ to show that $U_J \cap \pi_{\mathcal{K}}^{-1}(\mathcal{Q})$ is closed, we express it as the union of $Q_J^0 := U_J \cap \pi_{\mathcal{K}}^{-1}(\iota_{\mathcal{K}}(X) \smallsetminus \pi_{\mathcal{K}}(S_I))$ with the union over $H \subset I$ of

$$\begin{aligned} Q_{JH} \;:=\; U_J \cap \pi_{\mathcal{K}}^{-1}\big(\pi_{\mathcal{K}}(U_H) \smallsetminus \pi_{\mathcal{K}}(S_I)\big) &= U_J \cap \pi_{\mathcal{K}}^{-1}\big(\pi_{\mathcal{K}}(U_H \smallsetminus \varepsilon_H(S_I))\big) \\ &= \varepsilon_J\big(U_H \smallsetminus \phi_{HI}^{-1}(S_I)\big) \;=\; \varepsilon_J(C_H). \end{aligned}$$

Note here that $C_H := U_H \smallsetminus \phi_{HI}^{-1}(S_I) \subset U_H$ is closed since as above $\phi_{HI}^{-1}(S_I) \subset U_H$ is open. We moreover claim that this union can be simplified to

$$U_J \cap \pi_{\mathcal{K}}^{-1}(\mathcal{Q}) \;=\; Q_J^0 \,\cup \bigcup_{H \subset I \cap J} \varepsilon_J(C_H). \tag{2.3.8}$$

Indeed, in case $H \cap J = \emptyset$ we have $\varepsilon_J(C_H) \subset Q^0_J$ since Lemma 2.3.12 (d) gives

$$\begin{aligned}\varepsilon_J\big(U_H \smallsetminus \phi_{HI}^{-1}(S_I)\big) &= \psi_J^{-1}\big(\psi_H\big(\mathfrak{s}_H^{-1}(0_H) \smallsetminus \phi_{HI}^{-1}(S_I)\big)\big) \\ &= \psi_J^{-1}\big(F_H \smallsetminus \big(F_I \cap \pi_{\mathcal{K}}(S_I)\big)\big) \subset U_J \cap \pi_{\mathcal{K}}^{-1}(\iota_{\mathcal{K}}(X) \smallsetminus \pi_{\mathcal{K}}(S_I)) = Q^0_J.\end{aligned}$$

For $H \not\subset J$ with $H \cap J \neq \emptyset$ we have $\varepsilon_J(C_H) \subset \varepsilon_J(C_{H\cap J})$ since Lemma 2.3.12 (d) gives

$$\begin{aligned}&\varepsilon_J\big(U_H \smallsetminus \phi_{HI}^{-1}(S_I)\big) \\ &\qquad = \phi_{(H\cap J)J}\big(U_{(H\cap J)J} \cap \phi_{(H\cap J)H}^{-1}(U_H \smallsetminus \phi_{HI}^{-1}(S_I))\big) \subset U_J \cap \pi_{\mathcal{K}}^{-1}(C_{H\cap J}),\end{aligned}$$

where $\phi_{(H\cap J)H}^{-1}\big(U_H \smallsetminus \phi_{HI}^{-1}(S_I)\big) \subset U_{H\cap J} \smallsetminus \phi_{(H\cap J)I}^{-1}(S_I) = C_{H\cap J}$ by the cocycle condition. This confirms (2.3.8).

It remains to show that Q^0_J and $\varepsilon_J(C_H)$ for $H \subset I \cap J$ are closed. For the latter, Lemma 2.3.12 (d) gives $\varepsilon_J(U_H \smallsetminus \phi_{HI}^{-1}(S_I)) = \phi_{HJ}(U_{HJ} \cap C_H)$, which is closed in U_J since closedness of $C_H \subset U_H$ implies relative closedness of $U_{HJ} \cap C_H \subset U_{HJ}$, and $\phi_{HJ} : U_{HJ} \to U_J$ is a homeomorphism onto a closed subset of U_J by (2.2.7). Finally,

$$Q^0_J = U_J \cap \pi_{\mathcal{K}}^{-1}(\iota_{\mathcal{K}}(X) \smallsetminus \pi_{\mathcal{K}}(S_I)) \;=\; s_J^{-1}(0_J) \smallsetminus \psi_J^{-1}\big(F_J \cap \psi_I(\mathfrak{s}_I^{-1}(0_I) \cap S_I)\big)$$

is closed in $s_J^{-1}(0_J)$ and hence in U_J, since both F_J and $\psi_I(\mathfrak{s}_I^{-1}(0_I) \cap S_I)$ are open in X, and $\psi_J : s_J^{-1}(0_J) \to X$ is continuous. Thus $U_J \cap \pi_{\mathcal{K}}^{-1}(\mathcal{Q})$ is closed, as claimed, which finishes the proof of the homeomorphism property of $\pi_{\mathcal{K}}|_{U_I}$. □

Corollary 2.3.14. *If $\mathcal{K}$ is tame, then for every $I \in \mathcal{I}_{\mathcal{K}}$ we have*

$$\overline{\pi_{\mathcal{K}}(U_I)} \;\subset\; \iota(X) \cup {\textstyle\bigcup_{H\subset I}} \pi_{\mathcal{K}}(U_H).$$

Proof. It suffices to check that $\mathcal{Q}' := \iota(X) \cup \bigcup_{H\subset I} \pi_{\mathcal{K}}(U_H)$ is closed in $|\mathcal{K}|$. But $\mathcal{Q}'$ is the set $\mathcal{Q}$ of the above proof in the case when $S = \emptyset$. □

Proposition 2.3.15. *Proposition 2.2.5 holds.*

Proof. We proved part (i) just before stating Proposition 2.2.5, and part (ii) follows from Proposition 2.3.13 because by Lemma 2.3.8 every tame smooth atlas may also be considered as a tame topological atlas. Thus it remains to prove (iii), which asserts that in the smooth case the fibers of $|\operatorname{pr}_{\mathcal{K}}| : |\mathbf{E}_{\mathcal{K}}| \to |\mathcal{K}|$ have a unique linear structure such that for every $I \in \mathcal{I}_{\mathcal{K}}$ the embedding $\pi_{\mathcal{K}} : U_I \times E_I \to |\mathbf{E}_{\mathcal{K}}|$ is linear on the fibers.

To this end, for fixed $p \in |\mathcal{K}|$ denote the union of index sets for which $p \in \pi_{\mathcal{K}}(U_I)$ by

$$I_p := \bigcup_{I\in\mathcal{I}_{\mathcal{K}},\, p\in\pi_{\mathcal{K}}(U_I)} I \qquad \subset \{1,\dots,N\}.$$

To see that $I_p \in \mathcal{I}_{\mathcal{K}}$ we repeatedly use the observation that Lemma 2.3.12 (a) implies

$$p \in \pi_{\mathcal{K}}(U_I) \cap \pi_{\mathcal{K}}(U_J) \quad\Rightarrow\quad (I, \pi_{\mathcal{K}}^{-1}(p) \cap U_I) \sim (J, \pi_{\mathcal{K}}^{-1}(p) \cap U_J) \quad\Rightarrow\quad I \cup J \in \mathcal{I}_{\mathcal{K}}.$$

Moreover, $x_p := \pi_{\mathcal{K}}^{-1}(p) \cap U_{I_p}$ is unique by Lemma 2.3.12 (c). Next, any element in the fiber $[I, x, e] \in |\operatorname{pr}_{\mathcal{K}}|^{-1}(p)$ is represented by some vector over $(I, x) \in \pi_{\mathcal{K}}^{-1}(p)$, so we have $I \subset I_p$ and $\phi_{II_p}(x) = x_p$, and hence $(I, x, e) \sim (I_p, x_p, \widehat{\phi}_{II_p}(e))$. Thus $\pi_{\mathcal{K}} : \{x_p\} \times E_{I_p} \to |\operatorname{pr}_{\mathcal{K}}|^{-1}(p)$ is surjective, and by Lemma 2.3.12 (c) also injective. Thus the requirement of linearity for this bijection induces a unique linear structure on the fiber $|\mathrm{pr}_{\mathcal{K}}|^{-1}(p)$. To see that this is compatible with the injections $\pi_{\mathcal{K}}: \{x\} \times E_I \to |\mathrm{pr}_{\mathcal{K}}|^{-1}(p)$ for $(I,x) \sim (I_p, x_p)$ note again that $I \subset I_p$ since I_p was defined to be maximal, and hence by Lemma 2.3.12 (b) (ii) the embedding factors as $\pi_{\mathcal{K}}|_{\{x\}\times E_I} = \pi_{\mathcal{K}}|_{\{x_p\}\times E_{I_p}} \circ \widehat{\phi}_{I I_p}$, where $\widehat{\phi}_{I I_p}$ is linear by definition of coordinate changes. Thus $\pi_{\mathcal{K}}|_{\{x\}\times E_I}$ is linear as well. □

REMARK 2.3.16. (i) The above construction gives a rather nice picture of the virtual neighbourhood $|\mathcal{K}|$ for a tame topological atlas. By Corollary 2.3.14, it is a union of sets $\pi_{\mathcal{K}}(U_I)$, each of which has frontier $\overline{\pi_{\mathcal{K}}(U_I)}\smallsetminus\pi_{\mathcal{K}}(U_I)$ contained in the union of the zero set $\iota(X)$ with the sets $\bigcup_{H\subset I}\pi_{\mathcal{K}}(U_H)$. A pairwise intersection $\pi_{\mathcal{K}}(U_I)\cap\pi_{\mathcal{K}}(U_J)$ is nonempty only if the corresponding footprint intersection $F_I \cap F_J = F_{I\cup J}$ is nonempty, in which case Lemma 2.3.12 (a) implies that $\pi_{\mathcal{K}}(U_I)\cap \pi_{\mathcal{K}}(U_J) \subset \pi_{\mathcal{K}}(U_{I\cup J})$. If also $I\cap J \neq \emptyset$, then $\pi_{\mathcal{K}}(U_I)\cap\pi_{\mathcal{K}}(U_J)$ may be identified with the subset $\pi_{\mathcal{K}}\big(\mathfrak{s}_{I\cup J}^{-1}(\mathbb{E}_{(I\cap J)(I\cup J)})\big)$ of $\pi_{\mathcal{K}}(U_{I\cup J})$. However, if $I\cap J=\emptyset$ then these two sets intersect only along the zero set; in fact the intersection is $\iota_{\mathcal{K}}(F_{I\cup J})$, where $\iota_{\mathcal{K}}$ is as in (2.2.2). In the smooth case, each set $\pi_{\mathcal{K}}(U_I)$ is a homeomorphic image of a manifold, and it is easy to see from the index condition that if $I\cap J\neq\emptyset$ the intersection of $\pi_{\mathcal{K}}(U_I)$ with $\pi_{\mathcal{K}}(U_J)$ is transverse when considered inside the "submanifold" $\pi_{\mathcal{K}}(U_{I\cup J})$. For example, the images of the domains of two basic charts $\pi_{\mathcal{K}}(U_1)$ and $\pi_{\mathcal{K}}(U_2)$ will in general intersect nontransversally in their common footprint $\iota_{\mathcal{K}}(F_{12})$, while the two transition domains $\pi_{\mathcal{K}}(U_{12})$ and $\pi_{\mathcal{K}}(U_{23})$ intersect transversally in the submanifold $\pi_{\mathcal{K}}(U_2)\cap\pi_{\mathcal{K}}(U_{123})$ of $\pi_{\mathcal{K}}(U_{123})$.

(ii) The effect of the taming condition is to reduce the equivalence relation to a two step process: $(I,x)\sim(J,y)$ iff we can write $(I,x)\preceq(I\cup J,z)\succeq(J,y)$, or equivalently (if $s_I(x)\neq 0$) $(I,x)\succeq(I\cap J,w)\preceq(J,y)$. The reduction process described in §2.4 below will simplify the equivalence relation even further to a single step. In fact, this process discards all the elements in $U_I\smallsetminus V_I$, for suitable choice of open sets $V_I\subset U_I$, so that when $x\in V_I, y\in V_J$ we have $(I,x)\sim(J,y)$ only if $(I,x)\preceq(J,y)$ or $(I,x)\succeq(J,y)$.

(iii) See [**MW1**, Example 2.4.3] for a (non tame) atlas for which the map $\pi_{\mathcal{K}}$ is not injective on U_I.[15] It is also easy to construct non tame examples where the argument in Proposition 2.3.15 fails, so that some fibers of $|\mathrm{pr}|: |\mathbf{E}_{\mathcal{K}}|\to|\mathcal{K}|$ are not vector spaces. ◊

To complete the proof of the first part of Theorem 2.2.3 it remains to explain how to construct a metrizable tame shrinking of a weak atlas. To accomplish this, we again work in the context of topological atlases in order to be able to apply the argument when the isotropy is nontrivial. Note that the definition of a shrinking given in Definition 2.2.9 makes sense in this context without change. Moreover, since smooth atlases are automatically filtered by Lemma 2.3.8, we will suppose that the initial weak atlas $\mathcal{K}$ is filtered.

We define the filtration $(\mathbb{E}'_{IJ})_{I\subset J}$ on a shrinking $\mathcal{K}'$ by restriction $\mathbb{E}'_{IJ} := \mathbb{E}_{IJ}\cap\mathbb{E}'_J$ in the obvious way. Note that any shrinking of a weak Kuranishi atlas preserves the weak cocycle condition (since the latter only requires equality on overlaps). Moreover, a shrinking is determined by the choice of the domains $U'_I \sqsubset U_I$ of the

[15]In this example the atlas is not additive. However it can easily be made additive by replacing the charts $(U_i)_{1\le i\le 3}$ described there by charts $(U_{i4})_{1\le i\le 3}$ for a suitable chart $\mathbf{K}_4$ with $E_4=\mathbb{R}$ and where $E_i=\{0\}$.

transition charts (since condition (iii) then specifies the domains of the coordinate changes), and so can be considered as the restriction of $\mathcal{K}$ to the subset $\bigsqcup_{I\in\mathcal{I}_\mathcal{K}} U'_I \subset \mathrm{Obj}_{\mathbf{B}_\mathcal{K}}$. However, for a shrinking to satisfy a stronger form of the cocycle condition (such as tameness) the domains $U'_{IJ} := \phi_{IJ}^{-1}(U'_J)\cap U'_I$ of the coordinate changes must satisfy appropriate compatibility conditions, so that the domains U'_I can no longer be chosen independently of each other. Since the relevant conditions are expressed in terms of the U'_{IJ}, we next show that the construction of a tame shrinking can be achieved by iterative choice of these sets U'_{IJ}.

The following result is proved in [**MW1**, §3.3].

Proposition 2.3.17. *Every filtered weak topological Kuranishi atlas $\mathcal{K}$ has a tame shrinking $\mathcal{K}'$.*

Sketch of proof. Since X is compact and metrizable and the footprint open cover (F_i) is finite, it has a shrinking (F'_i) in the sense of Definition 2.2.8. In particular we can ensure that $F'_I \neq \emptyset$ whenever $F_I \neq \emptyset$ by choosing $\delta > 0$ so that every nonempty F_I contains some ball $B_\delta(x_I)$ and then choosing the F'_i to contain $B_{\delta/2}(x_I)$ for each $I \ni i$ (i.e. $F_I \subset F_i$). Then we obtain $F'_I \neq \emptyset$ for all $I \in \mathcal{I}_\mathcal{K}$ since $B_{\delta/2}(x_I) \subset \bigcap_{i\in I} F'_i = F'_I$.

In another preliminary step, we find precompact open subsets $U_I^{(0)} \sqsubset U_I$ and open sets $U_{IJ}^{(0)} \subset U_{IJ} \cap U_I^{(0)}$ for all $I, J \in \mathcal{I}_\mathcal{K}$ such that

$$U_I^{(0)} \cap \mathfrak{s}_I^{-1}(0) = \psi_I^{-1}(F'_I), \qquad U_{IJ}^{(0)} \cap \mathfrak{s}_I^{-1}(0) = \psi_I^{-1}(F'_I \cap F'_J). \tag{2.3.9}$$

Here we choose any suitable $U_I^{(0)}$ (which is possible by [**MW1**, Lemma 2.1.4]), and then define the $U_{IJ}^{(0)}$ by restriction:

$$U_{IJ}^{(0)} := U_I^{(0)} \cap \phi_{IJ}^{-1}(U_J^{(0)}) \subset U_{IJ}. \tag{2.3.10}$$

We then construct the required shrinking $\mathcal{K}'$ by choosing possibly smaller domains $U'_I \subset U_I^{(0)}$ and $U'_{IJ} \subset U_{IJ}^{(0)}$ with the same footprints F'_I. We also arrange $U'_{IJ} = U'_I \cap \phi_{IJ}^{-1}(U'_J)$, so that $\mathcal{K}'$ is a shrinking of the original $\mathcal{K}$. Therefore we just need to make sure that $\mathcal{K}'$ satisfies the tameness conditions (2.2.3) and (2.2.4).

We construct the domains U'_I, U'_{IJ} by a finite iteration, starting with $U_I^{(0)}, U_{IJ}^{(0)}$. Here we streamline the notation by setting $U_I^{(k)} := U_{II}^{(k)}$ and extend the notation to all pairs of nonempty subsets $I \subset J \subset \{1,\dots,N\}$ by setting $U_{IJ}^{(k)} = \emptyset$ if $J \notin \mathcal{I}_\mathcal{K}$. (Note that $J \in \mathcal{I}_\mathcal{K}$ and $I \subset J$ implies $I \in \mathcal{I}_\mathcal{K}$.) Then in the k-th step we construct open subsets $U_{IJ}^{(k)} \subset U_{IJ}^{(k-1)}$ for all $I \subset J \subset \{1,\dots,N\}$ so that the following holds.

(1) The zero set conditions $U_{IJ}^{(k)} \cap \mathfrak{s}_I^{-1}(0) = \psi_I^{-1}(F'_J)$ hold for all $I \subset J$.
(2) The first tameness condition (2.2.3) holds for all $I \subset J, K$ with $|I| \leq k$, that is
$$U_{IJ}^{(k)} \cap U_{IK}^{(k)} = U_{I(J\cup K)}^{(k)}.$$
In particular, we have $U_{IK}^{(k)} \subset U_{IJ}^{(k)}$ for $I \subset J \subset K$ with $|I| \leq k$.
(3) The second tameness condition (2.2.4) holds for all $I \subset J \subset K$ with $|I| \leq k$, that is
$$\phi_{IJ}(U_{IK}^{(k)}) = U_{JK}^{(k)} \cap \mathfrak{s}_J^{-1}(\mathbb{E}_{IJ}).$$
In particular we have $\phi_{IJ}(U_{IJ}^{(k)}) = U_J^{(k)} \cap \mathfrak{s}_J^{-1}(\mathbb{E}_{IJ})$ for all $I \subset J$ with $|I| \leq k$.

In other words, we need the tameness conditions to hold up to level k.

The above choice of the domains $U_{IJ}^{(0)}$ completes the 0-th step since conditions (ii) (iii) are vacuous. Now suppose that the $(k-1)$-th step is complete for some $k \geq 1$. We then define $U_{IJ}^{(k)} := U_{IJ}^{(k-1)}$ for all $I \subset J$ with $|I| \leq k-1$. For $|I| = k$ we also set $U_{II}^{(k)} := U_{II}^{(k-1)}$. This ensures that (i) and (ii) continue to hold for $|I| < k$. In order to preserve (iii) for triples $H \subset I \subset J$ with $|H| < k$ we then require that the intersection $U_{IJ}^{(k)} \cap \mathfrak{s}_I^{-1}(\mathbb{E}_{HI}) = U_{IJ}^{(k-1)} \cap \mathfrak{s}_I^{-1}(\mathbb{E}_{HI})$ is fixed. In case $H = \emptyset$, this is condition (i), and since $U_{IJ}^{(k)} \subset U_{IJ}^{(k-1)}$ it can generally be phrased as inclusion (i′) below. With that it remains to construct the open sets $U_{IJ}^{(k)} \subset U_{IJ}^{(k-1)}$ as follows.

(i′) For all $H \subsetneq I \subset J$ with $|H| < k$ and $|I| \geq k$ we have $U_{IJ}^{(k-1)} \cap \mathfrak{s}_I^{-1}(\mathbb{E}_{HI}) \subset U_{IJ}^{(k)}$. Here we include $H = \emptyset$, in which case the condition says that $U_{IJ}^{(k-1)} \cap \mathfrak{s}_I^{-1}(0) \subset U_{IJ}^{(k)}$ (which implies $U_{IJ}^{(k)} \cap \mathfrak{s}_I^{-1}(0) = \psi_I^{-1}(F_J')$, as explained above).

(ii′) For all $I \subset J, K$ with $|I| = k$ we have $U_{IJ}^{(k)} \cap U_{IK}^{(k)} = U_{I(J\cup K)}^{(k)}$.

(iii′) For all $I \subsetneq J \subset K$ with $|I| = k$ we have $\phi_{IJ}(U_{IK}^{(k)}) = U_{JK}^{(k)} \cap \mathfrak{s}_J^{-1}(\mathbb{E}_{IJ})$.

The construction is then completed in two steps.

Step A constructs $U_{IK}^{(k)}$ for $|I| = k$ and $I \subsetneq K$ satisfying (i′),(ii′) and

(iii″) $U_{IK}^{(k)} \subset \phi_{IJ}^{-1}(U_{JK}^{(k-1)})$ for all $I \subsetneq J \subset K$.

Step B constructs $U_{JK}^{(k)}$ for $|J| > k$ and $J \subset K$ satisfying (i′) and (iii′).

Step A assumes the existence of suitable sets $U_{IJ}^{(k-1)}$ for all $I \subset J$ and uses the following nontrivial result to show that the required sets $U_{IJ}^{(k)}$ with $|I| = k$ exist.

LEMMA 2.3.18 (Lemma 3.3.6 in [**MW1**]). *Let U be a locally compact metric space, $U' \sqsubset U$ a precompact open set, and $Z \subset U'$ a relatively closed subset. Suppose we are given a finite collection of relatively open subsets $Z_i \subset Z$ for $i = 1, \dots, N$ and open subsets $W_K \subset U'$ with*

$$W_K \cap Z = Z_K := \bigcap_{i \in K} Z_i$$

for all index sets $K \subset \{1, \dots, N\}$. Then there exist open subsets $U_K \subset W_K$ with $U_K \cap Z = Z_K$ and $U_J \cap U_K = U_{J\cup K}$ for all $J, K \subset \{1, \dots, N\}$.

Step B then completes the inductive step for $|I| = k$, modifying the sets $U_{JK}^{(k-1)}$ by removing the extra parts that contradict (iii′). In other words we define

$$U_{JK}^{(k)} := U_{JK}^{(k-1)} \setminus \bigcup_{I \subset J, |I| = k} \big(\mathfrak{s}_J^{-1}(\mathbb{E}_{IJ}) \setminus \phi_{IJ}(U_{IJ}^{(k)}) \big). \tag{2.3.11}$$

Note that $U_{JK}^{(k)}$ is open since the sets $\mathfrak{s}_J^{-1}(\mathbb{E}_{IJ}) \setminus \phi_{IJ}(U_{IJ}^{(k)})$ that we remove are closed, which holds because ϕ_{IJ} is a homeomorphism to its image which is open in $\mathfrak{s}_J^{-1}(\mathbb{E}_{IJ})$ by condition (iv) of Definition 2.3.6.

We illustrate how Step A works by explaining the construction of the level 1 sets $U_{IJ}^{(1)}$ with $|I| = 1$. Fix $I = \{i_0\}$, and choose an arbitrary order for the set

$$\{ i \mid i \neq i_0, \{i_0, i\} \in \mathcal{I} \} \cong \{1, \dots, N'\}.$$

Then the index sets $K \in \mathcal{I}$ containing $\{i_0\}$ as a proper subset are in one-to-one correspondence with nonempty index sets $K' \subset \{1, \dots, N'\}$ via $K = \{i_0\} \cup K'$. Apply Lemma 2.3.18 to the metric space $U := U_{i_0}$, its precompact open subset

$U' := U^{(0)}_{i_0}$, the relatively closed subset $Z := U^{(0)}_{i_0} \cap \mathfrak{s}^{-1}_{i_0}(0_{i_0}) = \psi^{-1}_{i_0}(F'_{i_0})$ of U', the relatively open subsets $Z_i := \psi^{-1}_{i_0}(F'_{i_0} \cap F'_i) \subset Z$ for all $i \in \{1, \dots, N'\}$, and the open subsets

$$W_{K'} := \bigcap_{\{i_0\} \subset J \subset K} \bigl(\phi^{-1}_{i_0 J}(U^{(0)}_{JK}) \cap U^{(0)}_{i_0 J}\bigr),$$

where $K = \{i_0\} \cup K'$ as above, and we write $U_{i_0 J} := U_{\{i_0\} J}$. With these choices, the assumptions of Lemma 2.3.18 are satisfied when K, N is replaced by K', N'. In particular, (2.3.9) implies that

$$\begin{aligned} W_{K'} \cap Z &= U^{(0)}_{i_0 K} \cap \mathfrak{s}^{-1}_{i_0}(0_{i_0}) = \psi^{-1}_{i_0}(F'_{i_0} \cap F'_K) \\ &= \psi^{-1}_{i_0}(F'_{i_0} \cap \bigcap_{i \in K'} F'_i) = \bigcap_{i \in K'} Z_i =: Z_{K'}. \end{aligned}$$

Thus the lemma provides open subsets $U^{(1)}_{i_0 K} := U_{K'} \subset W_{K'} = U^{(0)}_{i_0 K}$ which satisfy the conditions (i$'$), (ii$'$) and (iii$''$) as follows:

- (i$'$) $U^{(0)}_{i_0 J} \cap \mathfrak{s}^{-1}_{i_0}(0_{i_0}) \subset U^{(1)}_{i_0 J}$ for all $J \supset \{i_0\}$ translates to $U^{(0)}_{i_0 J} \cap Z \subset U_{J'}$, which holds because $U^{(0)}_{i_0 J} = W_{J'}$ and $W_{J'} \cap Z = Z_{J'} = U_{J'} \cap Z \subset U_{J'}$;
- (ii$'$) follows from the second property in Lemma 2.3.18;
- (iii$''$) $U^{(1)}_{i_0 K} \subset \phi^{-1}_{i_0 J}(U^{(0)}_{JK})$ for $\{i_0\} \subsetneq J \subset K$ translates to $U_{K'} \subset \phi^{-1}_{i_0 J}(U^{(0)}_{JK})$, which holds because (2.3.10) and the definition of $W_{K'}$ as an intersection imply

$$\phi^{-1}_{i_0 J}(U^{(0)}_{JK}) = \phi^{-1}_{i_0 J}\bigl(U^{(0)}_J \cap \phi^{-1}_{JK}(U^{(0)}_K)\bigr) = \phi^{-1}_{i_0 J}(U^{(0)}_J) \cap \phi^{-1}_{i_0 K}(U^{(0)}_K) \supset W_{K'},$$

while $U_{K'} \subset W_{K'}$ by construction.

If $k = 2$ the argument is similar. Given I wth $|I| = 2$ we apply Lemma 2.3.18 with $U = U_I$ and $U' = U^{(1)}_I$, but now defining Z, Z_i in terms of the level 1 sets $U^{(1)}_{HK}$ where $|H| = 1$. Note that the images of these sets in U_I lie in $\mathfrak{s}^{-1}_I(\mathbb{E}_{HI})$ for some $H \subsetneq I$. Thus we define

$$Z := \bigcup_{H \subsetneq I} \bigl(U^{(1)}_{II} \cap \mathfrak{s}^{-1}_I(\mathbb{E}_{HI})\bigr) \;=\; \bigcup_{H \subsetneq I} \phi_{HI}\bigl(U^{(1)}_{HI}\bigr) \;\subset\; U'$$

and then put $Z_i := U^{(1)}_{I(I \cup \{i\})} \cap Z$ for $i \in \{1, \dots, N\} \smallsetminus I$. Finally, in order to achieve property (iii$'$) we again define $W_{K'}$ to be an intersection, namely, if $K := I \cup K'$ we set

$$W_{K'} := \bigcap_{I \subset J \subset K} \bigl(\phi^{-1}_{IJ}(U^{(1)}_{JK}) \cap U^{(1)}_{IJ}\bigr).$$

For further details, see [**MW1**, §3.3]. □

Remark 2.3.19. As we explained in Remark 2.3.11 (i), the existence of the filtration on the initial weak atlas $\mathcal{K}$ is a crucial ingredient of this inductive proof. As a concrete example of this, consider a smooth weak atlas that contains just three charts $\mathbf{K}_1, \mathbf{K}_2$ and $\mathbf{K}_{12}$ each with obstruction space E so that $\widehat{\phi}_{i(12)} = \mathrm{id}$ for $i = 1, 2$. Then when $k = 1$ we must construct sets $U^{(1)}_{i(12)}$ for $i = 1, 2$ that both satisfy $\phi_{i(12)}\bigl(U^{(1)}_{i(12)}\bigr) = U^{(1)}_{12} \cap \mathfrak{s}^{-1}_{12}(E) = U^{(1)}_{12}$. Hence the choices of the two level one sets $U^{(1)}_{1(12)}$ and $U^{(1)}_{2(12)}$ are not independent. In an additive situation[16], one can only have $E_1 = E_{12} = E$ if $E_2 = \{0\}$. In this case we still need $\phi_{1(12)}\bigl(U^{(1)}_{1(12)}\bigr) = U^{(1)}_{12}$.

[16] i.e. one where $E_I = \prod_{i \in I} E_i$

However, the condition for $i = 2$ is $\phi_{2(12)}\big(U^{(1)}_{2(12)}\big) = \mathfrak{s}_2^{-1}(0)$, which has been arranged at level 0. $\Diamond$

Finally we establish the metrizability of preshrunk shrinkings.

LEMMA 2.3.20. *For every tame shrinking $\mathcal{K}'$ of a tame topological atlas $\mathcal{K}$*

(i) *the map $|\mathcal{K}'| \to |\mathcal{K}|$ of quotient spaces is injective and continuous, and*

(ii) *$\mathcal{K}'$ is metrizable, with a metric that induces the subspace topology.*

PROOF. We write U_I, U_{IJ} for the domains of the charts and coordinate changes of $\mathcal{K}$ and U'_I, U'_{IJ} for those of $\mathcal{K}'$, so that $U'_I \subset U_I, U'_{IJ} \subset U_{IJ}$ for all $I, J \in \mathcal{I}_{\mathcal{K}} = \mathcal{I}_{\mathcal{K}'}$. Suppose that $\pi_{\mathcal{K}}(I, x) = \pi_{\mathcal{K}}(J, y)$ where $x \in U'_I, y \in U'_J$. Then we must show that $\pi_{\mathcal{K}'}(I, x) = \pi_{\mathcal{K}'}(J, y)$. Since $\mathcal{K}$ is tame, Lemma 2.3.12 (a) implies that if $I \cap J \neq \emptyset$ there is $w \in U_{I\cap J}$ such that $\phi_{(I\cap J)I}(w)$ is defined and equal to x. Hence $x \in \mathfrak{s}_I^{-1}(\mathbb{E}_{I\cap J}) \cap U'_I = \phi_{(I\cap J)I}(U'_{(I\cap J)I})$ by the tameness equation (2.2.7) for $\mathcal{K}'$. Therefore $w \in U'_{(I\cap J)I}$. Similarly, because $\phi_{(I\cap J)J}(w)$ is defined and equal to y, we have $w \in U'_{(I\cap J)J}$. Then by definition of $\pi_{\mathcal{K}'}$ we deduce $\pi_{\mathcal{K}'}(I, x) = \pi_{\mathcal{K}'}(I \cap J, w) = \pi_{\mathcal{K}'}(J, y)$. On the other hand, if $I \cap J \neq \emptyset$ then Lemma 2.3.12 (a) implies that $\psi_I(x) = \psi_J(y)$ so that $\pi_{\mathcal{K}'}(I, x) = \pi_{\mathcal{K}'}(J, y)$ by the injectivity of $\iota_{\mathcal{K}} : X \to |\mathcal{K}|$ (see Lemma 2.2.2).

To prove (ii), notice that because U'_I is precompact in U_I for each I, the closure $\overline{\iota(|\mathcal{K}'|)}$ of $\iota(|\mathcal{K}'|)$ in $|\mathcal{K}|$ is compact. It follows that $\overline{\iota(|\mathcal{K}'|)}$ (with the subspace topology) is regular (i.e. points and closed sets have disjoint neighbourhoods) and second countable, hence metrizable by Urysohn's metrization theorem. (For details of these arguments see [**MW1**, Proposition 3.3.8].) Therefore one may obtain the desired metric on $|\mathcal{K}'|$ by restriction from $\overline{\iota(|\mathcal{K}'|)}$. $\square$

COROLLARY 2.3.21. *Proposition* 2.2.10 *holds.*

PROOF. By Proposition 2.3.17 each weak atlas has a tame shrinking $\mathcal{K}'$ with domains U'_I. Now apply Proposition 2.3.17 again to obtain a further tame shrinking $\mathcal{K}_{sh}$ of $\mathcal{K}'$. with domains $U_I \sqsubset U'_I$. The metric on $\mathcal{K}_{sh}$ provided by Lemma 2.3.20 pulls back to a metric on U_I that is compatible with the topology on U_I because the inclusion $U_I \stackrel{\pi_{\mathcal{K}_{sh}}}{\longrightarrow} |\mathcal{K}_{sh}| \subset |\mathcal{K}'|$ also factors as $U_I \to U'_I \stackrel{\pi_{\mathcal{K}'}}{\longrightarrow} |\mathcal{K}'|$ and hence is a homomorphism to its image when this is topologized as a subspace of $|\mathcal{K}'|$. Hence $\mathcal{K}_{sh}$ is metrizable in the sense of Definition 2.2.6. The other assertions of Proposition 2.2.10 hold by Lemma 2.3.20). $\square$

The above corollary, together with the results in §2.3, establish the key parts of the first set of statements in Theorem 2.2.3. The rest of this theorem concerns cobordisms. The proofs are fairly straightforward generalizations of those given above; for details see [**MW1**].

Finally, in order to work in §6 with atlases that are not filtered, we introduce the notion of a good atlas as follows.

DEFINITION 2.3.22. *A topological atlas $\mathcal{K} = \big(\mathbf{K}_I, \widehat{\Phi}_{IJ}\big)_{I\subset J, I, J\in\mathcal{I}_{\mathcal{K}}}$ is said to be* **good** *if*

(i) *the realization $|\mathcal{K}|$ of $\mathcal{K}$ is Hausdorff in the quotient topology;*

(ii) *for all $I \in \mathcal{I}_{\mathcal{K}}$ the map $\pi_{\mathcal{K}} : U_I \to |\mathcal{K}|$ is a homeomorphism to its image;*

(iii) *$\mathcal{K}$ is metrizable, i.e. there is a metric d on $|\mathcal{K}|$ whose pullback by $\pi_{\mathcal{K}}$ induces the given topology on U_I for each $I \in \mathcal{I}_{\mathcal{K}}$;*

(iv) *for all* $I \subset J$ *with* $I, J \in \mathcal{I}_{\mathcal{K}}$ *the image* $\phi_{IJ}(U_{IJ})$ *is a closed subset of* U_J.

LEMMA 2.3.23. *Every preshrunk tame topological atlas is good.*

PROOF. Proposition 2.3.13 shows that every tame topological atlas satisfies conditions (i) and (ii) above, while Lemma 2.3.20 shows that (iii) holds. Further (iv) holds by the tameness condition (2.3.4) with $J = K$. □

REMARK 2.3.24. Once we have constructed a well behaved virtual neighbourhood $|\mathcal{K}|$, condition (iv) in Definition 2.3.6 is all that we need of the filtration/tameness conditions in order to construct the VFC. It will be used in the proof that the zero set of a perturbed section is compact; cf Proposition 2.4.6 below. ◊

2.4. Reductions and the construction of perturbation sections. In this section, we explain how to construct the virtual moduli cycle $[X]^{vir}_{\mathcal{K}}$ for an oriented smooth good atlas (with trivial isotropy). The discussion has three main parts.

- We first develop the notion of a **reduction** $\mathcal{V}$ **of a good topological atlas**. This gives us a subcategory $\mathbf{B}_{\mathcal{K}}|_{\mathcal{V}}$ of $\mathbf{B}_{\mathcal{K}}$ whose footprints cover X and with the property that two of its objects $(I, x), (J, y)$ are equivalent if and only if either $(I, x) \preceq (J, y)$ or $(I, x) \succeq (J, y)$. Comparing with Lemma 2.3.12, we see that the equivalence relation $\sim_{\mathcal{K}}$ becomes significantly simpler when restricted to $\mathbf{B}_{\mathcal{K}}|_{\mathcal{V}}$.
- Secondly, we define the notion of a **perturbation section** $\nu : \mathbf{B}_{\mathcal{K}}|_{\mathcal{V}} \to \mathbf{E}_{\mathcal{K}}|_{\mathcal{V}}$ of a reduction of a smooth atlas (with trivial isotropy), and in Proposition 2.4.6 give conditions under which the local zero sets of $\mathfrak{s}|_{\mathcal{V}} + \nu$ fit together to form a closed oriented manifold $|\mathbf{Z}^{\nu}|$.
- We then explain how to construct suitable perturbation sections ν. Although the conditions formulated in Definition 2.4.9 are rather intricate, they give a great deal of control over ν, and in particular over the zero set of $\mathfrak{s}|_{\mathcal{V}} + \nu$. This will allow us both to construct cobordisms between admissible perturbations as in Proposition 2.4.11 and to adapt the construction to the case of nontrivial isotropy as explained at the end of §3.2.
- Finally we outline the proof of Theorem B in the smooth case with trivial isotropy, showing how to define $[X]^{vir}_{\mathcal{K}}$ from the zero sets $|\mathbf{Z}^{\nu}|$.

We begin by explaining why it is necessary to "reduce" the atlas $\mathcal{K}$. The cover of X by the footprints $(F_I)_{I \in \mathcal{I}_{\mathcal{K}}}$ of all the Kuranishi charts (both the basic charts and those that are part of the transitional data) is closed under intersection. This makes it easy to express compatibility of the charts, since the overlap of footprints of any two charts $\mathbf{K}_I$ and $\mathbf{K}_J$ is the footprint of a third chart $\mathbf{K}_{I \cup J}$. However, this yields so many compatibility conditions that a construction of compatible perturbations in the Kuranishi charts may not be possible. For example, a choice of perturbation (with values in E_I) in the chart $\mathbf{K}_I$ also fixes the perturbation in each chart $\mathbf{K}_J$ over $\phi^{-1}_{J(I \cup J)}(\text{im } \phi_{I(I \cup J)}) \subset U_J$, whenever $I \cup J \subset \mathcal{I}_{\mathcal{K}}$. Since we do not assume transversality of the coordinate changes, this subset of U_J need not be a submanifold, [17] and hence the perturbation may not extend smoothly to a map from U_J to E_J. Moreover, for such an extension to exist at all, the pushforward of

[17]As explained in Remark 2.3.16, it will be a submanifold if $I \cap J \neq \emptyset$, but usually not otherwise.

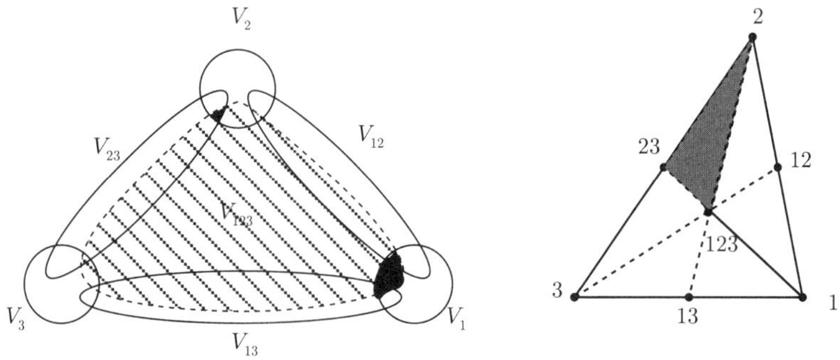

FIGURE 2.4.1. The right diagram shows the first barycentric subdivision of the triangle with vertices $1,2,3$. It has three new vertices labelled ij at the barycenters of the three edges and one vertex labelled 123 at the barycenter of the triangle. The left is a schematic picture of the cover (V_I) by the stars of the vertices of this barycentric subdivision. The black sets are examples of multiple intersections of the new cover, which correspond to the simplices in the barycentric subdivision. E.g. $V_2 \cap V_{23} \cap V_{123}$ corresponds to the triangle with vertices $2, 23, 123$, whereas $V_1 \cap V_{123}$ corresponds to the edge between 1 and 123. The cover (V_I) has the same intersection properties as the reduction of the original cover U_1, U_2, U_3, where U_i is the star of the vertex v_i.

the perturbation to $U_{I\cup J}$ would have to take values in the intersection

$$\widehat{\phi}_{I(I\cup J)}(E_I) \cap \widehat{\phi}_{J(I\cup J)}(E_J) \quad \subset \quad \widehat{\phi}_{(I\cap J)(I\cup J)}(E_{I\cap J}),$$

a very restrictive condition. In fact if $I \cap J = \emptyset$, this would mean that the perturbation would have to vanish over $F_{I\cup J}$. We will avoid these difficulties, and also make a first step towards compactness, by reducing the domains of the Kuranishi charts to precompact subsets $V_I \sqsubset U_I$ such that all compatibility conditions between $\mathbf{K}_I|_{V_I}$ and $\mathbf{K}_J|_{V_J}$ are given by direct coordinate changes $\widehat{\Phi}_{IJ}$ or $\widehat{\Phi}_{JI}$. As explained more fully in [**MW1**] the reduction process is analogous to replacing the star cover of a simplicial set by the star cover of its first barycentric subdivision; also see Figure 2.4.1.

In the following we work with good atlases (with properties as spelled out in Definition 2.3.22). Since all tame atlases are good, for the present purposes we could equally well work with tame atlases.

DEFINITION 2.4.1. *Let $\mathcal{K}$ be a good topological atlas. A* **reduction** *of $\mathcal{K}$ is an open subset $\mathcal{V} = \bigsqcup_{I\in\mathcal{I}_\mathcal{K}} V_I \subset \mathrm{Obj}_{\mathbf{B}_\mathcal{K}}$ i.e. a tuple of (possibly empty) open subsets $V_I \subset U_I$, satisfying the following conditions:*

(1) *$V_I \sqsubset U_I$ for all $I \in \mathcal{I}_\mathcal{K}$, and if $V_I \neq \emptyset$ then $V_I \cap \mathfrak{s}_I^{-1}(0) \neq \emptyset$;*
(2) *if $\pi_\mathcal{K}(\overline{V_I}) \cap \pi_\mathcal{K}(\overline{V_J}) \neq \emptyset$ then $I \subset J$ or $J \subset I$;*
(3) *the zero set $\iota_\mathcal{K}(X) = |\mathfrak{s}_\mathcal{K}|^{-1}(0)$ is contained in $\pi_\mathcal{K}(\mathcal{V}) = \bigcup_{I\in\mathcal{I}_\mathcal{K}} \pi_\mathcal{K}(V_I)$.*

Given a reduction $\mathcal{V}$, we define the **reduced domain category** *$\mathbf{B}_\mathcal{K}|_\mathcal{V}$ and the* **reduced obstruction category** *$\mathbf{E}_\mathcal{K}|_\mathcal{V}$ to be the full subcategories of $\mathbf{B}_\mathcal{K}$ and $\mathbf{E}_\mathcal{K}$ with objects $\bigsqcup_{I\in\mathcal{I}_\mathcal{K}} V_I$ resp. $\bigsqcup_{I\in\mathcal{I}_\mathcal{K}} V_I \times E_I$, and denote by $\mathfrak{s}|_\mathcal{V} : \mathbf{B}_\mathcal{K}|_\mathcal{V} \to \mathbf{E}_\mathcal{K}|_\mathcal{V}$ the section given by restriction of $\mathfrak{s}_\mathcal{K}$.*

There is a similar notion of good cobordism. It follows easily from condition (ii) above that the realization $|\mathbf{B}_{\mathcal{K}}|_{\mathcal{V}}|$ of the subcategory $\mathbf{B}_{\mathcal{K}}|_{\mathcal{V}}$ (i.e. its object space modulo the equivalence relation generated by its morphisms) injects into $|\mathbf{B}_{\mathcal{K}}| =: |\mathcal{K}|$. There is a related notion of cobordism reduction (see [**MW1**, §5.1]), which is just as one would imagine, keeping in mind that all sets have product form near the boundary.

Here is the main existence result. It is proved in [**MW1**, §5] by first constructing a reduction of the footprint cover (a process well understood in algebraic topology) as in Lemma 6.2.2 below, and then extending this suitably.

PROPOSITION 2.4.2. *The following statements hold.*

(a) *Every good topological Kuranishi atlas $\mathcal{K}$ has a reduction $\mathcal{V}$.*
(b) *Every good topological Kuranishi cobordism $\mathcal{K}^{[0,1]}$ has a cobordism reduction $\mathcal{V}^{[0,1]}$.*
(c) *Let $\mathcal{V}^0, \mathcal{V}^1$ be reductions of a good topological Kuranishi atlas $\mathcal{K}$. Then there exists a cobordism reduction $\mathcal{V}$ of $[0,1] \times \mathcal{K}$ such that $\partial^\alpha \mathcal{V} = \mathcal{V}^\alpha$ for $\alpha = 0, 1$.*

SKETCH PROOF OF (A). Define

$$\mathcal{C}(I) := \{J \in \mathcal{I}_{\mathcal{K}} \,|\, I \subset J \text{ or } J \subset I\},$$

We show below in Lemma 6.2.2 that the footprint cover $(F_I)_{I \in \mathcal{I}_{\mathcal{K}}}$ has a reduction $(Z_I)_{I \in \mathcal{I}_{\mathcal{K}}}$, i.e. there is an open cover of X such that $Z_I \sqsubset F_I$ and $\overline{Z_I} \cap \overline{Z_J} \neq \emptyset \Rightarrow J \in \mathcal{C}(I)$. For each I, choose an open subset $W_I \sqsubset U_I$ such that

$$W_I \cap \mathfrak{s}_I^{-1}(0_I) = \psi_I^{-1}(Z_I), \quad \overline{W_I} \cap \mathfrak{s}_I^{-1}(0_I) = \psi_I^{-1}(\overline{Z_I}),$$

and for each $J \notin \mathcal{C}(I)$ define

$$Y_{IJ} := \overline{W_I} \cap \pi_{\mathcal{K}}^{-1}(\pi_{\mathcal{K}}(\overline{W_J})) \;=\; U_I \cap \pi_{\mathcal{K}}^{-1}\big(\pi_{\mathcal{K}}(\overline{W_I}) \cap \pi_{\mathcal{K}}(\overline{W_J})\big).$$

Then Y_{IJ} is compact by the homeomorphism property of $\pi_{\mathcal{K}}$. Further, If $J \notin \mathcal{C}(I)$, then using $\psi_I^{-1}(\overline{Z_I}) = \mathfrak{s}_I^{-1}(0_I) \cap \overline{W_I}$ and the notation ε_I from (2.3.6) we have

$$\begin{aligned} \psi_I^{-1}(\overline{Z_I}) \cap Y_{IJ} &\subset \psi_I^{-1}(\overline{Z_I}) \cap \mathfrak{s}_I^{-1}(0_I) \cap \varepsilon_I(\overline{W_J}) \\ &= \psi_I^{-1}(\overline{Z_I}) \cap \varepsilon_I\big(\mathfrak{s}_J^{-1}(0_J) \cap \overline{W_J}\big) \\ &= \psi_I^{-1}(\overline{Z_I}) \cap \varepsilon_I\big(\psi_J^{-1}(\overline{Z_J})\big) \;=\; \psi_I^{-1}(\overline{Z_I}) \cap \psi_I^{-1}(\overline{Z_J}) \;=\; \emptyset, \end{aligned}$$

where the first equality holds because $\mathfrak{s}$ is compatible with the coordinate changes. Because the Y_{IJ} are compact, we can find closed neighbourhoods $\mathcal{N}(Y_{IJ}) \subset U_I$ of Y_{IJ} for each $J \notin \mathcal{C}(I)$ such that

$$\mathcal{N}(Y_{IJ}) \cap \psi_I^{-1}(\overline{Z_I}) = \emptyset.$$

It is now straightforward to check that the sets

$$V_I := W_I \smallsetminus {\textstyle\bigcup_{J \notin \mathcal{C}(I)}} \mathcal{N}(Y_{IJ}). \tag{2.4.1}$$

form the required reduction. The other parts of the proposition have similar proofs. □

EXAMPLE 2.4.3. (i) A reduction of the atlas $\mathcal{K}$ in Example 2.2.16 has three sets V_1, V_2, V_{12} that cover the zero set and have the property that $\pi_{\mathcal{K}}(\overline{V}_1) \cap \pi_{\mathcal{K}}(\overline{V}_2) = \emptyset$. For instance, we can take $V_1 = (-\infty, 2) \sqsubset U_1, V_2 = (3, \infty) \times \mathbb{R} \sqsubset U_2$ and $V_{12} = (1,3) \times \mathbb{R} \subset U_{12}(= U_2)$.

(ii) It is clear from the above proof that we can construct the reduction $(V_I)_{I\in\mathcal{I}_\mathcal{K}}$ in such a way that there is a coordinate change between V_I and V_J only if the corresponding footprints have nonempty intersection $Z_I \cap Z_J$. (After choosing the Z_I, simply replace $\mathcal{C}(I)$ in the above proof by the set $\mathcal{C}_Z(I) = \{J : Z_I \cap Z_J \neq \emptyset\}$.) For example, if $X = \{p_1, p_2, p_3\}$ and $\mathcal{K}$ has two basic charts U_1, U_2 with footprints $F_1 = \{p_1, p_3\}, F_2 = \{p_2, p_3\}$, there is one transition chart $\mathbf{K}_{12}$ with footprint $\{p_3\}$, and we may take $Z_1 = \{p_1\}, Z_2 = \{p_2\}, Z_{12} = \{p_3\}$. We can then choose $\mathcal{V}$ so that the sets $\big(\pi_\mathcal{K}(V_I)\big)_{I\in\mathcal{I}_\mathcal{K}}$ are disjoint, so that the only morphisms in the category $\mathbf{B}_\mathcal{K}|_\mathcal{V}$ are identity morphisms. In other words $V_I \cap \phi_{IJ}^{-1}(V_J) = \emptyset$ for all $I \subset J$, so that the compatibility condition between ν_I, ν_J in Definition 2.4.5 is automatically satisfied. $\Diamond$

REMARK 2.4.4. (i) The key property of a reduction $\mathcal{V}$ is that if any intersection $\pi_\mathcal{K}(\overline{V}_I)\cap\pi_\mathcal{K}(\overline{V}_J)$ is nonempty then $I \subset J$ or $J \subset I$ so that the induced equivalence relation on $\bigsqcup V_I$ has just one step. Contrast this with the equivalence relation for a tame atlas which, as explained in Remark 2.3.16 (ii), is given by a two step process. Another consequence in the smooth case is that the intersection $\pi_\mathcal{K}(V_I) \cap \pi_\mathcal{K}(V_J)$ is the image of a submanifold; see Remark 2.3.16 (i).

(ii) Another important property is that the inclusion of categories $\mathbf{B}_\mathcal{K}|_\mathcal{V} \to \mathbf{B}_\mathcal{K}$ induces an injection on their realizations, in other words the equivalence relation on $\bigsqcup_I V_I = \mathrm{Obj}_{\mathbf{B}_\mathcal{K}|_\mathcal{V}}$ induced by the morphisms in $\mathbf{B}_\mathcal{K}|_\mathcal{V}$ coincides with that induced by the morphisms in $\mathbf{B}_\mathcal{K}$. (To see that this is not true for arbitrary full subcategories of $\mathbf{B}_\mathcal{K}$ consider the full subcategory with objects $\bigcup_i U_i$ which only has identity morphisms.) It follows that the quotient topology on $|\mathcal{V}| := {}^{\bigcup V_I}/_{\sim_\mathcal{V}}$ induced by the morphisms in $\mathbf{B}_\mathcal{K}|_\mathcal{V}$ coincides with that induced by the morphisms in $\mathbf{B}_\mathcal{K}$, so that $|\mathcal{V}|$ can be identified with its image $\pi_\mathcal{K}(\mathcal{V})$ in $|\mathcal{K}|$ with the quotient topology. Further, because $\mathcal{V}$ is precompact in $\mathrm{Obj}_{\mathbf{B}_\mathcal{K}}$ one can show that the relative topology on $\pi_\mathcal{K}(\mathcal{V})$ as a subset of $|\mathcal{K}|$ equals its metric topology with respect to any admissible metric on $|\mathcal{K}|$; see [**MW1**, Lemma 5.2.5].

(iii) Subsets of $|\mathcal{K}|$ of the form $\pi_\mathcal{K}(\mathcal{V})$ contain the zero set $\iota_\mathcal{K}(X)$ and are the analog in the virtual neighbourhood $|\mathcal{K}|$ of "precompact neighbourhoods of the zero set". Since $|\mathcal{K}|$ is not locally compact even in the metric topology (see Example 2.2.16), there are no compact neighbourhoods of the zero set. On the other hand, because $V_I \sqsubset U_I$, the subset $\pi_\mathcal{K}(\mathcal{V})$ is precompact in $|\mathcal{K}|$, and is "open" to the extent that it is the image by $\pi_\mathcal{K}$ of the open set $\bigsqcup_I V_I$. (But, of course, it is almost never actually open in $|\mathcal{K}|$.) We can interpret Figure 2.4.1 as a schematic picture of the subsets $\pi_\mathcal{K}(V_I)$ in $|\mathcal{K}|$, though it is not very accurate since the dimensions of the V_I change. $\Diamond$

We now introduce the notion of a perturbation from [**MW2**, §7.2]. For the rest of this subsection we assume that our atlas is smooth as well as good in the sense of Definition 2.3.22.

DEFINITION 2.4.5. *Let $\mathcal{K}$ be a good smooth atlas with trivial isotropy. A* **reduced perturbation** *of $\mathcal{K}$ is a smooth functor $\nu : \mathbf{B}_\mathcal{K}|_\mathcal{V} \to \mathbf{E}_\mathcal{K}|_\mathcal{V}$ between the reduced domain and obstruction categories of some reduction $\mathcal{V}$ of $\mathcal{K}$, such that $\mathrm{pr}_\mathcal{K} \circ \nu$ is the identity functor. That is, $\nu = (\nu_I)_{I\in\mathcal{I}_\mathcal{K}}$ is given by a family of smooth* *maps* $\nu_I : V_I \to E_I$ *such that for each* $I \subsetneq J$ *we have a commuting diagram*

$$\begin{array}{ccc} V_I \cap \phi_{IJ}^{-1}(V_J) & \xrightarrow{\nu_I} & E_I \\ {\scriptstyle \phi_{IJ}}\downarrow & & \downarrow{\scriptstyle \widehat{\phi}_{IJ}} \\ V_J & \xrightarrow{\nu_J} & E_J. \end{array} \tag{2.4.2}$$

We say that a reduced perturbation ν *is* **admissible** *if*

$$\mathrm{d}_y\, \nu_J(\mathrm{T}_y\, V_J) \subset \operatorname{im} \widehat{\phi}_{IJ} \qquad \forall\, I \subsetneq J,\ y \in V_J \cap \phi_{IJ}(V_I \cap U_{IJ}). \tag{2.4.3}$$

Each reduced perturbation $\nu : \mathbf{B}_{\mathcal{K}}|_{\mathcal{V}} \to \mathbf{E}_{\mathcal{K}}|_{\mathcal{V}}$ induces a continuous map $|\nu| : |\mathcal{V}| \to |\mathbf{E}_{\mathcal{K}}|_{\mathcal{V}}|$ such that $|\operatorname{pr}_{\mathcal{K}}| \circ |\nu| = \operatorname{id}$, where $|\operatorname{pr}_{\mathcal{K}}|$ is as in Theorem 2.2.3. Each such map has the further property that $|\nu|\big|_{\pi_{\mathcal{K}}(V_I)}$ takes values in $\pi_{\mathcal{K}}(U_I \times E_I)$. Note that the zero section $0_{\mathcal{K}}$, given by $U_I \to 0 \in E_I$, restricts to an admissible perturbation $0_{\mathcal{V}} : \mathbf{B}_{\mathcal{K}}|_{\mathcal{V}} \to \mathbf{E}_{\mathcal{K}}|_{\mathcal{V}}$ in the sense of the above definition. Similarly, the canonical section $\mathfrak{s} := \mathfrak{s}_{\mathcal{K}}$ of the Kuranishi atlas restricts to a perturbation $\mathfrak{s}|_{\mathcal{V}} : \mathbf{B}_{\mathcal{K}}|_{\mathcal{V}} \to \mathbf{E}_{\mathcal{K}}|_{\mathcal{V}}$ of any reduction. However, the canonical section is generally not admissible. In fact, it follows from the index condition that for all $y \in V_J \cap \phi_{IJ}(V_I \cap U_{IJ})$ the map

$$\operatorname{pr}_{E_I}^{\perp} \circ \mathrm{d}_y\, s_J \;:\; {\mathrm{T}_y U_J}\big/{\mathrm{T}_y(\phi_{IJ}(U_{IJ}))} \;\longrightarrow\; {E_J}\big/{\widehat{\phi}_{IJ}(E_I)}$$

is an isomorphism, while for an admissible section it is identically zero. Thus for any reduction $\mathcal{V}$ and admissible perturbation ν, the sum

$$\mathfrak{s}|_{\mathcal{V}} + \nu := (s_I|_{V_I} + \nu_I)_{I \in \mathcal{I}_{\mathcal{K}}} \;:\; \mathbf{B}_{\mathcal{K}}|_{\mathcal{V}} \;\to\; \mathbf{E}_{\mathcal{K}}|_{\mathcal{V}} \tag{2.4.4}$$

is a reduced section that satisfies the index condition.

Here are some further definitions.

- We say that two reductions $\mathcal{C}, \mathcal{V}$ are **nested** and write $\mathcal{C} \sqsubset \mathcal{V}$ if $C_I \sqsubset V_I$ for all $I \in \mathcal{I}_{\mathcal{K}}$. One can show that any two such pairs $\mathcal{C}_0 \sqsubset \mathcal{V}_0, \mathcal{C}_1 \sqsubset \mathcal{V}_1$ are cobordant via a nested cobordism $\mathcal{C}^{01} \sqsubset \mathcal{V}^{01}$.
- A perturbation $\nu : \mathbf{B}_{\mathcal{K}}|_{\mathcal{V}} \to \mathbf{E}_{\mathcal{K}}|_{\mathcal{V}}$ is called **precompact** if there is a nested reduction $\mathcal{C} \sqsubset \mathcal{V}$ such that

$$\pi_{\mathcal{K}}\big((\mathfrak{s}|_{\mathcal{V}} + \nu)^{-1}(0)\big) \;\subset\; \pi_{\mathcal{K}}(\mathcal{C}). \tag{2.4.5}$$

- It is called **transverse** if for all $z \in V_I \cap (s_I|_{V_I} + \nu_I)^{-1}(0)$ the map $\mathrm{d}_z(s_I + \nu_I) : \mathrm{T}\, U_I \to E_I$ is surjective.

It is not hard to see if ν is admissible, transversality of the local sections ν_I is preserved under coordinate changes. More precisely, if $z \in V_I \cap U_{IJ}$ and $w \in V_J$ are such that $\phi_{IJ}(z) = w$, then z is a transverse zero of $s_I|_{V_I} + \nu_I$ if and only if w is a transverse zero of $s_J|_{V_J} + \nu_J$. (This holds because the sections $s_I|_{V_I} + \nu_I$ satisfy the index condition; see (2.4.4) above.) Moreover, if $\mathcal{K}$ is oriented then the local orientation of the zero set at $z \in U_I$ defined as in Lemma 3.3.9 is taken by $\mathrm{d}_z\, \phi_{IJ}$ to the orientation at w.

Here is the main result about the zero sets, from [**MW2**, Proposition 7.2.8]. In it, we show that the local zero sets $Z_I = (s_I|_{V_I} + \nu_I)^{-1}(0)$ can be assembled into a nonsingular ep groupoid[18] $\mathbf{Z}^{\nu}$ whose realization is a closed oriented manifold $|\mathbf{Z}^{\nu}|$ that injects into $|\mathcal{K}|$.

[18]For the definition of ep groupoid see the beginning of §5.1; nonsingular means that there is at most one morphism between any two points.

Proposition 2.4.6. *Let $\mathcal{K}$ be a good smooth d-dimensional Kuranishi atlas with trivial isotropy and reduction $\mathcal{V} \sqsubset \mathcal{K}$, and suppose that $\nu : \mathbf{B}_{\mathcal{K}}|_{\mathcal{V}} \to \mathbf{E}_{\mathcal{K}}|_{\mathcal{V}}$ is a precompact, admissible, transverse perturbation. Then $|\mathbf{Z}^\nu| = |(\mathfrak{s}|_{\mathcal{V}} + \nu)^{-1}(0)|$ is a smooth closed d-dimensional manifold, whose quotient topology agrees with the subspace topology on $|(\mathfrak{s}|_{\mathcal{V}} + \nu)^{-1}(0)| \subset |\mathcal{K}|$.*

Moreover, if $\mathcal{K}$ is oriented, so is $|\mathbf{Z}^\nu|$.

Proof. Since ν is transverse, each local zero set $Z_I := (s_I|_{V_I} + \nu_I)^{-1}(0)$ is a smooth manifold. Hence one may construct a groupoid $\mathbf{Z}^\nu$ with objects $\sqcup_I Z_I$ and morphisms generated by the restriction of the coordinate changes ϕ_{IJ} and their inverses. Since these are diffeomorphisms and there are a finite number of coordinate patches, the realization $|\mathbf{Z}^\nu|$ (with the quotient topology) is second countable. The space $|\mathbf{Z}^\nu|$ is also Hausdorff because its inclusion into the Hausdorff space $|\mathcal{K}|$ is continuous. Therefore it is a smooth d-dimensional manifold.

To check the compactness of $|\mathbf{Z}^\nu|$, it suffices to check sequential compactness. To this end, consider a sequence $(p_k)_{k\in\mathbb{N}} \subset |\mathbf{Z}^\nu|$. In the following we will index all subsequences by $k \in \mathbb{N}$ as well. By finiteness of $\mathcal{I}_{\mathcal{K}}$ there is a subsequence of (p_k) that has lifts in $(s_I|_{V_I} + \nu_I)^{-1}(0)$ for a fixed $I \in \mathcal{I}_{\mathcal{K}}$. In fact, by (2.4.5), and using the language of equation (2.3.6), the subsequence lies in

$$V_I \cap \pi_{\mathcal{K}}^{-1}\big(\pi_{\mathcal{K}}(\mathcal{C})\big) \;=\; V_I \cap \textstyle\bigcup_{J\in\mathcal{I}_{\mathcal{K}}} \varepsilon_I(C_J) \;\subset\; U_I.$$

Here $\varepsilon_I(C_J) = \emptyset$ unless $I \subset J$ or $J \subset I$, due to the intersection property (ii) of Definition 2.4.1 and the inclusion $C_J \subset V_J$. So we can choose another subsequence and lifts $(x_k)_{k\in\mathbb{N}} \subset V_I$ with $\pi_{\mathcal{K}}(x_k) = p_k$ such that either

$$(x_k)_{k\in\mathbb{N}} \subset V_I \cap \phi_{IJ}^{-1}(C_J) \qquad \text{or} \qquad (x_k)_{k\in\mathbb{N}} \subset V_I \cap \phi_{JI}(C_J \cap U_{JI})$$

for some fixed J satisfying either $I \subset J$ or $J \subset I$. In the first case, compatibility of the perturbations (2.4.2) implies that there are other lifts $\phi_{IJ}(x_k) \in (s_J|_{V_J} + \nu_J)^{-1}(0) \cap C_J$, which by precompactness $\overline{C_J} \sqsubset V_J$ have a convergent subsequence $\phi_{IJ}(x_k) \to y_\infty \in (s_J|_{V_J} + \nu_J)^{-1}(0)$. Thus we have found a limit point in the perturbed zero set $p_k = \pi_{\mathcal{K}}(\phi_{IJ}(x_k)) \to \pi_{\mathcal{K}}(y_\infty) \in |\mathbf{Z}^\nu|$, as required for sequential compactness.

On the other hand, if $J \subset I$ we use the relative closedness of $\phi_{JI}(U_{JI}) \subset U_I$ (which holds by condition (iii) for good atlases) and the precompactness $V_I \sqsubset U_I$ to find a convergent subsequence $x_k \to x_\infty \in \overline{V_I} \cap \phi_{JI}(U_{JI})$. Since ϕ_{JI} is a homeomorphism to its image, this implies convergence of the preimages $y_k := \phi_{JI}^{-1}(x_k) \to \phi_{JI}^{-1}(x_\infty) =: y_\infty \in U_{JI}$. By construction and compatibility of the perturbations (2.4.2), this subsequence (y_k) of lifts of $\pi_{\mathcal{K}}(y_k) = p_k$ moreover lies in $(s_J|_{V_J} + \nu_J)^{-1}(0) \cap C_J$. Now precompactness of $C_J \sqsubset V_J$ implies $y_\infty \in V_J$, and continuity of the section implies $y_\infty \in (s_J|_{V_J} + \nu_J)^{-1}(0)$. Thus we again have a limit point $p_k = \pi_{\mathcal{K}}(y_k) \to \pi_{\mathcal{K}}(y_\infty) \in |\mathbf{Z}^\nu|$. This proves that the perturbed zero set $|\mathbf{Z}^\nu|$ is sequentially compact, and hence compact. Therefore it is a closed manifold, as claimed.

So far we have considered $|\mathbf{Z}^\nu|$ with the quotient topology. By construction there is a continuous injection $|\mathbf{Z}^\nu| \to |\mathcal{K}|$. Because $|\mathcal{K}|$ is Hausdorff and $|\mathbf{Z}^\nu|$ is compact, this map is a homomorphism to its image, when this is given the subspace topology. This proves the last claim in the first paragraph. Finally note that if $\mathcal{K}$ is oriented, we may orient each Z_I as in Lemma 3.3.9. The fact that the orientation is defined by a bundle over $|\mathcal{K}|$ implies that its restriction to the Z_I is preserved by coordinate changes. Hence $\mathbf{Z}$ is oriented if $\mathcal{K}$ is. □

There are similar result for cobordisms. In particular, if $\mathcal{K}$ is an oriented cobordism (or concordance) from $\mathcal{K}^0$ to $\mathcal{K}^1$, then for appropriate perturbations ν the zero set $|\mathbf{Z}^\nu|$ provides an oriented cobordism (or concordance) between the zero set of $\nu|_{\mathcal{K}^0}$ and that of $\nu|_{\mathcal{K}^1}$.

Construction of the perturbation ν: The main remaining problem is to construct suitable perturbations ν. Even though the constructions are rather intricate, the statements of the main results in Propositions 2.4.10 and 2.4.11 below are very precise. Moreover, our language gives us great control over all aspects of the construction, so that it can be easily adapted to other situations.

In order that the construction apply to the case with isotropy, we will begin by working with a good topological atlas. By definition, we may therefore choose an **admissible metric** d on $|\mathcal{K}|$ as in Definition 2.3.22, i.e. a metric whose pullback d_I to each domain U_I is compatible with its topology. We denote the δ-neighbourhoods of subsets $Q \subset |\mathcal{K}|$ resp. $A \subset U_I$ for $\delta > 0$ by

$$B_\delta(Q) := \bigl\{w \in |\mathcal{K}| \mid \exists q \in Q : d(w,q) < \delta\bigr\},$$
$$B^I_\delta(A) := \bigl\{x \in U_I \mid \exists a \in A : d_I(x,a) < \delta\bigr\}.$$

Note that $\phi_{IJ}(B^I_\delta(A)) = B^J_\delta\bigl(\phi_{IJ}(A)\bigr)$ because all coordinate changes are isometries. Similarly $U_I \cap \pi_{\mathcal{K}}^{-1}(B_\delta(Q)) = B^I_\delta\bigl(U_I \cap \pi_{\mathcal{K}}^{-1}(Q)\bigr)$. Further, we define a decreasing sequence of nested reductions $\mathcal{V}^k := (V_I^k)_{I\in\mathcal{I}_\mathcal{K}} \sqsupset \mathcal{V}^{k+1} := (V_I^{k+1})_{I\in\mathcal{I}_\mathcal{K}}$, where

$$V_I^k := B^I_{2^{-k}\delta}(V_I) \sqsubset U_I \qquad \text{for } k \geq 0, \tag{2.4.6}$$

and $\delta > 0$ is chosen so that the following disjointness condition holds when $I \not\subset J, J \not\subset I$:

$$B_\delta\bigl(\pi_\mathcal{K}(V_I^k)\bigr) \cap B_\delta\bigl(\pi_\mathcal{K}(V_J^k)\bigr) \subset B_{\delta+2^{-k}\delta}\bigl(\pi_\mathcal{K}(V_I)\bigr) \cap B_{\delta+2^{-k}\delta}\bigl(\pi_\mathcal{K}(V_J)\bigr) = \emptyset. \tag{2.4.7}$$

This implies that when $I \subsetneq J$,

$$\begin{aligned} V_I^k \cap \pi_\mathcal{K}^{-1}(\pi_\mathcal{K}(V_J^k)) &= V_I^k \cap \phi_{IJ}^{-1}(V_J^k), \\ V_J^k \cap \pi_\mathcal{K}^{-1}(\pi_\mathcal{K}(V_I^k)) &= V_J^k \cap \phi_{IJ}(V_I^k \cap U_{IJ}) =: N_{JI}^k \end{aligned} \tag{2.4.8}$$

for the sets on which we will require compatibility of the perturbations ν_I and ν_J. Similarly, we have precompact inclusions for any $k' > k \geq 0$

$$N_{JI}^{k'} = V_J^{k'} \cap \phi_{IJ}(V_I^{k'} \cap U_{IJ}) \sqsubset V_J^k \cap \phi_{IJ}(V_I^k \cap U_{IJ}) = N_{JI}^k. \tag{2.4.9}$$

We abbreviate

$$N_J^k := \textstyle\bigcup_{J \supsetneq I} N_{JI}^k \subset V_J^k,$$

and call the union $N_J^{|J|}$ the **core** of $V_J^{|J|}$, since it is the part of this set on which we will prescribe ν_J by compatibility with the ν_I for $I \subsetneq J$. (When $|J| = k+1$ we also consider various **enlargements of the core** of the form $N_J^{k+\lambda} \supset N_J^{|J|}$ where $\lambda \in (0,1)$.)

In the iterative argument we work with quarter integers between 0 and $M := \max_{I\in\mathcal{I}_\mathcal{K}} |I|$, and need to introduce another constant $\eta_0 > 0$ that controls the intersection with $\operatorname{im} \phi_{IJ} = \phi_{IJ}(U_{IJ})$ for all $I \subsetneq J$ as in Figure 2.4.2,

$$\operatorname{im} \phi_{IJ} \cap B^J_{2^{-k-\frac12}\eta_0}\bigl(N_{JI}^{k+\frac34}\bigr) \subset N_{JI}^{k+\frac12} \qquad \forall\, k \in \{0,1,\dots,M\}. \tag{2.4.10}$$

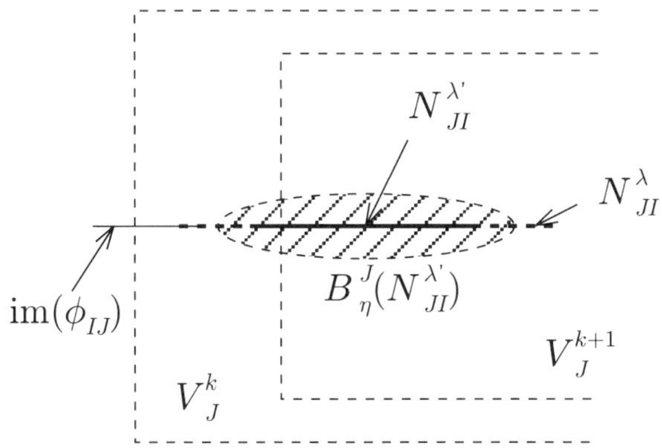

FIGURE 2.4.2. This figure illustrates the nested sets $V_J^{k+1} \sqsubset V_k^k$ and $N_{JI}^{\lambda'} \sqsubset N_{JI}^{\lambda} \subset \mathrm{im}\,(\phi_{IJ}) \cap V_J^k$ for $k+1 > \lambda' = k + \frac{3}{4} > \lambda = k + \frac{1}{2} > k$, the shaded neighbourhood $B^J_\eta(N^{\lambda'}_{JI})$ for $\eta = 2^{-\lambda}\eta_0$, and the inclusion given by (2.4.10).

Since ϕ_{IJ} is an isometric embedding, this inclusion holds whenever $2^{-k-\frac{1}{2}}\eta_0 + 2^{-k-\frac{3}{4}}\delta \le 2^{-k-\frac{1}{2}}\delta$ for all k. To minimize the number of choices in the construction of perturbations, we may thus simply fix η_0 in terms of δ by

$$\eta_0 := (1 - 2^{-\frac{1}{4}})\delta. \tag{2.4.11}$$

Then we also have $2^{-k}\eta_0 + 2^{-k-\frac{1}{2}}\delta < 2^{-k}\delta$, which provides the inclusions

$$B^I_{\eta_k}\Big(\overline{V_I^{k+\frac{1}{2}}}\Big) \sqsubset V_I^k \qquad \text{for } k \ge 0,\ \eta_k := 2^{-k}\eta_0. \tag{2.4.12}$$

We then define constants $\delta_{\mathcal{V}} > 0$ and $\sigma(\delta, \mathcal{V}, \mathcal{C}) > 0$ that depend only on the indicated data as follows.

DEFINITION 2.4.7. *Given a reduction $\mathcal{V}$ of a good topological atlas $(\mathcal{K}, d)$, we define $\delta_{\mathcal{V}} > 0$ to be the maximal constant such that any $\delta < \delta_{\mathcal{V}}$ satisfies*

$$B_{2\delta}(V_I) \sqsubset U_I \qquad \forall I \in \mathcal{I}_{\mathcal{K}}, \tag{2.4.13}$$

$$B_{2\delta}(\pi_{\mathcal{K}}(\overline{V_I})) \cap B_{2\delta}(\pi_{\mathcal{K}}(\overline{V_J})) \ne \emptyset \quad \Longrightarrow \quad I \subset J \text{ or } J \subset I. \tag{2.4.14}$$

Further, given a nested reduction $\mathcal{C} \sqsubset \mathcal{V}$ of $(\mathcal{K}, d)$ and $0 < \delta < \delta_{\mathcal{V}}$, we define

$$\eta_0 := (1 - 2^{-\frac{1}{4}})\delta, \qquad \eta_{|J|-\frac{1}{2}} := 2^{-|J|+\frac{1}{2}}\eta_0$$

and
(2.4.15)

$$\sigma(\delta, \mathcal{V}, \mathcal{C}) := \min_{J \in \mathcal{I}_{\mathcal{K}}} \inf \Big\{ \|s_J(x)\| \;\Big|\; x \in \overline{V_J^{|J|}} \smallsetminus \Big(\widetilde{C}_J \cup \bigcup_{I \subsetneq J} B^J_{\eta_{|J|-\frac{1}{2}}}\big(N_{JI}^{|J|-\frac{1}{4}}\big)\Big)\Big\},$$

where

$$\widetilde{C}_J := \bigcup_{K \supset J} \phi_{JK}^{-1}(C_K) \subset U_J.$$

LEMMA 2.4.8. $\sigma(\delta, \mathcal{V}, \mathcal{C}) > 0$.

SKETCH OF PROOF. Because $\mathcal{C} \sqsubset \mathcal{V} \sqsubset \mathcal{V}^1$ are nested reductions, $\pi_{\mathcal{K}}(\mathcal{C})$ contains the zero set $\iota_{\mathcal{K}}(X)$, and also

$$\pi_{\mathcal{K}}(V_J^1) \cap \pi_{\mathcal{K}}(\mathcal{C}) \subset \bigcup_{I \subsetneq J \text{ or } J \subset I} \pi_{\mathcal{K}}(C_I).$$

Hence the definition of $\widetilde{C}_J$ implies that

$$\overline{V_J^{|J|}} \cap s_J^{-1}(0) \;\subset\; \overline{V_J^{|J|}} \cap \pi_{\mathcal{K}}^{-1}(\pi_{\mathcal{K}}(\mathcal{C})) \;\subset\; \widetilde{C}_J \;\cup\; \textstyle\bigcup_{I \subsetneq J} \phi_{IJ}(C_I).$$

But when $I \subsetneq J$ we have $C_I \subset V_I \subset V_I^k$ for all k, so that $\phi_{IJ}(C_I)$ is a subset of the core N_J^k for all k. Hence $\sigma(\delta, \mathcal{V}, \mathcal{C})$ is the minimum of $\|s_J(x)\|$ as $x \in \overline{V_J^{|J|}}$ ranges over the complement of a neighbourhood of the zero set of s_I, and so is strictly positive. □

Now, let us suppose that $\mathcal{K}$ is smooth, good (e.g. metric and tame) and, as always in this section, with trivial isotropy. Here is a slightly shortened version of [**MW2**, Definition 7.3.5].

DEFINITION 2.4.9. *Given a nested reduction $\mathcal{C} \sqsubset \mathcal{V}$ of a smooth good Kuranishi atlas $(\mathcal{K}, d)$ and constants $0 < \delta < \delta_{\mathcal{V}}$ and $0 < \sigma \le \sigma(\delta, \mathcal{V}, \mathcal{C})$, we say that a perturbation ν of $s_{\mathcal{K}}|_{\mathcal{V}}$ is $(\mathcal{V}, \mathcal{C}, \delta, \sigma)$-**adapted** if the perturbations $\nu_I : V_I \to E_I$ extend to perturbations over $V_I^{|I|}$ (also denoted ν_I) so that the following conditions hold for every $k = 1, \dots, M$ where η_k is as in (2.4.12).*

a) *The perturbations are compatible on $\bigcup_{|I| \le k} V_I^k$, that is*

$$\nu_I \circ \phi_{HI}|_{V_H^k \cap \phi_{HI}^{-1}(V_I^k)} = \widehat{\phi}_{HI} \circ \nu_H|_{V_H^k \cap \phi_{HI}^{-1}(V_I^k)} \qquad \text{for all } H \subsetneq I, |I| \le k.$$

b) *The perturbed sections are transverse, that is $(s_I|_{V_I^k} + \nu_I) \pitchfork 0$ for each $|I| \le k$.*

c) *The perturbations are strongly admissible with radius η_k, that is for all $H \subsetneq I$ and $|I| \le k$ we have*

$$\nu_I(B_{\eta_k}^I(N_{IH}^k)) \subset \widehat{\phi}_{HI}(E_H) \qquad \text{with } N_{IH}^k = V_I^k \cap \phi_{HI}(V_H^k \cap U_{HI}).$$

d) *The perturbed zero sets are contained in $\pi_{\mathcal{K}}^{-1}(\pi_{\mathcal{K}}(\mathcal{C}))$; more precisely $s_I + \nu_I \ne 0$ on $V_I^k \smallsetminus \pi_{\mathcal{K}}^{-1}(\pi_{\mathcal{K}}(\mathcal{C}))$.*

e) *The perturbations are small, that is $\sup_{x \in V_I^k} \|\nu_I(x)\| < \sigma$ for $|I| \le k$.*

The above conditions are more than needed to ensure that every $(\mathcal{V}, \mathcal{C}, \delta, \sigma)$-adapted perturbation ν of $\mathfrak{s}_{\mathcal{K}}|_{\mathcal{V}}$ is an admissible, transverse perturbation with

$$\pi_{\mathcal{K}}((\mathfrak{s} + \nu)^{-1}(0)) \subset \pi_{\mathcal{K}}(\mathcal{C}).$$

Note that the role of the strong admissibility condition above is to allow us to deduce d) from e). In fact, the definition of σ in (2.4.15) and condition e) imply that the zero set of $s_I|_{V_I^{|I|}} + \nu_I$ must either lie in $\widetilde{C}_I$ and hence project to $\pi_{\mathcal{K}}(\mathcal{C})$ or lie in some neighbourhood $\bigcup_{H \subsetneq I} B_\eta^I(N_{IH}^k)$ of the enlarged core. But the strong admissibility condition c) implies that near the core N_{IH} the components in $I \smallsetminus H$ of $s_I + \nu_I$ equal those of s_I and hence only vanish on the core itself. It follows that any zero of $s_I + \nu_I$ that lies near the core must in fact lie in the core and hence project to $\pi_{\mathcal{K}}(\mathcal{C})$ by the inductive nature of the construction.

We now outline the argument that such perturbations ν exist. For full details see [**MW2**, Proposition 7.3.7].

PROPOSITION 2.4.10. *Let $(\mathcal{K}, d)$ be a smooth good Kuranishi atlas with nested reduction $\mathcal{C} \sqsubset \mathcal{V}$. Then for any $0 < \delta < \delta_{\mathcal{V}}$ and $0 < \sigma \le \sigma(\delta, \mathcal{V}, \mathcal{C})$ there exists a $(\mathcal{V}, \mathcal{C}, \delta, \sigma)$-adapted perturbation ν of $\mathfrak{s}_{\mathcal{K}}|_{\mathcal{V}}$.*

SKETCH OF PROOF. The construction is by an inductive process that constructs the required perturbations ν_I on sets larger than V_I. Namely, this proposition constructs functions $\nu_I : V_I^{|I|} \to E_I$ by an iteration over $k = 0, \ldots, M = \max_{I \in \mathcal{I}_{\mathcal{K}}} |I|$, where in step k we will define $\nu_I : V_I^k \to E_I$ for all $|I| = k$ that, together with the $\nu_I|_{V_I^k}$ for $|I| < k$ obtained by restriction from earlier steps, satisfy conditions a)-e) of Definition 2.4.9. Restriction to $V_I \subset V_I^{|I|}$ then yields a $(\mathcal{V}, \mathcal{C}, \delta, \sigma)$-adapted perturbation ν of $s_{\mathcal{K}}|_{\mathcal{V}}$. A key point in the construction is that because the different sets $V_I^{|I|}, |I| = k$, are disjoint (by (2.4.7)), at the kth step the needed functions ν_I can be constructed independently of each other.

Assume inductively that suitable $\nu_I : V_I^{|I|} \to E_I$ have been found for $|I| \le k$, and consider the construction of ν_J for some J with $|J| = k+1$. We construct ν_J as a sum $\widetilde{\nu}_J + \nu_{\pitchfork}$ where

- $\widetilde{\nu}_J|_{N_J^{k+1}} = \mu_J|_{N_J^{k+1}}$ where $\mu_J : N_J^k \to E_J$ is defined on the enlarged core N_J^k by the compatibility conditions, and
- $\nu_{\pitchfork}$ is a final perturbation chosen so as to achieve transversality.

We construct the extension $\widetilde{\nu}_J$ by extending each component $\mu_J^j, j \in J$, of μ_J separately. In turn, we construct each μ_J^j by an elaborate iterative process over the increasing family of sets $W_\ell, 1 \le \ell \le k$, defined in equation (7.3.25) of [**MW2**]. Here W_ℓ is a carefully chosen neighbourhood of the part $\bigcup_{|H| \le \ell} N_{JH}^{k+\frac{1}{2}}$ of the enlarged core $N_J^{k+\frac{1}{2}}$ defined by sets H with $|H| \le \ell$. In particular, when $\ell = k$ we define

$$W_k := B^J_{\eta_{k+\frac{1}{2}}}(N_J^{k+\frac{1}{2}}) =: W_J \subset \overline{W_J} \subset V_I^k, \tag{2.4.16}$$

where the last inclusion follows from (2.4.12). Thus, omitting J from the notation, we need to construct for $1 \le \ell \le k$ functions $\widetilde{\mu}_\ell^j : W_\ell \to E_j$ that extend μ_J^j and satisfy certain vanishing and size conditions that will guarantee (a-e). Again simplifying by omitting j from the notation, it turns out to suffice to construct the extension $\widetilde{\mu}_\ell$ on a certain set B'_ℓ, that is a union of disjoint sets B'_L, one for each $L \subset J$ with $|L| = \ell$. For each L, we localize the latter extension problem, reducing it to the construction of extensions $\widetilde{\mu}_z$ near each point z in the set B'_L of [**MW2**, Equation (7.3.27)], that we then sum up using a partition of unity. In most cases we can choose $\widetilde{\mu}_z$ either to be zero or to be given by the compatibility conditions. In fact, the only case in which this extension is nontrivial is when z is in the enlarged core, more precisely the case $z \in \overline{N_{JL}^{k+\frac{1}{2}}}$. In that case we define $\widetilde{\mu}_z$ by extending $\mu_J^j|_{B^J_{r_z}(z) \cap N^k_{JL}}$ to be "constant in the normal directions" as described in the third bullet point of the construction.

When these extensions have all been constructed for $k \le \ell$ and $j \in J$, we define

$$\widetilde{\nu}_J := \beta \cdot \left(\textstyle\sum_{j \in J} \widetilde{\mu}_k^j\right), \tag{2.4.17}$$

where $\beta : U_J \to [0,1]$ is a smooth cutoff function with $\beta|_{N_J^{k+\frac{1}{2}}} \equiv 1$ and $\operatorname{supp} \beta \subset W_J$, so that $\widetilde{\nu}_J$ extends trivially to $U_J \setminus W_J$. Here are some important points.

(A) By (2.4.12) the constants η_k are chosen so that $W_J \cap N_J^k$ is compactly contained in $N_J^{k+\frac{1}{4}}$. Further, (2.4.11) implies that $B^J_{\eta_{k+\frac{1}{2}}}(N_{JI}^{k+\frac{3}{4}} \cap N_{JI}^k) = N_{JI}^{k+\frac{1}{2}}$. Together with tameness this gives

$$s_J^{-1}(E_I) \cap B^J_{\eta_{k+\frac{1}{2}}}(N_{JI}^{k+\frac{3}{4}}) \subset N_{JI}^{k+\frac{1}{2}}$$

which means that at each point of the closure $cl\big(B^J_{\eta_{k+\frac{1}{2}}}(N_{JI}^{k+\frac{1}{2}})\big) \smallsetminus N_{JI}^k$ at least one component of $(s_J^j)_{j\in J\smallsetminus I}$ is nonzero. This gives control over zero sets as in (C), (D) below, for all sufficiently small perturbations $\widetilde{\nu}_J$ that are admissible as in (B).

(B) The following strong admissibility condition holds: if $I \subsetneq J$ and $j \in J\smallsetminus I$ then $\widetilde{\mu}_J^j = 0$ on $B^J_{\eta_{k+\frac{1}{2}}}(N_{JI}^{k+\frac{1}{2}}) \subset W_J$ and on N_{JI}^k.

(C) The section $s_J + \widetilde{\nu}_J$ is transverse to 0 on $B := B^J_{\eta_{k+\frac{1}{2}}}(N_J^{k+\frac{3}{4}}) \subset W_J$. (This holds because by (A),(B) all zeros in B must lie in the subset $N_J^{k+\frac{1}{2}}$ of B; but here the perturbation is pulled back from $\nu_I, I \subsetneq J$, so that zeros are transverse by the inductive hypothesis.)

(D) For any perturbation $\widetilde{\nu}_J$ with support in W_J, $\|\widetilde{\nu}_J\| < \sigma$, and satisfying the admissibility condition (B), we have

$$V_J^{k+1} \cap (s_J + \widetilde{\nu}_J)^{-1}(0) \subset N_J^{k+\frac{1}{4}} \cup (\widetilde{C}_J \smallsetminus \overline{W_J}).$$

The set B in (C) above compactly contains the neighbourhood $B' := B^J_{\eta_{k+1}}(N_J^{k+1})$ of the core $N := N_J^{|J|}$ on which compatibility requires $\nu_J|_N = \mu_J|_N = \widetilde{\nu}_J|_N$.

At this stage conditions a), c), d), e) hold, so that we only need work to achieve transversality b) while keeping $\nu_\pitchfork$ so small that e) and hence d) remain true. The difficulty here is to keep control of the zero set since it is not a compact subset of V_J^{k+1} though of course it is compact in $\overline{V_J^{k+1}}$. We first choose a relatively open neighbourhood W of $\overline{V_J^{k+1}} \smallsetminus B$ in $\overline{V_J^{k+1}}$ so that

$$(s_J + \widetilde{\nu}_J)^{-1}(0) \cap W \subset O := \overline{V_J^{k+1}} \cap \widetilde{C}_J, \quad \overline{B'} \cap W = \emptyset.$$

This is possible by the zero set condition (D) and the fact $B' \sqsubset B \subset W_J$. By (C), transversality holds in B and hence outside the compact subset $\overline{V_J^{k+1}} \smallsetminus B \subset W$. Hence there is an open precompact subset $P \sqsubset W$ (the complement of a slight shrinking of B) so that transversality holds on $W \smallsetminus P$. Finally we choose $\nu_\pitchfork$ to be a very small smooth function with support in P and values in E_J so that the $s_J + \widetilde{\nu}_J + \nu_\pitchfork$ is transverse on all of W and hence on all V_J^{k+1}, and is such that $(s_J + \widetilde{\nu}_J + \nu_\pitchfork)^{-1}(0) \subset O \subset \widetilde{C}_J$. This completes the inductive step, and hence the construction of ν. For more details see [**MW2**]. □

To show that different choices lead to cobordant zero sets we also need a relative version of this construction. The relevant constant $\sigma_{\rm rel}(\delta, \mathcal{V}, \mathcal{C})$ depends on the given data, and in particular on the constants $\sigma(\delta, \mathcal{V}^\alpha, \mathcal{C}^\alpha), \alpha = 0, 1$ that occur in Proposition 2.4.10. (See [**MW2**, Definition 7.3.8] for a precise formula.)

PROPOSITION 2.4.11. *Let $(\mathcal{K}, d)$ be a metric tame Kuranishi cobordism with nested cobordism reduction $\mathcal{C} \sqsubset \mathcal{V}$, let $0 < \delta < \min\{\varepsilon, \delta_{\mathcal{V}}\}$, where ε is the collar* *width of* $(\mathcal{K}, d)$ *and the reductions* $\mathcal{C}, \mathcal{V}$. *Then we have* $\sigma_{\rm rel}(\delta, \mathcal{V}, \mathcal{C}) > 0$ *and the following holds.*

(1) *Given any* $0 < \sigma \le \sigma_{\rm rel}(\delta, \mathcal{V}, \mathcal{C})$, *there exists an admissible, precompact, transverse cobordism perturbation* ν *of* $\mathfrak{s}_{\mathcal{K}}|_{\mathcal{V}}$ *with* $\pi_{\mathcal{K}}\big((\mathfrak{s}_{\mathcal{K}}|_{\mathcal{V}} + \nu)^{-1}(0)\big) \subset \pi_{\mathcal{K}}(\mathcal{C})$, *whose restrictions* $\nu|_{\partial^\alpha \mathcal{V}}$ *for* $\alpha = 0, 1$ *are* $(\partial^\alpha \mathcal{V}, \partial^\alpha \mathcal{C}, \delta, \sigma)$*-adapted perturbations of* $\mathfrak{s}_{\partial^\alpha \mathcal{K}}|_{\partial^\alpha \mathcal{V}}$.

(2) *Given any perturbations* ν^α *of* $\mathfrak{s}_{\partial^\alpha \mathcal{K}}|_{\partial^\alpha \mathcal{V}}$ *for* $\alpha = 0, 1$ *that are* $(\partial^\alpha \mathcal{V}, \partial^\alpha \mathcal{C}, \delta, \sigma)$*-adapted with* $\sigma \le \sigma_{\rm rel}(\delta, \mathcal{V}, \mathcal{C})$, *the perturbation* ν *of* $\mathfrak{s}_{\mathcal{K}}|_{\mathcal{V}}$ *in (i) can be constructed to have boundary values* $\nu|_{\partial^\alpha \mathcal{V}} = \nu^\alpha$ *for* $\alpha = 0, 1$.

This is proved by making minor modifications in the construction given above. If $\mathcal{K}$ is a metric tame Kuranishi atlas with nested reduction $\mathcal{C} \sqsubset \mathcal{V}$, then a perturbation ν of $\mathfrak{s}_{\mathcal{K}}|_{\mathcal{V}}$ is said to be **strongly** $(\mathcal{V}, \mathcal{C})$**-adapted** if it is $(\mathcal{V}, \mathcal{C}, \delta, \sigma)$-adapted for some $\delta < \delta_{\mathcal{V}}$ and $\sigma < \sigma_{\rm rel}(\delta, [0,1] \times \mathcal{V}, [0,1] \times \mathcal{C})$ where the constant $\sigma_{\rm rel}(\delta, \mathcal{V}, \mathcal{C})$ is defined with respect to the product cobordism $[0,1] \times \mathcal{K}$ and reductions $[0,1] \times \mathcal{C} \sqsubset [0,1] \times \mathcal{V}$. Note that the constant $\sigma_{\rm rel}(\delta, [0,1] \times \mathcal{V}, [0,1] \times \mathcal{C})$ in this definition depends only on $\mathcal{K}, \mathcal{V}, \mathcal{C}$ and the choice of δ. Using Proposition 2.4.11 one can then show the following.[19]

COROLLARY 2.4.12. *Given any two strongly* $(\mathcal{V}, \mathcal{C})$*-adapted perturbations* ν^0, ν^1 *of* $\mathcal{K}$, *there is an admissible, precompact, transverse cobordism perturbation* ν *of* $\mathfrak{s}_{[0,1]\times\mathcal{K}}|_{[0,1]\times\mathcal{V}}$ *that restricts to* ν^α *on* $\mathcal{K} \equiv \{\alpha\} \times \mathcal{K}$ *for* $\alpha = 0, 1$.

Sketch proof of Theorem B. By Theorem 2.2.3, each weak Kuranishi atlas has a tame shrinking $\mathcal{K}$ that is unique up to concordance, and by Proposition 2.4.2 $\mathcal{K}$ has a reduction $\mathcal{V}$ that is also unique up to concordance. Proposition 2.4.10 shows that there is a corresponding admissible, precompact, transverse perturbation ν, such that the corresponding zero set is an oriented manifold $|\mathbf{Z}^\nu|$ by Proposition 2.4.6. Moreover, there is a natural inclusion $\iota_Z : |\mathbf{Z}^\nu| \to |\mathcal{V}| \subset |\mathcal{K}|$, and using Corollary 2.4.12 one can show that if ν is $(\mathcal{V}, \mathcal{C})$-strongly adapted for some $\mathcal{C} \sqsubset \mathcal{V}$ the oriented bordism class of this inclusion is independent of choices. By construction $|\mathcal{V}|$ lies in a δ-neighbourhood of the zero set $|\mathfrak{s}|^{-1}(0) = \iota_{\mathcal{K}}(X)$ of $|\mathcal{K}|$. Hence, we may obtain a well defined element in the rational Čech homology group $\check{H}_d(X; \mathbb{Q})$ by taking an appropriate inverse limit as $\delta \to 0$. This is the basic idea of the proof. For (many) more details see [**MW2**, §8.2].

3. Kuranishi atlases with nontrivial isotropy

The main change in this case is that the domains of the charts are no longer smooth manifolds, but rather group quotients (U, Γ) where Γ is a finite group acting on U. We assume throughout that U is smooth. Our definitions are chosen so that the quotient of a Kuranishi chart $\mathbf{K}$ with isotropy group Γ is a topological Kuranishi chart $\underline{\mathbf{K}}$ (with trivial isotropy) that we call the **intermediate** chart. Similarly the quotient of a weak Kuranishi atlas by its finite isotropy groups is a weak topological Kuranishi atlas.[20] Hence we can apply the results of §2.3 to tame the intermediate atlas, and then lift this to a taming of $\mathcal{K}$.

[19] This corollary is not explicitly stated in [**MW2**] but can be extracted from the proof given there that the VFC is independent of choices; see [**MW2**, Remark 7.3.9].

[20] Note that we do not consider topological atlases with nontrivial isotropy.

The other key new idea is that the coordinate changes $\mathbf{K}_I \to \mathbf{K}_J$ should no longer be given by inclusions ϕ_{IJ} of an open subset $U_{IJ} \subset U_I$ into U_J. These inclusions exist on the intermediate level as $\underline{\phi}_{IJ} : \underline{U}_{IJ} \to \underline{U}_J$. However, it is the inverse

$$\underline{\phi}_{IJ}^{-1} : \underline{\phi}_{IJ}(\underline{U}_{IJ}) \to \underline{U}_{IJ}$$

that lifts to the charts themselves: there is a Γ_J-invariant submanifold $\widetilde{U}_{IJ} \subset s_J^{-1}(E_I)$ on which the kernel of the natural projection $\Gamma_J \to \Gamma_I$ acts freely with quotient homeomorphic to a Γ_I-invariant subset U_{IJ} of U_I. This gives a **covering map** $\rho_{IJ} : \widetilde{U}_{IJ} \to U_{IJ} \subset U_I$ that descends on the intermediate level to the inverse $\underline{\phi}_{IJ}^{-1}$ of $\underline{\phi}_{IJ}$. In the Gromov–Witten setting, these maps ρ_{IJ} occur very naturally as maps that forget certain sets of added marked points. (See the end of Lecture 2 in [**M5**], and §4 below.) Another important point is that the coordinate changes preserve stabilizer subgroups. Hence the construction equips each point of X with a stabilizer subgroup, well defined up to isomorphism. (In practice, one of course starts with a space X whose points do have well defined stabilizer subgroups, see for instance Example 3.4.8 below.)

Most of the proofs are routine generalizations of those in §2; the only real difficulty is to make appropriate definitions. This section therefore consists mostly of notation and definitions. The main reference is [**MW3**].

3.1. Kuranishi atlases.

DEFINITION 3.1.1. *A* **group quotient** *is a pair* (U, Γ) *consisting of a smooth manifold* U *and a finite group* Γ *together with a smooth action* $\Gamma \times U \to U$. *We will denote the quotient space by*

$$\underline{U} := {}^{U}\!/_{\Gamma},$$

giving it the quotient topology, and write $\pi : U \to \underline{U}$ *for the associated projection. Moreover, we denote the* **stabilizer** *of each* $x \in U$ *by*

$$\Gamma^x := \{\gamma \in \Gamma \,|\, \gamma x = x\} \subset \Gamma.$$

Both the basic and transition charts of Kuranishi atlases will be group quotients, related by coordinate changes that are composites of the following kinds of maps.

DEFINITION 3.1.2. *Let* $(U, \Gamma), (U', \Gamma')$ *be group quotients. A* **group embedding**

$$(\phi, \phi^\Gamma) : (U, \Gamma) \to (U', \Gamma')$$

is a smooth embedding $\phi : U \to U'$ *with the following properties:*

- *it is equivariant with respect to an injective group homomorphism* $\phi^\Gamma : \Gamma \to \Gamma'$;
- *for each* $x \in U$, ϕ^Γ *induces an isomorphism between the stabilizer subgroup* $\Gamma^x \subset \Gamma$ *and the stabilizer* $\Gamma'^{\phi(x)} \subset \Gamma'$;
- *it induces an injection* $\underline{\phi} : \underline{U} \to \underline{U}'$ *on the quotient spaces.*

Note that if $\phi^\Gamma : \Gamma \to \Gamma' := \Gamma$ is surjective (and hence an isomorphism) then the condition on the stabilizers is automatically satisfied. However, in general, this condition is needed to avoid the situation where Γ' contains elements that act trivially on U' and hence do not affect the quotient $\underline{U}'$ and yet are not in the image of Γ.[21] In a Kuranishi atlas we often consider embeddings $(\phi, \phi^\Gamma) : (U, \Gamma) \to (U', \Gamma)$ where $\dim U < \dim U'$ and $\phi^\Gamma : \Gamma \to \Gamma' := \Gamma$ is the identity map. On the other hand, group quotients of the same dimension are usually related either by restriction or by coverings as follows.

DEFINITION 3.1.3. *Let (U, Γ) be a group quotient and $\underline{V} \subset \underline{U}$ an open subset. Then the* **restriction of** (U, Γ) **to** $\underline{V}$ *is the group quotient $(\pi^{-1}(\underline{V}), \Gamma)$.*

Note that the inclusion $\pi^{-1}(\underline{V}) \to U$ induces an equidimensional group embedding $(\pi^{-1}(\underline{V}), \Gamma) \to (U, \Gamma)$ that covers the inclusion $\underline{V} \to \underline{U}$. The third kind of map that occurs in a coordinate change is a group covering. This notion is less routine; notice in particular the requirement in (ii) that $\ker \rho^\Gamma$ act freely. Further, the two domains $\widetilde{U}, U$ will necessarily have the same dimension since they are related by a regular covering ρ.

DEFINITION 3.1.4. *Let (U, Γ) be a group quotient. A* **group covering** *of (U, Γ) is a tuple $(\widetilde{U}, \widetilde{\Gamma}, \rho, \rho^\Gamma)$ consisting of*

(i) *a surjective group homomorphism $\rho^\Gamma : \widetilde{\Gamma} \to \Gamma$,*
(ii) *a group quotient $(\widetilde{U}, \widetilde{\Gamma})$ where $\ker \rho^\Gamma$ acts freely,*
(iii) *a regular covering $\rho : \widetilde{U} \to U$ that is the quotient map $\widetilde{U} \to {}^{\widetilde{U}}/_{\ker \rho^\Gamma}$ composed with a diffeomorphism ${}^{\widetilde{U}}/_{\ker \rho^\Gamma} \cong U$ that is equivariant with respect to the induced $\Gamma = \mathrm{im}\ \rho^\Gamma$ action on both spaces.*

Thus $\rho : \widetilde{U} \to U$ is a principal $\ker \rho^\Gamma$-bundle that is ρ^Γ-equivariant. We denote by $\underline{\rho} : \underline{\widetilde{U}} \to \underline{U}$ the induced map on quotients.

Here is an elementary but important lemma ([**MW3**, Lemma 2.1.5]).

LEMMA 3.1.5. *Let (U, Γ) be a group quotient.*

(i) *The projection $\pi : U \to \underline{U}$ is open and proper (i.e. the inverse image of a compact set is compact).*
(ii) *Every point $x \in U$ has a neighbourhood U_x that is invariant under Γ^x and is such that inclusion $U_x \hookrightarrow U$ induces a homeomorphism from ${}^{U_x}/_{\Gamma^x}$ to $\pi(U_x)$. In particular, the inclusion $(U_x, \Gamma^x) \to (U, \Gamma)$ is a group embedding.*
(iii) *If $(\widetilde{U}, \widetilde{\Gamma}, \rho, \rho^\Gamma)$ is a group covering of (U, Γ), then*
  - *for all $\widetilde{x} \in \widetilde{U}$, ρ^Γ induces an isomorphism $\Gamma^{\widetilde{x}} \to \Gamma^{\rho(\widetilde{x})}$;*
  - *$\underline{\rho} : \underline{\widetilde{U}} \to \underline{U}$ is a homeomorphism.*

SKETCH OF PROOF. Parts (i) and (ii) are well known. To prove (iii) notice first that because $\ker \rho^\Gamma$ acts freely on $\widetilde{U}$ we have $\Gamma^{\widetilde{x}} \cap \ker \rho^\Gamma = \{\mathrm{id}\}$. Hence ρ^Γ induces an injection from $\Gamma^{\widetilde{x}}$ onto a subgroup of Γ^x, where $x := \rho(\widetilde{x})$, and it suffices to show that $\Gamma^{\widetilde{x}}, \Gamma^x$ have the same order. But this holds because the subgroup $(\rho^\Gamma)^{-1}(\Gamma^x)$ has order equal to the product $|\Gamma^x|\,|\ker \rho^\Gamma|$ and acts transitively on the set $\rho^{-1}(x)$, which has $|\ker \rho^\Gamma|$ elements. The last claim then follows from (ii). □

DEFINITION 3.1.6. *A smooth Kuranishi chart for X is a tuple $\mathbf{K} = (U, E, \Gamma, s, \psi)$ made up of*

[21] One can also express this definition in terms of the topological category $\mathbf{C}_{U,\Gamma}$ with objects U and morphisms $U \times \Gamma$ associated to a group quotient (U, Γ). In this language we are simply saying that the functor $\mathbf{C}_{U,\Gamma} \to \mathbf{C}_{U',\Gamma'}$ defined by (ϕ, ϕ^Γ) is fully faithful.

- *the* **domain** U *which is a smooth finite dimensional manifold;*
- *a finite dimensional vector space* E *called the* **obstruction space***;*
- *a finite* **isotropy group** Γ *with a smooth action on* U *and a linear action on* E*;*
- *a smooth* Γ*-equivariant function* $s : U \to E$*, called the* **section***;*
- *a continuous map* $\psi : s^{-1}(0) \to X$ *that induces a homeomorphism*

$$\underline{\psi} : \underline{s^{-1}(0)} := {}^{s^{-1}(0)}/_{\Gamma} \to F$$

with open image $F \subset X$*, called the* **footprint** *of the chart.*

The **dimension** *of* $\mathbf{K}$ *is* $\dim(\mathbf{K}) := \dim U - \dim E$*, and we will say that the chart is*

- **minimal** *if there is a point* $x \in s^{-1}(0)$ *at which* $\underline{\psi}$ *is injective, i.e.* $x = \psi^{-1}(\psi(x))$*, or equivalently* $\Gamma x = x$*.*

In order to extend topological constructions from §2 to the case of nontrivial isotropy, we will also consider the following notion of intermediate Kuranishi charts which have trivial isotropy but less smooth structure.

DEFINITION 3.1.7. *We associate to each Kuranishi chart* $\mathbf{K} = (U, E, \Gamma, s, \psi)$ *the* **intermediate chart** $\underline{\mathbf{K}} := (\underline{U}, \underline{U \times E}, \underline{\mathfrak{s}}, \underline{\psi})$*, where* $\underline{U \times E}$ *is the quotient by the diagonal action of* Γ *and* $\underline{\mathfrak{s}}$ *is the section of the bundle* $\underline{\mathrm{pr}} : \underline{U \times E} \to \underline{U}$ *induced by* $\mathfrak{s} = \mathrm{id}_U \times s : U \to U \times E$*.*

Thus, as we explained in Remark 2.3.2, $\underline{\mathbf{K}}$ is a topological chart in the sense of Definition 2.3.1. We write $\pi : U \to \underline{U} := {}^{U}/_{\Gamma}$ for the projection from the Kuranishi domain to the intermediate domain, and will distinguish everything connected to the intermediate charts by underlines. Moreover if a chart $\mathbf{K}_I = (U_I, E_I, \Gamma_I, s_I, \psi_I)$ has the label I, the corresponding projection is denoted $\pi_I : U_I \to \underline{U}_I$.

We will find that many results (in particular the taming constructions) from §2 carry over to nontrivial isotropy via the intermediate charts, since precompact subsets of $\underline{U}$ lift to precompact subsets of U by Lemma 3.1.5 (i). An important exception is the construction of perturbations which must be done on the smooth spaces U rather than on their quotients $\underline{U}$.

DEFINITION 3.1.8. *Let* $\mathbf{K} = (U, E, \Gamma, s, \psi)$ *be a Kuranishi chart and* $F' \subset F$ *an open subset of its footprint. A* **restriction of K to** F' *is a Kuranishi chart of the form*

$$\mathbf{K}' = \mathbf{K}|_{\underline{U}'} := \big(U', E, \Gamma, s' = s|_{U'}, \psi' = \psi|_{s'^{-1}(0)}\big), \qquad U' = \pi^{-1}(\underline{U}')$$

given by a choice of open subset $\underline{U}' \subset \underline{U}$ *such that* $\underline{U}' \cap \underline{\psi}^{-1}(F) = \underline{\psi}^{-1}(F')$*. We call* $\underline{U}'$ *the* **intermediate domain** *of the restriction and* U' *its* **domain**.

Note that the restriction $\mathbf{K}'$ in the above definition has footprint $\psi'(s'^{-1}(0)) = F'$, and its domain group quotient (U', Γ) is the restriction of (U, Γ) to $\underline{U}'$ in the sense of Definition 3.1.3.

Moreover, because the restriction of a chart is determined by a subset of the intermediate domain $\underline{U}$, all results about restrictions are easy to generalize to the case of nontrivial isotropy. In particular the following result holds, where we use the notation $\sqsubset$ to denote a precompact inclusion and $\mathrm{cl}_V(V')$ to denote the closure of a subset $V' \subset V$ in the relative topology of V.

LEMMA 3.1.9. *Let* $\mathbf{K}$ *be a Kuranishi chart. Then for any open subset* $F' \subset F$ *there is a restriction* $\mathbf{K}'$ *to* F' *with domain* U' *such that* $\mathrm{cl}_U(U') \cap s^{-1}(0) = \psi^{-1}(\mathrm{cl}_X(F'))$. *Moreover, if* $F' \sqsubset F$ *is precompact, then* $U' \sqsubset U$ *can be chosen precompact.*

Most definitions in §2 extend with only minor changes to the case of nontrivial isotropy. However, the notion of coordinate change needs to be generalized significantly to include a covering map. We will again formulate the definition in the situation that is relevant to Kuranishi atlases. That is, we suppose that a finite set of Kuranishi charts $(\mathbf{K}_i)_{i\in\{1,\dots,N\}}$ is given such that for each $I \subset \{1,\dots,N\}$ with $F_I := \bigcap_{i\in I} F_i \neq \emptyset$ we have another Kuranishi chart $\mathbf{K}_I$ with[22]

- group $\Gamma_I = \prod_{i\in I}\Gamma_i$,
- obstruction space $E_I = \prod_{i\in I} E_i$ on which Γ_I acts with the product action,
- footprint $F_I := \bigcap_{i\in I} F_i$.

Then for $I \subset J$ note that the natural inclusion $\widehat{\phi}_{IJ} : E_I \to E_J$ is equivariant with respect to the inclusion $\Gamma_I \hookrightarrow \Gamma_I \times \{\mathrm{id}\} \subset \Gamma_J$, and we have a natural splitting $\Gamma_J \cong \Gamma_I \times \Gamma_{J\setminus I}$, so that the complement $\Gamma_{J\setminus I}$ of the inclusion acts trivially on the image $\widehat{\phi}_{IJ}(E_I) \subset E_J$.

DEFINITION 3.1.10. *Given* $I \subset J \subset \{1,\dots,N\}$ *let* $\mathbf{K}_I$ *and* $\mathbf{K}_J$ *be Kuranishi charts as above, so that* $F_I \supset F_J$. *A* **coordinate change** $\widehat{\Phi}_{IJ}$ *from* $\mathbf{K}_I$ *to* $\mathbf{K}_J$ *consists of*

- *a choice of domain* $\underline{U}_{IJ} \subset \underline{U}_I$ *such that* $\mathbf{K}_I|_{\underline{U}_{IJ}}$ *is a restriction of* $\mathbf{K}_I$ *to* F_J,
- *the splitting* $\Gamma_J \cong \Gamma_I \times \Gamma_{J\setminus I}$ *as above, and the induced inclusion* $\Gamma_I \hookrightarrow \Gamma_J$ *and projection* $\rho^\Gamma_{IJ} : \Gamma_J \to \Gamma_I$,
- *a* Γ_J*-invariant submanifold* $\widetilde{U}_{IJ} \subset U_J$ *on which* $\Gamma_{J\setminus I}$ *acts freely, and the induced* Γ_J*-equivariant inclusion* $\widetilde{\phi}_{IJ} : \widetilde{U}_{IJ} \to U_J$;
- *a group covering* $(\widetilde{U}_{IJ}, \Gamma_J, \rho_{IJ}, \rho^\Gamma_{IJ})$ *of* (U_{IJ}, Γ_I), *where* $U_{IJ} := \pi_I^{-1}(\underline{U}_{IJ}) \subset U_I$,
- *the linear* Γ_I*-equivariant injection* $\widehat{\phi}_{IJ} : E_I \to E_J$ *as above,*

such that the Γ_J*-equivariant inclusion* $\widetilde{\phi} : \widetilde{U}_{IJ} \hookrightarrow U_J$ *intertwines the sections and footprint maps,*

$$s_J \circ \widetilde{\phi}_{IJ} = \widehat{\phi}_{IJ} \circ s_I \circ \rho_{IJ} \quad \text{on } \widetilde{U}_{IJ}, \qquad \psi_J \circ \widetilde{\phi}_{IJ} = \psi_I \circ \rho_{IJ} \quad \text{on } \rho^{-1}(s_I^{-1}(0)). \tag{3.1.1}$$

Moreover, we denote $s_{IJ} := s_I \circ \rho_{IJ} : \widetilde{U}_{IJ} \to E_I$ *and require the* **index condition**:

(1) *the embedding* $\widetilde{\phi}_{IJ} : \widetilde{U}_{IJ} \hookrightarrow U_J$ *identifies the kernels,*

$$\mathrm{d}_u \widetilde{\phi}_{IJ}\big(\ker \mathrm{d}_u s_{IJ}\big) = \ker \mathrm{d}_{\widetilde{\phi}_{IJ}(u)} s_J \qquad \forall u \in \widetilde{U}_{IJ};$$

(2) *the linear embedding* $\widehat{\phi}_{IJ} : E_I \to E_J$ *identifies the cokernels,*

$$\forall u \in \widetilde{U}_{IJ} : \qquad E_I = \mathrm{im}\, \mathrm{d}_u s_{IJ} \oplus C_{u,I} \quad \Longrightarrow \quad E_J = \mathrm{im}\, \mathrm{d}_{\widetilde{\phi}_{IJ}(u)} s_J \oplus \widehat{\phi}(C_{u,I}).$$

The subset $\underline{U}_{IJ} \subset \underline{U}_I$ *is called the* **domain** *of the coordinate change, while* $\widetilde{U}_{IJ} \subset U_J$ *is its* **lifted domain**.

---

[22]Generalizations of these conditions will be considered in §6.

REMARK 3.1.11. (i) If the isotropy and covering $\rho_{IJ} =: \phi_{IJ}^{-1}$ are both trivial, this definition agrees with that in §2.1 with $\widetilde{U}_{IJ} = \phi_{IJ}(U_{IJ})$. Further, the index condition together with the condition that $\widetilde{U}_{IJ}$ is a submanifold of U_J implies that $\widetilde{U}_{IJ}$ is an open subset of $s_J^{-1}(E_I)$.

(ii) The following diagram of group embeddings and group coverings is associated to each coordinate change:

$$
\begin{array}{ccc}
 & (\widetilde{U}_{IJ}, \Gamma_J) & \stackrel{(\widetilde{\phi}_{IJ}, \mathrm{id})}{\longrightarrow} & (U_J, \Gamma_J) \\
 & \downarrow (\rho_{IJ}, \rho_{IJ}^{\Gamma}) & & \\
(U_I, \Gamma_I) \longleftarrow & (U_{IJ}, \Gamma_I) & &
\end{array} \tag{3.1.2}
$$

(iii) Since $\underline{\rho}_{IJ} : \underline{\widetilde{U}}_{IJ} \to \underline{U}_{IJ}$ is a homeomorphism by Lemma 3.1.5 (iii), each coordinate change $(\phi_{IJ}, \widehat{\phi}_{IJ}, \rho_{IJ}) : \mathbf{K}_I|_{\underline{U}_{IJ}} \to \mathbf{K}_J$ induces an injective map

$$\underline{\phi_{IJ}} := \underline{\widetilde{\phi}_{IJ}} \circ \underline{\rho}_{IJ}^{-1} : \underline{U}_{IJ} \to \underline{U}_J$$

on the domains of the intermediate charts. In fact, there is an induced orbifold coordinate change $\underline{\widehat{\Phi}_{IJ}} : \underline{\mathbf{K}}_I|_{\underline{U}_{IJ}} \to \underline{\mathbf{K}}_J$ on the level of the intermediate charts, given by the bundle map $\underline{\widehat{\Phi}_{IJ}} : \underline{U_{IJ} \times E_I} \to \underline{U_J \times E_J}$ which is induced by the multivalued map $(\widetilde{\phi}_{IJ} \circ \rho_{IJ}^{-1}) \times \widehat{\phi}_{IJ}$ and hence covers $\underline{\widetilde{\phi}_{IJ}} \circ \underline{\rho}_{IJ}^{-1} =: \underline{\phi}_{IJ}$. This is a topological coordinate change as in Definition 2.3.3 that respects the orbibundle structure of the projection $\underline{\mathrm{pr}}_I : \underline{U_I \times E_I} \to \underline{U}_I$.

(iv) Conversely, suppose we are given an orbifold coordinate change $\underline{\widehat{\Phi}_{IJ}} : \underline{\mathbf{K}}_I \to \underline{\mathbf{K}}_J$ with domain $\underline{U}_{IJ}$. Then any coordinate change from $\mathbf{K}_I$ to $\mathbf{K}_J$ that induces $\underline{\widehat{\Phi}_{IJ}}$ is determined by the Γ_J-invariant set $\widetilde{U}_{IJ} := \pi_J^{-1}(\underline{\phi}(\underline{U}_{IJ}))$ and a choice of Γ_I-equivariant diffeomorphism between ${}^{\widetilde{U}_{IJ}}/_{\Gamma_{J\setminus I}}$ and $U_{IJ} := \pi_I^{-1}(\underline{U}_{IJ})$. When constructing coordinate changes in the Gromov–Witten setting, we will see that there is a natural choice of this diffeomorphism since the covering maps ρ_{IJ} are given by forgetting certain added marked points.

(v) Because $\widetilde{U}_{IJ}$ is defined to be a subset of U_J it is sometimes convenient to think of an element $x \in \widetilde{U}_{IJ}$ as an element in U_J, omitting the notation for the inclusion map $\widetilde{\phi}_{IJ} : \widetilde{U}_{IJ} \to U_J$. ◊

The next step is to consider restrictions and composites of coordinate changes. Restrictions behave as before. Thus, for $I \subset J$, given any restrictions $\mathbf{K}'_I := \mathbf{K}_I|_{\underline{U}'_I}$ and $\mathbf{K}'_J := \mathbf{K}_J|_{\underline{U}'_J}$ whose footprints $F'_I \cap F'_J$ have nonempty intersection, and any coordinate change $\mathbf{K}_I|_{\underline{U}_{IJ}} \to \mathbf{K}_J$, there is an induced **restricted coordinate change** $\mathbf{K}'_I|_{\underline{U}'_{IJ}} \to \mathbf{K}'_J$ for any subset $\underline{U}'_{IJ} \subset \underline{U}_{IJ}$ satisfying the conditions

$$\underline{U}'_{IJ} \subset \underline{U}'_I \cap \underline{\phi}^{-1}(\underline{U}'_J), \qquad \underline{U}'_{IJ} \cap \underline{s}_I^{-1}(0) = \underline{\psi}_I^{-1}(F'_I \cap F'_J). \tag{3.1.3}$$

However, coordinate changes now do not directly compose due to the coverings involved. The induced coordinate changes on the intermediate charts still compose directly, but the analog of [**MW1**, Lemma 2.2.5] is the following. The proof is routine.

LEMMA 3.1.12. *Let $I \subset J \subset K$ (so that automatically $F_I \cap F_K = F_J$) and suppose that $\widehat{\Phi}_{IJ} : \mathbf{K}_I \to \mathbf{K}_J$ and $\widehat{\Phi}_{JK} : \mathbf{K}_J \to \mathbf{K}_K$ are coordinate changes with domains $\underline{U}_{IJ}$ and $\underline{U}_{JK}$ respectively. Then the following holds.*

(1) *The domain $\underline{U}_{IJK} := \underline{U}_{IJ} \cap \underline{\phi}_{IJ}^{-1}(\underline{U}_{JK}) \subset \underline{U}_I$ defines a restriction $\mathbf{K}_I|_{\underline{U}_{IJK}}$ to F_K.*

(2) *The composite $\rho_{IJK} := \rho_{IJ} \circ \rho_{JK} : \widetilde{U}_{IJK} \to U_{IJK} := \pi_I^{-1}(\underline{U}_{IJK})$ is defined on $\widetilde{U}_{IJK} := \pi_K^{-1}\big((\underline{\phi}_{JK} \circ \underline{\phi}_{IJ})(\underline{U}_{IJK})\big)$ via the natural identification of $\rho_{JK}(\widetilde{U}_{IJK}) \subset U_J$ with a subset of $\widetilde{U}_{IJ}$. Together with the natural projection $\rho^\Gamma_{IK} : \Gamma_K \to \Gamma_I$ with kernel $\Gamma_{K \smallsetminus I}$ (which factors $\rho^\Gamma_{IK} = \rho^\Gamma_{IJ} \circ \rho^\Gamma_{JK}$), this forms a group covering $(\widetilde{U}_{IJK}, \Gamma_K, \rho_{IJK}, \rho^\Gamma_{IK})$ of (U_{IJK}, Γ_I).*

(3) *The inclusion $\widetilde{\phi}_{IJK} : \widetilde{U}_{IJK} \hookrightarrow U_K$ together with the natural inclusion $\widehat{\phi}_{IK} : E_I \to E_K$ (which factors $\widehat{\phi}_{IK} = \widehat{\phi}_{JK} \circ \widehat{\phi}_{IJ}$) and ρ_{IJK} satisfies (3.1.1) and the index condition with respect to the indices I, K.*

Hence this defines a **composite coordinate change** $\widehat{\Phi}_{IJK} = (\widetilde{\phi}_{IJK}, \widehat{\phi}_{IJK}, \rho_{IJK})$ *from $\mathbf{K}_I$ to $\mathbf{K}_K$ with domain $\underline{U}_{IJK}$.*

REMARK 3.1.13. The induced orbifold coordinate change $\underline{\widehat{\Phi}}_{IJK} = (\underline{\phi}_{IJK}, \underline{\widehat{\phi}}_{IJK})$ between the intermediate charts $\underline{\mathbf{K}}_I$ and $\underline{\mathbf{K}}_K$ is the composite $\underline{\widehat{\Phi}}_{JK} \circ \underline{\widehat{\Phi}}_{IJ}$ as considered in §2.1. (For more detail, see [**MW1**, Lemma 2.2.5].) ◊

With these notions of Kuranishi charts and coordinate changes in place, we can now directly extend the definition of Kuranishi atlas from §2.1. The notions of a covering family $(\mathbf{K}_i)_{i=1,\dots,N}$ of basic charts for X and of transition data

$$(\mathbf{K}_J)_{J \in \mathcal{I}_\mathcal{K}, |J| \ge 2}, (\widehat{\Phi}_{IJ})_{I,J \in \mathcal{I}_\mathcal{K}, I \subsetneq J}$$

are as before. The cocycle conditions can now mostly be expressed in terms of the intermediate charts.

DEFINITION 3.1.14. *Let $\mathbf{K}_\alpha$ for $\alpha = I, J, K$ be Kuranishi charts with $I \subset J \subset K$ and let $\widehat{\Phi}_{\alpha\beta} : \mathbf{K}_\alpha|_{\underline{U}_{\alpha\beta}} \to \mathbf{K}_\beta$ for $(\alpha, \beta) \in \{(I,J), (J,K), (I,K)\}$ be coordinate changes. We say that this triple $\widehat{\Phi}_{IJ}, \widehat{\Phi}_{JK}, \widehat{\Phi}_{IK}$ satisfies the*

- **weak cocycle condition** *if $\widehat{\Phi}_{JK} \circ \widehat{\Phi}_{IJ} \approx \widehat{\Phi}_{IK}$ are equal on the overlap in the sense*

$$\rho_{IK} = \rho_{IJ} \circ \rho_{JK} \quad \text{on } \widetilde{U}_{IK} \cap \rho_{JK}^{-1}(\widetilde{U}_{IJ} \cap U_{JK}); \tag{3.1.4}$$

- **cocycle condition** *if $\widehat{\Phi}_{JK} \circ \widehat{\Phi}_{IJ} \subset \widehat{\Phi}_{IK}$, i.e. $\widehat{\Phi}_{IK}$ extends the composed coordinate change in the sense that (3.1.4) holds and*

$$\underline{U}_{IJ} \cap \underline{\phi}_{IJ}^{-1}(\underline{U}_{JK}) \subset \underline{U}_{IK}; \tag{3.1.5}$$

- **strong cocycle condition** *if $\widehat{\Phi}_{JK} \circ \widehat{\Phi}_{IJ} = \widehat{\Phi}_{IK}$ are equal as coordinate changes, that is if (3.1.4) holds and*

$$\underline{U}_{IJ} \cap \underline{\phi}_{IJ}^{-1}(\underline{U}_{JK}) = \underline{U}_{IK}. \tag{3.1.6}$$

As in [**MW3**, Lemma 2.2.13], it is straightforward to check that the cocycle condition (3.1.5) implies that

$$\rho_{IK} = \rho_{IJ} \circ \rho_{JK} \quad \text{on } \rho_{JK}^{-1}(\widetilde{U}_{IJ} \cap U_{JK}) \subset \widetilde{U}_{IK}. \tag{3.1.7}$$

Similarly, condition (3.1.4) implies

$$\underline{\phi}_{IK} = \underline{\phi}_{JK} \circ \underline{\phi}_{IJ} \quad \text{on } \underline{U}_{IK} \cap \big(\underline{U}_{IJ} \cap \underline{\phi}_{IJ}^{-1}(\underline{U}_{JK})\big). \tag{3.1.8}$$

Therefore we are led to the following definition. It is identical to the original Definition 2.1.9 of an atlas, except that now we allow more general charts and coordinate changes.

DEFINITION 3.1.15. *A* **(weak) smooth Kuranishi atlas of dimension d** *on a compact metrizable space X is a tuple*

$$\mathcal{K} = \big(\mathbf{K}_I, \widehat{\Phi}_{IJ}\big)_{I,J\in\mathcal{I}_\mathcal{K}, I\subsetneq J}$$

consisting of a covering family of basic charts $(\mathbf{K}_i)_{i=1,\dots,N}$ of dimension d and transition data $\big((\mathbf{K}_J)_{|J|\geq 2}, (\widehat{\Phi}_{IJ})_{I\subsetneq J}\big)$ for $(\mathbf{K}_i)_{i=1,\dots,N}$ that satisfies the (weak) cocycle condition for every triple $I, J, K \in \mathcal{I}_K$ with $I \subsetneq J \subsetneq K$.

The corresponding **intermediate atlas** *has charts $(\underline{\mathbf{K}}_I)_{I\in\mathcal{I}_\mathcal{K}}$ and coordinate changes $(\underline{\widehat{\Phi}}_{IJ})_{I\subset J, I,J\in\mathcal{I}_\mathcal{K}}$.*

LEMMA 3.1.16. *The intermediate atlas $\underline{\mathcal{K}}$ corresponding to a (weak) smooth Kuranishi atlas $\mathcal{K}$ is a (weak) filtered topological atlas with $\mathbb{E}_{IJ} = \underline{U_J \times \widehat{\phi}_{IJ}(E_I)}$ for all $I \subset J$.*

PROOF. This holds by the proof of Lemma 2.3.8 (a). □

EXAMPLE 3.1.17. (i) We show in Proposition 5.1.6 that every compact smooth orbifold has an atlas. As an example, consider the "football" $Y = S^2$ with two orbifold points, one at the north pole of order 2 and one at the south pole of order 3. Take charts $(U_1, \mathbb{Z}_2)$, $(U_2, \mathbb{Z}_3)$ about north/south pole with $\underline{U}_1 \cap \underline{U}_2 = \underline{U}_{\{12\}} = \underline{A}$ an annulus around the equator. Let $A_i = \pi_i^{-1}(\underline{A})$ where $\pi_i : U_i \to \underline{U}_i$ is the projection.[23] Then the restriction of $(U_1, \mathbb{Z}_2)$ over $\underline{A}$ is $(A_1, \mathbb{Z}_2)$, whereas the restriction of $(U_2, \mathbb{Z}_3)$ over $\underline{A}$ is $(A_2, \mathbb{Z}_3)$. There is no direct relation between these restrictions because the coverings $A_1 \to \underline{A}$ and $A_2 \to \underline{A}$ are incompatible. However, they do have a common free covering, namely the pullback defined by the diagram

$$\begin{array}{ccc} U_{12} & \longrightarrow & U_1 \\ \downarrow & & \downarrow{\scriptstyle \pi_1} \\ U_2 & \xrightarrow{\pi_2} & Y \end{array}$$

i.e. $U_{12} := \{(x, y) \in U_1 \times U_2 \,|\, \pi_1(x) = \pi_2(y)\}$ with group $\Gamma_{12} := \Gamma_1 \times \Gamma_2$. This defines an atlas with two basic charts and one transition chart. For a noneffective example with trivial obstruction spaces, see [**M6**, Example 3.6].

Note that this construction cannot be imitated for general orbifolds since the above fiber product U_{12} is not smooth if there are any points in the overlap $\pi_1(U_1) \cap \pi_2(U_2)$ with nontrivial stabilizers. However, it turns out that there is a very simple substitute construction for the domain of the transition chart. Namely, if the charts (U_i, Γ_i) inject into a groupoid representative for Y then we can take U_{12} to be the morphisms in this groupoid from U_1 to U_2; see the proof of Proposition 5.1.6.

(ii) For another example that illustrates the effect of a noneffective group action, consider an atlas over $X = S^1$ with two basic charts, where

[23] For simplicity, we here identify each $\underline{U}_I$ with its image $F_I \subset Y$.

- $\mathbf{K}_1$ has $U_1 = S^1, E_1 = \{0\}$, the group $\Gamma_1 = \mathbb{Z}_2$ acting trivially, and footprint map $\psi_1 = \mathrm{id} : U_1 \to S^1$, while
- $\mathbf{K}_2$ has $U_2 = S^1 \times \mathbb{R}^2$, $E_2 = \mathbb{R}^2$, $s_2(x,e) = e$, and $\Gamma_2 = \mathbb{Z}_2$ where $\gamma \neq \mathrm{id}$ acts on U_2 and E_2 by $(x,e) \mapsto (x,-e)$ and $e \mapsto -e$, so that $\psi_2 : s_2^{-1}(0) = S^1 \times \{0\} \to S^1$ is induced by the projection.

The transition chart $\mathbf{K}_{12}$ has $E_{12} = \widehat{\phi}_{2,12}(E_2) = \mathbb{R}^2$ and $\Gamma_{12} = \mathbb{Z}_2 \times \mathbb{Z}_2$. We must have $U_{1,12} = U_1$ and if we take $U_{2,12} = U_2$ (there is some choice here), then in order for Γ_i to act freely on $\widetilde{U}_{j,12}$ for $i \neq j$, we must take $U_{12} = \widetilde{U}_{2,12} = \sqcup_{\beta=\pm}(S^1 \times \mathbb{R}^2)_\beta$ to be two copies of $S^1 \times \mathbb{R}^2$ that are interchanged by the action of either group Γ_i. Thus the group $\Gamma_{12} = \mathbb{Z}_2 \times \mathbb{Z}_2$ acts as follows:

$$\text{on } U_{12} \text{ by } \ (\gamma_1,\gamma_2)(x,e)_\beta = \big((x,\gamma_2 e)\big)_{\gamma_1\gamma_2(\beta)} \ \text{ where } \ \gamma_i(\beta) = (-1)^{\gamma_i}\beta,$$
$$\text{on } E_{12} \text{ by } \ (\gamma_1,\gamma_2)(e) = \gamma_2 e,$$

and the coordinate changes are given by

$$\widetilde{U}_{1,12} = \{(x,0)_\beta \mid x \in S^1\}, \quad \rho_{1,12} : (x,0)_\beta \mapsto x,$$
$$\widetilde{U}_{2,12} = U_{12}, \quad \rho_{2,12} : (x,e)_\beta \mapsto (x,e).$$

Notice that although the action of Γ_1 is totally noneffective, the group Γ_{12} acts effectively on U_{12}. Moreover the stabilizer $\Gamma_{12}^{\widetilde{x}}$ of the point $\widetilde{x} = (x,0)_\beta \in \widetilde{U}_{1,12}$ is the diagonal subgroup $(\gamma,\gamma) \in \Gamma_{12}$. Thus although $\rho^\Gamma_{1,12} : \Gamma_{12} \to \Gamma_1$ does induce an isomorphism $\Gamma_{12}^{\widetilde{x}} \to \Gamma_1^x$ for all $\widetilde{x} \in \rho_{1,12}^{-1}(x)$ (cf. Lemma 3.1.5 (iii)), it does not preserve the property of acting noneffectively. See Figure 3.4.1 in Example 3.4.7 for an illustration of the fundamental cycle of this atlas. $\Diamond$

REMARK 3.1.18. The definition of atlas requires that each transition chart $\mathbf{K}_I$ has obstruction spaces and groups that are products over $i \in I$. We will generalize the requirements on E_I in §6. Assuming for now that the obstruction spaces are products, we can generalize the requirements on the groups Γ_I as follows. We assume given charts $(\mathbf{K}_I)_{I,J\in\mathcal{I}_\mathcal{K}}$ whose domains are group quotients (U_I,Γ_I) with coordinate changes given by group coverings $(\rho_{IJ},\rho^\Gamma_{IJ}) : \widetilde{U}_{IJ} \to U_{IJ}$ and linear embeddings $\widehat{\phi}_{IJ} : E_I \to E_J$ as before, where $\rho^\Gamma_{IJ} : \Gamma_J \to \Gamma_I$ is a family of split surjective homomorphisms with splittings $\iota^\Gamma_{IJ} : \Gamma_I \to \Gamma_J$ such that

- $\rho_{IJ} \circ \rho_{JK} = \rho_{IK}$, $\rho^\Gamma_{IJ} \circ \rho^\Gamma_{JK} = \rho^\Gamma_{IK}$ and $\iota^\Gamma_{JK} \circ \iota^\Gamma_{IJ} = \iota^\Gamma_{IK}$ whenever $I \subset J \subset K$;
- for all $I \subset J$
- the linear inclusion $\widehat{\phi}_{IJ} : E_I \to E_J$ is ι^Γ_{IJ}-equivariant and $\ker(\rho^\Gamma_{IJ})$ acts trivially on its image;

We will call such atlases **tameable**. They are very natural when one considers products; cf. §6.3. It is not hard to check that all the constructions in this section, go through as before with this more general kind of atlas. In particular, in the construction of $\widehat{\mathbf{Z}}^\nu$ in Proposition 3.4.4 the group $\Gamma_{I\setminus F}$ that appears in (3.4.3) should be replaced by $\ker\rho^\Gamma_{FI}$. $\Diamond$

REMARK 3.1.19. We now show how to obtain a Kuranishi structure in the sense of [**FOOO1**] from a Kuranishi atlas in the case when the isotropy groups act effectively on the domains U, the only case they consider. Recall that to define a Kuranishi structure one needs to specify a family of Kuranishi charts $(\mathbf{K}_p)_{p\in X}$ with footprints $F_p \ni p$, together with coordinate changes[24] $(\underline{\phi}_{qp} : \mathbf{K}_q \to \mathbf{K}_p)_{q\in F_p}$ that are defined on the level of the intermediate atlas and satisfy the weak cocycle condition stated below. Even though there are uncountably many charts $\mathbf{K}_p$, Fukaya et al. construct them from a finite covering family (called a *primary family* in [**FF**]) in much the same way that we now describe.

- First choose a precompact "shrinking" $\{G_i \sqsubset F_i\}_{i=1,\dots,N}$ of the footprints, and for $p \in X$, define $I_p := \{i \,|\, p \in \overline{G_i}\}$.
- For $p \in X$ define $\mathbf{K}_p$ by choosing a restriction of the chart $\mathbf{K}_{I_p}$ to
$$F_p := \big(\bigcap_{i\in I_p} F_i\big) \smallsetminus \big(\bigcup_{j\notin I_p} \overline{G_j}\big).$$
More precisely, since $p \in F_p$, we choose a lift $\mathfrak{o}_p$ of p to U_{I_p}, take Γ_p to be the stabilizer subgroup of $\mathfrak{o}_p$ (so that the chart is minimal as required by [**FOOO**]) and take the domain $U_p \subset U_{I_p}$ to be a Γ_p-invariant neighbourhood of $\mathfrak{o}_p$ such that the induced map ${}^{U_p}/_{\Gamma_p} \to {}^{U_{I_p}}/_{\Gamma_{I_p}}$ is injective. (Suitable (U_p, Γ_p) exist by Lemma 3.1.5 (ii).)
- For $q \in F_p$ we have $I_q \subset I_p$ since by construction $F_p \cap G_i = \emptyset$ for $i \notin I_p$. So we obtain a coordinate change $\widehat{\Phi}_{qp} : \mathbf{K}_q \to \mathbf{K}_p$ from a suitable restriction of $\widehat{\Phi}_{I_q I_p}$ to a $\Gamma_q^{x_q}$-invariant neighbourhood $U_{qp} \subset U_q$ of x_q. More precisely, we choose $U_{qp} \subset U_q$ small enough so that the projection $\rho_{I_q I_p} : U_p \cap \widetilde{U}_{I_q I_p} \to U_{I_q I_p}$ has a continuous section over U_{qp}. We denote its image by $\widetilde{U}_{qp}$ and thus obtain an embedding $\phi_{qp} := \rho_{I_q I_p}^{-1} : U_{qp} \to \widetilde{U}_{qp} \subset U_p \cap \widetilde{U}_{I_q I_p}$. Since the projection $\rho_{I_q I_p}$ induces an isomorphism on stabilizer subgroups by Lemma 3.1.5 (iii), this is equivariant with respect to a suitable injective homomorphism $h_{qp} : \Gamma_q \to \Gamma_p$ and induces an injection $\underline{\phi}_{qp} : \underline{U}_{qp} := {}^{U_{qp}}/_{\Gamma_q} \to \underline{U}_p := {}^{U_p}/_{\Gamma_p}$. Moreover, because the induced map $\underline{U}_p = {}^{U_p}/_{\Gamma_p} \to \underline{U}_{I_p} = {}^{U_{I_p}}/_{\Gamma_{I_p}}$ is a homeomorphism to its image by construction and similarly for q, we can identify the underlying map $\underline{\phi}_{qp}$ of the coordinate change with a suitable restriction of $\underline{\phi}_{I_q I_p}$. The coordinate change $\widehat{\Phi}_{qp} = (U_{qp}, \underline{\phi}_{qp})$ is then given by the domain U_{qp} and the restriction of $\underline{\phi}_{I_q I_p}$ to $\underline{U}_{qp} \subset \underline{U}_q$.

  The weak cocycle condition for $\mathcal{K}$ implies the required compatibility condition, namely for all triples $p, q, r \in X$ with $q \in F_p$ and $r \in \psi_q(U_{qp} \cap s_q^{-1}(0)) \subset F_q \cap F_p$, the equality $\underline{\phi}_{qp} \circ \underline{\phi}_{rq} = \underline{\phi}_{rp}$ holds on the common domain $\underline{\phi}_{rq}^{-1}(\underline{U}_{qp}) \cap \underline{U}_{rp}$ of the maps in this equation.

This process of passing to small charts loses information about isotropy, and seems a little inefficient, in that one needs to rebuild larger charts in order to get a "good coordinate system". In [**FOOO1**] this is done only on the level of the intermediate atlas, so that they need to assume the group actions are effective.[25] Further, although it is not clear how relevant the extra information contained in a Kuranishi atlas is to the question of how to define Gromov–Witten invariants for closed curves in smooth manifolds, it might prove useful in other situations, for

[24] For consistency with our conventions we write $\underline{\phi}_{qp}$ rather than $\underline{\phi}_{pq}$ here.

[25] Although this is almost always true in the Gromov–Witten setting and can always be arranged by adding marked points, there are a few situations where it does not hold; cf. Example 3.4.8.

example in the case of orbifold Gromov–Witten invariants, or in the recent work of Fukaya et al [**FOOO2**] where the authors consider a process that rebuilds a Kuranishi structure from a coordinate system. One way of formulating the approach in [**FOOO1**] is that they do taming and reduction simultaneously. It might be possible to do that in our context. However, it is actually very useful to have the intermediate object of a Kuranishi atlas since this captures all the needed information about the coordinate changes in the simplest possible form. $\Diamond$

REMARK 3.1.20. Although it seems that many choices are needed in order to construct a Kuranishi atlas, this is somewhat deceptive. For example, in the Gromov–Witten case considered in § 4.1 below, the choices involved in the construction of a family of basic charts $(\mathbf{K}_i)_{i=1,\dots,N}$ essentially induce the transition data as well. Namely, for each $I \subset \{1,\dots,N\}$ such that $F_I := \bigcap_{i\in I} F_i$ is nonempty, we will construct a "transition chart" $\mathbf{K}_I$ with group $\Gamma_I := \prod_{i\in I}\Gamma_i$ and obstruction space $E_I := \prod_{i\in I} E_i$. Moreover, each E_i is a product of the form $E_i = \prod_{\gamma\in\Gamma_i}(E_{0i})_\gamma$ of copies of a vector space E_{0i} that are permuted by the action of Γ_i, and Γ_I acts on E_I by the obvious product action.

More precisely, each basic chart $\mathbf{K}_i$ is constructed by adding a certain tuple $\mathbf{w}_i$ of marked points[26] to the domains of the stable maps $[\Sigma_f, \mathbf{z}, f]$, given by the preimages of a fixed hypersurface of M. When seen on spaces of equivalence classes of maps, the action of Γ_i is easy to understand, since it simply permutes this set of marked points $\mathbf{w}_i$; see (4.1.2). Similarly, elements of the domains U_J of the transition charts consist of certain maps $f : \Sigma \to M$ with the given marked points $\mathbf{z}$ together with $|J|$ sets of added tuples of marked points $(\mathbf{w}_j)_{j\in J}$, each taken by f to certain hypersurfaces in M. Each factor Γ_j of the group Γ_J acts by permuting the elements of the j-th tuple of points, leaving the others alone. Moreover, the covering map $\widetilde{U}_{IJ} \to U_I$ simply forgets the extra tuples $(\mathbf{w}_j)_{j\in J\smallsetminus I}$. Thus it is immediate from the construction that the group $\Gamma_{J\smallsetminus I}$ acts freely on the subset $\widetilde{U}_{IJ}$ of U_J, and that the covering map is equivariant in the appropriate sense. Further, when $I \subset J \subset K$ the compatibility condition $\rho_{IK} = \rho_{IJ}\circ\rho_{JK}$ holds whenever both sides are defined. Therefore, once given the basic charts and the choice of domains for the transition charts, the only new choice needed to construct an atlas is that of the domains $\underline{U}_{IJ}$ of the coordinate changes which are required to intersect the zero set $\underline{s}_I^{-1}(0)$ in $\underline{\psi}_I^{-1}(F_J)$. Since there is no obvious way to organize these choices so as to satisfy the cocycle condition, concrete constructions will usually only satisfy a weak cocycle condition. $\Diamond$

3.2. Categories and tamings. Just as in §2.2, we will associate to each Kuranishi atlas $\mathcal{K}$ two topological categories $\mathbf{B}_\mathcal{K}, \mathbf{E}_\mathcal{K}$ together with functors

$$\mathrm{pr}_\mathcal{K} : \mathbf{E}_\mathcal{K} \to \mathbf{B}_\mathcal{K}, \quad s_\mathcal{K} : \mathbf{B}_\mathcal{K} \to \mathbf{E}_\mathcal{K}, \quad \psi_\mathcal{K} : s_\mathcal{K}^{-1}(0) \to \mathbf{X},$$

where $\mathbf{X}$ is the category with objects X and only identity morphisms. Recall here that the morphism spaces will only be closed under composition (and thus generate an equivalence relation that defines the realization $|\mathcal{K}|$ as ambient space for X) if the cocycle condition holds. Thus for the following we assume that $\mathcal{K}$ is a Kuranishi

[26]One difference between our construction and that in [**FOOO1**, **FF**] is that here the added marked points $\mathbf{w}$ are labelled, while in [**FOOO1**, **FF**] they are unordered.

atlas. Then, as before, the domain category $\mathbf{B}_{\mathcal{K}}$ has objects

$$\mathrm{Obj}_{\mathbf{B}_{\mathcal{K}}} := \bigsqcup_{I \in \mathcal{I}_{\mathcal{K}}} U_I \;=\; \{(I,x) \,|\, I \in \mathcal{I}_{\mathcal{K}}, x \in U_I\},$$

where we identify $x \in U_I$ with $(I,x) \in \mathrm{Obj}_{\mathbf{B}_{\mathcal{K}}}$. The morphisms in $\mathbf{B}_{\mathcal{K}}$ are composites of morphisms of the following two types.

(a) For each $I \in \mathcal{I}_{\mathcal{K}}$ the action of Γ_I gives rise to morphisms between points in U_I. These form a space $U_I \times \Gamma_I$ with source and target maps

$$\begin{array}{rcccl} s \times t: & U_I \times \Gamma_I & \longrightarrow & U_I \times U_I & \subset \mathrm{Obj}_{\mathbf{B}_{\mathcal{K}}} \times \mathrm{Obj}_{\mathbf{B}_{\mathcal{K}}} \\ & (x,\gamma) & \longmapsto & \big((I,\gamma^{-1}x),(I,x)\big), & \end{array}$$

and inverses $(x,\gamma)^{-1} = (\gamma^{-1}x, \gamma^{-1})$.

(b) For each $I \subsetneq J$ the coordinate change $\widehat{\Phi}_{IJ}$ gives rise to non-invertible morphisms from points in U_I to points in U_J given by the space $\widetilde{U}_{IJ}$ with source and target maps

$$\begin{array}{rcccl} s \times t: & \widetilde{U}_{IJ} & \longrightarrow & U_I \times U_J & \subset \mathrm{Obj}_{\mathbf{B}_{\mathcal{K}}} \times \mathrm{Obj}_{\mathbf{B}_{\mathcal{K}}} \\ & y & \longmapsto & \big((I,\rho_{IJ}(y)),(J,\widetilde{\phi}_{IJ}(y))\big). & \end{array}$$

In order to determine the general morphisms in $\mathbf{B}_{\mathcal{K}}$ we will unify types (a) and (b) by allowing $I = J$, in which case we interpret $U_{II} := U_I$, $\widetilde{U}_{IJ} := U_J$, $\widetilde{\phi}_{II} = \rho_{II} := \mathrm{id}$. Thus the morphisms of type (b) are described by their targets $\widetilde{\phi}_{IJ}(y) \in U_J$ and the covering map ρ_{IJ}. In comparison, recall that in §2.1, we have no morphisms of type (a) and the morphisms of type (b) are described by their source $x \in U_{IJ} \subset U_I$ and the embedding $\phi_{IJ} : U_{IJ} \to U_J$. When the isotropy groups are all trivial, it makes no difference whether we use source or target since $\phi_{IJ} = \rho_{IJ}^{-1}$ and $\widetilde{\phi}_{IJ} = \mathrm{id}_{U_J}$. The corresponding isomorphism of categories is given in Lemma 3.2.2 below. For nontrivial isotropy, however, the only way to obtain a continuous description of the morphism spaces is to parametrize them essentially by their targets as follows.

LEMMA 3.2.1. *Let $\mathcal{K}$ be a Kuranishi atlas. Then the space of morphisms in $\mathbf{B}_{\mathcal{K}}$ is the disjoint union*

$$\mathrm{Mor}_{\mathbf{B}_{\mathcal{K}}} = \bigsqcup_{I \subset J} \widetilde{U}_{IJ} \times \Gamma_I = \{(I,J,y,\gamma) \,|\, I \subset J, y \in \widetilde{U}_{IJ}, \gamma \in \Gamma_I\},$$

with source and target maps given by

$$\begin{array}{rrcccl} (3.2.1) & s \times t: & \widetilde{U}_{IJ} \times \Gamma_I & \longrightarrow & U_I \times U_J & \subset \mathrm{Obj}_{\mathbf{B}_{\mathcal{K}}} \times \mathrm{Obj}_{\mathbf{B}_{\mathcal{K}}}, \\ & & (I,J,y,\gamma) & \longmapsto & \big((I,\gamma^{-1}\rho_{IJ}(y)),\ (J,\widetilde{\phi}_{IJ}(y))\big), & \end{array}$$

and composition given by

$$\widetilde{\phi}_{IJ}(x) = \delta^{-1}\rho_{JK}(y) \implies (I,J,x,\gamma) \circ (J,K,y,\delta) = \big(I,K,y,\rho^{\Gamma}_{IJ}(\delta)\gamma\big). \tag{3.2.2}$$

The **intermediate Kuranishi category** $\underline{\mathbf{B}}_{\mathcal{K}}$ is defined by the topological atlas $\underline{\mathcal{K}}$ as in §2.3. It has

$$\mathrm{Obj}_{\underline{\mathbf{B}}_{\mathcal{K}}} := \bigsqcup_{I \in \mathcal{I}_{\mathcal{K}}} \underline{U}_I, \quad \mathrm{Mor}_{\underline{\mathbf{B}}_{\mathcal{K}}} := \bigsqcup_{I,J \in \mathcal{I}_{\mathcal{K}}, I \subset J} \underline{U}_{IJ}$$

with source and target maps

$$s \times t: \underline{U}_{IJ} \to \underline{U}_I \times \underline{U}_J \subset \mathrm{Obj}_{\underline{\mathbf{B}}_{\mathcal{K}}} \times \mathrm{Obj}_{\underline{\mathbf{B}}_{\mathcal{K}}}, \quad (I, \underline{x}) \mapsto \big((I, \underline{x}), (J, \underline{\phi}_{IJ}(\underline{x}))\big).$$

The identity maps $\underline{\phi}_{II}$ on $\underline{U}_{II} = \underline{U}_I$ are included, giving rise to the identity morphisms.

As before, we denote by $|\mathcal{K}|$ the **realization** of the category $\mathbf{B}_{\mathcal{K}}$, i.e. the topological space obtained as the quotient of $\mathrm{Obj}_{\mathbf{B}_{\mathcal{K}}}$ by the equivalence relation generated by the morphisms in $\mathbf{B}_{\mathcal{K}}$. The quotient map $\pi_{\mathcal{K}} : \mathrm{Obj}_{\mathbf{B}_{\mathcal{K}}} \to |\mathcal{K}|$, $(I, x) \mapsto [I, x]$ now factors through the intermediate category,

$$\pi_{\mathcal{K}} : \mathrm{Obj}_{\mathbf{B}_{\mathcal{K}}} \to \mathrm{Obj}_{\underline{\mathbf{B}}_{\mathcal{K}}} \to |\mathcal{K}|.$$

In particular the two categories $\mathbf{B}_{\mathcal{K}}$ and $\underline{\mathbf{B}}_{\mathcal{K}}$ have the same realization. In the latter case, we denote the natural projection by $\underline{\pi}_{\mathcal{K}} : \mathrm{Obj}_{\underline{\mathbf{B}}_{\mathcal{K}}} \to |\mathcal{K}|$. More precisely, we can formulate this as follows.

LEMMA 3.2.2. *Let $\mathcal{K}$ be a Kuranishi atlas. Then there is a functor $\pi_\Gamma : \mathbf{B}_{\mathcal{K}} \to \underline{\mathbf{B}}_{\mathcal{K}}$ that is given on objects by the quotient maps $\pi_I : U_I \to \underline{U}_I$, and on morphisms by*

$$\widetilde{U}_{IJ} \times \Gamma_I \ \to \ \underline{U}_{IJ}, \qquad \big(I, J, y, \gamma\big) \ \mapsto \ \big(I, J, \underline{\rho_{IJ}(y)}\big).$$

If all isotropy groups $\Gamma_I = \{\mathrm{id}\}$ are trivial, then π_Γ is an isomorphism of categories with identical object spaces, and $\underline{\mathbf{B}}_{\mathcal{K}}$ is the category associated to the Kuranishi atlas in §2.1.

In general, the realization $|\mathcal{K}|$ of $\mathbf{B}_{\mathcal{K}}$ can be identified as topological space (with the quotient topology) with that of $\underline{\mathbf{B}}_{\mathcal{K}}$ via factoring the quotient map $\pi_{\mathcal{K}} = \underline{\pi}_{\mathcal{K}} \circ \pi_\Gamma$ into the functor π_Γ given on objects by quotienting by the group actions and the projection $\underline{\pi}_{\mathcal{K}} : \underline{\mathbf{B}}_{\mathcal{K}} \to |\mathcal{K}|$, that can be considered as a functor to a topological category with only identity morphisms. Moreover, π_Γ is proper, i.e. compact subsets of $\mathrm{Obj}_{\underline{\mathbf{B}}_{\mathcal{K}}}$ have compact preimage in $\mathrm{Obj}_{\mathbf{B}_{\mathcal{K}}}$.

There are similar obstruction space categories $\mathbf{E}_{\mathcal{K}}$ and $\underline{\mathbf{E}}_{\mathcal{K}}$. For example,

$$\mathrm{Obj}_{\mathbf{E}_{\mathcal{K}}} = \bigsqcup_I U_I \times E_I, \qquad \mathrm{Mor}_{\mathbf{E}_{\mathcal{K}}} = \bigsqcup_{I \subset J} \widetilde{U}_{IJ} \times \widehat{\phi}_{IJ}(E_I) \times \Gamma_I.$$

The projections pr_I and sections $\mathfrak{s}_I$ fit together to functors

$$\mathrm{pr}_{\mathcal{K}} : \mathbf{E}_{\mathcal{K}} \to \mathbf{B}_{\mathcal{K}}, \ \mathfrak{s}_{\mathcal{K}} : \mathbf{B}_{\mathcal{K}} \to \mathbf{E}_{\mathcal{K}}, \quad \underline{\mathrm{pr}}_{\mathcal{K}} : \underline{\mathbf{E}}_{\mathcal{K}} \to \underline{\mathbf{B}}_{\mathcal{K}}, \ \underline{\mathfrak{s}}_{\mathcal{K}} : \underline{\mathbf{B}}_{\mathcal{K}} \to \underline{\mathbf{E}}_{\mathcal{K}}.$$

PROPOSITION 3.2.3. *Let $\mathcal{K}$ be a Kuranishi atlas.*

(i) *The functors $\mathrm{pr}_{\mathcal{K}} : \mathbf{E}_{\mathcal{K}} \to \mathbf{B}_{\mathcal{K}}$ and $\underline{\mathrm{pr}}_{\mathcal{K}} : \underline{\mathbf{E}}_{\mathcal{K}} \to \underline{\mathbf{B}}_{\mathcal{K}}$ induce the same continuous map*

$$|\mathrm{pr}_{\mathcal{K}}| : |\mathbf{E}_{\mathcal{K}}| \to |\mathcal{K}|,$$

which we call the **obstruction bundle** *of $\mathcal{K}$, although its fibers generally do not have the structure of a vector space. However, it has a continuous zero section*

$$|0_{\mathcal{K}}| : \ |\mathcal{K}| \to |\mathbf{E}_{\mathcal{K}}|, \quad [I, x] \mapsto [I, x, 0].$$

(ii) *The sections $\mathfrak{s}_{\mathcal{K}} : \mathbf{B}_{\mathcal{K}} \to \mathbf{E}_{\mathcal{K}}$ and $\underline{\mathfrak{s}}_{\mathcal{K}} : \underline{\mathbf{B}}_{\mathcal{K}} \to \underline{\mathbf{E}}_{\mathcal{K}}$ descend to the same continuous section*

$$|\mathfrak{s}_{\mathcal{K}}| : |\mathcal{K}| \to |\mathbf{E}_{\mathcal{K}}|.$$

Both of these are sections in the sense that $|\,\mathrm{pr}_{\mathcal{K}}\,| \circ |\mathfrak{s}_{\mathcal{K}}| = |\,\mathrm{pr}_{\mathcal{K}}\,| \circ |0_{\mathcal{K}}| = \mathrm{id}_{|\mathcal{K}|}$.

(iii) *There is a natural homeomorphism from the realization of the subcategory $\mathfrak{s}_{\mathcal{K}}^{-1}(0)$ to the zero set of $|\mathfrak{s}_{\mathcal{K}}|$, with the relative topology induced from $|\mathcal{K}|$,*

$$\left|\mathfrak{s}_{\mathcal{K}}^{-1}(0)\right| \;=\; {}^{\mathfrak{s}_{\mathcal{K}}^{-1}(0)}\Big/_{\sim_{\mathfrak{s}_{\mathcal{K}}^{-1}(0)}} \;\xrightarrow{\cong}\; |s_{\mathcal{K}}|^{-1}(0) := \big\{[I,x]\,\big|\, s_I(x)=0\big\} \subset |\mathcal{K}|.$$

The proof is not difficult.

The next task is to establish the following analog of Theorem 2.2.3. Note that we say that $\mathcal{K}$ **is tame or metrizable** if its intermediate atlas $\underline{\mathcal{K}}$ has this property. In particular, we require that the metric d on $|\mathcal{K}|$ pulls back to a metric $\underline{d}_I := (\underline{\pi}_{\mathcal{K}}|_{\underline{U}_I})^* d$ on $\underline{U}_I = {}^{U_I}/_{\Gamma_I}$ that induces the quotient topology. (When $\Gamma_I \neq \mathrm{id}$ the pullback to U_I cannot be a metric.) Similarly, a **shrinking of** $\mathcal{K}$ is the pullback by the functor π_Γ in Lemma 3.2.2 of a shrinking of $\underline{\mathcal{K}}$.

THEOREM 3.2.4. *Let $\mathcal{K}$ be a weak Kuranishi atlas on a compact metrizable space X. Then there is a metrizable tame shrinking $\mathcal{K}'$ of $\mathcal{K}$ with domains $(U_I' \subset U_I)_{I\in\mathcal{I}_{\mathcal{K}'}}$ such that the realizations $|\mathcal{K}'|$ and $|\mathbf{E}_{\mathcal{K}'}|$ are Hausdorff in the quotient topology. Further, for each $I \in \mathcal{I}_{\mathcal{K}'} = \mathcal{I}_{\mathcal{K}}$ the projection maps $\underline{\pi}_{\mathcal{K}'} : \underline{U}_I' \to |\mathcal{K}'|$ and $\underline{\pi}_{\mathcal{K}'} : \underline{U_I' \times E_I} \to |\mathbf{E}_{\mathcal{K}'}|$ are homeomorphisms onto their images. In addition, these projections $\pi_{\mathcal{K}'}$ fit into a commutative diagram*

$$\begin{array}{ccc} U_I' \times E_I & \xrightarrow{\pi_{\mathcal{K}'}} & |\mathbf{E}_{\mathcal{K}'}| \\ \downarrow & & \downarrow |\,\mathrm{pr}_{\mathcal{K}'}\,| \\ U_I' & \xrightarrow{\pi_{\mathcal{K}'}} & |\mathcal{K}'| \end{array}$$

where the horizontal maps intertwine the linear structure on the fibers of $U_I' \times E_I \to U_I'$ with the induced orbibundle structure on the fibers of $|\,\mathrm{pr}_{\mathcal{K}'}\,|$. More precisely, for each $p \in |\mathcal{K}'|$ there is $I_p \in \mathcal{I}_{\mathcal{K}'}$ such that $p \in \pi_{\mathcal{K}'}(U_{I_p})$ and so that the composite map

$$\pi_{\mathcal{K}'} : \big(U_{I_p}' \cap \pi_{\mathcal{K}'}^{-1}(p)\big) \times E_{I_p} \;\to\; |\,\mathrm{pr}_{\mathcal{K}'}\,|^{-1}(p) \subset |\mathbf{E}_{\mathcal{K}'}|$$

induces an isomorphism from the quotient ${}^{(U_{I_p}' \cap \pi_{\mathcal{K}'}^{-1}(p))\times E_{I_p}}/_{\Gamma_{I_p}}$ onto the fiber $|\,\mathrm{pr}_{\mathcal{K}'}\,|^{-1}(p)$.

Moreover, any two such shrinkings are cobordant by a metrizable tame Kuranishi cobordism that also has the above Hausdorff, homeomorphism, and linearity properties.

PROOF. This is an immediate consequence of Theorem 2.2.3 applied to the intermediate atlas $\underline{\mathcal{K}}$, which is a filtered topological atlas by Lemma 3.1.16. The only nontrivial statement concerns the linear structure on the fibers of $|\,\mathrm{pr}_{\mathcal{K}'}\,|$. But this follows by arguing as in the proof of Proposition 2.3.15. □

3.3. Orientations. This section develops the notion of orientation. (See [**MW2**, §8.1] for the trivial isotropy case, and [**MW3**, §3.1] for the general case.) As in Definition 2.2.15, we define an orientation of $\mathcal{K}$ to be a nonzero section of its determinant line bundle. There are two possible definitions of this bundle, first as a bundle $\det(\mathcal{K})$ over $\mathcal{K}$ that restricts on each domain U_I to the local orientation bundle $\Lambda^{\max} TU_I \otimes (\Lambda^{\max} E_I)^*$ of Definition 2.2.15, and second as a bundle $\det(\mathfrak{s}_{\mathcal{K}})$ whose restriction to U_I is the standard determinant bundle of the (Fredholm) section s_I (see [**MS**, § A.2]). Although the local description of $\det(\mathfrak{s}_{\mathcal{K}})$ is more complicated because the dimension of $\ker s_I$ varies, its coordinate changes are more natural, being induced directly from the chart coordinate changes $\widehat{\Phi}_{IJ}$ with no auxiliary choices, while those for $\det(\mathcal{K})$ do involve the sections s_I as well as the choice of a normal bundle to $\widetilde{U}_{IJ}$ in U_J. Thus we first prove that $\det(\mathfrak{s}_{\mathcal{K}})$ is a bundle, and then show it is isomorphic to the bundle $\det(\mathcal{K})$. A similar argument shows that the restriction of $\det(\mathcal{K})$ to $\mathcal{K}|_{\mathcal{V}}$ is isomorphic to $\det(\mathfrak{s}_{\mathcal{K}}|_{\mathcal{V}} + \nu)$. But if $(s_{\mathcal{K}}|_{\mathcal{V}} + \nu) \pitchfork 0$, an orientation of $\det(\mathfrak{s}_{\mathcal{K}}|_{\mathcal{V}} + \nu)$ induces one on the zero set $\mathbf{Z}^\nu$; see Lemma 3.3.9. We do not attempt to give full proofs here, but at least define all the maps involved.

We begin by defining the notion of a bundle over a smooth atlas. Notice that in distinction to the "obstruction bundle" with fibers E_I over the domains U_I, we now assume that the fiber dimension is constant.

DEFINITION 3.3.1. *A* **vector bundle** $\Lambda = \big(\Lambda_I, \widetilde{\phi}_{IJ}\big)_{I,J\in\mathcal{I}_{\mathcal{K}}}$ **over a weak Kuranishi atlas** $\mathcal{K}$ *is a collection* $(\Lambda_I \to U_I)_{I\in\mathcal{I}_{\mathcal{K}}}$ *of vector bundles such that*

- *for each* $I \in \mathcal{I}_{\mathcal{K}}$ *the action of* Γ_I *on* U_I *lifts to* Λ_I*;*
- *there is an action of* Γ_J *on the pullback bundle* $\rho_{IJ}^*(\Lambda_I|_{U_{IJ}})$ *such that*
  - *the projection* $\rho_{IJ}^*(\Lambda_I|_{U_{IJ}}) \to \Lambda_I|_{U_{IJ}}$ *factors as the quotient by the action of* $\Gamma_{J\setminus I}$ *followed by an isomorphism, and*
  - *the inclusion* $\widetilde{\phi}_{IJ} : \widetilde{U}_{IJ} \to U_J$ *lifts to a* Γ_J*-equivariant embedding* $\widetilde{\phi}_{IJ}^\Lambda : \rho_{IJ}^*(\Lambda_I|_{U_{IJ}}) \to \Lambda_J$.
- *the weak cocycle condition holds*

$$\widetilde{\phi}_{IK}^\Lambda = \widetilde{\phi}_{JK}^\Lambda \circ \rho_{JK}^*(\widetilde{\phi}_{IJ}^\Lambda) \quad \text{on } \rho_{JK}^{-1}(\widetilde{\phi}_{IJ}(\widetilde{U}_{IJ})) \cap \widetilde{U}_{IK}, \quad \forall\, I \subset J \subset K.$$

A **section** *of a bundle* Λ *over* $\mathcal{K}$ *is a collection of smooth* Γ_I*-equivariant sections* $\sigma = \big(\sigma_I : U_I \to \Lambda_I\big)_{I\in\mathcal{I}_{\mathcal{K}}}$ *that are compatible with the pullbacks* ρ_{IJ}^* *and bundle maps* $\widetilde{\phi}_{IJ}^\Lambda$. *Thus there are commutative diagrams for each* $I \subset J$,

$$\begin{array}{ccccc}
\Lambda_I|_{U_{IJ}} & \xleftarrow{\ \rho_{IJ}\ } & \rho_{IJ}^*(\Lambda_I|_{U_{IJ}}) & \xrightarrow{\ \widetilde{\phi}_{IJ}^\Lambda\ } & \Lambda_J \\
\uparrow{\scriptstyle \sigma_I} & & \uparrow{\scriptstyle \rho_{IJ}^*(\sigma_I)} & & \uparrow{\scriptstyle \sigma_J} \\
U_{IJ} & \xleftarrow{\ \rho_{IJ}\ } & \widetilde{U}_{IJ} & \xrightarrow{\ \widetilde{\phi}_{IJ}\ } & U_J.
\end{array}$$

An **isomorphism** $\Psi : \Lambda \to \Lambda'$ *between vector bundles over* $\mathcal{K}$ *is a collection* $(\Psi_I : \Lambda_I \to \Lambda'_I)_{I\in\mathcal{I}_{\mathcal{K}}}$ *of* Γ_I*-equivariant bundle isomorphisms covering the identity on* U_I*, that intertwine the transition maps, i.e.* $\widetilde{\phi}_{IJ}^{\Lambda'} \circ \rho_{IJ}^*(\Psi_I) = \Psi_J \circ \widetilde{\phi}_{IJ}^\Lambda|_{\widetilde{U}_{IJ}}$ *for all* $I \subsetneq J$.

REMARK 3.3.2. (i) If $\mathcal{K}$ has trivial isotropy, this is just the definition in [**MW2**]. Since the smooth structure on U_I is not used, there is a corresponding notion for a bundle over the intermediate category $\underline{\mathcal{K}}$. Further any bundle $\underline{\Lambda} = (\underline{\Lambda}_I, \underline{\widetilde{\phi}}_{IJ}^\Lambda)_{I\in\mathcal{I}_{\mathcal{K}}}$ over $\underline{\mathcal{K}}$ pulls back to a bundle over $\mathcal{K}$ via the projections $\pi_I : U_I \to \underline{U}_I$.

(ii) There are similar definitions for bundles over a cobordism. As always, the key point is that the bundle has a chosen product structure (or collar) near its boundary. We will not need this notion in this generality here since we will only consider the case of a determinant bundle that is locally oriented and hence is pulled back from a line bundle over the intermediate (cobordism) atlas as described in Remark 3.3.6. Hence we can use the definitions and results from [**MW2**] that apply to cobordism atlases with trivial isotropy. $\Diamond$

The next task is to establish that each of the two forms of the determinant bundle gives rise to a line bundle over $\mathcal{K}$. Recall that we denote the top exterior power of a real vector space V by $\Lambda^{\max}V$.

We define the **determinant line of a linear map** $D : V \to W$ to be

$$\det(D) := \Lambda^{\max} \ker D \otimes \left(\Lambda^{\max}\left({}^{W}\!/_{\mathrm{im}\, D}\right)\right)^*,$$

and use the following conventions from [**MW2**]; see the discussion after [**MW2**, Definition 8.1.6]. Each isomorphism $F : Y \to Z$ between finite dimensional vector spaces induces an isomorphism

$$\Lambda_F : \Lambda^{\max} Y \xrightarrow{\cong} \Lambda^{\max} Z, \qquad y_1 \wedge \ldots \wedge y_k \mapsto F(y_1) \wedge \ldots \wedge F(y_k). \tag{3.3.1}$$

For example, if $I \subsetneq J$ and $\widetilde{x} \in \widetilde{U}_{IJ}$ is such that $\rho_{IJ}(\widetilde{x}) = x$, then because

$$s_I \circ \rho_{IJ} =: s_{IJ} : \widetilde{U}_{IJ} \to E_I$$

the derivative $\mathrm{d}_{\widetilde{x}}\, \rho_{IJ} : \ker \mathrm{d}\, s_{IJ} \to \ker \mathrm{d}\, s_I$ induces a isomorphism

$$\Lambda_{\mathrm{d}_{\widetilde{x}}\, \rho_{IJ}} \otimes (\Lambda_{\mathrm{id}})^* : \det(\mathrm{d}_{\widetilde{x}}\, s_{IJ}) = \rho_{IJ}^*(\det \mathrm{d}_x\, s_I) \to \det(\mathrm{d}_x\, s_I). \tag{3.3.2}$$

Further, it follows from the index condition in Definition 2.1.5 that with $y := \widetilde{\phi}_{IJ}(\widetilde{x})$ the map

$$\Lambda_{IJ}(\widetilde{x}) := \Lambda_{\mathrm{d}_{\widetilde{x}}\, \widetilde{\phi}_{IJ}} \otimes \left(\Lambda_{[\widehat{\phi}_{IJ}^{-1}]}\right)^* \ : \ \det(\mathrm{d}_{\widetilde{x}}\, s_{IJ}) \to \det(\mathrm{d}_y\, s_J) \tag{3.3.3}$$

is an isomorphism, induced by the isomorphisms $\mathrm{d}\, \widetilde{\phi}_{IJ} : \ker \mathrm{d}\, s_{IJ} \to \ker \mathrm{d}\, s_J$ and $[\widehat{\phi}_{IJ}] : {}^{E_I}\!/_{\mathrm{im}\, \mathrm{d}\, s_I} \to {}^{E_J}\!/_{\mathrm{im}\, \mathrm{d}\, s_J}$.

LEMMA 3.3.3. *There is a line bundle* $\det(\mathfrak{s}_{\mathcal{K}}) = \left(\det(\mathrm{d}\, s_I)\right)_{I \in \mathcal{I}_{\mathcal{K}}}$ *on* $\mathcal{K}$ *where*

$$\det(\mathrm{d}\, s_I) := \bigcup_{x \in U_I} \det(\mathrm{d}_x\, s_I) \ \to \ U_I \qquad \textit{for } I \in \mathcal{I}_{\mathcal{K}},$$

and the isomorphisms $\widetilde{\phi}_{IJ}^{\Lambda} := \Lambda_{IJ}(\widetilde{x})$ *in* (3.3.3) *for* $I \subsetneq J$ *and* $\widetilde{x} \in \widetilde{U}_{IJ}$.

PROOF. We showed in [**MW2**, Proposition 8.1.8] that the assignments $x \mapsto \det(\mathrm{d}_x\, s_I)$ and $\widetilde{x} \mapsto \Lambda_{IJ}(\widetilde{x})$ are smooth. Further, the weak cocycle condition holds by functoriality. It remains to check the conditions in Definition 3.3.1 pertaining to the group actions. Since $\gamma \in \Gamma_I$ acts by diffeomorphisms on U_I and by linear transformations on E_I and since s_I is Γ_I equivariant, the action of $\gamma \in \Gamma_I$ on U_I lifts to Λ_I by

$$\Lambda_{\mathrm{d}_x \gamma} \otimes \left(\Lambda_{\gamma^{-1}}\right)^* \ : \ \det(\mathrm{d}_x\, s_I) \to \det(\mathrm{d}_{\gamma x}\, s_I).$$

Further, because the image $\widehat{\phi}_{IJ}(E_I) \subset E_J$ is invariant under the Γ_J action, there is a similar action of Γ_J on the pullback $\rho_{IJ}^*(\det(\mathrm{d}\, s_I))$. Thus the required group actions exist. The map $\rho_{IJ}^*(\Lambda_I|_{U_{IJ}}) \to \Lambda_I|_{U_{IJ}}$ is the quotient by $\Gamma_{J \setminus I}$ followed by an isomorphism since this holds for ρ_{IJ}. Moreover, the natural lift of $\widetilde{\phi}_{IJ}$ to $\widetilde{\phi}_{IJ}^{\Lambda} := \Lambda_{IJ}(y)$ is Γ_J equivariant because the isomorphisms $\mathrm{d}\, \widetilde{\phi}_{IJ}$ and

$$\widetilde{\phi}_{IJ} \times [\widehat{\phi}_{IJ}] : \widetilde{U}_{IJ} \times {}^{E_I}\!/_{\mathrm{im}\, \mathrm{d}\, s_I} \to U_J \times {}^{E_J}\!/_{\mathrm{im}\, \mathrm{d}\, s_J}$$

are Γ_J-equivariant. $\square$

The hard part of the above proof is to show that the assignments $x \mapsto \det(\mathrm{d}_x s_I)$ and $\widetilde{x} \mapsto \Lambda_{IJ}(\widetilde{x})$ are smooth.[27] For this one constructs local trivializations that use the contraction isomorphisms defined in the next lemma. These maps will also be used in the definition of the second version of the determinant bundle: see Proposition 3.3.5 below.

LEMMA 3.3.4 ([**MW2**] Lemma 8.1.7). *Let V, W be finite dimensional vector spaces. Every linear map $F : V \to W$ together with an isomorphism $\phi : K \to \ker F$ induces an isomorphism*

$$\mathfrak{C}_F^{\phi} \; : \; \Lambda^{\max} V \otimes \left(\Lambda^{\max} W\right)^* \xrightarrow{\cong} \Lambda^{\max} K \otimes \left(\Lambda^{\max} \left({}^{W}\!/_{F(V)}\right)\right)^* \tag{3.3.4}$$

given by

$$(v_1 \wedge \dots v_n) \otimes (w_1 \wedge \dots w_m)^* \longmapsto \left(\phi^{-1}(v_1) \wedge \dots \phi^{-1}(v_k)\right) \otimes \left([w_1] \wedge \dots [w_{m-n+k}]\right)^*,$$

where $v_1, \dots, v_n$ is a basis for V with $\mathrm{span}(v_1, \dots, v_k) = \ker F$, and $w_1, \dots, w_m$ is a basis for W whose last $n-k$ vectors are $w_{m-n+i} = F(v_i)$ for $i = k+1, \dots, n$.

In particular, for every linear map $D : V \to W$ we may pick ϕ as the inclusion $K = \ker D \hookrightarrow V$ to obtain an isomorphism

$$\mathfrak{C}_D \; : \; \Lambda^{\max} V \otimes \left(\Lambda^{\max} W\right)^* \xrightarrow{\cong} \det(D). \tag{3.3.5}$$

Recall that the tangent bundle condition (2.1.3) implies that $\mathrm{d}\, s_J$ restricts at all $y = \widetilde{\phi}_{IJ}(\widetilde{x})$[28] to an isomorphism ${}^{\mathrm{T}_y U_J}\!/_{\mathrm{d}_{\widetilde{x}} \widetilde{\phi}_{IJ}(\mathrm{T}_{\widetilde{x}} \widetilde{U}_{IJ})} \stackrel{\cong}{\to} {}^{E_J}\!/_{\widehat{\phi}_{IJ}(E_I)}$. Therefore, if we choose a smooth normal bundle

$$N_{IJ} = \bigcup_{\widetilde{x} \in \widetilde{U}_{IJ}} N_{IJ,\widetilde{x}} \subset \widetilde{\phi}_{IJ}^*(\mathrm{T}\, U_J)$$

to the submanifold $\mathrm{im}\, \widetilde{\phi}_{IJ} \subset U_J$, the subspaces $\mathrm{d}_y\, s_J(N_{IJ,\widetilde{x}})$ form a smooth family of subspaces of E_J that are complements to $\widehat{\phi}_{IJ}(E_I)$. Hence letting $\mathrm{pr}_{N_{IJ}}(\widetilde{x}) : E_J \to \mathrm{d}_y\, s_J(N_{IJ,\widetilde{x}}) \subset E_J$ be the smooth family of projections with kernel $\widehat{\phi}_{IJ}(E_I)$, we obtain a smooth family of linear maps

$$F_{\widetilde{x}} := \mathrm{pr}_{N_{IJ}}(\widetilde{x}) \circ \mathrm{d}_y\, s_J \; : \; \mathrm{T}_y U_J \longrightarrow E_J \qquad \text{for } \widetilde{x} \in \widetilde{U}_{IJ}, \;\; y = \widetilde{\phi}_{IJ}(\widetilde{x})$$

with images $\mathrm{im}\, F_{\widetilde{x}} = \mathrm{d}_y\, s_J(N_{IJ,\widetilde{x}})$, and isomorphisms to their kernel

$$\phi_{\widetilde{x}} := \mathrm{d}_{\widetilde{x}} \widetilde{\phi}_{IJ} \; : \; \mathrm{T}_{\widetilde{x}} \widetilde{U}_I \xrightarrow{\cong} \ker F_{\widetilde{x}} = \mathrm{d}_{\widetilde{x}} \widetilde{\phi}_{IJ}(\mathrm{T}_{\widetilde{x}} \widetilde{U}_{IJ}) \; \subset \; \mathrm{T}_y U_J.$$

We may arrange that the maps $F_{\widetilde{x}}$ are Γ_J-equivariant by choosing the bundle N_{IJ} to be Γ_J-invariant, since s_J is Γ_J-equivariant, as is the subspace $\widehat{\phi}_{IJ}(E_I) \subset E_J$ (this time with trivial action of $\Gamma_{J \setminus I}$). Then the isomorphism $\phi_{\widetilde{x}}$ is also Γ_J-equivariant. By Lemma 3.3.4 the maps $F_{\widetilde{x}}, \phi_{\widetilde{x}}$ induce Γ_J-equivariant isomorphisms

$$\mathfrak{C}_{F_{\widetilde{x}}}^{\phi_{\widetilde{x}}} \; : \Lambda^{\max} \mathrm{T}_y U_J \otimes \left(\Lambda^{\max} E_J\right)^* \xrightarrow{\cong} \Lambda^{\max} \mathrm{T}_{\widetilde{x}} \widetilde{U}_{IJ} \otimes \left(\Lambda^{\max} \left({}^{E_J}\!/_{\mathrm{im}\, F_{\widetilde{x}}}\right)\right)^*.$$

We may combine this with the dual of the isomorphism $\Lambda^{\max}\left({}^{E_J}\!/_{\mathrm{d}_y\, s_J(N_{IJ,\widetilde{x}})}\right) \cong \Lambda^{\max} E_I$ induced via (3.3.1) by $\mathrm{pr}^{\perp}_{N_{IJ}}(\widetilde{x}) \circ \widehat{\phi}_{IJ} : E_I \to {}^{E_J}\!/_{\mathrm{d}_y\, s_J(N_{IJ,\widetilde{x}})}$ to obtain

[27] The local argument is essentially the same as that in [**MS**, Thm. A.2.2], but we choose a different order for the added basis vectors in order to get consistency of signs.

[28] To keep notation similar to that in [**MW3**] we always denote $y := \widetilde{\phi}_{IJ}(\widetilde{x})$ where $\widetilde{x} \in \widetilde{U}_{IJ}$.

Γ_J-equivariant isomorphisms

$$\mathfrak{C}_{IJ}(\widetilde{x}) \,:\, \Lambda^{\max}\, \mathrm{T}_y U_J \otimes \big(\Lambda^{\max} E_J\big)^* \;\xrightarrow{\cong}\; \Lambda^{\max}\, \mathrm{T}_{\widetilde{x}} \widetilde{U}_{IJ} \otimes \big(\Lambda^{\max} E_I\big)^* \tag{3.3.6}$$

given by $\mathfrak{C}_{IJ}(\widetilde{x}) := \big(\mathrm{id}_{\Lambda^{\max}\, \mathrm{T}_{\widetilde{x}} \widetilde{U}_{IJ}} \otimes \Lambda^*_{(\mathrm{pr}^{\perp}_{N_{IJ}}(\widetilde{x}) \circ \widehat{\phi}_{IJ})^{-1}}\big) \circ \mathfrak{C}^{\phi_{\widetilde{x}}}_{F_{\widetilde{x}}}$, where again $\Gamma_{J \setminus I}$ acts trivially on E_I.

Proposition 3.3.5 ([**MW3**] Proposition 3.1.13). *Let $\mathcal{K}$ be a weak Kuranishi atlas. Then there is a well defined line bundle $\det(\mathcal{K})$ over $\mathcal{K}$ given by the line bundles*

$$\Lambda^{\mathcal{K}}_I := \Lambda^{\max}\, \mathrm{T}\, U_I \otimes \big(\Lambda^{\max} E_I\big)^* \to U_I, \qquad I \in \mathcal{I}_{\mathcal{K}}$$

and the transition maps $\mathfrak{C}^{-1}_{IJ} : \rho^*_{IJ}\big(\Lambda^{\mathcal{K}}_I|_{U_{IJ}}\big) \to \Lambda^{\mathcal{K}}_J$ from (3.3.6) for $I \subsetneq J$. In particular, the latter isomorphisms are independent of the choice of normal bundle N_{IJ}.

Furthermore, the contractions $\mathfrak{C}_{\mathrm{d}\, s_I} : \Lambda^{\mathcal{K}}_I \to \det(\mathrm{d}\, s_I)$ from (3.3.5) with $D = \mathrm{d}\, s_I$ define an isomorphism $\Psi^{\mathfrak{s}_{\mathcal{K}}} := \big(\mathfrak{C}_{\mathrm{d}\, s_I}\big)_{I \in \mathcal{I}_{\mathcal{K}}}$ from $\det(\mathcal{K})$ to $\det(\mathfrak{s}_{\mathcal{K}})$.

Proof. The first step is to show that the maps $\mathfrak{C}_{\mathrm{d}_x s_I}$ are isomorphisms for each I, and in particular that they vary smoothly with $x \in U_I$. This holds just as before. Next, consider the following diagram:

$$\begin{array}{ccc}
\Lambda^{\mathcal{K}}_I = \Lambda^{\max}\, \mathrm{T}_x U_I \otimes \big(\Lambda^{\max} E_I\big)^* & \xrightarrow{\mathfrak{C}_{\mathrm{d}_x s_I}} & \det(\mathrm{d}_x s_I) \\
\Big\uparrow {\scriptstyle \Lambda_{\mathrm{d}_{\widetilde{x}} \rho_{IJ}} \otimes \Lambda_{\mathrm{id}}} & & \Big\uparrow {\scriptstyle \Lambda_{\mathrm{d}_{\widetilde{x}} \rho_{IJ}} \otimes \Lambda_{\mathrm{id}}} \\
\rho^*_{IJ}\big(\Lambda^{\mathcal{K}}_I\big) = \Lambda^{\max}\, \mathrm{T}_{\widetilde{x}} \widetilde{U}_{IJ} \otimes \big(\Lambda^{\max} E_I\big)^* & \xrightarrow{\mathfrak{C}_{\mathrm{d}_{\widetilde{x}} s_{IJ}}} & \det(\mathrm{d}_{\widetilde{x}} s_{IJ}) \\
\Big\uparrow {\scriptstyle \mathfrak{C}_{IJ}(\widetilde{x})} & & \Big\downarrow {\scriptstyle \Lambda_{IJ}(\widetilde{x})} \\
\Lambda^{\mathcal{K}}_J = \Lambda^{\max}\, \mathrm{T}_y U_J \otimes \big(\Lambda^{\max} E_J\big)^* & \xrightarrow{\mathfrak{C}_{\mathrm{d}_y s_J}} & \det(\mathrm{d}_y s_J).
\end{array} \tag{3.3.7}$$

The top square has right hand map given by (3.3.2), and commutes because $s_{IJ} = s_I \circ \rho_{IJ}$. Hence the maps $\mathfrak{C}_{\mathrm{d}_{\widetilde{x}} s_{IJ}}$ vary smoothly with $\widetilde{x} \in \widetilde{U}_{IJ}$, and we can define preliminary transition maps

$$(\widetilde{\phi}^{\Lambda}_{IJ})' := \mathfrak{C}^{-1}_{\mathrm{d}\, s_J} \circ \Lambda_{IJ} \circ \mathfrak{C}_{\mathrm{d}\, s_{IJ}} \,:\, \rho^*_{IJ}\big(\Lambda^{\mathcal{K}}_I|_{\widetilde{U}_{IJ}}\big) \to \Lambda^{\mathcal{K}}_J \qquad \text{for } I \subsetneq J \in \mathcal{I}_{\mathcal{K}} \tag{3.3.8}$$

by the transition maps (3.3.3) of $\det(\mathfrak{s}_{\mathcal{K}})$ and the isomorphisms $\mathfrak{C}_{\mathrm{d}\, s}$. These define a line bundle $\Lambda^{\mathcal{K}} := \big(\Lambda^{\mathcal{K}}_I, \widetilde{\phi}_{IJ}\big)_{I,J \in \mathcal{I}_{\mathcal{K}}}$ since the weak cocycle condition follows directly from that for the Λ_{IJ}. Moreover, this automatically makes the family of bundle isomorphisms $\Psi^{\mathcal{K}} := \big(\mathfrak{C}_{\mathrm{d}\, s_I}\big)_{I \in \mathcal{I}_{\mathcal{K}}}$ an isomorphism from $\Lambda^{\mathcal{K}}$ to $\det(\mathfrak{s}_{\mathcal{K}})$. It remains to see that $\Lambda^{\mathcal{K}} = \det(\mathcal{K})$ and $\Psi^{\mathcal{K}} = \Psi^{\mathfrak{s}_{\mathcal{K}}}$, i.e. we claim equality of transition maps $\widetilde{\phi}_{IJ} = \mathfrak{C}^{-1}_{IJ}$. This also shows that $\mathfrak{C}^{-1}_{IJ}$ and thus $\det(\mathcal{K})$ is independent of the choice of normal bundle N_{IJ} in (3.3.6). So to finish the proof it suffices to show that the lower square commutes. But this can be proved by slightly modifying the proof of the corresponding result in the trivial isotropy case. In fact, since this statement can be proved at each point separately, we can treat the map ρ_{IJ} as the identity, identifying x with $\widetilde{x}$ and hence reduce to the diagram [**MW2**, (8.1.18)] which is commutative by the proof of [**MW2**, Proposition 8.1.12]. □

This completes the construction of the two isomorphic line bundles $\det(\mathfrak{s}_{\mathcal{K}})$ and $\det(\mathcal{K})$ over $\mathcal{K}$. Similar arguments show that if $\mathcal{K}$ is a cobordism atlas as in

Definition 2.2.11 the corresponding bundles $\det(\mathcal{K})$ and $\det(\mathfrak{s}_{\mathcal{K}})$ are well defined and isomorphic.

REMARK 3.3.6. (i) If the bundle $\Lambda_I^{\mathcal{K}} := \Lambda^{\max} \mathrm{T} U_I \otimes (\Lambda^{\max} E_I)^* \to U_I$ has a nonvanishing Γ_I-equivariant section σ_I for each $I \in \mathcal{I}_{\mathcal{K}}$, then $\det(\mathcal{K})$ is the pullback of a line bundle on the intermediate atlas $\underline{\mathcal{K}}$. Indeed, the existence of σ_I implies that $\Lambda_I^{\mathcal{K}}$ is the pullback via the projection $\pi_I : U_I \to \underline{U}_I$ of some bundle $\underline{\Lambda}_I$ over $\underline{U}_I$ that is trivialized by σ_I. Further, the fact that the transition maps $\widetilde{\phi}_{IJ}^{\Lambda} : \rho_{IJ}^*(\Lambda_I^{\mathcal{K}}) \to \Lambda_J^{\mathcal{K}}$ are Γ_J-equivariant implies that they descend to well defined maps $\widetilde{\underline{\phi}}_{IJ}^{\Lambda} : \underline{\Lambda}_I|_{\underline{U}_{IJ}} \to \underline{\Lambda}_J$. Note that there is no need to assume here that the local sections σ_I are compatible. A similar remark applies to $\det(\mathfrak{s}_{\mathcal{K}})$.

(ii) In the case of Gromov–Witten atlases the obstruction space is a direct sum $E_I := \oplus_{\gamma \in \Gamma_I} E_I^0$ of copies of a fixed space E_I^0, and Γ_I acts by permuting its factors. If we assume that $\dim E_I^0$ is even then this action preserves an orientation of E_I so that the local section σ_I in (i) exists exactly if the group Γ_I acts by orientation preserving maps of U_I. $\Diamond$

Here are some basic definitions . For more detail about the isomorphisms $(\widetilde{\iota}_I^{\Lambda,\alpha})_{I \in \mathcal{I}_{\partial^\alpha \mathcal{K}}}$ see [**MW2**, equation (8.1.13)].

DEFINITION 3.3.7. *A weak Kuranishi atlas or Kuranishi cobordism $\mathcal{K}$ is* **orientable** *if there exists a nonvanishing section σ of the determinant bundle* $\det(\mathfrak{s}_{\mathcal{K}})$. *An* **orientation** *of $\mathcal{K}$ is a choice of nonvanishing section σ of* $\det(\mathfrak{s}_{\underline{\mathcal{K}}})$, *which pulls back to a section, also denoted σ, of* $\det(\mathfrak{s}_{\mathcal{K}})$. *An* **oriented Kuranishi atlas or cobordism** *is a pair $(\mathcal{K}, \sigma)$ consisting of a Kuranishi atlas or cobordism and an orientation σ of $\mathcal{K}$.*

For an oriented Kuranishi cobordism $(\mathcal{K}, \sigma)$, the **induced orientation of the boundary** $\partial^\alpha \mathcal{K}$ *for $\alpha = 0$ resp. $\alpha = 1$ is the orientation of $\partial^\alpha \mathcal{K}$,*

$$\partial^\alpha \sigma := \Big(\big((\widetilde{\iota}_I^{\Lambda,\alpha})^{-1} \circ \sigma_I \circ \iota_I^\alpha \big)|_{\partial^\alpha U_I \times \{\alpha\}} \Big)_{I \in \mathcal{I}_{\partial^\alpha \mathcal{K}}}$$

induced by the isomorphism $(\widetilde{\iota}_I^\alpha)_{I \in \mathcal{I}_{\partial^\alpha \mathcal{K}}}$ between a collar neighbourhood of the boundary in $\mathcal{K}$ and the product Kuranishi atlas $A_\varepsilon^\alpha \times \partial^\alpha \mathcal{K}$, followed by restriction to the boundary $\partial^\alpha \mathcal{K} = \partial^\alpha \big(A_\varepsilon^\alpha \times \partial^\alpha \mathcal{K} \big)$, where we identify $\partial^\alpha U_I \times \{\alpha\} \cong \partial^\alpha U_I$.

With that, we say that two oriented weak Kuranishi atlases $(\mathcal{K}^0, \sigma^0)$ and $(\mathcal{K}^1, \sigma^1)$ are **oriented cobordant** *(resp.* **oriented concordant***) if there exists a weak Kuranishi cobordism (resp. concordance) $\mathcal{K}$ from $\mathcal{K}^0$ to $\mathcal{K}^1$ and a section σ of* $\det(s_{\mathcal{K}})$ *such that $\partial^\alpha \sigma = \sigma^\alpha$ for $\alpha = 0, 1$.*

REMARK 3.3.8. (i) As we saw in Remark 3.3.6 the existence of a nonvanishing local section s_I of $\Lambda_I^{\mathcal{K}}$ (or $\det(\mathfrak{s}_I)$) implies that this bundle is pulled back from a trivial bundle over the intermediate chart. The orientability condition implies that one can choose compatible local orientations for all these charts.

(ii) As in [**MW2**], we define the induced orientation on the boundary $\partial^\alpha \mathcal{K}$ of a cobordism so that it is completed to an orientation of the collar by adding the positive unit vector 1 along $A_\varepsilon^\alpha \subset \mathbb{R}$ rather than the more usual outward normal vector. $\Diamond$

We show in [**MW2**, Proposition 8.1.13] that if $(\mathcal{K}, \sigma)$ is an oriented, tame Kuranishi atlas with trivial isotropy and with reduction $\mathcal{V}$, then the zero set $|\mathbf{Z}^\nu|$ of any admissible, precompact, transverse perturbation ν of $\mathfrak{s}_{\mathcal{K}}|_{\mathcal{V}}$ is an oriented closed manifold whose oriented cobordism class is independent of the choice of ν. The proof is based on the next lemma (a simplified version of [**MW3**, Lemma 3.1.14]) that describes how to use the orientation of $\mathcal{K}$ to orient the local zero sets Z_I. One then proves that these local orientations are compatible with coordinate changes by adapting the proof of Proposition 3.3.5 above.

LEMMA 3.3.9. *Let $\mathcal{K}$ be oriented and $F : U_I \to E_I$ be a smooth map that is transverse to 0. Then the zero set $Z_I := F^{-1}(0)$ inherits the structure of a smooth oriented manifold.*

PROOF. By transversality Z_I is a manifold and $\operatorname{im}(\mathrm{d}_z F) = E_I$ for each $z \in Z_I$. Hence $\Lambda^{\max}\left({}^{E_I}\!/_{\operatorname{im}(\mathrm{d}_z s_I + \mathrm{d}_z \nu_I)}\right) = \Lambda^{\max}\{0\} = \mathbb{R}$, which gives a natural isomorphism between the orientation bundle of Z_I and the restriction of the determinant line bundle of F:

$$\begin{aligned} \Lambda^{\max} \mathrm{T}\, Z_I &= \textstyle\bigcup_{z\in Z_I} \Lambda^{\max} \ker(\mathrm{d}_z F) \\ &\cong \textstyle\bigcup_{z\in Z_I} \Lambda^{\max} \ker(\mathrm{d}_z F) \otimes \mathbb{R} \;=\; \det(F)|_{Z_I}. \end{aligned}$$

By hypothesis, we are given a nonvanishing section of the bundle $\Lambda_I^{\mathcal{K}} := \Lambda^{\max}\big(\widetilde{U}_I\big) \otimes \Lambda^{\max}\big(E_I\big)^*$. The proof of Proposition 3.3.5 adapts to show that the contraction

$$\mathfrak{C}_{\mathrm{d}F} : \Lambda_I^{\mathcal{K}}(z) \longrightarrow \det(F)|_z = \Lambda^{\max} \mathrm{T}_z Z_I \qquad \text{for } z \in Z_I$$

is an isomorphism that varies continuously with $z \in Z_I$. Hence the given section of $\Lambda_I^{\mathcal{K}}$ pushes forward to a nonvanishing section of $\Lambda^{\max} \mathrm{T}\, Z_I$, i.e. an orientation of Z_I. □

When the isotropy groups of $\mathcal{K}$ are nontrivial, the above lemma shows how to orient the local zero set. However the global zero set is now a branched manifold. All the above work applies in this case, but we postpone further discussion until we have introduced the relevant language.

3.4. Perturbation sections and construction of the VFC. With the notion of orientation in hand, we are in a position to complete the construction of the VFC. Again, we represent this class via the zero set of a suitable reduced perturbation $\nu = (\nu_I)$. After making the relevant definitions, we explain how to construct a cycle from the local zero sets.

The notion of reduction is essentially the same as before. Recall from Lemma 3.1.16 that the intermediate atlas $\underline{\mathcal{K}}$ is a filtered topological atlas, and that we define $\mathcal{K}$ to be tame if $\underline{\mathcal{K}}$ is. Similarly, we define a reduction of a tame Kuranishi atlas $\mathcal{K}$ to be the pullback of a reduction $\underline{\mathcal{V}}$ of the intermediate atlas $\underline{\mathcal{K}}$. Here is a formal definition.

DEFINITION 3.4.1 (see Definition 2.4.1). *A* **reduction** $\mathcal{V} = \bigsqcup_{I\in\mathcal{I}_{\mathcal{K}}} V_I \subset \mathrm{Obj}_{\mathbf{B}_{\mathcal{K}}}$ *of a tame Kuranishi atlas $\mathcal{K}$ is a tuple of (possibly empty) precompact open subsets $V_I \sqsubset U_I$, satisfying the following conditions:*

(1) *$V_I = \pi_I^{-1}(\underline{V}_I)$ for each $I \in \mathcal{I}_{\mathcal{K}}$, i.e. V_I is pulled back from the intermediate category and so is Γ_I-invariant;*
(2) *$V_I \sqsubset U_I$ for all $I \in \mathcal{I}_{\mathcal{K}}$, and if $V_I \neq \emptyset$ then $V_I \cap s_I^{-1}(0) \neq \emptyset$;*
(3) *if $\pi_{\mathcal{K}}(\overline{V_I}) \cap \pi_{\mathcal{K}}(\overline{V_J}) \neq \emptyset$ then $I \subset J$ or $J \subset I$;*
(4) *the zero set $\iota_{\mathcal{K}}(X) = |\mathfrak{s}_{\mathcal{K}}|^{-1}(0)$ is contained in $\pi_{\mathcal{K}}(\mathcal{V}) = \bigcup_{I\in\mathcal{I}_{\mathcal{K}}} \pi_{\mathcal{K}}(V_I)$.*

Given a reduction $\mathcal{V}$, we define the **reduced domain category** $\mathbf{B}_{\mathcal{K}}|_{\mathcal{V}}$ *and the* **reduced obstruction category** $\mathbf{E}_{\mathcal{K}}|_{\mathcal{V}}$ *to be the full subcategories of* $\mathbf{B}_{\mathcal{K}}$ *and* $\mathbf{E}_{\mathcal{K}}$ *with objects* $\bigsqcup_{I\in\mathcal{I}_{\mathcal{K}}} V_I$ *resp.* $\bigsqcup_{I\in\mathcal{I}_{\mathcal{K}}} V_I\times E_I$*, and denote by* $\mathfrak{s}|_{\mathcal{V}}:\mathbf{B}_{\mathcal{K}}|_{\mathcal{V}}\to\mathbf{E}_{\mathcal{K}}|_{\mathcal{V}}$ *the section given by restriction of* $\mathfrak{s}_{\mathcal{K}}$.

It is crucial in this context that the quotient map Obj $\mathbf{B}_{\mathcal{K}}\to$ Obj $\underline{\mathbf{B}}_{\mathcal{K}}$ is proper (see Lemma 3.1.5), so that the pullback V_I of a precompact subset $\underline{V}_I\sqsubset\underline{U}_I$ is still precompact in U_I. Because of this, we can establish the existence and uniqueness of reductions modulo cobordism by working in the intermediate category, hence proving the analog of Proposition 2.4.2. Further, the results on **nested (cobordism) reductions** can be interpreted at the intermediate level, and hence go through as before. Here, we say that two reductions $\mathcal{C},\mathcal{V}$ are **nested** (written $\mathcal{C}\sqsubset\mathcal{V}$) if C_I is a precompact subset of V_I for all I. The only real change needed to the discussion in §2.4 above is that, to achieve transversality we should work with "multisections" rather than sections.[29] In our categorical framework, these can be defined very easily.

DEFINITION 3.4.2. *A* **reduced perturbation** ν *of* $\mathcal{K}$ *is a smooth map*

$$\nu:\mathcal{V}=\bigsqcup_{I\in\mathcal{I}_{\mathcal{K}}} V_I\longrightarrow \mathrm{Obj}_{\mathbf{E}_{\mathcal{K}}|_{\mathcal{V}}}$$

between the spaces of objects in the reduced domain and obstruction categories of some reduction $\mathcal{V}$ *of* $\mathcal{K}$*, such that* $\mathrm{pr}_{\mathcal{K}}\circ\nu$ *is the identity. Further, we require that* $\nu=(\nu_I)_{I\in\mathcal{I}_{\mathcal{K}}}$ *is given by a family of smooth maps* $\nu_I:V_I\to E_I$ *that are compatible with coordinate changes in the sense that*

$$\nu_J\big|_{\widetilde{V}_{IJ}}=\widehat{\phi}_{IJ}\circ\nu_I\circ\rho_{IJ}\big|_{\widetilde{V}_{IJ}}\qquad where\quad \widetilde{V}_{IJ}:=\widetilde{U}_{IJ}\cap V_J\cap\rho_{IJ}^{-1}(V_I).\tag{3.4.1}$$

We say that ν *is:*

- **admissible** *if*
  $$\mathrm{d}_y\nu_J(\mathrm{T}_yV_J)\subset\mathrm{im}\,\widehat{\phi}_{IJ}\qquad\forall\, I\subsetneq J,\ y\in\widetilde{U}_{IJ}\cap V_J\cap\rho_{IJ}^{-1}(V_I).$$
- **precompact** *if there is a nested reduction* $\mathcal{C}\sqsubset\mathcal{V}$ *such that*
  $$\bigcup_{I\in\mathcal{I}_{\mathcal{K}}}\pi_{\mathcal{K}}\big((s_I|_{V_I}+\nu_I)^{-1}(0)\big)\ \subset\ \pi_{\mathcal{K}}(\mathcal{C});$$
- **transverse** *(written* $\pitchfork 0$*) if* $s_I|_{V_I}+\nu_I:V_I\to E_I$ *is transverse to* 0 *for each* $I\in\mathcal{I}_{\mathcal{K}}$.

The above compatibility condition implies that when $I\subset J$ the perturbation ν_J is determined by ν_I on the part $\widetilde{V}_{IJ}$ of V_J that lies over V_I. In particular it takes values in $E_I\subset E_J$ and is invariant under the action of $\Gamma_{J\smallsetminus I}$, and this means that ν is compatible with morphisms of type (b) on page 55. However, ν is *not* in general a functor $\mathbf{B}_{\mathcal{K}}\big|_{\mathcal{V}}\to\mathbf{E}_{\mathcal{K}}\big|_{\mathcal{V}}$ since it is not required to be equivariant under the group actions. Hence it induces a multivalued map $|\nu|$ on the realization $|\mathcal{V}|$, given by

$$|\nu_I([I,x])|=\bigcup_{\gamma\in\Gamma_I}\pi_{\mathbf{E}_{\mathcal{K}}}([I,\nu(\gamma x)])\in|\,\mathrm{pr}\,|^{-1}([I,x]).$$

[29]To see that one cannot always achieve transversality by equivariant sections of an orbibundle, consider the bundle $\pi:(S^1\times\mathbb{R},\mathbb{Z}_2)\to(S^1,\mathrm{id})$ where the generator of $\mathbb{Z}_2$ acts on $S^1\times\mathbb{R}$ by $(x,e)\mapsto(x,-e)$. Then the only equivariant section is the constant section, which is not transverse to 0.

Thus one can think of ν as a "multisection", even though it is defined by a single valued function on $\mathcal{V}$.

REMARK 3.4.3. The reduced perturbation $\nu : \mathbf{B}_{\mathcal{K}}|_{\mathcal{V}} \to \mathbf{E}_{\mathcal{K}}|_{\mathcal{V}}$ is defined in [**MW3**] to be a functor between the **pruned** domain and obstruction categories $\mathbf{B}_{\mathcal{K}}|_{\mathcal{V}}^{\smallsetminus\Gamma}$ and $\mathbf{E}_{\mathcal{K}}|_{\mathcal{V}}^{\smallsetminus\Gamma}$ of [**MW3**, Lemma 3.2.3]. These categories have the same object spaces as $\mathbf{B}_{\mathcal{K}}|_{\mathcal{V}}$ and $\mathbf{E}_{\mathcal{K}}|_{\mathcal{V}}$ but fewer morphisms. For example in $\mathbf{B}_{\mathcal{K}}|_{\mathcal{V}}^{\smallsetminus\Gamma}$ one only keeps the morphisms of the form $(I,J,y) : (I,\rho_{IJ}(y)) \mapsto (J,y)$ given by the projections ρ_{IJ}. ◊

Because the perturbation section ν satisfies essentially the same compatibility conditions as does a perturbation section of an atlas with trivial isotropy, it may be constructed by the same method. Notice in particular that, although ν must be constructed on the space of objects $\bigsqcup_I V_I$ of $\mathbf{B}_{\mathcal{K}}|_{\mathcal{V}}$, the sets such as $V_J^{|J|}$ and $B_\delta^J(N_{JI}^{|J|})$ used in §2.4 to describe its inductive construction can all be pulled back from corresponding subsets of the domains of the intermediate category (which after all is where the metric lives). It follows that the construction of ν goes through without essential change.

However, because of the presence of nontrivial isotropy, the local zero sets $Z_I := (s_I|_{V_I} + \nu_I)^{-1}(0)$ no longer fit together to form a manifold. Rather we must use the notions of weighted nonsingular branched (wnb) groupoid (defined below) and weighted branched manifold developed in [**M3**]. Since this requires considerable preparation, we will first state the main result and show how to use it to construct the VFC.

PROPOSITION 3.4.4. *Let $(\mathcal{K},\sigma)$ be a tame, oriented, Kuranishi atlas of dimension d with nested reduction $\mathcal{C} \sqsubset \mathcal{V}$, and let $\nu : \mathbf{B}_{\mathcal{K}|_{\mathcal{V}}} \to \mathbf{E}_{\mathcal{K}|_{\mathcal{V}}}$ be an admissible reduced perturbation that is transverse and precompact with local zero sets*

$$Z_I := (s_I|_{V_I} + \nu_I)^{-1}(0) \subset V_I \cap \pi_{\mathcal{K}}^{-1}(\pi_{\mathcal{K}}(\mathcal{C})).$$

Let $\mathbf{Z}^\nu$ be the full subcategory of $\mathbf{B}_{\mathcal{K}|_{\mathcal{V}}}$ with object space $\bigcup_I Z_I$. Then $\mathbf{Z}^\nu$ can be completed to a wnb groupoid $(\widehat{\mathbf{Z}}^\nu, \Lambda^\nu)$ with the same realization $|\widehat{\mathbf{Z}}^\nu| = |\mathbf{Z}^\nu|$. Moreover, the maximal Hausdorff quotient $|\widehat{\mathbf{Z}}^\nu|_{\mathcal{H}}$ is compact and inclusion induces a map $|\iota_\nu|_{\mathcal{H}} : |\widehat{\mathbf{Z}}^\nu|_{\mathcal{H}} \to |\mathcal{K}|$ with image in $\pi_{\mathcal{K}}(\mathcal{C})$.

Sketch proof of Theorem B. By definition, the maximal Hausdorff quotient $|\widehat{\mathbf{Z}}|_{\mathcal{H}}$ (which is defined below) of a wnb groupoid $\widehat{\mathbf{Z}}$ is a weighted branched manifold. We show in [**M3**, Proposition 3.25] that when $Z := |\widehat{\mathbf{Z}}|_{\mathcal{H}}$ is compact, it carries a fundamental class $[Z] \in H_d(Z;\mathbb{Q})$, where d is the dimension of $\widehat{\mathbf{Z}}$. (See Examples 3.4.6 and 3.4.7 below.) With this information we can now repeat the arguments sketched at the end of §2 to establish the existence of the class $[X]_{\mathcal{K}}^{vir} \in \check{H}_d(X;\mathbb{Q})$. To prove that this class does not depend on our auxiliary choices, we first establish a version of Proposition 3.4.4 for cobordisms, and then argue as in the case with trivial isotropy. □

With this in hand, we now return to the discussion of wnb groupoids $\mathbf{G}$. Because these groupoids[30] are in general not proper, the realization $|\mathbf{G}|$ may not be Hausdorff, and we write $|\mathbf{G}|_{\mathcal{H}}$ for its **maximal Hausdorff quotient**. Thus $|\mathbf{G}|_{\mathcal{H}}$

[30]See the beginning of §5.1 for the definition of an étale groupoid; nonsingular means that there is at most one morphism between any two points. Note also that although the following

is a Hausdorff quotient of $|\mathbf{G}|$ that satisfies the following universal property: any continuous map from $|\mathbf{G}|$ to a Hausdorff space Y factors though the projection $|\mathbf{G}| \to |\mathbf{G}|_{\mathcal{H}}$. (The existence of such a quotient for any topological space is proved in [**M3**, Lemma 3.1]; see also [**MW3**, Lemma A.2].) We denote the natural maps by

$$\pi_{\mathbf{G}} : \mathrm{Obj}_{\mathbf{G}} \to |\mathbf{G}|, \qquad \pi^{\mathcal{H}}_{|\mathbf{G}|} : |\mathbf{G}| \longrightarrow |\mathbf{G}|_{\mathcal{H}}, \qquad \pi^{\mathcal{H}}_{\mathbf{G}} := \pi^{\mathcal{H}}_{|\mathbf{G}|} \circ \pi_{\mathbf{G}} : \mathrm{Obj}_{\mathbf{G}} \to |\mathbf{G}|_{\mathcal{H}}.$$

Moreover, for $U \subset \mathrm{Obj}_{\mathbf{G}}$ we write $|U| := \pi_{\mathbf{G}}(U) \subset |\mathbf{G}|$ and $|U|_{\mathcal{H}} := \pi_{\mathcal{H}}(U) \subset |\mathbf{G}|_{\mathcal{H}}$. The **branch locus** of $\mathbf{G}$ is defined to be the subset of $|\mathbf{G}|_{\mathcal{H}}$ consisting of points with more than one preimage in $|\mathbf{G}|$.

DEFINITION 3.4.5. *A* **weighted nonsingular branched groupoid** *(or* **wnb groupoid** *for short) of dimension d is a pair $(\mathbf{G}, \Lambda)$ consisting of an oriented, nonsingular étale groupoid $\mathbf{G}$ of dimension d, together with a rational weighting function $\Lambda : |\mathbf{G}|_{\mathcal{H}} \to \mathbb{Q}^+ := \mathbb{Q} \cap (0, \infty)$ that satisfies the following compatibility conditions. For each $p \in |\mathbf{G}|_{\mathcal{H}}$ there is an open neighbourhood $N \subset |\mathbf{G}|_{\mathcal{H}}$ of p, a collection $U_1, \dots, U_\ell$ of disjoint open subsets of $(\pi^{\mathcal{H}}_{\mathbf{G}})^{-1}(N) \subset \mathrm{Obj}_{\mathbf{G}}$ (called* **local branches***), and a set of positive rational weights $m_1, \dots, m_\ell$ such that the following holds:*

(Covering) $(\pi^{\mathcal{H}}_{|\mathbf{G}|})^{-1}(N) = |U_1| \cup \cdots \cup |U_\ell| \subset |\mathbf{G}|$;

(Local Regularity) *for each $i = 1, \dots, \ell$ the projection $\pi^{\mathcal{H}}_{\mathbf{G}}|_{U_i} : U_i \to |\mathbf{G}|_{\mathcal{H}}$ is a homeomorphism onto a relatively closed subset of N;*

(Weighting) *for all $q \in N$, the number $\Lambda(q)$ is the sum of the weights of the local branches whose image contains q:*

$$\Lambda(q) = \sum_{i: q \in |U_i|_{\mathcal{H}}} m_i.$$

Further we define a **weighted branched manifold** of dimension d to be a pair (Z, Λ_Z) consisting of a topological space Z together with a function $\Lambda_Z : Z \to \mathbb{Q}^+$ and an equivalence class[31] of wnb d-dimensional groupoids $(\mathbf{G}, \Lambda_{\mathbf{G}})$ and homeomorphisms $f : |\mathbf{G}|_{\mathcal{H}} \to Z$ that induce the function $\Lambda_Z = \Lambda_{\mathbf{G}} \circ f^{-1}$. Analogous definitions of a wnb cobordism groupoid (as always, assumed to have collared boundaries) and of a wnb cobordism are spelled out in [**MW3**, Appendix].

EXAMPLE 3.4.6. (i) A compact weighted branched manifold of dimension 0 consists of a finite set of points $Z = \{p_1, \dots, p_k\}$, each with a positive rational weight $m(p_i) \in \mathbb{Q}^+$ and orientation $\mathfrak{o}(p_i) \in \{\pm\}$. Its fundamental class pushes forward to the sum $\sum_i \mathfrak{o}(p_i) m(p_i) \in H_0(pt, \mathbb{Q}) = \mathbb{Q}$ under any map $Z \to \{pt\}$. Any representing groupoid $\mathbf{G}$ has as object space $\mathrm{Obj}_{\mathbf{G}}$ a set with the discrete topology, that is equipped with an orientation function $\mathfrak{o} : \mathrm{Obj}_{\mathbf{G}} \to \{\pm\}$. The morphism space $\mathrm{Mor}_{\mathbf{G}}$ is also a discrete set and, because we assume that $\mathbf{G}$ is oriented, defines an equivalence relation on $\mathrm{Obj}_{\mathbf{G}}$ such that $x \sim y \Longrightarrow \mathfrak{o}(x) = \mathfrak{o}(y)$. Moreover, because $|\mathbf{G}|$ is Hausdorff, we can identify $|\mathbf{G}| = |\mathbf{G}|_{\mathcal{H}}$ and hence conclude that $\mathrm{Obj}_{\mathbf{G}}$ has precisely k equivalence classes.

---

definition requires that $\mathbf{G}$ be oriented in order to be consistent with [**MW3**], the definitions of wnb groupoid and manifold work equally well in the nonoriented case.

[31]The precise notion of equivalence is given in [**M3**, Definition 3.12]. In particular it ensures that the induced function $\Lambda_Z := \Lambda_{\mathbf{G}} \circ f^{-1}$ and the dimension of $\mathrm{Obj}_{\mathbf{G}}$ are the same for equivalent structures $(\mathbf{G}, \Lambda_{\mathbf{G}}, f)$.

(ii) For the prototypical example of a 1-dimensional weighted branched cobordism $(|\mathbf{G}|_{\mathcal{H}}, \Lambda)$, take $\mathrm{Obj}(\mathbf{G}) = I \sqcup I'$ equal to two copies of the interval $I = I' = [0,1]$ with nonidentity morphisms from $x \in I$ to $x \in I'$ for $x \in [0, \frac{1}{2})$ and their inverses, where we suppose that I is oriented in the standard way. Then the realization and its Hausdorff quotient are

$$\begin{aligned} |\mathbf{G}| &= {}^{I \sqcup I'}\big/_{\{(I,x) \sim (I',x) \text{ iff } x \in [0,\frac{1}{2})\}}, \\ |\mathbf{G}|_{\mathcal{H}} &= {}^{I \sqcup I'}\big/_{\{(I,x) \sim (I',x) \text{ iff } x \in [0,\frac{1}{2}]\}}, \end{aligned}$$

and the branch locus is a single point $\mathrm{Br}(\mathbf{G}) = \{[I, \frac{1}{2}] = [I', \frac{1}{2}]\} \subset |\mathbf{G}|_{\mathcal{H}}$. The choice of weights $m, m' > 0$ on the two local branches I and I' determines the weighting function $\Lambda : |\mathbf{G}|_{\mathcal{H}} \to (0, \infty)$ as

$$\Lambda([I,x]) = \begin{cases} m + m' & \text{if } \ x \in [0, \frac{1}{2}], \\ m & \text{if } \ x \in (\frac{1}{2}, 1], \end{cases} \qquad \Lambda([I',x]) = \begin{cases} m + m' & \text{if } \ x \in [0, \frac{1}{2}], \\ m' & \text{if } \ x \in (\frac{1}{2}, 1]. \end{cases}$$

For example, if we give each branch I, I' the weight $m = m' = \frac{1}{2}$ and choose appropriate collar functors $\iota^\alpha_{\mathbf{G}}$, we obtain a weighted branched cobordism $(|\mathbf{G}|_{\mathcal{H}}, \iota^0_{\mathbf{G}}, \iota^1_{\mathbf{G}}, \Lambda)$ with $|\partial^0 \mathbf{G}|_{\mathcal{H}} = \{[I,0] = [I',0]\}$, which is a single point with weight 1, and $|\partial^1 \mathbf{G}|_{\mathcal{H}} = \{[I,1], [I',1]\}$, which consists of two points with weight $\frac{1}{2}$, all with positive orientation. The fundamental class $[Z]$ of $Z = |\mathbf{G}|_{\mathcal{H}}$ is then represented in $H_1(Z, \partial Z; \mathbb{Q})$ by the weighted sum of the two branches, thought of as singular 1-chains. Note that its image under the boundary map $H_1(Z, \partial Z; \mathbb{Q}) \to H_0(\partial Z)$ is $[\partial^1 Z] - [\partial^0 Z]$, where $\partial^\alpha Z := |\partial^\alpha \mathbf{G}|$.

Another choice of collar functors for the same weighted groupoid $(\mathbf{G}, \Lambda)$ might give rise to a different partition of the boundary into incoming $\partial^0 \mathbf{G}$ and outgoing $\partial^1 \mathbf{G}$, for example yielding a weighted branched cobordism with $|\partial^0 \mathbf{G}|_{\mathcal{H}} = \{[I,0] = [I',0], [I,1]\}$, consisting of two points with weights and orientations $(1,+)$ and $(\frac{1}{2}, -)$ and $|\partial^1 \mathbf{G}|_{\mathcal{H}} = \{[I',1]\}$, consisting of one point with weight $(\frac{1}{2}, +)$.

See Figure 3.4.1 for an illustration of the realization of a similar example. $\Diamond$

Sketch proof of Proposition 3.4.4. Note first that each set

$$\widetilde{V}_{IJ} := V_J \cap \rho_{IJ}^{-1}(V_I) \subset \widetilde{U}_{IJ} = V_J \cap \pi_{\mathcal{K}}^{-1}\big(\pi_{\mathcal{K}}(V_I)\big)$$

is invariant under the action of Γ_J, and is an open subset of the closed submanifold $\widetilde{U}_{IJ} = s_J^{-1}(E_I)$ of V_J, where the last equality holds by the tameness condition (2.2.4). Further if $F \subset I \subset J$,

$$V_J \cap \rho_{IJ}^{-1}(\widetilde{V}_{FI}) = \widetilde{V}_{FJ} \cap \widetilde{V}_{IJ} = V_J \cap \pi_{\mathcal{K}}^{-1}\big(\pi_{\mathcal{K}}(V_F) \cap \pi_{\mathcal{K}}(V_I)\big) \subset \widetilde{U}_{FJ}. \tag{3.4.2}$$

In fact, the second equality holds for any pair of subsets $F, I \subset J$. However, because $\mathcal{V}$ is a reduction, the intersection is empty unless F, I are nested, i.e. either $F \subset I$ or $I \subset F$. Finally, the group $\Gamma_{I \setminus F}$ acts freely on $\widetilde{U}_{FI}$ (by Definition 3.1.10 for a coordinate change) and hence also on $\widetilde{V}_{FI}$.

We begin by establishing the following properties of the local zero sets Z_I:

(1) each $Z_I := (s_I|_{V_I} + \nu_I)^{-1}(0) \subset V_I$ is a submanifold of dimension d;
(2) for all $I \subset J$, the intersection $\widetilde{V}_{IJ} \cap Z_J$ is an open subset of Z_J, and each $\alpha \in \Gamma_{J \setminus I}$ induces a diffeomorphism $Z_J \cap \widetilde{V}_{IJ} \to Z_J \cap \widetilde{V}_{IJ} : y \mapsto \alpha y$;
(3) for all $I \subset J$, $\rho_{IJ} : \widetilde{V}_{IJ} \cap Z_J \to Z_I$ is a submersion onto an open subset of Z_I;

(4) if $\mathcal{K}$ is oriented and each Z_I is oriented as in Lemma 3.3.9, then the maps in (i) and (ii) preserve orientation.

The proof that these properties hold is straightforward;. In particular, the first claim in (ii) holds by the index condition for the inclusion $\widetilde{\phi}_{IJ} : \widetilde{U}_{IJ} \to U_J$, while the second holds by the compatibility condition (3.4.1). Indeed, if $x \in \widetilde{V}_{IJ}$ and $\alpha \in \Gamma_{J \setminus I}$ then compatibility implies that $\nu_J(\alpha x) = \nu_J(x)$, while equivariance implies that $s_J(\alpha x) = \alpha s_J(x) = s_J(x)$ since $s_J|_{\widetilde{V}_{IJ}} = \widehat{\phi}_{IJ} \circ \rho_{IJ}|_{\widetilde{V}_{IJ}}$ takes values in E_I and $\Gamma_{J \setminus I}$ acts trivially on E_I. Hence $(s_J + \nu_J)(\alpha x) = 0$ if and only if $(s_J + \nu_J)(x) = 0$.

We then define the wnb groupoid $\widehat{\mathbf{Z}} := \widehat{\mathbf{Z}}^\nu$ to have objects $\mathrm{Obj}_{\widehat{\mathbf{Z}}} = \bigsqcup_I Z_I$, and morphisms that are generated by the covering maps $\rho_{IJ} : \widetilde{V}_{IJ} \to V_I$. In other words, we take all morphisms of this kind (which are denoted $(I, J, \mathrm{id}, y) \in \mathrm{Mor}_{\mathbf{B}_{\mathcal{K}}}$) and then complete to a groupoid by including all composites and inverses. As we sketch below, it turns out that the resulting space of morphisms with source in Z_I and target in Z_J for $I \subset J$ is given by

(3.4.3)

$$\begin{aligned}\mathrm{Mor}_{\widehat{\mathbf{Z}}}(Z_I, Z_J) &= \bigcup_{\emptyset \ne F \subset I} \big(Z_J \cap \widetilde{V}_{IJ} \cap \widetilde{V}_{FJ}\big) \times \Gamma_{I \setminus F},\\ &= \big\{(I, J, \alpha, y) \bigm| \alpha \in \Gamma_{I \setminus F}, y \in \widetilde{V}_{FJ} \cap \widetilde{V}_{IJ} \cap Z_J \text{ for some } F \subset I,\ F \ne \emptyset\big\},\end{aligned}$$

where we interpret $\Gamma_\emptyset = \mathrm{id}$; in particular the identity morphisms are given by taking $F = I = J$. Here we use the same notation as for the category $\mathbf{B}_{\mathcal{K}}$. In other words, source and target of these morphisms are given by

$$(I, J, \alpha, y) \in \mathrm{Mor}_{\mathbf{G}}\big((I, \alpha^{-1}\rho_{IJ}(y)),\ (J, y)\big). \tag{3.4.4}$$

The category $\widehat{\mathbf{Z}}$ also contains the inverses of these morphisms. Thus we obtain $\widehat{\mathbf{Z}}$ from the full subcategory of $\mathbf{B}_{\mathcal{K}}$ with objects $\bigsqcup_I Z_I$ by throwing away many of the morphisms, and then adding inverses. Note that a point $(J, x) \in Z_J$ that does not lie in any subset $\widetilde{V}_{IJ}$ for $I \subsetneq J$ is the target of exactly one morphism, namely the identity morphism, while if it lies in some $\widetilde{V}_{IJ}$ it is the target of morphisms with source in Z_I as well as morphisms with source in Z_J given by the (free) action of $\Gamma_{J \setminus I}$.

It is not too hard to see that this set of morphisms is closed under composition. For example, one can check that if $I \subset J \subset K$ and $\rho_{IJ}(y') = \rho_{IK}(y)$ so that $y' = \rho_{JK}(y)$, then the composite[32] on the left of the following identity is well defined and equals the morphism on the right:

$$(I, J, \mathrm{id}, y')^{-1} \circ (I, K, \mathrm{id}, y) = (J, K, \mathrm{id}, y).$$

On the other hand, if $I \subset K \subset J$ and there are $y' \in \widetilde{V}_{IJ} \cap V_J$, $y \in \widetilde{V}_{IK} \cap V_K$ with

$$\rho_{IJ}(y') = \rho_{IK}(\rho_{KJ}(y')) = \rho_{IK}(y) \in V_I,$$

then $\rho_{KJ}(y')$ and y lie in the same $\Gamma_{K \setminus I}$-orbit, so that there is $\beta \in \Gamma_{K \setminus I}$ such that $y = \beta\,\rho_{KJ}(y') = \rho_{KJ}(\beta y') \in V_K$, and we have

$$(I, J, \mathrm{id}, y')^{-1} \circ (I, K, \mathrm{id}, y) \;=\; (J, J, \beta, \beta\, y') \circ (K, J, \mathrm{id}, \beta\, y')^{-1}. \tag{3.4.5}$$

This shows that if we patch the local zero sets together by the projections ρ_{IJ} and then complete to a groupoid by adding inverses and composites, one necessarily

[32] as usual written in categorical order

must add the morphisms from Z_J to Z_J given by the actions of the groups $\Gamma_{J\setminus I}$ on the sets $\widetilde{V}_{IJ}$. The fact that these actions are free implies that $\widehat{\mathbf{Z}}$ is nonsingular. Further the category is étale by properties (ii) and (iii) above, and oriented by (iv). One might informally think that the manifolds Z_I form the local branches. However, this is not strictly true because the map $Z_J \to |\widehat{\mathbf{Z}}|$ is *not* injective but rather quotients out by the free actions of the subgroups $\Gamma_{J\setminus I}$ on the sets $\widetilde{V}_{IJ} \cap Z_J$.

Once $\widehat{\mathbf{Z}}$ is constructed, the next step is to understand its maximal Hausdorff quotient. By [**MW3**, Lemma 3.2.10], this is the realization of the groupoid $\widehat{\mathbf{Z}}_{\mathcal{H}}$ that is obtained from $\widehat{\mathbf{Z}}$ by closing the equivalence relation on $\mathrm{Obj}_{\widehat{\mathbf{Z}}}$ generated by $\mathrm{Mor}_{\mathbf{Z}}$. Thus $\widehat{\mathbf{Z}}_{\mathcal{H}}$ has extra morphisms from Z_J to Z_J given by

$$\begin{aligned}\mathrm{Mor}_{\widehat{\mathbf{Z}}_{\mathcal{H}}}(Z_J, Z_J) &= \bigcup_{F\subset J} \big(Z_J \cap \mathrm{Fr}(\widetilde{V}_{FJ})\big) \times \Gamma_{J\setminus F}, \\ &= \big\{(J, J, \alpha, y) \;\big|\; \exists F \subset J \;\alpha \in \Gamma_{J\setminus F} \; s.t. \; y \in \mathrm{Fr}(\widetilde{V}_{FJ}) \cap Z_J\big\},\end{aligned}$$

where $\mathrm{Fr}(A) := \overline{A}\setminus A$ denotes the frontier of the set A. As one can clearly see in the football example given below, these extra morphisms lie over the branch locus. Finally, one checks that the following formula gives a well-defined weighting function $\Lambda : |\widehat{\mathbf{Z}}_{\mathcal{H}}| \to \mathbb{Q}^+$:

$$\Lambda(p) := \sum_{x\in Z_J\cap \pi^{-1}_{\widehat{\mathbf{Z}}_{\mathcal{H}}}(p)} \frac{1}{|\Gamma_J|}.$$

Thus each local zero set Z_J is assigned the weight $\frac{1}{|\Gamma_J|}$, and the weight of a point $p \in |\widehat{\mathbf{Z}}_{\mathcal{H}}| := |\widehat{\mathbf{Z}}|_{\mathcal{H}}$ is obtained by multiplying this weight by the number of its inverse images in Z_I. This completes the sketch proof. □

Here are some examples to illustrate this construction of the perturbed zero set as weighted branched manifold..

EXAMPLE 3.4.7. We begin with the orbifold case, i.e. where the atlas $\mathcal{K}$ on X has trivial (i.e. zero dimensional) obstruction bundles so that $\mathfrak{s}_{\mathcal{K}}$ and ν are identically zero and $\iota_{\mathcal{K}} : X \to |\mathcal{K}|$ is a homeomorphism. Recall from Proposition 3.4.4 that $\mathbf{Z}$ is defined as a subcategory of $\mathbf{B}_{\mathcal{K}}|_{\mathcal{V}}$, while $\widehat{\mathbf{Z}}$ is its groupoid completion. Therefore to avoid confusion below we often designate groupoids with hats.

(i) First suppose that $X = {}^{M}/_{\Gamma}$ is the quotient of a compact oriented smooth manifold M by the action of a finite group Γ, and that $\mathcal{K}$ is the atlas with a single chart with domain M and $E = \{0\}$. Then $\mathbf{Z}$ is the category with objects M and only identity morphisms, because there are no pairs $I, J \in \mathcal{I}_{\mathcal{K}}$ such that $\emptyset \neq I \subsetneq J$. Therefore $\mathbf{Z} = \widehat{\mathbf{Z}}_{\mathcal{H}}$ has realization $|\widehat{\mathbf{Z}}_{\mathcal{H}}| = M$ and the weighting function $\Lambda : M \to \mathbb{Q}$ is given by $\Lambda(x) = \frac{1}{|\Gamma|}$. If the action of Γ is effective on every open subset of M then the pushforward of Λ by $\iota_{\widehat{\mathbf{Z}}_{\mathcal{H}}} : M \to X$, which is defined by

$$(\iota_{\widehat{\mathbf{Z}}_{\mathcal{H}}})_*\Lambda(p) := \sum_{x\in(\iota_{\widehat{\mathbf{Z}}_{\mathcal{H}}})^{-1}(p)} \Lambda(x)$$

takes the value 1 at every smooth point (i.e. point with trivial stabilizer) of the orbifold ${}^{M}/_{\Gamma}$. On the other hand, if Γ acts by the identity so that the action is totally noneffective then $\iota_{\widehat{\mathbf{Z}}_{\mathcal{H}}} : M \overset{\cong}{\to} X$ is the identity map and the weighting function $X \to \mathbb{Q}^+$ takes the constant value $\frac{1}{|\Gamma|}$.

We claim that this choice of weights gives a fundamental class that is consistent with standard conventions. Recall from [**M3**] that if the branch locus $\mathrm{Br}(|\widehat{\mathbf{Z}}|)$ (i.e. set of points with more than one inverse image in $|\widehat{\mathbf{Z}}|$) of a weighted branched manifold $|\widehat{\mathbf{Z}}|_{\mathcal{H}}$ is sufficiently nice then there is a triangulation of $|\widehat{\mathbf{Z}}|_{\mathcal{H}}$ such that $\mathrm{Br}(|\widehat{\mathbf{Z}}|)$ is contained in the codimension 1-skeleton, and each top dimensional simplex lies in a single branch. In this case, the compatibility condition on the weights implies that the fundamental class is represented by the sum of the top dimensional simplices, each provided with the appropriate orientation and weight. For example, if the action of Γ on M is effective and preserves an orientation then its singular set $M^{sing} = \{x : \Gamma^x \neq \mathrm{id}\}$ has codimension at least 2 and the fundamental class of the weighted branched manifold (M, Λ) pushes forward to the "usual" fundamental class of X, which is represented by the pseudocycle $X \setminus X^{sing}$, where $X^{sing} := {}^{M^{sing}}/_{\Gamma}$.

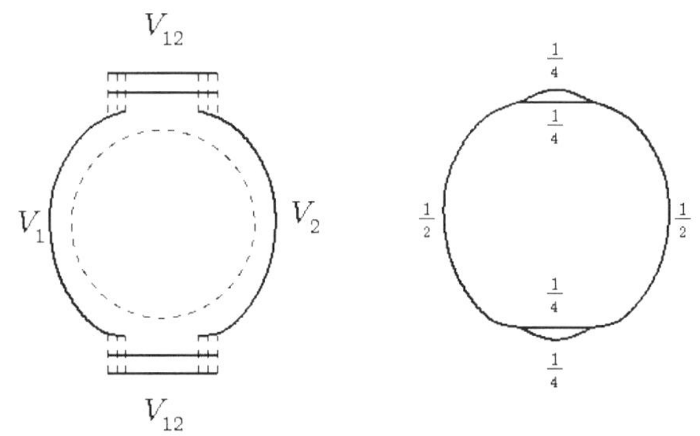

FIGURE 3.4.1. This illustrates the zero set $\mathbf{Z}$ in the case of the atlas on S^1 constructed in Example 3.1.17 (ii). The left diagram shows $\mathrm{Obj}_{\mathbf{Z}^\nu} := \sqcup V_I$, with vertical lines to denote the existence of morphisms, while the right shows the realization and weighting function.

(ii) If the action is noneffective, then one might wonder about the correct definition for the fundamental class. However, in dimension $d = 0$ we saw in Example 3.4.6 (i) that a branched manifold $Z = |\mathbf{Z}|_{\mathcal{H}}$ is a finite collection of points $\{p_1, \dots, p_k\}$, one for each equivalence class in $\mathrm{Obj}_{\mathbf{Z}}$, where the point p_i corresponding to an equivalence class with stabilizer Γ^i has weight $\frac{1}{|\Gamma^i|}$. If each point is positively oriented, then our construction counts the "number of elements" in $|\mathbf{Z}|_{\mathcal{H}}$ as the sum $\sum_{i=1}^k \frac{1}{|\Gamma^i|}$, which is the Euler characteristic of the groupoid; see Weinstein [**W**] for further discussion and references. Example 3.4.8 below gives another relevant example.

(iii) Again suppose that $X = {}^{M}/_{\Gamma}$ where $\Gamma = \mathbb{Z}_2$ acts trivially on M, and choose an atlas $\mathcal{K}$ with one chart

$$\big(U = M \times \mathbb{R}^2,\ E = \mathbb{R}^2,\ \Gamma := \mathbb{Z}_2,\ \mathfrak{s} := \mathrm{pr}_E : U \to E,\ \psi = \mathrm{pr}_M|_{M \times \{0\}}\big),$$

where the nonidentity element $\gamma \in \mathbb{Z}_2$ acts by $(x, e) \mapsto (x, -e)$ on U and by $e \mapsto -e$ on E. The reduction $\mathcal{V}$ is given by a choice of any open neighbourhood V of

$\mathfrak{s}^{-1}(0) = M \times \{0\} \subset U$. Further, any function $\nu : V \to \mathbb{R}$ of the form $\nu(x, e) = f(x)$ is an admissible transverse perturbation, and defines a zero groupoid $\widehat{\mathbf{Z}}^{\nu}$ whose space of objects $\{(x, e) | f(x) + e = 0\}$ can be identified with the graph of $-f : M \to \mathbb{R}^2$. As in (i), $\widehat{\mathbf{Z}}^{\nu}$ has only identity morphisms and its weighting function is the constant $\frac{1}{2}$.

(iv) For a more interesting example, consider the "football" discussed in Example 3.1.17, with reduction $\mathcal{V}$ given by two discs $V_1 \sqsubset U_1, V_2 \sqsubset U_2$ with disjoint images in X, together with an open annulus $V_{12} \sqsubset U_{12}$. For $j = 1, 2$ the sets $\widetilde{V}_{j,12} \subset V_{12}$ are disjoint open annuli that project into V_j by a covering map of degree 3 for $j = 1$ (that quotients out by $\Gamma_{(12)\smallsetminus 1} = \Gamma_2 = \mathbb{Z}_3$) and degree 2 for $j = 2$. Then $Z_I = V_I$ so that $\mathrm{Obj}_{\mathbf{Z}} = V_1 \sqcup V_2 \sqcup V_{12}$. For $j = 1, 2$ the category $\widehat{\mathbf{Z}}$ has the following morphisms (besides identities);

- morphisms $V_j \to V_{12}$ given by the projection $\rho_{j,12} : \widetilde{V}_{j,12} \to V_j$, together with their inverses;
- morphisms $V_{12} \to V_{12}$ given by the action of $\mathbb{Z}_3 = \Gamma_{(12)\smallsetminus 1}$ on $\widetilde{V}_{1,12}$, resp. of $\mathbb{Z}_2 = \Gamma_{(12)\smallsetminus 2}$ on $\widetilde{V}_{2,12}$.

To obtain $\widehat{\mathbf{Z}}_{\mathcal{H}}$ we add the morphisms given by the action of $\Gamma_{(12)\smallsetminus 1}$ on the boundary $\mathrm{Fr}(\widetilde{V}_{1,12}) := V_{12} \smallsetminus \widetilde{V}_{1,12}$ and the action of $\Gamma_{(12)\smallsetminus 2}$ on $\mathrm{Fr}(\widetilde{V}_{2,12}) := V_{12} \smallsetminus \widetilde{V}_{2,12}$. The weighting function Λ is given by:

$$
\begin{aligned}
\Lambda(p) \ &= \tfrac{1}{2} \text{ if } p \in \overline{\pi_{\widehat{\mathbf{Z}}_{\mathcal{H}}}(V_1)} = \pi_{\widehat{\mathbf{Z}}_{\mathcal{H}}}(V_1) \cup \pi_{\widehat{\mathbf{Z}}_{\mathcal{H}}}(cl(\widetilde{V}_{1,12})) \\
&= \tfrac{1}{3} \text{ if } p \in \overline{\pi_{\widehat{\mathbf{Z}}_{\mathcal{H}}}(V_2)} = \pi_{\widehat{\mathbf{Z}}_{\mathcal{H}}}(V_2) \cup \pi_{\widehat{\mathbf{Z}}_{\mathcal{H}}}(cl(\widetilde{V}_{2,12})) \\
&= \tfrac{1}{6} \text{ if } p \in \pi_{\widehat{\mathbf{Z}}_{\mathcal{H}}}(V_{12} \smallsetminus cl(\widetilde{V}_{1,12} \cup \widetilde{V}_{2,12}))
\end{aligned}
$$

Notice that for $j = 1, 2$ the weighting function does not change along the boundary of the intersection $\pi_{\widehat{\mathbf{Z}}_{\mathcal{H}}}(V_j) \cap \pi_{\widehat{\mathbf{Z}}_{\mathcal{H}}}(\widetilde{V}_{j,12})$, i.e. there is no branching there, while it does change along the internal boundaries $\pi_{\widehat{\mathbf{Z}}_{\mathcal{H}}}(\mathrm{Fr}(\widetilde{V}_{j,12}))$ in $\pi_{\widehat{\mathbf{Z}}_{\mathcal{H}}}(V_2)$. Note further that the pushforward of Λ by the map $|\iota_{\widehat{\mathbf{Z}}_{\mathcal{H}}}| : |\widehat{\mathbf{Z}}_{\mathcal{H}}| \to |\mathbf{B}_{\mathcal{K}}|_{\mathcal{V}}|$ takes the value 1 at all points except the two poles N, S:

$$
|\iota_{\widehat{\mathbf{Z}}_{\mathcal{H}}}|_*(\Lambda)(q) := \sum_{p \in |\iota_{\widehat{\mathbf{Z}}_{\mathcal{H}}}|^{-1}(q)} \Lambda(p) = 1, \quad \forall q \in X \smallsetminus \{N, S\}.
$$

(v) The above examples all consider the case when the bundle over the orbifold has dimension zero; for further discussion see §5.1. Example 3.2.12 in [**MW3**] builds Kuranishi atlases that model the tangent bundles of S^2 and the football, and then constructs an explicit perturbation that calculates the Euler class. Atlases for more general bundles over orbifolds are considered in §5.2 below. $\Diamond$

EXAMPLE 3.4.8. Here is an example from Gromov–Witten theory that further illustrates our choice of weighting function in the noneffective case. Consider the space X of equivalence classes of genus zero stable maps to S^2 of degree 2 with no marked points. An element $[\Sigma, f] \in X$ is represented either by a degree 2 map $f : \Sigma = S^2 \to S^2$ (classified modulo reparametrization by its two (unordered) critical values) or by a map $f : \Sigma = S^2 \sqcup S^2 \to S^2$ (classified by the image of the nodal point). Elements of the first kind form a stratum $\mathcal{S}_1$ diffeomorphic to ${}^{\{(x,y) \in S^2 \times S^2 | x \neq y\}}\!/_{\sim}$ where $(x, y) \sim (y, x)$, and one can show that elements of the second kind are Gromov limits of sequences in $\mathcal{S}_1$ in which the two critical values converge to points on the diagonal; e.g. see [**M2**, Lemma 3.7]. Thus, as a space we can consider X to be homeomorphic to $\mathbb{C}P^2 = {}^{S^2\times S^2}/_{\sim}$. However, each element $[\Sigma, f]$ has a nontrivial order two symmetry $\gamma_f : \Sigma \to \Sigma$ such that $f = f \circ \gamma_f$, so that in the current context we should consider X as an orbifold equipped with the identity action of the group $\Gamma = \mathbb{Z}_2$. Thus $[X]^{vir} = \frac{1}{2}[X]$. To justify this factor of $\frac{1}{2}$, consider the space $\widetilde{Y}$ of equivalence classes of genus zero, degree 2 stable maps to S^2 that are equipped with three marked points z_0, z_1, z_2 with the fixed images $0, 1, \infty$. Since the elements of $\widetilde{Y}$ have trivial isotropy, $\widetilde{Y}$ is a manifold so that $[\widetilde{Y}]^{vir}$ is its usual fundamental class $[\widetilde{Y}]$. As a map of manifolds, the forgetful map $\phi : \widetilde{Y} \to X$ has degree four: given a generic element $[\Sigma, f] \in X$ (i.e. one that does not branch at $\{0, 1, \infty\}$) there are two choices of position for each of the marked points, and hence eight choices overall, and but only four distinct lifts to $\widetilde{Y}$ because of the action of the Möbius group. (In fact, [**M2**, Lemma 3.7] shows that $\widetilde{Y} \cong \mathbb{C}P^2$ and identifies the forgetful map with the branched cover $\mathbb{C}P^2 \to \mathbb{C}P^2, [x, y, z] \mapsto [x^2, y^2, z^2]$.) On the other hand as a map of orbifolds $\phi : (Y, \mathrm{id}) \to (X, \Gamma)$ we should consider ϕ to have degree 8, and this is what is reflected in the identity $\phi_*([\widetilde{Y}]^{vir}) = 8[X]^{vir}$. $\Diamond$

4. Constructing atlases

Although the papers [**MW1**, **MW2**, **MW3**] are mostly concerned with developing the abstract theory of Kuranishi atlases, [**MW2**, §3,§4] explains some of the analytic difficulties that arise when attempting to construct atlases for Gromov–Witten moduli spaces using the tool of traditional Fredholm analysis on Banach spaces. The first subsection below outlines the construction of Gromov–Witten atlases, without going into the analytic proofs. The paper [**MW5**] gives many more details of this construction in the genus zero case, in particular giving precise formulas for the group actions and coordinate changes in terms of suitable local coordinates. However, it still uses the gluing theorem from [**MS**]. This may be used to equip the atlas with a weakly stratified smooth structure, in which the strata are smooth but there is not enough smoothness across strata for one to be able to construct the perturbation section needed to define $[X]^{vir}_{\mathcal{K}}$. This defect is remedied by Castellano in [**C1**, **C2**] who establishes the existence of $\mathcal{C}^1$-stratified smooth atlases in the genus zero case, and proves Theorem B (as well as the Gromov–Witten axioms) in this context.

The second brief subsection describes some general manipulations concerning atlases that will be useful in §5 and §6.

4.1. Sketch proof of Theorem A.

Consider a closed $2n$-dimensional symplectic manifold (M, ω) with ω-tame almost complex structure J, define $X := \overline{\mathcal{M}}_{g,k}(M, A, J)$ to be the space of equivalence classes of genus g, k-marked stable maps to M in class $A \in H_2(M; \mathbb{Z})$, and let

$$d := \mathrm{ind}\,(A) = (2n - 6)(1 - g) + 2c_1(A) + 2k.$$

In this section, we will sketch the construction of a d-dimensional weak Kuranishi atlas on X, omitting many details that (in the genus zero case at least) may be found in [**MW5**]. In particular, we do not discuss questions of smoothness.

A genus g stable map $(\Sigma, j, \mathbf{z}, f)$ is a genus g nodal Riemann surface (Σ, j), equipped with k labelled marked points $\mathbf{z} \subset (\Sigma \setminus \{nodes\})$ and a map $f : \Sigma \to M$ that satisfies the following conditions:

- the integral of $f^*(\omega)$ over each component of Σ is nonnegative;
- **(stability)** every genus 0 (resp. 1) component of (Σ, j) over which $f^*(\omega)$ has zero integral has at least three (resp. one) special (i.e. nodal or marked) points.

Two such maps are equivalent, denoted $(\Sigma, j, \mathbf{z}, f) \sim (\Sigma', j', \mathbf{z}', f')$, if there is a biholomorphic map $\phi : (\Sigma, j, \mathbf{z}) \to (\Sigma', j', \mathbf{z}')$ such that $f' \circ \phi = f$. The corresponding equivalence class is written $[\Sigma, j, \mathbf{z}, f]$. The space of equivalence classes of such maps[33] is denoted by $\overline{\mathcal{M}}_{g,k}(M, A)$. When completed with respect to suitable Banach norms, this space has a natural metrizable topology called the Gromov topology. However, in contrast to polyfold theory, our constructions do not use the properties of this ambient space in any serious way. The moduli space X is the compact subset $\overline{\mathcal{M}}_{g,k}(M, A, J) \subset \overline{\mathcal{M}}_{g,k}(M, A)$ consisting of all equivalence classes $[\Sigma, j, \mathbf{z}, f]$ of smooth J-holomorphic stable maps such that $f_*([\Sigma]) = A$. We provide X with the usual Gromov topology, so that it is a compact, metrizable space.

To define a basic chart in a neighbourhood of a point $\tau_0 = [\Sigma_0, j_0, \mathbf{z}_0, f_0] \in X$, we take Γ to be the group of self-equivalences of τ_0 and make the following choices:

- A **slicing manifold** Q that is transverse to $f_0(\Sigma_0)$, disjoint from $f_0(\mathbf{z}_0)$, orientable, and chosen so that the k points $\mathbf{z}_0$ together with the L points $\mathbf{w}_0$ in $f_0^{-1}(Q)$ stabilize the domain of f_0, and also add enough structure to it that the stabilized domain $[\Sigma_0, j_0, \mathbf{w}_0, \mathbf{z}_0]$ has no self-automorphisms, i.e. it is a smooth point in the orbifold $\overline{\mathcal{M}}_{g,k+L}$. We choose an order for the points $\mathbf{w}_0 = (w_0^i)_{1 \le i \le L}$, and note that the action of Γ on Σ_0 permutes these points. We show below that this action extends to a neighbourhood of $\widetilde{\tau}_0 := [\Sigma_0, j_0, \mathbf{w}_0, \mathbf{z}_0, f_0]$ in the space of equivalence classes of stable maps with $k + L$ marked points.
- An **obstruction space** E (which is a complex vector space) and a Γ-invariant complex linear map
$$\lambda : E \to \mathcal{C}^\infty\big(\mathrm{Hom}_J^{0,1}(\mathcal{C}|_\Delta \times M)\big),$$
where Δ is a small (manifold) neighbourhood of the stabilized domain $[\Sigma_0, j_0, \mathbf{w}_0, \mathbf{z}_0]$ in $\overline{\mathcal{M}}_{g,k+L}$. Here $\mathcal{C}|_\Delta \to \Delta$ is the restriction of the universal curve to Δ with fiber Σ_δ at $\delta \in \Delta$, and $\mathrm{Hom}_J^{0,1}(\mathcal{C}|_\Delta \times M)$ is the bundle over $\mathcal{C}|_\Delta \times M$ whose fiber at (z, δ, x) (where $z \in \Sigma_\delta, x \in M$) is the space $\mathrm{Hom}_J^{0,1}(T_z\Sigma_\delta, T_xM)$ is anti-J-holomorphic linear maps $T_z\Sigma_\delta \to T_xM$. As explained in more detail below, we choose E sufficiently large to cover the cokernel of the linearized Cauchy–Riemann operator at τ_0.

We now explain the group action in more detail. As above, we choose a lift $\widetilde{\tau}_0 := [\Sigma_0, j_0, \mathbf{w}_0, \mathbf{z}_0, f_0]$ of τ_0 to $\overline{\mathcal{M}}_{g,k+L}(M, A, J)$ and an ordering $\mathbf{w}_0 = (w_0^i)_{1 \le i \le L}$ of the points in $f_0^{-1}(Q)$. Since Γ is the isotropy group of τ_0, for each $\gamma \in \Gamma$ there is a biholomorphism $\phi_\gamma : (\Sigma_0, j_0, \mathbf{z}_0) \to (\Sigma_0, j_0, \mathbf{z}_0)$ such that $f_0 \circ \phi_\gamma = f_0$. Because the elements in $\mathbf{w}_0$ comprise the full inverse image $f_0^{-1}(Q)$, the tuple $\phi_\gamma^{-1}(\mathbf{w}_0)$ is a

[33]Many authors (e.g. [**HWZ4**]) call the elements of this space "stable curves" or "stable maps", and for simplicity we will sometimes do this as well. What is meant should be clear from the context and notation.

reordering of the tuple $\mathbf{w}_0$, so that we may define a permutation $\gamma \in \mathrm{Sym}_L$, $i \mapsto \gamma(i)$, by

$$w^{\gamma(i)} := (\gamma \cdot \mathbf{w}_0)^i, \ \forall i, \text{ where } \gamma \cdot \mathbf{w}_0 := \phi_\gamma^{-1}(\mathbf{w}_0).$$

Thus

$$\begin{aligned} &\phi_{\gamma_1\gamma_2} = \phi_{\gamma_2} \circ \phi_{\gamma_1}, \quad (\gamma_1\gamma_2) \cdot \mathbf{w}_0 = \gamma_2 \cdot (\gamma_1 \cdot \mathbf{w}_0), \\ &\gamma\big([\Sigma_0, j_0, \mathbf{w}_0, \mathbf{z}_0, f_0]\big) = [\Sigma_0, j_0, \gamma \cdot \mathbf{w}_0, \mathbf{z}_0, f_0] = [\Sigma_0, j_0, \mathbf{w}_0, \mathbf{z}_0, f_0], \end{aligned} \tag{4.1.1}$$

where the last equality holds because

$$(\Sigma_0, j_0, \gamma \cdot \mathbf{w}_0, \mathbf{z}_0, f_0) = (\Sigma_0, j_0, \phi_\gamma^{-1}(\mathbf{w}_0), \mathbf{z}_0, f_0 \circ \phi_\gamma) \sim (\Sigma_0, j_0, \mathbf{w}_0, \mathbf{z}_0, f_0).$$

We now define an action of Γ on a neighbourhood Δ of $[\Sigma_0, j_0, \mathbf{w}_0, \mathbf{z}_0]$ in $\overline{\mathcal{M}}_{g,k+L}$ by setting

$$\gamma\big([\Sigma, j, \mathbf{w}, \mathbf{z}]\big) = [\Sigma, j, \gamma \cdot \mathbf{w}, \mathbf{z}]. \tag{4.1.2}$$

Here we suppose that

(4.1.3)
Δ is so small that the points in $\mathbf{w}$ are small perturbations of those in $\mathbf{w}_0$.

More precisely, we assume that there is a metric on $\mathcal{C}|_\Delta$ so that w_0^i is the point in $\mathbf{w}_0$ closest to w^i. This implies that the points in $\mathbf{w}$ inherit a natural ordering from the chosen ordering of the added marked points $\mathbf{w}_0$. Any ordering of the points in $f^{-1}(Q)$ given by a tuple $\gamma \cdot \mathbf{w}$ for $\gamma \in \Gamma$ is said to be **admissible**. Thus the set $f^{-1}(Q)$ has $|\Gamma|$ admissible orderings. By replacing Δ by $\cap_\gamma \gamma(\Delta)$ if necessary, we may suppose that Δ is Γ-invariant. Note that, as in Example 4.1.2 below, although by (4.1.1) the action of Γ fixes the element $[\Sigma_0, j_0, \mathbf{w}_0, \mathbf{z}_0, f_0]$, it is often nontrivial.

This Γ action clearly extends to the bundle $\mathrm{Hom}_J^{0,1}(\mathcal{C}|_\Delta \times M)$. Further, it lifts to an action on a neighbourhood of $[\Sigma_0, j_0, \mathbf{w}_0, \mathbf{z}_0, f_0]$ in the space of stable maps given by

$$\gamma\big([\Sigma, j, \mathbf{w}, \mathbf{z}, f]\big) = [\Sigma, j, \gamma \cdot \mathbf{w}, \mathbf{z}, f], \quad \gamma \in \Gamma, \ [\Sigma, j, \mathbf{w}, \mathbf{z}] \in \Delta. \tag{4.1.4}$$

Note in particular that γ acts by relabelling the entries in $\mathbf{w}$, leaving the map f unchanged. One might be tempted to think that γ should act via some geometric action on the domain, but as we show in Example 4.1.2 below a suitable action ? exist.

LEMMA 4.1.1. *Given Δ as in (4.1.3), let $\mathcal{N} \subset \overline{\mathcal{M}}_{g,k+L}(M, A)$ be an open, Γ-invariant neighbourhood of $\widetilde{\tau}_0$ consisting of elements $[\Sigma, j, \mathbf{w}, \mathbf{z}, f]$ with $[\Sigma, j, \mathbf{w}, \mathbf{z}] \in \Delta$. Then the forgetful map*

$$\psi : \mathcal{N} \to \overline{\mathcal{M}}_{g,k}(M, A), \quad [\Sigma, j, \mathbf{w}, \mathbf{z}, f] \mapsto [\Sigma, j, \mathbf{z}, f]$$

is the composite of the quotient map $\mathcal{N} \to \mathcal{N}/\Gamma$ with an inclusion

$$\mathcal{N}/\Gamma \hookrightarrow \overline{\mathcal{M}}_{g,k}(M, A).$$

PROOF. Choose a representative $(\Sigma_0, j_0, \mathbf{z}_0, f_0)$ of τ_0. Because Γ is the stabilizer of τ_0, there are $|\Gamma|$ different lifts of $(\Sigma_0, j_0, \mathbf{z}_0, f_0)$ to a stable map of the form

$$(\Sigma_0, j_0, \mathbf{w}_0', \mathbf{z}_0, f_0) \sim (\Sigma_0, j_0, \mathbf{w}_0, \mathbf{z}_0, f_0) \ ;$$

namely those given by the ordered tuples $\mathbf{w}_0' = \gamma \cdot \mathbf{w}_0$ for $\gamma \in \Gamma$. Thus these lifts form a Γ-orbit, and by (4.1.3) this property extends to the set of stable maps lying over Δ.

Finally, we argue by contradiction that $\mathcal{N}$ can be chosen sufficiently small for the map $\mathcal{N}/\Gamma \to \overline{\mathcal{M}}_{g,k}(M,A)$ to be injective. For if this were not the case, we could find two convergent sequences of elements $(\widetilde{\tau}_k)_{k\geq 0}, (\widetilde{\tau}'_k)_{k\geq 0}$ both with limits in $\psi^{-1}(\tau_0)$ and such that

$$\psi(\widetilde{\tau}_k) = \psi(\widetilde{\tau}'_k), \qquad \widetilde{\tau}'_k \neq \gamma \cdot \widetilde{\tau}_k, \qquad \forall k,\ \gamma \in \Gamma.$$

But $\psi^{-1}(\tau_0)$ is the single point $\widetilde{\tau}_0$, and Γ is its automorphism group. Hence this is impossible. □

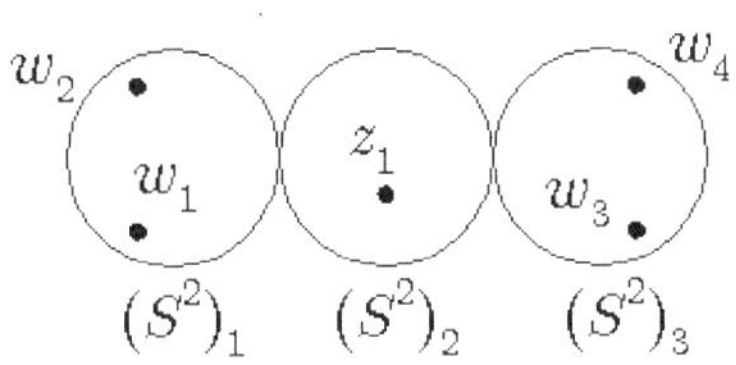

FIGURE 4.1.1. The domain of a map with nontrivial automorphism group.

EXAMPLE 4.1.2. Note that for elements near the center $\widetilde{\tau}_0$ the permutation $\mathbf{w} \mapsto \gamma \cdot \mathbf{w}$ may well not be induced by a biholomorphism or even a diffeomorphism of the underlying domain $(\Sigma, j, \mathbf{z})$. For example, suppose that $M = \mathbb{C}P^2$, that $\widetilde{\tau}_0$ is a chain of three spheres $S^2_1 \vee S^2_2 \vee S^2_3$, with one marked point on S^2_2, and that $f(S^2_1) = f(S^2_3)$ is a line ℓ_0, while $f(S^2_2) = pt$. To stabilize the domain we choose Q to consist of two small disjoint 2-discs that are transverse to ℓ_0 and disjoint from $f(S^2_2)$. Then $\mathbf{w}$ consists of four points (see Figure 4.1.1), and the generator γ of $\Gamma = \mathbb{Z}/2\mathbb{Z}$ takes the two points on S^2_1 to those on S^2_3; we can assume these labelled so that

$$\mathbf{w} = (w^1, w^2, w^3, w^4), \qquad \gamma \cdot \mathbf{w} = (w^3, w^4, w^1, w^2).$$

Now consider a nearby element in which only one node (say that between S^2_1, S^2_2) is glued. Its domain has two components, one with the three marked points w^1, w^2, z^1 and the other with w^3, w^4, and there is clearly no diffeomorphism that fixes z^1 while appropriately permuting the $\mathbf{w}$. ◊

The elements of the domain of a basic chart near $\tau_0 \in X$ are tuples of the form

(4.1.5)

$$(\vec{e}, [\Sigma, j, \mathbf{w}, \mathbf{z}, f]), \quad \vec{e} \in E, \delta := [\Sigma, \mathbf{w}, \mathbf{z}] \in \Delta, \ \overline{\partial}_{J,j} f = \lambda(\vec{e})|_{\mathrm{graph\, f}}, \ f(\mathbf{w}) \in Q.$$

Here we use the shorthand $\Sigma_\delta = [\Sigma, j, \mathbf{w}, \mathbf{z}]$ for the domain together with its marked points, that we identify with the fiber of the universal curve $\mathcal{C}|_\Delta \to \Delta \subset \overline{\mathcal{M}}_{k+L}$. For each $f: \Sigma_\delta \to M$, the image $\overline{\partial}_{J,j} f := \frac{1}{2}(df + J \circ df \circ j)$ of the Cauchy–Riemann operator is a section of the bundle of anti-(J,j)-holomorphic 1-forms on Σ_δ with values in $f^*(TM)$, and hence can be considered as a section of the restriction of $\mathrm{Hom}^{0,1}_J(\mathcal{C}|_\Delta, M)$ to

$$\mathrm{graph\, f} := \{(z, f(z)) | z \in \Sigma_\delta\} \subset \mathcal{C}|_\Delta \times M.$$

Therefore the equation $\overline{\partial}_{J,j} f = \lambda(e)|_{\mathrm{graph\, f}}$ in (4.1.5) makes sense.

In order to ensure that the set of its solutions near $(0, [\Sigma_0, j_0, \mathbf{w}_0, \mathbf{z}_0, f_0])$ forms a manifold of the right dimension, we must use the fact that its linearization is

Fredholm. If we restrict attention to the stable maps whose domain has the same number of nodes as $\widetilde{\tau}_0 = [\Sigma_0, j_0, \mathbf{w}_0, \mathbf{z}_0]$, then we can define the operator $\overline{\partial}_{J,j}$ as a section of a Banach space bundle over a suitable Banach base space, and then use the Fredholm property of its linearization. (If Σ has genus $g > 0$, this base space is determined by the choice of local coordinates on a neighbourhood of the fiber over $\widetilde{\tau}_0$ in the universal curve $\mathcal{C}|_\Delta$.) To deal with a full neighbourhood, we need a gluing theorem asserting that if the linearization at $\widetilde{\tau}_0$ is surjective then it remains surjective for all nearby stable maps. Here we will simply assume that this is true, and that the resulting solution space is smooth.[34]

It follows that we may choose a finite dimensional subspace E_0 of $\mathcal{C}^\infty$-sections of $\mathrm{Hom}_J^{0,1}(\mathcal{C}|_\Delta, M)$ such that the solution space

$$\big\{\widetilde{\tau} = [\Sigma, j, \mathbf{w}, \mathbf{z}, f] \;\big|\; \overline{\partial}_{J,j} f = e|_{\mathrm{graph\, f}}, e \in E_0,\ \widetilde{\tau} \approx \widetilde{\tau}_0\big\}$$

is a manifold of the "right" dimension, namely $d + 2L + \dim E_0$. We then define $E := \prod_{\gamma\in\Gamma} E_0$ with elements $\vec{e} := (e^\gamma)_{\gamma\in\Gamma}$. Then Γ acts on E by permutations:

$$\gamma \cdot \vec{e} = ((\gamma \cdot \vec{e})^\alpha)_{\alpha\in\Gamma} := (e^{\gamma\alpha}), \quad \alpha \in \Gamma,$$

and we define $\lambda : E \to \mathcal{C}^\infty(\mathrm{Hom}_J^{0,1}(\mathcal{C}|_\Delta, M))$ to be the Γ-equivariant extension of the inclusion $E_0 \subset \mathcal{C}^\infty(\mathrm{Hom}_J^{0,1}(\mathcal{C}|_\Delta, M))$.

With these definitions, the set

$$\text{(4.1.6)}\quad \widehat{U} := \big\{(\vec{e}, [\Sigma, j, \mathbf{w}, \mathbf{z}, f]) \;\big|\; \vec{e} \in E,\ \overline{\partial}_{J,j} f = \lambda(\vec{e})|_{\mathrm{graph\, f}},\ [\Sigma, j, \mathbf{w}, \mathbf{z}, f] \approx \widetilde{\tau}_0\big\}$$

is a manifold[35] of dimension $d + 2L + \dim E$ that supports a (local) action of Γ by Lemma 4.1.1. Because transversality of $f(\Sigma)$ to Q is an open condition, we may choose this neighbourhood so that $f(\Sigma) \pitchfork Q$ for all maps f in $\widehat{U}$, and then cut down by the slicing conditions $f(\mathbf{w}) \in Q$ and average over Γ to obtain the domain U of the chart:

$$\text{(4.1.7)}\quad U := \big\{(\vec{e}, [\Sigma, j, \mathbf{w}, \mathbf{z}, f]) \in \widehat{U} \;\big|\; \vec{e} \in E,\ \overline{\partial}_{J,j} f = \lambda(\vec{e})|_{\mathrm{graph\, f}},\ f(\mathbf{w}) \in Q\big\}.$$

By construction this is a manifold of dimension $d + \dim E$.

The other components of the basic chart are now easy to define. We have $\mathbf{K} = (U, E, \Gamma, s, \psi)$, where

- U, E are as above;
- Γ acts by $\gamma \cdot (\vec{e}, [\Sigma, j, \mathbf{w}, \mathbf{z}, f]) = (\gamma \cdot \vec{e}, [\Sigma, j, \gamma \cdot \mathbf{w}, \mathbf{z}, f])$;
- $s : U \to E$ is the projection $(\vec{e}, [\Sigma, j, \mathbf{w}, \mathbf{z}, f]) \to \vec{e}$;

[34] One certainly can manage with less smoothness here. However, to carry out our construction of transverse perturbation sections, it is useful to have at least $\mathcal{C}^1$-smoothness overall. The paper [**C1**] shows how to achieve this in the genus zero case, by improving the gluing theorem in [**MS**]. Pardon proves a C^0-version of gluing in [**P**], while [**HWZ1, HWZ2, HWZ3**] proves an analogous result in the polyfold context, i.e. for sc^∞-maps. Note also that [**FF**] describes the relevant setup in much more detail than is given here.

[35] The choice of E_0 guarantees that the linearization of the operator $(e, \widetilde{\tau}) \mapsto \overline{\partial}_{J,j} f - \lambda(\vec{e})|_{\mathrm{graph\, f}}$ is surjective. However, rather than trying to construct a Γ-invariant subspace of $\mathcal{C}^\infty\big(\mathrm{Hom}_J^{0,1}(\mathcal{C}|_\Delta, M)\big)$ that covers the cokernel, we simply impose Γ-invariance by defining E to be the product $\prod_{\gamma\in\Gamma} E_0$. Since E may not inject into $\mathcal{C}^\infty\big(\mathrm{Hom}_J^{0,1}(\mathcal{C}|_\Delta, M)\big)$, we incorporate it into the domain and add the linear map λ as part of the structure. This approach avoids the need for the injectivity requirement Sum Condition I' in [**MW2**, §4.3]. However, as explained in more detail below, these definitions make it essential to first consider the domain $\widehat{U}$ before imposing the slicing conditions.

- the footprint map $\psi : s^{-1}(0) \to X$ is the forgetful map
$$(0, [\Sigma, j, \mathbf{w}, \mathbf{z}, f]) \mapsto [\Sigma, j, \mathbf{z}, f] \in X.$$

It follows from Lemma 4.1.1 (and the choice of Δ) that the induced map $\underline{\psi} : \underline{U} \to X$ is injective, and from the surjectivity property of the gluing map that the footprint $F := \psi(U)$ is an open subset of X.

The sum of two charts: We now construct a transition chart $\mathbf{K}_I$ with $I = \{1, 2\}$ of length 2.[36] Suppose that we have chosen charts $\mathbf{K}_1, \mathbf{K}_2$ as above, whose elements have domains in the sets $\Delta_i \in \overline{\mathcal{M}}_{g,k+L_i}$ that both satisfy the condition in (4.1.3). Define

$$\Delta_{12} := \big\{[\Sigma, \mathbf{w}_1, \mathbf{w}_2, \mathbf{z}] \;\big|\; [\Sigma, \mathbf{w}_i, \mathbf{z}] \in \Delta_i\big\} \stackrel{(\iota_1, \iota_2)}{\longrightarrow} \overline{\mathcal{M}}_{g,k+L_1} \times \overline{\mathcal{M}}_{g,k+L_2}. \tag{4.1.8}$$

We define $E_{12} = E_1 \times E_2$ with elements $\underline{\vec{e}} = (\vec{e}_1, \vec{e}_2)$ and the obvious product action of $\Gamma_{12} := \Gamma_1 \times \Gamma_2$, and define the linear map λ_{12} from E_{12} to the appropriate space of forms over $\mathcal{C}|_{\Delta_{12}} \times M$ to be the sum of the pullbacks $\iota_1^*(\lambda_1) + \iota_2^*(\lambda_2)$. We then define U_{12} to be a suitable Γ_{12}-invariant open subset of the following space

$$W_{12} := \left\{ (\underline{\vec{e}}, [\Sigma, \mathbf{w}_1, \mathbf{w}_2, \mathbf{z}, f]) \;\middle|\; \begin{array}{ll} \underline{\vec{e}} \in E_{12}, & \delta_i := [\Sigma, \mathbf{w}_i, \mathbf{z}] \in \Delta_i, \\ f(\mathbf{w}_i) \in Q_i, & \overline{\partial}_{J,j} f = \lambda_{12}(\underline{\vec{e}})|_{\mathrm{graph} f} \end{array} \right\}. \tag{4.1.9}$$

Note that the equation $\overline{\partial}_{J,j} f = \lambda_{12}(\underline{\vec{e}})|_{\mathrm{graph} f}$ could also be written as the sum

$$\overline{\partial}_{J,j} f = \sum_{i=1,2} \lambda_i(\vec{e}_i)|_{\mathrm{graph}_i f}, \tag{4.1.10}$$

where $\mathrm{graph}_i f \subset \Sigma_{\delta_i} \times M$ is the graph of f considered as a map with domain $\Sigma_{\delta_i} = [\Sigma, \mathbf{w}_i, \mathbf{z}]$. The rest of the definition of the chart and coordinate changes $\Phi_{i,12} : \mathbf{K}_i \to \mathbf{K}_{12}$ is now very natural.

- Γ_{12} acts on U_{12} by permutation:
$$(\gamma_1, \gamma_2) \cdot \big(\vec{e}_1, \vec{e}_2, [\Sigma, \mathbf{w}_1, \mathbf{w}_2, \mathbf{z}, f]\big) = \big(\gamma_1 \cdot \vec{e}_1, \gamma_2 \cdot \vec{e}_2, [\Sigma, \gamma_1 \cdot \mathbf{w}_1, \gamma_2 \cdot \mathbf{w}_2, \mathbf{z}, f]\big).$$
- the section map s_{12} is the projection to E_{12};
- the footprint map ψ_{12} is the forgetful map
$$\big(0, [\Sigma, \mathbf{w}_1, \mathbf{w}_2, \mathbf{z}, f]\big) \to [\Sigma, \mathbf{z}, f] \in X,$$
with image F_{12};
- the group covering $\rho_{1,12} : \widetilde{U}_{1,12} \subset U_{12} \cap (s_{12}^{-1}(E_1)) \to U_{1,12} \subset U_1$ is given by the restriction of the forgetful map
$$\big((\vec{e}_1, 0), [\Sigma, \mathbf{w}_1, \mathbf{w}_2, \mathbf{z}, f]\big) \mapsto \big(\vec{e}_1, [\Sigma, \mathbf{w}_1, \mathbf{z}, f]\big) \in U_1;$$
and similarly for $\rho_{2,12}$.

It is mostly a routine matter to verify that one can choose the sets U_{12} and $\widetilde{U}_{1,12}$ so that the conditions in Definitions 3.1.6 and 3.1.10 for charts and coordinate changes are satisfied. Here are the main points.

- *We can choose U_{12} to have footprint F_{12}.*
  Each element $\tau = [\Sigma, \mathbf{z}, f] \in F_i$ has a Γ_i-orbit of lifts $(0, [\Sigma, \mathbf{w}_i, \mathbf{z}, f])$ to $s_i^{-1}(0) \subset U_i$, and hence a $\Gamma_1 \times \Gamma_2$-orbit of lifts $(0, [\Sigma, \mathbf{w}_1, \mathbf{w}_2, \mathbf{z}, f])$ to the space W_{12} defined in (4.1.9). This defines a subset $\psi_{12}^{-1}(F_{12})$ of

[36] Informally we often call $\mathbf{K}_{12}$ the "sum" of $\mathbf{K}_1, \mathbf{K}_2$.

$s_{12}^{-1}(0) \cap W_{12}$ along which the linearized Cauchy–Riemann operator is surjective. Hence $\psi_{12}^{-1}(F_{12})$ has an open neighbourhood on which this operator remains surjective; and we take U_{12} to be a Γ_{12}-invariant open subset of this.

- *We can choose U_{12} to be a smooth manifold of dimension* $d + \dim(E_{12})$.

  To check that U_{12} is smooth, we must first consider a space $\widehat{U}_{12}$ that is defined analogously to U_{12} but without the slicing conditions $f(\mathbf{w}_i) \in Q_i$; cf. (4.1.6). As explained in [**MW2**, §3,§4], the linear term $\lambda(\underline{\vec{e}})|_{\text{graph f}}$ is $\mathcal{C}^1$ when the points $\mathbf{w}_i$ can vary freely in the domain, but is not $\mathcal{C}^1$ when the $\mathbf{w}_i$ are constrained by the slicing conditions: see in particular the discussion just before [**MW2**, equation (4.3.11)]. Hence we can choose $\widehat{U}_{12}$ to be smooth, and we may then cut down by the slicing conditions (which are smooth when restricted to the finite dimensional manifold $\widehat{U}_{12}$).
- *$\rho_{1,12}$ is the quotient of $\widetilde{U}_{1,12} \subset s_{12}^{-1}(E_1) \subset U_{12}$ by a free action of Γ_2; and similarly for $\rho_{2,12}$.*

  We choose $\widetilde{U}_{1,12}$ to be any Γ_{12}-invariant neighbourhood of $\psi_{12}^{-1}(F_{12}) \cap s_{12}^{-1}(E_1)$, and then define $U_{1,12} := \rho_{1,12}(\widetilde{U}_{1,12})$. In particular the elements $(\vec{e}_1, \vec{e}_2, [\Sigma, \mathbf{w}_1, \mathbf{w}_2, \mathbf{z}, f]) \in \widetilde{U}_{1,12}$ have $\vec{e}_2 = 0$. Thus the group Γ_2 acts freely on $\widetilde{U}_{1,12}$. Indeed its action $[\Sigma, \mathbf{w}_1, \mathbf{w}_2, \mathbf{z}] \mapsto [\Sigma, \mathbf{w}_1, \gamma_2 \cdot \mathbf{w}_2, \mathbf{z}]$ on the set Δ_{12} defined in (4.1.8) is free and preserves the defining equation (4.1.10) (which is independent of $\mathbf{w}_2$ because $\vec{e}_2 = 0$).
- *The index condition holds (see Definition* 3.1.10*).* For points in the same stratum as $\widetilde{\tau}_0$ this holds by the openness of the set of regular points of a Fredholm operator. (A detailed proof in the genus zero case may be found in [**MW1**, Thm. 4.3.1].) In general, this step requires the gluing theorem.

This completes the definition of the sum of two charts and the corresponding coordinate changes.

REMARK 4.1.3. Above we have used the notion of "Fredholm stabilization" to incorporate the obstruction spaces into the domains; i.e. the elements of the domain of a chart are pairs $(\vec{e}, \widetilde{\tau})$ rather than simply stable maps $\widetilde{\tau} = [\Sigma, \mathbf{w}, \mathbf{z}, f]$ that solve the equation $\overline{\partial}_{J,j} f \in E$. This obviates the need for the injectivity and transversality requirements detailed in Sum Condition I′, II′ in [**MW2**, Section 4.3], and hence allows us to form the sum of any pair of charts with overlapping footprints.

Completion of the construction: Suppose given a collection $(\mathbf{K}_i)_{i \in I}$ of basic charts whose footprints $(F_i)_{1 \le i \le N}$ cover X. We aim to construct an atlas in the sense of Definition 3.1.15 in which the charts are indexed by $I \in \mathcal{I}_{\mathcal{K}}$ and have $E_I := \prod_{i \in I} E_i$, $\Gamma_I := \prod_{i \in I} \Gamma_i$. As before, we denote the elements of the obstruction space E_I by underlined tuples: $\underline{\vec{e}} := (\vec{e}_i)_{i \in I}$. Similarly $\underline{\mathbf{w}} := (\mathbf{w}_i)_{i \in I}$ are the tuples of added marked points. We define U_I to be a Γ_I-invariant open subset of the following space (where Γ_i acts on E_I via its action on E_i and permutes the elements in $\mathbf{w}_i$):

$$(4.1.11) \quad U_I \subset \left\{ \left(\underline{\vec{e}}, [\Sigma_\delta, \underline{\mathbf{w}}, \mathbf{z}, f]\right) \;\middle|\; \begin{array}{ll} \underline{\vec{e}} \in E_I, & \delta_i := [\Sigma, \mathbf{w}_i, \mathbf{z}] \in \Delta_i, \forall i \in I, \\ f(\mathbf{w}_i) \in Q_i, & \overline{\partial}_{J,j} f = \lambda_I(\underline{\vec{e}})|_{\text{graph} f} \end{array} \right\},$$

chosen so that the footprint is F_I. Here, we define λ_I to be a sum $\lambda_I(\underline{\vec{e}}) := \sum_{i\in I} \iota_i^*(\lambda_i)$ so that, as in (4.1.10), the defining equation can be written

$$\overline{\partial}_{J,j} f = \sum_{i\in I} \lambda_i(\vec{e}_i)|_{\mathrm{graph}_i f}.$$

As before, we define

$$s_I(\underline{\vec{e}}, [\Sigma, \underline{\mathbf{w}}, \mathbf{z}, f]) = \underline{\vec{e}}, \quad \psi_I(0, [\Sigma, \underline{\mathbf{w}}, \mathbf{z}, f]) = [\Sigma, \mathbf{z}, f].$$

Further, the atlas coordinate changes $\mathbf{K}_I \to \mathbf{K}_J$ are given by first choosing appropriate domains $\widetilde{U}_{IJ} \subset U_J$ and then simply forgetting the components $(\mathbf{w}_i)_{i\in(J\smallsetminus I)}$. One proves just as before that all the conditions for a weak atlas are satisfied.

Orientations: Finally we must check that the resulting weak atlas is oriented. To this end, recall that we chose the E_i, λ_i to be complex linear, and the slicing manifolds Q_i to be orientable. According to Definition 3.3.7, we must construct a nonvanishing section of $\det(\mathfrak{s}_{\mathcal{K}})$, or, equivalently, of the bundle $\det(\mathcal{K})$. For this we need compatible Γ_I invariant sections of the local bundles $\Lambda^{\max} \mathrm{T} U_I \otimes (\Lambda^{\max} E_i)^*$ over U_I. Since each E_I has a natural orientation because it is a complex vector space, it suffices to construct compatible orientations of the domains U_I. To this end, we now fix an orientation for each Q_i. It then suffices to orient the unsliced domains $\widehat{U}_I$. Since we did not set up the defining equation precisely, it is hard to give a very convincing reason why $\widehat{U}_I$ can be oriented. But this follows from the fact that the space $\mathcal{C}|_\Delta$ is oriented because it is a complex manifold, and the fact that the linearized Cauchy–Riemann operator is the sum of a complex linear first order operator with a compact perturbation, and hence has a canonical perturbation through Fredholm operators to a complex linear operator. (For more details of this last step, see the proof of [**MS**, Theorem 3.1.5] and also [**MS**, Appendix A].) Since the resulting orientation of $U_I \times E_I$ is preserved by the group actions and coordinate changes, we obtain an orientation of $\mathcal{K}$.

Constructing cobordisms: To prove that the VFC is independent of choices we need to build cobordisms between any two atlases on X. We defined Kuranishi cobordisms over $X \times [0,1]$ in Definition 2.2.11. These restricted cobordisms are also called **concordances**. The following notion is also useful.

DEFINITION 4.1.4. *Two (weak) atlases $\mathcal{K}, \mathcal{K}'$ on X are said to be* **directly commensurate** *if they are subatlases of a common (weak) atlas $\mathcal{K}''$ on X. They are commensurate if there is a sequence of (weak) atlases $\mathcal{K} =: \mathcal{K}_1, \dots, \mathcal{K}_\ell := \mathcal{K}'$ such that any consecutive pair $\mathcal{K}_i, \mathcal{K}_{i+1}$ are directly commensurate.*

One useful result is that any two commensurate (weak) atlases are concordant. The proof is sketched in Remark 6.3.5; for details see [**MW2**, §6.2]. Because we can construct the sum of any number of charts provided only that their footprints have nonempty intersection (see Remark 4.1.3), any two atlases constructed on X by the method described above with basic charts $(\mathbf{K}_i)_{1\le i\le N_1}$ and $(\mathbf{K}_i)_{N_1+1\le i\le N_2}$ are subatlases of a common atlas with basic charts $(\mathbf{K}_i)_{1\le i\le N_2}$. This implies the following.

COROLLARY 4.1.5. *Any two weak atlases on X constructed by the method described above are directly commensurate and hence concordant.*

Final comment on the proof of Theorem A: The above construction outlines the proof of Theorem A, though to fill in the details would require a gluing theorem that we did not even begin to state precisely. To do this, one must choose a local trivialization for the universal bundle $\mathcal{C}|_\Delta$, and work with stable maps $(\Sigma, j, \mathbf{w}, \mathbf{z}, f)$ rather than with equivalence classes of such. The paper [**MW5**] will set up these coordinates in the genus zero case, which is easier since the genus zero Deligne–Mumford space $\overline{\mathcal{M}}_{0,k+L}$ has natural coordinates given by cross ratios. One can then write down explicit formulas for all relevant maps, in preparation for checking that they have the required smoothness properties. As remarked earlier the gluing theorem in [**MS**] is not enough to give an overall smooth structure, though each stratum is smooth and there is enough information normal to the strata to prove that an analog of the index condition holds. Thus one gets a "weakly stratified smooth" atlas, which, as is not hard to see, suffices for the case of dimension 0 invariants with dimension 1 cobordisms. The paper [**C1**] completes the construction in the genus zero case by providing the analytic details of a $\mathcal{C}^1$-gluing theorem, thus establishing a $\mathcal{C}^1$ version of Theorem A. It is shown in [**C2**] that Theorem B still holds in this context.

REMARK 4.1.6. A similar argument shows that the cobordism class of the VFC is independent of the choice of almost complex structure J in the following sense. Suppose that J_0, J_1 are two ω-tame almost complex structures on the symplectic manifold (M, ω), join them by a path $(J_t)_{t\in[0,1]}$ of ω-tame almost complex structures (where $t \mapsto J_t$ is constant for t near $0, 1$), and define $X^{01} := \bigcup_{t\in[0,1]} \overline{\mathcal{M}}_{0,k}(M, A, J_t)$. In the same way that we build a cobordism atlas over $X \times [0, 1]$, we can build a cobordism atlas $\mathcal{K}^{01}$ over X^{01} since this has collared boundary. Moreover, we can arrange that its restrictions $\mathcal{K}^\alpha := \mathcal{K}^{01}|_\alpha$ at the end points $\alpha = 0, 1$ equal any given GW atlases $\mathcal{K}^\alpha$ for $X_\alpha := \overline{\mathcal{M}}_{0,k}(M, A, J_\alpha)$, and then prove that the two elements $\big([X_\alpha]^{vir}_{\mathcal{K}^\alpha}\big)_{\alpha=0,1}$ have the same image in $\check{H}_d(X^{01}; \mathbb{Q})$. It follows that all GW invariants calculated using $[X]^{vir}_{\mathcal{K}}$ are independent of the choice of J. This argument is not formally written anywhere; however its details are very similar to those in [**MW2**, §8.2] which deal with the case of concordances.

4.2. Manipulating atlases. In general, the notion of direct commensurability introduced in Definition 4.1.4 above is not an equivalence relation, although it is whenever the atlases under consideration are constructed by some method that permits the construction of a transition chart for any pair of charts. In the current context, we have no good language to describe this property. However, it is hard to imagine a case in which the theory of Kuranashi atlases is relevant, but this property does not hold. Later we will sometimes describe atlases with this property as **GW atlases**.

As we will see in §5.2 and §6.3 it is sometimes useful to perform abstract constructions on atlases. Hence it is useful to develop a notion of maps between atlases in order to compare different constructions.

DEFINITION 4.2.1. *Let $\mathcal{K}, \mathcal{K}'$ be smooth weak atlases on X, X' respectively, where X is a compact subset of X'. We say that a tuple $\mathfrak{f} := \big(\ell, (f_I, \widehat{f}_I, f^\Gamma_I)_{I\in\mathcal{I}_\mathcal{K}}\big)$ is a* **map from $\mathcal{K}$ to $\mathcal{K}'$** *if the following conditions hold:*

- *there is an injection $\ell : \{1, \dots, N\} \to \{1, \dots, N'\}$ such that $F_i \subset F'_{\ell(i)}$ for all $1 \le i \le N$; the associated injection $\mathcal{I}_\mathcal{K} \to \mathcal{I}_{\mathcal{K}'}$ is also denoted ℓ;*

- *for each* $I \in \mathcal{I}_{\mathcal{K}}$ *the triple* $(f_I, \widehat{f}_I, f_I^\Gamma)$ *consists of*
  - *a linear map* $\widehat{f}_I : E_I \to E'_{\ell(I)}$*;*
  - *an injective group homomorphism* $f_I^\Gamma : \Gamma_I \to \Gamma'_{\ell(I)}$*;*
  - *a* f_I^Γ*-equivariant smooth map* $f_I : U_I \to U'_{\ell(I)}$ *such that*
    - f_I^Γ *maps the stabilizer subgroup* $\mathrm{Stab}_x \subset \Gamma_I$ *onto the stabilizer* $\mathrm{Stab}'_{f_I(x)} \subset \Gamma'_{\ell(I)}$ *for each* $x \in U_I$*;*
    - f_I *is compatible with the section and footprint maps, i.e.* $s'_{\ell(I)} \circ f_I = \widehat{f}_I \circ s_I$ *and* $\psi'_{\ell(I)} \circ f_I = \psi_I$ *on* $s_I^{-1}(0)$*;*
- $(f_I, \widehat{f}_I, f_I^\Gamma)$ *are compatible with coordinate changes, i.e. for all* $I \subset J$*,*
  - $f_J(\widetilde{U}_{IJ}) \subset \widetilde{U}_{\ell(I)\ell(J)}$*;*
  - $\widehat{f}_J \circ \widehat{\phi}_{IJ} = \widehat{\phi}'_{\ell(I)\ell(J)} \circ \widehat{f}_I : E_I \to E'_{\ell(J)}$*;*
  - $(\rho^\Gamma_{\ell(I)\ell(J)})' \circ f_J^\Gamma = f_I^\Gamma \circ \rho^\Gamma_{JI} : \Gamma_J \to \Gamma'_{\ell(I)}$*; and*
  - $(\rho_{\ell(I)\ell(J)})' \circ f_J = f_I \circ \rho_{JI} : \widetilde{U}_{IJ} \to U'_{\ell(I)\ell(J)}$*.*

Further, we say that $\mathfrak{f}$ *is an* **embedding** *if each map* f_I *is an injection onto a smooth submanifold of* $U'_{\ell(I)}$ *and each* $\widehat{f}_I$ *is an isomorphism, and is an* **open embedding** *if in addition* $f_I(U_I)$ *is an open subset of* $U'_{\ell(I)}$*.*

For example, each shrinking $\mathcal{K}_{sh}$ of $\mathcal{K}$ gives rise to an open embedding $\mathcal{K}_{sh} \to \mathcal{K}$ with the maps $\ell, \widehat{f}_I, f_I^\Gamma$ all equal to the identity.

If $\mathcal{K}, \mathcal{K}'$ are atlases (rather than weak atlases), each map $\mathfrak{f}$ gives rise to functors $\mathfrak{f} : \mathbf{B}_{\mathcal{K}} \to \mathbf{B}_{\mathcal{K}'}$ and $\mathfrak{f}_{\mathbf{E}} : \mathbf{E}_{\mathcal{K}} \to \mathbf{E}_{\mathcal{K}'}$ that commute with the structural functors $\mathfrak{s}, \mathrm{pr}$ as well as being compatible with the homeomorphisms $\iota_{\mathcal{K}} : X \to |\mathfrak{s}_{\mathcal{K}}^{-1}(0)|$ and $\iota_{\mathcal{K}'} : X' \to |(\mathfrak{s}'_{\mathcal{K}'})^{-1}(0)|$.

The definition above is very general in order that it apply in a variety of circumstances. However to get a useful notion one should add enough conditions so that there is an appropriate relation between $[X]^{vir}_{\mathcal{K}}$ and $[X']^{vir}_{\mathcal{K}'}$; for work in this direction see [**C2**]. For now, we note the following easy result.

LEMMA 4.2.2. *If* $X = X'$ *and* $\mathfrak{f} : \mathcal{K} \to \mathcal{K}'$ *is an open embedding, then* $[X]^{vir}_{\mathcal{K}} = [X]^{vir}_{\mathcal{K}'}$*.*

PROOF. Any reduction $\mathcal{V} = (V_I)_{I \in \mathcal{I}_{\mathcal{K}}}$ of $\mathcal{K}$ pushes forward to a reduction $\mathfrak{f}(\mathcal{V})$ of $\mathcal{K}'$ for which $V'_{\ell(I)} = f_I(V_I)$ for $I \in \mathcal{I}_{\mathcal{K}}$, while $F'_J = \emptyset = V'_J$ for all $J \notin \ell(\mathcal{I}_{\mathcal{K}})$. Because the maps $\widehat{f}_I$ are linear isomorphisms, any (admissible transverse, precompact) perturbation section ν of $\mathcal{K}|_{\mathcal{V}}$ pushes forward to a similar perturbation section of $\mathcal{K}'$. Thus these two atlases define the same virtual class on X. □

We also note the following.

- Since we do not require the group homomorphisms f_I^Γ to be surjective, the image $f_I(U_I)$ need not be the full inverse image $(\pi'_I)^{-1}(\underline{f_I(U_I)})$. If f_I is an embedding then $(f_I, f_I^\Gamma) : (U_I, \Gamma_I) \to (U'_I, \Gamma'_I)$ is a group embedding in the sense of Definition 3.1.2.
- There are analogous definitions if $\mathcal{K}, \mathcal{K}'$ are (filtered) topological atlases.
- In §6.3 we will consider maps between atlases with more general indexing sets for which $\ell : \mathcal{I}_{\mathcal{K}} \to \mathcal{I}_{\mathcal{K}'}$ and the $\widehat{f}_I$ are not injective; see Remark 6.3.2.

5. Atlases for orbifolds and orbibundles

In this section, we explain how the theory of Kuranishi atlases applies when X is the zero set of an orbibundle. We begin with the case of orbifolds, showing that each orbifold has a **strict orbifold atlas**, that is a Kuranishi atlas with trivial obstruction spaces. We also show that each such atlas has a canonical completion to a groupoid that represents the orbifold. Hence the atlas is a stripped down model for the groupoid that captures all essential information. The zero set construction in Proposition 3.4.4 can then be interpreted as a simpler version of the resolution construction given in [**M3**]. We next show how to use atlases to compute Gromov–Witten invariants in some easy cases, for example if the moduli space is an orbifold with cokernels of constant rank so that the obstruction spaces form a bundle of constant rank.

5.1. Orbifolds. The aim of this section is to sketch the proof of Proposition C stated in §1, i.e. to show that every compact orbifold Y is the realization of a Kuranishi atlas with trivial obstruction spaces. We will take a naive approach to orbifolds, since that suffices for our current purposes. Thus, as in Moerdijk [**Mo**] and [**M3**], we consider them as equivalence classes of groupoids rather than as stacks as in Lerman [**L10**].

Recall that an **ep (étale proper) groupoid**[37] $\mathbf{G}$ is a topological category with (finite dimensional) smooth spaces of objects $\mathrm{Obj}_{\mathbf{G}}$ and morphisms $\mathrm{Mor}_{\mathbf{G}}$, such that

- all structural maps (i.e. source s, target t, identity, composition and inverse) are **étale** (i.e. local diffeomorphisms); and
- the map $s \times t : \mathrm{Mor}_{\mathbf{G}} \to \mathrm{Obj}_{\mathbf{G}} \times \mathrm{Obj}_{\mathbf{G}}$ given by taking a morphism to its source and target is **proper**.

The realization $|\mathbf{G}|$ of $\mathbf{G}$ is the quotient of the space of objects by the equivalence relation given by the morphisms: thus $x \sim y \Leftrightarrow \mathrm{Mor}_{\mathbf{G}}(x, y) \neq \emptyset$. When (as here) the domains are locally compact, the properness condition implies that $|\mathbf{G}|$ is Hausdorff.

Definition 5.1.1. *A d-dimensional* **orbifold structure** *on a paracompact Hausdorff space Y is a pair $(\mathbf{G}, f)$ consisting of a d-dimensional ep (étale proper) groupoid $\mathbf{G}$ together with a map $f : \mathrm{Obj}_{\mathbf{G}} \to Y$ that factors through a homeomorphism $|f| : |\mathbf{G}| \to Y$. Two orbifold structures $(\mathbf{G}, f)$ and $(\mathbf{G}', f')$ are* **Morita equivalent** *if they have a common refinement, i.e. if there is a third structure $(\mathbf{G}'', f'')$ on Y and functors $F : \mathbf{G}'' \to \mathbf{G}, F' : \mathbf{G}'' \to \mathbf{G}'$ such that $f'' = f \circ F = f' \circ F'$.*

An **orbifold** *is a paracompact Hausdorff space Y equipped with an equivalence class of orbifold structures. We say that Y is oriented if for each representing groupoid $\mathbf{G}$ the spaces $\mathrm{Obj}_{\mathbf{G}}$ and $\mathrm{Mor}_{\mathbf{G}}$ have orientations that are preserved by all structure maps and by the functors $F : \mathbf{G} \to \mathbf{G}'$ considered above.*

Definition 5.1.2. *We say that a compact Hausdorff space Y has a* **strict orbifold atlas** *$\mathcal{K}$ if Y has an open covering $Y = \bigcup_{i=1,\dots,N} F_i$ such that the following conditions hold with*

$$\mathcal{I}_Y := \bigl\{\emptyset \neq I \subset \{1, \dots, N\} : F_I := \textstyle\bigcap_{i \in I} F_i \neq \emptyset\bigr\}.$$

[37]There is a corresponding notion for polyfolds with a strengthened version of the properness condition; see [**H, HWZ7**].

(i) *For each* $I \in \mathcal{I}_Y$ *there is a manifold* W_I *on which the finite group* $\Gamma_I := \prod_{i \in I} \Gamma_i$ *acts and a map* $\psi_I : W_I \to F_I$ *that induces a homeomorphism* $\underline{\psi}_I : {}^{W_I}/_{\Gamma_I} \to F_I$*;*

(ii) *for all (nonempty) subsets* $I \subset J$ *there is a group covering*

$$(\rho_{IJ}, \rho^{\Gamma}_{IJ}) : (W_J, \Gamma_J) \to \big(W_{IJ} := (\psi_I)^{-1}(F_J), \Gamma_I\big)$$

in the sense of Definition 3.1.4*, where* $\rho^{\Gamma}_{IJ} : \Gamma_J \to \Gamma_I$ *is the projection and* ρ_{IJ} *is the composite of the quotient of* W_J *by a free action of* $\ker \rho^{\Gamma}_{IJ} = \Gamma_{J \setminus I}$ *with a* Γ^I*-equivariant homeomorphism* ${}^{W_J}/_{\Gamma_{J\setminus I}} \to W_{IJ} \subset U_I$*;*

(iii) $\psi_I \circ \rho_{IJ} = \psi_J$*, and* $\rho_{IJ} \circ \rho_{JK} = \rho_{IK}$ *for all* $I \subset J \subset K$*.*

Thus the charts of this atlas $\mathcal{K}$ *are the tuples* $\big(\mathbf{K}_I := (W_I, \Gamma_I, \psi_I)\big)_{I \in \mathcal{I}_Y}$ *with footprints* $(F_I)_{I \in \mathcal{I}_Y}$*, and the coordinate changes are induced by the covering maps* ρ_{IJ}*.*

REMARK 5.1.3. (i) Note that in an atlas with trivial obstruction spaces the section maps are identically zero so that the footprint maps $\psi_I : U_I \to F_I$ are defined on the whole domain and induce homeomorphisms $\underline{\psi}_I : \underline{U}_I \to X$. In particular, the lifted domains $\widetilde{U}_{IJ} \subset U_J$ of the coordinate changes equal the whole of U_J, which explains why we assume above that the covering map ρ_{IJ} is defined on the whole of W_J. Thus the above definition is precisely that of a Kuranishi atlas with trivial obstruction spaces.

(ii) One could also allow a more general choice of groups Γ_I as in Remark 3.1.18. However, the proof of Proposition 5.1.6 shows that the product groups considered above are natural to the problem. ◊

Such an atlas satisfies the strong cocycle condition, and is oriented if the domains are equipped with an orientation that is preserved by the group actions and other structure maps. The corresponding category $\mathbf{B}_{\mathcal{K}}$ has realization Y, since in this case the intermediate category $\underline{\mathbf{B}}_{\mathcal{K}}$ (see Lemma 3.2.2) is the category defined by the covering $(F_i)_{i=1,\dots,N}$ of Y, i.e. its objects are $\bigsqcup_{I \in I} F_I$ and there is a morphism $(I, x) \to (J, y)$ iff $I \subset J$ and $\psi_I(x) = \psi_J(y) \in F_J$. Although $\mathbf{B}_{\mathcal{K}}$ is not a groupoid since the nonidentity maps are not invertible, we now show that this category has a canonical **groupoid completion** $\mathbf{G}_{\mathcal{K}}$. (This justifies our language since it means that any compact Hausdorff space Y with a strict orbifold atlas is in fact an orbifold.)

DEFINITION 5.1.4. *Let* $\mathbf{M}$ *be an étale proper category with objects* $\bigsqcup_{I \in \mathcal{I}} W_I$ *and realization* $Y := \mathrm{Obj}_{\mathbf{M}} / \sim$ *such that*

- *for each* $I \in \mathcal{I}$ *the full subcategory of* $\mathbf{M}$ *with objects* W_I *can be identified with the group quotient* (W_I, Γ_I) *for some group* Γ_I*;*
- *for each* $I \in \mathcal{I}$ *the realization map* $\pi_{\mathcal{M}} : \mathrm{Obj}_{\mathbf{M}} \to Y$ *induces a homeomorphism* ${}^{W_I}/_{\Gamma_I} \to F_I \subset Y$*, where* F_I *is an open subset of* Y*.*

Then we say that an ep groupoid $\mathbf{G}$ *is a* **groupoid completion** *of* $\mathbf{M}$ *if there is an injective functor* $\iota : \mathbf{M} \to \mathbf{G}$ *that induces a bijection on objects and a homeomorphism on the realizations* $Y = |\mathbf{M}| \to |\mathbf{G}|$*.*

Thus for each component W_I of $\mathrm{Obj}_{\mathbf{M}}$ the groupoid completion (if it exists) has the same morphisms from W_I to W_I but (unless $\mathbf{M}$ is already a groupoid) will have more morphisms between the different components of $\mathrm{Obj}_{\mathbf{M}}$ that are obtained by adding inverses and composites.

PROPOSITION 5.1.5. *Let $\mathcal{K} = \big(W_I, \Gamma_I, \rho_{IJ}\big)_{I\subset J, I,J\in\mathcal{I}_\mathcal{K}}$ be a strict orbifold atlas as above. Then the category $\mathbf{B}_\mathcal{K}$ has a canonical completion to an ep groupoid $\mathbf{G}_\mathcal{K}$ with the same objects and realization as $\mathbf{B}_\mathcal{K}$ and morphisms*

$$\mathrm{Mor}_{\mathbf{G}_\mathcal{K}} = \bigsqcup_{I,J\in\mathcal{I}_\mathcal{K}: I\cup J\in\mathcal{I}_\mathcal{K}} \mathrm{Mor}_{\mathbf{G}_\mathcal{K}}(W_I, W_J), \quad \mathrm{Mor}_{\mathbf{G}_\mathcal{K}}(W_I, W_J) := W_{I\cup J}\times \Gamma_{I\cap J},$$

where $\Gamma_\emptyset := \mathrm{id}$, and with the following structural maps.

(i) *The source and target maps $s\times t: W_{I\cup J}\times\Gamma_{I\cap J}\to W_I\times W_J$ are*

$$(s\times t)(I,J,z,\gamma) \;=\; \Big(\big(I, \gamma^{-1}\rho_{I(I\cup J)}(z)\big),\,\big(J, \rho_{J(I\cup J)}(z)\big)\Big).$$

(ii) *Composition is given by*

$$m: \mathrm{Mor}_{\mathbf{G}_\mathcal{K}}(W_I, W_J)\,{}_t\times_s \mathrm{Mor}_{\mathbf{G}_\mathcal{K}}(W_J, W_K) \to \mathrm{Mor}_{\mathbf{G}_\mathcal{K}}(W_I, W_K),$$
$$\big((I,J,z,\gamma),(J,K,w,\delta)\big)\mapsto (I,K,v',\alpha\,\delta_{IJK}\gamma_{IJK})\in W_{I\cup K}\times\Gamma_{I\cap K},$$

where

$$v' := \rho_{I\cup K, I\cup J\cup K}(v)\in W_{I\cup K},$$
$$\gamma = (\gamma_{IJK}, \gamma_{IJ\smallsetminus K})\in \Gamma_{I\cap J\cap K}\times\Gamma_{(I\cap J)\smallsetminus K} = \Gamma_{I\cap J},$$
$$\delta = (\delta_{IJK}, \delta_{JK\smallsetminus I})\in \Gamma_{I\cap J\cap K}\times\Gamma_{(J\cap K)\smallsetminus I} = \Gamma_{J\cap K},$$

and $(v,\alpha)\in W_{I\cup J\cup K}\times\Gamma_{(I\cap K)\smallsetminus J}$ is the unique pair such that

$$\rho_{I\cup J, I\cup J\cup K}(v) = \gamma^{-1}_{IJ\smallsetminus K}\alpha\,\delta z, \quad \rho_{J\cup K, I\cup J\cup K}(v) = \gamma^{-1}_{IJ\smallsetminus K} w.$$

(iii) *The inverse is given by*

$$\iota: \mathrm{Mor}_{\mathbf{G}_\mathcal{K}}(W_I, W_J)\to \mathrm{Mor}_{\mathbf{G}_\mathcal{K}}(W_J, W_I), \quad (I,J,z,\gamma)\mapsto (J,I,\gamma^{-1}z,\gamma^{-1}).$$

SKETCH OF PROOF. The above formula for $\mathrm{Mor}_{\mathbf{G}_\mathcal{K}}(W_I, W_J)$ agrees with the definition of $\mathrm{Mor}_{\mathbf{B}_\mathcal{K}}(W_I, W_J)$ when $I\subset J$. We must extend this definition to all pairs I,J with $F_I\cap F_J\neq\emptyset$ (or equivalently $I\cup J\in\mathcal{I}_\mathcal{K}$) so as to be consistent with the footprint maps and the local group actions. In particular, we require that for all $x\in W_I, y\in W_J$

- $\mathrm{Mor}_{\mathbf{G}_\mathcal{K}}(x,y)\neq\emptyset$ iff $(I,x)\sim(J,y)$ in $\mathrm{Obj}_{\mathbf{B}_\mathcal{K}}$ iff $\psi_I(x) = \psi_J(y)\in F_{I\cup J}$.

To check this, note first that $\mathrm{Mor}_{\mathbf{G}_\mathcal{K}}(x,y)\neq\emptyset$ implies that x,y have the same image in Y. Conversely, suppose given $x\in W_I, y\in W_J$ with $I\not\subset J$ and such that $\psi_I(x) = \psi_J(y)$. Since $\rho_{J(I\cup J)}: W_{I\cup J}\to W_{J(I\cup J)} := \psi^{-1}(F_{I\cup J})$ is surjective and factors out by the free action of $\Gamma_{I\smallsetminus J}$ we may choose $z\in W_{I\cup J}$ so that $\rho_{J(I\cup J)}(z) = y$. Then $\rho_{I(I\cup J)}(z)$ lies in the Γ_I-orbit of x because $\psi_{I\cup J}(z) = \psi_I(x)$, so that by replacing z by δz for some $\delta\in\Gamma_{I\smallsetminus J}$ we may arrange that $\rho_{I(I\cup J)}(z)$ lies in the $\Gamma_{I\cap J}$-orbit of x, where $\Gamma_{I\cap J} := \mathrm{id}$ if $I\cap J = \emptyset$. Therefore there is a pair $(z,\gamma)\in(W_{I\cup J},\Gamma_{I\cap J})$ with $\rho_{I(I\cup J)}(z) = \gamma x$, $\rho_{J(I\cup J)}(z) = y$ and hence with source x and target y, as required. Hence the equivalence relation on $\mathrm{Obj}_{\mathbf{B}_\mathcal{K}}$ generated by the morphisms in $\mathbf{B}_\mathcal{K}$ agrees with that induced by the morphisms in $\mathbf{G}_\mathcal{K}$, which implies that the induced map $|\mathbf{B}_\mathcal{K}|\to|\mathbf{G}_\mathcal{K}|$ is a homeomorphism.

To complete the proof that $\mathbf{G}_\mathcal{K}$ is a category we must check that the composition operation defined on $\mathrm{Mor}_{\mathbf{B}_\mathcal{K}}$ extends to a well defined and associative multiplication operation. Notice that each morphism $(I, J, z, \gamma) \in W_{I\cup J} \times W_{I\cap J}$ may be written as a composite $m_1 \circ (m_2)^{-1}$:

$$(I, \gamma^{-1}x) \overset{m_1}{\mapsto} (I \cup J, z) \overset{(m_2)^{-1}}{\mapsto} (J, \rho_{(I\cup J)J}(z)),$$

where $m_1 = (I, I\cup J, \gamma, z)$ and $m_2 = (J, I\cup J, \mathrm{id}, z)$ are morphisms in $\mathbf{B}_{\mathcal{K}}$. Therefore to define the composite of two such morphisms we must find an element $v \in W_{I\cup J\cup K}$ together with an appropriate group element that completes the following diagram, where for simplicity we omit the subscripts for the projections ρ and label the morphisms by the relevant group elements:

$$\begin{array}{ccccc}
(I, \gamma^{-1}x) & \xrightarrow{\gamma} & (I \cup J, z) & \dashrightarrow & (I \cup J \cup K, v) \\
 & & \uparrow & & \uparrow \\
 & & \big(J, \rho(z) = \delta^{-1}\rho(w)\big) & \xrightarrow{\delta} & \big(J \cup K, w = \rho(v)\big) \\
 & & & & \uparrow \\
 & & & & (K, \rho(w))
\end{array}$$

It is not hard to check that the formula for v given in (ii) has the required properties. Details are in [**M6**]. Moreover, this definition implies that composition is a local diffeomorphism that is compatible with the inverse operation in (iii).

If the group actions are effective, we can now check indirectly that m is associative as follows. (The proof in the general case is given in [**M6**].) For if not, there are objects x, y and composable morphisms f, g, h such that $f \circ (g \circ h) \neq (f \circ g) \circ h \in \mathrm{Mor}(x, y)$. Since m is a local diffeomorphism and the sets $\mathrm{Mor}(x, y)$ are finite, this identity must continue to hold as x and hence f, g, h, y vary. Therefore we can identify $f \circ (g \circ h) \circ \big((f \circ g) \circ h\big)^{-1} \in \mathrm{Mor}(x, x)$ with a group element $\gamma \in \Gamma_I^x$ that must remain fixed as x varies. Therefore γ must belong to a component of morphisms that are not equal to identity morphisms but yet have $s = t$. Since the existence of such γ contradicts the effectivity hypothesis, it follows that m is associative. □

We now sketch the proof that every orbifold has a strict orbifold atlas. Recall from Definition 4.1.4 that two atlases $\mathcal{K}, \mathcal{K}'$ are said to be **directly commensurate** if they are subatlases of a common atlas $\mathcal{K}''$ on X, and **commensurate** if there is a finite sequence of pairwise directly commensurate atlases starting with $\mathcal{K}$ and ending at $\mathcal{K}'$.

PROPOSITION 5.1.6. *Every compact orbifold Y has a strict orbifold atlas $\mathcal{K}$ whose associated groupoid $\mathbf{G}_{\mathcal{K}}$ is an orbifold structure on Y. Moreover, there is a bijective correspondence between commensurability classes of such atlases and Morita equivalence classes of ep groupoids.*

PROOF. Let $\mathbf{G}$ be an ep groupoid with footprint map $f : \mathrm{Obj}_{\mathbf{G}} \to Y$. Our first aim is to construct an atlas $\mathcal{K}$ on Y together with a functor $\mathcal{F} : \mathbf{B}_{\mathcal{K}} \to \mathbf{G}$ that covers the identity map on Y and hence extends to an equivalence from the groupoid completion $\mathbf{G}_{\mathcal{K}}$ to $\mathbf{G}$.

By Moerdijk [**Mo**], each point in Y is the image of a group quotient that embeds into $\mathbf{G}$. Therefore since Y is compact we can find a finite set of basic charts $\mathbf{K}_i := \big(W_i, \Gamma_i, \psi_i\big)_{1\le i\le N}$ on Y whose footprints $(F_i)_{1\le i\le N}$ cover Y, together with embeddings

$$\sigma : \textstyle\bigsqcup_i W_i \hookrightarrow \mathrm{Obj}_{\mathbf{G}}, \quad \widetilde{\sigma} : \bigsqcup_i W_i \times \Gamma_i \hookrightarrow \mathrm{Mor}_{\mathbf{G}}$$

that are compatible in the sense that the following diagrams commute:

$$\begin{array}{ccc} W_i \times \Gamma_i & \xrightarrow{\widetilde{\sigma}_i} & \mathrm{Mor}_{\mathbf{G}} \\ {\scriptstyle s\times t}\downarrow & & \downarrow{\scriptstyle s\times t} \\ W_i \times W_i & \xrightarrow{\sigma_i} & \mathrm{Obj}_{\mathbf{G}} \times \mathrm{Obj}_{\mathbf{G}}, \end{array} \qquad \begin{array}{ccc} W_i & \xrightarrow{\sigma_i} & \mathrm{Obj}_{\mathbf{G}} \\ {\scriptstyle \psi_i}\downarrow & & \downarrow{\scriptstyle f} \\ Y & \xrightarrow{\mathrm{id}} & Y, \end{array}$$

We claim that there is a Kuranishi atlas $\mathcal{K}$ with these basic charts whose footprint maps ψ_I extend $f\circ\sigma : \bigsqcup_i W_i \to Y$.

To see this, we first consider the sum of two charts. Given $I := \{i_0, i_1\}$ with $F_I \ne \emptyset$, order its elements so that $i_0 < i_1$ and consider the set

$$W_I := W_{\{i_1,i_0\}} := \mathrm{Mor}_{\mathbf{G}}(\sigma(W_{i_0}), \sigma(W_{i_1})) := (s\times t)^{-1}\big(\sigma(W_{i_0}) \times \sigma(W_{i_1})\big)$$

of morphisms in $\mathbf{G}$ from $\sigma(W_{i_0})$ to $\sigma(W_{i_1})$. Then W_I is the inverse image of an open subset of $\mathrm{Obj}_{\mathbf{G}} \times \mathrm{Obj}_{\mathbf{G}}$, hence open in $\mathrm{Mor}_{\mathbf{G}}$, and thus a smooth manifold. Since the points in $f^{-1}(F_I)\cap\sigma(W_{i_0})$ are identified with points in $f^{-1}(F_I)\cap\sigma(W_{i_1})$ by morphisms in $\mathbf{G}$, the restrictions of s, t to W_I have images

$$s(W_I) = f^{-1}(F_I)\cap\sigma(W_{i_0}), \quad t(W_I) = f^{-1}(F_I)\cap\sigma(W_{i_1}).$$

Moreover, for any $x\in s(W_I)$ and $\alpha\in\mathrm{Mor}_{\mathbf{G}}(x,y)\in W_I$, we have

$$s^{-1}(x)\cap W_I \cong \mathrm{Mor}_{\mathbf{G}}\big(t(\alpha), \sigma(W_{i_1})\big) \cong \Gamma_{i_1},$$

where the second isomorphism holds because by assumption $f\circ\sigma_{i_1}$ is the footprint map $\psi_{i_1} : W_{i_1} \mapsto {}^{W_{i_1}}\!/_{\Gamma_{i_1}} \cong F_{i_1}$. Rephrasing this in terms of the action of the group $\Gamma_I := \Gamma_{i_1}\times\Gamma_{i_0}$ on $\alpha\in W_I$ by

$$(\gamma_{i_1}, \gamma_{i_0})\cdot\alpha = \widetilde{\sigma}(\gamma_{i_1})\circ\alpha\circ\widetilde{\sigma}(\gamma_{i_0}^{-1}),$$

one finds that Γ_{i_1} acts freely on W_I and that the source map $s : W_I \to \sigma(W_{i_0})$ induces a diffeomorphism ${}^{W_I}\!/_{\Gamma_{i_1}} \to \sigma(W_{i_0})\cap f^{-1}(F_I)$. Similarly, Γ_{i_0} acts freely, and the target map $t : W_I \to \sigma(W_{i_1})$ induces a diffeomorphism ${}^{W_I}\!/_{\Gamma_{i_0}} \to \sigma(W_{i_1})\cap f^{-1}(F_I)$. Since the footprint map for the chart W_i factors out by the action of Γ_i, the same is true for this sum chart: in other words the footprint map $\psi_I : W_I \to Y, \alpha\mapsto f\big(s(\alpha)\big) = f\big(t(\alpha)\big)$ induces a homeomorphism ${}^{W_I}\!/_{\Gamma_I} \xrightarrow{\cong} F_I$. Therefore W_I satisfies all the requirements of a sum of two charts.

To define a transition chart for general $I\in\mathcal{I}_Y$, enumerate its elements as $i_0 < i_1 \cdots < i_k$, where $k+1 := |I| \ge 2$ and define W_I to be the set of composable k-tuples of morphisms $(\alpha_{i_k}, \cdots, \alpha_{i_1})$, where $(s\times t)(\alpha_{i_\ell}) \in \big(\sigma(W_{i_{\ell-1}}), \sigma(W_{i_\ell})\big)$. Then W_I supports an action of Γ_I given by

$$\gamma\cdot(\alpha_{i_k}, \cdots, \alpha_{i_1}) = (\alpha_{i_k}, \cdots, \alpha_{i_{\ell+1}}\widetilde{\sigma}(\gamma)^{-1}, \widetilde{\sigma}(\gamma)\alpha_{i_\ell}, \cdots, \alpha_{i_1}), \quad \gamma\in\Gamma_{i_\ell}.$$

Notice that for any $H \subsetneq I$ the subgroup $\Gamma_{I\setminus H}$ acts freely, and the quotient can be identified with W_H by means of the appropriate partial compositions and forgetful maps. Moreover, we define the functor $\mathcal{F} : \mathbf{B}_{\mathcal{K}} \to \mathbf{G}$ on objects by

$$W_I \to \mathrm{Obj}_{\mathbf{G}}, \quad \begin{cases} x\mapsto\sigma(x), & \text{if } I = \{i_0\}, x\in W_{i_0}, \\ (\alpha_{i_k}, \cdots, \alpha_{i_1}) \mapsto t(\alpha_{i_k}) \in \sigma(W_{i_k}) & \text{if } |I| > 1. \end{cases}$$

Further details of this construction are given in [**M6**].

This sketches why every orbifold has a Kuranishi atlas of the required type. To see that this atlas is unique up to commensurability, note first that because we can start from any covering set of basic charts any two atlases constructed in this way from the same groupoid are directly commensurate. Further, given groupoid structures $(\mathbf{G}, f), (\mathbf{G}', f')$ on Y with common refinement $F : (\mathbf{G}'', f'') \to (\mathbf{G}, f), F' : (\mathbf{G}'', f'') \to (\mathbf{G}', f')$, all atlases on Y that are constructed from $\mathbf{G}$ or from $\mathbf{G}'$ are commensurate to an atlas whose basic charts are pushed forward from $\mathbf{G}''$, and hence they all belong to the same commensurability class.

Conversely, we must show that if $\mathcal{K}, \mathcal{K}'$ are commensurate, the groupoids $\mathbf{G}_{\mathcal{K}}$ and $\mathbf{G}_{\mathcal{K}'}$ are equivalent. It suffices to consider the case when $\mathcal{K}, \mathcal{K}'$ are directly commensurate. But then they are contained in a common atlas $\mathcal{K}''$ on Y that defines a groupoid $\mathbf{G}_{\mathcal{K}''}$ that contains both $\mathbf{G}_{\mathcal{K}}$ and $\mathbf{G}_{\mathcal{K}'}$ as subgroupoids with the same realization Y. Thus the inclusions $\mathbf{G}_{\mathcal{K}} \to \mathbf{G}_{\mathcal{K}''}$ and $\mathbf{G}_{\mathcal{K}'} \to \mathbf{G}_{\mathcal{K}''}$ are equivalences. This completes the proof. □

EXAMPLE 5.1.7. (i) Example 3.1.17 explains the "football", which is a simple example of this construction. In [**M6**, Example 3.6] we also discuss in detail the different possible choices of strict orbifold atlas for the noneffective but nontrivial orbifold structure $\mathbf{G}_{S^2}$ on S^2 with group $\Gamma := \mathbb{Z}/2\mathbb{Z}$: the nontriviality can appear either in the nontriviality of the individual covering maps $\rho_{IJ} : W_J \to W_I$ (if, for example, W_I is an annulus), or, if all components of the W_I are contractible, in the relation between the different free actions of the groups $\Gamma_{J \setminus I}$ on the sets of these components.

(ii) If one instead describes an orbifold via a Kuranishi structure as in Remark 3.1.19, then the coordinate changes between the local charts (F_p, Γ_p) are related by injections $\phi_{qp} : (F_q, \Gamma_q) \to (F_p, \Gamma_p)$ that satisfy a compatibility condition of the form $\phi_{qp} \circ \phi_{rq} = \gamma_{pqr}\phi_{rp}$ for some locally constant map $\gamma_{pqr} : F_{pqr} \to \Gamma_p$. In this language, all the local charts for the nontrivial groupoid $\mathbf{G}_{S^2}$ in (i) above have the form (F, Γ), where $\Gamma = \mathbb{Z}/2$ acts trivially on $F \subset S^2$, and the twisting is encoded in a suitable choice for the γ_{pqr}. For example, one could describe a structure of this type with three basic charts (F_i, Γ_i) with $(F_i)_{i=1,2,3}$ as above, where the coordinate changes $\phi_{ij} : (W_{ij}, \Gamma_i) \to (W_j, \Gamma_j)$ satisfy the relation $\phi_{13} = \gamma\phi_{23} \circ \phi_{12}$ where $\gamma = \mathrm{id}$ on $F_S \subset F_{123}$ and $\gamma \neq \mathrm{id}$ on $F_N \subset F_{123}$. ◊

REMARK 5.1.8. (i) The construction in Proposition 5.1.6 is reminiscent of that given in [**M3**, §4] for the "resolution" of an orbifold. However, the two constructions have different aims: here we want to build a simple model for the orbifold Y, while there we wanted to construct a *resolution*, i.e. a corresponding branched manifold with the same fundamental class. In fact, we can simplify the construction in [**M3**] of such a resolution by first constructing the atlas $\mathcal{K}_Y$ and its corresponding category $\mathbf{B}_{\mathcal{K}}$ and then removing morphisms from $\mathbf{B}_{\mathcal{K}}$ by the recipe used in Proposition 3.4.4 to construct the zero set.

To this end, consider a strict orbifold atlas $\mathcal{K}$ with realization Y and choose a reduction $\mathcal{V}$ of the footprint covering. Since there is no obstruction bundle, the perturbation ν is identically zero, so that the "local zero set" Z_I equals V_I for all I. Thus the wnb category that we will call $(\mathbf{V}, \Lambda_V)$ constructed in Proposition 3.4.4 has objects $\sqcup_I V_I$ and morphisms and weighting function as described in §3.4. It is a nonsingular subcategory of the groupoid completion $\mathbf{G}_{\mathcal{K}}$ of $\mathbf{B}_{\mathcal{K}}$, such that inclusion induces a map

$$|\iota_{\mathbf{V}}| : |\mathbf{V}|_{\mathcal{H}} \to |\mathbf{G}_{\mathcal{K}}| = Y, \qquad |\iota_{\mathbf{V}}|_*(\Lambda_V) = \Lambda_Y, \qquad \text{where } \Lambda_Y(y) = \tfrac{1}{|\Gamma^y|}.$$

Note that this statement holds also in the noneffective case (although [**M3**] mostly considered the effective case).[38] By [**M3**, Proposition 4.19], the fundamental class of $(|\mathbf{V}|_{\mathcal{H}}, \Lambda_V)$ is constructed in such a way that $|\iota_{\mathbf{V}}|$ pushes it forward to the usual fundamental class of Y.

(ii) The relations between the notions of directly commensurate, commensurate and concordant are not clear even for atlases with trivial obstruction bundles. For example, the concordance relation between atlases $\mathcal{K}^0, \mathcal{K}^1$ on X requires there to be an atlas $\mathcal{K}^{01}$ over $[0,1] \times X$ with prescribed collar isomorphisms between the restrictions of $\mathcal{K}^{01}$ to the collars $[0,\varepsilon) \times X$ and $(1-\varepsilon, 1] \times X$ and the product atlases $[0,\varepsilon) \times \mathcal{K}^0, (1-\varepsilon, 1] \times \mathcal{K}^1$. However, there is no requirement on the interior charts (i.e. those whose footprint does not intersect $\{0,1\} \times X$) that they are in any way compatible with the product structure, for example the local action of the stabilizer group of a point need not decompose as a product. Hence even when the obstruction bundles are trivial so that the footprint maps ψ_I are defined over the whole of the domains U_I, it is not immediately clear that the relation of commensurability for atlases on X with trivial obstruction spaces is the same as the notion of concordance, though it could well be true. Since we are using concordance (or more generally cobordism) simply for convenience, we will not pursue this question further here. Moreover, because orbifolds are most properly described in terms of stacks, the "correct" equivalence relation would no doubt involve the notion of 2-category as in the work of Joyce [**J15**]. ◊

5.2. Nontrivial obstruction bundles.

When calculating Gromov–Witten invariants one often starts with moduli spaces X that have nice geometric structure, though they are not regular. For example, X might be a manifold (or more generally orbifold) of solutions to the Cauchy–Riemann equations, such that the cokernels form a bundle over X. In this case the VFC should be the Euler class of the (orbi)bundle. We now explain some simple examples of this type, both in the abstract and as applied in the GW setting.

There are two possible ways of incorporating nontrivial obstruction bundles into our framework. We can trivialize the bundle either by adding a complementary bundle or by using local trivializations. The first method is simpler, but may not adapt well to more complicated situations. The second method abstracts the procedure used to construct GW atlases.

Method 1: We explain this method in the case when the isotropy is trivial. It generalizes to cases when the obstruction bundle is a global quotient. With more complicated isotropy, one would need more charts and so should use Method 2.

LEMMA 5.2.1. *Suppose that $\pi_X : T \to X$ is a nontrivial k-dimensional oriented dimensional bundle over a compact $d+k$ dimensional oriented manifold X. Then*

[38] In fact, in the orbifold setting it does suffice for these purposes to restrict to the effective case, since, as is well known, each connected component of Y has an effective quotient (see for example [**M3**, Lemma 2.15]). However, as explained in Example 3.4.8 – and in contrast to the approach in [**M3**] – the weighting function should take into account the number of group elements that act trivially.

there is an oriented Kuranishi atlas $\mathcal{K}$ on X whose VFC equals the Euler class $\chi(T) \in H_d(X)$.

PROOF. Choose an oriented complementary bundle $\pi_X^\perp : T^\perp \to X$ such that $T \oplus T^\perp \cong X \times \mathbb{R}^m$. Denote by $\widehat{\iota} : T^\perp \to \mathbb{R}^m$ the composite of the inclusion $T^\perp \to X \times \mathbb{R}^m$ with the projection. Then define an atlas $\mathcal{K}$ with a single chart

$$\mathbf{K} = \big(U = T^\perp,\ E = \mathbb{R}^m,\ s = \widehat{\iota},\ \psi\big),$$

where ψ identifies the zero section of $T^\perp$ with X. It has no nontrivial coordinate changes. Any section $\nu^X : X \to T$ gives rise to a perturbation section $\iota_T \circ \nu^X \circ \pi_X^\perp : U \to \mathbb{R}^m$, where $\iota_T : T \to \mathbb{R}^m$ is the inclusion. Then $s + \iota_T \circ \nu(u) = 0$ only if $s(u) = 0 \in T^\perp$ and $\nu(u) = 0 \in E$. Hence the sections ν^X and $s + \iota_T \circ \nu^X \circ \pi_X^\perp$ have the same zero set; moreover one is transverse to zero if and only if the other is. □

REMARK 5.2.2. Suppose that $X = \overline{\mathcal{M}}_{0,k}(M, A, J)$ is a manifold consisting of equivalence classes of stable maps with domains of constant topological type. If the cokernels of the linearized Cauchy–Riemann operator form a bundle over X of constant rank, then one might be able to carry out this construction in the GW setting since, at least locally, one can always find a suitable embedding λ of $E^\perp$; see [**MS**, Example 7.2.4]. However there may be no a global stabilization for the domains of the curves. The next method is more local, and so more adaptable. ◊

Method 2: We begin with the case of trivial isotropy. The first step is to build an oriented (standard) Kuranishi atlas that models the nontrivial bundle $T \to X$. To this end, choose a finite open cover $(F_i)_{i=1,\dots,N}$ of X together with trivializations $\tau_i : T_0 \times F_i \to T|_{F_i}$, where T_0 is the fiber of T. We will define an atlas with indexing set $\mathcal{I}_\mathcal{K} = \big\{ I \subset \{1, \dots, N\} : F_I \neq \emptyset \big\}$, and basic charts

(5.2.1) $$\mathbf{K}_i := \big(F_i, E_i := T_0, s_i \equiv 0, \psi_i = \mathrm{id}\big).$$

The transition charts are: $\mathbf{K}_I := \big(U_I, E_I := \prod_{i \in I} E_i, s_I, \psi_I\big)$ where

(5.2.2)
$$U_I = \Big\{ (\vec{e}, x) \in E_I \times F_I \mid \textstyle\sum_{i \in I} \tau_i(e_i, x) = 0 \Big\}, \quad s_I(\vec{e}, x) = \vec{e}, \quad \psi_I(\vec{0}, x) = x \in F_I.$$

The coordinate changes have domains $U_{IJ} := (E_I \times F_J) \cap U_I$ and are induced by the obvious inclusions

$$\widehat{\phi}_{IJ} : E_I \to E_J, \quad (e_i)_{i \in I} \mapsto \big((e_i)_{i \in I}, (0_j)_{j \in J \setminus I}\big),$$
$$\phi_{IJ}(\vec{e}, x) = \big(\widehat{\phi}_{IJ}(\vec{e}), x\big).$$

Thus each chart has dimension $\dim X - \dim T_0$. The cocycle condition is immediate. Moreover the index condition holds because the inclusion $\widehat{\phi}_{IJ} : E_I|_0 \to E_J|_0$ and $\widehat{\phi}_{IJ} : E_I \to E_J$ have isomorphic cokernels, where $E|_0 = \{\vec{e} \in E : \sum \vec{e} = 0\}$. Therefore this set $\mathcal{K} = \big(\mathbf{K}_I, \widehat{\Phi}_{IJ}\big)_{I \subset J, I, J \in \mathcal{I}_\mathcal{K}}$ of charts and coordinate changes defines an additive atlas, which is tame by construction. Note the commutative diagram

(5.2.3)
$$\begin{array}{ccccc}
\mathbf{E}_\mathcal{K} & \xrightarrow{\pi_\mathcal{K}} & |\mathbf{E}_\mathcal{K}| & \xrightarrow{\tau} & T \\
\downarrow{\scriptstyle \mathrm{pr}} & & \downarrow{\scriptstyle \mathrm{pr}} & & \downarrow{\scriptstyle \pi_X} \\
\mathbf{B}_\mathcal{K} & \xrightarrow{\pi_\mathcal{K}} & |\mathcal{K}| & \xrightarrow{\pi} & X,
\end{array}$$

where $\tau : |\mathbf{E}_{\mathcal{K}}| \to T$ is induced by $\tau\big((e_i)_{i\in I}\big) = \sum_{i\in I} \tau_i(e_i)$ and $\pi : |\mathcal{K}| \to X$ by the projection $(\vec{e}, x) \mapsto x$.

By Lemma 6.2.2, the footprint covering $(F_i)_{i=1,\dots,N}$ has a reduction $(Z_I)_{I\in\mathcal{I}_{\mathcal{K}}}$. This defines a reduction $(V_I)_{I\in\mathcal{I}_{\mathcal{K}}}$ of $\mathcal{K}$ with $V_I := \big\{(\vec{e}, x) \in U_I \,|x \in Z_I\big\}$. We now show that each section $\nu^X : X \to T$ that is transverse to the zero section lifts to an admissible perturbation section of $\mathrm{pr} : \mathbf{E}_{\mathcal{K}}|_{\mathcal{V}} \to \mathbf{B}_{\mathcal{K}}|_{\mathcal{V}}$.

PROPOSITION 5.2.3. *Let $\pi_X : T \to X$ be a nontrivial bundle over a compact manifold X with atlas $\mathcal{K}$ as above. Let $(Z_I^X \sqsubset F_I)_{I\in\mathcal{I}_{\mathcal{K}}}$ be any reduction of the footprint cover and $\mathcal{V}$ the associated reduction of $\mathcal{K}$ as above. Then any section $\nu^X : X \to T$ of π_X lifts to a functor $\nu : \mathbf{B}_{\mathcal{K}}|_{\mathcal{V}} \to \mathbf{E}_{\mathcal{K}}|_{\mathcal{V}}$ with local zero sets $(\mathfrak{s}|_{V_I} + \nu_I)^{-1}(0) = \psi_I^{-1}\big(F_I \cap (\nu^X)^{-1}(0)\big)$. Moreover, if ν^X is transverse to 0, so is its lift. Therefore $[X]^{vir}_{\mathcal{K}} = \chi(T)$.*

PROOF. The first step is to choose a smooth partition of unity $(\beta_i)_{i=1,\dots,N}$ on X such that

$$x \in Z_J^X \Longrightarrow \sum_{i\in J} \beta_i(x) = 1. \tag{5.2.4}$$

To do this, fix a metric d on X, and define

$$\rho_i(x) := d(x, \textstyle\bigcup_{J:i\notin J} \overline{Z}_J^X).$$

Thus $\rho_i(x) = 0$ if and only if $x \in \bigcup_{J:i\notin J} \overline{Z}_J^X$. We claim that for all x there is i such that $\rho_i(x) \neq 0$. To see this, note that for each x the set of J such that $x \in \overline{Z}_J^X$ is totally ordered and so can be written as a chain $J_1^x \subsetneq J_2^x \subsetneq \cdots \subsetneq J_q^x$ for some $q \geq 1$. Therefore if $i \in J_1^x$ and $x \in \overline{Z}_J^X$ we must have $i \in J$. Hence $\rho_i(x) \neq 0$ for all $i \in J_1^x$. Thus the function $\sum_{i=1}^N \rho_i$ does not vanish anywhere on X, which implies that the function $\beta_i(x) := \frac{1}{\sum_j \rho_j(x)} \rho_i(x)$ is well defined, and gives a partition of unity on X. Moreover (5.2.4) holds because

$$x \in Z_J^X,\ i \notin J \implies \beta_i(x) = 0.$$

Next define $V_I := \big\{(\vec{e}, x) \in U_I \mid x \in Z_I^X\big\}$. These sets $(V_I)_{I\in\mathcal{I}_{\mathcal{K}}}$ form a reduction of $\mathcal{K}$. Further, given a section $\nu^X : X \to T$ there is an associated functor $\nu : \mathbf{B}_{\mathcal{K}}|_{\mathcal{V}} \to \mathbf{E}_{\mathcal{K}}|_{\mathcal{V}}$ defined by

$$\nu_I(x) := \big(\nu_I^i(x)\big)_{i\in I} = \Big(\beta_i(x)\tau_i^{-1}\big(\nu^X(x)\big)\Big)_{i\in I} \in E_I.$$

These perturbations are compatible with the coordinate transformations and have the property that for each $x \in Z_I^X$ we have $\sum_i \tau_i(\nu_I^i(x)) = \nu^X(x)$. On the other hand, by definition of U_I and s_I the elements $\vec{e} = (e_i)_{i\in I}$ in $\mathrm{im}\, s_I(x) \subset E_I$ have the property that $\sum_i \tau_i(e_i, x) = 0$. Therefore $(s_I|_{V_I} + \nu_I)(x) = 0$ precisely if $\nu^X(x) = 0$. Further, the fact that ν^X is transverse to the zero section easily implies that $s_I|_{V_I} + \nu_I$ is also transverse to zero. Similarly, one can check that the orientation of $T \to X$ induces an orientation of $\mathcal{K}$ and that the induced orientations (defined as in Lemma 3.3.9) on the zero sets agree. Since the zero set is compact, this completes the proof. □

REMARK 5.2.4. If the manifold X is a GW moduli space (with trivial isotropy) and T is isomorphic to the bundle of cokernels of the linearized Cauchy–Riemann operator, then it is not hard to build an atlas with basic charts $(\mathbf{K}_i)_{1,\dots,N}$ as above, provided that we choose suitably small footprints F_i whose elements can be stabilized by a slicing manifold Q_i. Note that the section s_i is zero since there are no solutions of the equation $\overline{\partial}_J f = \lambda(\vec{e})|_{\text{graph} f}$ for f sufficiently close to f_0. To see that the transition charts provided by the construction in §4.1 have the form described above, we must compare the defining equation $\sum_{i\in I} \tau_i(e_i, x) = 0$ for the subset $U_I \subset X \times E_I$ with the equation $\overline{\partial}_{J,j} f = \lambda_I(\underline{\vec{e}})|_{\text{graph} f}$ that defines the transition chart in (4.1.11). Here the linear map λ_I plays the role of the local trivialization, and is defined as a sum of terms $\lambda_i(\vec{e}_i)|_{\text{graph}_i f}$ as in (4.1.10). Therefore these equations are the same; the only difference in the construction is that in equation (5.2.2) we take U_I to consist of *all* solutions to the equation $\sum_{i\in I} \tau_i(e_i, x) = 0$ while in the GW construction we restrict to a neighbourhood of the zero set of s_I. Hence, in this situation there is an open embedding in the sense of Definition 4.2.1 of the GW atlas into the atlas constructed in Proposition 5.2.3. $\Diamond$

The case with isotropy. Now suppose given an oriented orbibundle $\pi : T \to X$ over a smooth compact orbifold X with fiber T_0. We first explain the standard way of defining its Euler class, using the strict orbifold atlases for X, T constructed in §5.1.

Choose a covering of X by group quotients $\big((W_i, \Gamma_i)\big)_i$ that is compatible with the bundle in the sense that for each i the orbibundle $T|_{F_i}$ pulls back to a trivial bundle $(\psi_i^X)^*(T|_{F_i})$ on which Γ_i acts by a product action. By Proposition 5.1.6 we may extend this family of basic charts to an orbifold atlas, $\mathcal{K}_X^{orb}$ on X with charts $(W_I, \Gamma_I, \psi_I^X)$ and footprint cover (F_I). Moreover the orbifold T has a corresponding atlas $\mathcal{K}_T^{orb}$ with charts $(T|_{W_I}, \Gamma_I, \psi_I^T)$ and footprint cover $(T|_{F_I}) \subset T$, where for simplicity we denote the pullback $(\psi_I^X)^*(T|_{F_I})$ of T to W_I simply by $T|_{W_I}$. By Proposition 5.1.5 the categories

$$\mathbf{B}_X^{orb} := \mathbf{B}_{\mathcal{K}_X^{orb}}, \quad \mathbf{B}_T^{orb} := \mathbf{B}_{\mathcal{K}_T^{orb}},$$

corresponding to these orbifold atlases have completions to ep groupoids that we will denote $\mathbf{G}_T^{orb}, \mathbf{G}_X^{orb}$ and the projection π induces a functor $\pi : \mathbf{G}_T^{orb} \to \mathbf{G}_X^{orb}$ that restricts on the object spaces to the bundle projection $\bigsqcup_I T|_{W_I} \to \bigsqcup_I W_I$.

One way to define the Euler class of $\pi : T \to X$ is to consider a "nonsingular resolution" of the groupoid $\mathbf{G}_X^{orb}$, pull the bundle $T \to X$ back to this resolution and then push forward to X the (weighted) zero set of a generic section ν^X of this bundle. As we explained in Remark 5.1.8 (i), we can take the resolution of $\mathbf{G}_X^{orb}$ to be the wnb groupoid $\mathbf{V}_X^{orb}$ formed as in Proposition 3.4.4 from a reduction of $\mathcal{K}_X^{orb}$. Indeed,if (G_I^X) is a reduction of the footprint cover $(F_I = \psi_I^X(W_I))$ of $\mathcal{K}_X^{orb}$, the sets $\big(V_I^{orb} = (\psi_I^X)^{-1}(Z_I^X)\big) \sqsubset W_I$ form a reduction of $\mathcal{K}_X^{orb}$. The corresponding wnb groupoid $\mathbf{V}_X^{orb}$ has

- objects $\mathrm{Obj}_{\mathbf{V}^{orb}} = \bigsqcup_I V_I^{orb}$,
- morphisms generated by the projections $\rho_{IJ}^X : \widetilde{V}_{IJ}^{orb} \to V_I^{orb}$ for $I \subsetneq J$.
- weighting function given by assigning the weight $\frac{1}{|\Gamma_I|}$ to each component V_I^{orb}.

The pullback of T to $\mathbf{V}_X^{orb}$ is the corresponding wnb groupoid with objects $\bigsqcup_I T|_{V_I^{orb}}$. Therefore the section $\nu^X = (\nu_I^X)$ is given by a compatible family of sections

$$\nu_I^X : V_I^{orb} \to T|_{V_I^{orb}}, \quad \nu_J^X|_{\widetilde{V}_{IJ}^{orb}} = \nu_I^X \circ \rho_{IJ}^X. \tag{5.2.5}$$

If $\nu \pitchfork 0$, there is a full subcategory $\mathbf{Z}_X^{orb}$ of $\mathbf{V}_X^{orb}$ with objects $\nu_I^{-1}(0) \subset V_I^{orb}$. Again this forms a wnb groupoid, with the induced weighting function. It follows from

[**M3**, Proposition 4.19] [39] that its pushforward to X represents the Euler class of $T \to X$.

Let us now compute this Euler class by using a Kuranishi atlas. The first step is to build a Kuranishi atlas $\mathcal{K}$ such that the bundle $\mathrm{pr} : \mathbf{E}_{\mathcal{K}} \to \mathbf{B}_{\mathcal{K}}$ is a model for the original bundle $T \to X$, in the sense that there is an analog of the commutative diagram (5.2.3). To this end, choose Γ_i-equivariant trivializations

$$\tau_i : T_i \times W_i \overset{\cong}{\to} (\psi_i^X)^*(T|_{F_i}),$$

where we denote the fiber (which is isomorphic to T_0) by T_i to emphasize that it supports an action of Γ_i. We then define a Kuranishi atlas essentially as before, incorporating the groups in the natural way. Thus

$$\mathbf{K}_I := \left(U_I,\ E_I := \textstyle\prod_{i\in I} T_i,\ \Gamma_I := \prod_{i\in I}\Gamma_i,\ s_I,\ \psi_I\right) \tag{5.2.6}$$

where

$$U_I = \Big\{(\vec{e}, x) \in E_I \times W_I \mid \sum_{i\in I} \tau_i(e_i, x) = 0\Big\},$$

$$s_I(\vec{e}, x) = \vec{e}, \quad \psi_I(\vec{0}, x) = \psi_I^X(x) \in F_I.$$

The coordinate changes have lifted domains $\widetilde{U}_{IJ} := \widehat{\phi}_{IJ}(E_I) \times W_J \subset U_J$ and are induced by the obvious projections

$$\rho_{IJ} : \big((e_i)_{i\in I}, (0_j)_{j\in J\setminus I}, x\big) \mapsto \big((e_i)_{i\in I}, \rho_{IJ}^X(x)\big) \in U_I,$$

where $\rho_{IJ}^X : W_J \to W_I$ is part of the corresponding coordinate change in the orbifold atlas $\mathcal{K}_X^{orb}$.

As before, $\mathcal{K}$ is tame so that the theory developed in §3 applies. In particular, the reduction (Z_I^X) of the footprint cover chosen above induces a compatible reduction $\mathcal{V} \subset \mathcal{K}$ with $V_I := U_I \cap (E_I \times V_I^{orb})$. Further, the proof of Proposition 5.2.3 carries through to show that every section (ν_I^X) as in (5.2.5) may be lifted via a partition of unity to a perturbation $\nu : \mathbf{B}_{\mathcal{K}}|_{\mathcal{V}} \to \mathbf{E}_{\mathcal{K}}|_{\mathcal{V}}$ in the sense of Definition 3.4.2. Hence, as before, the VFC defined by the Kuranishi atlas $\mathcal{K}$ is the Euler class of $T \to X$. Further details are left to the interested reader. Thus we conclude:

- *The analog of Proposition* 5.2.3 *holds for general oriented orbibundles.*

REMARK 5.2.5. If X is more complicated, for example a union of strata each of which has fixed dimension and cokernel bundle of constant rank, then one should first build local atlases that model each stratum separately, and then put them together via the gluing parameters. Suppose, for example, that X is a compact $2k$-dimensional manifold that contains a codimension 2 submanifold Y consisting of curves with one node, and that the cokernels have constant rank $2r$ so that they form bundles $T^Y \to Y$ and $T^X \to X \setminus Y$. The complex line bundle L_Y over Y formed by the gluing parameter at the node is the normal bundle to Y in X. Let $\mathcal{N}(Y) \subset X$ be a neighbourhood of Y that forms a disc bundle $\pi_Y : \mathcal{N}(Y) \to Y$. Notice that the restriction of T^X to $\mathcal{N}(Y)$ may not simply be the pullback $\pi_Y^*(T^Y)$ because when one glues with the family of parameters $a = \varepsilon e^{2\pi i\theta}, \theta \in S^1$, one of the components twists by $2\pi\theta$ relative to the other. In many situations the bundles T^X, T^Y have a natural complex structure that is preserved by the S^1 action in the fibers of $\partial\mathcal{N}(Y) \to Y$, and T^Y decomposes into a finite sum $\oplus_k T_k^Y$ so that

[39] Warning: In [**M3**] the word "atlas" has a different meaning from our current usage; see [**M3**, Definition 2.11].

$T^X|_{\mathcal{N}(Y)\setminus Y} = \oplus_k \pi_Y^*\big(T_k^Y \otimes_{\mathbb{C}} (L_Y)^{\otimes k}\big)$. One should then build the atlas over X in stages, with one atlas over a neighbourhood $\mathcal{N}_1(Y)$ of Y with obstruction bundle T^Y and gluing parameters in L_Y, another over $X \setminus \mathcal{N}_2(Y)$ with obstruction bundle T^X, and appropriate transition charts over the deleted neighbourhood $\mathcal{N}_1(Y)\setminus\mathcal{N}_2(Y)$. The interaction of the gluing parameters and the obstruction bundles will be seen in the structure of these transition charts. Of course, if one really wanted to carry out such computations, rather than using the rather rigid notion of atlases one probably should first generalize the previous discussion to show that the fundamental cycle can be calculated using more general atlases, in which, for example, the bundles $\mathrm{pr}: U_I \times E_I \to U_I$ are allowed to be nontrivial. $\Diamond$

6. Order structures and products

Atlases are defined by a rather rigid process from a set of basic charts. In particular, by (2.1.2) our definition requires an atlas to be **additive** in the sense that it satisfies the following condition for the obstruction spaces of the transition charts:

$$E_I = \textstyle\prod_{i\in I} E_i. \tag{6.0.1}$$

On the one hand, this condition is essential for the taming process; on the other it does not hold for product atlases which are relevant to various geometrically interesting situations. In this section, we define a notion of semi-additive atlas, which does apply to products.[40] Since these atlases are not filtered, they cannot be tamed; see Remark 6.1.10. However, if one builds them starting from tame or good atlases (as in Definition 2.3.22), then they have nice properties.

Here are our main results, stated somewhat informally.

- A good semi-additive atlas has a well defined VFC; see Proposition 6.2.5.
- A product of good atlases $\mathcal{K}_1 \times \mathcal{K}_2$ is semi-additive and good; see Lemma 6.1.9. By Proposition 6.2.1, the corresponding virtual class $[X_1 \times X_2]^{vir}_{\mathcal{K}_1\times\mathcal{K}_2}$ equals the product $[X_1]^{vir}_{\mathcal{K}_1} \times [X_2]^{vir}_{\mathcal{K}_2}$ in $\check{H}_*(X_1 \times X_2)$.
- We show that in Proposition 6.3.3 that every semi-additive atlas is concordant to an atlas that can be tamed. We then sketch how this might be applied in Hamiltonian Floer theory, where one needs a coherent system of atlases for a family of spaces X whose boundary is a union of products $X_1 \times \cdots \times X_k$.

We begin by with a general discussion of posets (i.e. partially ordered sets) since the category $\mathbf{B}_{\mathcal{K}}$ determined by an atlas with trivial isotropy (or, more generally, by a topological atlas) is a poset; see Remark 2.2.17 (ii).

6.1. Semi-additive atlases. In our approach, an atlas (of whatever kind) on X always consists of a family of basic charts $(\mathbf{K}_i)_{i=1,\dots,N}$ with footprints $F_i \subset X$, together with transition charts $(\mathbf{K}_I)_{I\in\mathcal{I}_{\mathcal{K}}}$ that are indexed by a poset $\mathcal{I}_{\mathcal{K}} \subset \mathcal{P}^*\big(\{1,\dots,N\}\big)$ that is determined by the intersection pattern of the footprints $F_i \subset X$, where $\mathcal{P}^*(\mathcal{A})$ denotes the poset of nonempty subsets of the set $\mathcal{A}$. Indeed there is a transition chart with footprint $F_I = \cap_{i\in I} F_i$ for every (nonempty) collection

[40] This turns out to be fairly close to the notion of a "MW weak good coordinate system" in Joyce [**J14**, Definition A.18]. Since Joyce keeps only infinitesimal information along the zero set of a chart, the charts in his atlases inherit enough compatibility from the zero sets that they do not need to be tamed in our sense. However, the taming process is essential in our approach, since we use open neighbourhoods of the zero set.

of charts whose footprints have nonempty intersection. In some cases (such as the product of two atlases) *different* sets of basic charts have the *same* footprint intersection. One might therefore consider permitting an atlas to have just one transition chart for each such intersection. The corresponding indexing set $\mathcal{I}$ would no longer be a subset of $\mathcal{P}^*(\{1,\dots,N\})$. Nevertheless, it is minimally generated in the following sense.

DEFINITION 6.1.1. *The finite poset $\mathcal{I}$ is said to be* **minimally generated** *if it satisfies the following assumptions:*

(a) *every subset $\{I_1,\dots,I_k\}$ of $\mathcal{I}$ with an upper bound has a unique least upper bound $I_1\vee\cdots\vee I_k\in\mathcal{I}$;*
(b) *if $\mathcal{M}(\mathcal{I})$ denotes the set of minimal elements in $\mathcal{I}$, then each $J\in\mathcal{I}$ is the least upper bound of the set $m(J)=\{H\in\mathcal{M}(\mathcal{I})\mid H\le J\}$ of minimal elements it dominates.*

Further, we say that $\mathcal{I}$ is **minimally** $(\mathcal{A},\tau)$**-generated** *if in addition there is an order preserving injection $\tau:\mathcal{I}\to\mathcal{P}^*(\mathcal{A})$ such that*

(c) *for all $I,J\in\mathcal{I}$, $\tau(I\vee J)=\tau(I)\cup\tau(J)$,*
(d) *if $A\in\operatorname{im}(\tau), H\in\mathcal{M}(\mathcal{I})$ then*

$$\bigl\{B\in\mathcal{P}^*(\mathcal{A}):\tau(H)\subset B\subset A\bigr\}\ \subset\ \operatorname{im}(\tau).$$

LEMMA 6.1.2. *If $\mathcal{I}$ is minimally $(\mathcal{A},\tau)$-generated, then $\tau(J)=\bigcup_{H\in m(J)}\tau(H)$.*

PROOF. This follows from conditions (b) and (c). □

Condition (d) is a completeness condition that asserts the existence of "enough" elements in $\mathcal{I}$.

EXAMPLE 6.1.3. (i) A subset $\mathcal{I}$ of $\mathcal{P}^*(\{1,\dots,N\})$ (with the induced partial order) is minimally generated if

- it contains all the one point subsets, and
- $A\in\mathcal{I},\ \ \emptyset\ne B\subset A\Longrightarrow B\in\mathcal{I}$.

In fact, in this case it is minimally $(\mathcal{A},\tau)$-generated, with $\mathcal{A}=\{1,\dots,N\}$ and $\tau=\mathrm{id}$. The index sets of a Kuranishi atlas have this form.

(ii) Let $\mathcal{A}:=\mathcal{A}_1\sqcup\mathcal{A}_2$ where $\mathcal{A}_t:=\{1,\dots,N_t\}$ for $t=1,2$. The set of subsets $\mathcal{P}(\mathcal{A})$ of $\mathcal{A}$ consists of all pairs $(I_1,I_2)\in\mathcal{P}(\mathcal{A}_1)\times\mathcal{P}(\mathcal{A}_2)$, where one or both of I_1,I_2 may be empty. The set $\mathcal{I}$ of all pairs (I_1,I_2) with $I_t\ne\emptyset$ for $t=1,2$ is minimally $(\mathcal{A},\tau)$-generated, where $\tau:\mathcal{I}\to\mathcal{P}^*(\mathcal{A})$ is given by $\tau(I_1,I_2)=I_1\cup I_2\subset\mathcal{A}$. Note that the minimal subsets of $\mathcal{I}$ are the pairs of one point subsets $(\{i_1\},\{i_2\})$.

(iii) In our examples of minimally $(\mathcal{A},\tau)$-generated posets, the elements of $\mathcal{A}$ have the following interpretation. Each $\alpha\in\mathcal{A}$ indexes a subset $F_\alpha\subset X$, and we associate to each $A\in\mathcal{P}^*(\mathcal{A})$ the set $F_A:=\bigcap_{\alpha\in A}F_\alpha$. Further, the image of the map $\tau:\mathcal{I}\to\mathcal{P}^*(\mathcal{A})$ is contained in $\{A\in\mathcal{P}^*(\mathcal{A}):F_A\ne\emptyset\}$, and typically consists of all such sets A that contain the image $\tau(H)$ of a minimal element in $\mathcal{I}$. For example, with $\mathcal{A}=\mathcal{A}_1\sqcup\mathcal{A}_2$ as in (ii), suppose that $X=X_1\times X_2$, and that, for $t=1,2$, $(F_{t\,i})_{i\in\mathcal{I}_t}$ is an open covering of X_t. For each $I\subset\{1,\dots,N_t\}$ define

$$F_{t\,I}:=\bigcap\nolimits_{i\in I}F_{t\,i},\quad F_{t\,\emptyset}:=X_t.$$

Then the product covering $\big(F_{1I_1}\times F_{2I_2}\big)$ of $X_1\times X_2$ is indexed by the pairs $(I_1,I_2)\in\mathcal{I}_1\times\mathcal{I}_2)$ with the product order

$$(I_1,I_2)\le(J_1,J_2)\iff I_1\le J_1,\ I_2\le J_2.$$

The injection $\tau:\mathcal{I}_1\times\mathcal{I}_2\to\mathcal{P}^*(\mathcal{A})$ is given by $(I_1,I_2)\mapsto(I_1\cup I_2)\subset\mathcal{P}^*(\mathcal{A})$. It clearly satisfies conditions (c),(d) above. Note that $\mathcal{I}_1\times\mathcal{I}_2$ contains no elements of the form $(\emptyset,I_2)$ or $(I_1,\emptyset)$,[41] and hence its image by τ contains none of the minimal elements of $\mathcal{P}^*(\mathcal{A})$.

(iv) In the Gromov–Witten context it is appropriate to use semi-additive atlases in situations where different collections of basic charts have the *same* footprint intersection. For example, in the case of products the two pairs $\mathbf{K}_{1i_1}\times\mathbf{K}_{2j_3},\mathbf{K}_{1i_2}\times\mathbf{K}_{2j_4}$ and $\mathbf{K}_{1i_1}\times\mathbf{K}_{2j_4},\mathbf{K}_{1i_2}\times\mathbf{K}_{2j_3}$ have the same footprint intersection, namely

$$F_{1(i_1i_2)}\times F_{2(j_3j_4)}:=\big(F_{1i_1}\cap F_{1i_2}\big)\times\big(F_{2j_3}\cap F_{2j_4}\big).$$

The product atlas has a *single* transition chart $\mathbf{K}_{1(i_1i_2)}\times\mathbf{K}_{2(j_3j_4)}$ with this footprint rather than many different ones. In Remark 6.3.8 and the subsequent discussion we indicate how one might show that in the GW context the product atlas is commensurate (in an appropriate sense) to the additive atlas with the same basic charts. $\Diamond$

REMARK 6.1.4. Property (ii) in Definition 6.1.1 implies that the map

$$\iota_m:\mathcal{I}\mapsto\mathcal{P}^*(\mathcal{M}(\mathcal{I})),\quad J\mapsto m(J),$$

is injective. Further, the greatest lower bound $I\wedge J$ exists iff and only if there is $H\in\mathcal{M}(\mathcal{I})$ such that $H\le I$ and $H\le J$, and in this case $H\le I\wedge J$ iff $H\le I$ and $H\le J$. Hence we have

$$\iota_m(I\wedge J)=\iota_m(I)\cap\iota_m(J)\quad\forall I,J,\ I\wedge J\in\mathcal{I}.\tag{6.1.1}$$

However, as in the case of products, ι_m may not satisfy the condition $\iota_m(I\vee J)=\iota_m(I)\cup\iota_m(J)$. $\Diamond$

We now define the notion of semi-additive atlas. This involves generalizing the definition of coordinate change $\widehat{\Phi}_{IJ}$, since the groups Γ_I,Γ_J will in general no longer be products of groups indexed by the elements of I,J. However, their main characteristic remains unchanged, namely, they are defined by group covering maps $(\rho_{IJ},\rho^\Gamma_{IJ}):(\widetilde{U}_{IJ},\Gamma_J)\to(U_{IJ},\Gamma_I)$ as in Definition 3.1.4, that are compatible in the sense of Remark 3.1.18.

REMARK 6.1.5. Unless explicitly mentioned otherwise, we will assume throughout this section that all charts are smooth. One might attempt similar definitions in the topological case. However, when it comes to the notion of tameable atlas (see §6.3), one would need to check that the corresponding filtration satisfies all the required properties. (The potential problem here is that we use the index condition for smooth coordinate changes in order to establish the filtration condition

[41] With our definitions, if $(F_{t\ i})_{i\in\mathcal{A}_t}$ is the footprint covering of an atlas on X_t, then the products $(F_{1I_1}\times F_{2I_2})_{(I_1,I_2)\in\mathcal{I}_1\times\mathcal{I}_2}$ form the footprint covering of the product atlas, see Definition 6.1.8 below. Pardon [**P**] works with a more general definition of atlas that allows charts with footprints such as $F_{I\emptyset}$ even though such charts may contain no smooth points. Hence his treatment of products does not meet the particular difficulties discussed here, though it has some technicalities of its own.

(iv) in Definition 2.3.6. To deal with this, one might need to rephrase/strengthen condition (iv) in Definition 2.3.22 of good atlas.) ◊

DEFINITION 6.1.6. *Consider a family* $\mathcal{K} := (\mathbf{K}_I, \widehat{\Phi}_{IJ})_{I\le J, I,J\in\mathcal{I}}$ *of (smooth) Kuranishi charts and coordinate changes on* X *whose charts are indexed by a set* $\mathcal{I}$ *that is minimally* $(\mathcal{A},\tau)$*-generated, where the minimal elements in* $\mathcal{I}$ *correspond to the basic charts, and* $\mathcal{A}$ *is related to the footprint covering as follows:*

- *for each* $\alpha\in\mathcal{A}$ *there is a subset* $F_\alpha\subset X$ *such that the chart* $\mathbf{K}_I$ *has footprint* $F_I := F_{\tau(I)}$*, where for* $A\subset\mathcal{A}$ *we define*

$$F_A := \bigcap\nolimits_{\alpha\in A} F_\alpha,\quad A\subset\mathcal{P}^*(\mathcal{A}). \tag{6.1.2}$$

We say that $\mathcal{K}$ *is* **semi-additive** *if in addition the following conditions hold on the charts and coordinate changes.*

- *There are tuples* $(E_\alpha)_{\alpha\in\mathcal{A}}$, $(\Gamma_\alpha)_{\alpha\in\mathcal{A}}$ *where* E_α *is a finite dimensional vector space and* Γ_α *is a finite group that acts on* E_α*. For each subset* $A\subset\mathcal{A}$ *we define* $E_A := \prod_{\alpha\in A} E_\alpha$*, so that* $\Gamma_A := \prod_{\alpha\in A}\Gamma_\alpha$ *acts on* E_A *by the product action.*
- *The map* $\tau:\mathcal{I}\to\mathcal{P}^*(\mathcal{A})$ *satisfies the following conditions with respect to the groups* Γ_I *and obstruction spaces* E_I *of the transition charts* $\mathbf{K}_I$*.*
  - $\Gamma_I := \Gamma_{\tau(I)}$ *for each* I*, and the surjections* $\rho^\Gamma_{IJ}:\Gamma_J\to\Gamma_I$ *in the coordinate changes are given by the projections* $\prod_{\alpha\in\tau(J)}\Gamma_\alpha\to\prod_{\alpha\in\tau(I)}\Gamma_\alpha$ *and hence have kernel* $\prod_{\alpha\in\tau(J)\smallsetminus\tau(I)}\Gamma_\alpha$*; in particular, by definition of coordinate change each group* $\prod_{\alpha\in\tau(J)\smallsetminus\tau(I)}\Gamma_\alpha$ *acts freely on the set* $\widetilde{U}_{IJ}$ *with quotient* Γ_I*-equivariantly isomorphic to* (U_{IJ},Γ_I)*.*
  - $E_I = \prod_{\alpha\in\tau(I)} E_\alpha$ *and* $\widehat{\phi}_{IJ}: E_I\to E_J$ *is the natural inclusion for all* $I\le J$*.*

We say that $\mathcal{K}$ *is a* **(weak) semi-additive atlas** *if in addition the tangent bundle condition and (weak) cocycle condition hold. Such an atlas is* **good** *if in addition the conditions of Definition 2.3.22 hold.*
For short we will denote a semi-additive atlas by the tuple $(\mathbf{K}_I,\widehat{\Phi}_{IJ})_{\mathcal{I},\mathcal{A},\tau}$*.*

EXAMPLE 6.1.7. If $\mathcal{K}$ is an atlas in which τ is induced by an isomorphism $\mathcal{M}(\mathcal{I})\cong\mathcal{A}$, then the above notion of semi-additivity reduces to the notion of additivity in [**MW2**], and is already assumed in the definitions of atlas given here. In this case we say that $\mathcal{K}$ is **standard**. ◊

Our main example is that of products where $\mathcal{A}=\mathcal{A}_1\sqcup\mathcal{A}_2$ as in Example 6.1.3.

DEFINITION 6.1.8. *Let* $(\mathcal{K}_t)_{t=1,2}$ *be semi-additive atlases*

$$\big((\mathbf{K}_I,\widehat{\Phi}_{IJ})_{\mathcal{I}_t,\mathcal{A}_t,\tau_t}\big)_{t=1,2}.$$

The product atlas $\mathcal{K}_1\times\mathcal{K}_2$ *is an atlas on* $X_1\times X_2$ *whose basic charts are products* $\mathbf{K}_{i_1,i_2}$ *of the basic charts in* $\mathcal{K}_1,\mathcal{K}_2$*, where for* $I_t\in\mathcal{I}_t$ *we define*

$$\mathbf{K}_{I_1,I_2} := \big(U_{I_1}\times U_{I_2}, E_{I_1}\times E_{I_2}, \Gamma_{I_1}\times\Gamma_{I_2}, s_{I_1}\times s_{I_2}, \psi_{I_1}\times\psi_{I_2}\big).$$

The indexing set of $\mathcal{K}_1\times\mathcal{K}_2$ *is* $\mathcal{I}=\mathcal{I}_1\times\mathcal{I}_2$ *with the product order*

$$(I_1,I_2)\le(J_1,J_2)\iff I_t\le J_t,\ \ t=1,2.$$

Further $\mathcal{A} := \mathcal{A}_1 \sqcup \mathcal{A}_2$ *and* τ *is given by*

$$\tau:\ \mathcal{M}(\mathcal{I}_1\times\mathcal{I}_2)=\mathcal{M}(\mathcal{I}_1)\sqcup\mathcal{M}(\mathcal{I}_2)\to\mathcal{P}^*(\mathcal{A})=\mathcal{P}^*(\mathcal{A}_1\sqcup\mathcal{A}_2),$$
$$\tau((i_1,i_2)):=\tau_1(i_1)\sqcup\tau_2(i_2),\qquad \text{where } i_t\in\mathcal{M}(\mathcal{I}_t).$$

Thus its transition charts are the products $\mathbf{K}_{I_1,I_2}$, *and we require that the coordinate changes are also given by product maps.*

LEMMA 6.1.9. *The product* $\mathcal{K}_1\times\mathcal{K}_2$ *of two semi-additive (weak) atlases* $\mathcal{K}_1,\mathcal{K}_2$ *is a semi-additive (weak) atlas. Further if* $\mathcal{K}_1,\mathcal{K}_2$ *are good atlases (see Definition* 2.3.22*), then so is* $\mathcal{K}_1\times\mathcal{K}_2$.

PROOF. As in Example 6.1.3, one readily checks that the given tuple $\mathcal{I},\tau,\mathcal{A}$ and footprints satisfy the conditions in Definition 6.1.1 as well as (6.1.2), while the product charts and coordinate changes have the form required by the semi-additivity conditions. Further, if $\mathcal{K}_1,\mathcal{K}_2$ satisfy the cocycle condition, so does the product $\mathcal{K}_1\times\mathcal{K}_2$. This proves the first statement.

If $\mathcal{K}_1,\mathcal{K}_2$ are atlases, then the quotient space $|\mathcal{K}_1\times\mathcal{K}_2|$ equals the product $|\mathcal{K}_1|\times|\mathcal{K}_2|$ as a point set because

$$(I_1,I_2,(x_1,x_2))\sim(J_1,J_2,(y_1,y_2))\iff(I_t,x_t)\sim(J_t,y_t),\ \ t=1,2.$$

Therefore the map

$$\mathrm{Obj}_{\mathcal{K}_1\times\mathcal{K}_2}=\bigsqcup U_{I_1}\times U_{I_2}\to|\mathcal{K}_1|\times|\mathcal{K}_2|$$

descends to a continuous injection $|\mathcal{K}_1\times\mathcal{K}_2|\to|\mathcal{K}_1|\times|\mathcal{K}_2|$. If the $\mathcal{K}_t$ are good, the spaces $|\mathcal{K}_t|$ are Hausdorff, which implies that $|\mathcal{K}_1\times\mathcal{K}_2|$ is also. We must check that $\pi_{\mathcal{K}_1\times\mathcal{K}_2}$ is a homeomorphism when restricted to $U_{I_1}\times U_{I_2}$. Because the topology on $U_{I_1}\times U_{I_2}$ has a basis consisting of product sets $S_1\times S_2$, it suffices to check that for each such set there is an open set $\mathcal{W}\subset|\mathcal{K}_1\times\mathcal{K}_2|$ that intersects $\pi_{\mathcal{K}_1\times\mathcal{K}_2}(U_{I_1}\times U_{I_2})$ in $\pi_{\mathcal{K}_1\times\mathcal{K}_2}(S_1\times S_2)$. But since for $t=1,2$ the atlas $\mathcal{K}_t$ is good, there is an open subset $\mathcal{W}_t\subset|\mathcal{K}_t|$ such that $\mathcal{W}_t\cap\pi_{\mathcal{K}_t}(U_{I_t})=S_t$. Hence we may take $\mathcal{W}$ to be the pullback of $\mathcal{W}_1\times\mathcal{W}_2$ to $|\mathcal{K}_1\times\mathcal{K}_2|$. The other two conditions in the definition of "good", namely the existence of a suitable metric, and the closedness of the sets $\phi_{IJ}(U_{IJ})$, also are preserved under taking products. This completes the proof. □

REMARK 6.1.10. We will see in the next section that for many purposes one can work just as well with semi-additive atlases as with additive atlases. The main difference is that in the semi-additive case the taming construction does not work because these atlases do not have a filtration satisfying condition (iii) in Definition 2.3.6. This implies that the taming construction in Proposition 2.3.17 does not go through. (See also Remarks 2.3.11 and 2.3.19.) Moreover, even if the atlas were tame, the proof in Proposition 2.3.13 that the atlas is therefore good does not go through because the key identity (2.3.7) in Lemma 2.3.12 also fails. However, if we know for other reasons (e.g. from its construction) that the atlas is good, then we show in the next section that it defines a VFC as before. ◊

6.2. From a good semi-additive atlas to the VFC. We now show that every cobordism class of good, oriented, semi-additive atlases $\mathcal{K}$ on X defines a unique homology class $[X]^{vir}_{\mathcal{K}}\in H_*(X;\mathbb{Q})$. As an illustration, we prove the following result. Note that the statement makes sense because the product atlas of Definition 6.1.8 is good by Lemma 6.1.9.

Proposition 6.2.1. *If $\mathcal{K}_t$ is a good oriented, semi-additive Kuranishi atlas on X_t of dimension d_t for $t = 1, 2$, then the fundamental class $[X_1 \times X_2]^{vir}_{\mathcal{K}_1 \times \mathcal{K}_2}$ of the product atlas $\mathcal{K}_1 \times \mathcal{K}_2$ on $X = X_1 \times X_2$ is the product*

$$[X_1]^{vir}_{\mathcal{K}_1} \times [X_2]^{vir}_{\mathcal{K}_2} \in \check{H}_{d_1+d_2}(X_1 \times X_2; \mathbb{Q}).$$

We construct the VFC by the method explained in §2.4. Thus the first step is to construct a reduction as in Definition 2.4.1. First we prove a general result about coverings.

Lemma 6.2.2. *Let $\mathcal{I}$ be a poset that is minimally generated in the sense of Definition 6.1.1, and suppose given a finite open cover of a compact metrizable Hausdorff space $X = \bigcup_{I \in \mathcal{I}} P_I$ such that*

$$P_I \cap P_J \subset P_{I \vee J} \quad \forall I, J \in \mathcal{I}.$$

Then there exists a **cover reduction** $(Z_I)_{I \in \mathcal{I}}$ *where $Z_I \subset X$ is a (possibly empty) open subset satisfying*

(1) *$Z_I \sqsubset P_I$ for all I;*
(2) *if $\overline{Z_I} \cap \overline{Z_J} \neq \emptyset$ then $I \leq J$ or $J \leq I$;*
(3) *$X = \bigcup_I Z_I$.*

Proof. Since X is compact Hausdorff, we may choose precompact open subsets $Q'_I \sqsubset P_I$ that still cover X. We first claim we may enlarge these sets to Q_I with $Q'_I \subset Q_I \sqsubset P_I$ so that $Q_I \cap Q_J \subset Q_{I \vee J}$ for all I, J. For this, we define $Q_I = Q'_I$ if I is minimal, and then define Q_I by induction over $|I| := |m(I)|$ by setting

$$Q_I = Q'_I \cup \bigcup_{S \in \mathcal{S}(I)} \bigcap\nolimits_{j \in S} Q_{I_j},$$

where $\mathcal{S}(I)$ is the set of all collections $S := (I_j)_{j \in S}$ such that $\vee_{j \in S} I_j = I$ and $I_j \neq I$. Note that for each such collection S, the induction hypothesis implies that

$$\bigcap\nolimits_{j \in S} Q_{I_j} \sqsubset \bigcap\nolimits_{j \in S} P_{I_j} \subset P_I.$$

Therefore $Q_I \sqsubset P_I$ since it is a finite union of precompact subsets of P_I.

Repeating this procedure $2M$ times where $M = \max_{I \in \mathcal{I}} |m(I)|$, we obtain families of nested sets

$$Q_I^0 \sqsubset P_I^1 \sqsubset Q_I^1 \sqsubset P_I^2 \sqsubset \ldots \sqsubset Q_I^M := P_I, \tag{6.2.1}$$

such that $P_I^k \cap P_J^k \subset P_{I \vee J}^k$, and $Q_I^k \cap Q_J^k \subset Q_{I \vee J}^k$ for all k. For $I \in \mathcal{I}$, define

$$Z_I := P_I^{|I|} \smallsetminus \bigcup\nolimits_{|J| > |I|} \overline{Q_J^{|I|}}, \qquad |I| := |m(I)|. \tag{6.2.2}$$

These sets are open since they are the complement of a finite union of closed sets in the open set $P_I^{|I|}$. The precompact inclusion $Z_I \sqsubset P_I$ in (i) holds since $P_I^{|I|} \sqsubset P_I$.

To prove the covering property (iii), let $x \in X$ be given. Then we claim that $x \in Z_{I_x}$ for

$$I_x := \bigvee_{I : x \in P_I^{|I|}} I \quad \subset \quad \{1, \ldots, N\}.$$

To see this, denote by S_x the collection of $I \in \mathcal{I}$ such that $x \in P_I^{|I|}$. Then $I_x = \vee_{I \in S_x} I$, which implies $|I| \leq |I_x|$ for all $I \in S_x$. Hence

$$x \in \bigcap\nolimits_{I \in S_x} P_I^{|I|} \subset \bigcap\nolimits_{I \in S_x} P_I^{|I_x|} \subset P_{I_x}^{|I_x|},$$

where the last step holds by assumption on the covering $(P_I^k)_{I\in\mathcal{I}}$. On the other hand, if $x \in \overline{Q_J^{|I_x|}}$ for some J with $|J| > |I_x|$, then $x \in \overline{Q_J^{|I_x|}} \subset P_J^{|J|}$, which contradicts the definition of I_x. Hence $x \in Z_{I_x}$ as claimed.

To prove the intersection property (ii), suppose to the contrary that $x \in \overline{Z_I} \cap \overline{Z_J}$ where $|I| \le |J|$ but $m(I) \smallsetminus m(J) \ne \emptyset$ so that $I \not\le J$. Then $I \vee J > J$ and we have $x \in \overline{Z_I} \cap \overline{Z_J} \subset Q_I^{|I|} \cap Q_J^{|J|} \subset Q_{J\vee I}^{|J|}$, which is impossible because $Q_{J\vee I}^{|J|}$ has been removed from $\overline{Z_J}$. Thus the sets Z_I form a cover reduction. $\square$

REMARK 6.2.3. This is a version of [**MW1**, Lemma 5.3.1], modified to apply to a covering that is indexed by a minimally generated poset $\mathcal{I}$, rather than just by the poset $\mathcal{P}^*(\mathcal{M}(\mathcal{I}))$. $\diamondsuit$

We define the **reduction** of a good semi-positive atlas just as in Definition 2.4.1. Thus it is a collection of open sets $V_I \sqsubset U_I$ such that

$$\iota_{\mathcal{K}}(X) \subset \textstyle\bigcup_{I\in\mathcal{I}} \pi_{\mathcal{K}}(V_I \cap s_I^{-1}(0)); \quad \text{and}$$
$$\pi_{\mathcal{K}}(\overline{V_I}) \cap \pi_{\mathcal{K}}(\overline{V_J}) \ne \emptyset \implies I \le J \text{ or } J \le I.$$

The next result is the semi-additive analog of Proposition 2.4.2.

LEMMA 6.2.4. *Suppose that $\mathcal{K}$ is a good semi-additive atlas with footprint cover $(F_I)_{I\in\mathcal{I}}$, and let $P_I \subset F_I$ be any open sets that cover X and are such that $P_I \cap P_J \subset P_{I\vee J}$. Suppose further that $W_I \subset U_I$ are Γ_I-invariant sets such that $W_I \cap s_I^{-1}(0) = \psi_I^{-1}(P_I)$. Then $\mathcal{K}$ has a reduction $\mathcal{V} := (V_I)_{I\in\mathcal{I}}$ such that $V_I \subset W_I$ for all I. It is unique up to concordance.*

PROOF. Suppose first that the isotropy is trivial. Choose a cover reduction $(Z_I \sqsubset P_I)_{I\in\mathcal{I}}$ of X as in the previous lemma, and then for each I choose an open set $W_I' \sqsubset W_I$ such that $W_I' \cap s_I^{-1}(0) = \psi_I^{-1}(Z_I)$. For each I, let $\mathcal{C}(I) = \{J \in \mathcal{I} : I \le J, \text{ or } J \le I\}$ and then define $Y_I := \bigcup_{J\notin\mathcal{C}(I)} \overline{W_I'} \cap \pi_{\mathcal{K}}^{-1}(\pi_{\mathcal{K}}(\overline{W_J'}))$. Then Y_I is closed because $\pi_{\mathcal{K}} : U_J \to |\mathcal{K}|$ is homeomorphism for each J by goodness. Further, by construction

$$Y_I \cap s_I^{-1}(0) \subset \bigcup_{J\notin\mathcal{C}(I)} \psi_I^{-1}(F_I \cap F_J) = \emptyset.$$

Hence we may choose an open neighbourhood $\mathcal{N}(Y_I)$ of Y_I in $U_I \smallsetminus s_I^{-1}(0)$, and then set $V_I := W_I' \smallsetminus \mathcal{N}(Y_I)$. This has the same zero set of W_I' and satisfies the other conditions by construction. The statement about concordance follows as in [**MW1**].

If the isotropy is nontrivial, the argument is the essentially same except that we construct the corresponding sets $\underline{W}_I \subset \underline{U}_I$ on the intermediate level and then define $W_I := \pi_I^{-1}(\underline{W}_I)$, $\square$

The notions of Kuranishi concordance, cobordism and of orientation bundle extend immediately to the semi-additive case. The only difference is that because the atlases that we now consider cannot be tamed, we must assume from the outset that all atlases under consideration are good. Thus the relevant relation is that of **good cobordism**. We can now proceed to construct representatives for the fundamental class $[X]^{vir}_{\mathcal{K}}$ as in §3.4. Thus, given a reduction $\mathcal{V}$ of a good atlas, we use the method of Proposition 2.4.10 to construct a reduced perturbation $\nu := \big(\nu_I :$

$V_I \to E_I\big)_{I\in\mathcal{I}}$ that satisfies the conditions of Definition 3.4.2 whose zero set

$$\mathbf{Z}^\nu := \bigcup_{I\in\mathcal{I},\gamma\in\Gamma_I} (\mathfrak{s}_I|_{V_I} + \nu)^{-1}(0)$$

can be completed to a wnb groupoid $\widehat{\mathbf{Z}}^\nu$ as in Proposition 3.4.4. This construction is done by induction over $|m(I)|$, which is permissible because $I < J \Longrightarrow |m(I)| < |m(J)|$. It follows that $\widehat{\mathbf{Z}}^\nu$ has a fundamental class that is represented in the rational singular homology of a small neighbourhood in $|\mathcal{K}|$ of the zero set $\iota_{\mathcal{K}}(X)$. Taking a sequence ν_k of admissible perturbations with norm converging to 0, one obtains an element in the Čech homology of X. It is unique because, as in Proposition 2.4.11, it is possible to join any two admissible, precompact, transverse perturbations of $\mathcal{K}$ by an admissible, precompact, transverse perturbation over the product $[0,1]\times\mathcal{K}$. Therefore, finally we obtain the following result.

PROPOSITION 6.2.5. *Every oriented, good, semi-additive atlas $\mathcal{K}$ on X of dimension d determines a Čech homology class $[X]^{vir}_{\mathcal{K}} \in \check{H}_d(X;\mathbb{Q})$ that depends only on the good oriented cobordism class of $\mathcal{K}$.*

Proof of Proposition 6.2.1. Again, let us first suppose that the isotropy is trivial. Since each $\mathcal{K}_i$ is good, we may choose reductions $\mathcal{V}_i$ of $\mathcal{K}_i$ and admissible perturbations $\nu_i : \mathbf{B}_{\mathcal{K}_i}\big|_{\mathcal{V}_i} \to \mathbf{E}_{\mathcal{K}_i}\big|_{\mathcal{V}_i}$. For each chart $\mathbf{K}_{I_i}\times\mathbf{K}_{I_2}$ of the product atlas, define $P_{I_1,I_2} = V_{I_1}\times V_{I_2} \subset U_{I_1}\times U_{I_2}$. Since $P_{I_1,I_2}\cap P_{J_1,J_2} = P_{I_1\vee J_1, I_2\vee J_2}$, we can apply Lemma 6.2.4 to construct a reduction $\mathcal{V} = (V_{I_1,I_2})_{(I_1,I_2)\in\mathcal{I}_{\mathcal{K}_1}\times\mathcal{I}_{\mathcal{K}_2}}$ with $V_{I_1,I_2}\subset P_{I_1,I_2} := V_{I_1}\times V_{I_2}$.[42]

Now define

$$\nu : \mathbf{B}_{\mathcal{K}_1\times\mathcal{K}_2}\big|_{\mathcal{V}} \to \mathbf{E}_{\mathcal{K}_1\times\mathcal{K}_2}\big|_{\mathcal{V}},$$

by setting

$$\nu_{I_1,I_2} : V_{I_1,I_2} \to E_{I_1}\times E_{I_2}, \quad (u_1,u_2) = \big(\nu_1(u_1),\nu_2(u_2)\big).$$

It is easy to check that ν satisfies the necessary conditions. In particular, $\mathfrak{s}|_{\mathcal{V}}+\nu$ is transverse to 0 on each V_{I_1,I_2} with zero set equal to the restriction to V_{I_1,I_2} of the product of the zero sets of ν_{I_1} and ν_{I_2}. It follows that the realization $|\mathbf{Z}^\nu|$ of the zero set is the product $|\mathbf{Z}^{\nu_1}|\times|\mathbf{Z}^{\nu_2}|$. Hence the resulting homology class $[X_1\times X_2]^{vir}_{\mathcal{K}_1\times\mathcal{K}_2}$ in $\check{H}_{d_1+d_2}(X_1\times X_2)$ is the product $[X_1]^{vir}_{\mathcal{K}_1}\times[X_2]^{vir}_{\mathcal{K}_2}$.

If the isotropy is nontrivial, the argument is again the essentially same. By Proposition 3.4.4, the zero set can now be constructed as a wnb groupoid $\widehat{\mathbf{Z}}^\nu$. The inclusions $V_{I_1,I_2}\to V_{I_1}\times V_{I_2}$ induce a functor from $\widehat{\mathbf{Z}}^\nu$ to the product wnb groupoid $\widehat{\mathbf{Z}}^{\nu_1}\times\widehat{\mathbf{Z}}^{\nu_2}$ that takes each local zero set $Z_{I_1,I_2}\subset V_{I_1,I_2}$ diffeomorphically onto the subset $(Z_{I_1}\times Z_{I_2})\cap V_{I_1,I_2}$ of $V_{I_1}\times V_{I_2}$, preserving its weight $\frac{1}{|\Gamma_{I_1}|\,|\Gamma_{I_2}|}$. Because the weighting function on $|\widehat{\mathbf{Z}}^\nu|_{\mathcal{H}}$ is induced by the weights of the components Z_{I_1,I_2}, it follows that this functor is a layered covering in the sense of [**M3**], and hence pushes the fundamental class of $\widehat{\mathbf{Z}}^\nu$ forward to that of the product. □

[42]Note that the sets P_{I_1,I_2} themselves do *not* satisfy the requirements of a reduction: for example if $I_1 \lneq J_1$ and $I_2 \gneq J_2$ then $(V_{I_1}\times V_{I_2})\cap(V_{J_1}\times V_{J_2}) = V_{J_1}\times V_{I_2}$ but (I_1,I_2) is not comparable to (J_1,J_2) in the product order.

6.3. From semi-additive to tameable atlases. In this section we first show that every semi-additive atlas is cobordant to a tameable atlas, and then discuss the relevance of this result to Hamiltonian Floer theory.

Remark 3.1.18 defines the notion of a (smooth) **tameable** atlas that we will call $\mathcal{K}'$ to avoid confusion. The transition charts of such an atlas have the usual indexing set $\mathcal{I}_{\mathcal{K}'} \subset \mathcal{P}^*(\mathcal{M}(\mathcal{I}_{\mathcal{K}'}))$ (where $\mathcal{M}(\mathcal{I}_{\mathcal{K}'})$ is the set of minimal elements in $\mathcal{I}_{\mathcal{K}'}$), and the obstruction spaces $E'_I = \prod_{i\in I} E'_i$ are products. However the isotropy groups $(\Gamma'_I)_{I\in\mathcal{I}_{\mathcal{K}'}}$ are no longer required to be the products of the corresponding basic groups. For example, suppose given a tuple $(\mathcal{I}, \mathcal{A}, \tau)$ with associated spaces and groups E_α, Γ_α as in Definition 6.1.6, where $\mathcal{I} \subset \mathcal{P}^*(\mathcal{M}(\mathcal{I}))$. Then the transition charts might be indexed by the subsets $K \in \mathcal{P}^*(\mathcal{M}(\mathcal{I}))$ and have obstruction spaces $E'_K := \prod_{i\in K}\prod_{\alpha\in\tau(i)} E_\alpha$ as usual, but the groups Γ'_K could have the form $\prod_{\alpha\in\tau(K)}\Gamma_\alpha$. Then E'_K might contain several factors of E_α, while Γ'_K contains only one, acting diagonally on the E_α factors in E'_K, and trivially on the rest. (For an explicit example, see Proposition 6.3.1 below.)

Since the corresponding intermediate atlas with domains ${}^{U'_K}/_{\Gamma'_K}$ is a filtered topological atlas, all taming results apply to these atlases.

Here is the first main result of this section.

PROPOSITION 6.3.1. *Every semi-additive (weak) atlas $\mathcal{K} = (\mathbf{K}_I, \widehat{\Phi}_{IJ})_{\mathcal{I},\mathcal{A},\tau}$ has a canonical extension to a tameable (weak) atlas $\mathcal{K}'$ with the same basic charts as $\mathcal{K}$.*

PROOF. Define

$$\mathcal{I}_{\mathcal{K}'} := \{K \subset \mathcal{P}^*(\mathcal{M}(\mathcal{I})) \mid F_K := \textstyle\bigcap_{i\in K} F_{\tau(i)} \neq \emptyset\},$$

where $\tau : \mathcal{M}(\mathcal{I}) \to \mathcal{A}$ is part of the defining structure for $\mathcal{K}$. By assumption on $\mathcal{I}$, the least upper bound function

$$\ell : \mathcal{P}^*(\mathcal{M}(\mathcal{I})) \to \mathcal{I}, \quad K \mapsto \ell(K) := \mathrm{l.u.b.}(K) = \vee_{i\in K}\, i \tag{6.3.1}$$

defines a map $\ell : \mathcal{I}_{\mathcal{K}'} \to \mathcal{I}$ such that $F_K = F_{\ell(K)}$ for all $K \in \mathcal{I}_{\mathcal{K}'}$. The weak atlas $\mathcal{K}'$ has charts indexed by $K \in \mathcal{I}_{\mathcal{K}'}$.

For simplicity, let us first consider the case when all isotropy groups are trivial. For $i \in \mathcal{M}(\mathcal{I})$ define $E'_i := E_i = \prod_{\alpha\in\tau(i)} E_\alpha$. Then the obstruction space E'_K for the chart $\mathbf{K}'_K$ is $E'_K = \prod_{i\in K} E'_i$. This may also be written as the product

$$E'_K := \prod_{\alpha\in\tau(\ell(K))} E_\alpha^{\;m_{\alpha,K}},$$

where the multiplicities $m_{\alpha,K} \geq 1$ are defined as follows:

$$m_{\alpha,K} = \big|\{i \in K \mid \alpha \in \tau(i)\}\big|.$$

Therefore we can write the elements e'_K of E'_K as tuples $\big(\vec{e}_\alpha\big)_{\alpha\in\tau(\ell(K))}$ where $\vec{e}_\alpha = (e^k_\alpha)_{1\leq k\leq m_{\alpha,K}}$ is an $m_{\alpha,K}$-tuple of vectors in E_α. With this notation the map $\widehat{\phi}'_{KL} : E'_K \to E'_L$ is the obvious inclusion with image equal to

$$\widehat{\phi}'_{KL}(E'_K) = \Big\{\big(\vec{e}_\alpha\big)_{\alpha\in\tau(\ell(L))} \mid e^k_\alpha = 0,\ m_{\alpha,K} < k \leq m_{\alpha,L}\Big\} \subset E'_L. \tag{6.3.2}$$

Further, there is a projection:

$$\sigma_K : E'_K \to E_{\ell(K)}, \quad \big(\vec{e}_\alpha\big)_{\alpha\in\tau(\ell(K))} \mapsto \big(\sigma_\alpha(\vec{e}_\alpha)\big)_{\alpha\in\tau(\ell(K))}, \tag{6.3.3}$$

where we define $\sigma_\alpha(\vec{e}_\alpha) := \sum_{k=1}^{m_{\alpha,K}} e_\alpha^k \in E_\alpha$. This map satisfies the compatibility condition:

$$\sigma_L \circ \widehat{\phi}'_{KL} = \widehat{\phi}_{KL} \circ \sigma_K, \quad E'_K \to E_L.$$

Now define the domains U'_K of the charts of $\mathcal{K}'$ and the section s'_K by setting:

$$U'_K := \Big\{(e'_K, u) \in E'_K \times U_{\ell(K)} \mid s_{\ell(K)}(u) = \sigma_K(e'_K)\Big\}, \quad s'_K(e'_K, u) := e'_K. \tag{6.3.4}$$

Thus, since $\ell(i) = i$, the basic chart $\mathbf{K}_i = (U'_i, E'_i = E_i, s'_i, \psi'_i)$ has domain U'_i consisting of all pairs $(e_i, u) \in E_i \times U_i$ with $s_i(u) = e_i \in E_i$. Thus we can identify U'_i with U_i, and take the footprint map $\psi'_i := \psi_i$ so that $\mathbf{K}'_i \cong \mathbf{K}_i$. On the other hand, if $|K| > 1$ there is a fibration

$$f_K : U'_K \to U_{\ell(K)}, \quad (e'_K, u) \mapsto (\sigma'_K(e'_K), u) \mapsto u \in U_{\ell(K)},$$

where the second map is a diffeomorphism since $\sigma'_K(e'_K) \in E_{\ell(K)} = s_{\ell(K)}(u)$ is uniquely determined by u. Thus the fiber of f_K at $u \in U_K$ is the affine space $\{e \in E'_K \mid \sigma_K(e) = s_{\ell(K)}(u)\}$. Further f_K restricts to a diffeomorphism from $(s'_K)^{-1}(0) = \{(e'_K, u) \in U'_K \mid e'_K = 0\}$ to $s^{-1}_{\ell(K)}(0)$. Therefore the footprint map

$$\psi'_K := \psi_{\ell(K)} \circ f_K : (s'_K)^{-1}(0) \to F_K$$

can be identified with $\psi_{\ell(K)} : (s_{\ell(K)})^{-1}(0) \to F_{\ell(K)} = F_K$ and so induces a homeomorphism $(s'_K)^{-1}(0) \to F_K$. Therefore the chart

$$\mathbf{K}'_K := \big(U'_K, E'_K, s'_K, \psi'_K\big),$$

has the footprint $F_K = F_{\ell(K)}$. If $K \subset L$ define

$$U'_{KL} := \big\{(e'_K, u) \in U'_K \mid \ u \in U_{\ell(K),\ell(L)}\big\} = f_K^{-1}(U_{\ell(K),\ell(L)}), \tag{6.3.5}$$

and define the coordinate change $\phi'_{KL} : U_{KL} \to U_L$ to be the product $\widehat{\phi}'_{KL} \times \phi_{KL}$. This is compatible with the section s'_K and footprint maps, and satisfies the weak cocycle condition. Since $\mathcal{K}'$ is tameable by construction, this completes the proof of (i) in the case when the isotropy is trivial.

In the general case, we use the same definition of E'_K and U'_K, and take $\Gamma'_K := \Gamma_{\ell(K)} = \prod_{\alpha \in \ell(K)} \Gamma_\alpha$. It acts on E'_K in the obvious way, with each factor Γ_α acting diagonally on the tuple $\vec{e}_\alpha \in E_\alpha^{m_\alpha}$, and trivially on the other components. By taking the product of this action with its action on $u \in U_{\ell(K)}$, we obtain an action of $\Gamma'_K = \Gamma_{\ell(K)}$ on the domain U'_K in (6.3.4) that lifts its action on U_K. Hence it is compatible with the footprint maps, the projections σ_K and the fibration f_K. Moreover, because $(s'_K)^{-1}(0)$ can again be identified with $s^{-1}_{\ell(K)}(0)$, the fact that the footprint map induces a homeomorphism ${}^{(s'_K)^{-1}(0)}/_{\Gamma_K} \to F_K = F_{\ell(K)}$ can be deduced from the corresponding result for the chart $\mathbf{K}_{\ell(K)}$. As for coordinate changes, we define $\rho^\Gamma_{KL} : \Gamma'_L \to \Gamma'_K$ to equal $\rho^\Gamma_{\ell(K)\ell(L)}$, and define

$$\widetilde{U}_{KL} = \big\{(\widehat{\phi}_{JK}(e'_K), u) \in U'_L \bigm| e'_K \in E'_K, u \in \widetilde{U}_{\ell(K)\ell(L)}\big\} \tag{6.3.6}$$

Then $f_L : \widetilde{U}_{KL} \to \widetilde{U}_{\ell(K)\ell(L)}$ is surjective, so that $\ker(\rho^\Gamma_{KL})$ acts freely on $\widetilde{U}_{KL}$ with quotient U'_{KL} as defined in (6.3.5). Thus we may take

$$\rho'_{KL} = \mathrm{id}_{E'_K} \times \rho_{\ell(K)\ell(L)} : \widetilde{U}_{KL} \to U_{KL}.$$

It is then easy to see that all the compatibility requirements in Remark 3.1.18 are satisfied.

It remains to check that if $\mathcal{K}$ is an atlas (rather than a weak atlas), then so is $\mathcal{K}'$. This follows because the domains U'_{KL} of the coordinate changes in $\mathcal{K}'$ fiber over the corresponding domains U_{KL} in $\mathcal{K}$. Hence if $\mathcal{K}$ satisfies the cocycle condition then $\mathcal{K}'$ does as well. Hence there is a well defined category $\mathbf{B}_{\mathcal{K}'}$. Further the fibrations (f_K) fit together to provide a functor $\mathfrak{f}: \mathbf{B}_{\mathcal{K}'} \to \mathbf{B}_{\mathcal{K}}$ that intertwines the section and footprint functors. □

REMARK 6.3.2. Definition 4.2.1 spells out a notion of map between (weak) atlases that readily extends to the semi-additive atlases considered here; all we need is a suitable map ℓ between the indexing sets. In particular, the maps $f_K : U'_K \to U_{\ell(K)}$ defined above extend to a map $\mathfrak{f} : \mathcal{K}' \to \mathcal{K}$ with $\ell : \mathcal{I}_{\mathcal{K}'} \to \mathcal{I}$ given by (6.3.1), $f^{\Gamma}_K : \Gamma'_K \to \Gamma_{\ell(K)}$ equal to the identity, and $\widehat{f}_K : E'_K \to E_{\ell(K)}$ equal to the projection σ_K of (6.3.3). ◊

We next show that the atlas $\mathcal{K}'$ constructed above is concordant to $\mathcal{K}$.

PROPOSITION 6.3.3. *Let $\mathcal{K}$ and $\mathcal{K}'$ be as in Proposition* 6.3.1. *Then there is a (weak) cobordism atlas $\mathcal{K}^{01}$ on $[0,1] \times X$ from $\partial^0\mathcal{K}^{01} = \mathcal{K}$ to $\partial^1\mathcal{K}^{01} = \mathcal{K}'$ and with basic charts*

$$[0, \tfrac{2}{3}) \times \mathbf{K}_i, \quad (\tfrac{1}{3}, 1] \times \mathbf{K}_i, \quad i \in \mathcal{M}(\mathcal{I}).$$

PROOF. We define the cobordism (weak) atlas to have an indexing set $\mathcal{I}^{01}$ that is a hybrid of the given order set $\mathcal{I}$ with the standard order set $\mathcal{I}'$ of the atlas $\mathcal{K}'$. Thus

- $\mathcal{M}(\mathcal{I}^{01})$ is the union of two copies of $\mathcal{M}(\mathcal{I})$ that for clarity we denote $\mathcal{M}(\mathcal{I}^0) := \mathcal{M}(\mathcal{I})$ and $\mathcal{M}(\mathcal{I}^1) := \mathcal{M}(\mathcal{I}')$;
- we take $\mathcal{A}^{01} = \mathcal{A} \sqcup \mathcal{M}(\mathcal{I}^1)$, and
- we define

$$\begin{aligned} \tau^{01} : \mathcal{M}(\mathcal{I}^0) \sqcup \mathcal{M}(\mathcal{I}^1) &\to \mathcal{P}^*(\mathcal{A}^{01}) \subset \mathcal{P}(\mathcal{A}) \times \mathcal{P}(\mathcal{M}(\mathcal{I}^1)), \\ (I, K) &\mapsto (\tau(I), m(K)) \end{aligned}$$

The transition charts are of three types:

- $[0, \frac{2}{3}) \times \mathbf{K}_{(I,\emptyset)}$ where $\mathbf{K}_{(I,\emptyset)} = \mathbf{K}_I$ is a transition chart in $\mathcal{K}$;
- $(\frac{1}{3}, 1] \times \mathbf{K}_{(\emptyset,K)}$ where $\mathbf{K}_{(\emptyset,K)} = \mathbf{K}_K$ is a transition chart in $\mathcal{K}'$;
- "transition" charts $(\frac{1}{3}, \frac{2}{3}) \times \mathbf{K}_{(I,K)}$ where $\mathbf{K}_{(I,K)}$ is a chart on X with footprint $F_{I\vee\ell(K)}$, obstruction space $E_I \times E'_K$, group $\Gamma_{I,K} = \Gamma_{I\vee\ell(K)}$, and domain

$$U_{(I,K)} = \big\{(e_I, e'_K, u) \;\big|\; u \in U_{I\vee\ell(K)}, \widehat{\phi}_{I(I\vee\ell(K))}(e_I) + \sigma_K(e'_K) = s_{I\vee\ell(K)}(u)\big\}.$$

The section $s_{(I,K)}$ is given by $(e_I, e'_K, u) \mapsto (e_I, e'_K)$.

If the isotropy groups are trivial, then the coordinate changes for the cobordism charts involve the obvious inclusions on the obstruction spaces together with the following maps on the domains:

$$\begin{aligned} &(\tfrac{1}{3}, \tfrac{2}{3}) \times U_{(I,\emptyset)} \to (\tfrac{1}{3}, \tfrac{2}{3}) \times U_{(J,K)} : (t, u) \mapsto \big(t, \widehat{\phi}_{IJ}(s_I(u)), 0, \phi_{I(J\vee\ell(K))}(u)\big), \\ &(\tfrac{1}{3}, \tfrac{2}{3}) \times U_{(\emptyset,K)} \to (\tfrac{1}{3}, \tfrac{2}{3}) \times U_{(J,L)} : (t, 0, e'_K, u) \mapsto \big(t, 0, \widehat{\phi}_{KL}(e'_K), \phi_{\ell(K)(J\vee\ell(L))}(u)\big), \\ &(\tfrac{1}{3}, \tfrac{2}{3}) \times U_{(I,K)} \to (\tfrac{1}{3}, \tfrac{2}{3}) \times U_{(J,L)} : (t, e_I, e'_K, u) \mapsto \\ &\qquad\qquad \big(t, \widehat{\phi}_{IJ}(e_I), \widehat{\phi}'_{KL}(e'_K), \phi_{(I\vee\ell(K))\ (J\vee\ell(L))}(u)\big). \end{aligned}$$

It is straightforward to check that the cocycle condition holds, so that this is a cobordism atlas.

If the isotropy is nontrivial, then we define the domains

$$(\tfrac{1}{3}, \tfrac{2}{3}) \times \widetilde{U}_{(I,K)(J,L)} \subset (\tfrac{1}{3}, \tfrac{2}{3}) \times U_{(J,L)}$$

of the coordinate changes as in (6.3.6), with elements

$$\Big\{\big(t, \widehat{\phi}_{IJ}(e_I), \widehat{\phi}'_{KL}(e'_K), u\big) \mid e_I \in E_I, e'_K \in E'_K, u \in \widetilde{U}_{(I\vee\ell(K))\,(J\vee\ell(L))}, t \in (\tfrac{1}{3}, \tfrac{2}{3})\Big\},$$

where $\widetilde{U}_{(I\vee\ell(K))\,(J\vee\ell(L))}$ is the corresponding domain in the atlas $\mathcal{K}$. We must check that $\widetilde{U}_{(I,K)(J,L)}$ supports a free action of the kernel of the group homomorphism $\Gamma_{(J,L)} \to \Gamma_{(I,K)}$. But $\Gamma_{(J,L)} = \Gamma_{J\cup\ell(L)}$ is the isotropy group for the chart $\mathbf{K}_{J\vee\ell(L)}$ in $\mathcal{K}$. Hence this kernel acts freely on $\widetilde{U}_{(I,K)(J,L)}$ because it does on the corresponding set $\widetilde{U}_{(I\vee\ell(K))\,(J\vee\ell(L))}$ in $\mathcal{K}$. The rest of the details are straightforward, and are left to the interested reader. □

COROLLARY 6.3.4. *If both $\mathcal{K}$ and the cobordism $\mathcal{K}^{01}$ constructed above are good, oriented atlases, then $[X]^{vir}_{\mathcal{K}} = [X]^{vir}_{\mathcal{K}'}$.*

PROOF. This follows by the arguments that establish Proposition 6.2.5. □

REMARK 6.3.5. (i) The proof of Proposition 6.3.3 builds a cobordism atlas over $[0,1] \times X$ that has a natural restriction over $\{\frac{1}{2}\} \times X$ to an atlas that contains both $\mathcal{K}$ and $\mathcal{K}'$. Hence these two atlases are directly commensurate. Conversely, given two standard atlases that are directly commensurate (or, more generally, commensurate) in the sense of Definition 4.1.4, one can build a cobordism atlas on $[0,1] \times X$ as above and hence deduce that the two atlases are concordant.

(ii) Formally, when the isotropy is trivial the construction of $\mathcal{K}'$ from $\mathcal{K}$ given above is very similar to the construction in Proposition 5.2.3 that builds an atlas from a nontrivial bundle over a manifold. Indeed, if the basic charts of $\mathcal{K}$ are GW charts, the corresponding standard GW atlas with these basic charts can be mapped into $\mathcal{K}'$ by an open embedding in the sense of Definition 4.2.1; the only real difference between the constructions is that in the present context we include the full fiber of f_I into the domain U_I. Just after Remark 5.2.4 we sketched how to obtain a (standard) atlas in the corresponding orbifold case. This construction was possible because in this case the isotropy is determined by its action on the orbifold E so that we could use the results of §5.1 to build the needed transition charts. We explain briefly below how in the Gromov–Witten situation it is also possible to promote the tameable atlas $\mathcal{K}'$ to a standard atlas. For general atlases, it is not immediately clear how we might do this. However, this does not matter since all our constructions work for tameable atlases just as well as they do for standard atlases. ◊

LEMMA 6.3.6. *Suppose that $\mathcal{K}$ is a good semi-additive atlas with trivial isotropy. Then the atlas $\mathcal{K}'$ constructed in Proposition 6.3.1 and the cobordism atlas $\mathcal{K}^{01}$ constructed in Proposition 6.3.3 are also good.*

PROOF. Define the vector space $E'_{\mathcal{M}} := \prod_{i\in\mathcal{M}(\mathcal{I})} E'_i$, and consider the product category $E'_{\mathcal{M}} \times \mathbf{B}_{\mathcal{K}}$ with objects $\bigsqcup_K E'_{\mathcal{M}} \times U'_K$ and morphisms $\mathrm{id}_{E'_{\mathcal{M}}} \times \mathrm{Mor}_{\mathcal{B}'_{\mathcal{K}}}$. Then there is an injective functor $\iota_{\mathcal{M}} : \mathcal{K}' \to E'_{\mathcal{M}} \times \mathcal{K}$ defined on objects by

$$\iota_{\mathcal{M}} :\ (K, (e'_K, u))) \mapsto \big(\iota^E_K(e'_K),\ f_K(K, (e'_K, u))\big),$$

that intertwines the coordinate change maps ϕ'_{KL} of $\mathcal{K}'$ with the products $\mathrm{id} \times \phi_{\ell(K)\ell(L)}$. It follows that $\iota_{\mathcal{M}}$ descends to give continuous injections

$$|\iota_{\mathcal{M}}| : |\mathcal{K}'| \to |E'_{\mathcal{M}} \times \mathbf{B}_{\mathcal{K}}| \to E'_{\mathcal{M}} \times |\mathcal{K}|,$$

where the last space has the product topology. Then the composite

$$|\iota_{\mathcal{M}}| \circ \pi_{\mathcal{K}'} : U'_I \to E'_{\mathcal{M}} \times |\mathcal{K}|$$

is a homeomorphism to its image, which implies that $\pi'_{\mathcal{K}}$ is. The Hausdorff and metrizability conditions in Definition 2.3.22 follow similarly, while $\operatorname{im} \phi'_{IJ}$ is closed in U'_J since it fibers over $\operatorname{im} \phi_{IJ}$ with fibers that are closed in E'_J. The proof that $\mathcal{K}^{01}$ is good is similar. □

REMARK 6.3.7. If $\mathcal{K}$ has nontrivial isotropy, then the above argument extends to show there is a continuous map $|\mathcal{K}'| \to {}^{E'_{\mathcal{M}}}/_{\Gamma'_{\mathcal{M}}} \times |\mathcal{K}|$, where $\Gamma'_{\mathcal{M}} = \prod_{\alpha \in \Gamma} \Gamma_\alpha$ acts diagonally on the factors of E_α in $E'_{\mathcal{M}}$. However, in general this map is no longer injective and so does not quite give what we need. When $\mathcal{K}$ is good, one should nevertheless be able to prove directly that the atlases $\mathcal{K}'$ and $\mathcal{K}^{01}$ constructed above satisfies all the conditions in Definition 2.3.22 except perhaps for metrizability. The problem here is that we assumed so little about the metric topology $(|\mathcal{K}|, d)$ that it is not clear how to proceed. However, the original proof of metrizability in Lemma 2.3.20 uses the fact that the atlas under consideration is "preshrunk". This implies that its domains can be coherently compactified, so that $|\mathcal{K}|$ injects into a compact space $|\overline{\mathcal{K}}|$ that can be directly shown to be metrizable. The good semi-additive atlases that we encounter in practice are constructed from preshrunk tame atlases, for example, by taking products. Hence this argument should adapt to prove the analog of Lemma 6.3.6 in all cases of interest. Since we do not fully develop any such applications here, we will leave this point for now. ◊

Propositions 6.3.1 and 6.3.3 describe abstract constructions that apply to all semi-additive atlases. We now explain how one can do better in the Gromov–Witten context, promoting the atlas $\mathcal{K}'$ from a tameable atlas to a standard atlas $\mathcal{K}''$, because in this case the isotropy groups are controlled explicitly via the added marked points $\mathbf{w}$.

REMARK 6.3.8. In §4.1 we gave a recipe for constructing basic and transition charts using stabilizations that are determined by the choice of a slicing manifold $Q_i \subset M$ for each basic chart $\mathbf{K}_i$. The isotropy group Γ_i acts on the tuples $(\vec{e}_i, [\Sigma, \mathbf{w}, \mathbf{z}, f])$ in U_i by permuting the corresponding admissibly ordered set $\mathbf{w}_i$ of marked points and also the components $\vec{e}_i$ of the obstruction space $E_i := \prod_{\gamma \in \Gamma_i} E_i^0$. The elements of the transition chart $\mathbf{K}_K$ contain tuples $\underline{\vec{e}} := (\vec{e}_i)_{i \in K}, \underline{\mathbf{w}} := (\mathbf{w}_i)_{i \in K}$ of such elements that are again permuted by the components of the groups $\Gamma_K := \prod_{i \in K} \Gamma_i$. The construction of a standard atlas treats the extra data E, Q for each basic chart as distinct. For example, if we choose the same obstruction space $E_1 = E_2$ and slicing manifold Q for two different basic charts $\mathbf{K}_1, \mathbf{K}_2$ and if the two groups Γ_1, Γ_2 are isomorphic, then the standard construction adds two copies of the marked points $\mathbf{w}_1, \mathbf{w}_2$ (agreeing modulo admissible labellings; see the discussion after (4.1.3)) to the elements in the domain of the transition chart U_{12} as well as two copies $E_{12} = E_1 \times E_2$ of the obstruction space thus giving rise to an action of $\Gamma_1 \times \Gamma_2$ on the domain. However, we could instead simply add one set of marked points and take the group to be Γ_1 with the diagonal action on $E_{12} = E_1 \times E_2$, thereby getting a simplified chart whose domain embeds as an open subset of the usual transition chart $\mathbf{K}_{12}$. Note that this simplification of the structure of the chart does not affect which stable maps $[\Sigma, \underline{\mathbf{w}}, \mathbf{z}, f]$ lie in the domain of this chart in any essential way: the map f satisfies the equation $\overline{\partial}_J f = \lambda(\underline{\vec{e}})|_{\mathrm{graph} f}$ where the map λ sums over the action of the relevant group Γ_I over all components of E_I, and one can check that having repeated copies of one of the components E_i, Γ_i does not increase the image of λ.

We explain below how the tameable atlas $\mathcal{K}'$ constructed in Proposition 6.3.1 can similarly be considered as a simplified version of the standard GW atlas with the same basic charts and obstruction spaces. $\diamondsuit$

The variations on the construction described in Remark 6.3.8 above, though possible, are not very useful when the moduli spaces X consist of closed curves, since in this case one can always build a standard atlas. However, as we now explain they are relevant in the context of **Hamiltonian Floer theory**.[43]

REMARK 6.3.9. In this situation we have a family of (noncompact) moduli spaces $X(p, r)$ consisting of unbroken trajectories from p to r (where p, r are periodic orbits of the Hamiltonian flow) whose boundary decomposes into products of the form $X(p, q_1) \times \cdots \times X(q_k, r)$ consisting of trajectories that are broken at k interior orbits $q_1, \ldots, q_k$. For example if the spaces $X(p, q), X(q, r)$ are compact, then $X(p, r)$ will have a boundary component given by the product $X(p, q) \times X(q, r)$. To be consistent with our approach to cobordism, we will thicken up $X(p, r)$ near its boundary by adding appropriate collars (in general, cornered). In particular, we will assume that this component of the boundary of $X(p, r)$ has a neighborhood that can be identified with $[0, \varepsilon) \times X(p, q) \times X(q, r)$. In order to define a chain complex one needs to build a coherent family of atlases, reductions, and sections. By induction, one can assume that atlases $\mathcal{K}_{p,q}$ for $X(p, q)$ and $\mathcal{K}_{q,r}$ for $X(q, r)$ are already constructed, and need to extend the product $\mathcal{K}_{p,q} \times \mathcal{K}_{q,r}$ to a good atlas over $X(p, r)$. Once this is done, the arguments in Proposition 6.2.1 would allow us to extend a product perturbation section $\nu_{p,q} \times \nu_{q,r}$ over the interior of the atlas.

Given a basic chart $\mathbf{K}_1$ at $\tau_1 \in X(p, q)$ and $\mathbf{K}_2$ at $\tau_2 \in X(q, r)$, the elements of the domain $U_1 \times U_2$ of the product chart $\mathbf{K}_1 \times \mathbf{K}_2$ are tuples whose map component $f = (f_1, f_2)$ is a broken trajectory from p to r with first component f_1 (going from p to q) stabilized by points in $f_1^{-1}(Q_1)$ and second component f_2(going from q to r) stabilized by points in $f_2^{-1}(Q_2)$. By gluing at q, we can build a chart with domain $[0, \varepsilon) \times U_1 \times U_2$ that models a neighbourhood of (τ_1, τ_2) in $\overline{X}(p, r)$. Note that this chart has the essentially the same form as a basic GW chart that would be constructed for trajectories going straight from p to r; the only difference is that the slicing manifold is the union $Q_1 \cup Q_2$ of two submanifolds of the ambient manifold M which we do not assume to be disjoint. However, because the associated marked points $\mathbf{w}_t := f^{-1}(Q_t)$ belong to distinct parts of the domain of the curve it is not hard to extend the definition to allow this variation. Therefore we can use the GW construction to build a standard atlas $\mathcal{K}_{p,r}$ over $X(p, r)$ that restricts over $[0, \varepsilon) \times X(p, q) \times X(q, r)$ to $[0, \varepsilon) \times \mathcal{K}''$, where $\mathcal{K}''$ is the standard atlas over $X(p, q) \times X(q, r)$ with these product basic charts. On the other hand, the product $[0, \varepsilon) \times \mathcal{K}_{p,q} \times \mathcal{K}_{q,r}$

[43] The general set-up is described and analyzed in Pardon [**P**], and we do not attempt a complete treatment here. However, what we have done above is enough to deal formally with trajectories that are once broken; to deal with more breaks one would need to introduce a (collared) corner structure.

is another (semi-additive) atlas over the neighbourhood $[0,\varepsilon) \times X(p,q) \times X(q,r)$, which, by the inductive hypothesis and the construction in Proposition 6.2.1 already has a reduction and a perturbation section. To be able to extend these structures over $\mathcal{K}_{p,r}$ it suffices to check that the product atlas $\mathcal{K}_{p,q} \times \mathcal{K}_{q,r}$ is concordant to the standard atlas $\mathcal{K}''$. Note that we do need the concordance to be good (in particular metrizable) in order to be able to apply the results about the extension of perturbation sections; see Proposition 2.4.11 and Remark 6.3.7.

If the isotropy is trivial, the arguments we gave above prove what we need. Indeed the construction of $\mathcal{K}'$ from the product semi-additive $\mathcal{K}$ given in Proposition 6.3.1 is mirrored in the GW set up: the only difference in the constructions is that we usually do not include the whole fiber of f_K into the chart U'_K. Thus the category of the GW atlas embeds as an open subcategory of $\mathbf{B}_{\mathcal{K}'}$. Hence, if we similarly restrict the domains of the transition charts over $(\frac{1}{3}, \frac{2}{3})$ we can define the cobordism as in Proposition 6.3.3. Moreover, it is good by Lemma 6.3.6.

In the case of nontrivial isotropy, the two atlases $\mathcal{K}'$ and $\mathcal{K}''$ are different because the chart with label K has isotropy group $\Gamma'_K := \prod_{\alpha \in \ell(K)} \Gamma_\alpha$ in the first case and $\Gamma''_K := \prod_{i\in K} \Gamma_i = \prod_{\alpha\in\ell(K)} \Gamma_\alpha^{m_{\alpha,K}}$ in the second. As in Remark 6.3.8 one can obtain $\mathcal{K}'$ from $\mathcal{K}''$ by simplifying the charts in $\mathcal{K}''$. (This can be accomplished by either by forgetting all repetitions of the added marked points $\mathbf{w}_\alpha$, so that the points in the domain only contain one such set instead of $m_{\alpha,K}$ of them, or, equivalently, by restricting to those elements in the domain U''_K of the standard GW chart for which the repetitions of $\mathbf{w}_\alpha$ are identical (rather than being identical modulo order).) However, more relevant here is the fact that one can build a concordance from the product to the standard atlas $\mathcal{K}''$ much as in Proposition 6.3.3, using the indexing set $\mathcal{I}^{01}$ and function τ^{01} described there, but this time using the set $\tau^{01}\big((I,K)\big)$ to define the group as well as the obstruction space for the transition chart $\mathbf{K}_{(I,K)}$. Thus $\Gamma''_{(I,K)} = \Gamma_I \times \Gamma''_K$ instead of $\Gamma_{I\vee\ell(K)}$. To make this possible the tuples of added marked points in the elements of the domain $U''_{(I,K)}$ should also be indexed by the elements of $\tau^{01}\big((I,K)\big)$. One would then have to check directly that the concordance can be constructed to be good. As suggested in Remark 6.3.7, this should be possible if one assumes that the initial atlases $\mathcal{K}_{p,q}, \mathcal{K}_{q,r}$ are preshrunk. Further details are left for later. $\Diamond$

Acknowledgements I wish to thank Jingchen Niu, Robert Castellano and Yasha Savelyev for making some useful comments on earlier drafts of these notes, and the referee for a careful reading. I also benefitted by conversations with John Pardon, Dominic Joyce and Eugene Lerman about orbifolds. Finally I wish to thank Katrin Wehrheim for contributing so many ideas to this collaboration and patiently helping to clarify many arguments. Sections §2 and §3 of these notes are taken very directly from the joint papers with Katrin Wehrheim. However I am solely responsible for the exposition of these results as well as for the rest of these notes. I wish to thank the Simons Center for Geometry and Physics for its hospitality during Spring 2014 while these notes were being prepared.

References

[Bbk] N. Bourbaki, *General topology. Chapters 1–4*, Elements of Mathematics (Berlin), Springer-Verlag, Berlin, 1998. Translated from the French; Reprint of the 1989 English translation. MR1726779

[C1] R. Castellano, Smoothness of Kuranishi atlases on Gromov–Witten moduli spaces, arXiv:1511.04350.

[C2] R. Castellano, Genus zero Gromov-Witten axioms via Kuranishi atlases, arXiv:1601.04048.

[FF] M. Farajzadeh Tehrani and K. Fukaya, Gromov–Witten theory via Kuranishi Structures, this volume.

[FO] K. Fukaya and K. Ono, *Arnold conjecture and Gromov-Witten invariant*, Topology **38** (1999), no. 5, 933–1048, DOI 10.1016/S0040-9383(98)00042-1. MR1688434

[FOOO] K. Fukaya, Y.-G. Oh, H. Ohta, and K. Ono, *Lagrangian intersection Floer theory: anomaly and obstruction. Part II*, AMS/IP Studies in Advanced Mathematics, vol. 46, American Mathematical Society, Providence, RI; International Press, Somerville, MA, 2009. MR2548482

[FOOO1] K. Fukaya, Y.-G. Oh, H. Ohta, and K. Ono, Technical detail on Kuranishi structure and Virtual Fundamental Chain, arXiv:1209.4410.

[FOOO2] K. Fukaya, Y.-G. Oh, H. Ohta, and K. Ono, Kuranishi structure, Psuedoholomorphic curve and and Virtual Fundamental Chain: Part I, arXiv:1503.07631.

[H] H. Hofer, *Polyfolds and Fredholm theory*, Lectures on geometry, Clay Lect. Notes, Oxford Univ. Press, Oxford, 2017, pp. 87–158. MR3676594

[HWZ1] H. Hofer, K. Wysocki, and E. Zehnder, *A general Fredholm theory. I. A splicing-based differential geometry*, J. Eur. Math. Soc. (JEMS) **9** (2007), no. 4, 841–876, DOI 10.4171/JEMS/99. MR2341834

[HWZ2] H. Hofer, K. Wysocki, and E. Zehnder, *A general Fredholm theory. II. Implicit function theorems*, Geom. Funct. Anal. **19** (2009), no. 1, 206–293, DOI 10.1007/s00039-009-0715-x. MR2507223

[HWZ3] H. Hofer, K. Wysocki, and E. Zehnder, *A general Fredholm theory. III. Fredholm functors and polyfolds*, Geom. Topol. **13** (2009), no. 4, 2279–2387, DOI 10.2140/gt.2009.13.2279. MR2515707

[HWZ4] H. Hofer, K. Wysocki, and E. Zehnder, *Applications of polyfold theory I: The polyfolds of Gromov-Witten theory*, Mem. Amer. Math. Soc. **248** (2017), no. 1179, v+218, DOI 10.1090/memo/1179. MR3683060

[HWZ6] H. Hofer, K. Wysocki, and E. Zehnder, Deligne-Mumford type spaces with a View Towards Symplectic Field Theory, lecture notes in preparation.

[HWZ7] H. Hofer, K. Wysocki, and E. Zehnder, Polyfold and Fredholm theory, arXiv:1707.08941.

[J14] D. Joyce, A new definition of Kuranishi space, arXiv:1409.6908.

[J15] D. Joyce, Kuranishi spaces as a 2-category, this volume.

[L10] E. Lerman, *Orbifolds as stacks?*, Enseign. Math. (2) **56** (2010), no. 3-4, 315–363, DOI 10.4171/LEM/56-3-4. MR2778793

[M2] D. McDuff, *Almost complex structures on $S^2 \times S^2$*, Duke Math. J. **101** (2000), no. 1, 135–177, DOI 10.1215/S0012-7094-00-10116-0. MR1733733

[M3] D. McDuff, *Groupoids, branched manifolds and multisections*, J. Symplectic Geom. **4** (2006), no. 3, 259–315. MR2314215

[M4] D. McDuff, Smooth Kuranishi atlases, Lecture at Mathematical Congress of the Americas, Guanajuato, Aug 2013, https://math.barnard.edu/sites/default/files/mexsmoothlectv2.pdf

[M5] D. McDuff, Lectures on Kuranishi atlases, videotaped at SCGP Workshop, March 2014. Lecture 1: http://media.scgp.stonybrook.edu/presentations/20140317_McDuff.pdf Lecture 2: http://media.scgp.stonybrook.edu/presentations/20140318_McDuff.pdf

[M6] D. McDuff, *Strict orbifold atlases and weighted branched manifolds*, J. Symplectic Geom. **15** (2017), no. 2, 507–540, DOI 10.4310/JSG.2017.v15.n2.a3. MR3683790

[M7] D. McDuff, Constructing the virtual fundamental class of a Kuranishi atlas, arXiv:1708.01127, accepted by *Algebraic Geometry and Topology*

[MS] D. McDuff and D. Salamon, *J-holomorphic curves and symplectic topology*, American Mathematical Society Colloquium Publications, vol. 52, American Mathematical Society, Providence, RI, 2004. MR2045629

[MW0] D. McDuff and K. Wehrheim, *Smooth Kuranishi atlases with isotropy*, Geom. Topol. **21** (2017), no. 5, 2725–2809, DOI 10.2140/gt.2017.21.2725. MR3687107

[MW1] D. McDuff and K. Wehrheim, *The topology of Kuranishi atlases*, Proc. Lond. Math. Soc. (3) **115** (2017), no. 2, 221–292, DOI 10.1112/plms.12032. MR3684105

[MW2] D. McDuff and K. Wehrheim, *The fundamental class of smooth Kuranishi atlases with trivial isotropy*, J. Topol. Anal. **10** (2018), no. 1, 71–243, DOI 10.1142/S1793525318500048. MR3737511

[MW3] D. McDuff and K. Wehrheim, *Smooth Kuranishi atlases with isotropy*, Geom. Topol. **21** (2017), no. 5, 2725–2809, DOI 10.2140/gt.2017.21.2725. MR3687107

[MW4] D. McDuff and K. Wehrheim, Finite dimensional models of polyfold-Fredholm sections, work in progress.

[MW5] D. McDuff and K. Wehrheim, Constructing Kuranishi atlases: the genus zero case, in preparation.

[Mi] J. Milnor, *Collected papers of John Milnor. IV*, American Mathematical Society, Providence, RI, 2009. Homotopy, homology and manifolds; Edited by John McCleary. MR2590677

[Mo] I. Moerdijk, *Orbifolds as groupoids: an introduction*, Orbifolds in mathematics and physics (Madison, WI, 2001), Contemp. Math., vol. 310, Amer. Math. Soc., Providence, RI, 2002, pp. 205–222, DOI 10.1090/conm/310/05405. MR1950948

[P] J. Pardon, *An algebraic approach to virtual fundamental cycles on moduli spaces of pseudo-holomorphic curves*, Geom. Topol. **20** (2016), no. 2, 779–1034, DOI 10.2140/gt.2016.20.779. MR3493097

[RS] J. W. Robbin and D. A. Salamon, *A construction of the Deligne-Mumford orbifold*, J. Eur. Math. Soc. (JEMS) **8** (2006), no. 4, 611–699, DOI 10.4171/JEMS/101. MR2262197

[TW] H.-H. Tseng and D. Wang, Seidel representation for symplectic orbifolds, arXiv:1207.4246.

[W] A. Weinstein, *The volume of a differentiable stack*, Lett. Math. Phys. **90** (2009), no. 1-3, 353–371, DOI 10.1007/s11005-009-0343-2. MR2565044

[Y14] D. Yang, *A Choice-independent Theory of Kuranishi Structures and the Polyfold–Kuranishi Correspondence*, ProQuest LLC, Ann Arbor, MI, 2014. Thesis (Ph.D.)–New York University. MR3218117

Department of Mathematics, Barnard College, Columbia University
Email address: dusa@math.columbia.edu

Gromov-Witten Theory via Kuranishi Structures

Mohammad F. Tehrani and Kenji Fukaya

Preface

In the past few decades, moduli spaces have been a major tool to study families of geometric objects and to define geometric invariants of various kind for these spaces. Among various moduli spaces that have been extensively studied, moduli spaces of pseudoholomorphic maps have been a corner stone of various subjects in symplectic geometry and string theory. For example, through these moduli spaces we can define Gromov-Witten invariants; these are rational numbers that, in certain situations, count pseudoholomorphic curves meeting prescribed conditions in a given symplectic manifold.

In physics, more specifically in topological sigma models, GW invariants were first studied by Witten [**55**]. Conjectural predictions of mirror symmetry [**5**] relating the GW invariants of a symplectic manifold on one side, and the solutions of Picard-Fuchs equation over some "mirror" family of complex manifolds on the other side, raised the efforts and interests to rigorously define these numbers mathematically. Naively, GW invariants are obtained by integrating cohomology classes against the fundamental class of the moduli space; equally, they are given by intersecting cycles on the moduli space.

However, a priori, every moduli space of pseudoholomorphic maps of some fixed topological type does not carry the structure of a (compact) differentiable orbifold in general; this is mainly due to the transversality issues of the corresponding Fredholm operators. Nevertheless, one would like to at least associate a "virtual" fundamental class to the moduli space in question.

In past twenty years, several different[1] approaches to the construction of virtual fundamental class have emerged, each of which has its own advantages, drawbacks, and motivations. Among all proposed analytical/topological approaches, we can mention the works of Li-Tian [**35**], Fukaya-Ono [**20**], Siebert [**51**], Hofer [**28**], and many other recent works.

Construction of virtual fundamental class for moduli spaces via Kuranishi models is a traditional technique for describing the local structure of moduli spaces cut out by non-linear equations whose linearization is Fredholm. First in the "semi-positive" case, the transversality issue was handled by Ruan-Tian [**48**] via global

The authors were partially supported by NSF Grant No. 1406423 and Simons Collaboration on Homological Mirror Symmetry.

[1] But to some extent similar.

inhomogeneous perturbations of the Cauchy-Riemann equation. The Kuranishi method can be regarded as a local and multi-valued version of the perturbation method.

A more elaborate version of Kuranishi structure is used by Fukaya, Oh, Ohta, and Ono [**17**] over moduli spaces of pseudoholomorphic discs and strips to define and study the Lagrangian intersection Floer homology of two Lagrangian submanifolds. Such moduli spaces can also be used to define GW-type invariants counting pseudoholomorphic discs (and higher genus analogues of that) with boundary on Lagrangian submanifolds.

In this manuscript, we review the construction of GW virtual fundamental class via Kuranishi structures for moduli spaces of pseudoholomorphic maps defined on closed Riemann surfaces. We consider constraints coming from the ambient space and Deligne-Mumford moduli, called primary insertions, as well as intrinsic classes such as ψ-classes and Hodge classes. Readers interested in the Floer theoretic aspects of Kuranishi structures should consult [**17**].

The first part of this article, i.e. Sections 2-4, is about abstract and topological aspects of Kuranishi structures and is fairly self-contained. The second part, i.e. Sections 5 and 6, is about construction of a natural class of Kuranishi structures for moduli spaces of interest. In part 2, some of the main analytical steps, such as the gluing theorem, are stated without proof. For the proof we refer to [**16**] and [**14**].

Disclaimer: The construction provided in this article is a repackaging of the relevant materials in [**14–16**, **20**], and the seminar series delivered by Fukaya at the Simons center during Spring semester of 2014, which are available at SCGP's video portal [**13**]. It also relies on our discussions in the past two years to simplify or rephrase some of the statements and definitions. For example, we use "dimensionally graded systems" in place of "good coordinate systems" to simplify[2] the notation and avoid complications of working with partially ordered sets. Our main priority has been to collect all the concepts and ideas which are relevant to Gromov-Witten theory in the work of FOOO in one relatively short article. There is no doubt that some of the techniques, details, and statements in this note do not directly come from the original article of Fukaya and Ono [**20**] or the book of Fukaya-Oh-Ohta-Ono [**17**]. In [**16**], motivated by feedback to and questions about [**20**] and [**17**], FOOO provided more details and further arguments to explain their construction in [**20**] and [**17**]. We have adapted some of this material here. It is beyond the scope and goal of this text to cover the long history and all the contributions of various people that have led to our current understanding of virtual fundamental classes. For that, we refer the interested readers to [**16**] and [**39**]. In Sections 1-3 of [**39**], McDuff and Wehrheim give a detailed description of their approach and the motivations behind it. They compare their approach to that of FOOO and various others, and list some issues that they believe have not been addressed properly before. In [**16**, Part 6], FOOO discuss the questions that were raised[3] about the details of their approach, provide some history, and address these questions with details. An interested reader may read both to get a sense of how things have evolved to their current form. Last but not least, some of the arguments in this note have benefited from many helpful and exciting discussions Mohammad

[2]A dimensionally graded system is a good coordinate system where the index set is just $\mathbb{Z}$.

[3]Originally, as a Google group discussion.

Tehrani had with Dusa McDuff, John Morgan, and Gang Tian. We thank them especially for their interest and support.

1. Introduction

Enumerative geometry, originally a branch of algebraic geometry with a history of more than a century, concerns certain counts of algebraic objects, mainly algebraic curves, in generic situations. This has been a challenging field of study for various reasons; the category of algebraic (more generally holomorphic) functions does not admit partitions of unity, algebraic objects are fairly rigid, and a precise formulation of the necessary intersection theory needs elaborate techniques (etc.). On the other hand, these tools are easily accessible and play a fundamental role in topology and smooth category.

Symplectic manifolds are a softened version of Kähler manifolds, such as complex projective varieties, where we drop the integrality condition of the complex structure and mainly focus on the underlying non-degenerate two form which we call the symplectic form. At a quick glance, symplectic manifolds and complex projective varieties seem very far apart. Surprisingly, in [**21**], Gromov combined the rigidity of algebraic geometry with the flexibility of the smooth category and initiated the use of "pseudoholomorphic curves" as a generalization of holomorphic curves, or better said "parametrized holomorphic curves". These allow the formulation of symplectic analogues of enumerative questions from algebraic geometry as well-defined invariants of symplectic manifolds. Moreover, from a purely symplectic perspective, these invariants give us a tool to distinguish isotopy classes of symplectic manifolds which are indistinguishable as smooth manifolds, see [**46**].

Other important applications of pseudoholomorphic curves include (now classical) Floer homology for giving lower bounds for the number of fixed points of Hamiltonian diffeomorphisms, initiated by Floer [**12**] and further enhanced to prove Arnold conjecture in some cases; e.g. see [**27**] and [**42**]. This idea was also extended to study the intersection of a pair of Lagrangians, known as Lagrangian intersection Floer theory [**17**], in an arbitrary symplectic manifold.

The invariants of symplectic manifolds obtained by counting the number of pseudoholomorphic maps has also relations to topological sigma models in string theory physics. Among various works in this direction, the 1988 paper of Witten [**55**] laid a background for further studies. Since then, the invariants obtained by the counts of pseudoholomorphic maps into symplectic manifolds are called **Gromov-Witten** invariants.

Let us begin with a review of "classical" Gromov-Witten theory, as a theory of counting pseudoholomorphic curves; we refer interested reader unfamiliar with the basics of this subject to [**37**]. Let (X^{2n},ω) be a symplectic manifold, which we assume to be closed (i.e. compact without boundary) throughout this article. An almost complex structure J on X is an endmorphism $J\colon TX \longrightarrow TX$ such that $J^2=-\mathrm{id}$. We say J is ω-compatible if $g_J\equiv\omega(\cdot,J\cdot)$ is a metric; in this case the action of J is an isometry (with respect to g_J). More generally, we say J is ω-tame if

$$\omega(v,Jv)>0 \qquad \forall\, v\in TX,\ v\neq 0.$$

Let $\mathcal{J}(X,\omega)$ and $\mathcal{J}_\tau(X,\omega)$ denote the space[4] of smooth almost complex structures that are ω-compatible and ω-tame, respectively. Fortunately, each of these spaces is

[4] With its natural topology.

contractible; therefore, any number associated to a tuple (X, ω, J) which is invariant under the deformation of J within these spaces is an invariant of (deformation equivalence class of) (X, ω).

In the Kähler category, the complex structure gives an almost complex structure on the underlying real manifold which is compatible with the Kähler symplectic form. In a general symplectic manifold, the Nijenhuis tensor

$$\begin{aligned} &N_J \in \Gamma(X, \Omega^2_X \otimes TX), \\ &N_J(u,v) \equiv [u,v] + J[u, Jv] + J[Ju, v] - [Ju, Jv] \quad \forall u,v \in TX, \end{aligned} \tag{1.1}$$

calculates the deviation of a compatible triple (X, ω, J) from being Kähler; i.e. (X, ω, J) is Kähler if and only if $N_J \equiv 0$; see [**37**, Section 2.1].

1.1. Moduli space of pseudoholomorphic maps. Given a fixed $J \in \mathcal{J}_\tau(X, \omega)$, $A \in H_2(X, \mathbb{Z})$, and $g, k \in \mathbb{N}$, a k-marked genus g degree A (i.e. the image in X represents the homology class A) **pseudoholomorphic** map, or J-holomorphic map, is a differentiable map[5] $u\colon \Sigma \longrightarrow X$ from a closed genus g Riemann surface (complex curve) $(\Sigma, \mathfrak{j})$ with k distinct ordered points

$$\vec{z} = (z^1, \dots, z^k) \subset \Sigma$$

to X that satisfies the **Cauchy-Riemann equation**

$$\Gamma\big(\Sigma, \Omega^{0,1}_{\Sigma,\mathfrak{j}} \otimes_{\mathbb{C}} u^*TX\big) \ni \bar{\partial} u = \frac{1}{2}\,(\mathrm{d}u + J \circ \mathrm{d}u \circ \mathfrak{j}) \equiv 0. \tag{1.2}$$

It follows from an elliptic bootstrapping argument, cf. [**37**, Theorem B.4.1], that every such map is automatically smooth (more precisely, it is as smooth as the almost complex structure). For every positive integer N let

$$[N] \equiv \{1, 2, \dots, N\}.$$

Two k-marked[6] J-holomorphic maps $u_i : (\Sigma_i, \mathfrak{j}_i, \vec{z}_i) \longrightarrow X$, with $i = 1, 2$, are said to be **equivalent** if there is a holomorphic diffeomorphism

$$\varphi\colon (\Sigma_1, \mathfrak{j}_1) \longrightarrow (\Sigma_2, \mathfrak{j}_2) \text{ s.t. } u_2 = u_1 \circ \varphi \quad \text{and} \quad \varphi(z_1^a) = z_2^a \quad \forall a \in [k]. \tag{1.3}$$

Let $\mathfrak{Aut}(\Sigma, \mathfrak{j})$ be the group of biholomorphisms of $(\Sigma, \mathfrak{j})$, $\mathfrak{Aut}(\Sigma, \mathfrak{j}, \vec{z})$ be the subgroup fixing the marked points, and $\mathfrak{Aut}(u, \Sigma, \mathfrak{j}, \vec{z})$ be the subgroup that does not change the map and the marked points; we have

$$\mathfrak{Aut}(u, \Sigma, \mathfrak{j}, \vec{z}) \subset \mathfrak{Aut}(\Sigma, \mathfrak{j}, \vec{z}) \subset \mathfrak{Aut}(\Sigma, \mathfrak{j}).$$

We denote by $\mathcal{M}_{g,k}(X, A, J)$, or $\mathcal{M}_{g,k}(X, A)$ for short, to be the space of equivalence classes of k-marked genus g degree A J-holomorphic maps into X. This space carries a natural Haussdorf topology and is in fact metrizable[7]. However, it is normally not compact and usually very singular. A priori, the Cauchy-Riemann equation (1.2) is independent of the symplectic structure. Nevertheless, the symplectic form plays an

[5] More precisely, a $W^{\ell,p}$-smooth map for some $p > 1$ with $\ell p > 2$, where $W^{\ell,p}(\Sigma, X)$ is the space of ℓ-times differentiable maps whose ℓ-th derivative has finite L_p-norm.

[6] We write an ordered set of marked points as $\vec{z} \equiv (z^1, \dots, z^k) \subset \Sigma$. Thus, in our notion, upper indices indicate the ordering of the marked points and lower indices, if present, take care of other labelings of the marked points.

[7] The Gromov topology is paracompact, Hausdorff, and locally metrizable; hence, by Smirnov's theorem, it is metrizable. See Remark 3.4.6 for a comparison of various metrizability theorems.

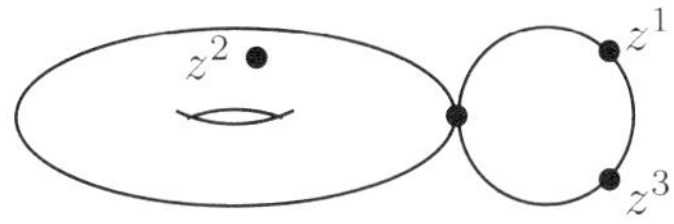

FIGURE 1. A 3-marked 1-nodal curve of genus 1.

important role; it provides an energy function which is essential for compactifying $\mathcal{M}_{g,k}(X,A)$.

In order to compactify the moduli space, we need to add "proper" limits of sequences of J-holomorphic maps. By the celebrated compactness theorem of Gromov [**21**, Theorem 15.B], a sequence of J-holomorphic maps in an almost complex manifold with a uniform area bound (with respect to some J-invariant metric g) has a subsequence which limits to a "stable" connected set of J-holomorphic maps attached to each other along some nodes; see Theorem 5.3.4.

A marked (smooth) complex curve $C=(\Sigma,\mathfrak{j},\vec{z})$ is said to be **stable** if $\mathfrak{Aut}(C)$ is finite. This is the case whenever $2g+k\geq 3$. Similarly, a marked J-holomorphic map $f\equiv(u,C)$ is said to be stable if $\mathfrak{Aut}(f)$ is finite. A **nodal** marked complex curve (or Riemann surface) is a connected space obtained by considering a disjoint union of smooth curves and identifying them along some pairs of (unordered) distinguished points disjoint from the marked points; i.e.

$$C=(\Sigma,\mathfrak{j},\vec{z})\equiv\Big(\coprod_{i=1}^{N}\big(\Sigma_i,\mathfrak{j}_i,\vec{z}_i,\{w_i^a\}_{a\in[k_i]}\big)\Big)/\sim,\qquad w_i^a\sim w_{\sigma(i,a)}^{\delta(i,a)},$$

where $(i,a)\longrightarrow(\sigma(i,a),\delta(i,a))$ is a fixed point free involution on the set of pairs

$$\{(i,a)\colon i\in[N],\ a\in[k_i]\};$$

see Section 5.1 or [**29**, Section 22.2]. The ordered set of marked points on the union corresponds to an ordering on

$$\vec{z}=\bigcup_{i=1}^{N}\vec{z}_i$$

which respects the ordering of each $\vec{z}_i$. The genus of a nodal curve is the sum of the genera of the connected components and the genus of its dual graph; see Equation 5.1.

Figure 1 illustrates a 3-marked 1-nodal curve of genus 1 made of two smooth curves (a torus and a sphere). A nodal J-holomorphic map is union of J-holomorphic maps on $\{\Sigma_i\}_{i\in[N]}$ that have identical values on each pair of identified points. Similarly, a nodal J-holomorphic map is said to be stable if its automorphism group is finite; see (5.5) and (5.6). The degree of a nodal map is the sum of degrees of its components. Then, we denote by $\overline{\mathcal{M}}_{g,k}(X,A)$ to be the space of equivalence classes of all k-marked genus g degree A J-holomorphic maps into X. The Deligne-Mumford space of stable curves, denoted by $\overline{\mathcal{M}}_{g,k}$, corresponds to the special case where X is just a point. In order to make our notation consistent, in the unstable range $2g+k<3$, we define $\overline{\mathcal{M}}_{g,k}$ to be a point![8]

[8] In the original approach of [**20**], $\overline{\mathcal{M}}_{1,0}$ is taken to be the j-line; however, for the sake of consistency in Theorem 1.4.1, we take it to be a point in this article.

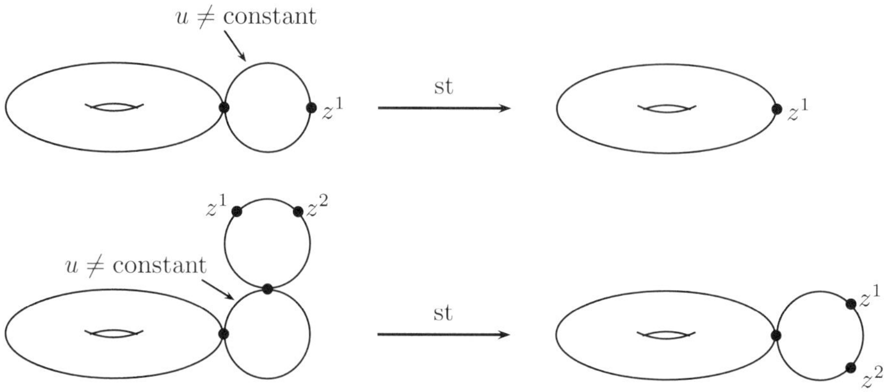

FIGURE 2. Examples of the stabilization process.

Let

$$\text{ev}_i\colon \overline{\mathcal{M}}_{g,k}(X,A) \longrightarrow X, \quad \text{ev}_i\big([u,\Sigma,\mathfrak{j},\vec{z}]\big) = u(z^i) \in X, \tag{1.4}$$

be the natural **evaluation** map at the marked points. For every (equivalence class of) stable map $[u,\Sigma,\mathfrak{j},\vec{z}]$, forgetting about u, we obtain a nodal curve which a priori may not be stable; it may have spherical components with at most two distinguished and marked points, or Σ could be a smooth genus 1 curve with no marked point. If $2g+k\geq 3$, we can collapse every unstable component into a node or a marked point and obtain a stable curve $[\bar{\Sigma},\bar{\mathfrak{j}},\vec{z}]$ with possibly fewer components but the same number of marked points; see [**29**, Section 26.3] for a more through description. If $2g+k<3$, we follow the convention before (1.4). This process gives us a well-defined **forgetful** (or stabilization) map from the space of stable maps into the Deligne-Mumford space,

$$\text{st}\colon \overline{\mathcal{M}}_{g,k}(X,A) \longrightarrow \overline{\mathcal{M}}_{g,k}, \quad \text{st}\big([u,\Sigma,\mathfrak{j},\vec{z}]\big) = [\bar{\Sigma},\bar{\mathfrak{j}},\vec{z}]. \tag{1.5}$$

Figure 2-top shows an example of stabilization process where an spherical component with one marked point and one distinguished point is collapsed into a marked point on the next component. Figure 2-bottom shows an example where an spherical component with two distinguished points is collapsed into a node. By a similar procedure, for every $I\subset[k]$ we obtain similar forgetful maps (this time we remove some of the marked points)

$$\begin{aligned} &\pi_I\colon \overline{\mathcal{M}}_{g,k}(X,A) \longrightarrow \overline{\mathcal{M}}_{g,k-|I|}(X,A), \\ &\pi_I([u,\Sigma,\mathfrak{j},(z^j)_{j\in[k]}]) = [u,\bar{\Sigma},\bar{\mathfrak{j}},(z^j)_{j\in[k]-I}], \end{aligned} \tag{1.6}$$

that removes $(z^j)_{j\in I}$, collapses the possible unstable components, and relabels the remaining points by $1,\ldots,k-|I|$, preserving the ordering; cf. [**29**, Section 23.4].

With this setup, the Gromov's Compactness Theorem 5.3.4, states that every sequence in $\overline{\mathcal{M}}_{g,k}(X,A)$ has a subsequence with a well-defined limit in $\overline{\mathcal{M}}_{g,k}(X,A)$ and this sequential convergence provides a compact Hausdorff topology on $\overline{\mathcal{M}}_{g,k}(X,A)$. Moreover, with respect to this topology (1.4), (1.5), and (1.6) are continuous.

1.2. GW invariants. If $\overline{\mathcal{M}}_{g,k}(X,A)$ has the structure of a smooth (in some suitable sense) oriented orbifold of the expected real dimension

$$d = 2\big((n-3)(1-g)+c_1(A)+k\big) \tag{1.7}$$

and the evaluation and forgetful maps are smooth, Gromov-Witten invariants are defined via integrals of the form

$$\mathrm{GW}^X_{g,A}(\theta_1,\dots,\theta_k;\kappa) = \int_{[\overline{\mathcal{M}}_{g,k}(X,A)]^{\mathrm{FC}}} \mathrm{ev}_1^*\theta_1\wedge\cdots\wedge\mathrm{ev}_k^*\theta_k\wedge\mathrm{st}^*\kappa, \tag{1.8}$$

where θ_i are de Rham cohomology classes on X, κ is a cohomology class on $\overline{\mathcal{M}}_{g,k}$, and $[\overline{\mathcal{M}}_{g,k}(X,A)]^{\mathrm{FC}}$ is the fundamental class of the oriented orbifold structure (see Remark 2.1.11). Similarly, we may consider the push-forward of the singular homology fundamental class $[\overline{\mathcal{M}}_{g,k}(X,A)]^{\mathrm{FC}} \in H_*(\overline{\mathcal{M}}_{g,k}(X,A),\mathbb{Q})$ under the map

$$\mathrm{ev}\times\mathrm{st} \equiv \mathrm{ev}_1\times\cdots\times\mathrm{ev}_k\times\mathrm{st}\colon \overline{\mathcal{M}}_{g,k}(X,A) \longrightarrow X^k\times\overline{\mathcal{M}}_{g,k} \tag{1.9}$$

and define the image

$$\mathrm{GW}^X_{g,A} = (\mathrm{ev}\times\mathrm{st})_*\big([\overline{\mathcal{M}}_{g,k}(X,A)]^{\mathrm{FC}}\big) \in H_d(X^k\times\overline{\mathcal{M}}_{g,k},\mathbb{Q})$$

to be the **Gromov-Witten fundamental class** of $\overline{\mathcal{M}}_{g,k}(X,A)$. In this sense, GW invariants are defined via intersection of homology/cohomology classes in $X^k\times\overline{\mathcal{M}}_{g,k}$ with the GW fundamental class.

Except in some very special cases (e.g. $X\cong\mathbb{CP}^n$ and $g=0$) $\overline{\mathcal{M}}_{g,k}(X,A)$ does not carry a nice orbifold structure of the right dimension; it may have many components which are singular, intersect non-trivially, and have different dimensions! Treating these difficulties is the main point of this manuscript.

1.3. Semi-positive case. There is still one case, the case of semi-positive manifolds, where the GW invariants can be defined by a direct count of geometric objects. Let us review this case; see the original article of Ruan-Tian [**49**] or the book of McDuff-Salamon [**37**] for more details. A marked J-holomorphic map with smooth domain

$$f = [u,\Sigma,\mathfrak{j},\vec{z}] \in \mathcal{M}_{g,k}(X,A)$$

is said to be **somewhere injective** if there is a point $z\in\Sigma$ such that

$$du(z)\neq 0, \qquad u^{-1}(u(z)) = \{z\}.$$

Let $\mathcal{M}^*_{g,k}(X,A)\subset\mathcal{M}_{g,k}(X,A)$ denote the subspace of somewhere injective maps. By [**37**, Theorem A], over a dense subset of second category

$$\mathcal{J}^{\mathrm{reg}}(X,\omega,g,A)\subset\mathcal{J}_\tau(X,\omega)$$

(such J is called **regular** with respect to g and A), $\mathcal{M}^*_{g,k}(X,A)$ is a naturally oriented smooth manifold of the expected real dimension (1.7).

Definition 1.3.1. *A closed $2n$-dimensional symplectic manifold (X,ω) is called* ***semi-positive*** *if $c_1(A)\geq 0$ for all $A\in H_2(M)$ such that*

$$\omega(A)>0 \quad\text{and}\quad c_1(A)\geq 3-n. \tag{1.10}$$

It is called ***positive***[9] *if $c_1(A)>0$ for all $A\in H_2(M)$ such that (1.10) holds.*

[9] Other terms such as monotone, strongly semi-positive, convex, etc. have also been used in the literature for this notion.

In particular, every symplectic manifold of complex dimension 3 or less is semi-positive. If $g=0$ and (X,ω) is positive, for generic $J\in\mathcal{J}^{\mathrm{reg}}(X,\omega,0,A)$, a deliberate dimension counting argument shows that the image of the complement

$$\partial\overline{\mathcal{M}}_{0,k}(X,A,J)=\overline{\mathcal{M}}_{0,k}(X,A,J)\setminus\mathcal{M}^*_{0,k}(X,A,J)$$

under ev$\times$st in $X^k\times\overline{\mathcal{M}}_{0,k}$ is a set of at least real codimension 2 and can be ignored; i.e. the inclusion $\mathcal{M}^*_{0,k}(X,A)\subset\overline{\mathcal{M}}_{0,k}(X,A)$ gives rise to a GW **pseudocycle**

$$\mathrm{GW}^X_{0,A}\subset H_d(X^k\times\overline{\mathcal{M}}_{0,k},\mathbb{Q})$$

independent of the choice of the generic J and the particular choice of ω in its deformation equivalence class; see [**37**, Theorem 6.6.1] for more details. This way, we can extend the definition of GW fundamental class to the case of positive manifolds.

In the higher genus case the same argument does not work and one needs to perturb the Cauchy- Riemann equation to take care of multiple cover maps. Given a (sufficiently) smooth map $u\colon(\Sigma,\mathfrak{j})\longrightarrow(X,J)$, let

$$\mathcal{E}_f=\Gamma\big(\Sigma,\Omega^{0,1}_{\Sigma,\mathfrak{j}}\otimes_{\mathbb{C}}u^*TX\big),\quad\text{with } f=\big(u,(\Sigma,\mathfrak{j})\big),\tag{1.11}$$

be the space of (sufficiently) smooth u^*TX-valued $(0,1)$-forms with respect to $\mathfrak{j}$ on Σ and J on TX. Given a **perturbation** term $\nu\in\mathcal{E}_f$, we say u is (J,ν)-holomorphic if it satisfies the perturbed Cauchy-Riemann equation

$$\bar{\partial}u=\nu.\tag{1.12}$$

Given a "smooth" family of perturbations over the Deligne-Mumford space (see [**49**] for the precise description) we define $\overline{\mathcal{M}}_{g,k}(X,A,J,\nu)$ to be the space of equivalence classes of perturbed pseudoholomorphic maps, in a similar way. If (X,ω) is semi-positive, by [**49**, Corollary 3.9], the space $\overline{\mathcal{M}}_{g,k}(X,A,J,\nu)$ is Hausdorff and compact with respect to a similarly defined Gromov convergence topology. By [**49**, Theorem 3.16], for generic J and sufficiently small ν, the main stratum $\mathcal{M}^*_{g,k}(X,A,J,\nu)$ consisting of somewhere-injective maps is a smooth manifold of the expected dimension (1.7). By [**49**, Theorem 3.11], it has a canonical orientation. By [**49**, Proposition 3.21], the image of $\partial\overline{\mathcal{M}}_{g,k}(X,A,J,\nu)$ under ev $\times$ st is contained in images of maps from smooth even-dimensional manifolds of at least 2 real dimension less than the main stratum. Thus, similar to the positive case, the inclusion

$$\mathcal{M}^*_{g,k}(X,A,J,\nu)\subset\overline{\mathcal{M}}_{g,k}(X,A,J,\nu)$$

provides a GW pseudocycle independent of the choice of the admissible almost complex structure J or the perturbation ν.

In the positive/semi-positive situations, the resulting GW invariants are **enumerative**, in the sense that they can be interpreted as a finite $\mathbb{Q}$-weighted count of (J,ν)-holomorphic maps of fixed degree and genus meeting some prescribed cycles at the marked points.

1.4. Virtual Fundamental Class. Extending the definition of GW invariants to more general symplectic manifolds occupies the rest this manuscript. Let us briefly indicate how we shall proceed. Locally, the argument is analogous to the semi-positive case. Roughly speaking, we consider local perturbations in a consistent way and patch the solution sets of the perturbed equations (1.12) to obtain a so called **Virtual Fundamental Class**, or **VFC** for short.

To this end, we construct a **Kuranishi structure** on the moduli space. Abstractly speaking, a Kuranishi structure on a compact metrizable space M is made of **Kuranishi charts** on open sets of M (similar to manifolds) which are "compatible" along intersections. An abstract Kuranishi chart of dimension d over an open subset $F \subset M$ (called the support or footprint of the chart) is made of an orbifold vector bundle $\mathrm{pr}\colon \mathcal{U} \longrightarrow \mathcal{V}$, an orbifold section $s\colon \mathcal{V} \longrightarrow \mathcal{U}$, and a homeomorphism from the underlying topological space of the zero set of s onto F, $\overline{\psi}\colon \overline{s}^{-1}(0) \longrightarrow F \subset M$, such that

$$\dim \mathcal{V} - \operatorname{rank} \mathcal{U} = d.$$

See Section 2 for the basics of orbifolds and Section 3 for the precise definition of Kuranishi charts.

One can put a Kuranishi structure on a fixed topological space M in many different ways. However, there is a natural way of constructing Kuranishi charts over moduli spaces of stable maps. With notation as in (1.11), given a (family of) J-holomorphic map

$$u\colon (\Sigma, \mathfrak{j}) \longrightarrow (X, J),$$

instead of considering a fixed perturbation $\nu \in \mathcal{E}_f$ as in (1.12), we consider a finite dimensional $\mathfrak{Aut}(f)$-invariant subspace $E_f \subset \mathcal{E}_f$ of perturbations and look at the space of all maps

$$V_f = \{u'\colon (\Sigma, \mathfrak{j}) \longrightarrow (X, J) \colon \exists\ \nu \in E_f \text{ s.t } \bar\partial u' = \nu \text{ and } \mathfrak{d}(u, u') < \epsilon\};$$

here $\epsilon > 0$ is a small positive number and $\mathfrak{d}(u, u') < \epsilon$ means u' is ϵ-close to u with respect to some specified distance function $\mathfrak{d}$. For an "appropriate" and sufficiently large choice of E_f (see Definition 6.2.2), V_f is $\mathfrak{Aut}(f)$-invariant, $[V_f/\mathfrak{Aut}(f)]$ is a smooth orbifold chart,

$$\mathcal{U} = [U_f = (V_f \times E_f)/\mathfrak{Aut}(f)] \longrightarrow \mathcal{V} = [V_u/\mathfrak{Aut}(f)]$$

is an orbibundle chart, and the map

$$s\colon \mathcal{V}_f \longrightarrow \mathcal{U}_f, \quad s([u']) = [u', \bar\partial u']$$

is an orbifold section. The zero set of this section is the set of actual J-holomorphic maps (on the fixed domain $(\Sigma, \mathfrak{j})$) sufficiently close to u. More generally, in Section 6 we will consider the case where the complex structure $\mathfrak{j}$ on Σ can change and the domain has marked points. We extend this construction to nodal maps and cover the moduli space with Kuranishi charts of this specific type. In Section 4, we construct a VFC for every abstract oriented Kuranishi structure. In the case of natural Kuranishi structures on a moduli space of pseudoholomorphic maps, this gives us a VFC which we use to define GW invariants. The following theorem is the main result of this construction in its simplest form; see Theorem 6.5.1 for the precise statement.

THEOREM 1.4.1 ([**20**, Theorem 1.3]). *Let (X^{2n}, ω) be a closed symplectic manifold, $J \in \mathcal{J}(X, \omega)$ be an arbitrary compatible*[10] *almost complex structure, $A \in H_2(X, \mathbb{Z})$, and $g, k \in \mathbb{N}$. With the dimension d as in (1.7), there exist a class of natural oriented cobordant d-dimensional Kuranishi structures on $\overline{\mathcal{M}}_{g,k}(X, A, J)$ and an associated rational homology class*

$$[\overline{\mathcal{M}}_{g,k}(X, A, J)]^{\mathrm{VFC}} \in H_d(X^k \times \overline{\mathcal{M}}_{g,k}, \mathbb{Q})$$

which is independent of J and which only depends on the isotopy class of (X, ω).

[10] One may consider tame almost complex structures, instead.

The VFC provided by Theorem 1.4.1 only allows us to define GW invariants with **primary insertions**; i.e. constraints coming from the ambient space X or the Deligne-Mumford space $\overline{\mathcal{M}}_{g,k}$. A more complete list of GW invariants include invariants involving ψ-classes, Hodge classes, and other similarly defined intrinsic classes. First, let us briefly recall the definition of these cohomology classes.

For every genus g k-marked degree A J-holomorphic map $f \equiv (u, \Sigma, \mathfrak{j}, \vec{z})$ (nodal or not), let $\mathcal{L}_i|_f$ be the holomorphic cotangent space of $(\Sigma, \mathfrak{j})$ at the i-th marked point,

$$\mathcal{L}_i|_f = \mathcal{T}^*_{z^i}\Sigma.$$

Since every z^i is fixed with respect to $\mathfrak{Aut}(f)$, $\mathfrak{Aut}(f)$ acts on $\mathcal{L}_i|_f$. If $\overline{\mathcal{M}}_{g,k}(X, A)$ is an orbifold, then the quotient spaces $[\mathcal{L}_i|_f/\mathfrak{Aut}(f)]$ for different $[f] \in \overline{\mathcal{M}}_{g,k}(X, A)$ form an orbifold complex line bundle $\mathcal{L}_i$ over $\overline{\mathcal{M}}_{g,k}(X, A)$. This line bundle is called the i-th **tautological** line bundle. In this situation, the i-th ψ-class, denoted by ψ_i, is the orbifold first Chern class of the i-th tautological line bundle; see [**29**, Definition 25.2.1]. In the general case, for every natural Kuranishi structure on $\overline{\mathcal{M}}_{g,k}(X, A)$, these one-dimensional complex vector spaces form a "Kuranishi complex line bundle" in the sense of Definition 3.9.1; also see [**29**, Section 25.2]. In this case, ψ_i is the "virtual" first Chern class of $\mathcal{L}_i$ against VFC of $\overline{\mathcal{M}}_{g,k}(X, A)$; see the next paragraph. Similarly, for every $g \geq 1$, we can define a rank g Kuranishi vector bundle $\mathcal{E}_g$, called **Hodge bundle**, whose fiber at every $[f] \equiv [u, \Sigma, \mathfrak{j}, \vec{z}]$ is equal to

$$[H^0(\Sigma, \omega_{\Sigma,\mathfrak{j}})/\mathfrak{Aut}(f)], \tag{1.13}$$

where $\omega_{\Sigma,\mathfrak{j}}$ is the dualizing sheaf of $(\Sigma, \mathfrak{j})$; see [**29**, Section 25.3.1]. For smooth Σ, $\omega_{\Sigma,\mathfrak{j}}$ is simply the sheaf of antiholomorphic 1-forms $\Omega^{0,1}_{\Sigma,\mathfrak{j}}$. Then, the Hodge class λ_g is the "virtual" top Chern class of $\mathcal{E}_g$ against VFC of $\overline{\mathcal{M}}_{g,k}(X, A)$.

In order to define a VFC involving the top Chern class of some Kuranishi vector bundle $\mathcal{L}$, we change the original Kuranishi structure of Theorem 1.4.1 into some augmented Kuranishi structure by increasing the obstruction bundle. For example, in the case of GW invariants, at every J-holomorphic map f, we replace the obstruction bundle E_f with $E_f \oplus L_f$, where L_f is the fiber of $\mathcal{L}$ (before taking quotient w.r.t $\mathfrak{Aut}(f)$) at f, and extend the Kuranishi map by zero on the second factor. Then, we define the "virtual" top Chern class of $\mathcal{L}$ against VFC of $\overline{\mathcal{M}}_{g,k}(X, A)$ to be the VFC of the augmented system. We can then pair the VFC of the augmented system with constraints from X and $\overline{\mathcal{M}}_{g,k}$ to define invariants involving a mixture of primary and non-primary insertions. See Section 3.9 and Section 6.7 for more details.

REMARK 1.4.2. Given a compact smooth oriented submanifold $M_{\mathcal{J}} \subset \mathcal{J}(X, \omega)$ of real dimension $d_{\mathcal{J}}$, let

$$\overline{\mathcal{M}}_{g,k}(X, A, M_{\mathcal{J}}) = \bigcup_{J \in M_{\mathcal{J}}} \overline{\mathcal{M}}_{g,k}(X, A, J) \tag{1.14}$$

be the family of moduli spaces over $M_{\mathcal{J}}$ (with the Gromov topology) and

$$\pi_{\mathcal{J}} : \overline{\mathcal{M}}_{g,k}(X, A, M_{\mathcal{J}}) \longrightarrow M_{\mathcal{J}}$$

be the projection map. Then a family version of Theorem 1.4.1 holds: there exists a natural family of oriented $(d+d_{\mathcal{J}})$-dimensional Kuranishi structures on (1.14) and an associated relative homology class (obtained via $\mathrm{ev}\times\mathrm{st}\times\pi_{\mathcal{J}}$)

$$[\overline{\mathcal{M}}_{g,k}(X,A,M_{\mathcal{J}})]^{\mathrm{VFC}} \in H_{d+d_{\mathcal{J}}}(X^k\times\overline{\mathcal{M}}_{g,k}\times M_{\mathcal{J}}, X^k\times\overline{\mathcal{M}}_{g,k}\times\partial M_{\mathcal{J}},\mathbb{Q})$$

which only depends on the isotopy class of (X,ω) and $M_{\mathcal{J}}$. For example, this family version of Theorem 1.4.1, with $M_{\mathcal{J}}\cong S^2$, has been used to define GW-type invariants of certain S^1-family of K3-surfaces; see [**56**].

REMARK 1.4.3. It is important to mention that unlike in algebraic geometry, all we need in the symplectic approach to Gromov-Witten theory is a space on which we can do intersection theory. So the main functorial aspects of studying moduli spaces in algebraic geometry, e.g. the existence of universal families, are absent in this approach. The result of this flexibility is a spectrum of different analytical/topological approaches [**7, 20, 28, 31, 35, 38**] to formalizing such intersection theory in symplectic geometry.

1.5. Outline. In Sections 2.1 and 2.2, we will review some basic definitions and facts about orbifolds, orbibundles, and sections of orbibundles. We will mostly focus on effective orbifolds, however, at certain points throughout the paper, we will consider some basic non-effective orbifolds as well. In Section 2.3, we introduce the notion of a multisection. A multisection $\mathfrak{s}$ of an orbibundle $\mathrm{pr}\colon \mathcal{E}\longrightarrow\mathcal{M}$ is locally given by a finite family of sections of an orbifold chart invariant under the group action. As we will show in Section 2.4, multisections give us a way of deforming possibly non-transversal orbifold sections into a compatible collection of transversal multisections. Otherwise, it is not always feasible to perturb an arbitrary orbifold section into a transversal one. In Section 2.5, we introduce the notion of a resolution of a multisection. Given a multisection $\mathfrak{s}$, we decompose the base orbifold into a finite union of open sets labeled by positive integers $\ell\in\mathbb{N}$, such that in the region with label ℓ, $\mathfrak{s}$ is given by a set of ℓ distinct branches (possibly with different multiplicities). Then, a resolution is a compatible collection of *branched orbifold ℓ-covering spaces* over the corresponding regions such that on each piece $\mathfrak{s}$ is a weighted push-forward of an orbifold section s_ℓ of the pull back bundle. In Section 2.6, we use these resolutions to define the orbifold Euler class of the corresponding relatively oriented orbibundle. This construction lays out a foundation for the more complicated construction of VFC in Section 4.

In Sections 3.1 and 3.2, we introduce an abstract notion of a Kuranishi structure $\mathcal{K}$ for a compact metrizable space M. It consists of a set of compatible Kuranishi charts $\mathcal{U}_p$ indexed by $p\in M$. Each $\mathcal{U}_p$ is a local orbibundle with a section and a homeomorphism, called footprint map, from the zero set of the section onto an open subset F_p of M. The open sets $\{F_p\}_{p\in M}$ cover M, and the Kuranishi charts are compatible on the overlaps in a suitable sense. In Section 3.3, we assemble this uncountable data into a finite set $\mathfrak{L}$ of compatible orbibundles with sections and footprint maps

$$\begin{aligned} &\mathrm{pr}_i\colon \mathcal{E}(i)\longrightarrow\mathcal{Y}(i), \quad s_i\colon\mathcal{Y}(i)\longrightarrow\mathcal{E}(i), \quad \mathrm{rank}\ \mathcal{E}(i)=i,\\ &\psi\colon s_i^{-1}(0)\longrightarrow F(i)\subset M, \quad \forall i\in\mathfrak{B}\subset\mathbb{N}, \end{aligned}$$

called a **dimensionally graded system** (or **DGS**), such that the set of footprints $\{F(i)\}$ gives us a "nice" covering of M. The compatibility conditions are given by orbibundle embeddings $\mathcal{E}(i)|_{\mathcal{Y}(i,j)}\longrightarrow\mathcal{E}(j)$ from the restriction to open sub-orbifolds of the orbibundles of the lower rank into the orbibundles of the higher rank. In Section 3.4, we introduce the notation of a shrinking of a dimensionally graded system. A shrinking allows us to replace orbifold pieces of an DGS with smaller orbifolds to accommodate certain perturbations and gluing-pasting arguments. Our construction of a VFC for $(M, \mathcal{K})$ in this article depends on many auxiliary choices, including a particular choice of a Kuranishi structure itself. For this reason, in Section 3.5, we briefly review the notion of a cobordism between two Kuranishi structures, and more generally, between two sets of auxiliary data. In Section 3.7, we extend the perturbation results of Section 2.4 to dimensionally graded systems. Finally, in Section 4, we give two different constructions of a virtual fundamental class associated to a cobordism class of Kuranishi structures. In the simplest case of an DGS made of a single orbibundle, the resulting VFC is simply the Euler class of the orbibundle. We give three examples of this situation in Section 6.7.

In Section 5, we will review some well-known facts about the structure of the Deligne-Mumford space [**9**, **45**] of stable curves and moduli spaces of stable pseudoholomorphic maps.

In Section 6, we show that the moduli spaces of stable pseudoholomorphic maps have a natural cobordism class of Kuranishi structures. In order to construct Kuranishi charts around different types of maps in such a moduli space, in Section 6.2, we start with the simplest case of pseudoholomorphic maps with smooth stable domain. In Section 6.3, we extend this construction to maps with stable nodal domain, and in Section 6.4, we address the case of maps with unstable domain. The collection of Kuranishi charts constructed in these three sections, which we call **primary charts**, are not necessarily compatible along the overlaps. For this reason, in Section 6.5, we modify the construction and construct a new collection of Kuranishi charts, which we call **induced charts**, satisfying all the requirements of a Kuranishi structure. A collection of induced charts, and thus the resulting Kuranishi structure, depends on a finite collection of primary charts. Different such collections result in different but cobordant Kuranishi structures. Together with the results of Section 4, this gives the GW virtual fundamental class of Theorem 1.4.1.

The most difficult step in the construction of Kuranishi charts around maps with nodal domain is establishing the gluing argument of smoothing the nodes and its exponential decay properties. Due to the lack of space, we will state and use the gluing theorem as a black box and refer the interested reader to the dedicated article of FOOO [**14**]. The main purpose of this article is to give a fairly detailed, yet efficient, description of Kuranishi structures and of the resulting VFC, and to show that how they can be used to define GW invariants.

2. Preliminaries

In [**50**], Satake introduced the notion of orbifold (which he called V-manifold) and orbibundle to generalize the notion of manifold and vector bundle, in order to include local quotients of smooth manifolds and vector bundles by finite group actions; see also the book of Thurston [**53**]. In this chapter, we briefly recall the definitions and prove necessary statements for the construction of VFC. We mainly consider effective orbifolds. More general case of non-effective orbifolds can be better studied in the context of stacks or groupoids; see [**11**] and [**40**] for an introduction to stacks, and [**1**, Section 1.4], [**36**, Section 2], [**41**], and the works of Haefliger [**22**–**24**] to learn about groupoids.

2.1. Orbifolds.

DEFINITION 2.1.1. *An **effective orbifold chart** of real dimension n on a Haussdorff topological space M is a tuple $\mathcal{V}=(V,G,\psi)$ such that V is a connected open manifold of real dimension n, G is a finite group acting effectively and smoothly on V, and $\psi\colon V\longrightarrow F\subset M$ is a continuous map of the form $\psi=\overline{\psi}\circ\pi$, where $\pi\colon V\longrightarrow V/G$ is the quotient map and $\overline{\psi}\colon V/G\longrightarrow F$ is a homeomorphism (with respect to the quotient topology on V/G) onto an open subset of M.*

Given an effective orbifold chart (V,G,ψ), for every $x\in V$ let

$$G_x=\{g\in G\colon g\cdot x=x\}$$

be the **isotropy** group at x.

DEFINITION 2.1.2. *Given an n-dimensional effective orbifold chart $\mathcal{V}=(V,G,\psi)$ on M, a **refinement** of $\mathcal{V}$ is another n-dimensional effective orbifold chart $\mathcal{V}'=(V',G',\psi')$ together with a group homomorphism $h\colon G'\longrightarrow G$ and an h-equivariant smooth embedding $\phi\colon V'\longrightarrow V$ such that $\psi'=\psi\circ\phi$. Two charts $\mathcal{V}_1$ and $\mathcal{V}_2$ on M are said to be **compatible** if they admit a common refinement around every point of $\psi_1(V_1)\cap\psi_2(V_2)$.*

For the group homomorphism of refinements, it follows from the effectiveness of the group action that h is an injective homomorphism.

More generally, if $\overline{\phi}\colon M'\longrightarrow M$ is a continuous map between two topological spaces, given two effective orbifold charts (V',G',ψ') and (V,G,ψ) on M' and M, respectively, a **smooth lift**

$$(\phi,h)\colon(V',G',\psi')\longrightarrow(V,G,\psi) \tag{2.1}$$

of $\overline{\phi}|_{\psi'(V')}\colon\psi'(V')\longrightarrow\psi(V)$ consists of a group homomorphism $h\colon G'\longrightarrow G$ and an h-equivariant smooth map $\phi\colon V'\longrightarrow V$ such that

$$\psi\circ\phi=\overline{\phi}\circ\psi'. \tag{2.2}$$

In this situation, for every $x\in V'$ we get an induced homomorphism

$$h_x\colon G'_x\longrightarrow G_{\phi(x)}. \tag{2.3}$$

With this definition, a refinement is a smooth lift of the inclusion map over an open embedding $F'\subset F\subset M$. The group homomorphism of a general smooth lift may not be injective or surjective. In the case of refinements, every two such smooth embedding lifts $(\phi_i,h_i)\colon\mathcal{V}'\longrightarrow\mathcal{V}$, with $i=1,2$, differ by an element of G, i.e. there exists $g\in G$ such that $\phi_2=g\circ\phi_1$. If $g\big(\mathrm{Im}(\phi_1)\big)\cap\mathrm{Im}(\phi_2)\big)\neq\emptyset$, then $g\in\mathrm{Im}(h)$; therefore, in this case, (2.3) is an isomorphism and the conjugacy class of the isotropy group for different lifts of $\overline{x}=\psi(x)\in M$ is independent of the particular choice. We denote by $\mathrm{I}_{\overline{x}}$ to be the conjugacy class of the isotropy groups over different lifts of $\overline{x}\in M$ (w.r.t. the orbifold structure $\mathcal{M}$) and define

$$\mathfrak{I}_{\mathcal{M}}\colon M\longrightarrow\mathbb{Z},\quad \mathfrak{I}_{\mathcal{M}}(\overline{x})=|\mathrm{I}_{\overline{x}}|. \tag{2.4}$$

We use the function $\mathfrak{I}_{\mathcal{M}}$ in Section 2.5 and Section 2.6 in the Definition of Euler class of an orbibundle.

DEFINITION 2.1.3. *Given a metrizable topological space M, an n-dimensional smooth **effective orbifold atlas** on M is a countable collection*

$$\mathcal{A}\equiv\{\mathcal{V}_\alpha=(V_\alpha,G_\alpha,\psi_\alpha)\}_{\alpha\in S} \tag{2.5}$$

of compatible n-dimensional orbifold charts covering M. Two effective orbifold atlases $\mathcal{A}$ and $\mathcal{A}'$ are said to be ***equivalent*** *if they admit a common refinement*[11]. *An* ***effective orbifold structure*** $\mathcal{M}$ *on M is an equivalence class of such atlases.*

It actually requires a little bit of work to show that "refinement" defines an equivalence relation on the set of orbifold atlases. We refer to [**1**, Section 1] for a more detailed discussion of orbifold atlases, their refinements, and the refinement equivalence relation between atlases. By replacing diffeomorphisms with homeomorphisms in the defining equation of the orbifold charts and the lift maps, we can define topological orbifolds, similarly. By Definition 2.1.1, the orbifolds considered in this article, unless specifically mentioned (e.g. in Section 3.5), do not have boundary.

NOTATION 2.1.4. *In the rest of this article, we let calligraphic letters $\mathcal{M}$, $\mathcal{Y}$, $\mathcal{Z}$, and $\mathcal{W}$ to denote for orbifold structures on topological spaces (or manifolds) M, Y, Z, and W, respectively. We also use $\mathcal{M}$ for moduli spaces (as they are orbifolds in some nice cases). Similarly, after the introduction of orbibundles in Section 2.2, we let $\mathcal{E}$ denote an orbibundle with underlying topological space E.*

REMARK 2.1.5. *Unless explicitly mentioned, e.g. in comparison of our results and framework with similar works and in the construction of Euler class, the orbifolds considered in this manuscript are effective. Therefore, we omit the word "effective" from our terminology for simplicity. We refer to* [**18**, Section 4] *for some discussion on the complexity of non-effective orbifold quotients.*

The simplest examples of orbifolds are global quotients of manifolds by discrete groups (which indeed can be infinite). If M is a smooth manifold and G is a discrete group acting effectively and properly discontinuously on M, then M/G has the structure of an orbifold which we denote by $[M/G]$. More generally, if G is a compact Lie group acting smoothly, effectively, and almost freely (i.e. with finite stabilizers) on a smooth manifold M, the quotient space M/G has the structure of an orbifold. For example, all the weighted projective spaces are orbifolds of this sort with $M=\mathbb{C}^N-\{0\}$ and $G=\mathbb{C}^*$. In the terminology of [**1**, Definition 1.8], the former is called an **effective global quotient** and the latter is called an **effective quotient**.

As another example, if Σ is a Riemann surface of genus g, hence a manifold, by removing finitely many disjoint open discs $\{D_i\}_{i=1}^k$ from Σ and gluing back copies of $[D_i/\mathbb{Z}_{m_i}]$, where $\mathbb{Z}_{m_i}=\mathbb{Z}/(m_i\mathbb{Z})$ acts by multiplication with m_i-th roots of unity, we obtain new orbifold structures on the underlying space Σ. Thurston [**53**, Theorem 13.3.6 and p. 304] shows that the result is an effective global quotient unless $g=0$ and $k=1$ or $k=2$ with $m_1\neq m_2$. Examples of this form naturally arise in the study of moduli spaces; see Section 7.2. On the other hand, by [**1**, Theorem 1.23], every (effective) orbifold is an effective quotient of its frame bundle.

As the previous example shows, a fixed topological space can carry many different orbifold structures. It is possible to adopt most of the invariants of manifolds, e.g. fundamental group and Euler characteristic, to orbifolds and these invariants are usually different from the corresponding invariants of the underlying topological space.

---

[11] An atlas $\mathcal{A}'$ is a refinement of another atlas $\mathcal{A}$, if every chart of $\mathcal{A}'$ is the refinement of some chart of $\mathcal{A}$.

EXAMPLE 2.1.6. Among various examples of orbifolds, we are particularly interested in those arising from moduli problems. For example, the moduli space $\mathcal{M}_{1,1}$ of isomorphism classes of smooth elliptic curves with 1 marked point is isomorphic to $[\mathbb{H}/\mathrm{SL}_2(\mathbb{Z})]$, where

$$\mathbb{H} = \{\tau \in \mathbb{C}\colon \mathrm{im}(\tau) > 0\}, \quad A\tau = \frac{a\tau + b}{c\tau + d} \quad \forall A = \begin{bmatrix} a & b \\ c & d \end{bmatrix} \in \mathrm{SL}_2(\mathbb{Z}).$$

Since the action of $-\mathrm{Id}$ is trivial, $\mathcal{M}_{1,1}$ is not an effective orbifold. In order to get an effective orbifold, we need to reduce the group action to

$$\mathrm{PSL}_2(\mathbb{Z}) = \mathrm{SL}_2(\mathbb{Z}) / \langle -\mathrm{Id} \rangle .$$

Then, the reduced orbifold structure

$$\mathcal{M}_{1,1}^{\mathrm{red}} \equiv [\mathbb{H}/\mathrm{PSL}_2(\mathbb{Z})]$$

is the parameter space of isomorphism classes of smooth elliptic curves with 1 marked point and $\mathcal{M}_{1,1}$ is some (stacky) $\mathbb{Z}_2$-quotient of that. An alternative way to approach the issue is to increase the number of marked points. For example, the moduli space $\mathcal{M}_{1,2}$ of isomorphism classes of smooth elliptic curves with 2 marked points is an open subset of the effective global quotient

$$[(\mathbb{H} \times \mathbb{C})/(\mathrm{SL}_2(\mathbb{Z}) \ltimes \mathbb{Z}^2)]$$

where

$$(\tau, z) \sim \Big(A\tau, (c\tau + d)^{-1}(z + m + \tau n) \Big) \quad \forall A = \begin{bmatrix} a & b \\ c & d \end{bmatrix} \in \mathrm{SL}_2(\mathbb{Z}), \ (m, n) \in \mathbb{Z}^2.$$

In our construction of VFC, we often use the trick of adding more marked points to avoid encountering stacks and more complicated groupoids.

DEFINITION 2.1.7. *Given an orbifold atlas $\mathcal{A}$ on M as in (2.5), we say $\mathcal{A}$ is* ***oriented*** *if each V_α is oriented, the action of G_α is orientation-preserving, and the orientations agree on the overlaps. An* ***oriented*** *orbifold structure on M is a maximal oriented atlas.*

DEFINITION 2.1.8. *An* ***orbifold smooth map***, *$\phi\colon \mathcal{M}_1 \longrightarrow \mathcal{M}_2$, between two orbifolds $\mathcal{M}_1$ and $\mathcal{M}_2$ is an underlying continuous map $\overline{\phi}\colon M_1 \longrightarrow M_2$ that admits local smooth lifts as in (2.1). An* ***orbifold morphism*** *is an orbifold smooth map for which the induced homomorphisms $h_x\colon G_x \longrightarrow G_{\phi(x)}$ as in (2.3) are isomorphism. An orbifold* ***diffeomorphism*** *is an orbifold morphism with an inverse. An* ***orbifold embedding*** *is an orbifold morphism such that the lift maps (2.1) are smooth embeddings. Finally, a* ***sub-orbifold*** *$\mathcal{M}'$ of an orbifold structure $\mathcal{M}$ on M is a subspace $M' \subset M$ together with an orbifold structure $\mathcal{M}'$ on M' such that the inclusion map $\bar{\iota}\colon M' \longrightarrow M$ gives an orbifold embedding $\iota\colon \mathcal{M}' \longrightarrow \mathcal{M}$.*

The isomorphism condition on (2.3) for orbifold embeddings is not standard across the literature but is the one we use in our construction of Kuranishi structures in this article.

NOTATION 2.1.9. *In Definition 2.1.8, the local group homomorphisms h of (2.1) are embedded in the notation ϕ, i.e. in Definition 2.1.8 ϕ denotes a global map between orbifolds and at the level of charts the map ϕ also comes with a group homomorphism h as in (2.1).*

EXAMPLE 2.1.10. Let $\mathcal{M}=[\mathbb{R}^2/\mathbb{Z}_2]$ where $\mathbb{Z}_2$ acts by $(x,y)\longrightarrow(-x,y)$. Then the map $\phi\colon \mathbb{R}\longrightarrow\mathcal{M}$, given by $\phi(y)=(0,y)$, is a smooth orbifold map which is *not* an orbifold embedding.

REMARK 2.1.11. We conclude this section with some final remarks on the structure and properties of orbifolds. In what follows, $\mathcal{M}$ is a closed (compact without boundary) orbifold.

(1) If $\mathcal{M}$ is oriented, then the singular locus

$$M^{\text{sing}}\equiv\{\overline{x}\in M\colon \mathrm{I}_{\overline{x}}\neq 1\}$$

of $\mathcal{M}$ has real codimension greater than or equal to 2. Let M^{sm} be the complement of M^{sing} in M; M^{sm} has the structure of an oriented smooth manifold. Then, the inclusion map $M^{\text{sm}}\longrightarrow M$ provides a pseudocycle, and thus a fundamental class for M; see [**58**] for the definition and properties of pseudocycles. In the case of semi-positive symplectic manifolds in Section 1.3, we use this idea to define GW fundamental class of $\overline{\mathcal{M}}_{g,k}(X,A)$.

(2) More generally, given a group H, let $M^H\subset M$ be the set of points where the isotropy group is isomorphic to H. For example, for the trivial group $\langle 1\rangle$, $M^{\langle 1\rangle}=M^{\text{sm}}$. By the argument in [**18**, Section 5], every M^H has the structure of a smooth manifold (induced from the orbifold structure on $\mathcal{M}$). Let $\mathcal{H}$ be the set of all H such that $M^H\neq\emptyset$; if M is compact, then $\mathcal{H}$ is finite and $M=\bigcup_{H\in\mathcal{H}}M^H$ gives a stratification of M into smooth manifolds.

(3) For every point $\overline{x}\in M$, there exists an orbifold chart $\mathcal{V}=(V,G,\psi)$ **centered at** $\overline{x}$ in the sense that $\psi^{-1}(\overline{x})=\{x\}$ and $G=G_x$.

(4) An orbifold chart (V,G,ψ) is said to be **linear**, whenever $V\subset\mathbb{R}^n$ is a connected open set and the (restriction to V of the) action of G is linear. As [**1**, Page 2] indicates, since smooth actions are locally linear, any orbifold has an atlas consisting of linear charts. Moreover, we can choose the linear charts so that $G\subset O(n)$, and if $\mathcal{M}$ is orientable, we can choose them so that $G\subset SO(n)$. This is the point of view in the original approach of [**20**]; all the local structures taken there are linear. In the case of orbifold charts centered at $\overline{x}$, we can choose the linear chart so that the unique preimage x of $\overline{x}$ is the origin.

(5) As we mentioned earlier, orbifolds can be approached via the frame work of **Lie groupoids**. A Lie groupoid is a category in which all morphisms are invertible (i.e. they are isomorphisms), the space of objects $\widetilde{M}$ and the space of morphisms R (relations) are smooth manifolds (possibly with infinitely many, but countable, connected components), and the following set of structure maps are all smooth. For $x,y\in\widetilde{M}$, let $R(x,y)$ be the set of morphisms from x to y. Since all morphisms are invertible, there is an **inversion** involution $\iota\colon R\longrightarrow R$ which takes $R(x,y)$ to $R(y,x)$. From this set-up, we get the **source** and **target** maps

$$s\colon R\longrightarrow\widetilde{M}\times\widetilde{M}\ \text{and}\ t\colon R\longrightarrow\widetilde{M}\times\widetilde{M}$$

which are defined on every element $\gamma\in R(x,y)$ by $s(\gamma)=x$ and $t(\gamma)=y$, respectively. Note that $s\circ\iota=t$. By composing morphisms, whenever possible, we get a multiplication map

$$m\colon R_t\times_s R\longrightarrow R,\qquad m(\gamma_1,\gamma_2)=\gamma_2\circ\gamma_1\quad\forall\gamma_1\times\gamma_2\in R(x,y)\times R(y,z).$$

The **orbit space** of $(\widetilde{M},R)$, denoted by $M\equiv\widetilde{M}/R$, is the quotient space of equivalence classes of objects in $\widetilde{M}$ with respect to relations in R.

A **Lie groupoid orbifold** is a lie groupoid where

$$s \times t \colon R \to \widetilde{M} \times \widetilde{M}$$

is proper and $s, t\colon R \longrightarrow \widetilde{M}$ are local diffeomorphisms; see [**1**, Example 1.33] for more details.

Let $\mathcal{A}$ be an orbifold atlas on M as in (2.5). Associated to $\mathcal{A}$, there is a Lie groupoid orbifold with orbit space M given by

$$\widetilde{M} = \coprod_{\alpha \in S} V_\alpha, \quad R = \coprod_{\alpha,\beta \in S} \Gamma_{\alpha\beta},$$

where $\Gamma_{\alpha\beta}$ is the disjoint union of the irreducible components of the fiber product $V_\alpha \times_M V_\beta$. Therefore, all (effective) orbifolds are special cases of Lie groupoid orbifolds but the latter includes non-effective orbifolds as well. The most relevant example to us is the Deligne-Mumford space of stable marked curves which is worked out in details in [**45**]. By [**45**, Theorem 6.5], for every $g, k \in \mathbb{N}$ such that $2g+k \geq 3$, $\overline{\mathcal{M}}_{g,k}$ has the structure of a Lie groupoid orbifold. If $(g,k) \neq (1,1)$ or $(2,0)$, then the Lie groupoid orbifold structure comes from an effective orbifold as above, but in those two special cases, one either has to mildly generalize the notion of orbifold or to use the language of Lie groupoid orbifold.

2.2. Orbibundles. Orbibundles are the analogue of vector bundles for orbifolds. They show up as obstruction bundles, tangent bundle of orbifolds, and normal bundle of orbifold embeddings in the construction of VFC.

DEFINITION 2.2.1. *An* ***orbibundle chart*** *of rank r over an n-dimensional orbifold chart $\mathcal{V} = (V, G, \psi)$ is a triple*

$$\mathcal{U} = (\mathrm{pr}\colon U \longrightarrow V, G, \psi) \tag{2.6}$$

where $\mathrm{pr}\colon U \longrightarrow V$ *is a smooth G-equivariant*[12] *vector bundle of rank r over V.*

Similarly, given an orbibundle chart (2.6), a **refinement** of $\mathcal{U}$ is another orbibundle chart $\mathcal{U}' = (\mathrm{pr}'\colon U' \longrightarrow V', G', \psi')$ of the same rank over a refinement $\mathcal{V}' = (V', G', \psi')$ of $\mathcal{V}$ which is defined via an embedding (ϕ, h), and an h-equivariant fiber-wise linear smooth embedding $\mathfrak{D}\phi\colon U' \longrightarrow U$ such that the following diagram commutes,

$$\begin{array}{ccc} U' & \xrightarrow{\mathfrak{D}\phi} & U \\ \downarrow{\scriptstyle \mathrm{pr}'} & & \downarrow{\scriptstyle \mathrm{pr}} \\ V' & \xrightarrow{\phi} & V\,. \end{array} \tag{2.7}$$

NOTATION 2.2.2. *We use the notation $\mathfrak{D}\phi$ for the map between vector bundles of orbibundle charts to emphasize on its linearity and compatibility with ϕ. In the example of tangent bundle below and few other instances, $\mathfrak{D}\phi$ is actually the derivative ϕ, but otherwise, the letter $\mathfrak{D}$ behind ϕ is just a notation.*

Two orbibundle chart enhancements $\mathcal{U}_1$ and $\mathcal{U}_2$ of the orbifold charts $\mathcal{V}_1$ and $\mathcal{V}_2$ are said to be **compatible**, whenever they admit a common refinement around every point of $\psi_1(V_1) \cap \psi_2(V_2)$.

[12] i.e. the G-action on U and V commutes with the projection map, and for every $g \in G$ and $x \in V$, $g\colon U_x \longrightarrow U_{g\cdot x}$ is linear.

DEFINITION 2.2.3. *Given an orbifold atlas $\mathcal{A}_{\mathcal{M}}$ on M as in (2.5), an* ***orbibundle atlas*** *of rank r over $\mathcal{A}_{\mathcal{M}}$ is a compatible enhancement of the orbifold charts of $\mathcal{A}_{\mathcal{M}}$ to orbibundle charts of rank r,*

$$\mathcal{A}_{\mathcal{E}} \equiv \{\mathcal{U}_\alpha = (\mathrm{pr}_\alpha : U_\alpha \longrightarrow V_\alpha, G_\alpha, \psi_\alpha)\}_{\alpha\in S}. \tag{2.8}$$

We say two orbibundle atlases $\mathcal{A}_{\mathcal{E}}$ and $\mathcal{A}'_{\mathcal{E}}$ over $\mathcal{M}$ are ***equivalent*** *if they admit a common refinement. An* ***orbibundle*** *$\mathcal{E}$ on $\mathcal{M}$ is an equivalence class of such atlases.*

Equivalently, we can think of the total space of an orbibundle as an orbifold, itself, which admits certain type of orbifold projection map.

DEFINITION 2.2.4. *(alternative definition) Let $\mathcal{E}$ and $\mathcal{M}$ be two orbifolds, and*

$$\mathrm{pr}\colon \mathcal{E} \longrightarrow \mathcal{M} \tag{2.9}$$

be a surjective orbifold smooth map[13]. *We say (2.9) is an* orbibundle *of rank r, if for every sufficiently small orbifold chart $\mathcal{V}=(V,G,\psi_V)$ of $\mathcal{M}$ there is an orbifold chart $\mathcal{U}=(U,G,\psi_U)$ over $\overline{\mathrm{pr}}^{-1}(\psi_V(V)) \subset E$ such that $\mathrm{pr}\colon U \longrightarrow V$ is a smooth (G-equivariant by definition) vector bundle of rank r over V and for every $g\in G$ and $x\in V$, $g\colon U_x \longrightarrow U_{g\cdot x}$ is linear.*

EXERCISE 2.2.5. Prove that Definition 2.2.3 and Definition 2.2.4 are equivalent.

REMARK 2.2.6. Similar to Remark 2.1.11.(4), for every $\overline{x}\in M$, we can take a sufficiently small orbibundle chart $\mathcal{U}$ **centered at** $\overline{x}$ which is linear and trivial in the sense that

$$V\subset\mathbb{R}^n,\ x=\psi^{-1}(\overline{x})=0\in V,\ U=V\times\mathbb{R}^r,$$

the group G acts linearly on both V and $\mathbb{R}^r$, and pr is the projection map onto the first component. This is the point of view in the original approach of [**20**].

EXAMPLE 2.2.7. Let $\mathcal{M}$ be a smooth orbifold of dimension n, then its tangent bundle is a well defined orbibundle of rank n. Given a smooth orbifold chart $\mathcal{V}=(V,G,\psi)$ for $\mathcal{M}$, the action of G on V lifts to $U=TV$ by

$$\rho\colon G\times U \longrightarrow U, \quad \rho(g,(x,v_x)) = (g\cdot x, (\mathrm{d}g)_x(v_x)).$$

REMARK 2.2.8. In general, if $\phi\colon \mathcal{M}' \longrightarrow \mathcal{M}$ is an orbifold smooth map and $\mathrm{pr}\colon \mathcal{E} \longrightarrow \mathcal{M}$ is an orbibundle, then there is no well-defined notion of pull-back orbibundle similar to the manifold case. For example, with $\mathcal{M}$ and ϕ as in Example 2.1.10 and $\mathcal{E} = T\mathcal{M}$, the fiber of underlying space E over $[(0,y)]$ is a $\mathbb{Z}_2$-quotient of $\mathbb{R}^2$, but every orbibundle over $\mathbb{R}$ is a vector bundle. Nevertheless, the pull-back bundle is well-defined whenever the homomorphisms h_x that come with ϕ as in Definition 2.1.8 are monomorphism; see [**1**, Theorem 2.43]. We will use this fact in Section 2.5.

EXAMPLE 2.2.9. Let $\phi\colon\mathcal{M}' \longrightarrow\mathcal{M}$ be a smooth embedding of orbifolds as in Definition 2.2.3. Then there exists a well-defined normal orbibundle

$$\mathcal{N}_{\mathcal{M}'}\mathcal{M} \cong \frac{T\mathcal{M}|_{M'}}{T\mathcal{M}'} \longrightarrow \mathcal{M}'$$

generalizing the notion of normal bundle for smooth embedding of manifolds.

[13]Surjective means the underlying continuous map $\overline{\mathrm{pr}}\colon E \longrightarrow M$ is surjective.

DEFINITION 2.2.10. *An* ***orbibundle smooth map***, $(\mathfrak{D}\phi,\phi)\colon(\mathcal{E}_1,\mathcal{M}_1)\longrightarrow(\mathcal{E}_2,\mathcal{M}_2)$, *between two orbibundles* $\mathcal{E}_1$ *and* $\mathcal{E}_2$ *over* $\mathcal{M}_1$ *and* $\mathcal{M}_2$, *respectively, is given by an underlying pair of continuous maps* $(\overline{\mathfrak{D}\phi},\overline{\phi})\colon(E_1,M_1)\longrightarrow(E_2,M_2)$ *that commute with the projection maps* $\overline{\mathrm{pr}}_1$ *and* $\overline{\mathrm{pr}}_2$, *admit local smooth lifts as in (2.7), and* $\mathfrak{D}\phi$ *is linear on each fiber. An* ***orbibundle morphism*** *is an orbibundle smooth map for which the underlying orbifold smooth map* $\phi\colon\mathcal{M}_1\longrightarrow\mathcal{M}_2$ *is a morphism. An* ***orbibundle embedding*** *is an orbibundle morphism such that the lift maps (2.7) are smooth embeddings. Finally, a* ***sub-orbibundle*** $\mathcal{E}'$ *of an orbibundle* $\mathcal{E}$ *over* $\mathcal{M}$ *is a subspace* $(E',M')\subset(E,M)$ *together with an orbibundle structure* $\mathcal{E}'\longrightarrow\mathcal{M}'$ *on* (E',M') *such that the inclusion map* $\overline{\iota}\colon(E',M')\longrightarrow(E,M)$ *gives an orbibundle embedding* $\iota\colon(\mathcal{E}',\mathcal{M}')\longrightarrow(\mathcal{E},\mathcal{M})$.

We can similarly extend the notions of quotient bundle, tensor product, pullback bundle, etc. to the context of orbibundles.

From the perspective of Definition 2.2.3, an **orbifold section** s of an orbibundle $\mathrm{pr}\colon\mathcal{E}\longrightarrow\mathcal{M}$ is an orbifold map $s\colon\mathcal{M}\longrightarrow\mathcal{E}$ such that $\mathrm{pr}\circ s=\mathrm{id}$. The simplest orbifold section is the zero section. It is clear from the definition of orbibundle that the zero section is an orbifold embedding whose normal bundle is canonically isomorphic to $\mathcal{E}$ itself. We say that a section is **transverse** if all local lifts are transverse to the zero section.

EXAMPLE 2.2.11. For $m,n>0$, with $\gcd(m,n)=1$, consider $M=S^2$ and let $\mathcal{M}=\mathbb{P}^1_{m,n}$ denote the orbifold structure of type (m,n) obtained by replacing neighborhoods of 0 and ∞ in $\mathbb{P}^1\cong S^2$ with $\mathbb{Z}_m$ and $\mathbb{Z}_n$ quotients of a disk, respectively. We can cover $\mathbb{P}^1_{m,n}$ with two orbifold charts $\mathcal{V}_0$ and $\mathcal{V}_\infty$ where $V_0\cong V_\infty\cong\mathbb{C}$, $\mathbb{Z}_m$ and $\mathbb{Z}_n$ act by the corresponding roots of unity, $\psi_0(z)=z^m$, and $\psi_\infty(w)=w^n$ ($w^{-n}=z^m$ on the overlap). Let $\mathcal{E}=T\mathcal{M}$. Then it is easy to see that

$$s_0\colon V_0\longrightarrow TV_0,\ s_0(z)=z\frac{\partial}{\partial z},\ \text{ and } s_\infty\colon V_\infty\longrightarrow TV_\infty,\ s_\infty(w)=\frac{-m}{n}w\frac{\partial}{\partial w},$$

define a holomorphic orbifold section of the orbibundle $\mathcal{E}$ which is transverse to the zero section.

For an orbibundle $\mathcal{E}\longrightarrow\mathcal{M}$, let

$$\det\mathcal{E}\longrightarrow\mathcal{M}$$

be the determinant (top exterior power) orbi-line bundle of $\mathcal{E}$.

DEFINITION 2.2.12. *We say an orbibundle* $\mathcal{E}\longrightarrow\mathcal{M}$ *is* ***relatively orientable*** *if*

$$\det\mathcal{E}\otimes\det T\mathcal{M}\cong\mathcal{M}\times\mathbb{R}. \tag{2.10}$$

A relative orientation is a choice of trivialization in (2.10).

If s is a transverse section of a relatively oriented orbibundle $\mathcal{E}\longrightarrow\mathcal{M}$, then

$$Z(s)=\overline{s}^{-1}(0)\subset M$$

inherits a possibly non-effective but oriented orbifold structure $\mathcal{Z}(s)$ from the pair $(\mathcal{E},\mathcal{M})$. In the case of Example 2.2.11, $\mathcal{Z}(s)$ is a disjoint union of the two points 0 and ∞ with trivial (thus non-effective) actions of $\mathbb{Z}_m$ and $\mathbb{Z}_n$, respectively. For simplicity, we will allow these basic non-effective orbifold structures in Section 2.6.

2.3. Multisections. In this section, we introduce the notion of multisection. We are forced to consider multisections in order to achieve transversality. Let $s\colon \mathcal{M} \longrightarrow \mathcal{E}$ be an orbifold section, $\mathcal{U}$ as in (2.6) be an orbibundle chart of $\mathcal{E}$, $x \in V$ such that $G_x \neq 1$, and assume G_x acts non-trivially on U_x. Then $s(x) \in U_x^{G_x}$ and if this invariant subspace is zero, every section has to vanish at that point. Therefore, it is quite likely (e.g. if the set of such points has larger dimension than $\dim V - \text{rank } U$) that s does not admit any transversal deformation.

This leads us to consider multisections, i.e. a G-invariant set of possibly non G-equivariant sections in place of a single equivariant section. Given a vector bundle $\text{pr}\colon U \longrightarrow V$, we denote by $S^\ell(U)$ to be the quotient of $U^{\times \ell}$ (ℓ-times fiber-product of U over V) with respect to the action of the symmetric group of order ℓ,

$$v^1 \times \cdots \times v^\ell \sim v^{\sigma(1)} \times \cdots \times v^{\sigma(\ell)} \qquad \forall x \in V,\ v^1, \ldots, v^\ell \in U_x, \text{ and } \sigma \in S_\ell.$$

The projection map pr extends to a similarly denoted map $\text{pr}\colon S^\ell(U) \longrightarrow V$; however, the fibers are -usually- singular. An **ℓ-section** is then a (continuous) section of $S^\ell(U)$.

Similarly, given an orbibundle chart $\mathcal{U}$ as in (2.6), we can lift the G-action to a component-wise G-action on $U^{\times \ell}$ and it descends to a G-action on $S^\ell(U)$ commuting with pr. Then, by definition, an **orbifold ℓ-section** for $\mathcal{U}$ is a G-equivariant (continuous) section of $S^\ell(U)$. If $G = \{g_1, \ldots, g_\ell\}$, every arbitrary section $s\colon V \longrightarrow U$ can be naturally enhanced to an orbifold ℓ-section $\mathfrak{s} = [G \cdot s] \in S^\ell(U)$ defined by

$$\mathfrak{s} = [g_1 \cdot s, \ldots, g_\ell \cdot s], \quad (g_i \cdot s)(x) = g_i(s(g_i^{-1}(x)))\ \forall i \in [\ell],\ x \in V. \tag{2.11}$$

Every ℓ-section can be canonically extended to orbifold ℓm-sections, with $m \in \mathbb{N}$: we replace $[v^1, \ldots, v^\ell]$ with

$$m \cdot [v^1, \ldots, v^\ell] = [\underbrace{v^1, \ldots, v^\ell\ , \ldots,\ v^1, \ldots, v^\ell}_{\text{repeated m times}}].$$

We say two orbifold multisections $\mathfrak{s}_1 \in S^{\ell_1}(U)$ and $\mathfrak{s}_2 \in S^{\ell_2}(U)$ are **equivalent**, and write $\mathfrak{s}_1 \cong \mathfrak{s}_2$, if there exists some common multiple ℓ of ℓ_1 and ℓ_2 such that

$$\frac{\ell}{\ell_2} \cdot \mathfrak{s}_1 = \frac{\ell}{\ell_1} \cdot \mathfrak{s}_2; \tag{2.12}$$

the equality means that the un-ordered set of of values of $\frac{\ell}{\ell_2} \cdot \mathfrak{s}_1$ and $\frac{\ell}{\ell_1} \cdot \mathfrak{s}_2$ at every point $x \in V$ are the same in U_x^ℓ. The map

$$S^{\ell_1}(U) \times S^{\ell_2}(U) \longrightarrow S^{\ell_1 \ell_2}(U), \qquad \left([v^i]_{i \in [\ell_1]}, [u^j]_{j \in [\ell_2]}\right) \longrightarrow [v^i + u^j]_{\substack{i \in [\ell_1] \\ j \in [\ell_2]}} \tag{2.13}$$

extends the definition of addition to the equivalence classes of orbifold multisections. The extended sum operation is commutative and associative but does not have an inverse; hence, the space of equivalence classes of multisections together with this addition is just a commutative monoid.

Given an orbibundle $\text{pr}\colon \mathcal{E} \longrightarrow \mathcal{M}$ with an orbibundle atlas $\mathcal{A}_{\mathcal{E}}$ as in (2.8), an **orbifold multisection** with respect to $\mathcal{A}_{\mathcal{E}}$ is a family $\mathfrak{s} \equiv \{\mathfrak{s}_\alpha\}_{\alpha \in S}$ of (continuous) G_α-equivariant ℓ_α-sections $\mathfrak{s}_\alpha \in S^{\ell_\alpha}(U_\alpha)$ which are equivalent along intersections; cf. [**36**, Definition 4.13] for a global (coordinate independent) functorial definition. Note that if S is finite, e.g. whenever M is compact, we can take a large enough ℓ such that all the local representatives $\mathfrak{s}_\alpha$ are orbifold ℓ-sections and $\mathfrak{s}_\alpha = \mathfrak{s}_\beta$ on the overlap. A multisection with respect to $\mathcal{A}_{\mathcal{E}}$ induces a multisection with respect to any refinement of $\mathcal{A}_\mathcal{E}$. Given two multisections $\mathfrak{s}$ and $\mathfrak{s}'$ over $\mathcal{A}_\mathcal{E}$ and $\mathcal{A}'_\mathcal{E}$, respectively, $\mathfrak{s}+\mathfrak{s}'$ is a multisection defined over any common refinement of $\mathcal{A}_\mathcal{E}$ and $\mathcal{A}'_\mathcal{E}$.

DEFINITION 2.3.1. *Let $\mathcal{U}=(\mathrm{pr}\colon U\longrightarrow V, G, \psi)$ be an orbibundle chart and $\mathfrak{s}$ be an orbifold ℓ-section of that. We say $\mathfrak{s}$ is* ***liftable*** *(over this chart) if there is a G-invariant family of ℓ (not necessarily G-equivariant) sections*

$$s^i\colon V\longrightarrow U, \quad \forall i\in\{1,\dots,\ell\},$$

such that $\mathfrak{s}=[s^1,\dots,s^\ell]$. Similarly, with notation as in the paragraph before Example 2.2.11, let $\mathrm{pr}\colon\mathcal{E}\longrightarrow\mathcal{M}$ be an orbibundle and $\mathfrak{s}=\{\mathfrak{s}_\alpha\}_{\alpha\in S}$ be an orbifold multisection with respect to some orbibundle atlas $\mathcal{A}_\mathcal{E}$. We say $\mathfrak{s}$ is liftable over $\mathcal{A}_\mathcal{E}$ if every $\mathfrak{s}_\alpha$ is liftable. A multisection is called ***locally liftable*** *if it is liftable over some orbibundle atlas.*

Given two multisections $\mathfrak{s}$ and $\mathfrak{s}'$ which are liftable over $\mathcal{A}_\mathcal{E}$ and $\mathcal{A}'_\mathcal{E}$, respectively, the multisection sum $\mathfrak{s}+\mathfrak{s}'$ is liftable over any common refinement of $\mathcal{A}_\mathcal{E}$ and $\mathcal{A}'_\mathcal{E}$.

EXAMPLE 2.3.2. With $\mathcal{M}=\mathbb{P}^1_{m,n}$ as in Example 2.2.11, consider the trivial line bundle

$$\mathcal{E}=\mathbb{P}^1_{m,n}\times\mathbb{C}.$$

Over $\mathcal{V}_0$, consider the section $s_0(z)=\rho_0(|z|)z$ where ρ_0 is a bump function supported near $z=0$. The extension of this section to $\mathcal{V}_\infty$ is a multisection of the form

$$\mathfrak{s}_0(w)=[\rho_0(|w|^{-n/m})w^{-n/m}]\in S^m(\mathcal{V}_\infty\times\mathbb{C}).$$

Similarly, consider the section $s_\infty(w)=\rho_\infty(|w|)w$ and extend it to a multisection of the form

$$\mathfrak{s}_\infty(z)=[\rho_\infty(|z|^{-m/n})z^{-m/n}]\in S^n(\mathcal{V}_0\times\mathbb{C})$$

over $\mathcal{V}_0$. Let $\mathfrak{s}=\mathfrak{s}_0+\mathfrak{s}_\infty$. Since both $\mathfrak{s}_0$ and $\mathfrak{s}_\infty$ are liftable on their defining charts, $\mathfrak{s}$ is a locally liftable multisection; however, it is not liftable over the orbibundle atlas made of $\mathcal{V}_0\times\mathbb{C}$ and $\mathcal{V}_\infty\times\mathbb{C}$, i.e. $\mathcal{V}_0\times\mathbb{C}$ and $\mathcal{V}_\infty\times\mathbb{C}$ do not form a sufficiently refined atlas for $\mathfrak{s}$.

REMARK 2.3.3. If $\mathfrak{s}$ is a liftable orbifold ℓ-section of an orbibundle chart $\mathcal{U}$, the decomposition $\mathfrak{s}=[s^1,\dots,s^\ell]$ into a G-invariant set of sections of U is not unique. For example, assume $U=\mathbb{R}\times\mathbb{R}$, $V=\mathbb{R}$, pr is the projection onto the first factor, and $G=\mathrm{id}$. For each pair $(\epsilon_1,\epsilon_2)=(\pm,\pm)$, define

$$s_{\epsilon_1,\epsilon_2}(t)=\begin{cases}\epsilon_1 e^{-1/|t|} & \text{if } t\le 0\\ \epsilon_2 e^{-1/|t|} & \text{if } t\ge 0\end{cases}$$

to be the corresponding section of U. Then, the equations

$$\mathfrak{s}_1=[s_{+,+},s_{-,-}], \quad \mathfrak{s}_2=[s_{+,-},s_{-,+}],$$

describe different decompositions of the same 2-section $\mathfrak{s}=(\mathfrak{s}_1\cong\mathfrak{s}_2)$ into a set of two smooth branches.

REMARK 2.3.4. In light of the preceding remark, it will be important in the construction of Euler class of an orbibundle in Section 2.6, and its generalization to the construction of VFC of a Kuranishi structure in Section 4, that we consider a consistent choice of the liftings of a multisection over different local charts. For this reason, we consider lifted multisections instead of just liftable multisections; see definition below. This change of perspective, compared to the original definitions in [**16**, **17**, **20**], simplifies some of the steps in the construction of perturbations and VFC in Sections 3 and 4.

DEFINITION 2.3.5. *Let $\mathcal{E} \longrightarrow \mathcal{M}$ be an orbibundle and $\mathfrak{s}$ be a locally liftable multisection of $\mathcal{E}$. A **lifting** of $\mathfrak{s}$ consists of an orbibundle atlas $\mathcal{A}_{\mathcal{E}}$ as in (2.8) with respect to which $\mathfrak{s} = \{\mathfrak{s}_\alpha\}_{\alpha\in S}$ is liftable and a choice of lifting of every $\mathfrak{s}_\alpha$ into an unordered G_α-invariant set of sections (called **branches**),*

$$\mathfrak{s}_\alpha = [s_\alpha^1, \ldots, s_\alpha^{\ell_\alpha}],$$

which are compatible on the overlaps in the following sense. For every common refinement $\Phi_\alpha, \Phi_\beta \colon \mathcal{U}_{\alpha,\beta} \longrightarrow \mathcal{U}_\alpha, \mathcal{U}_\beta$, as in the paragraph following Definition 2.2.1, the pull-back branches

$$\mathfrak{D}\phi_\alpha^*(\ell_\beta \cdot \{s_\alpha^1, \ldots, s_\alpha^{\ell_\alpha}\}) \text{ and } \mathfrak{D}\phi_\beta^*(\ell_\alpha \cdot \{s_\beta^1, \ldots, s_\beta^{\ell_\beta}\})$$

*are equal as an unordered set with multiplicities. By a **lifted** multisection, we mean a locally liftable multisection together with a choice of lifting.*

Given a lifting over some orbibundle atlas $\mathcal{A}_{\mathcal{E}}$, by restricting to sub-charts, we obtain a lifting over any refinement $\mathcal{A}'_{\mathcal{E}}$ of $\mathcal{A}_{\mathcal{E}}$. Therefore, the multisection sum of two lifted multisections is "lifted" as well. As we showed in Example 2.3.3, a multisection can have many different liftings; thus, the notion of "equivalence" between lifted multisections should be enhanced as well.

DEFINITION 2.3.6. *We say two lifted multisections $\mathfrak{s}$ and $\mathfrak{s}'$ are **equivalent**, and write $\mathfrak{s} \cong \mathfrak{s}'$, if they are equivalent as in (2.12) and up to some multiple have the same germ of branches around every point.*

DEFINITION 2.3.7. *An orbifold multisection is said to be **smooth** (resp. C^m, with $m \geq 0$), if it is lifted*[14] *and each local branch of the chosen lift is smooth (resp. C^m). An orbifold smooth (or C^m for some $m>0$) multisection is said to be **transversal** if each local branch of the chosen smooth lift is transversal to the zero section.*

We denote the set of C^m-smooth (hence lifted) orbifold multisections of $\mathcal{E}$ by $C^m_{\text{multi}}(\mathcal{E})$, the set of C^m-smooth orbifold sections of $\mathcal{E}$ by $C^m(\mathcal{E})$, and the set of C^m-smooth functions on $\mathcal{M}$ by $C^m(\mathcal{M})$.

DEFINITION 2.3.8. *Let $\mathcal{E} \longrightarrow \mathcal{M}$ be an orbibundle, $(\mathfrak{s}_i)_{i=1}^\infty$ be a sequence of lifted orbifold multisections, and $||\cdot||$ be some specified norm. We say this sequence $||\cdot||$-**converges** to the lifted orbifold multisection $\mathfrak{s}_\infty$, and write*

$$\lim_{i\longrightarrow\infty} \mathfrak{s}_i = \mathfrak{s}_\infty,$$

if there exist a finite atlas $\mathcal{A}_{\mathcal{E}}$ as in (2.8) and a family of positive integers $\{\ell_\alpha\}_{\alpha\in S}$ such that the following conditions hold.

- *For every $i \in \mathbb{N} \cup \{\infty\}$, the given lifting of $\mathfrak{s}_i$ is equivalent to a lifting $\{\mathfrak{s}_{i,\alpha}\}_{\alpha\in S}$ of the form*

$$\mathfrak{s}_{i,\alpha} = [s_{i,\alpha}^1, \ldots, s_{i,\alpha}^{\ell_\alpha}], \quad s_{i,\alpha}^j \colon V_\alpha \longrightarrow U_\alpha \ \forall \alpha\in S,\ j\in[\ell_\alpha]. \tag{2.14}$$

- *Every $s_{i,\alpha}^j$ has finite $||\cdot||$-norm, and (for some choice of ordering of (2.14))*

$$\lim_{i\longrightarrow\infty} ||s_{i,\alpha}^j - s_{\infty,\alpha}^j|| = 0, \quad \forall \alpha\in S,\ j\in[\ell_\alpha]. \tag{2.15}$$

[14] This is a prerequisite.

2.4. Perturbations. The purpose of this section is to introduce a method which allows us to perturb any smooth orbifold section to a smooth (and hence lifted) multisection transverse to the zero section.

Assume $\mathcal{E}\longrightarrow\mathcal{M}$ is an orbibundle and s is an orbifold C^m-section of $\mathcal{E}$. Let

$$K_1\subset W_1\subset K_2\subset M \tag{2.16}$$

such that K_1 and K_2 are compact and W_1 is open. Let

$$\mathfrak{t}_1\in C^m_{\text{multi}}(\mathcal{E}|_{W_1})$$

be a given lifted multisection of $\mathcal{E}$ over W_1 such that $s|_{W_1}+\mathfrak{t}_1$ is a transversal orbifold multisection of $\mathcal{E}|_{W_1}$.

Let K_1^c and W_1^c be the complements of K_1 and W_1 in M, respectively. By (2.16), W_1^c is closed and K_1^c is an open neighborhood of W_1^c. By compactness of K_2 and Remark 2.2.6, there exists a finite orbibundle atlas $\mathcal{A}_{\mathcal{E}}$ as in (2.8) for $\mathcal{E}$ over a neighborhood W_2 of K_2 with the following properties.

- *For every* $\alpha\in S$, V_α *is isomorphic to some small ball in* $\mathbb{R}^{n+n_\alpha}$ *and* U_α *is trivial; we fix a choice of smooth trivialization* $U_\alpha\cong V_\alpha\times\mathbb{R}^{n_\alpha}$ *and an identification*

$$V_\alpha\cong B_{\delta_\alpha}(0)\subset\mathbb{R}^{n+n_\alpha}. \tag{2.17}$$

- *There exists a decomposition* $S=S_1\sqcup S_2$ *such that*

$$K_1\subset\Big(O_1=\bigcup_{\alpha\in S_1}\psi_\alpha(V_\alpha)\Big)\subset W_1,\qquad W_1^c\cap K_2\subset\Big(O_2=\bigcup_{\alpha\in S_2}\psi_\alpha(V_\alpha)\Big)\subset K_1^c\cap W_2. \tag{2.18}$$

- *Restricted to* O_1, *the given lifting of* $\mathfrak{t}_1$ *is equivalent to a lifting* $\{\mathfrak{t}_{1,\alpha}\}_{\alpha\in S_1}$ *of the form*

$$\mathfrak{t}_{1,\alpha}=[t^1_\alpha,\ldots,t^{\ell_\alpha}_\alpha],\quad t^j_\alpha\colon V_\alpha\longrightarrow U_\alpha\quad\forall j\in[\ell_\alpha],\ \alpha\in S_1. \tag{2.19}$$

- *The underlying orbifold atlas*

$$\mathcal{A}_{W_2}\equiv\{\mathcal{V}_\alpha=(V_\alpha,G_\alpha,\psi_\alpha)\}_{\alpha\in S}$$

covering W_2 *admits an orbifold smooth* ***partition of unity*** $\{\chi_\alpha\}_{\alpha\in S}$ *with respect to* K_2 *(cf.* [**50**, p. 361]*), i.e. each* $\chi_\alpha\colon V_\alpha\longrightarrow[0,1]$ *is a compactly supported* G_α*-invariant smooth function on* V_α *and*

$$\overline{\chi}(x)=\sum_{\alpha\in S:\,\overline{x}\in\psi_\alpha(V_\alpha)}\overline{\chi}_\alpha(\overline{x})=1\quad\forall\overline{x}\in K_2.$$

Let

$$W_3=\{\overline{x}\in W_2:\overline{\chi}(\overline{x})\neq0\},\quad K_2\subset W_3\subset W_2.$$

We fix a choice of such atlas. Define

$$\overline{\eta}\colon W_2\longrightarrow\mathbb{R},\qquad\overline{\eta}(\overline{x})=\sum_{\alpha\in S_1}\overline{\chi}_\alpha(\overline{x});$$

this is the underlying continuous function of an orbifold smooth function supported in O_1 with $\overline{\eta}|_{K_1}\equiv1$. Via the cut-off function $\overline{\eta}$, we extend $\mathfrak{t}_1$ to the well-defined lifted C^m-multisection $\eta\mathfrak{t}_1$ over entire W_2.

Let ν be an arbitrary set of not necessarily equivariant local C^m-sections

$$\nu\equiv\{\nu_\alpha\colon V_\alpha\longrightarrow U_\alpha\}_{\alpha\in S_2}.$$

With respect to the chosen set of trivializations of each U_α and the identifications (2.17), we define the C^m-norm of ν by

$$||\nu||^2_{C^m} = \sum_{\alpha\in S_2} ||\nu_\alpha||^2_{C^m}.$$

Let $\mathrm{Pert}(\mathcal{A}_\mathcal{E})$ be the Banach space of such ν with finite norm. Corresponding to every $\nu \in \mathrm{Pert}(\mathcal{A}_\mathcal{E})$, we build a C^m-smooth lifted multisection $\mathfrak{t}(\nu)$ of $\mathcal{E}|_{W_2}$ such that $\mathfrak{t}(\nu)|_{K_1} = \mathfrak{t}_1|_{K_1}$ in the following way.

For every $\beta\in S_2$, if $G_\beta = \{g_1, \ldots, g_\ell\}$, let

$$\mathfrak{t}_\beta = [g_1 \cdot \nu_\beta, \ldots, g_\ell \cdot \nu_\beta]$$

be the corresponding natural lifted multisection in (2.11). Let

$$\mathfrak{t}_2 = \sum_{\beta\in S_2} \chi_\beta \mathfrak{t}_\beta. \tag{2.20}$$

Since χ_β is supported in V_β, $\chi_\beta\mathfrak{t}_\beta$ naturally extends to a lifted multisection over W_2. The sum considered in (2.20) is the sum of multisections as in (2.13). Since the multisection sum of a finite set of lifted multisections is lifted, $\mathfrak{t}_2$ is a lifted C^m-multisection supported in O_2 (hence it is supported away from K_1). Finally, we put

$$\mathfrak{t}(\nu) = \eta\mathfrak{t}_1 + \mathfrak{t}_2;$$

this is again a sum of multisections as in (2.13). It is lifted over an appropriate finite refinement of $\mathcal{A}_\mathcal{E}$. More precisely, for every $\overline{x} \in W_3$, let

$$S(\overline{x}) = \{\alpha \in S \colon \overline{x} \in \psi_\alpha(V_\alpha)\},$$

and $\mathcal{U}_x = (U_x, V_x, G_x, \psi_x, \mathrm{pr}_x)$ be an orbibundle chart centered at $\overline{x}$ which is a refinement of $\mathcal{U}_\alpha$, for every $\alpha\in S(\overline{x})$, and

$$\psi_x(V_x) \subset W_3, \quad \overline{\chi}_\alpha|_{\psi_x(V_x)} \equiv 0 \qquad \forall \alpha \notin S(\overline{x}).$$

The latter condition can be satisfied because support of $\overline{\chi}_\alpha$, for every $\alpha \notin S(\overline{x})$, is a compact subset of V_α and $\overline{x} \notin \psi_\alpha(V_\alpha)$. By the choice of $\mathcal{U}_x$, for every $\overline{x}\in W_3$, the multisection $\mathfrak{t}(\nu)$ and hence $s + \mathfrak{t}(\nu)$ is lifted over $\mathcal{U}_x$; moreover, the set of branches is unique up to ordering. Let

$$(s + \mathfrak{t}(\nu))|_{V_x} = [s + t(\nu)^1, \ldots, s + t(\nu)^{\ell_x}] \tag{2.21}$$

be the set of branches. By the compactness of K_2, there exists a finite set T of such charts covering a neighborhood $W_4 \subset W_3$ of K_2; moreover, fixing an ordering of S, an ordering of G_β, for every $\beta\in S_2$, and an ordering of the set branches of $\mathfrak{t}_{1,\alpha}$, for all $\alpha\in S_1$, determines[15] an ordering of branches in (2.21), for all $\overline{x} \in T$. Then, for every $\nu\in\mathrm{Pert}(\mathcal{A}_\mathcal{E})$ let

$$\mathcal{Z}(\nu) \equiv \coprod_{\overline{x}\in T} \coprod_{i=1}^{\ell_x} \{y \in V_x \colon (s + t(\nu)^i)(y) = 0\} \subset \coprod_{\overline{x}\in T} \coprod_{i=1}^{\ell_x} V_x.$$

This completes the construction of extension and its zero set. Every $\nu\in\mathrm{Pert}(\mathcal{A}_\mathcal{E})$ is a **perturbation term** and the construction above gives us a systematic way of constructing a multisection $\mathfrak{t}(\nu)$ out of a perturbation term. Next, we are going to show that for generic ν, the perturbed multisection $s + \mathfrak{t}(\nu)$ is transverse to the

[15] Although it is clear from the definition of $\mathfrak{t}(\nu)$, the precise argument for this claim is a bit tedious to write.

zero section on W_4. Note that by definition, $\mathfrak{t}(\nu)$ is already transverse to the zero section along K_1.

A set

$$\Omega = \{\nu_1, \ldots, \nu_N\} \subset \mathrm{Pert}(\mathcal{A}_{\mathcal{E}})$$

is called a **globally generating set relative to** K_1, if for every $\overline{x} \in T$, every $i \in [\ell_x]$, and every $y \in V_x$, either $\overline{\eta}(\overline{y}) = 1$ or for every $u \in U_y$ there exists a linear combination $\nu = \sum_{j \in [N]} a_j \nu_j$ such that $t(\nu)^i(y) = u$.

EXAMPLE 2.4.1. With n_α as in (2.17), let

$$I = \{j \equiv (j_\alpha)_{\alpha \in S_2} : j_\alpha \in [n_\alpha]\}.$$

For every $j \in I$, let

$$\nu_j \equiv ((\nu_j)_\alpha)_{\alpha \in S_2}, \ (\nu_j)_\alpha : V_\alpha \longrightarrow U_\alpha \cong V_\alpha \times \mathbb{R}^{n_\alpha}, \ (\nu_j)_\alpha = \underbrace{(0, \ldots, 0, 1, 0, \ldots, 0)}_{\text{only 1 at the } j_\alpha\text{-th position}} .$$

Then the set $\Omega = \{\nu_j\}_{j \in I}$ is a globally generating set relative to K_1.

Given a globally generating set $\Omega = \{\nu_1, \ldots, \nu_N\}$, let

$$\mathbb{R}^\Omega = \{\nu = \sum_{j=1}^{N} a_j \nu_j, \ a_j \in \mathbb{R}\}$$

and

$$\mathcal{Z} = \bigcup_{\nu \in \mathbb{R}^\Omega} \mathcal{Z}(\nu) \times \{\nu\}, \quad \mathcal{Z} \subset \Big(\coprod_{\overline{x} \in T} \coprod_{i=1}^{\ell_x} V_x \Big) \times \mathbb{R}^\Omega.$$

Let $\pi : \mathcal{Z} \longrightarrow \mathbb{R}^\Omega$ be the projection map. By Sard's theorem and transversality of $s + \mathfrak{t}(\nu)$ on the complement of support of $(1 - \overline{\eta})$, for large enough m (i.e. if the sections are differentiable enough), generic $\nu \in \mathbb{R}^\Omega$ is a regular value of π. This implies that for generic $\nu \in \mathbb{R}^\Omega$, $s + \mathfrak{t}(\nu)$ is a transverse multisection on the complement of $\eta \equiv 1$. Since $s + \mathfrak{t}(\nu)$ is already transverse on O_1, this implies that $s + \mathfrak{t}(\nu)$ is transverse everywhere on W_4. We summarize the outcome of this argument as the following proposition. This proposition is a relative (and more general) version of [**20**, Lemma 3.14] which is used (but not explicitly stated) in [**16**, Section 6].

PROPOSITION 2.4.2. *Assume $\mathcal{E} \longrightarrow \mathcal{M}$ is an orbibundle and s is an orbifold C^m-section of $\mathcal{E}$, for some sufficiently large $m \in \mathbb{N}$. Let*

$$K_1 \subset W_1 \subset K_2 \subset M$$

such that K_1 and K_2 are compact and W_1 is open. Let

$$(\mathfrak{t}_{1,i})_{i \in \mathbb{N}} \subset C^m_{\mathrm{multi}}(\mathcal{E}|_{W_1})$$

be a given sequence of multisections of $\mathcal{E}$ over W_1 such that $(s|_{W_1} + \mathfrak{t}_{1,i})_{i \in \mathbb{N}}$ is a sequence of transversal orbifold multisections of $\mathcal{E}|_{W_1}$ and $\lim_{i \to \infty} \mathfrak{t}_{1,i} = 0$ in C^m-convergence norm over W_1. Then, there exists a neighborhood[16] *W of K_2 in M and a sequence of a C^m-multisections $(\mathfrak{t}_i)_{i \in \mathbb{N}} \subset C^m_{\mathrm{multi}}(\mathcal{E}|_W)$ such that $\mathfrak{t}_i|_{K_1} \cong \mathfrak{t}_{1,i}|_{K_1}$, $(s + \mathfrak{t}_i)_{i \in \mathbb{N}}$ is a sequence of transversal orbifold multisections of $\mathcal{E}|_W$, and $\lim_{i \to \infty} \mathfrak{t}_i = 0$ in C^m-convergence norm over W.*

[16]This is W_4 of the construction above.

REMARK 2.4.3. The orbifolds, Kuranishi structures, and gluing maps considered in this article are assumed to be (sufficiently) smooth, only to satisfy of the smoothness condition in Sard's theorem we used above. Nevertheless, C^1-smoothness is enough for the rest of arguments throughout the article. In genus 0 (and assumably higher genus), a middle ground between the C^1-smoothness and smoothness condition is the C^1SS (C^1 Stratified Smooth) condition considered in [**6**]. According to [**6**, Lemma 5.2.2], Sard's theorem holds in the category of C^1SS-smooth manifolds as well.

2.5. Resolution of multisections. Let $E \longrightarrow M$ be a relatively oriented vector bundle and $s\colon M \longrightarrow E$ be a transversal section with compact support. Then the zero set of s is a smooth oriented manifold which defines the Euler class of E as a singular homology class in M; see the beginning of Section 2.6. Similarly, we can define the Euler class of orbibundles, if we can find a transversal orbifold section.

As we argued at the beginning of Section 2.3, it is quite likely that an orbibundle does not admit any transversal orbifold section. In the light of Proposition 2.4.2, we consider a transversal multisection, in place of an orbifold section, and construct the Euler class from the "zero set" of the multisection. To this end, we introduce the notion of "resolution" of multisections which allows us, after passing to some covering spaces, to reduce the problem back to the of case of orbifold sections.

Throughout this section, assume $\mathrm{pr}\colon \mathcal{E} \longrightarrow \mathcal{M}$ is a smooth (or C^m for some $m>0$) relatively oriented orbibundle and $\mathfrak{s}$ is a lifted orbifold multisection of $\mathcal{E}$ with compact support. By assumption, there is a finite orbibundle atlas $\mathcal{A}_{\mathcal{E}}$ as in (2.8) covering a neighborhood of the zero set of $\mathfrak{s}$ such that the given lifting of $\mathfrak{s}$ is equivalent to a lifting $\{\mathfrak{s}_\alpha\}_{\alpha\in S}$ of the form

$$\mathfrak{s}_\alpha = [s_\alpha^1,\ldots,s_\alpha^{\ell_\alpha}], \quad s_\alpha^j\colon V_\alpha \longrightarrow U_\alpha \quad \forall \alpha\in S,\ j\in[\ell_\alpha]. \tag{2.22}$$

Let

$$Z(\mathfrak{s}) = \bigcup_{\substack{\alpha\in S\\ j\in[\ell_\alpha]}} \psi_\alpha((s_\alpha^j)^{-1}(0)) \subset M \tag{2.23}$$

be the zero set of $\mathfrak{s}$.

In general, $Z(\mathfrak{s})$ can be a complicated subspace of M, i.e. it may not be a sub-orbifold or it may not even admit a triangulation. In order to resolve this issue, we introduce the notion of **resolution** of $\mathfrak{s}$. In short, a resolution of $\mathfrak{s}$ consist of a covering of a neighborhood W of $Z(\mathfrak{s})$ by a finite number of open sets, and a compatible system of orbifold covering spaces over these open sets, such that the multisection $\mathfrak{s}$ on each piece is the push forward of some orbifold section on the covering. In the next section, we start with a resolution of $\mathfrak{s}$, compatibly triangulate the zero sets of the associated orbifold sections, and push forward these triangulations (with some $\mathbb{Q}$-coefficients) into $Z(\mathfrak{s})$ to define the Euler class of $\mathcal{E}$. We begin the discussion by recalling the notion of covering space for orbifolds; see [**53**] for more details.

DEFINITION 2.5.1. *An orbifold smooth map* $q\colon \mathcal{M}_1 \longrightarrow \mathcal{M}_2$ *is called a c-**covering map**, if every* $\overline{x}\in M_2$ *has an orbifold chart* (V,G,ψ) *centered around* *that, over which the orbifold structure on* $\overline{q}^{-1}(\psi(V))$ *is of the form*

$$\coprod_{a=1}^{N} V \longrightarrow \coprod_{a=1}^{N} V/G_a, \tag{2.24}$$

where G_a *(for every* $a\in[N]$*) is a subgroup of* G*,*

$$\sum_{a=1}^{N} |G/G_a| = c, \tag{2.25}$$

and the map q *on each component of* (2.24) *is given by the identity map on* V *as in the following diagram,*

$$\begin{array}{ccc} V & \xrightarrow[\cong]{q} & V \\ \big\downarrow \pi & & \big\downarrow \pi \\ V/G_a & \xrightarrow{\overline{q}} & V/G\,. \end{array}$$

For example, if $\mathcal{M}_2$ is an effective global quotient $[M/G]$, then the manifold $\mathcal{M}_1 = M$ is a $|G|$-covering of $\mathcal{M}_2$. In fact, for every $x \in M$ and some sufficiently small G_x-invariant open neighborhood V of x, $[V/G_x]$ is an orbifold chart of $[M/G]$ centered at $\overline{x}$, $\{[g]\cdot V\}_{[g]\in G/G_x}$ are the manifold charts around different preimages of $[V/G_x]$, and (2.25) is simply equal to

$$\sum_{i=1}^{|G/G_x|} |G_x| = |G|.$$

If $\mathcal{M}_2$ is a manifold, then $\mathcal{M}_1$ is a covering space in the usual sense. It is clear from Definition 2.5.1 that every covering map is an orbifold map such that all the homomorphisms h_x as in Definition 2.1.8 are monomorphism. Therefore, by Remark 2.2.8, if $q\colon \mathcal{M}_1 \longrightarrow \mathcal{M}_2$ is a c-covering map, every orbibundle $\mathcal{E}_2$ over $\mathcal{M}_2$ naturally pulls back to an orbibundle $\mathcal{E}_1 = q^*\mathcal{E}_2$ over $\mathcal{M}_1$. If $\mathcal{E}_2 \longrightarrow \mathcal{M}_2$ is relatively oriented, then so is $\mathcal{E}_1 \longrightarrow \mathcal{M}_1$.

Let $\mathcal{M}_{1,o}$, with $1 \le o \le t$, be the connected components of $\mathcal{M}_1$ and suppose $q|_{\mathcal{M}_{1,o}}$ is a c_o-covering; we have

$$\sum_{o=1}^{t} c_o = c.$$

Let $\{w(o)\}_{o=1}^{t}$ be a set of positive integers, called **weights** below, and set

$$\ell = \sum_{o=1}^{t} c_o\, w(o).$$

Then, every orbifold section s of $\mathcal{E}_1$ descends to a lifted ℓ-section $\mathfrak{s}$ of $\mathcal{E}_2$, called the **weighted push forward multisection**, in the following way. With notation as in Definition 2.5.1, let $\mathrm{pr}\colon U \longrightarrow V$ be an orbibundle chart of $\mathcal{E}_2$ and

$$s^a\colon V \longrightarrow U, \quad 1 \le a \le N,$$

be the family of local G_a-equivariant representatives of s over different components of $q^{-1}[V/G]$ in (2.24). Let $c'_a = |G/G_a|$, for all $1 \le a \le N$. Each orbifold chart $[V/G_a]$

for $\mathcal{M}_1$ belongs to some connected component $\mathcal{M}_{1,o(a)}$; we define $w(a)\in\mathbb{Z}^{>0}$ to be $w(o(a))$. We define the weighted push forward multisection

$$\mathfrak{s}=q_*^w s$$

to be the multisection given by the following set of branches on the orbifold chart $[V/G]$ for $\mathcal{M}_2$,

$$\mathfrak{s}|_V = [\underbrace{g_{11}s^1,\ldots,g_{1c_1'}s^1,\ldots}_{\text{repeated } w(1) \text{ times}},\ldots\ldots,\underbrace{g_{N1}s^N,\ldots,g_{Nc_N'}s^N,\ldots}_{\text{repeated } w(N) \text{ times}}], \tag{2.26}$$

where $\{g_{a1},\ldots,g_{ac_a'}\}$ forms a set of representatives of G/G_a. In simple words, $\mathfrak{s}$ is a lifted multisection whose branches over V are give by the section s over different lifts of V in (2.24), and the number of times each branch appears is equal to the weight of the component it comes from.

REMARK 2.5.2. Replacing the weight function w with a uniform multiple of that does not change the equivalence class of the resulting weighted push forward multisection. We define the **weight ratio** by

$$\overline{w}(o) = \frac{w(o)}{\ell} = \frac{w(o)}{\sum_{o'=1}^{t} c_{o'}\ w(o')}. \tag{2.27}$$

We can think of (2.27) as a locally constant rational function

$$\overline{w}\colon M_1 \longrightarrow \mathbb{Q}.$$

We use the weight ratio function (2.27) in the formulation of Euler class singular cycle in (2.43). Note that $\overline{w}\equiv\frac{1}{c}$, if all the weights are equal to 1.

Let us return to the construction of resolution. Given an orbibundle chart $\mathcal{U}=(\mathrm{pr}\colon U\longrightarrow V,G,\psi)$ and a lifted ℓ-section $\mathfrak{s}=[s^1,\ldots,s^\ell]$ on $\mathcal{U}$, we define the **multiplicity** of every s^i **over** V to be the number of branches s^j which are equal to s^i over V. For a push forward multisection of the form (2.26), for generic choices of $s^1,\ldots,s^N$, $\mathfrak{s}$ has $c=\sum_{a=1}^N c_a'$ distinct branches where each $g_{ij}s^i$ has multiplicity $w(i)$.

For every $\overline{x}\in M$, let

$$\mathrm{val}_{\mathfrak{s}}(\overline{x})\in\mathbb{N} \tag{2.28}$$

be the number of distinct germs of branches of the given transversal multisection $\mathfrak{s}$ around $\overline{x}$. In other words, if $\mathrm{val}_{\mathfrak{s}}(\overline{x})=c$, then there exists some open set $W\subset M$ around $\overline{x}$ such that for every orbibundle chart $\mathcal{U}$ with $\psi(V)\subset W$, $\mathfrak{s}|V$ has exactly c distinct branches. The function $\mathrm{val}_{\mathfrak{s}}$ is well-defined and upper semi-continuous; see Figure 3. Therefore, for every $c\in\mathbb{N}$, the set

$$M(\geq c)=\{\overline{x}\in M\colon \mathrm{val}_{\mathfrak{s}}(\overline{x})\geq c\}$$

is closed. Let

$$M(c)=\{\overline{x}\in M\colon \mathrm{val}_{\mathfrak{s}}(\overline{x})=c\}\quad \forall c\in\mathbb{N}.$$

DEFINITION 2.5.3. *For any $c\in\mathbb{N}$ and any compact set $K\subset M(c)$, by a **resolution of $\mathfrak{s}$ around** K we mean an open neighborhood W of K and an orbifold $\widetilde{\mathcal{W}}$ with a c-covering map $q\colon\widetilde{\mathcal{W}}\longrightarrow\mathcal{W}=\mathcal{M}|_W$, such that the lifted multisection $\mathfrak{s}$ restricted to W is equivalent to the weighted push forward of some orbifold section s of $\widetilde{\mathcal{E}}=q^*(\mathcal{E}|_W)$.*

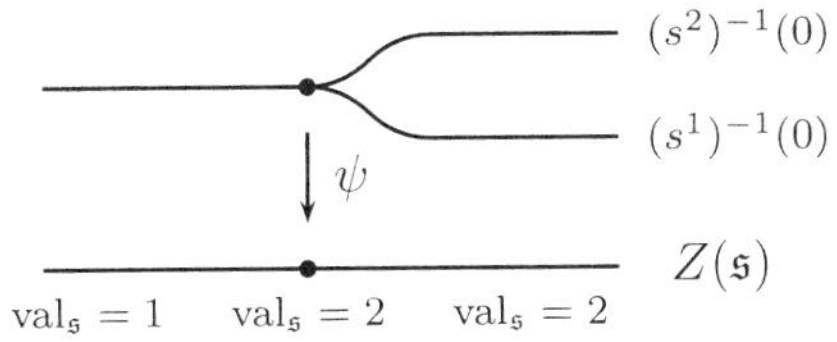

FIGURE 3. The $\mathrm{val}_{\mathfrak{s}}$ function for a multisection with two branches.

LEMMA 2.5.4. *For every $c\in\mathbb{N}$ and every compact set $K\subset M(c)$ such that*

$$K\cap M(>c)=\emptyset,$$

there exists a resolution of $\mathfrak{s}$ around K.

PROOF. For every $\overline{x}\in K$ choose an orbibundle chart $\mathcal{U}_x=(\mathrm{pr}_x\colon U_x\longrightarrow V_x, G_x, \psi_x)$ centered at $\overline{x}$ such that

$$\mathfrak{s}|_{V_x}=[s^1,\ldots,s^{\ell_x}],$$

for some $\ell_x\geq c$, has exactly c distinct branches and $\psi_x(V_x)\cap M(>c)=\emptyset$. Since K is compact, there exists a finite subset of points $I\subset K$ such that

$$W=\bigcup_{\overline{x}\in I}\psi_x(V_x)\subset M$$

is an open neighborhood of K in M disjoint from $M(>c)$. Over each V_x, if $s^{j_1},\ldots,s^{j_c}$ forms a set of distinct branches of $\mathfrak{s}|_{V_x}$, let

$$J_x=\{j_1,\ldots,j_c\}\subset[\ell_x]$$

and define the weight $w(j_a)$ of j_a to be the multiplicity of s^{j_a}. Since the set of branches of $\mathfrak{s}|_{V_x}$ is G_x-invariant, G_x acts on J_x and hence on the product

$$\widetilde{V}_x=V_x\times J_x.$$

The projection map into the first component, $q_x\colon\widetilde{V}_x\longrightarrow V_x$, is a G_x-equivariant c-covering map. If $|I|=1$, say $I=\{\overline{x}\}$, the effective global quotient orbifold $\widetilde{\mathcal{W}}=[\widetilde{V}_x/G_x]$ has the desired property. In fact, every connected component of $\widetilde{V}_x/G_x$ is of the form

$$(V_x\times o)/G_x\cong V_x/G_{x,o},\tag{2.29}$$

where $o=G\cdot j_a$ is some G_x-orbit in J_x and $G_{x,o}$ is the isotropy group of j_a (Changing the choice of j_a in o changes everything by conjugation). Since the weight function defined above is constant in each orbit, we define the weight $w(o)$ over the connected component (2.29) to be the weight of j_a. Restricted to such component, the $G_{x,o}$-equivariant orbifold section $s_o\colon V_x\longrightarrow U_x$ can be chosen to be s^{j_a}. If $|I|>1$, the conclusion follows from Lemma 2.5.5 below and induction on $|I|$. □

LEMMA 2.5.5. *For some $c\in\mathbb{N}$, let $K_1,K_2\subset M(c)$ be two compact subsets such that*

$$K_1\cap M(>c)=K_2\cap M(>c)=\emptyset$$

and both K_1 and K_2 admit a resolution of $\mathfrak{s}$ around them. Then $K=K_1\cup K_2$ admits a resolution of $\mathfrak{s}$ around it.

PROOF. By assumption, for $j=1,2$, there exist a neighborhood W_j of K_j, disjoint from $M(>c)$, and an orbifold $\widetilde{\mathcal{W}}_j$ with a c-covering map

$$q_j\colon \widetilde{\mathcal{W}}_j \longrightarrow \mathcal{W}_j = \mathcal{M}|_{W_j},$$

such that the multisection $\mathfrak{s}$ restricted to W_j is equivalent to the weighted push forward, with weight w_j, of an orbifold section s_j of $\widetilde{\mathcal{E}}_j = q_j^*(\mathcal{E}|_{W_j})$.

For every $\overline{x}\in K_{12}=K_1\cap K_2$, we can choose an orbibundle chart $\mathcal{U}_x$ centered at $\overline{x}$, such that $\mathfrak{s}|_{V_x}$ is lifted and has exactly c distinct branches, $\psi_x(V_x)$ is a subset of $W_1\cap W_2$, the covering map q_j (for $j=1,2$) lifts over V_x as in Definition 2.5.1, and $\mathfrak{s}=q_{j*}^{w_j}s_j$ as in (2.26). Moreover, we can choose these charts sufficiently small so that the following two conditions hold.

(1) If $\psi_x(V_x)\cap\psi_y(V_y)\neq\emptyset$, there exists some orbibundle chart $\mathcal{U}=(\mathrm{pr}\colon U\longrightarrow V,G,\psi)$ with the same properties such that

$$\psi_x(V_x)\cup\psi_y(V_y)\subset\psi(V). \tag{2.30}$$

(2) For every $\overline{x},\overline{y}\in K_{12}$,

$$\psi_x(V_x)\cap\psi_y(V_y)\cap M(c)\neq\emptyset,\ \text{whenever}\ \psi_x(V_x)\cap\psi_y(V_y)\neq\emptyset. \tag{2.31}$$

Since K_{12} is compact, a finite set I_{12} of such charts covers K_{12}. Let

$$W_{12}=\bigcup_{\overline{x}\in I_{12}}\psi_x(V_x),\qquad K_{12}\subset W_{12}\subset W_1\cap W_2.$$

Since $K_1\backslash W_{12}$ and $K_2\backslash W_{12}$ are compact and

$$(K_1\setminus W_{12})\cap(K_2\setminus W_{12})=\emptyset,$$

by possibly reducing W_1 and W_2 to smaller open sets, we can assume that

$$W_1\cap W_2=W_{12}.$$

For every $\overline{x}\in I_{12}$, let

$$q_x\colon\widetilde{V}_x=V_x\times J_x\longrightarrow V_x$$

be as in the proof of Lemma 2.5.4. On the other hand, for any of $j=1,2$ and by Definition 2.5.1, $q_j^{-1}[V_x/G_x]$ is also of the form $(V_x\times J_x^j)/G_x$, where

$$J_x^j=\coprod_{a=1}^{N^j}G_x/G_{x,a}^j,\quad |J_x^j|=c\ .$$

Here $\{G_{x,a}^j\}_{a=1}^N$ is a set of subgroups of G_x and G_x acts on J_x^j by acting on each coset $G_x/G_{x,a}^j$ from the left. For either of $j=1,2$, by definition, both J_x and J_x^j correspond to the set of distinct branches of the multisection $\mathfrak{s}$ on V_x; therefore, there is a unique identification between the set of orbits of the action of G_x on J_x^1 and J_x^2. This gives an identification of $\widetilde{\mathcal{W}}_1$ and $\widetilde{\mathcal{W}}_2$ along $q_1^{-1}[V_x/G_x]$ and $q_2^{-1}[V_x/G_x]$. The natural identifications corresponding to distinct $\overline{x},\overline{y}\in I_{12}$ are compatible along the intersections for the following two reasons. By (2.31), we can identify the set of distinct branches over V_x and V_y by their agreement near a point

$$\overline{z}\in\psi_x(V_x)\cap\psi_y(V_y)\cap M(c). \tag{2.32}$$

By (2.30), every two intersecting charts lie in a bigger chart over which there are exactly c distinct branches; this condition assures us that the previous set of identifications is independent of the choice of the point $\overline{z}$ in (2.32). We conclude that,

$\overline{q}_1^{-1}W_{12}$ and $\overline{q}_2^{-1}W_{12}$ are isomorphic as topological spaces and the orbifold structure induced on $\overline{q}_1^{-1}W_{12}$ by $\widetilde{\mathcal{W}}_1$ is the same as the orbifold structure induced on $\overline{q}_2^{-1}W_{12}$ by $\widetilde{\mathcal{W}}_2$. Identifying $\widetilde{\mathcal{W}}_1$ and $\widetilde{\mathcal{W}}_2$ along this common intersection, we obtain a c-covering $\widetilde{\mathcal{W}}$ of $\mathcal{W}|_{W_1\cup W_2}$. Finally, notice that the weight functions w_1 and w_2, up to multiplication by a constant, are determined by the multiplicity of distinct branches of the multisection $\mathfrak{s}$. Therefore, after possibly multiplying each of w_1 and w_2 by a constant, $q_{1*}^{w_1}s_1$ and $q_{2*}^{w_2}s_2$ are equal along W_{12} and we obtain a weight function w on the union such that $\mathfrak{s}|_{W_1\cup W_2} = q_*^w s$. □

REMARK 2.5.6. Let K be a compact subspace of M. More generally, we can define the notion of the **germ of an orbifold c-covering** of $\mathcal{M}$ along K: simply the direct limit of orbifold c-coverings of neighborhoods of K under restriction. In order to define such a germ, it suffices for each point of $\overline{x}$ in K to admit an orbifold chart centered at $\overline{x}$ and a c-covering as in Definition 2.5.1 with the property that if $\overline{y}\in\psi_x(V_x)$, then restricted to some smaller open set $V_y'\subset V_y$ such that V_y' is a refinement of V_x, the orbifold covering of V_y restricted to V_y' is "naturally" identified with the restriction of the orbifold covering of V_x restricted to V_y'. For example, in the case of Lemma 2.5.4, the natural identifications are given by the set of distinct branches of $\mathfrak{s}$.

Going back to the main discussion, by assumption, $Z(\mathfrak{s})$ is compact; therefore, $\mathrm{val}_{\mathfrak{s}}|_{Z(\mathfrak{s})}$ takes only finitely many values

$$c_1 < \cdots < c_N.$$

We use Lemma 2.5.4 to build a sequence of resolutions of $\mathfrak{s}$ around some compact subsets of

$$Z(\mathfrak{s})_i = M(c_i)\cap Z(\mathfrak{s}) \quad \forall 1\le i\le N.$$

We then use these coverings to build the overall resolution of $\mathfrak{s}$.

Starting with $i=N$, since $M(\ge c_N)=M(c_N)$ is closed and $Z(\mathfrak{s})$ is compact, we conclude that $Z(\mathfrak{s})_N = M(c_N)$ is compact. Therefore, by Lemma 2.5.4, there exist a neighborhood W_N of $Z(\mathfrak{s})_N$, and an orbifold $\widetilde{\mathcal{W}}_N$ with a c_N-covering map

$$q_N\colon \widetilde{\mathcal{W}}_N \longrightarrow \mathcal{W}_N = \mathcal{M}|_{W_N},$$

such that the multisection $\mathfrak{s}$, restricted to W_N, is equivalent to the weighted push forward, with weight w_N, of some orbifold section s_N of

$$\widetilde{\mathcal{E}}_N = q_N^*(\mathcal{E}|_{W_N}).$$

We proceed inductively. Having constructed $q_i\colon\widetilde{\mathcal{W}}_i \longrightarrow \mathcal{W}_i$, for all $N\ge i\ge t$, such that

$$W_{\ge t} = \bigcup_{t\le i\le N} W_i$$

is a neighborhood of $Z(\mathfrak{s})\cap M(\ge c_t)$, let

$$K = (Z(\mathfrak{s})\cap M(c_{t-1}))\setminus W_{\ge t}\,.$$

By the upper semicontinuity of $\mathrm{val}_{\mathfrak{s}}$ and compactness of $Z(\mathfrak{s})$, K is a compact subset of $M(c_{t-1})$ disjoint from $M(c_i)$, for all $i\ge t$. Therefore, again by Lemma 2.5.4, there exist a neighborhood W_{t-1} of K disjoint from $M(\ge c_t)$, and an orbifold $\widetilde{\mathcal{W}}_{t-1}$ with a c_{t-1}-covering map

$$q_{t-1}\colon \widetilde{\mathcal{W}}_{t-1} \longrightarrow \mathcal{W}_{t-1} = \mathcal{M}|_{W_{t-1}}$$

such that the multisection $\mathfrak{s}$, restricted to W_{t-1}, is equivalent to the weighted push forward, with weight w_{t-1}, of some orbifold section s_{t-1} of

$$\widetilde{\mathcal{E}}_{t-1} = q_{t-1}^*(\mathcal{E}|_{W_{t-1}}).$$

We continue this induction process until we exhaust $Z(\mathfrak{s})$ at c_1.

It also follows from the proof of Lemma 2.5.5 that on the overlaps

$$W_{i,j} = W_i \cap W_j \quad \forall 1 \le j < i \le N,$$

we have $q_i = q_j \circ q_{i,j}$, where $q_{i,j}$ is a component-wise[17] covering map

$$q_{i,j} : \widetilde{\mathcal{W}}_i|_{\overline{q}_i^{-1}(W_{i,j})} \longrightarrow \widetilde{\mathcal{W}}_j|_{\overline{q}_j^{-1}(W_{i,j})}.$$

Finally, the weight ration functions (2.27) are related to each other on the overlaps in the following way.

LEMMA 2.5.7. *For every* $i \in [N]$, *let* $\mathfrak{I}_i = \mathfrak{I}_{\widetilde{\mathcal{W}}_i}$ *be the integer valued function*[18] *in* (2.4) *and* $\overline{w}_i$ *be the weight ratio rational function in* (2.27). *For every* $1 \le j < i \le N$, *the push forward via* $\overline{q}_{i,j}$ *of the twisted weight ratio function* $(\overline{w}_i/\mathfrak{I}_i)_{\overline{q}_i^{-1}(W_{i,j})}$ *is equal to* $(\overline{w}_j/\mathfrak{I}_j)_{\overline{q}_j^{-1}(W_{i,j})}$, *i.e.*

$$\overline{q}_{i,j*}(\overline{w}_i/\mathfrak{I}_i)|_{\overline{q}_i^{-1}(W_{i,j})} = (\overline{w}_j/\mathfrak{I}_j)|_{\overline{q}_j^{-1}(W_{i,j})}.$$

In other words, for every $\overline{x} \in \overline{q}_j^{-1}(W_{i,j}) \subset \widetilde{W}_j$, *we have*

$$\frac{\overline{w}_j(\overline{x})}{\mathfrak{I}_j(\overline{x})} = \sum_{\overline{y} \in \overline{q}_{i,j}^{-1}(\overline{x}) \subset \overline{q}_i^{-1}(W_{i,j})} \frac{\overline{w}_i(\overline{y})}{\mathfrak{I}_i(\overline{y})}. \tag{2.33}$$

PROOF. The equality (2.33) is a consequence of the following explicit local calculation. Let $\mathcal{U} = (\mathrm{pr}: U \longrightarrow V, G, \psi)$ be an orbibundle chart of $\mathcal{E}$ over which $\mathfrak{s}$ is equal to a lifted ℓ-section

$$\mathfrak{s} \cong [s^1, \ldots, s^\ell].$$

Suppose $\mathfrak{s}$ restricted to V has c distinct branches given by the index set

$$J = \{j_1, \ldots, j_c\} \subset [\ell].$$

For every $\overline{x} \in \psi(V)$ with $\mathrm{val}_{\mathfrak{s}}(\overline{x}) = c_x \le c$, let $\mathcal{U}_x = (\mathrm{pr}|_{U_x}: U_x \longrightarrow V_x, G_x, \psi|_{V_x})$ be a sub-chart of $\mathcal{U}$ centered around $\overline{x}$ over which $\mathfrak{s}$ has c_x distinct branches given by the index set

$$J_x = \{i_1, \ldots, i_{c_x}\} \subset [\ell].$$

There exists an obvious projection map $\pi: J \longrightarrow J_x$ which sends j_a to i_b whenever

$$s^{j_a}|_{V_x} = s^{i_b}|_{V_x}.$$

This projection map commutes with the subgroup action of G_x on J and the action of G_x on J_x. In this projection, a G_x-orbit O in J can be larger than its image G_x-orbit $o = \pi(O)$ in J_x. As in the proof of Lemma 2.5.4, corresponding to $\mathcal{U}$ we obtain a c-covering

$$q_2 : \mathcal{W}_2 = [(V \times J)/G] \longrightarrow [V/G]$$

whose connected components are of the form

$$(V \times \mathfrak{O})/G \cong V/G_{\mathfrak{O}},$$

[17]For example if $c_i = 3$ and $c_j = 2$, $q_{i,j}$ is a 2-covering map on some connected components and an isomorphism everywhere else.

[18]Which is equal to the size of isotropy group at each point.

where $\mathfrak{O}$ is an orbit of the action of G on J and $G_{\mathfrak{O}} < G$ is the isotropy group of some element of $\mathfrak{O}$. Let $\mathfrak{I}_2 \colon W_2 \longrightarrow \mathbb{Z}$ be the corresponding function as in (2.4). Similarly, corresponding to $\mathcal{U}_x$ we obtain a c_x-covering

$$q_1 \colon \mathcal{W}_1 = [(V_x \times J_x)/G_x] \longrightarrow [V_x/G_x]$$

whose connected components are of the form

$$(V_x \times o)/G_x \cong V_x/(G_x)_o,$$

where o is an orbit of the action of G_x on J_x and $(G_x)_o < G_x$ is the isotropy group of some element of o. The function $\mathfrak{I}_1 \colon W_1 \longrightarrow \mathbb{Z}$ is defined similarly. Let w_1 and w_2 be the weight functions of these two coverings; $w_2(\mathfrak{O})$ is the multiplicity of s^{j_a} in $\{s^1, \dots, s^\ell\}$ for any $j_a \in \mathfrak{O}$, and $w_1(o)$ is the multiplicity of $s^{i_b}|_{V_x}$ in $\{s^1|_{V_x}, \dots, s^\ell|_{V_x}\}$ for any $i_b \in o$. Let $q_{2,1}$ be the projection map from $q_2^{-1}([V_x/G_x])$ to $\mathcal{W}_1$. We have the following commutative diagram

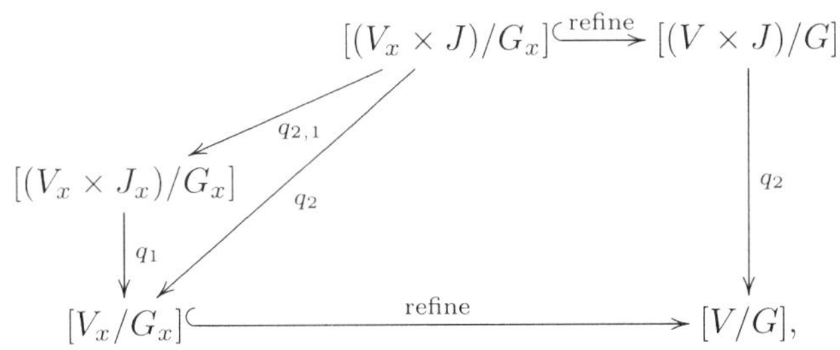

where $q_{2,1}$ sends $[y \times O]_{G_x}$ to $[y \times \pi(O)]_{G_x}$. Fix a G_x-orbit o in J_x, say $o = G_x \cdot \{i_1\}$, and let $\overline{x}_o = [x \times o]_{G_x}$ be the unique element in $\overline{q}_1^{-1}(\overline{x})$ which belongs to the connected component $(V_x \times o)/G_x$. Let $\{O_a\}_{a=1}^t$ be the set of all G_x-orbits in J where $\pi(O_a) = o$, and denote by $\overline{x}_a = [x \times O_a]_{G_x}$ to be the unique element in $\overline{q}_2^{-1}(\overline{x})$ which belongs to $(V_x \times O_a)/G_x$. We have

$$\overline{q}_{2,1}^{-1}(\overline{x}_o) = \{\overline{x}_a\}_{a=1}^t,$$

and

$$\left((\overline{q}_{2,1})_* \left(\frac{\overline{w}_2}{\mathfrak{I}_2} \right) \right) (\overline{x}_o) = \sum_{a=1}^{t} \frac{\overline{w}_2(\overline{x}_a)}{|G_{x,a}|}, \tag{2.34}$$

where $G_{x,a} < G_x$ is the isotropy subgroup of some element of O_a. By definition, if $O_a = G_x \cdot \{j_a\}$, where $s^{j_a}|_{V_x} = s^{i_1}|_{V_x}$, then

$$\begin{aligned} \overline{w}_1(\overline{x}_o) = \frac{\text{multiplicity}|_{V_x}(s^{i_1})}{\ell} = \sum_{a=1}^{t} \frac{\text{multiplicity}|_V(s^{j_a})}{\ell} \frac{|O_a|}{|o|} = \\ \sum_{a=1}^{t} \frac{\text{multiplicity}|_V(s^{j_a})}{\ell} \frac{|G_{x,o}|}{|G_{x,a}|} = \sum_{a=1}^{t} \overline{w}_2(\overline{x}_a) \frac{|G_{x,o}|}{|G_{x,a}|}, \end{aligned} \tag{2.35}$$

where $G_{x,o} < G_x$ is the isotropy subgroup of some element of o. Moving $|G_{x,o}|$ in (2.35) to the left-hand side, together with (2.34), we obtain

$$\frac{\overline{w}_1(\overline{x}_o)}{\mathfrak{I}_1(\overline{x}_o)} = \frac{\overline{w}_1(\overline{x}_o)}{|G_{x,o}|} = \sum_{a=1}^{t} \frac{\overline{w}_2(\overline{x}_a)}{|G_{x,a}|} = \left((\overline{q}_{2,1})_* \left(\frac{\overline{w}_2}{\mathfrak{I}_2} \right) \right) (\overline{x}_o);$$

this finishes the proof of (2.33). □

We have thus proved that every multisection $\mathfrak{s}$ with compact support admits a resolution in the following sense.

DEFINITION 2.5.8. *Let $\mathcal{E} \longrightarrow \mathcal{M}$ be an orbibundle and $\mathfrak{s}$ be a lifted multisection with compact support. A resolution of $\mathfrak{s}$ consist of a set of c_i-covering maps*

$$q_i : \widetilde{\mathcal{W}}_i \longrightarrow \mathcal{W}_i, \quad i = 1, \ldots, N, \tag{2.36}$$

over open subsets of $\mathcal{M}$ such that the following properties hold.

(1) *We have*

$$Z(\mathfrak{s}) \subset \bigcup_{i=1}^{N} W_i, \quad W_t \cap M(>c_t) = \emptyset, \quad \text{and} \quad M(\geq c_t) \subset \bigcup_{i=t}^{N} W_i \quad \forall t \in [N].$$

(2) *For every $i \in [N]$, the multisection $\mathfrak{s}$ restricted to W_i is equivalent to the weighted push forward with weight w_i of some orbifold section s_i of $\widetilde{\mathcal{E}}_i$.*

(3) *On the overlaps*

$$W_{i,j} = W_{j,i} = W_i \cap W_j \quad \forall 1 \leq j < i \leq N,$$

we have $q_i = q_j \circ q_{i,j}$, where $q_{i,j}$ is a component-wise covering map

$$q_{i,j} : \widetilde{\mathcal{W}}_i|_{\overline{q}_i^{-1}(W_{i,j})} \longrightarrow \widetilde{\mathcal{W}}_j|_{\overline{q}_j^{-1}(W_{i,j})};$$

moreover,

$$q_{i,j} \circ q_{j,k} = q_{i,k} \quad on \quad W_i \cap W_j \cap W_k \quad \forall 1 \leq k < j < i \leq N. \tag{2.37}$$

(4) *For every $1 \leq j < i \leq N$,*

$$(\overline{w}_j/\mathfrak{I}_{\mathcal{W}_j})|_{\overline{q}_j^{-1}(W_{i,j})} = (\overline{q}_{i,j})_*(\overline{w}_i/\mathfrak{I}_{\mathcal{W}_i})|_{\overline{q}_i^{-1}(W_{i,j})}. \tag{2.38}$$

REMARK 2.5.9. For a generic choice of $\mathfrak{s}$, the orbifolds $\widetilde{\mathcal{W}}_i$ are all manifolds. In fact, this happens if for every sufficiently small orbibundle chart $(\mathrm{pr}: U \longrightarrow V, G, \psi)$, in the local presentation

$$\mathfrak{s}|_V = [s^1, \ldots, s^\ell], \quad s^i : V \longrightarrow U \ \forall\, i \in [\ell],$$

none of the s^i is equivariant with respect to some subgroup of G; see Equation 2.29. This is the case if we choose ℓ large enough and the branches s^i generically[19]. In this situation, the triangulation argument in the next section simply involves manifolds, vector bundles, and triangulation of smooth submanifolds, which is standard. For the specific type of transverse perturbations constructed in Section 2.4, if we choose the perturbation terms ν_α generically, the resulting multisection has this property. We continue without this assumption so that our formula for the resulting rational cycle applies to every arbitrary choice of transversal multisection in the upcoming examples.

[19]We need to consider a sufficiently large ℓ, with respect to $|G|$, to achieve this.

2.6. Euler class. Let $E \longrightarrow M$ be a relatively oriented vector bundle over a manifold without boundary and $s\colon M \longrightarrow E$ be a transverse section with compact support. Then $s^{-1}(0) \subset M$ is a compact oriented submanifold of M; thus it defines a singular homology class in

$$H_{\dim M - \mathrm{rank} E}(M, \mathbb{Z}).$$

Two such sections s and s' are called deformation equivalent if there exists a one-parameter family of sections $(s_t)_{t\in[0,1]}$ with compact support such that $s = s_0$ and $s' = s_1$. Let $\hbar$ be a deformation equivalence class of sections of E with compact support and $s \in \hbar$ be transverse. Then the homology class of $s^{-1}(0)$, which we call the Euler class of the pair $(E, \hbar)$ and denote it by $e(E, \hbar)$, is a homological invariant of the pair $(E, \hbar)$. If M is oriented, this homology class is Poincaré dual to a cohomology class with compact support

$$\mathrm{PD}(e(E, \hbar)) \in H_c^{\dim M - \mathrm{rank} E}(M).$$

EXAMPLE 2.6.1. Let $E \longrightarrow M$ be the trivial complex line bundle over the open unit disk in the plane. For every $k \geq 0$, let $\hbar_k$ be the homotopy class of the sections with compact support containing $s(z) = z^k$. Then

$$\langle e_k, M \rangle = k, \quad e_k = \mathrm{PD}(e(E, \hbar_k)) \in H_c^2(M).$$

Similarly, after some simple generalization of this argument, if $s\colon \mathcal{M} \longrightarrow \mathcal{E}$ is a transverse orbifold section of a relatively oriented orbibundle, $\overline{s}^{-1}(0) \subset M$ inherits a possibly non-effective oriented orbifold structure from the pair $(\mathcal{E}, \mathcal{M})$ and its homology class, with some $\mathbb{Q}$-coefficients that come from the orbifold structure, defines a homology class in $H_{\dim M - \mathrm{rank} E}(M, \mathbb{Q})$ invariant under deformation of s.

As we argued before, it is quite likely that an orbibundle does not admit any transverse orbifold section; thus this construction does now work in all cases. In the light of Proposition 2.4.2, we consider a transversal multisection deformation $\mathfrak{s}$ of s and construct an Euler class via the zero set $Z(\mathfrak{s})$. To this end, we use a resolution of the multisection to reduce the construction back to the case of orbibundles and orbifold sections.

In the context of Gromov-Witten theory, an enhanced version of the Euler class for the obstruction bundle of an associated Kuranishi structure defines the Gromov-Witten VFC of the underlying moduli space.

Recall that a triangulation of some topological subspace $Z \subset M$ is a simplicial complex Δ, together with a homeomorphism $\sigma\colon \Delta \longrightarrow Z$. A pure simplicial n-complex Δ is a simplicial complex where every simplex of dimension less than n is a face of some n-simplex in Δ. If Δ is a cycle and if the top simplexes of Δ are compatibly oriented, then $\sigma(\Delta)$ defines a class in $H_n(M, \mathbb{Z})$. In general, the zero set $Z(\mathfrak{s})$ (2.23) of a transverse multisection $\mathfrak{s}$ may not admit any triangulation because different branches may intersect wildly. However, in what follows, we consider a "triangulation" of the zero set of the sections of some resolution of $\mathfrak{s}$ and push that into $Z(\mathfrak{s})$, with some rational weights, to obtain the desired singular n-cycle with $\mathbb{Q}$-coefficients. Our construction is inductive, explicit, but somewhat sketchy at certain points; for a more functorial and detailed construction, we refer to the definition and construction of branched coverings in [**36**].

For the rest of this section, we fix a relatively oriented orbibundle $\mathcal{E} \longrightarrow \mathcal{M}$, a transverse orbifold multisection with compact support $\mathfrak{s}$, and a resolution

$$\{q_i : \widetilde{\mathcal{W}}_i \longrightarrow \mathcal{W}_i, \quad s_i : \widetilde{\mathcal{W}}_i \longrightarrow \widetilde{\mathcal{E}}_i\}_{i\in[N]}$$

of $\mathfrak{s}$ as in Definition 2.5.8. Since $\mathfrak{s}$ is assumed to be transverse, for each $i=1,\ldots,N$, the orbifold section s_i is transverse to the zero section. Let

$$\widetilde{Z}_i = Z(s_i) \subset \widetilde{W}_i \tag{2.39}$$

be the zero set of s_i. If $\widetilde{\mathcal{W}}_i = \widetilde{W}_i$ is a manifold, see Remark 2.5.9, $\widetilde{Z}_i$ is an oriented submanifold of $\widetilde{W}_i$ and can be triangulated. More generally, in the orbifold case, over each local chart (V, G, ψ) of $\widetilde{\mathcal{W}}_i$, since

$$s_{i,V} = s_i|_V : V \longrightarrow U$$

is G-equivariant and transverse to the zero section, $s_{i,V}^{-1}(0)$ is a G-invariant n-dimensional oriented submanifold of V. The G-action on (any component of) $s_{i,V}^{-1}(0)$ can be non-effective, i.e. the orbifold structure $\widetilde{\mathcal{W}}_i$ induces a possibly non-effective orbifold structure $\widetilde{\mathcal{Z}}_i$ on $\widetilde{Z}_i$. More precisely, for each connected component $s_{i,V}^{-1}(0)_o$ of $s_{i,V}^{-1}(0)$, let G_o be the isotropy group of a generic point[20] $y \in s_{i,V}^{-1}(0)_o$. This is well-defined because by the orientability of the action on $s_{i,V}^{-1}(0)_o$, the isotropy group is fixed outside a set of codimension 2 in $s_{i,V}^{-1}(0)_o$; see Remark 2.1.11(1). In other words, $G_o < G$ is the largest subgroup acting as identity on $s_{i,V}^{-1}(0)_o$. Thus, $G_o < G$ is normal and the action of G/G_o on $s_{i,V}^{-1}(0)_o$ is effective. Therefore,

$$[s_{i,V}^{-1}(0)_o/(G/G_o)] \tag{2.40}$$

is an effective orbifold chart on $\widetilde{Z}_i$. We denote this reduced effective orbifold structure by $\widetilde{\mathcal{Z}}_i^{\mathrm{red}}$. Globally, for each connected component $\widetilde{Z}_{i,o}$ of $\widetilde{Z}_i$, let I_o be the conjugacy class of the isotropy group of a generic point of $\widetilde{Z}_{i,o}$. We conclude that $\widetilde{\mathcal{Z}}_{i,o}$ is possibly a non-effective orbifold with global isotropy group I_o such that

$$\widetilde{\mathcal{Z}}_{i,o}^{\mathrm{red}} = \widetilde{\mathcal{Z}}_{i,o}/\mathrm{I}_o. \tag{2.41}$$

For this reason, in the construction of Euler class rational cycle below, the isotropy group I_o contributes by $\frac{1}{|\mathrm{I}_o|}$ to the weight of each simplex mapped into $\widetilde{Z}_{i,o}$.

Similarly to case of manifolds (see [**54**] for the manifold case) every orbifold admits a "good" triangulation in the following sense[21].

Definition 2.6.2. *A **good** triangulation of an n-dimensional orbifold[22] $\mathcal{M}$ is a triangulation $\overline{\sigma}\colon \Delta \longrightarrow M$ with a pure simplicial n-complex $\Delta = \{\Delta^b\}_{b\in\Omega}$ such that the following conditions hold.*

(1) *For every n-simplex $\Delta^b \in \Delta$, there exists an orbifold chart (V, G, ψ) of $\mathcal{M}$ such that $\overline{\sigma}(\Delta^b) \subset \psi(V)$ and $\overline{\sigma}|_{\Delta^b}$ lifts to a continous embedding $\sigma\colon \Delta^b \longrightarrow V$.*

---

[20] If $s_{i,V}^{-1}(0)_o$ is a single point $\{y\}$, then $G_o = G_y$.

[21] We do not prove the existence of such triangulation in this article. We do not know of a reference that discusses the construction either. However, existence of such triangulation is a straightforward generalization of [**54**]. As Remark 2.5.9 indicates, we can avoid orbifolds by considering a generic choice of transverse multisection.

[22] For this definition, the orbifold structure does not really need to be effective.

(2) *The interior of every subface δ of every simplex Δ^b is mapped into a strata M^H of M; see Remark* 2.1.11(2).

If $\mathcal{M}$ is oriented, by the second condition in Definition 2.6.2, the interior of every n-dimensional simplex and the interior of every $(n-1)$-dimensional face of that are mapped into the smooth locus $M^{\rm sm}$ (which is a manifold); see Remark 2.1.11(1). Therefore, up to codimension 2 faces, a triangulation of $\mathcal{M}$ gives a triangulation of the smooth manifold $M^{\rm sm}$ and this is enough to conclude that $\overline{\sigma}\colon \Delta \longrightarrow M$ defines a cycle. This cycle is the fundamental cycle of $\mathcal{M}$. If $q\colon \mathcal{M}_1 \longrightarrow \mathcal{M}_2$ is a c-covering, it is clear from Definition 2.5.1 that a good triangulation of $\mathcal{M}_2$, after possibly subdividing into smaller triangles[23], canonically lifts to a good triangulation of $\mathcal{M}_1$.

DEFINITION 2.6.3. *With notation as above, given a resolution*

$$\mathfrak{R} \equiv \Big(\{q_i\colon \widetilde{\mathcal{W}}_i \longrightarrow \mathcal{W}_i\}_{i\in[N]}, \{q_{i,j}\colon \widetilde{\mathcal{W}}_i|_{\overline{q}_i^{-1}(W_{i,j})} \longrightarrow \widetilde{\mathcal{W}}_j|_{\overline{q}_j^{-1}(W_{i,j})}\}_{1\le j<i\le N}\Big)$$

of $\mathfrak{s}$ *as in Definition* 2.5.8, *an* ***admissible*** *triangulation of* $\mathfrak{R}$ *consists of a good triangulation*

$$\overline{\sigma}_i\colon \Delta_i \longrightarrow T_i$$

of some compact subset $T_i \subset \widetilde{Z}_i$, *for every* $i=1,\dots,N$, *where* Δ_i *is a pure simplicial* n-*complex* $\Delta_i = \{\Delta_i^b\}_{b\in\Omega_i}$, *such that the following conditions hold.*

(1) *The union of images of* T_j *in each* $\widetilde{Z}_i$ *cover* $\widetilde{Z}_i$, *i.e. for every* $1\le i\le N$,

$$T_i \cup \bigcup_{j>i} \overline{q}_{j,i}(T_j\cap \overline{q}_j^{-1}(W_{i,j})) \cup \bigcup_{j<i} \overline{q}_{i,j}^{-1}(T_j) = \widetilde{Z}_i.$$

(2) *For every* $i,j\in[N]$, *with* $i\neq j$, *there exist a sub-triangulation*

$$\overline{\sigma}_i\colon \Delta_{i,j} \longrightarrow T_{i,j} \subset T_i$$

for some pure simplicial n-*complex* $\Delta_{i,j} = \{\Delta_{i,j}^b\}_{b\in\Omega_{i,j}}$ *such that*

(1) $T_{i,j} \subset \overline{q}_i^{-1}(W_{i,j})$,

(2) $\overline{q}_{i,j}^{-1}(T_{j,i}) = T_{i,j}$, *i.e.* $T_{i,j}$ *is the lift of* $T_{j,i}$ *via* $\overline{q}_{i,j}$, *whenever* $i>j$,

(3) *and*

$$\overline{q}_{i,j}(T_i \cap \overline{q}_j^{-1}(W_{i,j})) \cap T_j = T_{j,i} \quad \forall i>j,$$

i.e. the overlap of T_i *and* T_j *via* $\overline{q}_{i,j}$ *is equal to* $T_{i,j}$ *and* $T_{j,i}$ *in the corresponding spaces.*

Note that we are considering a triangulation of $\widetilde{Z}_i$ and not the whole space $\widetilde{W}_i$. However, we can assume that the triangulation of each $\widetilde{Z}_i$ is the restriction of some good triangulation of $\widetilde{W}_i$ and these extended triangulations are compatible on the overlaps as well (as in Definition 2.6.3). We do not go into the details of existence of such an admissible triangulation in this article. The proof of existence is by induction on $i=1,\dots,N$, and uses the fact that every $\widetilde{Z}_i$ is triangulatable and M is metrizable. For the induction, we would start from $i=1$ (the smallest value) and consider a triangulation of $\widetilde{Z}_1$. After passing to some barycentric subdivision of the triangulation, we can compatibly extend it to the next level $\widetilde{Z}_2$ and continue until $i=N$.

[23]So that the image of each simplex sits inside an orbifold chart as in Definition 2.5.1.

Given an admissible triangulation of a resolution of $\mathfrak{s}$, as above, we construct the Euler class singular n-cycle of $(\mathcal{E}, \mathfrak{s})$ in the following way.

For every $i=1,\ldots,N$, and every n-simplex $\Delta_i^b \in \Delta_i$, we say Δ_i^b is **primary**, if there is no $j<i$ such that $\Delta_i^b \in T_{i,j}$. Let $\Delta_i^{\text{prim}} \subset \Delta_i$ be the subset of primary n-simplexes and

$$\Delta^{\text{prim}} = \coprod_{i=1}^{N} \Delta_i^{\text{prim}}.$$

For every $\Delta_i^b \in \Delta_i^{\text{prim}}$, by Definition 2.6.2, there exists an orbibundle chart

$$(U, V, G, \psi, \text{pr})$$

of $\widetilde{\mathcal{E}}_i \longrightarrow \widetilde{\mathcal{W}}_i$ such that $\overline{\sigma}_i(\Delta_i^b) \subset \psi(V)$ and $\overline{\sigma}_i|_{\Delta_i^b}$ lifts to

$$\sigma_i \colon \Delta_i^b \longrightarrow V \tag{2.42}$$

By assumption, the restriction $s_{i,V} \colon V \longrightarrow U$ of the orbifold section s_i on $\widetilde{\mathcal{W}}_i$ to V is transverse to the zero section. Since the bundle $U \longrightarrow V$ is relatively oriented, $s_{i,V}^{-1}(0)$ is oriented. We orient Δ_i^b such that (2.42) is orientation preserving. Let $\mathfrak{I}_i(\Delta_i^b)$ be equal to $\mathfrak{I}_i(\overline{x})$ for some generic $\overline{x} \in \overline{\sigma}_i(\Delta_i^b)$. In other words, $\mathfrak{I}_i(\Delta_i^b)$ is equal to the order of isotropy group of the connected component $\widetilde{Z}_{i,o}$ containing $\overline{\sigma}_i(\Delta_i^b)$; see (2.40), (2.41), and the argument there. Let $\overline{w}_i(\Delta_i^b)$ be the value of the weight ratio function $\overline{w}_i \colon \widetilde{W}_i \longrightarrow \mathbb{Q}$ along $\overline{\sigma}_i(\Delta_i^b)$.

PROPOSITION 2.6.4. *With notation as in Lemma 2.5.7 and Definition 2.6.3, the singular n-chain*

$$e(\mathcal{E}, \mathfrak{s}) = \sum_{\Delta_i^b \in \Delta^{\text{prim}}} \frac{\overline{w}_i(\Delta_i^b)}{\mathfrak{I}_i(\Delta_i^b)} \, \overline{q}_i(\overline{\sigma}_i(\Delta_i^b)) \subset Z(\mathfrak{s}) \subset M \tag{2.43}$$

is a rational n-cycle. Moreover, its homology class is independent of the choice of resolution, admissible triangulation, and deformation equivalence class of the transversal multisection $\mathfrak{s}$; thus we denote it by $e(\mathcal{E}, \mathfrak{s})$.

EXAMPLE 2.6.5. Let M be a connected closed manifold, $G = \{g_1, \ldots, g_c\}$ be a finite subgroup of diffeomorphisms of M, and $\mathcal{M} = [M/G]$ be the resulting effective global quotient orbifold. In this case, the manifold M is a c-covering of the orbifold $\mathcal{M}$ and $T\mathcal{M} \longrightarrow \mathcal{M}$ is relatively oriented. A generic[24] transverse section s of $T\mathcal{M}$ gives a push forward transverse multisection

$$\mathfrak{s} = [g_1 \cdot s, \ldots, g_c \cdot s]$$

of $T\mathcal{M}$ for which $\text{val}_{\mathfrak{s}} \equiv c$. Let $s^{-1}(0) = \{p_1, \cdots, p_N\}$. For every $i \in [N]$, let $\epsilon_i \in \{\pm 1\}$ be the intersection number of p_i. The weight ratio of every p_i is simply $\frac{1}{|G|}$ and $\mathfrak{I}_M \colon M \longrightarrow \mathbb{Q}$ is the constant function 1. Then, by (2.43), the orbifold Euler characteristic (class) of $\mathcal{M}$ is equal to

$$\chi(\mathcal{M}) = e(T\mathcal{M}) = \sum_{i=1}^{\ell} \frac{\epsilon_i}{|G|} = \frac{\chi(M)}{|G|} \in \mathbb{Q} \cong H_0(M/G, \mathbb{Q}).$$

---

[24] If s is not generic, $\text{val}_{\mathfrak{s}}$ may not be a constant function in which case M is not a resolution of $\mathfrak{s}$ as we built above.

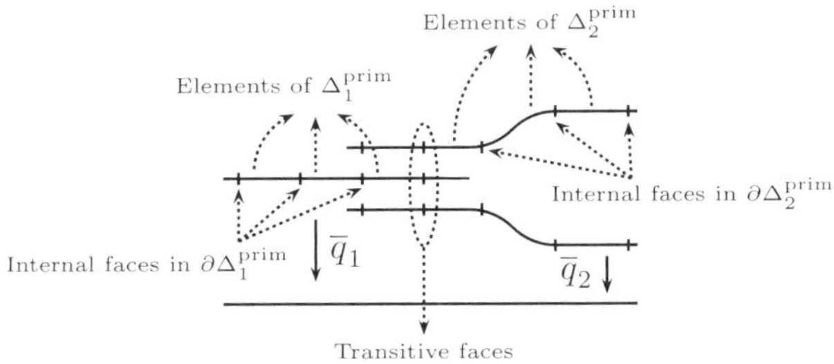

FIGURE 4. A schematic description of primary simplexes, internal faces, and transitive faces.

EXAMPLE 2.6.6. With notation as in Example 2.2.11, since s is already a transverse orbifold section, we do not need to pass to any non-trivial cover. We have $Z(s) = \{0, \infty\}$, isotropy group of 0 has order m, and isotropy group of ∞ has order n; therefore, by (2.43), $\chi(T\mathbb{P}^1_{m,n}) = \frac{1}{m} + \frac{1}{n}$.

PROOF OF PROPOSITION 2.6.4. We just show that (2.43) is a cycle. Proof of independence of its homology class from various choices is by constructing a cobordism between different choices and we skip that.

Assume $n = \dim \mathcal{M} - \text{rank}\ \mathcal{E} > 0$, otherwise, $\partial e(\mathcal{E}, \mathfrak{s}) = 0$ for dimensional reason. Let $\partial\Delta^{\text{prim}}$ and $\partial\Delta_i^{\text{prim}}$ be the set of all $(n-1)$-dimensional faces -or simply faces- of primary simplexes in Δ^{prim} and Δ_i^{prim}, respectively.

DEFINITION 2.6.7. *We call $\delta \in \partial\Delta_i^{\text{prim}}$ **internal** if there exists another $\delta' \in \partial\Delta_i^{\text{prim}}$ such that $\overline{\sigma}_i(\delta) = \overline{\sigma}_i(\delta')$. We call $\delta \in \partial\Delta_i^{\text{prim}}$ **transitive** if there exists $j \neq i$ and $\delta' \in \partial\Delta_j^{\text{prim}}$ such that*

$$\overline{\sigma}_i(\delta) = \overline{q}_{j,i}\overline{\sigma}_j(\delta') \text{ if } j > i, \text{ or } \quad \overline{\sigma}_j(\delta) = \overline{q}_{i,j}\overline{\sigma}_i(\delta') \text{ if } i > j; \tag{2.44}$$

*see Figure 4. In this situation, $j \in [N]$ is called an **associated** index to δ.*

LEMMA 2.6.8. *Every $\delta \in \partial\Delta^{\text{prim}}$ is either internal or transitive, but not both. For every internal δ, the corresponding face δ' is unique. For every transitive face $\delta \in \partial\Delta_i^{\text{prim}}$ there is only one index $j \in [N]$ associated to δ.*

PROOF. For every $\Delta_i^b \in \Delta_i^{\text{prim}}$ and every $\delta \in \partial\Delta_i^b$, let δ^o be the interior of δ. By the orientability of $\widetilde{\mathcal{Z}}_i$ and the second condition in Definition 2.6.2, $\overline{\sigma}_i(\delta^o)$ lies in the smooth locus $\widetilde{Z}_i^{\text{sm}}$ of the reduced orbifold structure $\widetilde{\mathcal{Z}}_i^{\text{red}}$. Therefore, as in the manifold case, either there exists another $\delta' \in \partial\Delta_i^{\text{prim}}$ (which could be a different face of the same simplex Δ_i^b) such that $\overline{\sigma}_i(\delta') = \overline{\sigma}_i(\delta)$, or $\overline{\sigma}_i(\Delta_i^{\text{prim}})$ does not cover any sufficiently small enough neighborhood of any point $\overline{x} \in \overline{\sigma}_i(\delta^o) \in \widetilde{Z}_i^{\text{sm}}$. In the first case, δ is an internal face. Moreover, similar to the manifold case, since $\widetilde{\mathcal{Z}}_i$ is oriented, the contribution of δ and δ' to $\partial e(\mathcal{E}, \mathfrak{s})$ have equal absolute values and different orientations (the weight ratio and the order of isotropy group over the simplexes contacting δ and δ' are the same, because they belong to the same connected component of $\widetilde{Z}_i$); therefore, they cancel out.

In the second case, for every $\overline{x} \in \overline{\sigma}_i(\delta^o)$, it follows from the first condition in Definition 2.6.3 that there exists another $\Delta_j^a \in \Delta_j^{\text{prim}}$, for some $j \in [N]$ different from

i, and $\delta' \in \partial\Delta_j^a$ such that

$$\begin{cases} \overline{q}_{i,j}(\overline{x}) \in \overline{\sigma}_j(\delta') & \text{if } i > j, \\ \overline{x} \in \overline{q}_{j,i}(\overline{\sigma}_j(\delta')) & \text{if } j > i. \end{cases}$$

In either case, it follows from Definition 2.6.3.2(c) that $\overline{x} \in T_{i,j}$. Finally, it follows from Condition 2(b) that

$$\begin{cases} \overline{q}_{i,j}(\overline{\sigma}_i(\delta)) = \overline{\sigma}_j(\delta') & \text{if } i > j, \\ \overline{q}_{j,i}(\overline{\sigma}_j(\delta')) = \overline{\sigma}_i(\delta) & \text{if } j > i. \end{cases}$$

An internal face can not be transitive because by Definition 2.6.3.2(c), if

$$\delta, \delta' \in \partial\Delta_i^{\text{prim}}, \ \delta \in \partial\Delta_i^b, \ \delta' \in \partial\Delta_i^{b'}, \ \overline{\sigma}_i(\delta) = \overline{\sigma}_i(\delta'), \text{ and } \overline{\sigma}_i(\delta) = \overline{q}_{j,i}\overline{\sigma}_j(\delta''),$$

for some $\delta'' \in \partial\Delta_j^{b''}$, where $\Delta_j^{b''} \in \Delta_j^{\text{prim}}$ and $j < i$, since $\overline{\sigma}_i(\delta)$ locally divides $\widetilde{Z}_i^{\text{sm}}$ into two components, either $\overline{\sigma}_i(\Delta_i^b) = \overline{q}_{j,i}(\overline{\sigma}_j(\Delta_j^{b''}))$ or $\overline{\sigma}_i(\Delta_i^{b'}) = \overline{q}_{j,i}(\overline{\sigma}_j(\Delta_j^{b''}))$. This is a contradiction because it implies that $\Delta_j^{b''}$ is not primary. The case of $j < i$ is similar.

Suppose there exist $i < j < k$ and

$$\delta \in \partial\Delta_i^b, \quad \delta' \in \partial\Delta_j^{b'}, \quad \text{and } \delta \in \partial\Delta_k^{b''}$$

where Δ_i^b, $\Delta_j^{b'}$, and $\Delta_k^{b''}$ are primary and

$$\overline{\sigma}_i(\delta) = \overline{q}_{j,i}(\overline{\sigma}_j(\delta')), \quad \overline{\sigma}_i(\delta) = \overline{q}_{k,i}(\overline{\sigma}_k(\delta'')).$$

First, it follows from (2.37) that

$$\overline{\sigma}_j(\delta) = \overline{q}_{k,j}(\overline{\sigma}_k(\delta'')). \tag{2.45}$$

Since $\overline{\sigma}_i(\delta)$ locally divides $\widetilde{Z}_i^{\text{sm}}$ into two components, both

$$\overline{q}_{j,i}(\overline{\sigma}_j(\Delta_j^{b'})) \quad \text{and} \quad \overline{q}_{k,i}(\overline{\sigma}_k(\Delta_k^{b''}))$$

are mapped to the same side of $\overline{\sigma}_i(\delta)$. It follows from (2.45) and Definition 2.6.3.2(c) that $\overline{\sigma}_j(\Delta_j^{b'}) = \overline{q}_{k,j}(\overline{\sigma}_k(\Delta_k^{b''}))$. This is a contradiction to the assumption that both $\Delta_j^{b'}$ and $\Delta_k^{b''}$ are primary. This establishes the last claim of Lemma 2.6.8. □

From Lemma 2.6.8, we conclude that internal faces come in pairs with identical images and that the contribution of each pair to $\partial e(\mathcal{E}, \mathfrak{s})$ is zero. As for transitive faces, proof of Lemma 2.6.8 shows that the elements of $\partial\Delta^{\text{prim}}$ can be grouped into tuples of the form

$$(\delta; \delta'_1, \ldots, \delta'_t), \quad \delta \in \partial\Delta_i^{\text{prim}}, \quad \delta'_1, \ldots, \delta'_t \in \partial\Delta_j^{\text{prim}} \quad \text{for some } j > i, \tag{2.46}$$

where j is the unique associated index to δ and $\delta'_1, \ldots, \delta'_t$ are the set of all faces where (2.44) holds. Finally, by (2.33), the contribution of δ and the sum of contributions of $\delta'_1, \ldots, \delta'_t$ to $\partial e(\mathcal{E}, \mathfrak{s})$ cancel out each other, i.e.

$$\sum_{a=1}^{t} \frac{\overline{w}_j(\Delta_j^a)}{\mathfrak{I}_j(\Delta_j^a)} = \frac{\overline{w}_i(\Delta_i^b)}{\mathfrak{I}_i(\Delta_i^b)}, \tag{2.47}$$

where Δ_i^b is the simplex containing δ and Δ_j^a, for each $a=1,\ldots,t$, is the simplex containing δ'_a. More precisely, by choosing a generic point $\overline{x} \in \overline{\sigma}_i(\delta^o)$ such that $\mathfrak{I}_i(\Delta_i^b) = \mathfrak{I}_i(\overline{x})$ (this is possible because $\widetilde{Z}_i \setminus \widetilde{Z}_i^{\mathrm{sm}}$ is of codimension at least 2),

$$\frac{\overline{w}_i(\Delta_i^b)}{\mathfrak{I}_i(\Delta_i^b)} = \frac{\overline{w}_i(\overline{x})}{\mathfrak{I}_i(\overline{x})} = \sum_{\overline{y} \in \overline{q}_{j,i}^{-1}(\overline{x})} \frac{\overline{w}_j(\overline{y})}{\mathfrak{I}_j(\overline{y})} = \sum_{a=1}^{t} \frac{\overline{w}_j(\Delta_j^a)}{\mathfrak{I}_j(\Delta_j^a)}.$$

□

3. Abstract Kuranishi structures

3.1. Introductory remarks. In differential geometry, various moduli spaces appear as the set of solutions of nonlinear partial differential equations. The zero set of the defining equation of the moduli space can be very singular. The idea of using obstruction bundle to resolve this issue goes back to Kuranishi [**33**]: Let (X, J) be a closed holomorphic manifold with the group of holomorphic diffeomorphisms $\mathfrak{Aut}(X, J)$. In order to describe the local structure of the moduli space of complex structures on X near J, Kuranishi proved that there exist

- a finite dimensional complex manifold V on which $\mathfrak{Aut}(X, J)$ acts homomorphically,
- a complex representation E of $\mathfrak{Aut}(X, J)$,
- and an $\mathfrak{Aut}(X, J)$-equivariant holomorphic map $s\colon V \longrightarrow E$,

such that the set of complex manifold (X, J') with J' sufficiently close to J has a canonical bijection to the quotient space $s^{-1}(0)/\mathfrak{Aut}(X, J)$. Roughly speaking, we can exchange the possibly singular moduli space $s^{-1}(0)$ with a larger smooth space V and an obstruction space E such that the moduli space is the zero set of a section of the vector bundle $V \times E$. After his work, such a local structure is called a "Kuranishi model" or a "Kuranishi chart". For example, motivated by this construction, in [**52**], Taubes used this idea to study the space of solutions of the Yang-Mills equations near bubbling in way that led to the modern use.

In 1980's, first by Donaldson in gauge theory [**10**] and then by Gromov [**21**] in the theory of pseudoholomorphic curves, the idea of using the fundamental homology class of moduli spaces to obtain invariants in various topological contexts was introduced. In the theory of pseudoholomorphic curves, whenever the moduli space in question has a nice orbifold structure, Gromov-Witten invariants were first obtained in this way. For example, such a theory was first rigorously built in the case of semi-positive symplectic manifolds by Ruan [**47**] and Ruan-Tian [**48**] and is now a straight-forward tool in these cases, see McDuff-Salamon [**37**].

However, moduli spaces of pseudoholomorphic maps of arbitrary genus over arbitrary symplectic manifolds can be quite singular and may have components of larger than expected dimension. In this situation, the moduli space does not have a fundamental homology class and a precise definition of Gromov-Witten invariants, as some sort of intersection theory on the moduli space, needs more elaborated tools. In order to solve the issue, we use the idea of Kuranishi to define a well-defined intersection theory on the compactified moduli space of pseudoholomorphic maps. We cover the moduli space with Kuranishi charts as above; these charts are obtained via finite dimensional reductions of the Cauchy-Riemann equation defining the moduli space and generalize the orbifold structure of the moduli space in the semi-positive cases. A suitably defined "compatible" covering of the moduli space with Kuranishi charts is called a Kuranishi structure. After adding some extra assumptions to the definition of Kuranishi structure[25], we define an intersection theory which results in Gromov-Witten invariants and generalizes the definition of these invariants in the semi-positive case.

In Section 3.2, we introduce the notion of abstract Kuranishi chart and Kuranishi structure over a compact metrizable topological space so that a version of basic concepts like tangent space, orientability, and smoothness make sense. In Section 3.3, in order to simplify the calculations, we assemble the uncountable data of Kuranishi structure into a finite set of triples of orbifolds, orbibundles, and orbifold sections, with transition maps from lower rank orbibundles to higher ones. The result is called a dimensionally graded system. In Section 3.7, similar to Section 2.4, we consistently perturb the orbifold sections of the dimensionally graded system into a consistent set of transversal but multivalued sections of the obstruction bundles.

REMARK 3.1.1. It is yet unknown how to build a category from the set of spaces with Kuranishi structure; thus, a so called Kuranishi category is the dream of this theory. The problem is that we do not know how to define morphisms. By a Kuranishi category, we mean a category that includes morphisms between Kuranishi structures and such that basic operations like fiber product are well-defined. For some attempts in this direction see [**31**]. In algebraic geometry, theory of 2-categories with 2-morphisms and 2-fiber products is one such big category that includes all moduli spaces of holomorphic curves in complex projective varieties.

3.2. Kuranishi structures. In this section, we continue to use the notation conventions of Notation 2.1.4, 2.1.9, and 2.2.2.

DEFINITION 3.2.1. *A **Kuranishi chart** of real dimension n on a compact metrizable topological space M is a tuple $\mathcal{U}\equiv(\mathrm{pr}\colon U\longrightarrow V, G, s, \psi)$ made up of*

(1) *an orbibundle chart $(\mathrm{pr}\colon U\longrightarrow V, G)$ of rank $\dim V-n$ as[26] in Definition 2.2.1,*
(2) *a smooth G-equivariant section $s\colon V\longrightarrow U$, called **Kuranishi map** or section,*
(3) *and a continuous map $\psi\colon s^{-1}(0)\longrightarrow M$ defined by $\psi=\overline{\psi}\circ\pi$, where*

$$\pi\colon s^{-1}(0)\longrightarrow \overline{s}^{-1}(0)=s^{-1}(0)/G$$

*is the quotient map and $\overline{\psi}\colon\overline{s}^{-1}(0)\longrightarrow F\subset M$ is a homeomorphism into an open subset F, called **footprint**, of M.*

Similar to Section 2.1, for $x\in s^{-1}(0)$ and $\overline{x}=\psi(x)\in M$, let G_x and $\mathrm{I}_{\overline{x}}$ denote the isotropy group and the conjugacy class of the isotropy group, respectively. We say $\mathcal{U}$ is a Kuranishi neighborhood **centered at** $\overline{x}\in M$ if $G_x=G$; in this case there is a unique $x\in s^{-1}(0)$ such that $\psi(x)=\overline{x}$.

Given a Kuranishi chart $\mathcal{U}$ as above, a subgroup $G'<G$, and a G'-invariant connected open subset $V'\subset V$ such that

$$g(V')\cap V'=\emptyset \qquad \forall g\in G\backslash G',$$

by restricting to V', we obtain a Kuranishi chart

$$\mathcal{U}'=\big(\mathrm{pr}\colon U'=\mathrm{pr}^{-1}(V')\longrightarrow V', G', s'=s|_{V'}, \psi'=\psi|_{s^{-1}(0)\cap V'},\big) \tag{3.1}$$

[25] e.g. existence of "tangent bundle", Hausdorffness, ...

[26] With the exception of the map ψ which in this case is only defined on the zero set of Kuranishi map.

which we call a **sub-chart**. For every $\overline{x} \in \psi(s^{-1}(0))$, by reducing to a sub-chart, we can always obtain a Kuranishi neighborhood centered at $\overline{x}$. Similar to the argument before Example 2.2.7, every sufficiently small Kuranishi (sub-) chart centered at any $\overline{x}\in M$ is isomorphic to a linear chart where U is a trivial bundle; this is the point of view in the original approach of [**20**].

DEFINITION 3.2.2. *Let $\mathcal{U}_i=(\mathrm{pr}_i\colon U_i\longrightarrow V_i, G_i, s_i, \psi_i)$, with $i=1,2$, be a pair of n-dimensional Kuranishi charts with intersecting footprints F_i, such that*

$$\overline{p}\in F_{12} = F_1\cap F_2 \text{ and } \dim V_1 \le \dim V_2.$$

*A **coordinate change** from $\mathcal{U}_1$ to $\mathcal{U}_2$ **centered at** $\overline{p}$ consists of a sub-chart*[27] *$\mathcal{U}_{1,2}\subset\mathcal{U}_1$ centered at $\overline{p}$ and an orbibundle embedding*

$$\Phi_{12} \equiv (\mathfrak{D}\phi_{12}, \phi_{12})\colon \mathcal{U}_{1,2} \longrightarrow \mathcal{U}_2$$

as in Definition 2.2.10 such that

$$\mathfrak{D}\phi_{12}\circ s_{1,2} = s_2\circ\phi_{12}, \qquad \psi_{1,2} = \psi_2\circ\phi_{12}. \tag{3.2}$$

Recall that by definition of orbibundle embedding, the group homomorphism

$$h_{12}\colon G_{1,2}=(G_{1,2})_p \longrightarrow (G_2)_{\phi(p)} \tag{3.3}$$

that comes with ϕ_{12} (see Notation 2.1.9) is an isomorphism.

EXAMPLE 3.2.3. Let $M=(0,3)$ and cover M with two 1-dimensional Kuranishi charts $\mathcal{U}_1$ and $\mathcal{U}_2$ such that $U_2=(0,2)\times(-1,1)\times\mathbb{R}$, $V_2=(0,2)\times(-1,1)$, pr_2 is the projection onto the first two components, $s_2(x,y) = y$, $V_1 = (1,3)$, U_1 is trivial, both isotropy groups are trivial, and ψ_1 and ψ_2 are the obvious maps to $(0,2)$ and $(1,3)$, respectively. In this situation, for any $x\in(1,2)$, by restricting $\mathcal{U}_1$ to $(1,2)$, we obtain a coordinate change centered at x from U_1 to U_2.

If we replace the trivial group G_2 with $G_2=\mathbb{Z}_2$, acting on U_2 by $(x,y,z)\longrightarrow(x,-y,-z)$, the resulting charts do not admit a coordinate change anymore; the isomorphism condition (3.3) can never be satisfied.

Given a coordinate change as above, since (3.3) is an isomorphism and ϕ_{12} is embedding, we conclude that for every $x_1\in s_{1,2}^{-1}(0)$ and $x_2\in s_2^{-1}(0)$ with the same image $\overline{x}$ in M, h_{12} induces an isomorphism between the isotropy groups $(G_1)_{x_1}$ and $(G_2)_{x_2}$. Therefore, similarly to orbifolds, the conjugacy class of the isotropy group $\mathrm{I}_{\overline{x}}$ is well-defined.

DEFINITION 3.2.4. *Let $\mathcal{U}_i$, with $i=1,2,3$, be a triple of n-dimensional Kuranishi charts centered at $\overline{p}_i\in M$ with intersecting footprints F_i, such that*

$$\overline{p}_1\in F_{123}=F_1\cap F_2\cap F_3, \quad \overline{p}_2\in F_{23}=F_2\cap F_3, \quad \text{and } \dim V_1\le\dim V_2\le\dim V_3.$$

*Given a triple set of coordinate change maps $\Phi_{ij}\colon\mathcal{U}_{i,j}\longrightarrow\mathcal{U}_j$ centered at $\overline{p}_i$, with $1\le i<j\le 3$, we say they satisfy the **cocycle condition** if*

$$(\Phi_{13}=\Phi_{23}\circ\Phi_{12})|_{\overline{V}_{1,2,3}}, \quad \text{where } \overline{V}_{1,2,3}=\overline{\phi}_{12}^{-1}(\overline{V}_{2,3})\cap\overline{V}_{1,3}\subset\overline{V}_{1,2}. \tag{3.4}$$

More explicitly, the equality (3.4) means that for every connected component $V_{1,2,3;\alpha}$ of $V_{1,2,3}=\phi_{12}^{-1}(V_{2,3})\cap V_{1,3}\subset V_{1,2}$, there exists $g_\alpha\in G_3$ such that

(1) $\phi_{13}|_{V_{1,2,3;\alpha}}=g_\alpha\circ\phi_{23}\circ\phi_{12}|_{V_{1,2,3;\alpha}}$,
(2) $\mathfrak{D}\phi_{13}|_{U_{1,2,3;\alpha}}=g_\alpha\circ\mathfrak{D}\phi_{23}\circ\mathfrak{D}\phi_{12}|_{U_{1,2,3;\alpha}}$, where $U_{1,2,3;\alpha}=\mathrm{pr}_1^{-1}(V_{1,2,3;\alpha})$,

[27] Any object labeled by $1,2$ is the restriction to $\mathcal{U}_{1,2}$ of the corresponding object in $\mathcal{U}_1$.

(3) and $\mathrm{ad}(g_\alpha)\circ h_{13}=h_{23}\circ h_{12}$, where for every $g\in G_3$, $\mathrm{ad}(g)\colon G_3\longrightarrow G_3$ is the adjoint action $g'\longrightarrow gg'g^{-1}$.

REMARK 3.2.5. Since the coordinate change maps commute with the section and footprint maps, the composite maps on the right-hand sides of the first and second condition and the coordinate change map Φ_{13} automatically satisfy the cocycle condition along the graph of $s_1|_{V_{1,2,3}}$. Otherwise, the cocycle condition away from the graph of s_1 has to be considered as a separate condition[28]. In the definition of orbifolds, where there is no Kuranishi map, the cocycle condition is automatic. Also, in the natural construction of Kuranishi structure on moduli space of pseudoholomorphic maps in Section 6.5, the cocycle condition naturally follows from the construction.

DEFINITION 3.2.6. *A Kuranishi structure $\mathcal{K}$ of real dimension n on a compact metrizable topological space M consists of a Kuranishi chart $\mathcal{U}_p$ of real dimension n centered at $\overline{p}$ with footprint F_p, for every $\overline{p}\in M$, and a coordinate change map $\Phi_{qp}\colon\mathcal{U}_{q,p}\longrightarrow\mathcal{U}_p$, for every $\overline{q}\in F_p$, such that whenever $\overline{r}\in F_p\cap F_q$ and $\overline{q}\in F_p$, then Φ_{qp}, Φ_{rq}, and Φ_{rp} satisfy the cocycle condition.*

NOTATION 3.2.7. *Ideally, we should label the coordinate charts $\mathcal{U}_p$ as $\mathcal{U}_{\overline{p}}$. However, since by definition the preimage $p=\psi^{-1}(\overline{p})$ is unique, we label the Kuranishi charts and the coordinate change maps between them by p,q,r, etc, instead of $\overline{p},\overline{q},\overline{r}$, etc, respectively.*

Given a Kuranishi structure

$$\mathcal{K}=\big\{(\mathcal{U}_p)_{\overline{p}\in M},(\Phi_{qp}\colon\mathcal{U}_{q,p}\longrightarrow\mathcal{U}_p)_{\overline{p}\in M,\overline{q}\in F_p}\big\}$$

as in Definition 3.2.6, define the **level map**

$$L\colon M\longrightarrow\mathbb{N},\qquad L(\overline{p})=\operatorname{rank}U_p\,. \tag{3.5}$$

This is an upper semi-continuous function and gives us a stratification

$$M=\coprod_{k\in\mathbb{N}}M(k),\qquad M(k)=L^{-1}(k), \tag{3.6}$$

where each **level set** $M(k)$ is an open subset of $M(\geq k)=\bigcup_{\ell\geq k}M(\ell)$. Let

$$\mathfrak{B}=\mathfrak{B}(M,\mathcal{K})=\{k\in\mathbb{N}\colon M(k)\neq\emptyset\}. \tag{3.7}$$

It follows from the compactness of M and the upper semi-continuity of L that $\mathfrak{B}$ is a finite set and $M(\geq k)$ is a closed, hence compact, subset of M.

EXAMPLE 3.2.8. The simplest examples of (abstract) Kuranishi structures come from orbibundles over orbifolds. Given a non-negative integer n, let $\mathrm{pr}\colon\mathcal{E}\longrightarrow\mathcal{Y}$ be an orbibundle of rank $\dim\mathcal{Y}-n$ and $s\colon\mathcal{Y}\longrightarrow\mathcal{E}$ be an orbifold section with compact zero locus $M=\overline{s}^{-1}(0)\subset Y$. For every $\overline{x}\in M$, we can take the n-dimensional Kuranishi neighborhood $\mathcal{U}_x$ centered at $\overline{x}$ to be an orbibundle chart of $\mathcal{E}$ centered at $\overline{x}$ together with the Kuranishi map s. The coordinate change maps are simply the refinements and the cocycle condition is a consequence of the compatibility condition of orbibundle charts. In this situation, we say M has a **pure orbibundle**

[28]Because we will perturb the Kuranishi maps and we need to preserve their compatibility throughout the deformation process.

Kuranishi structure. We may also consider a disjoint union

$$\coprod_{i=1}^{m} (\mathrm{pr}_i : \mathcal{E}_i \longrightarrow \mathcal{Y}_i)$$

of orbibundles of rank

$$\operatorname{rank} \mathcal{E}_i = \dim \mathcal{M}_i - n.$$

Given a set of orbifold sections $\{s_i\}_{i=1}^m$ as above, the zero set

$$M = \coprod_{i=1}^{m} \left(M_i \equiv \overline{s}_i^{-1}(0)\right)$$

inherits a Kuranishi structure as above that we still denote by a pure orbibundle Kuranishi structure. The examples in Section 7 are moduli spaces of pseudoholomorphic maps which admit Kuranishi structures of this type.

In the next section, using the level function L above, we assemble the local charts of an arbitrary Kuranishi structure into a finite union of pure orbibundle Kuranishi structures compatible along intersections. We call the result a **dimensionally graded system**, or **DGS** for short, associated to the Kuranishi structure. In [**20**] and the related followups, this is denoted by "good coordinate system" or "dimensionally graded good coordinate system". In an DGS, the pure orbibundles are indexed by their rank, while in good coordinate systems, they are indexed by some partially ordered set; thus, as D. Joyce [**32**, Definition 3.5] names it, an DGS is an *excellent* good coordinate system. This notion of "reduction" in [**39**] plays the same role as DGS.

For every arbitrary smooth vector bundle $\mathrm{pr}\colon U \longrightarrow V$ and a smooth section s, we can canonically decompose the tangent bundle of U along the zero set of s to

$$TU|_{s^{-1}(0)} = U|_{s^{-1}(0)} \oplus TV|_{s^{-1}(0)}. \tag{3.8}$$

With respect to this decomposition, and by projection onto the the first factor in the right-hand side of (3.8), we obtain the linearization map

$$\mathrm{d}s\colon TV|_{s^{-1}(0)} \longrightarrow U_{s^{-1}(0)}.$$

Therefore, for a coordinate change embedding map as in Definition 3.2.2, we obtain a commutative diagram

(3.9)

$$\begin{array}{ccccccccc}
0 & \longrightarrow & TV_{1,2}|_{s_{1,2}^{-1}(0)} & \xhookrightarrow{d\phi_{12}} & TV_2|_{\phi_{12}(s_{1,2}^{-1}(0))} & \longrightarrow & \mathcal{N}_{\phi_{12}(V_{1,2})}V_2|_{\phi_{12}(s_{1,2}^{-1}(0))} & \longrightarrow & 0 \\
 & & \Big\downarrow {\scriptstyle \mathrm{d}s_{1,2}} & & \Big\downarrow {\scriptstyle \mathrm{d}s_2} & & \Big\downarrow {\scriptstyle \mathrm{d}s_{2/1} = \mathrm{d}s_2/\mathrm{d}s_{1,2}} & & \\
0 & \longrightarrow & U_{1,2}|_{s_{1,2}^{-1}(0)} & \xhookrightarrow{\mathfrak{D}\phi_{12}} & U_2|_{\phi_{12}(s_{1,2}^{-1}(0))} & \longrightarrow & (U_2/\mathfrak{D}\phi_{12}(U_{1,2}))|_{\phi_{12}(s_{1,2}^{-1}(0))} & \longrightarrow & 0.
\end{array}$$

where the last vertical map, denoted by $\mathrm{d}s_{2/1}$, is defined in the following way. For every $v \in TV_2|_{\phi_{12}(s_{1,2}^{-1}(0))}$,

$$\mathrm{d}s_2(v) \in U_2|_{\phi_{12}(s_{1,2}^{-1}(0))}.$$

By the left-hand side of (3.2), if $v \in \mathrm{d}\phi_{12}(TV_{1,2}) \subset TV_2$, then

$$\mathrm{d}s_2(v) \in \mathfrak{D}\phi_{12}(U_{1,2}) \subset U_2;$$

therefore, we obtain a well-defined quotient map $ds_{2/1}$ from the normal bundle of the embedding into the quotient bundle (restricted to $\phi_{12}(s_{1,2}^{-1}(0))$). Note that $G_{1,2} \cong h_{12}(G_{1,2}) \subset G_2$ acts on both

$$\mathcal{N}_{\phi_{12}(V_{1,2})}V_2|_{\phi_{12}(s_{1,2}^{-1}(0))} \text{ and } (U_2/\mathfrak{D}\phi_{12}(U_{1,2}))|_{\phi_{12}(s_{1,2}^{-1}(0))},$$

and $ds_{2/1}$ is equivariant with respect to this action.

DEFINITION 3.2.9. *We say that a Kuranishi structure $\mathcal{K}$ on a topological space M has* ***tangent bundle*** *if for every coordinate change map $\Phi_{qp}\colon \mathcal{U}_{q,p} \longrightarrow \mathcal{U}_p$ in $\mathcal{K}$ the map*

$$ds_{p/q}\colon \mathcal{N}_{\phi_{qp}(V_{q,p})}V_p|_{\phi_{qp}(s_{q,p}^{-1}(0))} \longrightarrow \big(U_p/\mathfrak{D}\phi_{qp}(U_{q,p})\big)|_{\phi_{qp}(s_{q,p}^{-1}(0))}, \tag{3.10}$$

given by (3.9) is an orbibundle isomorphism. In other words, we require the restriction of s_p to $\phi_{qp}(s_{q,p}^{-1}(0))$ to be transverse in the normal direction modulo $U_{q,p}$. By a ***Kuranishi space****, we mean a pair $(M,\mathcal{K})$ where $\mathcal{K}$ is a Kuranishi structure with tangent bundle on M.*

REMARK 3.2.10. If the two charts $\mathcal{U}_1$ and $\mathcal{U}_2$ used in (3.9) have the same dimension, $ds_{2/1}$ is trivial and the tangent bundle condition is automatically satisfied. Therefore, an equidimensional set of Kuranishi neighborhoods automatically satisfies the tangent bundle condition. This is for example the case when $\mathcal{K}$ is a pure orbibundle Kuranishi structure.

We say a Kuranishi chart $\mathcal{U} = (\mathrm{pr}\colon U \longrightarrow V, G, s, \psi)$ is **orientable** if the orbibundle $[U/G] \longrightarrow [V/G]$ is relatively orientable; i.e. there is a G-equivariant isomorphism

$$\det U^* \otimes \det TV \cong V \times \mathbb{R}, \tag{3.11}$$

where the action of G on $\mathbb{R}$ is the trivial action. An orientation is a choice of (homotopy class of) such trivialization. Given a $G_{q,p}$-equivariant isomorphism (3.10), it induces a canonical $G_{q,p}$-equivariant isomorphism

$$\det ds_{p/q}\colon \big(\det U_{q,p}^* \otimes \det TV_{q,p}\big)|_{s_{q,p}^{-1}(0)} \longrightarrow \big(\det U_p^* \otimes \det TV_p\big)|_{\phi_{qp}(s_{q,p}^{-1}(0))}. \tag{3.12}$$

DEFINITION 3.2.11. *We say that a Kuranishi space $(M,\mathcal{K})$ is* ***orientable*** *if every Kuranishi chart $\mathcal{U}_p$ can be oriented in a way that following conditions hold. For every coordinate change map $\Phi_{qp}\colon \mathcal{U}_{q,p} \longrightarrow \mathcal{U}_p$ in $\mathcal{K}$, the canonical isomorphism (3.12) is orientation-preserving. Moreover, if $L(\overline{p}) = L(\overline{q})$, then the isomorphism*

$$\big(\det U_{q,p}^* \otimes \det TV_{q,p}\big) \xrightarrow{\cong} \big(\det U_p^* \otimes \det TV_p\big)|_{\phi_{qp}(V_{q,p})}$$

is orientation preserving.

REMARK 3.2.12. A pure orbibundle Kuranishi space as in Example 3.2.8 is orientable if and only if each orbibundle $\mathcal{E}_i \longrightarrow \mathcal{Y}_i$ is relatively orientable.

3.3. Dimensionally graded systems. Similar to the orbifold case of Section 2.6, our idea for constructing a VFC for an arbitrary Kuranishi space is to perturb the Kuranishi maps s_p, in a compatible way with the coordinate change maps, and make them transverse to the zero section. We construct these perturbations inductively; we start from the points with the lowest value of L in (3.5) and move up. For this induction to work properly, we build a pure orbibundle Kuranishi structure -in the sense of Example 3.2.8- around a compact subset of each level set $M(k)$ with orbibundle embeddings from (sub-orbibundles of) lower levels to higher

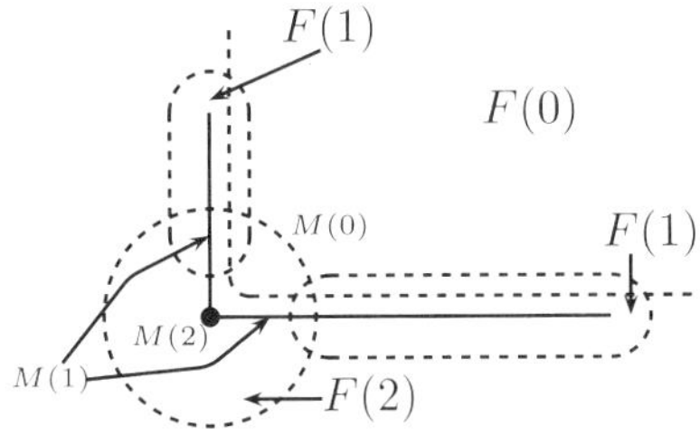

FIGURE 5. A possible configuration of footprints of different levels.

levels. In this section, we introduce the notion of dimensionally graded system for Kuranishi structures which makes sense of this outline. This notion is a special case of the notion of "good coordinate system" in [**20**] where the orbibundles are graded by their rank instead of an arbitrary partially ordered set.

NOTATION 3.3.1. *For a topological subspace $A\subset B$, let $\mathrm{cl}(A)\subset B$ denote the closure of A inside B. We are adopting this notation, instead of the common notion $\overline{A}$, in order to avoid confusion with the quotient space of group actions. However, there is one special case, the case of compactified moduli spaces of maps and curves, $\overline{\mathcal{M}}_{g,k}(X,A)$ and $\overline{\mathcal{M}}_{g,k}$, respectively, where the over-line has its usual meaning.*

DEFINITION 3.3.2. *Let $\mathcal{K}$ be an n-dimensional Kuranishi structure on M. A* ***dimensionally graded system (or DGS)*** *for $(M,\mathcal{K})$ consist of an orbibundle (possibly empty!)*

$$\mathrm{pr}_i\colon \mathcal{E}(i)\longrightarrow \mathcal{Y}(i)$$

and an orbifold section $s_i\colon\mathcal{Y}(i)\longrightarrow\mathcal{E}(i)$, for every $i\in\mathfrak{B}=\mathfrak{B}(M,\mathcal{K})\subset\mathbb{Z}$ with notation as in (3.7), such that the following conditions hold (See the notation convention in Notation 2.1.4).

(1) *For every $i\in\mathfrak{B}$, (if $Y(i)\neq\emptyset$) $\mathrm{rank}\ \mathcal{E}(i)=i$ and $\dim\mathcal{Y}(i)=n+i$.*
(2) *There are open embeddings $\overline{\psi}_i\colon\, \overline{s}_i^{-1}(0)\longrightarrow M$ such that the set of footprints*

$$\{F(i)=\overline{\psi}_i(\overline{s}_i^{-1}(0))\}_{i\in\mathfrak{B}}$$

is an open covering of M and

$$\mathrm{cl}(F(k))\cap M(>k)=\emptyset\quad \forall k\in\mathfrak{B};$$

see Figure 5.
(3) *For every $i,j\in\mathfrak{B}$, with $i<j$, there exists an open subset $Y(i,j)\subset Y(i)$ and an embedding of orbibundles commuting with the section and footprint maps*

$$\Phi_{ij}=(\mathfrak{D}\phi_{ij},\phi_{ij})\colon(\mathcal{E}(i,j),\mathcal{Y}(i,j))\longrightarrow(\mathcal{E}(j),\mathcal{Y}(j)),$$
$$\mathcal{Y}(i,j)=\mathcal{Y}(i)|_{Y(i,j)},\ \ \mathcal{E}(i,j)=\mathcal{E}(i)|_{\mathcal{Y}(i,j)},$$

as in Definition 2.2.10, such that[29]

$$F(i,j)=\overline{\psi}_{i,j}(\overline{s}_{i,j}^{-1}(0))=F(i)\cap F(j),\quad \text{where } s_{i,j}=s_i|_{\mathcal{Y}(i,j)}. \tag{3.13}$$

(4) *For $i,j,k\in\mathfrak{B}$, with $i<j<k$,*

$$\overline{\phi}_{ij}^{-1}(Y(j,k))=Y(i,k)\cap Y(i,j).$$

[29]Anything indexed by i,j denotes the restriction of the corresponding object indexed by i to $\mathcal{E}(i,j)\longrightarrow\mathcal{Y}(i,j)$.

We denote the intersection on right by $Y(i,j,k)$ *and the restriction of* $\mathcal{E}(i,j)$ *to* $Y(i,j,k)$ *by* $\mathcal{E}(i,j,k)$.

(5) *For* $i,j,k \in \mathfrak{B}$, *with* $i<j<k$,

$$\overline{\phi}_{ik}(Y(i,k)) \cap \overline{\phi}_{jk}(Y(j,k)) = \overline{\phi}_{ik}(Y(i,j,k)).$$

(6) *For every* $i,j,k \in \mathfrak{B}$, *with* $i<j<k$, *the cocycle condition*

$$(\Phi_{jk} \circ \Phi_{ij})|_{\mathcal{Y}(i,j,k)} = \Phi_{ik}|_{\mathcal{Y}(i,j,k)} \tag{3.14}$$

of Definition 3.2.4 *holds.*

(7) *For every* $i \in \mathfrak{B}$ *and* $\overline{p} \in F(i)$, *there exists a sub-chart* $\mathcal{U}_{p,i} \subset \mathcal{U}_p$ *centered at* $\overline{p}$ *and an orbibundle embedding*

$$\Phi_{pi} = (\mathfrak{D}\phi_{pi}, \phi_{pi})\colon ([U_{p,i}/G_{p,i}], [V_{p,i}/G_{p,i}]) \longrightarrow (\mathcal{E}(i), \mathcal{Y}(i)) \tag{3.15}$$

that commutes with the projection, section, and footprint maps, and the following compatibility conditions (with $\mathcal{K}$*) hold.*

(a) *Whenever* $\overline{q} \in F_p \cap F(i)$ *and* $\overline{p} \in F(i)$, *restricted to* $\overline{\phi}_{qp}^{-1}(\overline{V}_{p,i}) \cap \overline{V}_{q,i}$ *we have* $\overline{\Phi}_{pi} \circ \overline{\Phi}_{qp} = \overline{\Phi}_{qi}$, *where* $\overline{V}_{p,i}$ *is the underlying topological space of the orbifold* $[V_{p,i}/G_{p,i}]$.

(b) *Whenever* $\overline{p} \in F(i) \cap F(j)$ *and* $i<j$, *restricted to* $\overline{\phi}_{pi}^{-1}(Y_{i,j}) \cap \overline{V}_{p,j}$ *we have* $\overline{\Phi}_{ij} \circ \overline{\Phi}_{pi} = \overline{\Phi}_{pj}$.

(8) *If in addition* $(\mathcal{K}, M)$ *is a Kuranishi space, i.e. if* $\mathcal{K}$ *has tangent bundle, then for every* $i \in \mathfrak{B}$ *and every* $\overline{p} \in F(i)$, *with* Φ_{pi} *as in* (3.15), *the quotient map* $\mathrm{d}s_{i/p}$ *as in* (3.9) *is an equivariant isomorphism of vector bundles.*

REMARK 3.3.3. It follows from Definition 3.3.2.(2)(7) that

$$M(\geq k) \subset \bigcup_{i \in \mathfrak{B}(\geq k)} F(i), \qquad F(k) \subset \bigcup_{\overline{p} \in M(k)} F_p.$$

REMARK 3.3.4. Note that by Definition 3.3.2.(2), the necessary condition

$$\operatorname{rank} \mathcal{U}_{p,i} \leq \operatorname{rank} \mathcal{U}_i$$

for (7) to be plausible is automatically satisfied. The first and second conditions in Definition 3.3.2 together with (7) simply assert that $(\mathcal{E}(i), \mathcal{Y}(i))$ is made of Kuranishi charts centered at points of level i. Conditions (3) and (6) allow us to glue different $(\mathcal{E}(i), \mathcal{Y}(i))$ along their common intersections and form a "thickening" of the moduli space; see Lemma 3.3.8 and the argument preceding that. The equality (3.13) in (3) insures that every coordinate change sub-orbibundle covers the whole intersection of the corresponding footprints. Conditions (3) and (4) are necessary for (6) to make sense. We need (5) as well as (6) for the relation defined in Lemma 3.3.8 to be an equivalence relation. Condition (7) indicates how the level structure is related to the underlying Kuranishi structure. Finally, Condition (8) (as stated) needs some further clarification. The orbibundle $\mathcal{E}(i)$ itself, by definition, is made of local orbibundle charts. Then (8) means that for each $\overline{p} \in F(i)$, with Φ_{pi} as in (3.15), there exist a sub-chart $\mathcal{U}_1 = (\mathrm{pr}_1 \colon U_1 \longrightarrow V_1, G_1, s_1, \psi_1)$ of $\mathcal{U}_p$ centered at $\overline{p}$ and a defining orbibundle chart (together with the restriction of the footprint map and section maps) $\mathcal{U}_2 = (\mathrm{pr}_2 \colon U_2 \longrightarrow V_2, G_2, s_2, \psi_2)$ of $\mathcal{E}(i)$ centered at

$$\overline{p} \cong \overline{\psi}_i^{-1}(\overline{p}) \in \overline{s}_i^{-1}(0) \subset Y(i),$$

over which Φ_{pi} is given by an equivariant embedding of vector bundles

$$(\mathfrak{D}\phi,\phi)\colon \mathcal{U}_1 \longrightarrow \mathcal{U}_2, \quad \mathfrak{D}\phi\circ s_1 = s_2\circ\phi, \quad \overline{\phi}_1\circ\psi = \psi_2\circ\phi,$$

and the quotient map $\mathrm{d}s_{2/1}$ as in (3.9) is an equivariant isomorphism of vector bundles.

REMARK 3.3.5. In the case of pure orbibundle Kuranishi structures introduced in Example 3.2.8, we simply set

$$\big(M(i),\mathcal{E}(i),\mathcal{Y}(i)\big) = \coprod_{j\in[m]\colon \mathrm{rank}\ \mathcal{E}_j=i} (M_j,\mathcal{E}_j,\mathcal{Y}_j)$$
$$\text{and } \big(\mathcal{E}(i,j),\mathcal{Y}(i,j)\big)=\emptyset \qquad \forall i,j\in\mathfrak{B},\ i<j.$$

Therefore, there are no overlap conditions for pure orbibundle Kuranishi structures and we are essentially in the realm of orbifold and orbibundles. This also illuminates the idea behind dimensionally graded systems; we assemble the uncountable data of Kuranishi structure into a finite set of orbibundles, graded by their ranks, with transition embeddings from lower grades to higher grades. We then use this finite set to inductively build up the VFC.

LEMMA 3.3.6. *Assume*

$$\mathfrak{L}=\mathfrak{L}(\mathcal{K})\equiv\big(\{\mathcal{E}(i),\mathcal{Y}(i),s_i,\psi_i\}_{i\in\mathfrak{B}},\{\mathcal{E}(i,j),\mathcal{Y}(i,j),\Phi_{ij}\}_{i<j,\ i,j\in\mathfrak{B}}\big) \tag{3.16}$$

is a dimensionally graded system for the Kuranishi space $(M,\mathcal{K})$. Then $\mathfrak{L}$ also has tangent bundle in the sense that for every $i,j\in\mathfrak{B}$, with $i<j$, the map

$$\mathrm{d}s_{j/i}\colon \mathcal{N}_{\phi_{ij}(\mathcal{Y}(i,j))}\mathcal{Y}(j)|_{\overline{\phi}_{ij}(\overline{s}_{i,j}^{-1}(0))} \longrightarrow \big(\mathcal{E}(j)/\mathfrak{D}\phi_{ij}\mathcal{E}(i,j)\big)|_{\overline{\phi}_{ij}(\overline{s}_{i,j}^{-1}(0))}. \tag{3.17}$$

defined locally by the right column of (3.9) is an isomorphism of orbibundles.

PROOF. The point is that the condition (3.17) is local, and by the compatibility conditions (8) and (7) of Definition 3.3.2, local charts of $(\mathcal{E}(i),\mathcal{Y}(i))$ and that of Kuranishi structure are essentially the same. More precisely, for every $\overline{p}\in F(i,j)$, let $\mathcal{U}_0=(\mathrm{pr}_0\colon U_0\longrightarrow V_0,G_0,s_0,\psi_0)$ be a sub-chart of $\mathcal{U}_p$ centered at $\overline{p}$, $\mathcal{U}_1=(\mathrm{pr}_1\colon U_1\longrightarrow V_1,G_1,s_1,\psi_1)$ be a sub-chart of $\mathcal{E}(i)$ centered at

$$\overline{p}\cong\overline{\psi}_i^{-1}(\overline{p})\in\overline{s}_i^{-1}(0)\subset Y(i),$$

and $\mathcal{U}_2=(\mathrm{pr}_2\colon U_2\longrightarrow V_2,G_2,s_2,\psi_2)$ be a sub-chart of $\mathcal{E}(j)$ centered at $\overline{p}$, such that Φ_{pi}, Φ_{pj} and Φ_{ij} are given by equivariant embeddings of vector bundles

$$(\mathfrak{D}\phi_{ab},\phi_{ab})\colon \mathcal{U}_a\longrightarrow\mathcal{U}_b \qquad \forall a=0,1,\ b=1,2,\ a<b,$$

and the quotient maps $\mathrm{d}s_{1/0}$ and $\mathrm{d}s_{2/0}$ are as in (3.9). To distinguish between preimages of $\overline{p}$ in V_0, V_1, and V_2, we denote them by p_0, p_1, and p_2, respectively.

We then obtain a commutative diagram

$$
\begin{array}{ccc}
0 & & 0 \\
\downarrow & & \downarrow \\
\mathcal{N}_{\phi_{01}(V_0)}V_1|_{p_1=\phi_{01}(p_0)} & \xrightarrow{\mathrm{d}s_{1/0}} & \big(U_1/\mathfrak{D}\phi_{01}(U_0)\big)|_{p_1} \\
\downarrow{\scriptstyle \mathrm{d}\phi_{12}} & & \downarrow{\scriptstyle \mathrm{d}\mathfrak{D}\phi_{12}} \\
\mathcal{N}_{\phi_{02}(V_0)}V_2|_{p_2} & \xrightarrow{\mathrm{d}s_{2/0}} & \big(U_2/\mathfrak{D}\phi_{02}(U_0)\big)|_{p_2} \\
\downarrow{\scriptstyle \pi} & & \downarrow{\scriptstyle \pi} \\
\mathcal{N}_{\phi_{12}(V_1)}V_2|_{p_2} & \xrightarrow{\mathrm{d}s_{2/1}} & \big(U_2/\mathfrak{D}\phi_{12}(U_1)\big)|_{p_2} \\
\downarrow & & \downarrow \\
0 & & 0
\end{array}
$$

where π is the obvious projection map. Since the columns are exact and the first two rows are isomorphism, we conclude that the bottom row is also an isomorphism. $\square$

DEFINITION 3.3.7. *Let $\mathfrak{L}$ be an DGS with tangent bundle as in Lemma 3.3.6. We say $\mathfrak{L}$ is **orientable** (resp. oriented) if every orbibundle $\mathcal{E}(i)\longrightarrow\mathcal{Y}(i)$ is relatively orientable (resp. oriented) and there exists a (resp. the given) set of trivializations*

$$\det\mathcal{E}(i)^*\otimes\det T\mathcal{Y}(i)\cong\mathcal{Y}(i)\times\mathbb{R}\quad\forall i\in\mathfrak{B} \tag{3.18}$$

(resp. are) compatible with isomorphisms

(3.19)
$$\det \mathrm{d}s_{j/i}\colon \big(\det\mathcal{E}(i,j)^*\otimes\det T\mathcal{Y}(i,j)\big)|_{\overline{s}_{i,j}^{-1}(0)}\longrightarrow\big(\det\mathcal{E}(j)^*\otimes\det T\mathcal{Y}(j)\big)|_{\overline{\phi}_{ij}(\overline{s}_{i,j}^{-1}(0))}$$

induced by (3.17).

Similarly to Lemma 3.3.6, if $(M,\mathcal{K})$ is orientable (resp. oriented) and $\mathfrak{L}$ is an DGS for $(M,\mathcal{K})$, then a shrinking of $\mathfrak{L}$, which we will introduce in Section 3.4, is orientable (resp. oriented).

In the rest of this section, we introduce an equivalence relation on the disjoint union

$$\coprod_{i\in\mathfrak{B}}Y(i) \tag{3.20}$$

such that the quotient space $Y(\mathfrak{L})$ can be seen as a global **thickening** of the space M in which a perturbation of the zero set $M\subset Y(\mathfrak{L})$ will reside.

LEMMA 3.3.8. *The relation $\sim$ on (3.20) defined by*

$$\overline{x}\in Y(i)\sim\overline{y}\in Y(j)\Leftrightarrow\begin{cases} i=j \text{ and } \overline{x}=\overline{y}, \text{ or}\\ i<j,\ \overline{x}\in Y(i,j), \text{ and } \overline{\phi}_{ij}(\overline{x})=\overline{y}, \text{ or}\\ j<i,\ \overline{y}\in Y(j,i), \text{ and } \overline{\phi}_{ji}(\overline{y})=\overline{x},\end{cases} \tag{3.21}$$

is an equivalence relation.

PROOF. Among reflexivity, symmetry, and transitivity, only the last one is somewhat non-trivial. Suppose $\overline{x}_1 \sim \overline{x}_2$ and $\overline{x}_2 \sim \overline{x}_3$, $\overline{x}_a \in Y(j_a)$. If two of j_a are equal, the desired conclusion is trivial. If $j_1 < j_2 < j_3$, $\overline{x}_1 \sim \overline{x}_3$ follows from Definition 3.3.2.(6). If $j_2 < j_1, j_3$, we may assume $j_1 < j_3$, then again

$$\overline{x}_2 \in \overline{\phi}_{j_2 j_1}^{-1}(Y(j_1, j_3)) = Y(j_2, j_1) \cap Y(j_2, j_3)$$

and the result follows from the cocycle condition. If $j_1, j_3 < j_2$, we may assume $j_1 < j_3$, then by Definition 3.3.2.(5),

$$\overline{x}_2 \in \overline{\phi}_{j_1,j_2}(Y(j_1, j_2)) \cap \overline{\phi}_{j_3,j_1} Y(j_3, j_2) = \overline{\phi}_{j_1,j_2} Y(j_1, j_3, j_2).$$

By the uniqueness of $\overline{x}_1 \in Y(j_1, j_3)$ and $\overline{x}_3 \in Y(j_3, j_2)$, we conclude $\overline{x}_1 \in Y(j_1, j_3, j_2)$ and $\overline{\phi}_{j_1 j_3}(x_1) = \overline{x}_3$. □

Let $\wp_i : Y(i) \longrightarrow Y(\mathfrak{L})$ be the obvious projection maps. By (3.21) and Definition 3.3.2.(3), the inverse footprint maps $\overline{\psi}_i^{-1} : F(i) \longrightarrow \overline{s}_i^{-1}(0)$ followed by $\wp_i$ can be glued on the overlaps into a global continuous one-to-one map

$$\iota \colon M \longrightarrow Y(\mathfrak{L}).$$

DEFINITION 3.3.9. *A dimensionally graded system $\mathfrak{L}$ as in (3.16) is called **Hausdorff** if the quotient topology on the thickening*

$$Y(\mathfrak{L}) \equiv \Big(\coprod_{i \in \mathfrak{B}} Y(i) \Big) / \sim \tag{3.22}$$

is Hausdorff.

LEMMA 3.3.10. *A dimensionally graded system $\mathfrak{L}$ as in (3.16) is Hausdorff if and only if for every $i, j \in \mathfrak{B}$, with $i > j$, the map*

$$Y(i,j) \xrightarrow{\overline{\iota}_{ij} \times \overline{\phi}_{ij}} Y(i) \times Y(j), \tag{3.23}$$

where $\overline{\iota}_{ij}$ is the inclusion map, is proper.

PROOF. Assume $\overline{\iota}_{ij} \times \overline{\phi}_{ij}$ is not proper but $Y(\mathfrak{L})$ is Hausdorff. Then there exists a compact set $K \subset Y(i) \times Y(j)$ such that $\widetilde{K} \equiv (\overline{\iota}_{ij} \times \overline{\phi}_{ij})^{-1}(K)$ is not compact. Let $(\overline{x}_k)_{k \in \mathbb{N}} \subset \widetilde{K}$ be a sequence[30] with no subsequence limit in $\widetilde{K}$ and $\big((\overline{x}_k, \overline{y}_k)\big)_{k \in \mathbb{N}} \subset K$ be the image of this sequence under (3.23). By assumption, and after possibly restricting to a subsequence, there exists $(\overline{x}, \overline{y}) \in K$ such that

$$\lim_{k \longrightarrow \infty} (\overline{x}_k, \overline{y}_k) = (\overline{x}, \overline{y}) \quad \text{and} \quad \overline{x} \in Y(i) \backslash Y(i,j).$$

By definition of quotient topology on $Y(\mathfrak{L})$ and since $\overline{x}_k \sim \overline{y}_k$, for all $k \in \mathbb{N}$, every two open sets around $[\overline{x}] = \wp_i(\overline{x})$ and $[\overline{y}] \in \wp_j(\overline{y})$ in $Y(\mathfrak{L})$ intersect non-trivially; thus, from Hausdorffness we deduce $[\overline{x}] = [\overline{y}]$. This is a contradiction because $\overline{x} \notin Y(i,j)$. The other direction is similar. □

[30] Since $Y(i,j)$, $Y(i)$, and $Y(j)$ are metrizable, compactness is equal to sequential compactness.

3.4. Shrinking. Given an DGS for $(M,\mathcal{K})$ as in Definition 3.3.2, we often need to shrink the orbibundle pieces to accommodate certain perturbations and gluing-pasting arguments. For example, we use shrinking to achieve Hausdorffness and metrizability in the inductive construction of DGS. For the general case of good coordinate systems we refer to [**19**].

For the construction of VFC in Section 4, we need the thickening $Y(\mathfrak{L})$ of (3.22) to be metrizable; cf. Definition 3.3.9. Even if the quotient topology on $Y(\mathfrak{L})$ is Hausdorff, it is often the case that this topology is not paracompact, thus not metrizable. An example of this issue is discussed in [**39**, Example 6.1.11]; we recall this example here. For simplicity, the underlying topological space M is taken to be the non-compact manifold $\mathbb{R}$. Assume $\mathfrak{L}$ is an DGS made of two vector bundles, $Y(1)=\mathbb{R}$ together with the trivial bundle, section, and footprint map, and

$$Y(2)=(0,\infty)\times\mathbb{R},\quad E(2)=Y(2)\times\mathbb{R},\quad s_2(x,y)=y,\quad \psi_2(x,y)=x.$$

The embedding Φ_{12} is the obvious inclusion over $Y(1,2)=(0,\infty)\subset Y(1)$. Let $[0]$ be the image of $0\in Y(1)$ in $Y(\mathfrak{L})$. Then the quotient topology on $Y(\mathfrak{L})$ near $[0]$ is strictly stronger than the subspace topology $Y(\mathfrak{L})\subset\mathbb{R}^2$. In fact, for every $\epsilon>0$ and every continuous function $f\colon(0,\epsilon)\longrightarrow(0,\infty)$, the set

$$U_{f,\epsilon}=\big\{[x]\colon x\in Y(1),\ |x|<\epsilon\big\}\cup\big\{[(x,y)]\colon (x,y)\in Y(2),\ |x|<\epsilon,\ y<f(x)\big\}$$

is open in the quotient topology. Moreover, these sets form a basis around $[0]$ in the quotient topology.

On the otherhand, by Proposition 3.4.4 below, if $Y(\mathfrak{L})$ is Hausdorff, then the induced topology on $Y(\widetilde{\mathfrak{L}})$, where $\widetilde{\mathfrak{L}}$ is an arbitrary shrinking of $\mathfrak{L}$, is metrizable.

DEFINITION 3.4.1. *With the notation as in Definition 3.3.2, given a dimensionally graded system $\mathfrak{L}$, a **shrinking** of $\mathfrak{L}$ consist of relatively compact*[31] *open sub-orbifolds $\widetilde{\mathcal{Y}}(i)\subset\mathcal{Y}(i)$, for each $i\in\mathfrak{B}$, such that*

$$M(\geq k)\subset\bigcup_{i\in\mathfrak{B}(\geq k)}\widetilde{F}(i)\quad\forall k\geq 0,\quad\text{with } \widetilde{F}(i)=\overline{\psi}_i(\overline{s}_i^{-1}(0)\cap\widetilde{Y}(i))\quad\forall i\in\mathfrak{B}. \tag{3.24}$$

In this situation, we define $\widetilde{\mathcal{E}}(i)$ to be the restriction to $\widetilde{\mathcal{Y}}(i)$ of $\mathcal{E}(i)$. For each $i\in\mathfrak{B}$, the section and footprint maps s_i and ψ_i, respectively, restrict to section and footprint maps on $\widetilde{\mathcal{Y}}(i)$; for simplicity, we will often denote them by the same notation.

LEMMA 3.4.2. *Let $\mathfrak{L}$ be an DGS as in Definition 3.3.2. Given a shrinking $\{\widetilde{\mathcal{Y}}(i)\}_{i\in\mathfrak{B}}$ of $\mathfrak{L}$ as in Definition 3.4.1, with*

$$\widetilde{\mathcal{Y}}(i,j)=\widetilde{\mathcal{Y}}(i)\cap\mathcal{Y}(i,j)\cap\phi_{ij}^{-1}(\widetilde{\mathcal{Y}}(j)),\quad \widetilde{\mathcal{E}}(i,j)=\mathcal{E}(i,j)|_{\widetilde{\mathcal{Y}}(i,j)}\quad\forall i,j\in\mathfrak{B},\ i<j, \tag{3.25}$$

the shrunk collection

$$\widetilde{\mathfrak{L}}=\big(\{\widetilde{\mathcal{E}}(i),\widetilde{\mathcal{Y}}(i),s_i,\psi_i\}_{i\in\mathfrak{B}},\{\widetilde{\mathcal{E}}(i,j),\widetilde{\mathcal{Y}}(i,j),\Phi_{ij}|_{\widetilde{\mathcal{Y}}(i,j)}\}_{i<j,\ i,j\in\mathfrak{B}}\big) \tag{3.26}$$

is also an DGS for $(M,\mathcal{K})$.

PROOF. It immediately follows from the definition of $\widetilde{\mathcal{Y}}(i,j)$ and (3.24) that the new collection satisfies Conditions (1), (2), (3), (8) of Definition 3.3.2. For

[31] i.e. the underlying topological inclusions are relatively compact.

Condition (4), we have

$$\begin{aligned}\overline{\phi}_{ij}^{-1}(\widetilde{Y}(j,k))\cap\widetilde{Y}(i) =& \overline{\phi}_{ij}^{-1}\Big(\widetilde{Y}(j)\cap Y(j,k)\cap\overline{\phi}_{jk}^{-1}(\widetilde{Y}(k))\Big)\cap\widetilde{Y}(i) = \\ & \overline{\phi}_{ij}^{-1}(\widetilde{Y}(j))\cap\overline{\phi}_{ij}^{-1}\Big(Y(j,k)\cap\overline{\phi}_{jk}^{-1}(\widetilde{Y}(k))\Big)\cap\widetilde{Y}(i) = \\ & \overline{\phi}_{ij}^{-1}(\widetilde{Y}(j))\cap Y(i,j)\cap Y(i,k)\cap\overline{\phi}_{ik}^{-1}(\widetilde{Y}(k))\cap\widetilde{Y}(i) = \\ & \widetilde{Y}(i,k)\cap\widetilde{Y}(i,j).\end{aligned}$$

For Condition (5), if

$$\begin{aligned}\overline{x}\in\widetilde{Y}(i,j,k) =& \overline{\phi}_{ij}^{-1}(\widetilde{Y}(j))\cap Y(i,j)\cap Y(i,k)\cap\overline{\phi}_{ik}^{-1}(\widetilde{Y}(k))\cap\widetilde{Y}(i) = \\ & \overline{\phi}_{ij}^{-1}(\widetilde{Y}(j)\cap Y(j,k))\cap\overline{\phi}_{ik}^{-1}(\widetilde{Y}(k))\cap\widetilde{Y}(i),\end{aligned} \tag{3.27}$$

then, by the first line of (3.27),

$$\overline{\phi}_{ik}(\overline{x})\in\overline{\phi}_{ik}\Big(\widetilde{Y}(i)\cap Y(i,j)\Big)\cap\widetilde{Y}(k)=\overline{\phi}_{ik}(\widetilde{Y}(i,k)),$$

and, by the second line of (3.27),

$$\overline{\phi}_{jk}(\overline{y})\in\overline{\phi}_{jk}\Big(\widetilde{Y}(j)\cap Y(j,k)\Big)\cap\widetilde{Y}(k)=\overline{\phi}_{jk}(\widetilde{Y}(j,k)),\quad \overline{y}=\overline{\phi}_{ij}(\overline{x}).$$

Therefore, $\overline{\phi}_{ik}(\overline{x})\in\overline{\phi}_{ik}(\widetilde{Y}(i,k))\cap\overline{\phi}_{jk}(\widetilde{Y}(j,k))$. Conversely, if

$$\overline{z}\in\overline{\phi}_{ik}(\widetilde{Y}(i,k))\cap\overline{\phi}_{jk}(\widetilde{Y}(j,k))\subset\overline{\phi}_{ik}(Y(i,k))\cap\overline{\phi}_{jk}(Y(j,k))=\overline{\phi}_{ik}(Y(i,j,k)),$$

then

$$\overline{x}=\overline{\phi}_{ik}^{-1}(\overline{z})\in Y(i,j,k),\ \widetilde{Y}(i),\ \text{and}\ \phi_{ik}^{-1}(\widetilde{Y}(k)),$$

and

$$\overline{x}=\overline{\phi}_{ij}^{-1}(\overline{y}),\quad \overline{y}=\overline{\phi}_{jk}^{-1}(\overline{z})\in\widetilde{Y}(j);$$

i.e. $\overline{x}\in\overline{Y}(i,j,k)$. Condition (6) is just the restriction to the shrunk sub-orbifolds of the original cocycle condition. Finally, for Condition (7), we can replace $\mathcal{U}_{p,i}$ with a sub-chart centered at $\overline{p}$ of $\mathcal{U}_{p,i}\cap\Phi_{pi}^{-1}(\widetilde{\mathcal{Y}}(i))$. □

By (3.25) and Lemma 3.3.8, the composite projection maps

$$\widetilde{Y}(i)\hookrightarrow Y(i)\xrightarrow{\wp_i}Y(\mathfrak{L})\quad\forall i\in\mathfrak{B}$$

descend to a continuous (both with quotient topologies) one-to-one map

$$Y(\widetilde{\mathfrak{L}})\hookrightarrow Y(\mathfrak{L}). \tag{3.28}$$

In what follows, by $Y(\widetilde{\mathfrak{L}})\subset Y(\mathfrak{L})$ we mean the image of (3.28), and we denote by $Y_{\mathfrak{L}}(\widetilde{\mathfrak{L}})$ the induced topology given by this inclusion. The quotient topology on $Y(\widetilde{\mathfrak{L}})$ could be different than $Y_{\mathfrak{L}}(\widetilde{\mathfrak{L}})$.

Lemma 3.4.3. *Let $\widetilde{\mathfrak{L}}$ be a shrinking of $\mathfrak{L}$ as in Lemma 3.4.2. If $Y(\mathfrak{L})$ is Hausdorff, then $\mathrm{cl}(Y(\widetilde{\mathfrak{L}}))\subset Y(\mathfrak{L})$ is compact and is equal to*

$$\bigcup_{i\in\mathfrak{B}}\wp_i(\mathrm{cl}(\widetilde{Y}(i))).$$

PROOF. Let W be the complement of $\mathrm{cl}(Y(\widetilde{\mathfrak{L}})) \subset Y(\mathfrak{L})$ and

$$W_i = \wp_i^{-1}(W) \quad \forall i \in \mathfrak{B}.$$

Every $W_i \subset Y(i)$ is open and is disjoint from $\widetilde{Y}(i)$; thus, it is disjoint from $\mathrm{cl}(\widetilde{Y}(i))$. We conclude that

$$\bigcup_{i\in\mathfrak{B}} \wp_i(\mathrm{cl}(\widetilde{Y}(i))) \subset \mathrm{cl}(Y(\widetilde{\mathfrak{L}})). \tag{3.29}$$

Since every $\mathrm{cl}(\widetilde{Y}(i))$ is compact and $\wp_i$ is continuous, we conclude that the left-hand side of (3.29) is a compact subset of $Y(\mathfrak{L})$. Therefore, if (3.29) is proper, there exists $[\overline{x}] \in Y(\widetilde{\mathfrak{L}})$ such that

$$[\overline{x}] \notin \bigcup_{i\in\mathfrak{B}} \wp_i(\mathrm{cl}(\widetilde{Y}(i))).$$

By Haussdorfness of $Y(\mathfrak{L})$, there exists an open neighborhood W of $[\overline{x}]$ in $Y(\mathfrak{L})$ which is disjoint from $\wp_i(\mathrm{cl}(\widetilde{Y}(i)))$, for all $i \in \mathfrak{B}$. Therefore, $\wp_i^{-1}(W)$ is disjoint from $\widetilde{Y}(i)$, for all $i \in \mathfrak{B}$. This is a contradiction because by definition of $Y(\widetilde{\mathfrak{L}})$, there exists some $i \in \mathfrak{B}$ and $\overline{x}_i \in \widetilde{Y}(i)$ such that $[\overline{x}] = \wp_i(\overline{x}_i)$. □

PROPOSITION 3.4.4. *Let $\mathfrak{L}$ be an DGS as in Definition 3.3.2 and $\widetilde{\mathfrak{L}}$ be a shrinking of that as in (3.26). If $Y(\mathfrak{L})$ with the quotient topology is Hausdorff, then $Y(\widetilde{\mathfrak{L}})$ with the quotient topology is Hausdorff and with the induced topology from $Y(\mathfrak{L})$ (i.e. $Y_{\mathfrak{L}}(\widetilde{\mathfrak{L}})$) is metrizable.*

PROOF. By definition

$$\widetilde{Y}(i,j) = Y(i,j) \cap \widetilde{Y}(i) \cap \overline{\phi}_{ij}^{-1}\widetilde{Y}(j) = (\overline{\iota}_{ij} \times \overline{\phi}_{ij})^{-1}(\widetilde{Y}(i) \times \widetilde{Y}(j)).$$

Therefore, from Lemma 3.3.10 we conclude that if $Y(\mathfrak{L})$ with the quotient topology is Hausdorff, then so is $Y(\widetilde{\mathfrak{L}})$. This establishes the first claim.

In order to prove that $Y_{\mathfrak{L}}(\widetilde{\mathfrak{L}})$ is metrizable, first, we show that $Y_{\mathfrak{L}}(\widetilde{\mathfrak{L}})$ is regular. Let $\mathrm{pt} = [\overline{x}] \in Y_{\mathfrak{L}}(\widetilde{\mathfrak{L}})$ be a point and $C \subset Y_{\mathfrak{L}}(\widetilde{\mathfrak{L}})$ be a non-empty closed set that does not include pt. By Lemma 3.4.3, as subsets of $Y(\mathfrak{L})$,

$$\mathrm{cl}(C) \subset \mathrm{cl}(Y(\widetilde{\mathfrak{L}})) \subset Y(\mathfrak{L})$$

is compact and $\mathrm{pt} \notin \mathrm{cl}(C)$. Since $Y(\mathfrak{L})$ is Hausdorff, for every other point $\mathrm{pt}' = [\overline{y}] \in \mathrm{cl}(C)$, there exist open disjoint neighborhoods

$$\mathrm{pt} \in W_{\mathrm{pt}'} \quad \text{and} \quad \mathrm{pt}' \in W'_{\mathrm{pt}'},$$

separating pt and pt′ from each other. Since $\mathrm{cl}(C)$ is compact, there exists a finite set of such points

$$J = \{\mathrm{pt}_1, \cdots, \mathrm{pt}_\ell\} \subset \mathrm{cl}(C)$$

such that $W_C = \bigcup_{i=1}^{\ell} W'_{\mathrm{pt}_i}$ covers $\mathrm{cl}(C)$. In the induced topology, the open sets

$$W = \bigcap_{i=1}^{\ell} W_{\mathrm{pt}_i} \quad \text{and} \quad W_C$$

separate pt and C. Next, we show that $Y_{\mathfrak{L}}(\widetilde{\mathfrak{L}})$ is second countable. Since each $Y(i)$ is metrizable, $Y(i)$ is second-countable and we can take a countable basis

$\{W_{i,\alpha}\}_{\alpha\in\Omega_i}$ for it, including the empty set. For every $i\in\mathfrak{B}$, let

$$Y'(i) = Y(i) \setminus \bigcup_{j>i} \mathrm{cl}(\widetilde{Y}(i,j)), \tag{3.30}$$

and denote by $\Omega_i' \subset \Omega_i$ to be the set of those $\alpha \in \Omega_i$ such that $W_{i,\alpha} \subset Y'(i)$. For every $i\in\mathfrak{B}$ and every $\alpha\in\Omega_i'$, let

$$[W_{i,\alpha}] = \wp_i(W_{i,\alpha}); \tag{3.31}$$

this may not a priori be an open set in $Y(\mathfrak{L})$, but by (3.30), its restriction to $Y(\widetilde{\mathfrak{L}})$ is open. Then, we claim that the set

$$\{[W_{i,\alpha}] \cap Y(\widetilde{\mathfrak{L}})\}_{i\in\mathfrak{B},\alpha\in\Omega_i'}$$

is a basis for the induced topology. In fact, every open set of $Y_{\mathfrak{L}}(\widetilde{\mathfrak{L}})$ is the restriction of some open set W in $Y(\mathfrak{L})$. Let

$$W_i = \wp_i^{-1}(W) \subset Y(i).$$

By (3.30),

$$Y(\widetilde{\mathfrak{L}}) \cap W = Y(\widetilde{\mathfrak{L}}) \cap \bigcup_{i\in\mathfrak{B}} \wp_i(W_i) = Y(\widetilde{\mathfrak{L}}) \cap \bigcup_{i\in\mathfrak{B}} \wp_i(W_i \cap Y'(i)).$$

For every $i\in\mathfrak{B}$, we have

$$W_i \cap Y'(i) = \bigcup_{\alpha\in I_i} W_{i,\alpha},$$

for some $I_i \subset \Omega_i'$. We conclude that

$$Y(\widetilde{\mathfrak{L}}) \cap W = \bigcup_{i\in\mathfrak{B}} \bigcup_{\alpha\in I_i} [W_{i,\alpha}].$$

Since $Y_{\mathfrak{L}}(\widetilde{\mathfrak{L}})$ is already Hausdorff, it follows from Nagata-Smirnov's enhancement of Urysohn's theorem, see Remark 3.4.6 below, that $Y_{\mathfrak{L}}(\widetilde{\mathfrak{L}})$ is separable and metrizable. □

REMARK 3.4.5. In Proposition 3.4.4, a similar argument shows that the induced topology on $\mathrm{cl}(Y(\widetilde{\mathfrak{L}})) \subset Y(\mathfrak{L})$ is also metrizable. For example, we can consider a shrinking $\mathfrak{L}'$ of $\mathfrak{L}$ such that $\widetilde{\mathfrak{L}}$ is a shrinking of $\mathfrak{L}'$. Then the conclusion follows from, first, applying Proposition 3.4.4 to $Y(\mathfrak{L}')$ and then restricting to $\mathrm{cl}(Y(\widetilde{\mathfrak{L}}))$. Moreover, in this case, the quotient topology and the induced topology on $\mathrm{cl}(Y(\widetilde{\mathfrak{L}}))$ are the same.

REMARK 3.4.6. There are some widely-recognized metrizability theorems. One of the first ones, due to Urysohn, states that every Hausdorff second-countable regular space is metrizable. The converse does not hold; there exist metric spaces that are not second countable. The Nagata-Smirnov theorem, provides a more specific theorem where the converse does hold. Urysohn's Theorem can be restated as: A topological space is separable and metrizable if and only if it is regular, Hausdorff, and second-countable. The Nagata-Smirnov metrizability theorem extends this to the non-separable case. It states that a topological space is metrizable if and only if it is regular, Hausdorff, and has a σ-locally finite base. A σ-locally finite base is a base which is a union of countably many locally finite collections of open sets. A space is said to be locally metrizable if every point has a metrizable neighborhood. Smirnov proved that a locally metrizable space is metrizable if and only if it is

Hausdorff and paracompact. Such spaces admit partitions of unity subordinate to any open cover.

REMARK 3.4.7. We have not yet shown that an arbitrary dimensionally graded system $\mathfrak{L}$ admits a Haussdorf shrinking $\widetilde{\mathfrak{L}}$ as in (3.26). We will do this in Theorem 3.5.3. Proposition 3.4.4 only shows that once we build a Haussdorf DGS, the simple shrinking process of Lemma 3.4.2 preserves this property.

3.5. Cobordism. Our construction of a VFC in Section 4 depends on a choice of a dimensionally graded system $\mathfrak{L}$, deformation of Kuranishi maps, etc. Before all these other auxiliary data, it depends on a choice of a Kuranishi structure $\mathcal{K}$ on the topological space in question M. For example, in Section 6.5, we build a class of natural Kuranishi structures on the moduli space of pseudoholomorphic maps; thus, it is important to show that the resulting VFC does not depend on the particular choice of a natural Kuranishi structure or the defining almost complex structure J. For this purpose we sketch a notion of a cobordism between two sets of choices.

The definition of an orbifold in Section 2.1 does not include charts with boundary and corners. In order to define a cobordism, we need to at least include orbifold charts with boundary. A bordered orbifold chart can be simply taken to be a product chart of the form

$$\mathcal{V}=(V\times[0,a),G,\psi); \tag{3.32}$$

where (V,G,ψ) is an orbifold chart in the sense of Definition 2.1.1, a is some positive integer, and G acts trivially on $[0,a)$. Then

$$\partial\mathcal{V}=(V\times\{0\},G,\psi)\cong(V,G,\psi)\subset\mathcal{V} \quad\text{and}\quad \mathcal{V}\backslash\partial\mathcal{V}=(V\times(0,a),G,\psi)\subset\mathcal{V}$$

are orbifold charts in the sense of Definition 2.1.1. A refinement of such $\mathcal{V}$ at an interior point is a refinement of $\mathcal{V}\backslash\partial\mathcal{V}$ as in Definition 2.1.2, and a refinement at a boundary point is another orbifold chart with boundary $\mathcal{V}'=(V'\times[0,a'),G',\psi')$ (of the same dimension) together with a group homomorphism $h\colon G'\longrightarrow G$ and an h-equivariant smooth embedding

$$\phi\colon V'\times[0,a')\longrightarrow V\times[0,a) \tag{3.33}$$

which takes $\partial\mathcal{V}'$ to $\partial\mathcal{V}$. Eventually, for a metrizable topological space M with "boundary" $\partial M\subset M$, an n-dimensional smooth (effective) orbifold atlas on $(M,\partial M)$ is a countable collection

$$\mathcal{A}=\big\{\mathcal{V}_\alpha=(V_\alpha,G_\alpha,\psi_\alpha)\big\}_{\alpha\in S}\cup\big\{\mathcal{V}_\beta=(V_\beta\times[0,a_\beta),G_\beta,\psi_\beta)\big\}_{\beta\in S'} \tag{3.34}$$

of compatible n-dimensional orbifold charts covering M such that

$$\psi_\alpha(V_\alpha)\cap\partial M=\emptyset \quad \forall\alpha\in S \quad\text{and}\quad \psi_\beta^{-1}(\partial M)=V_\beta\times\{0\} \quad \forall\beta\in S'.$$

With this definition, the restriction

$$\mathcal{A}|_{\partial M}=\big\{\mathcal{V}_\beta=(V_\beta,G_\beta,\psi_\beta)\big\}_{\beta\in S'}$$

defines an orbifold structure $\partial\mathcal{M}$ on ∂M in the sense of Definition 2.1.3. Oriented orbifold structures are defined similarly to Definition 2.1.7 and if an orbifold structure $\mathcal{A}$ on $(M,\partial M)$ is oriented, by (3.33), it induces an orientation of the induced orbifold structure $\partial\mathcal{M}$. The latter depends on a sign convention; we consider the induced orientation on the boundary of (3.32) such that

$$T(V\times[0,a))|_{V\times\{0\}} = TV \oplus \mathbb{R}\cdot\frac{\partial}{\partial t},$$

where t is parametrizing $[0,a)$, is an isomorphism of oriented vector spaces. We can define other notions like orbifold morphisms, embeddings, orbibundles, multisections, etc, on bordered orbifolds, similarly.

Given an orbibundle $\mathcal{E}$ over a bordered orbifold $(\mathcal{M},\partial\mathcal{M})$ and a section $s\colon \mathcal{M}\longrightarrow\mathcal{E}$ with compact support, the analogue of Proposition 2.6.4 gives us a rational homology class $e(\mathcal{E},s)$ in the relative singular homology group $H_n(M,\partial M,\mathbb{Q})$, where $n=\dim(\mathcal{M})-\operatorname{rank}(\mathcal{E})$. The image of $e(\mathcal{E},s)$ under

$$H_n(M,\partial M,\mathbb{Q})\longrightarrow H_{n-1}(\partial M,\mathbb{Q})$$

is the Euler class of the restriction $s|_{\partial\mathcal{M}}$.

The definition of a Kuranishi structure, an DGS for it, and other related notions analogously extend to the case of compact metrizable spaces with "boundary".

DEFINITION 3.5.1. *For $i=0,1$, let $\mathcal{K}_i$ be a Kuranishi structure on the compact metrizable space M_i and $\mathfrak{L}_i$ be an DGS for $\mathcal{K}_i$ as in Definition 3.3.2. A* ***cobordism*** *between $(M_0,\mathcal{K}_0,\mathfrak{L}_0)$ and $(M_1,\mathcal{K}_1,\mathfrak{L}_1)$ consists of a compact metrizable space $\widetilde{M}$ with boundary*

$$\widetilde{M}\supset\partial\widetilde{M}\cong M_0\sqcup M_1,$$

a bordered Kuranishi structure $\widetilde{\mathcal{K}}$ on $(\widetilde{M},\partial\widetilde{M})$, and an DGS $\widetilde{\mathfrak{L}}$ for $\widetilde{\mathcal{K}}$, such the restriction of $(\widetilde{\mathcal{K}},\widetilde{\mathfrak{L}})$ to M_i, with $i=0,1$, coincides with $(\mathcal{K}_i,\mathfrak{L}_i)$. In the oriented case, we require the orientation induced by the orientation of $\widetilde{\mathcal{K}}$ on $\mathcal{K}_0$ to be equal to its given orientation and the one induced on $\mathcal{K}_1$ to be the opposite orientation.

In Definition 3.5.1, we may forget about the DGS to define a cobordism between two Kuranishi structures. In particular, let M be a fixed metrizable space as in the statement of Definition 3.2.6, and $\mathcal{K}_0$ and $\mathcal{K}_1$ be two Kuranishi structures on M. Then, by a **deformation equivalence** between $\mathcal{K}_1$ and $\mathcal{K}_2$, we mean a cobordism for which $\widetilde{M}=M\times[0,1]$. Similarly, let $\mathfrak{L}_0$ and $\mathfrak{L}_1$ be two dimensionally graded systems for a Kuranishi structure $(M,\mathcal{K})$. By $\mathcal{K}\times[0,1]$ we mean the product bordered Kuranishi structure on $M\times[0,1]$ obtained by taking products of Kuranishi charts of $\mathcal{K}$ with[32] $[0,1]$. Then, by a deformation equivalence between $\mathfrak{L}_1$ and $\mathfrak{L}_2$, we mean a cobordism for which $\widetilde{\mathcal{M}}=\mathcal{M}\times[0,1]$ and $\widetilde{\mathcal{K}}=\mathcal{K}\times[0,1]$.

REMARK 3.5.2. In [**39**, Section 6.2], a deformation equivalence between two Kuranishi structures is called a "concordance". McDuff-Wehrheim also use the notion of "commensurability" which is an equivalence relation generated by calling $\mathcal{K}_0$ and $\mathcal{K}_1$ directly commensurate if they are both contained in the same Kuranishi structure $\mathcal{K}_{01}$ on M. By [**39**, Lemma 6.2.16], commensurability implies concordance (deformation equivalence), so it is a stronger notion. For a more detailed study of various notions of cobordism of Kuranishi structures we refer to [**39**, Section 6.2].

We are now ready to state the existence result for dimensionally graded systems.

---

[32] More precisely, to comply with (3.33), by $[0,a)$ and $(b,1]\cong[0,1-b)$, where $b<a$ are some fixed real numbers.

THEOREM 3.5.3. *Let $(M,\mathcal{K})$ be a Kuranishi space*[33] *as in Definition 3.2.9. Then $(M,\mathcal{K})$ admits a plethora of "natural" Hausdorff*[34] *dimensionally graded systems. If $(M,\mathcal{K})$ is oriented, then the resulting natural DGS are also oriented. Moreover, every two of such DGS are (orientably) "deformation equivalent".*

We prove this theorem in the next section. As the proof shows, the same statements[35] hold without the tangent bundle assumption. The word "natural" in the statement means that the orbibundles $\{\mathcal{E}(i) \longrightarrow \mathcal{Y}(i)\}_{i\in\mathfrak{B}}$ are built by gluing Kuranishi charts of the same dimension along the overlaps. For the more general case of good coordinate systems, the construction of Hausdorff thickenings via the shrinking process is done in [**19**].

3.6. Existence of DGS (proof of Theorem 3.5.3). Our construction of natural dimensionally graded systems, out of a given Kuranishi structure or space $(M,\mathcal{K})$, is by reverse induction on $i\in\mathfrak{B}$, where $\mathfrak{B}\subset\mathbb{Z}$ is the index set of (3.7). With notation as in (3.6), starting from the maximal element $i_{\max}\in\mathfrak{B}$, we first construct an orbibundle $\mathcal{E}(i_{\max})\longrightarrow\mathcal{Y}(i_{\max})$ such that $M(i_{\max})\subset F(i_{\max})$ -recall that $M(i_{\max})\subset M$ is compact- and then proceed through lower values of $i\in\mathfrak{B}$ inductively.

LEMMA 3.6.1. *Let $\mathcal{K}$ be an n-dimensional Kuranishi structure on M. For every $k\in\mathfrak{B}$ and every compact subset $K\subset M(k)$, there exists an orbibundle* $\mathrm{pr}\colon\mathcal{E}\longrightarrow\mathcal{Y}$ *together with a section $s\colon\mathcal{Y}\longrightarrow\mathcal{E}$ and an open embedding $\overline{\psi}\colon\overline{s}^{-1}(0)\longrightarrow M$ with footprint F such that $K\subset F$,* $\operatorname{rank}\mathcal{E}=k$, *$\dim\mathcal{Y}=n+k$, compatibility conditions*[36] *(7) of Definition 3.3.2 hold, and*

$$F\subset\bigcup_{\overline{p}\in K}F_p.$$

If $(M,\mathcal{K})$ is a Kuranishi space then Condition (8) also holds.

We prove this lemma by induction, via the following lemma. In the light of Example 3.2.8, we call a tuple $(\mathrm{pr}\colon\mathcal{E}\longrightarrow\mathcal{Y},s,\psi)$ as in Lemma 3.6.1 an n-dimensional **pure orbibundle structure** around K. In [**16**, Definition 7.3], this is called a "pure orbifold structure" around K.

LEMMA 3.6.2. *Let $\mathcal{K}$ be an n-dimensional Kuranishi structure on M. Give $k\in\mathfrak{B}$ and compact subsets $K_1,K_2\subset M(k)$, for $a=1,2$, suppose* $(\mathrm{pr}_a\colon\mathcal{E}_a\longrightarrow\mathcal{Y}_a,s_a,\psi_a)$ *is an n-dimensional pure orbibundle structure of rank k with footprint F_a around K_a as described in Lemma 3.6.1. Then $K=K_1\cup K_2$ admits a pure orbibundle structure around it.*

PROOF OF LEMMA 3.6.1 VIA LEMMA 3.6.2. We cover K with finitely many Kuranishi charts of $\mathcal{K}$ cantered at points of K,

$$\{\mathcal{U}_{p_i}=(\mathrm{pr}_{p_i}\colon U_{p_i}\longrightarrow V_{p_i},G_{p_i},s_{p_i},\psi_{p_i})\}_{i\in[N]},\qquad \overline{p}_i\in K\ \forall i\in[N].$$

---

[33] i.e. $\mathcal{K}$ is a Kuranishi structure on M with tangent bundle.

[34] With respect to the quotient topology. Unless we specifically mention, the topology considered on $Y(\mathfrak{L})$ is the quotient topology.

[35] i.e. existence of Hausdorff DGS and their deformation equivalence.

[36] Between Conditions (7)(a) and (7)(b), only (7)(a) is relevant in this case.

If $N=1$, we can simply take $\mathcal{E}=\mathcal{U}_{p_1}$. If $N>1$, we fix a set of relatively compact open subsets $K^{\circ}_{p_i}\subset F_{p_i}$ such that

$$K\subset\bigcup_{i\in[N]}K^{\circ}_{p_i}.$$

Let K_{p_i} be the closure of $K^{\circ}_{p_i}$ in F_{p_i}. We then proceed by induction. By the $N=1$ case discussed above, $K\cap K_{p_1}$ admits such a pure orbibundle structure around it. Assume $K\cap\big(\cup_{i=1}^{\ell-1}K_{p_i}\big)$ admits such a pure orbibundle structure around it. Then by Lemma 3.6.2,

$$K\cap\Big(\bigcup_{i=1}^{\ell}K_{p_i}\Big)=\Big(K\cap\Big(\bigcup_{i=1}^{\ell-1}K_{p_i}\Big)\Big)\cup\big(K\cap K_{p_\ell}\big)$$

admits such an orbibundle structure around it. □

PROOF OF LEMMA 3.6.2. We assume $\mathcal{K}$ has tangent bundle, i.e. $(M,\mathcal{K})$ is a Kuranishi space; the general case is a simpler version of this. For every $\overline{p}\in F_1\cap F_2$, by definition, there exist sub-charts $p\in\mathcal{U}_{p,i}\subset\mathcal{U}_p$, with $i=1,2$, such that the compatibility conditions (7) and (8) of Definition 3.3.2 for these sub-charts hold. Since both $V_{p,1}$ and $V_{p,2}$ are G_p-invariant neighborhoods of $p\in V_p$, $V_{p,12}=V_{p,1}\cap V_{p,2}$ is G_p-invariant. If $V_{p,12}$ is disconnected, we replace it with the connected component containing the unique point p. Then $\mathcal{U}_{p,12}=\mathcal{U}_p|_{V_{p,12}}$ is a sub-chart of $\mathcal{U}_p$ which can be embedded into both $\mathcal{Y}_1$ and $\mathcal{Y}_2$.

Shrink each $\mathcal{U}_{p,12}$ into some relatively compact sub-chart[37] $\mathcal{U}^m_{p,12}$ (i.e. $V^m_{p,12}\subset V_{p,12}$ is relatively compact), and take finitely many such points in

$$K_{12}=K_1\cap K_2\subset M(k),$$

say $\{\overline{p}_\alpha\}_{\alpha\in I}\subset K_{12}$, such that

$$K_{12}\subset\bigcup_{\alpha\in I}F^m_{p_\alpha,12}. \tag{3.35}$$

Claim 1. *For every $\overline{q}\in K_{12}$, there exists a sub-chart[38] $\mathcal{U}^t_{q,12}\subset\mathcal{U}_{q,12}$ such that for $j=1,2$ and $\alpha\in I$, if*

$$\overline{\phi}_{qj}(\overline{V}^t_{q,12})\cap\overline{\phi}_{p_\alpha j}(\overline{V}^m_{p_\alpha,12})\neq\emptyset, \tag{3.36}$$

then $\overline{q}\in F_{p_\alpha}$ and $\overline{V}^t_{q,12}\subset\overline{\phi}^{-1}_{qp_\alpha}(\overline{V}_{p_\alpha,12})$.

PROOF. Decompose I into disjoint unions $I=I^{\text{in}}\sqcup I^{\text{out}}$ such that

$$I^{\text{in}}=\big\{\alpha\in I\colon\overline{q}\in\text{cl}(F^m_{p_\alpha,12})\big\}.$$

Note that $\overline{\psi}_{p_\alpha,12}\colon\overline{s}^{-1}_{p_\alpha,12}(0)\longrightarrow F_{p_\alpha,12}$ is a homeomorphism, therefore we can think of $\overline{q}\in F_{p_\alpha,12}$ as a point in

$$\overline{s}^{-1}_{p_\alpha,12}(0)\subset\overline{V}_{p_\alpha,12}=V_{p_\alpha,12}/G_{p_\alpha,12}.$$

Moreover, if $\overline{q}\in\text{cl}(F^m_{p_\alpha,12})$, since $\text{cl}(F^m_{p_\alpha,12})\subset F_{p_\alpha}$, the coordinate change map Φ_{qp_α} of Kuranishi structure is defined. Then we take $\mathcal{U}^t_{q,12}\subset\mathcal{U}_{q,12}$ to be the sub-chart

[37] The superscript "m" used for these sub-charts denotes for "main".

[38] The superscript "t" used for these sub-charts denotes for "transition".

over the preimage[39] of

$$\overline{V}^t_{q,12} = \bigcap_{j=1,2}\left(\bigcap_{\alpha\in I^{\mathrm{in}}}\left(\overline{\phi}^{-1}_{qp_\alpha}(\overline{V}_{p_\alpha,12})\right)\cap\bigcap_{\alpha\in I^{\mathrm{out}}}\left(\overline{\phi}^{-1}_{qj}\left(Y_j\setminus \mathrm{cl}\left(\overline{\phi}_{p_\alpha j}(\overline{V}^m_{p_\alpha,12})\right)\right)\right)\right).$$

It is easy to see that $\mathcal{U}^t_{q,12}$ has the desired property. □

We take finitely many such points $\overline{q}$ in K_{12}, say $\{\overline{q}_\beta\}_{\beta\in J}\subset K_{12}$, such that

$$K_{12}\subset\bigcup_{\beta\in J}F^t_{q_\beta,12}. \tag{3.37}$$

Then, for $(i,j)=(1,2)$ and $(2,1)$ define

$$Y_{i,j}=\bigcup_{\alpha\in I}\bigcup_{\beta\in J}\left(\overline{\phi}_{q_\beta i}(\overline{V}^t_{q_\beta,12})\cap\overline{\phi}_{p_\alpha i}(\overline{V}^m_{p_\alpha,12})\right)\subset Y_i. \tag{3.38}$$

These are open subsets of Y_1 and Y_2. By restricting the orbibundles $\mathcal{E}_1\longrightarrow\mathcal{Y}_1$ and $\mathcal{E}_2\longrightarrow\mathcal{Y}_2$ to $Y_{1,2}$ and $Y_{2,1}$, we obtain sub-orbibundles $\mathcal{E}_{1,2}\longrightarrow\mathcal{Y}_{1,2}$ and $\mathcal{E}_{2,1}\longrightarrow\mathcal{Y}_{2,1}$, respectively. Let $s_{1,2}$, $s_{2,1}$, and $F_{1,2}\cong F_{2,1}$ be the restriction of corresponding objects, respectively. By (3.35) and (3.37), $F_{1,2}\cong F_{2,1}$ is an open neighborhood of K_{12}.

Claim 2. *The tuples $(\mathcal{E}_{1,2}\longrightarrow\mathcal{Y}_{1,2},s_{1,2},\psi_{1,2})$ and $(\mathcal{E}_{2,1}\longrightarrow\mathcal{Y}_{2,1},s_{2,1},\psi_{2,1})$ are naturally isomorphic.*

Proof. We show that the maps

$$\begin{array}{ccc} E_{1,2} & \xrightarrow{\ \overline{\mathfrak{D}\varphi}\ } & E_{2,1}\\ \Big\downarrow{\scriptstyle \mathrm{pr}_{1,2}} & & \Big\downarrow{\scriptstyle \mathrm{pr}_{2,1}}\\ Y_{1,2} & \xrightarrow{\ \overline{\varphi}\ } & Y_{2,1}\end{array}$$

which send

$$\overline{x}=\overline{\phi}_{q_\beta 1}(\overline{x}_\beta)=\overline{\phi}_{p_\alpha 1}(\overline{x}_\alpha),\quad \overline{x}_\beta\in\overline{V}^t_{q_\beta,12},\ \overline{x}_\alpha\in\overline{V}^m_{p_\alpha,12}, \tag{3.39}$$

to $\overline{y}=\overline{\phi}_{q_\beta 2}(\overline{x}_\beta)$ and

$$\overline{v}=\overline{\mathfrak{D}\phi}_{q_\beta 1}(\overline{v}_\beta)=\overline{\mathfrak{D}\phi}_{p_\alpha 1}(\overline{v}_\alpha),\quad \overline{v}_\beta\in\overline{U}^t_{q_\beta,12},\ \overline{v}_\alpha\in\overline{U}^m_{p_\alpha,12}, \tag{3.40}$$

to $\overline{u}=\overline{\mathfrak{D}\phi}_{q_\beta 2}(\overline{v}_\beta)$ are well-defined continuous maps and lift to an orbibundle isomorphism $(\mathfrak{D}\varphi,\varphi)$ commuting with the section and footprint maps.

Given $\overline{x}\in Y_{1,2}$, suppose there are two different pairs of indices $(\alpha,\beta),(\alpha',\beta')\in I\times J$ such that

$$\begin{aligned}\overline{x}&=\overline{\phi}_{q_\beta 1}(\overline{x}_\beta)=\overline{\phi}_{p_\alpha 1}(\overline{x}_\alpha),\quad \overline{x}_\beta\in\overline{V}^t_{q_\beta,12},\ \overline{x}_\alpha\in\overline{V}^m_{p_\alpha,12},\\ \overline{x}&=\overline{\phi}_{q_{\beta'} 1}(\overline{x}_{\beta'})=\overline{\phi}_{p_{\alpha'} 1}(\overline{x}_{\alpha'}),\quad \overline{x}_{\beta'}\in\overline{V}^t_{q_{\beta'},12},\ \overline{x}_{\alpha'}\in\overline{V}^m_{p_{\alpha'},12}.\end{aligned}$$

It follows from (3.36) that

$$\begin{aligned}\overline{x}&\in\overline{\phi}_{q_\beta 1}(\overline{V}^t_{q_\beta,12})\cap\overline{\phi}_{p_{\alpha'} 1}(\overline{V}^m_{p_{\alpha'},12})\neq\emptyset,\\ \overline{x}&\in\overline{\phi}_{q_{\beta'} 1}(\overline{V}^t_{q_{\beta'},12})\cap\overline{\phi}_{p_\alpha 1}(\overline{V}^m_{p_\alpha,12})\neq\emptyset;\end{aligned}$$

[39] We would only consider the connected component containing q, if it is disconnected.

therefore,

$$\overline{q}_\beta \in F_{p_{\alpha'}},\ \overline{q}_{\beta'} \in F_{p_\alpha},\ \overline{V}^t_{q_\beta,12} \subset \overline{\phi}^{-1}_{q_\beta p_{\alpha'}}(\overline{V}_{p_{\alpha'},12}),\ \text{and } \overline{V}^t_{q_{\beta'},12} \subset \overline{\phi}^{-1}_{q_{\beta'} p_\alpha}(\overline{V}_{p_\alpha,12}).$$

By Definition 3.3.2.(7), for $i=1,2$, restricted to

$$\overline{V}^t_{q_\beta,12} \subset \overline{\phi}^{-1}_{q_\beta p_{\alpha'}}(\overline{V}_{p_{\alpha'},12}) \cap \overline{V}_{q_\beta,12} \subset \overline{\phi}^{-1}_{q_\beta p_{\alpha'}}(\overline{V}_{p_{\alpha'},i}) \cap \overline{V}_{q_\beta,i}$$

we have

$$\overline{\phi}_{p_{\alpha'}1} \circ \overline{\phi}_{q_\beta p_{\alpha'}} = \overline{\phi}_{q_\beta 1} \quad \text{and} \quad \overline{\phi}_{p_{\alpha'}2} \circ \overline{\phi}_{q_\beta p_{\alpha'}} = \overline{\phi}_{q_\beta 2}. \tag{3.41}$$

From the first identity in (3.41) and the fact that $\overline{\phi}_{p_{\alpha'}1}$ is one-to-one we conclude that

$$\overline{\phi}_{q_\beta p_{\alpha'}}(\overline{x}_\beta) = \overline{x}_{\alpha'}.$$

Then from the second identity in (3.41) we get

$$\overline{\phi}_{q_\beta 2}(\overline{x}_\beta) = \overline{\phi}_{p_{\alpha'}2}(\overline{\phi}_{q_\beta p_{\alpha'}}(\overline{x}_\beta)) = \overline{\phi}_{p_{\alpha'}2}(\overline{x}_{\alpha'}).$$

Therefore, (3.39) is independent of the choice of (α,β) or (α',β'). The case of (3.40) is similar.

By definition, over

$$\overline{\phi}_{q_\beta 1}(\overline{V}^t_{q_\beta,12}) \cap \overline{\phi}_{p_\alpha 1}(\overline{V}^m_{p_\alpha,12}) \subset Y_1,$$

$(\overline{\mathfrak{D}\varphi},\overline{\varphi})$ is equal to the restriction of $\overline{\Phi}_{p_\alpha 2}\circ\overline{\Phi}^{-1}_{p_\alpha 1}$; therefore, $(\overline{\mathfrak{D}\varphi},\overline{\varphi})$ is locally liftable to smooth orbibundle isomorphisms $\Phi_{p_\alpha 2} \circ \Phi^{-1}_{p_\alpha 1}$ between equal rank orbibundle charts. Compatibility of $(\mathfrak{D}\varphi,\varphi)$ with the section and footprint maps follow from the corresponding set of assumptions in Definition 3.3.2.(7). □

In order to finish the proof of Lemma 3.6.2, we identify $\mathrm{pr}_1\colon \mathcal{E}_1 \longrightarrow \mathcal{Y}_1$ and $\mathrm{pr}_2\colon \mathcal{E}_2 \longrightarrow \mathcal{Y}_2$ along the isomorphic sub-orbibundles $\mathrm{pr}_{i,j}\colon \mathcal{E}_{i,j} \longrightarrow \mathcal{Y}_{i,j}$, with $(i,j)=(1,2)$ and $(2,1)$, to obtain an orbibundle $\mathrm{pr}\colon \mathcal{E} \longrightarrow \mathcal{Y}$ together with a section $s\colon \mathcal{Y} \longrightarrow \mathcal{E}$ and an open embedding $\overline{\psi}\colon \overline{s}^{-1}(0) \longrightarrow M$ with footprint $F=F_1\cup F_2$ such that $K=K_1\cup K_2 \subset F$, rank $\mathcal{E}=k$, $\dim \mathcal{Y}=n+k$, and the compatibility conditions (7) and (8) of Definition 3.3.2 hold. However, without further restrictions on Y_1, Y_2, and $Y_{1,2}$, the resulting quotient space

$$Y = \Big(Y_1 \coprod Y_2\Big)/Y_{1,2} \sim Y_{2,1}$$

may not be a Hausdorff topological space, or the resulting continuous map $\overline{\psi}$ may not be injective. In order to obtain a pure orbibundle structure around K as above, we further shrink Y_1, Y_2, and $Y_{1,2} \cong Y_{2,1}$ in several steps to get the right orbibundle structure.

First shrinking to make $\overline{\psi}$ injective. For $i=1,2$, choose open sets $F_i^{(1)} \subset F_i$ such that

$$K_i \subset F_i^{(1)} \quad \text{and} \quad F_{12}^{(1)} = F_1^{(1)} \cap F_2^{(1)} \subset F_{1,2} = F_{2,1}.$$

Take open sets $Y_i^{(1)} \subset Y_i$ such that

$$Y_i^{(1)} \cap \overline{s}_i^{-1}(0) = \overline{\psi}_i^{-1}(F_i^{(1)}).$$

Let

$$Y_{1,2}^{(1)} = Y_{1,2} \cap Y_1^{(1)} \cap \overline{\varphi}^{-1}(Y_{2,1} \cap Y_2^{(1)}) \text{ and } Y_{2,1}^{(1)} = Y_{2,1} \cap Y_2^{(1)} \cap \overline{\varphi}(Y_{1,2} \cap Y_1^{(1)}).$$

Let $\mathcal{E}_1^{(1)}$, $\mathcal{E}_2^{(1)}$, and $\mathcal{E}_{1,2}^{(1)} \overset{\mathfrak{D}\varphi}{\cong} \mathcal{E}_{2,1}^{(1)}$, together with the similarly denoted sections and footprint maps, be restrictions to $Y_1^{(1)}$, $Y_2^{(1)}$, and $Y_{1,2}^{(1)} \overset{\overline{\varphi}}{\cong} Y_{2,1}^{(1)}$ of the corresponding orbibundles, respectively. Since

$$F_1^{(1)} \cap F_2^{(1)} = F_{1,2}^{(1)} = F_{2,1}^{(1)},$$

we conclude that the the induced map $\overline{\psi}$ on

$$\overline{s}^{-1}(0) \cap \left(Y^{(1)} = \left(Y_1^0 \cup_{Y_{1,2}^{(1)} \cong Y_{2,1}^{(1)}} Y_2^{(1)} \right) \right)$$

is injective.

Second shrinking to make the glued space Hausdorff.

Claim 3. *For $i=1,2$, there exist relatively compact open subsets $Y_i^{(2)} \subset Y_i^{(1)}$ and relatively compact open subsets $W_1 \subset Y_{1,2}^{(1)}$ and $W_2 \subset Y_{2,1}^{(1)}$ such that $K_i \subset F_i^{(2)}$,*

$$\text{(3.42)} \quad \mathrm{cl}\Big(Y_1^{(2)} \cap \overline{\varphi}^{-1}(Y_{2,1}^{(1)} \cap Y_2^{(2)})\Big) \subset W_1, \quad \mathrm{cl}\Big(Y_2^{(2)} \cap \overline{\varphi}(Y_{1,2}^{(1)} \cap Y_1^{(2)})\Big) \subset W_2,$$

and $\overline{\varphi}(W_1) = W_2$.

Proof. We start from an arbitrary pair of relatively compact open subsets $\widetilde{Y}_i^{(2)} \subset Y_i^{(1)}$ and find W_i such that the following pair of weaker conditions

$$\text{(3.43)} \quad \begin{aligned} &\overline{s}_1^{-1}(0) \cap \mathrm{cl}\Big(\widetilde{Y}_1^{(2)} \cap \overline{\varphi}^{-1}(Y_{2,1}^{(1)} \cap \widetilde{Y}_2^{(2)})\Big) \subset W_1, \\ &\overline{s}_2^{-1}(0) \cap \mathrm{cl}\Big(\widetilde{Y}_2^{(2)} \cap \overline{\varphi}(Y_{1,2}^{(1)} \cap \widetilde{Y}_1^{(2)})\Big) \subset W_2, \end{aligned}$$

hold. Then, we further shrink $\widetilde{Y}_i^{(2)}$ such that the original inclusions of (3.42) hold.

Note that the left-hand side terms in (3.43) are compact subsets of $Y_1^{(1)}$ and $Y_2^{(1)}$, respectively. The existence of such W_1 and W_2 follows, if we they are also subsets of $Y_{1,2}^{(1)}$ and $Y_{2,1}^{(1)}$, respectively. Since $\widetilde{Y}_1^{(2)} \subset Y_1^{(1)}$ is relatively compact and $\overline{s}_1^{-1}(0) \cap Y_1^{(1)}$ is closed

$$\overline{s}_1^{-1}(0) \cap \mathrm{cl}\Big(\widetilde{Y}_1^{(2)} \cap \overline{\varphi}^{-1}(Y_{2,1}^{(1)} \cap \widetilde{Y}_2^{(2)})\Big) \subset \mathrm{cl}\Big(\widetilde{Y}_1^{(2)} \cap Y_{1,2}^{(1)}\Big) \subset Y_1^{(1)}.$$

For every

$$\overline{x} \in \overline{s}_1^{-1}(0) \cap \mathrm{cl}\Big(\widetilde{Y}_1^{(2)} \cap \overline{\varphi}^{-1}(Y_{2,1}^{(1)} \cap \widetilde{Y}_2^{(2)})\Big)$$

there exists a sequence

$$\text{(3.44)} \quad (\overline{x}_k)_{k=1}^{\infty} \subset \overline{s}_1^{-1}(0) \cap \overline{\varphi}^{-1}(Y_{2,1}^{(1)} \cap \widetilde{Y}_2^{(2)}) \subset Y_{1,2}^{(1)}$$

such that $\lim_{k \longrightarrow \infty} \overline{x}_k = \overline{x}$. Since $\widetilde{Y}_2^{(2)} \subset Y_2$ is relatively compact, there exist $\overline{y} \in Y_2^{(1)}$ such that the sequence

$$(\overline{\varphi}(\overline{x}_k))_{k=1}^{\infty} \subset Y_{2,1}^{(1)} \cap \widetilde{Y}_2^{(2)}$$

converges to $\overline{y}$; moreover, $s_2(\overline{y}) = 0$. Since $F_1^{(1)} \cap F_2^{(1)} = F_{1,2}^{(1)}$ and $\overline{\psi}_1(\overline{x}) = \overline{\psi}_2(\overline{y})$, we conclude that $\overline{x} \in Y_{1,2}^{(1)}$. The argument for the inclusion of

$$\overline{s}_2^{-1}(0) \cap \mathrm{cl}\Big(\widetilde{Y}_2^{(2)} \cap \overline{\varphi}(Y_{1,2}^{(1)} \cap \widetilde{Y}_1^{(2)})\Big)$$

in $Y_{2,1}^{(1)}$ is similar. Fix such W_1, W_2.

We now shrink $\widetilde{Y}_1^{(2)}$ and $\widetilde{Y}_2^{(2)}$ into relatively compact open subsets $Y_1^{(2)}$ and $Y_2^{(2)}$, still covering K_1 and K_2, respectively, such that (3.42) hold.

For every $\overline{x}\in\overline{s}_1^{-1}(0)\cap\widetilde{Y}_1^{(2)}$ such that $\overline{\psi}_1(\overline{x})\in K_1$, we can find a sufficiently small open neighborhood $W_{\overline{x}}\subset\widetilde{Y}_1^{(2)}$ such that

(1) $\overline{x}\notin \mathrm{cl}\Big(\widetilde{Y}_1^{(2)}\cap\overline{\varphi}^{-1}(Y_{2,1}^{(1)}\cap\widetilde{Y}_2^{(2)})\Big)\Longrightarrow \mathrm{cl}(W_{\overline{x}})\cap\mathrm{cl}\Big(\widetilde{Y}_1^{(2)}\cap\overline{\varphi}^{-1}(Y_{2,1}^{(1)}\cap\widetilde{Y}_2^{(2)})\Big)=\emptyset$,

(2) $\overline{x}\in \mathrm{cl}\Big(\widetilde{Y}_1^{(2)}\cap\overline{\varphi}^{-1}(Y_{2,1}^{(1)}\cap\widetilde{Y}_2^{(2)})\Big)\Longrightarrow \mathrm{cl}(W_{\overline{x}})\subset W_1$.

Cover K_1 by finitely many such neighborhoods and let $Y_1^{(2)}$ be their union. We construct $Y_2^{(2)}$ similarly. If

$$\overline{y}\in\mathrm{cl}\Big(Y_1^{(2)}\cap\overline{\varphi}^{-1}(Y_{2,1}^{(1)}\cap Y_2^{(2)})\Big),$$

then there is one of those finitely many $\overline{x}$ such that

$$\overline{y}\in\mathrm{cl}\Big(W_{\overline{x}}\cap\overline{\varphi}^{-1}(Y_{2,1}^{(1)}\cap Y_2^{(2)})\Big)\subset\mathrm{cl}(W_{\overline{x}})\cap\mathrm{cl}\Big(\widetilde{Y}_1^{(2)}\cap\overline{\varphi}^{-1}(Y_{2,1}^{(1)}\cap\widetilde{Y}_2^{(2)})\Big).$$

In this case, $\overline{x}$ should be of type (2), hence, $\overline{y}\in W_1$. □

Let

$$\begin{aligned}Y^{(2)}=\Big(Y_1^{\mathrm{out}}&=\Big(Y_1^{(2)}\setminus\mathrm{cl}\Big(Y_1^{(2)}\cap\overline{\varphi}^{-1}(Y_{2,1}^{(1)}\cap Y_2^{(2)})\Big)\Big)\coprod\\ Y_2^{\mathrm{out}}&=\Big(Y_2^{(2)}\setminus\mathrm{cl}\Big(Y_2^{(2)}\cap\overline{\varphi}(Y_{1,2}^{(1)}\cap Y_1^{(2)})\Big)\Big)\coprod\Big((W_1\overset{\overline{\varphi}}{\cong}W_2)\Big)\Big)/\sim,\end{aligned}$$

where the equivalence relation is given by the intersections

$$Y_1^{\mathrm{out}}\supset W_{11}=Y_1^{\mathrm{out}}\cap W_1\subset W_1,\quad Y_2^{\mathrm{out}}\supset W_{22}=Y_2^{\mathrm{out}}\cap W_2\subset W_2. \tag{3.45}$$

Note that $\overline{\varphi}(W_{11})\cap W_{22}=\emptyset$. In fact,

$$\overline{\varphi}(W_{11})=\overline{\varphi}\left((Y_1^{(2)}\cap W_1)\setminus\mathrm{cl}\Big(Y_1^{(2)}\cap\overline{\varphi}^{-1}(Y_{2,1}^{(1)}\cap Y_2^{(2)})\Big)\right)\subset\overline{\varphi}\left(Y_{1,2}^{(1)}\cap Y_1^{(2)}\right);$$

therefore, $\overline{\varphi}(W_{11})\cap Y_2^{\mathrm{out}}=\emptyset$. Since W_1 and Y_1^{out} are open subsets of the same Hausdorff space Y_1, the quotient topology with respect to only the first relation in (3.45) is Hausdorff. Similarly, since W_2 and Y_2^{out} are open subsets of the same Hausdorff space Y_2, the quotient topology with respect to only the second relation in (3.45) is Hausdorff. Finally, since the intersections W_{11} and W_{22} in $W_1\overset{\overline{\varphi}}{\cong}W_2$ are disjoint, the quotient topology on $Y^{(2)}$ with respect to both relations in (3.45) is Hausdorff. Moreover, since both Y_1 and Y_2 are metrizable, $Y^{(2)}$ is metrizable; see Remark 3.4.6.

We identify the sub-orbibundles $\mathcal{E}_1|_{Y_1^{\mathrm{out}}}$, $\mathcal{E}_2|_{Y_2^{\mathrm{out}}}$, and $\mathcal{E}_1|_{W_1}\cong\mathcal{E}_2|_{W_2}$ to obtain an open orbibundle $\mathrm{pr}\colon\mathcal{E}^{(2)}\longrightarrow\mathcal{Y}^{(2)}$ together with a section $s^{(2)}\colon\mathcal{Y}^{(2)}\longrightarrow\mathcal{E}^{(2)}$ and an open embedding $\overline{\psi}^{(2)}\colon\overline{s}^{(2)^{-1}}(0)\longrightarrow M$ with footprint $F^{(2)}$ that contains $K=K_1\cup K_2$, rank $\mathcal{E}^{(2)}=k$, $\dim\mathcal{Y}^{(2)}=n+k$, and the compatibility conditions of Definition 3.3.2.(8) hold. This finishes the proof of Lemma 3.6.2.

Proof of Theorem 3.5.3. Proof of existence is by reverse induction on the index set $\mathfrak{B}$. By the argument after Definition 3.2.6, for the largest index $i_{\max}\in\mathfrak{B}$, $M(i_{\max})\subset M$ is compact. Therefore, by Lemma 3.6.1, there exists a pure orbibundle structure around $K=M(i_{\max})$.

For every $i \in \mathfrak{B}$, let

$$\mathfrak{B}(>i) = \{j \in \mathfrak{B}\colon j > i\}.$$

For some fixed $i \in \mathfrak{B}$ and every $j \in \mathfrak{B}(>i)$, suppose we have built an orbibundle $\mathrm{pr}_j\colon \mathcal{E}(j) \longrightarrow \mathcal{Y}(j)$ together with a section $s_j\colon \mathcal{Y}(j) \longrightarrow \mathcal{E}(j)$ and an open embedding $\overline{\psi}_j\colon \overline{s}_j^{-1}(0) \longrightarrow M$ with footprint $F(j)$ such that the conditions of Definition 3.3.2 and Definition 3.3.9, with $\mathfrak{B}(>i)$ instead of $\mathfrak{B}$, are satisfied; i.e.

$$\mathfrak{L}(>i) \equiv \left(\{\mathcal{E}(j), \mathcal{Y}(j)\}_{j\in\mathfrak{B}(>i)}, \{\mathcal{E}(j_1,j_2), \mathcal{Y}(j_1,j_2), \Phi_{j_1j_2}\}_{\substack{j_1,j_2\in\mathfrak{B}(>i)\\ j_1<j_2}}\right)$$

is an DGS on

$$M(>i) \subset F(>i) = \bigcup_{j\in\mathfrak{B}(>i)} F(j) \subset M$$

such that corresponding thickening $Y(\mathfrak{L}(>i))$ as in (3.22) is Hausdorff.

Let K_i be a compact subset[40] of $M(i)$ such that

$$\mathrm{cl}(M(\geq i)\backslash K_i) \subset F(>i). \tag{3.46}$$

In fact, we can start from a relatively compact open subset $F' \subset F(>i)$ such that $M(>i) \subset F'$ and set $K_i = M(i)\backslash F'$. Let

$$\mathcal{E}(i) \underset{s_i}{\overset{\mathrm{pr}_i}{\rightleftarrows}} \mathcal{Y}(i)\,, \quad \overline{\psi}_i\colon \overline{s}_i^{-1}(0) \longrightarrow F(i) \subset M,$$

be a pure orbibundle structure as in Lemma 3.6.1 around $K_i \subset M(i)$. After possibly shrinking $F(i)$, we may assume

$$\mathrm{cl}(F(i)) \cap M(>i) = \emptyset;$$

cf. Definition 3.3.2.(2). In what follows, we shrink each each $Y(j)$, for $j \in \mathfrak{B}(>i)$, and build the required transition sub-charts $\mathcal{E}(i,j) \longrightarrow \mathcal{Y}(i,j)$, such that the conditions of Definition 3.3.2 and Definition 3.3.9 for the extended DGS over $\mathfrak{B}(\geq i)$ are satisfied.

We first shrink $\{Y(j)\}_{j\in\mathfrak{B}(>i)}$, respectively $Y(i)$, as in Definition 3.4.1, into relatively compact open subsets $\{Y^{(1)}(j)\}_{i\in\mathfrak{B}(>i)}$, respectively $Y_0^{(1)}(i)$, such that the shrunk system

$$\mathfrak{L}^{(1)}(>i) \equiv \left(\{\mathcal{E}^{(1)}(j), \mathcal{Y}^{(1)}(j)\}_{j\in\mathfrak{B}(>i)}, \{\mathcal{E}^{(1)}(j_1,j_2), \mathcal{Y}^{(1)}(j_1,j_2), \Phi_{j_1j_2}\}_{\substack{j_1,j_2\in\mathfrak{B}(>i)\\ j_1<j_2}}\right)$$

as in Lemma 3.4.2 still covers $M(>i)$, $K_i \subset F_0^{(1)}(i)$, all the conditions of Definition 3.3.2 and Definition 3.3.9, with $\mathfrak{B}(>i)$ instead of $\mathfrak{B}$, are still satisfied, and the inclusion condition (3.46) is still valid; i.e. the shrunk system still has all the properties of last three paragraphs.

Let

$$O = M(i) \cap F_0^{(1)}(i) \cap F^{(1)}(>i) \subset M.$$

For every $\overline{p} \in \mathrm{cl}(O) \subset F(i) \cap F(>i)$, let $\mathcal{U}_{p,O} \subset \mathcal{U}_p$ be a sufficiently small sub-chart centered at $\overline{p}$ such that the following conditions hold.

(1) For every $j \in \mathfrak{B}(>i)$, if $\overline{p} \in F(j)$, then $\mathcal{U}_{p,O} \subset \mathcal{U}_{p,j}$, where $\mathcal{U}_{p,j}$ are the compatibility sub-charts with $\mathfrak{L}(>i)$ of Definition 3.3.2.(7) for $\mathcal{E}(j) \longrightarrow \mathcal{Y}(j)$.

(2) For every $j \in \mathfrak{B}(>i)$, if $\overline{p} \notin \mathrm{cl}\big(F^{(1)}(j)\big)$, then $F_{p,O} \cap F^{(1)}(j) = \emptyset$; moreover, if $\overline{p} \in F(j)\backslash\mathrm{cl}\big(F^{(1)}(j)\big)$, then

$$\overline{\phi}_{pj}(\overline{V}_{p,O}) \cap Y^{(1)}(j) = \emptyset. \tag{3.47}$$

---

[40]Possibly empty, in which case we proceed to lower values of $\mathfrak{B}$.

(3) For every $j_1, j_2 \in \mathfrak{B}(>i)$, if $j_1 < j_2$ and $\overline{p} \in F(j_1) \cap F(j_2)$, then

$$\overline{\phi}_{pj_1}(\overline{V}_{p,O}) \subset Y(j_1, j_2); \tag{3.48}$$

(4) For every $j, k \in \mathfrak{B}(>i)$, if $\overline{p} \in F(j)$ but $\overline{p} \notin \mathrm{cl}\big(F^{(1)}(k)\big)$, then

$$\begin{cases} \overline{\phi}_{pj}(\overline{V}_{p,O}) \cap \overline{\phi}_{kj}(Y^{(1)}(j,k)) = \emptyset & \text{if } j > k, \\ \overline{\phi}_{pj}(\overline{V}_{p,O}) \cap Y^{(1)}(j,k) = \emptyset & \text{if } j < k. \end{cases} \tag{3.49}$$

(5) Finally, $\mathcal{U}_{p,O} \subset \mathcal{U}_{p,i}$.

For every $\overline{p} \in \mathrm{cl}(O) \subset M$, shrink $\mathcal{U}_{p,O}$ into a relatively compact sub-chart $\mathcal{U}^m_{p,O} \subset \mathcal{U}_{p,O}$. For every finite set of points $S = \{\overline{p}_\alpha\}_{\alpha \in I} \subset \mathrm{cl}(O)$ and every $j \in \mathfrak{B}(>i)$ let

$$I_j = \{\alpha \in I \colon \overline{p}_\alpha \in S \cap \mathrm{cl}\big(F^{(1)}(j)\big)\}.$$

Since $\mathrm{cl}(O)$ is compact, we can choose a finite set S as above such that

$$\mathrm{cl}(O) \cap \mathrm{cl}\big(F^{(1)}(j)\big) \subset \bigcup_{\alpha \in I_j} F^m_{p_\alpha,O} \quad \forall j \in \mathfrak{B}(>i). \tag{3.50}$$

Similar to Claim 1 in the proof of Lemma 3.6.2, for every $\overline{q} \in \mathrm{cl}(O)$ there also exists a sub-chart $\mathcal{U}^t_{q,O} \subset \mathcal{U}_{q,O}$ such that for every $j \in \mathfrak{B}(>i)$, every $\alpha \in I_j$, and $a = i$ or j,

$$\overline{\phi}_{qa}(\overline{V}^t_{q,O}) \cap \overline{\phi}_{p_\alpha a}(\overline{V}^m_{p_\alpha,O}) \neq \emptyset \Longrightarrow \overline{q} \in F_{p_\alpha} \text{ and } \overline{V}^t_{q,O} \subset \overline{\phi}^{-1}_{qp_\alpha}(\overline{V}_{p_\alpha,O}). \tag{3.51}$$

Choose a finite set of points $T = \{\overline{q}_\beta\}_{\beta \in J} \subset \mathrm{cl}(O)$ such that

$$\mathrm{cl}(O) \cap \mathrm{cl}\big(F^{(1)}(j)\big) \subset \bigcup_{\beta \in J_j} F^t_{q_\beta,O} \quad \forall j \in \mathfrak{B}(>i). \tag{3.52}$$

Then, similar to (3.38), for every $j \in \mathfrak{B}(>i)$, let

$$Y(i,j) = \bigcup_{\beta \in J_j, \alpha \in I_j} \Big(Y_{\alpha,\beta} = \Big(\overline{\phi}_{q_\beta i}(\overline{V}^t_{q_\beta,O}) \cap \overline{\phi}_{p_\alpha i}(\overline{V}^m_{p_\alpha,O})\Big)\Big). \tag{3.53}$$

For every $j_1, j_2 \in \mathfrak{B}(>i)$, by (3.53) and Condition (2) in Page 174,

$$F(i,j_1) \cap F^{(1)}(j_2) = \bigcup_{\substack{\beta \in J_{j_1} \cap J_{j_2} \\ \alpha \in I_{j_1} \cap I_{j_2}}} (F^t_{q_\beta,O} \cap F^m_{p_\alpha,O} \cap F^{(1)}(j_2)) \subset F(i,j_2) \cap F^{(1)}(j_2). \tag{3.54}$$

Let $\mathcal{E}(i,j) \longrightarrow \mathcal{Y}(i,j)$ (with similarly denoted section and footprint maps) be sub-orbibundles of $\mathcal{E}(i) \longrightarrow \mathcal{Y}(i)$ obtained via restriction to $Y(i,j) \subset Y(i)$. By (3.50), (3.52), and (3.53), we have

$$\mathrm{cl}(O) \cap \mathrm{cl}\big(F^{(1)}(j)\big) \subset F(i,j) \quad \forall j \in \mathfrak{B}(>i).$$

Claim 1. *For every $j \in \mathfrak{B}(>i)$, there exists a natural orbibundle embedding*

$$\Phi_{ij} \equiv (\mathfrak{D}\phi_{ij}, \phi_{ij}) \colon (\mathcal{E}(i,j), \mathcal{Y}(i,j)) \longrightarrow (\mathcal{E}(j), \mathcal{Y}(j))$$

that commutes with the section and footprint maps. Moreover, for $j_1, j_2 \in \mathfrak{B}(>i)$, with $j_1 < j_2$, restricted to $\overline{\phi}^{-1}_{ij_1}(Y(j_1,j_2)) \cap Y(i,j_2)$ we have

$$\overline{\Phi}_{ij_2} = \overline{\Phi}_{j_1 j_2} \circ \overline{\Phi}_{ij_1}$$

and Definition 3.3.2.(7) holds.

PROOF. The proof is similar to the proof of Claim 2 in the proof of Lemma 3.6.2. For $j \in \mathfrak{B}(>i)$, $\alpha \in I_j$, and $\beta \in J_j$, we show that the maps

$$(\overline{\mathfrak{D}\phi}_{ij}^{\alpha,\beta}, \overline{\phi}_{ij}^{\alpha,\beta})\colon (E(i,j), Y(i,j))|_{Y_{\alpha,\beta}} \longrightarrow (E(j), Y(j))$$

which send

$$\overline{x} = \overline{\phi}_{q_\beta i}(\overline{x}_\beta) = \overline{\phi}_{p_\alpha i}(\overline{x}_\alpha), \quad \overline{x}_\beta \in \overline{V}^t_{q_\beta,O}, \ \overline{x}_\alpha \in \overline{V}^m_{p_\alpha,O}, \tag{3.55}$$

to $\overline{y} = \overline{\phi}_{q_\beta j}(\overline{x}_\beta)$, and

$$\overline{v} = \overline{\mathfrak{D}\phi}_{q_\beta i}(\overline{v}_\beta) = \overline{\mathfrak{D}\phi}_{p_\alpha i}(\overline{v}_\alpha), \quad \overline{v}_\beta \in \overline{U}^t_{q_\beta,O}, \ \overline{v}_\alpha \in \overline{U}^m_{p_\alpha,O}, \tag{3.56}$$

to $\overline{u} = \overline{\mathfrak{D}\phi}_{q_\beta j}(\overline{v}_\beta)$ are compatible on the overlaps and define a well-defined natural continuous map $\overline{\Phi}_{ij}$ as stated which lifts to a bundle isomorphism

$$\Phi_{ij} = (\mathfrak{D}\phi_{ij}, \phi_{ij})$$

with the required properties.

Given $\overline{x} \in Y(i,j)$, suppose there are two different pairs of indices $(\alpha,\beta), (\alpha',\beta') \in I_j \times J_j$ such that

$$\overline{x} = \overline{\phi}_{q_\beta i}(\overline{x}_\beta) = \overline{\phi}_{p_\alpha i}(\overline{x}_\alpha), \quad \overline{x}_\beta \in \overline{V}^t_{q_\beta,O}, \ \overline{x}_\alpha \in \overline{V}^m_{p_\alpha,O},$$
$$\overline{x} = \overline{\phi}_{q_{\beta'} i}(\overline{x}_{\beta'}) = \overline{\phi}_{p_{\alpha'} i}(\overline{x}_{\alpha'}), \quad \overline{x}_{\beta'} \in \overline{V}^t_{q_{\beta'},O}, \ \overline{x}_{\alpha'} \in \overline{V}^m_{p_{\alpha'},O}.$$

By (3.51), it follows from

$$\overline{x} \in \overline{\phi}_{q_\beta i}(\overline{V}^t_{q_\beta,O}) \cap \mathrm{cl}\Big(\overline{\phi}_{p_{\alpha'} i}(\overline{V}^m_{p_{\alpha'},O})\Big) \neq \emptyset,$$
$$\overline{x} \in \overline{\phi}_{q_{\beta'} i}(\overline{V}^t_{q_{\beta'},O}) \cap \mathrm{cl}\Big(\overline{\phi}_{p_\alpha i}(\overline{V}^m_{p_\alpha,O})\Big) \neq \emptyset;$$

that

$$\overline{q}_\beta \in F_{p_{\alpha'}}, \ \overline{q}_{\beta'} \in F_{p_\alpha}, \ \overline{V}^t_{q_\beta,O} \subset \overline{\phi}^{-1}_{q_\beta p_{\alpha'}}(\overline{V}_{p_{\alpha'},O}), \text{ and } \overline{V}^t_{q_{\beta'},O} \subset \overline{\phi}^{-1}_{q_{\beta'} p_\alpha}(\overline{V}_{p_\alpha,O}).$$

By Definition 3.3.2.(7).(a), with $a = i$ or j, restricted to

$$\overline{V}^t_{q_\beta,O} \subset \overline{\phi}^{-1}_{q_\beta p_{\alpha'}}(\overline{V}_{p_{\alpha'},O}) \cap \overline{V}_{q_\beta,O} \subset \overline{\phi}^{-1}_{q_\beta p_{\alpha'}}(\overline{V}_{p_{\alpha'},a}) \cap \overline{V}_{q_\beta,a},$$

we have

$$\overline{\Phi}_{p_{\alpha'} a} \circ \overline{\Phi}_{q_\beta p_{\alpha'}} = \overline{\Phi}_{q_\beta a}.$$

Therefore, with $a = i$ and since $\overline{\Phi}_{p_{\alpha'} i}$ is one-to-one, we get

$$\overline{x}_{\alpha'} = \overline{\Phi}_{q_\beta p_{\alpha'}}(\overline{x}_\beta);$$

and then, with $a = j$, we get

$$\overline{\phi}_{q_\beta j}(\overline{x}_\beta) = \overline{\phi}_{p_{\alpha'} j}(\overline{x}_{\alpha'}).$$

We conclude that (3.55) is independent of the choice of (α,β) or (α',β'). The case of (3.56) is similar.

By definition, restricted to $Y_{\alpha,\beta}$,

$$(\overline{\mathfrak{D}\phi}_{ij}, \overline{\phi}_{ij}) = \overline{\Phi}_{p_\alpha j} \circ \overline{\Phi}^{-1}_{p_\alpha i}; \tag{3.57}$$

therefore, it is locally liftable to a smooth orbibundle embedding between orbibundle charts. The cocycle condition $\overline{\Phi}_{ij_2} = \overline{\Phi}_{j_1 j_2} \circ \overline{\Phi}_{ij_1}$ is a consequence of (3.57). Compatibility with the section and footprint maps and Definition 3.3.2.(7) follow from the corresponding compatibility conditions of Definition 3.3.2 for $\mathfrak{L}(>i)$. □

The extended system of pure orbibundle structures

$$\mathfrak{L}(\geq i) \equiv \left(\{\mathcal{E}(j), \mathcal{Y}(j), s_j, \psi_j\}_{j\in\mathfrak{B}(\geq i)}, \{\mathcal{E}(j_1,j_2), \mathcal{Y}(j_1,j_2), \Phi_{j_1j_2}\}_{\substack{j_1,j_2\in\mathfrak{B}(\geq i)\\ j_1<j_2}} \right)$$

constructed so far does not satisfy all the requirements of Definition 3.3.2. Also, the corresponding thickening may not be Hausdorff. In the remaining part of the proof, we shrink these orbibundles in a way that the reduced system satisfies all the conditions of Definition 3.3.2 and Definition 3.3.9, as well.

First, for $j > i$, the intersection condition (3.13) does not necessarily hold. In fact,

$$\mathrm{cl}\left(M(i) \cap F_0^{(1)}(i) \cap F^{(1)}(j)\right) \subset F(i,j) \subset F(i) \cap F(j)$$

and the second inclusion could be proper. Let

$$K = K_i \setminus \bigcup_{j\in\mathfrak{B}(>i)} F(i,j) \subset M(i);$$

this is a compact set which has empty intersection with $\mathrm{cl}(F^{(1)}(>i))$. For $j>i$, we replace $Y(j)$ with $Y^{(1)}(j)$. For $i<j_1<j_2$, we replace $Y(j_1,j_2)$ with $Y^{(1)}(j_1,j_2)$. We replace $Y(i)$ with

$$Y^{(1)}(i) = W \cup \bigcup_{j\in\mathfrak{B}(>i)} Y(i,j) \tag{3.58}$$

such that W has the following properties:

$$W \subset Y(i), \quad K \subset \overline{\psi}_i(W), \text{ and } W \cap \bigcup_{j\in\mathfrak{B}(>i)} \overline{\phi}_{ij}^{-1}(Y^{(1)}(j)) = \emptyset. \tag{3.59}$$

For every $j \in \mathfrak{B}(>i)$, we replace $Y(i,j)$ with

$$Y^{(1)}(i,j) = \overline{\phi}_{ij}^{-1}(Y^{(1)}(j)) \subset Y(i,j) \subset Y^{(1)}(i). \tag{3.60}$$

It is immediate that the new collection

$$\mathfrak{L}^{(1)}(\geq i) \equiv \Big(\{\mathcal{E}^{(1)}(j), \mathcal{Y}^{(1)}(j), s_j, \psi_j\}_{j\in\mathfrak{B}(\geq i)},$$
$$\{\mathcal{E}^{(1)}(j_1,j_2), \mathcal{Y}^{(1)}(j_1,j_2), \Phi_{j_1j_2}\}_{\substack{j_1,j_2\in\mathfrak{B}(\geq i)\\ j_1<j_2}} \Big)$$

satisfies Conditions (1), (2), (7), and (8) of Definition 3.3.2. It is a bit tedious, but we show that $\mathfrak{L}^{(1)}(\geq i)$ also satisfies Conditions (3), (4), and (5). Then it follows that it also satisfies (6). Eventually, we achieve Hausdorffness by another appropriate shrinking.

Condition (3) **of Definition 3.3.2.** For every $j \in \mathfrak{B}(>i)$, by (3.54), (3.58), and (3.59), we have

$$\begin{aligned} &F^{(1)}(i) \cap F^{(1)}(j) = \big(\overline{\psi}_i(W) \cup \bigcup_{k\in\mathfrak{B}(>i)} F(i,k)\big) \cap F^{(1)}(j) = \\ &\bigcup_{k\in\mathfrak{B}(>i)} (F(i,k) \cap F^{(1)}(j)) = F(i,j) \cap F^{(1)}(j) = F^{(1)}(i,j). \end{aligned} \tag{3.61}$$

Therefore, $\mathfrak{L}^{(1)}(\geq i)$ satisfies (3.13), i.e. Definition 3.3.2(3) holds.

Claim 2. *For* $j_1, j_2 \in \mathfrak{B}(>i)$,

$$Y^{(1)}(i,j_1) \cap Y^{(1)}(i,j_2) = \overline{\phi}_{ij_1}^{-1}(Y^{(1)}(j_1)) \cap \overline{\phi}_{ij_2}^{-1}(Y^{(1)}(j_2)) \cap \bigcup_{\substack{\beta \in J_{j_1} \cap J_{j_2} \\ \alpha \in I_{j_1} \cap I_{j_2}}} Y_{\alpha,\beta}$$

Proof. The inclusion of right-hand side in the left-hand side is clear from (3.53) and (3.60); moreover,

$$Y^{(1)}(i,j_1) \cap Y^{(1)}(i,j_2) = \phi_{ij_1}^{-1}(Y^{(1)}(j_1)) \cap \phi_{ij_2}^{-1}(Y^{(1)}(j_2)) \cap \bigcup_{\substack{\beta_1 \in J_{j_1}, \beta_2 \in J_{j_2} \\ \alpha_1 \in I_{j_1}, \alpha_2 \in I_{j_2}}} \Big(Y_{\alpha_1,\beta_1} \cap Y_{\alpha_2,\beta_2} \Big).$$

We may assume $j_1 < j_2$. For some $\beta_1 \in J_{j_1}$, $\beta_2 \in J_{j_2}$, $\alpha_1 \in I_{j_1}$, and $\alpha_2 \in I_{j_2}$, if

$$\phi_{ij_1}^{-1}(Y^{(1)}(j_1)) \cap \phi_{ij_2}^{-1}(Y^{(1)}(j_2)) \cap Y_{\alpha_1,\beta_1} \cap Y_{\alpha_2,\beta_2} \neq \emptyset, \tag{3.62}$$

it follows from (3.51) and (3.48) that

$$\overline{q}_{\beta_1} \in F^{(1)}(j_1) \cap F(j_2), \quad \overline{q}_{\beta_2} \in F(j_1) \cap F^{(1)}(j_2),$$

and

$$\overline{\phi}_{ij_1}(Y_{\alpha_1,\beta_1}), \overline{\phi}_{ij_1}(Y_{\alpha_2,\beta_2}) \subset Y(j_1,j_2). \tag{3.63}$$

If $\overline{q}_{\beta_1} \notin F^{(1)}(j_2)$, by (3.47),

$$\overline{\phi}_{ij_2}(Y_{\alpha_1,\beta_1}) \cap Y^{(1)}(j_2) \subset \overline{\phi}_{q_{\beta_1} j_2}(\overline{V}^t_{q_{\beta_1},O}) \cap Y^{(1)}(j_2) = \emptyset;$$

which contradicts (3.62). Thus, $\overline{q}_{\beta_1} \in F^{(1)}(j_2)$. Similarly, we conclude $\overline{q}_{\beta_2} \in F^{(1)}(j_1)$. If $\overline{p}_{\alpha_1} \notin F^{(1)}(j_2)$, by (3.49),

$$\overline{\phi}_{ij_1}(Y_{\alpha_1,\beta_1}) \cap Y^{(1)}(j_1,j_2) \subset \overline{\phi}_{p_{\alpha_1} j_1}(\overline{V}^m_{p_{\alpha_1},O}) \cap Y^{(1)}(j_1,j_2) = \emptyset.$$

Therefore, by (3.63),

$$Y_{\alpha_1,\beta_1} \cap \phi_{ij_1}^{-1}(Y^{(1)}(j_1)) \cap \phi_{ij_2}^{-1}(Y^{(1)}(j_2)) = \emptyset;$$

which contradicts (3.62). Thus, $\overline{p}_{\alpha_1} \in F^{(1)}(j_2)$. Similarly, we conclude $\overline{p}_{\alpha_2} \in F^{(1)}(j_1)$. This finishes the proof of Claim 2. □

Condition (4) **of Definition 3.3.2.** For $j_1, j_2 \in \mathfrak{B}(>i)$, with $j_1 < j_2$, by Claim 2 and (3.48),

$$Y^{(1)}(i,j_1) \cap Y^{(1)}(i,j_2) \subset \overline{\phi}_{ij_1}^{-1}(Y^{(1)}(j_1,j_2)). \tag{3.64}$$

On the other hand, for every $\alpha \in I_{j_1}$ and $\beta \in J_{j_1}$, if

$$\overline{p}_\alpha \notin \mathrm{cl}(F^{(1)}(j_2)) \quad \text{or} \quad \overline{q}_\beta \notin \mathrm{cl}(F^{(1)}(j_2)),$$

by the second case of (3.49) we get

$$\overline{\phi}_{ij_1}(Y_{\alpha,\beta}) \cap Y^{(1)}(j_1,j_2) = \emptyset; \tag{3.65}$$

i.e.

$$\overline{\phi}_{ij_1}^{-1}(Y^{(1)}(j_1,j_2)) \cap Y_{\alpha,\beta} = \emptyset. \tag{3.66}$$

Therefore, by (3.66) and the definition of $Y^{(1)}(j_1,j_2)$ in the proof Lemma 3.4.2,

$$\begin{aligned}&\overline{\phi}_{ij_1}^{-1}(Y^{(1)}(j_1,j_2)) = \overline{\phi}_{ij_1}^{-1}\big(Y^{(1)}(j_1)\cap \overline{\phi}_{j_1j_2}^{-1}(Y^{(1)}(j_2))\cap Y(j_1,j_2)\big) \subset \\ &\phi_{ij_1}^{-1}(Y^{(1)}(j_1))\cap \phi_{ij_2}^{-1}(Y^{(1)}(j_2))\cap \bigcup_{\substack{\beta\in J_{j_1}\cap J_{j_2}\\ \alpha\in I_{j_1}\cap I_{j_2}}} Y_{\alpha,\beta} = Y^{(1)}(i,j_1)\cap Y^{(1)}(i,j_2).\end{aligned}$$

Together with (3.64), we conclude that the equality in Definition 3.3.2.(4) is satisfied.

Let

$$Y(i,j_1,j_2) = \overline{\phi}_{ij_1}^{-1}(Y^{(1)}(j_1,j_2)) = Y^{(1)}(i,j_1)\cap Y^{(1)}(i,j_2)$$

as in Definition 3.3.2.(4).

Condition (5) **of Definition 3.3.2.** For $j_1,j_2\in\mathfrak{B}(>i)$, with $j_1<j_2$, by (3.53),

$$\begin{aligned}&\overline{\phi}_{ij_2}(Y^{(1)}(i,j_2)) = Y^{(1)}(j_2)\cap \bigcup_{\beta\in J_{j_2},\alpha\in I_{j_2}} \left(\overline{\phi}_{q_\beta j_2}(\overline{V}^t_{q_\beta,O})\cap \overline{\phi}_{p_\alpha j_2}(\overline{V}^m_{p_\alpha,O})\right),\\ &\overline{\phi}_{j_1j_2}(Y^{(1)}(i,j_2)) = Y^{(1)}(j_2)\cap \overline{\phi}_{j_1,j_2}(Y^{(1)}(j_1)\cap Y(j_1,j_2)),\end{aligned} \tag{3.67}$$

By Claim 2 above,

$$\begin{aligned}\overline{\phi}_{ij_2}(Y^{(1)}(i,j_1,j_2)) =&Y^{(1)}(j_2)\cap \overline{\phi}_{j_1,j_2}(Y^{(1)}(j_1)\cap Y(j_1,j_2))\cap\\ &\bigcup_{\substack{\beta\in J_{j_1}\cap J_{j_2}\\ \alpha\in I_{j_1}\cap I_{j_2}}} \left(\overline{\phi}_{q_\beta j_2}(\overline{V}^t_{q_\beta,O})\cap \overline{\phi}_{p_\alpha j_2}(\overline{V}^m_{p_\alpha,O})\right)\end{aligned} \tag{3.68}$$

For every $\alpha\in I_{j_2}$ and $\beta\in J_{j_2}$, if

$$Y^{(1)}(j_2)\cap \overline{\phi}_{q_\beta j_2}(\overline{V}^t_{q_\beta,O})\cap \overline{\phi}_{p_\alpha j_2}(\overline{V}^m_{p_\alpha,O})\cap \overline{\phi}_{j_1,j_2}(Y^{(1)}(j_1)\cap Y(j_1,j_2))\neq\emptyset, \tag{3.69}$$

by (3.49), $\overline{q}_\beta,\overline{p}_\alpha\in \mathrm{cl}\left(F^{(1)}(j_1)\right)$. It immediately follows from (3.69), (3.67), and (3.68), that

$$\overline{\phi}_{ij_2}(Y^{(1)}(i,j_1,j_2)) = \overline{\phi}_{ij_2}(Y^{(1)}(i,j_2))\cap \overline{\phi}_{j_1j_2}(Y^{(1)}(j_1,j_2)).$$

Hence, Definition 3.3.2.(5) is also satisfied.

Achieving of Hausdorffness. The thickening $Y(\mathfrak{L}^{(1)}(\geq i))$ has the form

$$Y(\mathfrak{L}^{(1)}(\geq i)) = \left(Y^{(1)}(i)\coprod Y(\mathfrak{L}^{(1)}(>i))\right)/\sim,$$

where, by induction, $Y(\mathfrak{L}^{(1)}(>i))$ is Haussdorf and

$$Y^{(1)}(i)\ni \overline{x}\sim[\overline{y}]\in Y(\mathfrak{L}^{(1)}(>i))$$

whenever there exists $j\in\mathfrak{B}(>i)$ such that $\overline{y}\in Y^{(1)}(j)$, $[\overline{y}]=\wp_j(\overline{y})$, and $\overline{y}=\overline{\phi}_{ij}(\overline{x})$. In order to achieve Hausdorffness, we shrink $Y^{(1)}(i)$ and $Y(\mathfrak{L}^{(1)}(>i))$ into relatively compact open sets $Y^{(2)}(i)$ and $\widetilde{Y}(\mathfrak{L}^{(1)}(>i))$, respectively, similarly to the argument after the proof of Claim 3 in page 172. For $j\in\mathfrak{B}(>i)$, let

$$Y^{(2)}(j) = \wp_j^{-1}(\widetilde{Y}(\mathfrak{L}^{(1)}(>i))),$$

where $\wp_i\colon Y^{(1)}(j) \longrightarrow Y(\mathfrak{L}^{(1)}(\geq i))$ is the inclusion map as in the paragraph before Definition 3.3.9. The shrunk DGS

$$\mathfrak{L}^{(2)}(\geq i) \equiv$$
$$\left(\{\mathcal{E}^{(2)}(j), \mathcal{Y}^{(2)}(j), s_j, \psi_j\}_{j\in\mathfrak{B}(\geq i)}, \{\mathcal{E}^{(2)}(j_1, j_2), \mathcal{Y}^{(2)}(j_1, j_2), \Phi_{j_1j_2}\}_{\substack{j_1,j_2\in\mathfrak{B}(\geq i)\\ j_1<j_2}}\right)$$

covers $M(\geq i)$, is Hausdorff , and satisfies all the conditions of Definition 3.3.2. This completes the induction step and the proof of the first statement in Theorem 3.5.3.

For every $i\in\mathfrak{B}$, the orbibundle $\mathcal{E}(i) \longrightarrow \mathcal{Y}(i)$ is made of Kuranishi charts of the same rank i. Therefore, if $(M, \mathcal{K})$ is oriented, by the last condition in Definition 3.2.11, every $\mathcal{E}(i) \longrightarrow \mathcal{Y}(i)$ is relatively oriented. Compatibility of (3.18) and (3.19) in Definition 3.3.7 follow from the corresponding condition in Definition 3.2.11.

It is just left to show that every two such dimensionally graded systems are deformation equivalent. In fact, it follows from the construction of the proof that every two of such DGS admit a common refinement[41] in the following sense: for every two dimensionally graded systems $\mathfrak{L}_0$ and $\mathfrak{L}_1$ constructed as above, there exist shrinkings $\widetilde{\mathfrak{L}}_0$ and $\widetilde{\mathfrak{L}}_1$ of $\mathfrak{L}_0$ and $\mathfrak{L}_1$, respectively, which are isomorphic in the following sense. There are orbibundle isomorphisms

$$\Phi_{01;i}\colon : \widetilde{\mathcal{E}}_0(i) \longrightarrow \widetilde{\mathcal{E}}_1(i) \quad \forall i\in\mathfrak{B}$$

such that

- each Φ_i commutes with the section and footprint maps,
- for every $\overline{p}\in F_0(i)=F_1(i)$, the compatibility embeddings of Definition 3.3.2.(7) (taken over a sub-chart in the intersection of their domains)

$$\Phi_{0;pi}, \Phi_{1;pi}\colon ([U_{p,i}/G_{p,i}], [V_{p,i}/G_{p,i}]) \longrightarrow (\mathcal{E}_0(i), \mathcal{Y}_0(i)), (\mathcal{E}_1(i), \mathcal{Y}_1(i))$$

  satisfy $\Phi_{1;pi} = \Phi_{01;i}\circ\Phi_{0;pi}$.

Such a common refinement gives us a deformation equivalence[42] dimensionally graded system $\widetilde{\mathfrak{L}}$ for $(M\times[0,1], \mathcal{K}\times[0,1])$ (see the argument after Definition 3.5.1) with orbibundle pieces of the form

$$\begin{aligned}(\widetilde{\mathcal{E}}(i), \widetilde{\mathcal{Y}}(i)) = (\mathcal{E}_0(i), \mathcal{Y}_0(i))\times[0,1/3) \coprod (\widetilde{\mathcal{E}}_0(i), \widetilde{\mathcal{Y}}_0(i))\times(0,1) \coprod \\ (\widetilde{\mathcal{E}}_1(i), \widetilde{\mathcal{Y}}_1(i))\times(0,1) \coprod (\mathcal{E}_1(i), \mathcal{Y}_1(i))\times(2/3,1]/\sim \quad \forall i\in\mathfrak{B},\end{aligned}$$

where the gluing relation $\sim$ between $(\widetilde{\mathcal{E}}_0(i), \widetilde{\mathcal{Y}}_0(i))\times(0,1)$ and $(\widetilde{\mathcal{E}}_1(i), \widetilde{\mathcal{Y}}_1(i))\times(0,1)$ is by $\Phi_{01;i}\times\mathrm{id}$ and the gluing relations in the first and second lines are given by obvious inclusions restricted to $(0,1/3)$ and $(2/3,1)$, respectively. The footprint map and section on each piece are the obvious product footprint map and section. This finishes the proof of Theorem 3.5.3.

□

[41] This is analogue of commensurability in [**39**].

[42] This is analogue of [**39**, Lemma 6.2.16].

3.7. Deformations of Kuranishi maps. Let $\mathfrak{L}$ be an oriented Hausdorff DGS as in Definition 3.3.2 for an n-dimensional oriented Kuranishi structure $(\mathcal{K}, M)$ as in Section 3.3. In Section 4, with an approach similar to the orbibundle case of Section 2.6, we will build a virtual fundamental class for M with respect to $\mathfrak{L}$. More precisely, after shrinking $\mathfrak{L}$ finitely many times into a sufficiently small $\widetilde{\mathfrak{L}}$, we build a rational singular homology n-cycle

$$\mathrm{VFC}(\widetilde{\mathfrak{L}}) \in H_n(Y(\widetilde{\mathfrak{L}}), \mathbb{Q})$$

which will play the role of fundamental class of M in calculations. If $Y(\widetilde{\mathfrak{L}})$ admits a continuous map into some topological manifold, say $f\colon Y(\widetilde{\mathfrak{L}}) \longrightarrow X$, we can push-forward this homology class to a homology class in X. This is for instance the case in the construction of Gromov-Witten VFC. In this particular application, the evaluation and forgetful maps, (1.4) and (1.5), naturally extend to the thickening $Y(\widetilde{\mathfrak{L}})$ and the Gromov-Witten VFC of $\overline{\mathcal{M}}_{g,k}(X, A)$, as we define it in this article, is the image under $\mathrm{ev} \times \mathrm{st}$ of $\mathrm{VFC}(\widetilde{\mathfrak{L}})$ in $H_*(X^k \times \overline{\mathcal{M}}_{g,k}, \mathbb{Q})$. In the case of abstract Kuranishi structures, up to cobordism, VFC is independent of the particular choice of natural $\mathfrak{L}$ in Theorem 3.5.3; thus we may denote it by $\mathrm{VFC}(\mathcal{K})$. In the particular case of GW theory, $\mathrm{VFC}(\mathcal{K})$ does not depend on the particular choice of natural $\mathcal{K}$ in Theorem 1.4.1, either; thus we denote it by $\mathrm{VFC}(\overline{\mathcal{M}}_{g,k}(X, A, J))$ or $[\overline{\mathcal{M}}_{g,k}(X, A, J)]^{\mathrm{VFC}}$.

With notation as in (3.16), in order to construct $\mathrm{VFC}(\mathfrak{L})$, we start from the smallest value $i_1 \in \mathfrak{B}$, shrink $\mathfrak{L}$ into $\mathfrak{L}^{(1)}$, perturb the kuranishi map s_{i_1} into a sequence of transversal lifted multisections $(\mathfrak{s}_{i_1;a} = s_{i_1} + \mathfrak{t}_{i_1;a})_{a \in \mathbb{N}}$ on $\mathrm{cl}(Y^{(1)}(i_1))$, where $\lim_{a \longrightarrow \infty} \mathfrak{t}_{i_1;a} \equiv 0$ as in Proposition 2.4.2, and inductively extend the multisections $\mathfrak{t}_{i_1;a}$ to $\mathcal{E}(i) \longrightarrow \mathcal{Y}(i)$ for higher values of $i \in \mathfrak{B}$, again via Proposition 2.4.2. The result of this process is a compatible set of sequences of deformations of Kuranishi maps,

$$(\mathfrak{s}_{i;a} = s_i + \mathfrak{t}_{i;a})_{i \in \mathfrak{B}, a \in \mathbb{N}},$$

described in Definition 3.7.12 below.

For this induction process, we need to "compatibly" extend every sub-orbibundle

$$\mathfrak{D}\phi_{ij}(\mathcal{E}(i,j)) \subset \mathcal{E}(j)$$

to a sub-orbibundle

$$\mathcal{E}_{\mathrm{ext}}(i,j) \subset \mathcal{E}(j)$$

over a tubular neighborhood $\mathcal{N}(i,j)$ of $\phi_{ij}(\mathcal{Y}(i,j)) \subset \mathcal{Y}(j)$. We need some compatibility conditions between these extensions in order for the resulting deformations of Kuranishi maps to be equal one the overlaps. Following definitions abstractly characterize such extensions.

For a sub-orbifold $\mathcal{M}_1 \subset \mathcal{M}_2$ as in Definition 3.7.2, let

$$\mathrm{pr}\colon \mathcal{N}_{\mathcal{M}_1}\mathcal{M}_2 \longrightarrow \mathcal{M}_1$$

be the orbifold normal bundle of that as in Example 2.2.9. Similarly to the case of manifolds, the restriction of orbifold tangent bundles $T\mathcal{M}_2$ and $T\mathcal{N}_{\mathcal{M}_1}\mathcal{M}_2$ to $\mathcal{M}_1$ are naturally isomorphic to each other. We make use of this identification in (3.70). In what follows, by a **tubular neighborhood** of $\mathcal{M}_1$ in $\mathcal{N}_{\mathcal{M}_1}\mathcal{M}_2$ we mean the restriction $\mathcal{N}'_{\mathcal{M}_1}\mathcal{M}_2$ of $\mathcal{N}_{\mathcal{M}_1}\mathcal{M}_2$ to some open set $N'_{M_1}M_2$ including M_1. Here, by convention of Notation 2.1.4, we are thinking about $\mathcal{N}_{\mathcal{M}_1}\mathcal{M}_2$ as an orbifold with underlying topological space $N_{M_1}M_2$.

REMARK 3.7.1. Our definition of tubular neighborhood is weaker than its usual meaning in the literature. In our definition, $N'_{M_1}M_2$ does not even have to be connected; it is just an open set in $N_{M_1}M_2$ containing M_1.

DEFINITION 3.7.2. *Given a sub-orbifold $\mathcal{M}_1 \subset \mathcal{M}_2$, a* ***tubular neighborhood*** *of $\mathcal{M}_1$ in $\mathcal{M}_2$ is an open sub-orbifold $\mathcal{W} \subset \mathcal{M}_2$ containing $\mathcal{M}_1$ with an orbifold smooth map $\pi\colon \mathcal{W} \longrightarrow \mathcal{M}_1$ such that there exists a tubular neighborhood $\mathcal{N}'_{\mathcal{M}_1}\mathcal{M}_2$ of the zero section $\mathcal{M}_1 \subset \mathcal{N}_{\mathcal{M}_1}\mathcal{M}_2$ and an orbifold diffeomorphism*

$$f\colon \mathcal{N}'_{\mathcal{M}_1}\mathcal{M}_2 \longrightarrow \mathcal{W}$$

satisfying

$$\mathrm{pr} = \pi \circ f \text{ and } \mathrm{d}f|_{T\mathcal{M}_2|\mathcal{M}_1} = \mathrm{id}. \tag{3.70}$$

In the context of Definition 3.7.2, we call such f a **regularization** of the tubular neighborhood $\pi\colon \mathcal{W} \longrightarrow \mathcal{M}_1$.

DEFINITION 3.7.3. *Given a tubular neighborhood $\pi\colon \mathcal{W} \longrightarrow \mathcal{M}_1$ for $\mathcal{M}_1 \subset \mathcal{M}_2$, a* ***sub-tubular neighborhood*** *of $\mathcal{W}$ is some open*[43] *sub-orbifold $\mathcal{W}' \subset \mathcal{W}$ containing $\mathcal{M}_1$.*

For a tubular neighborhood $\mathcal{W}$ as in Definition 3.7.2 and any $\mathcal{M}'_1 \subset \mathcal{M}_1$, let

$$\mathcal{W}|_{\mathcal{M}'_1} = \pi^{-1}(\mathcal{M}'_1) \subset \mathcal{W};$$

if $\mathcal{M}'_1$ is an open sub-orbifold of $\mathcal{M}_1$, then $\mathcal{W}|_{\mathcal{M}'_1}$ is a tubular neighborhood of $\mathcal{M}'_1 \subset \mathcal{M}_2$.

DEFINITION 3.7.4. *Let $\mathfrak{L}$ be an DGS as in Definition 3.3.2 and $\ell \in \mathbb{N}$. An* ***extension of the obstruction bundles (EOB)*** *$\mathcal{N}$ of* ***width*** *ℓ for $\mathfrak{L}$ consists of a set of tubular neighborhoods*

$$\pi_{ji}\colon \mathcal{N}(i,j) \longrightarrow \phi_{ij}(\mathcal{Y}(i,j))$$

for $\phi_{ij}(\mathcal{Y}(i,j)) \subset \mathcal{Y}(j)$ and orbibundle embeddings

$$\iota_{ij}\colon \pi^*_{ji}(\mathfrak{D}\phi_{ij}\mathcal{E}(i,j)) \longrightarrow \mathcal{E}(j), \quad \iota_{ij}|_{\phi_{ij}(\mathcal{Y}(i,j))} = \mathrm{id}, \tag{3.71}$$

for all $i,j \in \mathfrak{B}$, with $0 < j-i \le \ell$, such that the following ***compatibility*** *conditions, for every $i,j,k \in \mathfrak{B}$, with $i<j<k$ and $k-i \le \ell$, hold.*

(1) *With $\mathcal{N}(i,j,k) = \mathcal{N}(i,k) \cap \mathcal{N}(j,k)$, we have*

$$\phi^{-1}_{jk}(\pi_{kj}(\mathcal{N}(i,j,k))) \subset \mathcal{N}(i,j). \tag{3.72}$$

(2) *Restricted to $\mathcal{N}(i,j,k)$, we have*

$$(\phi^{-1}_{ik} \circ \pi_{ki}) = (\phi^{-1}_{ij} \circ \pi_{ji}) \circ (\phi^{-1}_{jk} \circ \pi_{kj}). \tag{3.73}$$

(3) *For every $x \in \mathcal{N}(i,j,k)$, with $y = \phi^{-1}_{jk}(\pi_{kj}(x))$ and $z = \phi^{-1}_{ki}(\pi_{ki}(x))$,*

$$\iota_{ik}\left(x, \mathfrak{D}\phi_{ik}(v)\right) = \iota_{jk}\left(x, \mathfrak{D}\phi_{jk}\left(\iota_{ij}\left(y, \mathfrak{D}\phi_{ij}(v)\right)\right)\right) \quad \forall v \in \mathcal{E}(i,k)_z.$$

Here $(x, \mathfrak{D}\phi_{ik}(v))$, and the similar notation on the right-hand side of the equation, denotes the pull-back of vector $\mathfrak{D}\phi_{ik}(v) \in \mathfrak{D}\phi_{ik}(\mathcal{E}(i,k))$ to the fiber of $\pi^*_{ki}\mathfrak{D}\phi_{ik}(\mathcal{E}(i,k))$ at $x \in \pi^{-1}_{ki}(\phi_{ik}(z))$.

[43] Together with the restriction of the corresponding projection map.

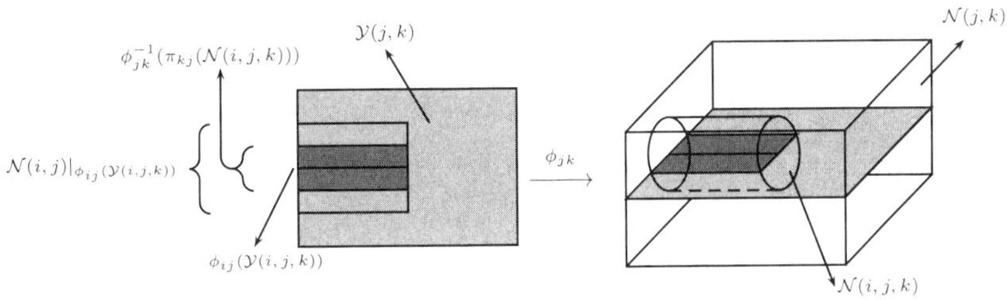

FIGURE 6. Overlap of tubular neighborhoods.

REMARK 3.7.5. With $\mathcal{Y}(i,j,k)$ as in Definition 3.3.2(4), (3.72) and (3.73) imply that $\mathcal{N}(i,j,k)$ is a sub-tubular neighborhood of $\mathcal{N}(i,k)|_{\phi_{ik}(\mathcal{Y}(i,j,k))}$; i.e. the following diagram is well-defined and commutes:

$$\begin{array}{ccc} \mathcal{N}(i,j,k) & \xrightarrow{\phi_{jk}^{-1}\circ\pi_{kj}} & \mathcal{N}(i,j)|_{\phi_{ij}(\mathcal{Y}(i,j,k))} \\ & \searrow^{\phi_{ik}^{-1}\circ\pi_{ki}} & \downarrow \phi_{ij}^{-1}\circ\pi_{ji} \\ & & \mathcal{Y}(i,j,k). \end{array}$$

Figure 6 illustrates a sequence of compatible neighborhoods and their overlaps.

In the context of Definition 3.7.4, for every $i,j\in\mathfrak{B}$, with $0<j-i\leq\ell$, let

$$\mathcal{E}_{\text{ext}}(i,j)=\iota_{ij}(\pi_{ji}^{*}(\mathfrak{D}\phi_{ij}\mathcal{E}(i,j)))\subset\mathcal{E}(j)|_{\mathcal{N}(i,j)}.$$

For simplicity, by an EOB, i.e. whenever we do not specify its width, we mean an EOB of the width infinity (or any number larger than the width of $\mathfrak{B}$). The case of EOB with bounded width appears in the inductive proof Theorem 3.7.11; otherwise, we only consider EOB of the width infinity. The case of $\ell=1$ corresponds to a set of tubular neighborhoods and extension of the corresponding orbibundles without any compatibility condition on the overlaps.

DEFINITION 3.7.6. *Let $\mathfrak{L}$ be an DGS as in Definition 3.3.2. Let $\mathcal{N}$ be an EOB of width ℓ for $\mathfrak{L}$ as in Definition 3.7.4. A shrinking of $(\mathfrak{L},\mathcal{N})$ consist of a shrinking $\widetilde{\mathfrak{L}}$ of $\mathfrak{L}$ as in Lemma 3.4.2 and an EOB $\widetilde{\mathcal{N}}$ of width ℓ for $\widetilde{\mathfrak{L}}$, such that for all $i,j\in\mathfrak{B}$, with $0<j-i\leq\ell$, $\widetilde{\mathcal{N}}(i,j)$ is a sub-tubular neighborhood of $\mathcal{N}(i,j)|_{\phi_{ij}(\widetilde{\mathcal{Y}}(i,j))}$.*

The extensions $\widetilde{\mathcal{E}}_{\text{ext}}(i,j)$ in this situation are simply the restrictions of the corresponding bundles to $\widetilde{\mathcal{N}}(i,j)$.

LEMMA 3.7.7. *Let $(\mathfrak{L},\mathcal{N})$ be a pair as in Definition 3.7.6. For every shrinking $\widetilde{\mathfrak{L}}$ of $\mathfrak{L}$ as in Lemma 3.4.2, the sub-tubular neighborhoods*

$$\widetilde{\mathcal{N}}(i,j)=\mathcal{N}(i,j)|_{\phi_{ij}(\widetilde{\mathcal{Y}}(i,j))}\cap\widetilde{\mathcal{Y}}(j)\quad\forall i,j\in\mathfrak{B},\ \text{with } i<j, \tag{3.74}$$

give rise to a shrinking $(\widetilde{\mathfrak{L}},\widetilde{\mathcal{N}})$ of $(\mathfrak{L},\mathcal{N})$ as in Definition 3.7.6.

PROOF. We just need to verify (3.72), the rest follows by restriction. By definition we have

$$\widetilde{\mathcal{N}}(i,j,k)=\mathcal{N}(i,k)|_{\phi_{ik}(\widetilde{\mathcal{Y}}(i,k))}\cap\mathcal{N}(j,k)|_{\phi_{jk}(\widetilde{\mathcal{Y}}(j,k))}\cap\widetilde{\mathcal{Y}}(k). \tag{3.75}$$

Suppose there exists $x \in \widetilde{\mathcal{N}}(i,j,k)$ such that

$$y = \phi_{jk}^{-1}(\pi_{kj}(x)) \notin \widetilde{\mathcal{N}}(i,j) = \mathcal{N}(i,j)|_{\phi_{ij}(\widetilde{\mathcal{Y}}(i,j))} \cap \widetilde{\mathcal{Y}}(j).$$

By (3.75),

$$y \in \widetilde{\mathcal{Y}}(j,k) \subset \widetilde{\mathcal{Y}}(j). \tag{3.76}$$

Therefore,

$$y \notin \mathcal{N}(i,j)|_{\phi_{ij}(\widetilde{\mathcal{Y}}(i,j))}.$$

Since $y \in \mathcal{N}(i,j)$ (because (3.72) holds for $\mathcal{N}$), we conclude that

$$z = \phi_{ij}^{-1}(\pi_{ji}(y)) \notin \widetilde{\mathcal{Y}}(i,j) = \phi_{jk}^{-1}(\widetilde{\mathcal{Y}}(j)) \cap \mathcal{Y}(i,j) \cap \widetilde{\mathcal{Y}}(i).$$

It follows from (3.76) that $z \notin \mathcal{Y}(i,j) \cap \widetilde{\mathcal{Y}}(i)$. On the other hand, it follows from (3.75), (3.73), and Remark 3.7.5 that

$$z \in \mathcal{Y}(i,j,k) \cap \widetilde{\mathcal{Y}}(i) \cap \phi_{ik}^{-1}(\widetilde{\mathcal{Y}}(k)) \subset \mathcal{Y}(i,j) \cap \widetilde{\mathcal{Y}}(i);$$

this is a contradiction. □

The following lemma is one of the reasons we consider EOB. More importantly, we will make use of EOB to consistently deform Kuranishi maps into traversal orbifolds multisections.

LEMMA 3.7.8. *Let $\mathfrak{L}$ be an DGS as in Definition 3.3.2 and $\mathcal{N}$ be an EOB for $\mathfrak{L}$ as in Definition 3.7.4. Via $\mathcal{N}$, for every $i, j \in \mathfrak{B}$, with $i < j$, the normal direction derivative map $\mathrm{d}s_{j/i}$ in (3.9), originally defined along the zero locus, extends to a similarly denoted map*

$$\mathrm{d}s_{j/i} \colon \mathcal{N}_{\phi_{ij}(\mathcal{Y}(i,j))}\mathcal{Y}(j) \longrightarrow \mathcal{E}(j)/\mathfrak{D}\phi_{ij}\mathcal{E}(i,j). \tag{3.77}$$

PROOF. The argument is local. For simplicity we forget about the orbifold structure and consider manifolds and vector bundles instead. The argument readily extends to local charts with group actions. Thus, let $M_1 \subset M_2$ be a submanifold, $\pi \colon E_2 \longrightarrow M_2$ be a vector bundle, $\pi \colon E_1 \longrightarrow M_1$ be a sub-bundle of $E_2|_{M_1}$, and E_1' be an extension of E_1 to a sub-bundle of E_2 over a neighborhood M_{12} of $M_1 \subset M_2$. Let s_2 be a section of E_2 such that restricted to M_1, $s_1 = s_2|_{M_1}$, is a section of E_1. Fix some $x \in M_1$. Let $U_{12} \subset M_{12}$ be a sufficiently small chart around x such that

$$E_1'|_{U_{12}} \cong E_1|_x \times U_{12}, \quad E_2|_{U_{12}} \cong E_2|_x \times U_{12}. \tag{3.78}$$

Given a choice of such trivialization, it determines a normal direction derivative map

$$\mathrm{d}s_{2/1} \colon N_{M_1}M_2|_x \longrightarrow (E_2/E_1)|_x. \tag{3.79}$$

A different choice of trivialization in (3.78) corresponds to some

$$g \colon M_{12} \longrightarrow \mathrm{End}(E_2|_x, E_2|_x)$$

such that

$$g(y)(E_1|_x) = E_1|_x \quad \forall\, y \in M_{12}.$$

For the new choice of trivialization, the map (3.79) changes to

$$\mathrm{d}^{\mathrm{new}}s_{2/1}([v]) = [g(x)][\mathrm{d}s_{2/1}(v)] + [\mathrm{d}_x g(v) s_2(x)].$$

Since g preserves E_1' and $s_2(x) \in E_1$, the second term on the left-hand side is zero; therefore, $\mathrm{d}s_{2/1}$ is independent of the choice of trivialization. □

By definition, if $\mathfrak{L}$ has tangent bundle, $\mathrm{d}s_{j/i}$ restricted to the zero set of Kuranishi map is an isomorphism. Therefore, after possibly shrinking $(\mathfrak{L},\mathcal{N})$, we may assume that (3.77) is an isomorphism along entire $\phi_{ij}(\mathcal{Y}(i,j))$.

For $i,j\in\mathfrak{B}$, with $i<j$, let

$$f\colon \mathcal{N}'_{\phi_{ij}(\mathcal{Y}(i,j))}\mathcal{Y}(j)\longrightarrow \mathcal{N}(i,j)$$

be a regularization of $\mathcal{N}(i,j)$ as in Definition 3.7.2. Fix metrics g on $\mathcal{N}_{\phi_{ij}(\mathcal{Y}(i,j))}\mathcal{Y}(j)$ and h on $\mathcal{E}(j)$. If $\mathrm{d}s_{j/i}$ is an isomorphism on the entire $\phi_{ij}(\mathcal{Y}(i,j))$, by the second assumption in (3.71) and after restricting to a sufficiently small sub-neighborhood $\mathcal{N}''_{\phi_{ij}(\mathcal{Y}(i,j))}\mathcal{Y}(j)$, there exists a positive continuous function

$$c\colon \phi_{ij}(\mathcal{Y}(i,j))\longrightarrow \mathbb{R}^+$$

such that

$$|[s_j(x)]|\geq c|v|\quad \forall\ x=f(v)\in\mathcal{N}(i,j),\ v\in\mathcal{N}''_{\phi_{ij}(\mathcal{Y}(i,j))}\mathcal{Y}(j). \tag{3.80}$$

Here, $|v|$ is the norm of v with repeat to g, $[s_j(x)]$ is the class of $s_j(x)$ in $\mathcal{E}(j)/\mathcal{E}_{\mathrm{ext}}(i,j)$, and $|[s_j(x)]|$ is the norm of $[s_j(x)]$ with respect to the metric induced by h. Changing the regularization f or the metrics g and h changes the function c but the existence of a lower bound (3.80) is a property of the tubular neighborhood independent of such choices.

DEFINITION 3.7.9. *Let $\mathfrak{L}$ be an DGS as in Definition 3.3.2 with tangent bundle. A **perfect** EOB $\mathcal{N}$ for $\mathfrak{L}$ is an EOB such that for every $i,j\in\mathfrak{B}$, with $i<j$, the normal direction derivative map $\mathrm{d}s_{j/i}$ along entire $\phi_{ij}(\mathcal{Y}(i,j))$ is an isomorphism and (3.80) holds.*

REMARK 3.7.10. Via a perfect EOB, and after possibly some shrinking so that $s_{i,j}^{-1}(0)$ over each connected component of $\mathcal{Y}(i,j)$ is non-empty, we can extend the isomorphism (3.19) to an isomorphism over entire $\phi_{ij}(\mathcal{Y}(i,j))$.

THEOREM 3.7.11. *Let $\mathfrak{L}$ be an DGS as in Definition 3.3.2 with tangent bundle. There exists a shrinking of $\mathfrak{L}$ that admits a perfect EOB.*

We postpone the proof to Section 3.8. Given such an EOB, in the rest of this section, we introduce the notion of deformation of Kuranishi map. Note that further shrinkings of an DGS with a perfect EOB as in Lemma 3.7.7 remain perfect.

Similar to Section 2.3, in order to deform the Kuranishi maps into something transversal, we substitute the orbifold sections with transversal orbifold multisections. We first need to extend the notion of normal direction derivative map in (3.17) to the case of multisections. Since the argument is local, suppose

$$\mathcal{U}_i=(\mathrm{pr}_i\colon U_i\longrightarrow V_i,G_i,s_i,\psi_i),\quad i=1,2,$$

are two Kuranishi charts of dimension $n=\dim V_i-\operatorname{rank} U_i$ centered at $\overline{x}$, and

$$\Phi=(\mathfrak{D}\phi,\phi,h)\colon \mathcal{U}_1\longrightarrow\mathcal{U}_2$$

is a coordinate change map as in Definition 3.2.2. Since $\mathcal{U}_1$ and $\mathcal{U}_2$ are local charts centered at $\overline{x}$, we have

$$(G_1)_x\cong G_1\cong h(G_1)\cong G_2\cong (G_2)_x$$

and we simply denote these isomorphic groups by G. We also identify $\phi(V_1)$ and $\mathfrak{D}\phi(U_1)$ with their images and think of V_1 as an G-invariant sub-manifold of V_2.

Assume U_1^{ext} is an extension of U_1 to a G-invariant sub-bundle of U_2. Suppose $\mathfrak{s}_2=[s_2^1,\ldots,s_2^k]$ is a smooth lifted orbifold k-section of $\mathcal{U}_2$ such that

$$s_1^i = s_2^i|_{V_1} \in U_1 \quad \forall i=1,\ldots,k.$$

Similar to Lemma 3.7.8, the extension U_1^{ext} results in a set of well-defined normal direction derivative maps

$$\text{(3.81)} \qquad \mathrm{d}s^i_{2/1}\colon N_{V_1}V_2 \longrightarrow U_2/U_1 \quad \forall i=1,\ldots,k.$$

While every individual map in (3.81) may not be G-invariant, the set $\{\mathrm{d}s^i_{2/1}\}_{i\in[k]}$ is G-invariant. Putting together, we obtain a well-defined map

$$\text{(3.82)} \qquad \mathrm{d}\mathfrak{s}_{2/1}\colon S^k(N_{V_1}V_2) \longrightarrow S^k(U_2/U_1), \quad \mathfrak{s}_1=\mathfrak{s}_2|_{V_1},$$

which we call the **normal direction derivative map** for lifted multisections. Then $\mathrm{d}\mathfrak{s}_{2/1}$ is an isomorphism if and only if all branches in (3.81) are isomorphisms.

DEFINITION 3.7.12. *Let $\mathcal{N}$ be an EOB for an DGS $\mathfrak{L}$ as in Definition* 3.7.4 *and $m\in\mathbb{N}$. A set of* ***compatible sequences of deformations*** *of Kuranishi maps $\{s_i\}_{i\in\mathfrak{B}}$, is a set of sequences of transversal (thus lifted) orbifold multisections,*

$$\mathfrak{s}_{i;a} \in C^m_{\text{multi}}(\mathcal{E}(i)) \quad \forall i\in\mathfrak{B},\ a\in\mathbb{N},$$

satisfying the following conditions.

(1) *For every $i,j\in\mathfrak{B}$, with $i<j$, and every $a\in\mathbb{N}$,*

$$\mathfrak{D}\phi_{ij}(\mathfrak{s}_{i;a}|_{\mathcal{Y}(i,j)}) \cong \mathfrak{s}_{j;a}|_{\phi_{ij}(\mathcal{Y}(i,j))}$$

*as lifted multisections, and the normal derivative map $\mathrm{d}\mathfrak{s}_{j;a/i;a}$ (*3.82*) along $\phi_{ij}(\mathcal{Y}(i,j))$, defined via $\mathcal{N}$, is an isomorphism.*

(2) *For every $i,j\in\mathfrak{B}$, with $i<j$, and every $a\in\mathbb{N}$,*

$$\mathfrak{s}_{j;a}^{-1}(0)\cap\mathcal{N}(i,j) = \phi_{ij}(\mathfrak{s}_{i;a}^{-1}(0)\cap\mathcal{Y}(i,j)).$$

(3) *With the function* $\mathrm{val}_{\mathfrak{s}}$ *as in (*2.28*)*

$$\mathrm{val}_{\mathfrak{s}_{i;a}}(\overline{x}) = \mathrm{val}_{\mathfrak{s}_{j;a}}(\overline{\phi}_{ij}(\overline{x})) \quad \forall i,j\in\mathfrak{B}, i<j, a\in\mathbb{N}, \overline{x}\in Y(i,j).$$

(4) *For every $i\in\mathfrak{B}$, as $a\longrightarrow\infty$, $\mathfrak{s}_{i;a}$ C^m-converges to s_i.*

THEOREM 3.7.13. *Let $\mathfrak{L}$ be an DGS with tangent bundle as in Definition* 3.3.2 *and $m\in\mathbb{N}$ is sufficiently large. Then there exists a shrinking $\mathfrak{L}'$ of $\mathfrak{L}$, as in Lemma* 3.4.2*, and a prefect EOB $\mathcal{N}$ for $\mathfrak{L}'$ such that $(\mathfrak{L}',\mathcal{N})$ admits a compatible set of sequences of deformations as in Definition* 3.7.12.

PROOF. After possibly shrinking $\mathfrak{L}$, we equip $\mathfrak{L}$ with a perfect EOB $\mathcal{N}$ as in Theorem 3.7.11. Let i_1 be the smallest value in $\mathfrak{B}$. Let $(\mathfrak{L}^{(1)},\mathcal{N}^{(1)})$ be a shrinking of $(\mathfrak{L},\mathcal{N})$ as in Lemma 3.7.7; thus, $\mathrm{cl}(Y^{(1)}(i_1))\subset Y(i_1)$ is compact. Then, by Proposition 2.4.2, with $K_1=W_1=\emptyset$ and $K_2=\mathrm{cl}(Y^{(1)}(i_1))$ in $M=Y(i_1)$, there exists a sequence of lifted multisections

$$\left(\mathfrak{t}_{i_1;a}\in C^m_{\text{multi}}(\mathcal{E}^{(1)}(i_1))\right)_{a\in\mathbb{N}}$$

such that

$$\left(\mathfrak{s}_{i_1;a} = s_{i_1}|_{\mathcal{Y}^{(1)}(i_1)} + \mathfrak{t}_{i_1;a}\right)_{a\in\mathbb{N}}$$

is a sequence of transversal orbifold multisections C^m-converging to $s_{i_1}|_{\mathcal{Y}^{(1)}(i_1)}$.

Next, for every $j\in\mathfrak{B}(>i_1)$ and $a\in\mathbb{N}$, restricted to $\mathcal{N}^{(1)}(i_1,j)$, define

$$\begin{aligned}&\mathfrak{t}_{j,i_1;a}=\iota_{i_1j}(\pi^*_{ji_1}(\mathfrak{D}\phi_{i_1j}(\mathfrak{t}_{i_1;a}))),\quad \mathfrak{s}_{j,i_1;a}=s_j|_{\mathcal{N}^{(1)}(i_1,j)}+\mathfrak{t}_{j,i_1;a},\\ &\mathfrak{t}_{j,i_1;a},\mathfrak{s}_{j,i_1;a}\in C^m_{\text{multi}}(\mathcal{E}^{(1)}(j)|_{\mathcal{N}^{(1)}(i_1,j)}).\end{aligned}\tag{3.83}$$

By the second condition in (3.71), restricted to $\phi_{i_1j}(\mathcal{Y}^{(1)}(i_1,j))$ we have

$$\mathfrak{s}_{j,i_1;a}=\mathfrak{D}\phi_{i_1j}\mathfrak{s}_{i_1;a}\quad \forall j\in\mathfrak{B}(>i_1).$$

Moreover, the pull-back of every lifted multisection is lifted; therefore, $\mathfrak{t}_{j,i_1;a}$ and $\mathfrak{s}_{j,i_1;a}$ are lifted as well. It is also clear from (3.83) that

$$\text{val}_{\mathfrak{s}_{i_1;a}}(\overline{x})=\text{val}_{\mathfrak{s}_{j,i_1;a}}(\overline{\phi}_{i_1j}(\overline{x}))\quad \forall j\in\mathfrak{B}(>i_1),a\in\mathbb{N},\overline{x}\in Y^{(1)}(i_1,j).$$

Claim 1. *For every $j\in\mathfrak{B}(>i_1)$, $(\mathfrak{s}_{j,i_1;a})_{a\in\mathbb{N}}$ is a of sequence transversal multisections over $\mathcal{N}^{(1)}(i_1,j)$ converging in C^m-norm to $s_j|_{\mathcal{N}^{(1)}(i_1,j)}$. Moreover,*

$$\mathfrak{s}^{-1}_{j,i_1;a}(0)=\phi_{i_1j}(\mathfrak{s}^{-1}_{i_1;a}(0)\cap\mathcal{Y}^{(1)}(i_1,j)).\tag{3.84}$$

Proof. By (3.83), the $\mathcal{E}^{(1)}_{\text{ext}}(i_1,j)^{\perp}$ component of $\mathfrak{s}_{j,i_1;a}$ is equal to the $\mathcal{E}^{(1)}_{\text{ext}}(i_1,j)^{\perp}$ component of s_j; therefore, (3.84) follows from (3.80).

Consider an arbitrary point $x\in\mathfrak{s}^{-1}_{j,i_1;a}(0)$. The $\mathcal{E}^{(1)}_{\text{ext}}(i_1,j)^{\perp}$ component

$$\mathrm{d}^{\perp}\mathfrak{s}_{j,i_1;a/i_1;a}(x)$$

of $\mathrm{d}\mathfrak{s}_{j,i_1;a/i_1;a}(x)$ is equal to $\mathrm{d}s_{j/i_1}(x)$, which surjects onto $\mathcal{E}^{(1)}_{\text{ext}}(i_1,j)^{\perp}_x$ by the tangent bundle condition. The $\mathcal{E}^{(1)}_{\text{ext}}(i_1,j)$ component

$$\mathrm{d}^{i_1}\mathfrak{s}_{j,i_1;a/i_1;a}(x)$$

of $\mathrm{d}\mathfrak{s}_{j,i_1;a/i_1;a}(x)$ is equal to $\mathrm{d}(s_{i_1}+\mathfrak{t}_{i_1;a})(x)$, which surjects onto $\mathcal{E}^{(1)}_{\text{ext}}(i_1,j)_x$, because $(s_{i_1}+\mathfrak{t}_{i_1;a})_{a\in\mathbb{N}}$ is a sequence of transversal orbifold multisections for $\mathcal{E}^{(1)}(i_1)$. $\square$

We proceed by induction. For $k\geq 1$, let $\mathfrak{B}_k$ be the set of first k values in $\mathfrak{B}$. Suppose $(\mathfrak{L}^{(k)},\mathcal{N}^{(k)})$ is a shrinking of $(\mathfrak{L},\mathcal{N})$, as in Lemma 3.7.7, admitting a compatible set of sequences of deformations

$$\mathfrak{s}_{i;a}=s_i|_{\mathcal{Y}^{(k)}}+\mathfrak{t}_{i;a}\in C^m_{\text{multi}}(\mathcal{E}^{(k)}(i))\quad \forall i\in\mathfrak{B}_k,\ a\in\mathbb{N},$$

for a family of lifted multisections

$$\mathfrak{t}_{i;a}\in C^m_{\text{multi}}(\mathcal{E}^{(k)}(i))\quad \forall i\in\mathfrak{B}_k,\ a\in\mathbb{N},$$

such that for every $i,i'\in\mathfrak{B}_k$, with $i<i'$ and $\mathfrak{t}_{i',i;a}$ defined over $\mathcal{N}^{(k)}(i,i')$ as in (3.83),

$$\mathfrak{t}_{i';a}|_{\mathcal{N}^{(k)}(i,i')}=\mathfrak{t}_{i',i;a}\quad \forall a\in\mathbb{N}.$$

Let i_{k+1} be the smallest number in $\mathfrak{B}\setminus\mathfrak{B}_k$ and $(\mathfrak{L}^{(k+1)},\mathcal{N}^{(k+1)})$ be a shrinking of $(\mathfrak{L}^{(k)},\mathcal{N}^{(k)})$ as in Lemma 3.7.7. Set

$$K_1^{\circ}=\bigcup_{i\in\mathfrak{B}_k}N^{(k+1)}(i,i_{k+1})\subset Y^{(k)}(i_{k+1}),\quad W_1=\bigcup_{i\in\mathfrak{B}_k}N^{(k)}(i,i_{k+1})\subset Y^{(k)}(i_{k+1}),$$

and $K_1=\text{cl}(K_1^{\circ})\subset W_1$. For every $a\in\mathbb{N}$, by Definition 3.7.4.(3), the multisection

$$\mathfrak{t}'_{i_{k+1};a}\in C^m_{\text{multi}}(\mathcal{E}^{(k)}(i_{k+1})|_{W_1}),\quad \mathfrak{t}'_{i_{k+1};a}|_{\mathcal{N}^{(k)}(i,i_{k+1})}=\mathfrak{t}_{i_{k+1},i;a}\quad \forall i\in\mathfrak{B}_k,$$

is well-defined and lifted. By an argument similar to Claim 1 above and the paragraph before that, for every $a\in\mathbb{N}$, the lifted multisection

$$\mathfrak{s}'_{i_{k+1};a}\in C^m_{\text{multi}}(\mathcal{E}^{(\mathrm{k})}(i_{k+1})|_{W_1}),\quad \mathfrak{s}'_{i_{k+1};a}=s_{i_{k+1}}+\mathfrak{t}'_{i_{k+1};a},$$

is transverse to the zero-section,

$$(\mathfrak{s}'_{i_{k+1};a})^{-1}(0)\cap\mathcal{N}^{(\mathrm{k})}(i,i_{k+1})=\phi_{ii_{k+1}}(\mathfrak{s}^{-1}_{i;a}(0)\cap\mathcal{Y}^{(\mathrm{k})}(i,i_{k+1}))\quad\forall i\in\mathfrak{B}_k,$$

and

$$\mathrm{val}_{\mathfrak{s}_{i;a}}(\overline{x})=\mathrm{val}_{\mathfrak{s}'_{i_{k+1},i;a}}(\overline{\phi}_{ii_{k+1}}(\overline{x}))\quad\forall i\in\mathfrak{B}_k,a\in\mathbb{N},\overline{x}\in Y^{(\mathrm{k})}(i,i_{k+1}).$$

Therefore, by Proposition 2.4.2, with K_1 and W_1 above and

$$K_2=\mathrm{cl}(Y^{(\mathrm{k})}(i_{k+1}))\subset M=Y(i_{k+1}),$$

there exists a sequence of lifted multisections

$$\big(\mathfrak{t}_{i_{k+1};a}\in C^m_{\text{multi}}(\mathcal{E}^{(\mathrm{k})}(i_{k+1}))\big)_{a\in\mathbb{N}}$$

such that

$$(\mathfrak{s}_{i_{k+1};a}=s_{i_{k+1}}|_{\mathcal{Y}^{(\mathrm{k})}(i_{k+1})}+\mathfrak{t}_{i_{k+1};a})_{a\in\mathbb{N}}$$

is a sequence of transversal orbifold multisections C^m-converging to $s_{i_{k+1}}|_{\mathcal{Y}^{(\mathrm{k})}(i_{k+1})}$ and

$$\mathfrak{t}_{i_{k+1};a}|_{\mathcal{N}^{(\mathrm{k}+1)}(i,i_{k+1})}=\mathfrak{t}'_{i_{k+1};a}|_{\mathcal{N}^{(\mathrm{k}+1)}(i,i_{k+1})}=\mathfrak{t}_{i_{k+1},i;a}\quad\forall a\in\mathbb{N},\ i\in\mathfrak{B}_k.$$

After the restriction to $(\mathfrak{L}^{(\mathrm{k}+1)},\mathcal{N}^{(\mathrm{k}+1)})$, this establishes the induction step and thus completes the proof of Theorem 3.7.13. □

3.8. Construction of perfect EOB (proof of Theorem 3.7.11). Let $\mathfrak{L}$ be an DGS with tangent bundle as in Definition 3.3.2. For $\ell=1$, an EOB of width 1 is simply a set of tubular neighborhoods $\mathcal{N}(i,i+1)$ and extensions of obstruction bundles $\mathcal{E}(i,i+1)$, for every $i\in\mathfrak{B}$ with $i+1\in\mathfrak{B}$, without any compatibility condition. Construction of such extensions, via exponential map and parallel extension of sub-orbibundle to a neighborhood can be done in the following way.

We equip every $T\mathcal{Y}(i)$ with an orbifold Riemannian metric g_i. Let $\nabla^{\mathcal{Y}(i)}$ the Levi-Civita connection of g_i. Also, we equip every obstruction bundle $\mathcal{E}(i)$ with a metric h_i and a compatible orbibundle connection $\nabla^{\mathcal{E}(i)}$. For every $i,j\in\mathfrak{B}$, with $j=i+1$, identify $\mathcal{N}_{\phi_i(\mathcal{Y}(i,j))}\mathcal{Y}(j)$ with the orthogonal complement $T\mathcal{Y}(i,j)^\perp$ of the sub-orbibundle

$$\mathrm{d}\phi_{ij}(T\mathcal{Y}(i,j))\subset T\mathcal{Y}(j)|_{\phi_{ij}(\mathcal{Y}(i,j))}.$$

Then, via the exponential map of g_j, restricted to $T\mathcal{Y}(i,j)^\perp$, we obtain an isomorphism

$$\exp^{ij}\colon\mathcal{N}^{\delta_{ij}}_{\phi_{ij}(\mathcal{Y}(i,j))}\mathcal{Y}(j)\longrightarrow\mathcal{N}(i,j)\tag{3.85}$$

from a tubular neighborhood of the zero section in $\mathcal{N}_{\phi_{ij}(\mathcal{Y}(i,j))}\mathcal{Y}(j)$ to a neighborhood of $\phi_{ij}(\mathcal{Y}(i,j))$ in $\mathcal{Y}(j)$. In (3.85), $\delta_{ij}\colon Y(i,j)\longrightarrow\mathbb{R}^+$ is a continuous positive function and $\mathcal{N}^{\delta_{ij}}_{\phi_{ij}(\mathcal{Y}(i,j))}\mathcal{Y}(j)$ is the set of vectors of the length less than δ_{ij} with respect to the metric induced by g_j. Let

$$\mathcal{N}^{\delta_{ij}}_{\phi_{ij}(\mathcal{Y}(i,j))}\mathcal{Y}(j)\xrightarrow{\pi_{ji}}\phi_{ij}(\mathcal{Y}(i,j)),$$

be the projection map. Via the orbifold diffeomorphism $\exp^{ij}$, π_{ji} gives rise to a similarly denoted projection map

$$\pi_{ji}\colon \mathcal{N}(i,j) \longrightarrow \phi_{ij}(\mathcal{Y}(i,j)).$$

Then the orbibundle embedding

$$\iota_{ij}\colon \pi_{ji}^*(\mathfrak{D}\phi_{ij}\mathcal{E}(i,j)) \xrightarrow{\cong} \mathcal{E}_{\rm ext}(i,j) \subset \mathcal{E}(j)|_{\mathcal{N}(i,j)}$$

at

$$\pi_{ji}^*(\mathfrak{D}\phi_{ij}\mathcal{E}(i,j))|_y, \quad y=\exp_x^{ij}(v),\ x\in\phi_{ij}(\mathcal{Y}(i,j)),\ v\in\mathcal{N}^{\delta_{ij}}_{\phi_{ij}(\mathcal{Y}(i,j))}\mathcal{Y}(j)|_x,$$

is given by parallel-wise extension of $\mathfrak{D}\phi_{ij}\mathcal{E}(i,j)|_x$, with respect to $\nabla^{\mathcal{E}(j)}$, along the geodesic ray $\{\exp_x^{ij}(tv)\}_{t\in[0,1]}$.

Let $\mathcal{N}$ be an EOB of width $\ell-1$ for $\mathfrak{L}$ for some $\ell>1$, $\widetilde{\mathfrak{L}}$ be a shrinking of $\mathfrak{L}$ as in Lemma 3.4.2, and $\widetilde{\mathcal{N}}$ be a the shrinking of $\mathcal{N}$ associated to $\widetilde{\mathfrak{L}}$ as in (3.74). Let $i,k\in\mathfrak{B}$, with $k-i=\ell$. For every $j\in\mathfrak{B}$, with $i<j<k$, the tubular neighborhoods $\mathcal{N}(i,j)$ and $\mathcal{N}(j,k)$ and the extension of orbibundles $\mathcal{E}^{\rm ext}(i,j)$ and $\mathcal{E}^{\rm ext}(j,k)$ are given by assumption. Choose

$$\mathcal{W}_{ik;j} \subset \mathcal{N}(i,j)|_{\phi_{ij}(\mathcal{Y}(i,j,k))}$$

to be sub-tubular neighborhood of

$$\phi_{ij}(\mathcal{Y}(i,j,k)) \subset \mathcal{N}(i,j)\cap\mathcal{Y}(j,k)$$

and set

$$\mathcal{N}(i,k)_j = \pi_{kj}^{-1}\big(\phi_{jk}(\mathcal{W}_{ik;j})\big) \subset \mathcal{N}(j,k).$$

Let

$$\pi_{ki;j}\colon \mathcal{N}(i,k)_j \longrightarrow \phi_{ik}(\mathcal{Y}(i,j,k)), \quad \pi_{ki;j} = \phi_{jk}\circ\pi_{ji}\circ\phi_{jk}^{-1}\circ\pi_{kj}.$$

Then, for $j,j'\in\mathfrak{B}$, with $i<j<j'<k$, by (3.73) applied to j,j',k, we get

$$\pi_{ki;j}(x) = \pi_{ki;j'}(x) \quad \forall x\in\mathcal{N}(i,k)_j\cap\mathcal{N}(i,k)_{j'}. \tag{3.86}$$

By (3.86), the projection maps $\{\pi_{ki;j}\}_{i<j<k}$ give rise to a projection map

$$\pi'_{ki}\colon \mathcal{N}'(i,k) = \bigcup_{\substack{j\in\mathfrak{B}\\ i<j<k}} \mathcal{N}(i,k)_j \longrightarrow \phi_{ik}(\mathcal{Y}'(i,k)), \quad \mathcal{Y}'(i,k) = \bigcup_{\substack{j\in\mathfrak{B}\\ i<j<k}} \mathcal{Y}(i,j,k).$$

Similarly, by Definition 3.7.4.(3), with

$$\mathcal{E}(i,k)_j = \pi_{kj}^*\mathfrak{D}\phi_{jk}(\iota_{ij}(\pi_{ji}^*(\mathfrak{D}\phi_{ij}(\mathcal{E}(i,j)|_{\mathcal{Y}(i,j,k)})))) \subset \pi_{kj}^*(\mathfrak{D}\phi_{jk}(\mathcal{E}(j,k)))|_{\mathcal{N}(i,k)_j},$$

the orbibundle embedding

$$\iota'_{ik}\colon (\pi'_{ki})^*\mathfrak{D}\phi_{ik}(\mathcal{E}(i,k)|_{\mathcal{Y}'(i,k)}) \longrightarrow \mathcal{E}(k)|_{\mathcal{N}'(i,k)}, \quad \iota'_{ij}|_{\mathcal{E}(i,k)_j} = \iota_{jk}|_{\mathcal{E}(i,k)_j},$$

is well-defined.

LEMMA 3.8.1. *There exists a tubular neighborhood*

$$\pi_{ki}\colon \mathcal{N}(i,k) \longrightarrow \phi_{ik}(\mathcal{Y}(i,k))$$

with an orbibundle embedding

$$\iota_{ik}\colon \pi_{ki}^*(\mathfrak{D}\phi_{ik}\mathcal{E}(i,k)) \longrightarrow \mathcal{E}(k), \quad \iota_{ik}|_{\phi_{ik}(\mathcal{Y}(i,k))} = \mathrm{id},$$

such that $\big(\pi_{ki}, \mathcal{N}(i,k)|_{\phi_{ik}(\widetilde{\mathcal{Y}}(i,j,k))}\big)$ *is a sub-tubular neighborhood of*

$$\big(\pi'_{ki}, \mathcal{N}'(i,k)|_{\phi_{ik}(\widetilde{\mathcal{Y}}(i,j,k))}\big),$$

for all $i<j<k$, *and*

$$(\iota_{ik} = \iota'_{ik})|_{\pi^*_{ki}(\mathfrak{D}\phi_{ik}\widetilde{\mathcal{E}}(i,j,k))}.$$

PROOF. For every $j \in \mathfrak{B}$, with $i < j < k$, $\phi_{ik}(\widetilde{\mathcal{Y}}(i,j,k)) \subset \mathcal{Y}(k)$ is relatively compact. With

$$\widetilde{\mathcal{M}}_1 = \bigcup_{\substack{j\in\mathfrak{B}\\ i<j<k}} \phi_{ik}(\widetilde{\mathcal{Y}}(i,j,k)), \quad \mathcal{M}'_1 = \phi_{ik}(\mathcal{Y}'(i,k)), \quad \mathcal{W}' = \mathcal{N}'(i,k), \quad \mathcal{M}_1 = \phi_{ik}(\mathcal{Y}(i,k)),$$

and $\mathcal{M}_2 = \mathcal{Y}(k)$, the existence of $\mathcal{N}(i,k)$ and ι_{ik} follows from Proposition 3.8.2 below. □

After replacing $\mathcal{N}(i,k)$ with $\mathcal{N}(i,k) \cap \widetilde{\mathcal{Y}}(k)$ and restricting everything to $\widetilde{\mathfrak{L}}$ and $\widetilde{\mathcal{N}}$, we obtain a set of tubular neighborhoods $\widetilde{\mathcal{N}}(i,j)$ for orbifold embeddings $\phi_{ij}(\widetilde{\mathcal{Y}}(i,j)) \subset \widetilde{\mathcal{Y}}(j)$ and orbibundle embeddings ι_{ij} as in (3.71), for all $i,j \in \mathfrak{B}$ with $j-i \leq \ell$. For every $i,j,k \in \mathfrak{B}$, with $i<j<k$, Definition 3.7.4.(1)-(3) hold by induction, if $k-i \leq \ell - 1$, and they hold by Lemma 3.8.1, if $k-i=\ell$. This finishes the inductive step and thus the proof of Theorem 3.7.11. □

PROPOSITION 3.8.2. *Let* $\mathcal{M}_1 \subset \mathcal{M}_2$ *be an orbifold embedding. For* $i=1,2$, *let* $\mathrm{pr}_i\colon \mathcal{E}_i \to \mathcal{M}_i$ *be orbibundles such that* $\mathcal{E}_1$ *is a sub-orbibundle of* $\mathcal{E}_2|_{\mathcal{M}_1}$. *Let* $\mathcal{M}'_1$ *be an open sub-orbifold of* $\mathcal{M}_1$ *and* $\widetilde{\mathcal{M}}_1$ *be a relatively compact open sub-orbifold of* $\mathcal{M}'_1$. *Let* $\pi'\colon \mathcal{W}' \longrightarrow \mathcal{M}'_1$ *be a tubular neighborhood of the orbifold embedding* $\mathcal{M}'_1 \subset \mathcal{M}_2$ *and*

$$\iota'\colon (\pi')^*(\mathcal{E}_1|_{\mathcal{M}'_1}) \longrightarrow \mathcal{E}_2|_{\mathcal{W}'}, \quad \iota'|_{\mathcal{M}'_1} = \mathrm{id},$$

be an embedding of the orbibundles. Then there exists a tubular neighborhood $\pi\colon \mathcal{W} \longrightarrow \mathcal{M}_1$ *and an orbibundle embedding*

$$\iota\colon \pi^*\mathcal{E}_1 \longrightarrow \mathcal{E}_2|_{\mathcal{W}}, \quad \iota|_{\mathcal{M}_1} = \mathrm{id},$$

such that $(\pi|_{\widetilde{\mathcal{M}}_1}, \mathcal{W}|_{\widetilde{\mathcal{M}}_1})$ *is a sub-tubular neighborhood of* $(\pi'|_{\widetilde{\mathcal{M}}_1}, \mathcal{W}'|_{\widetilde{\mathcal{M}}_1})$ *and*

$$(\iota = \iota')|_{\pi^*(\mathcal{E}_1|_{\widetilde{\mathcal{M}}_1})}.$$

PROOF. . Let

$$\pi_0\colon \mathcal{W}_0 \longrightarrow \mathcal{M}_1$$

be a sufficiently small tubular neighborhood of $\mathcal{M}_1 \subset \mathcal{M}_2$ and

$$\iota_0\colon \pi_0^*\mathcal{E}_1 \longrightarrow \mathcal{E}_2|_{\mathcal{W}_0}, \quad \iota_0|_{\mathcal{M}_1} = \mathrm{id},$$

be an embedding of orbibundles such that

$$\mathcal{W}_0|_{\mathrm{cl}(\widetilde{\mathcal{M}}_1)} \subset \mathcal{W}'. \tag{3.87}$$

Such tubular neighborhood and orbibundle embedding exist by the exponential map construction of the beginning of the section. The inclusions of open orbifolds in (3.87) is not necessarily compatible with π_0 and π'.

LEMMA 3.8.3. *There exists a sufficiently small sub-neighborhood* $\mathcal{W} \subset \mathcal{W}_0$ *of* $\mathcal{M}_1$ *in* $\mathcal{M}_2$ *with an orbifold embedding*

$$\mathcal{F}\colon \mathcal{W} \longrightarrow \mathcal{W}_0, \quad \mathrm{d}\mathcal{F}|_{T\mathcal{W}|_{\mathcal{M}_1}} = \mathrm{id},$$

such that

$$(\pi_0 \circ \mathcal{F} = \pi')|_{\pi'^{-1}(\widetilde{\mathcal{M}}_1)\cap\mathcal{W}}. \tag{3.88}$$

For such $\mathcal{W}$ and $\mathcal{F}$, let

$$\pi = \pi_0 \circ \mathcal{F} \colon \mathcal{W} \longrightarrow \mathcal{M}_1, \quad \iota = \mathcal{F}^* \iota_0 \colon \pi^* \mathcal{E}_1 \longrightarrow \mathcal{E}_2|_{\mathcal{W}}. \tag{3.89}$$

The projection map and orbibundle embedding (3.89) have the required properties of Proposition 3.8.2. □

We finish this section by proving Lemma 3.8.3 which follows from an orbifold version of the isotropy extension theorem.

PROOF OF LEMMA 3.8.3. Let

$$f' \colon \mathcal{N}'_{\mathcal{M}'_1}\mathcal{M}_2 \longrightarrow \mathcal{W}' \quad \text{and} \quad f_0 \colon \mathcal{N}^0_{\mathcal{M}_1}\mathcal{M}_2 \longrightarrow \mathcal{W}_0$$

be arbitrary regularizations of $\mathcal{W}'$ and $\mathcal{W}_0$ as in Definition 3.7.2, respectively. Let

$$\widehat{\mathcal{W}} \subset \mathcal{W}'_0 = (\mathcal{W}' \cap \mathcal{W}_0)$$

be a sufficiently small relatively compact open sub-orbifold such that $\mathrm{cl}(\widetilde{\mathcal{M}}_1) \subset \widehat{\mathcal{W}}$ and

$$\widehat{\mathcal{F}} = f' \circ f_0^{-1} \colon \mathrm{cl}(\widehat{\mathcal{W}}) \longrightarrow \mathcal{W}'_0 \tag{3.90}$$

is defined.

Claim. *For sufficiently small $\widehat{\mathcal{W}}$, there exist a diffeotopy (i.e. a one-parameter family of orbifold diffeomorphisms)*

$$\mathcal{F}_t \colon \mathcal{W}'_0 \longrightarrow \mathcal{W}'_0, \quad \mathrm{d}\mathcal{F}_t|_{T\mathcal{M}_2|_{\mathcal{W}'_0 \cap \mathcal{M}_1}} = \mathrm{id}, \quad \forall t \in [0,1], \tag{3.91}$$

supported in $\mathcal{W}'_0$ such that

$$\mathcal{F}_0 = \mathrm{id} \quad \text{and} \quad \mathcal{F}_1|_{\widetilde{\mathcal{W}}} = \widehat{\mathcal{F}}|_{\widetilde{\mathcal{W}}}, \tag{3.92}$$

where $\widetilde{\mathcal{W}} \subset \widehat{\mathcal{W}}$ is some smaller open neighborhood of $\mathrm{cl}(\widetilde{\mathcal{M}})$.

PROOF. In the case of manifolds, this follows from the isotopy extension theorem [**26**, Theorem 1.4, Ch 8]. More precisely, by definition of regularization and (3.90) we have

$$\mathrm{d}\widehat{\mathcal{F}}|_{T\mathcal{M}_2|_{\mathrm{cl}(\widehat{\mathcal{W}} \cap \mathcal{M}_1)}} = \mathrm{id};$$

in particular, $\widehat{\mathcal{F}}|_{\mathrm{cl}(\widehat{\mathcal{W}} \cap \mathcal{M}_1)} = \mathrm{id}$. Therefore, given a Riemannian metric g on $\mathcal{M}_2$, for sufficiently small $\widehat{\mathcal{W}}$, there is a unique vector field

$$\xi \in \Gamma(\mathcal{M}_2|_{\mathrm{cl}(\widehat{\mathcal{W}})}, T\mathcal{M}_2|_{\mathrm{cl}(\widehat{\mathcal{W}})}), \quad \xi|_{\mathrm{cl}(\widehat{\mathcal{W}} \cap \mathcal{M}_1)} \equiv 0, \quad \nabla\xi|_{\mathrm{cl}(\widehat{\mathcal{W}} \cap \mathcal{M}_1)} \equiv 0, \tag{3.93}$$

such that

$$\widehat{\mathcal{F}}(x) = \exp_x(\xi(x)) \quad \text{and} \quad \mathcal{F}_t(x) = \exp_x(t\xi(x)) \in \mathcal{W}'_0 \quad \forall t \in [0,1], \ x \in \mathrm{cl}(\widehat{\mathcal{W}}). \tag{3.94}$$

For $\widehat{\mathcal{W}}$ sufficiently small, by the last two conditions of (3.93) and the implicit function theorem, $(\mathcal{F}_t)_{t\in[0,1]}$ is one-parameter family of embeddings of $\widehat{\mathcal{W}}$ in $\mathcal{W}'_0$. Then by [**26**, Theorem 1.4, Ch 8], we can extend $(\mathcal{F}_t)_{t\in[0,1]}$ to a diffeotopy supported in $\mathcal{W}'_0$ such that (3.91) over some smaller open neighborhood $\widetilde{\mathcal{W}} \subset \widehat{\mathcal{W}}$ of $\mathrm{cl}(\widetilde{\mathcal{M}})$ holds. The proof is by cutting off the time dependent vector field of the isotopy (3.94) away from a neighborhood of $\mathrm{cl}(\widetilde{\mathcal{M}})$ to obtain a time dependent vector field on entire $\mathcal{W}'_0$ (such that the last two conditions of (3.93) remain valid). In the orbifold case, the same argument applies. We start with an orbifold Riemannian metric and define (3.94) similarly. We perform the same operation on the resulting time dependent vector field and consider the orbifold ODE flow of that to get (3.92). □

Going back to the proof of Lemma 3.8.3, choose a sufficiently small sub-neighborhood $\mathcal{W}\subset\mathcal{W}_0$ such that

$$\mathcal{W}|_{\widetilde{\mathcal{M}}_1}\subset\widetilde{\mathcal{W}}.$$

Define

$$\mathcal{F}\colon \mathcal{W}\longrightarrow\mathcal{W}_0,\quad \mathcal{F}(x)=\mathcal{F}_1(x)\quad \forall x\in\mathcal{W}\cap\mathcal{W}_0',\quad \mathcal{F}(x)=x\quad \forall x\in\mathcal{W}\setminus\mathcal{W}_0'. \tag{3.95}$$

Since $\mathcal{F}_1$ is supported in $\mathcal{W}_0'$, (3.95) is a well-defined orbifold embedding. Finally, it follows from (3.92) and (3.90) that (3.88) holds. This finishes the proof of Lemma 3.8.3.

3.9. Kuranishi vector bundles. A Kuranishi structure, as we defined in Section 3.2, is locally given by a pair of an orbibundle and a section, such that the difference of the dimension of the base and the rank of orbibundle is some fixed constant. In this section, we define the notation of Kuranishi vector bundle. It consists of an additional orbibundle of some fixed rank (unlike obstruction bundle) on the base orbifold of every Kuranishi chart. As we show below, we can incorporate this extra bundle into the obstruction bundle and obtain a new augmented Kuranishi structure with a bigger obstruction bundle for each chart. Then, by the results of Section 4, we define the Euler class of such Kuranishi vector bundle to be the VFC of corresponding augmented Kuranishi structure. In Section 6.7, we use the notion of Kuranishi vector bundle to define GW invariants involving ψ-classes.

DEFINITION 3.9.1. *For a Kuranishi structure $\mathcal{K}$ of real dimension n on a compact metrizable topological space M as in Definition* 3.2.6, *with notation as in Definition* 3.2.1, *a* ***Kuranishi vector bundle*** $\mathfrak{E}$ *of rank m on $\mathcal{K}$ consists of an additional rank m G_p-equivariant vector bundle $\mathrm{pr}_p'\colon U_p'\longrightarrow V_p$, for every Kuranishi chart $\mathcal{U}_p=(\mathrm{pr}_p\colon U_p\longrightarrow V_p, G_p, s_p, \psi_p)$, such the following conditions hold.*

(1) *For every $\overline{q}\in F_p$, the coordinate change map ϕ_{qp} lifts to an equivariant vector bundle isomorphism*

$$\mathfrak{D}\phi_{qp}'\colon U_{q,p}'\equiv U_q'|_{V_{q,p}}\longrightarrow U_p'|_{\phi_{qp}(V_{q,p})}. \tag{3.96}$$

(2) *Whenever $\overline{r}\in F_p\cap F_q$ and $\overline{q}\in F_p$, then $\mathfrak{D}\phi_{qp}'$, $\mathfrak{D}\phi_{qp}'$, and $\mathfrak{D}\phi_{rp}'$ satisfy the cocycle condition of Definition* 3.2.4.

Similarly to Section 3.6, the local defining bundles of a Kuranishi vector bundle $\mathfrak{E}$ can be glued into a **dimensionally graded bundle** $\mathfrak{E}_{\mathfrak{L}}$, or **DGB** for short, over different orbifolds $\mathcal{Y}(i)$, with $i\in\mathfrak{B}$, of an associated natural DGS $\mathfrak{L}$. In other words, $\mathfrak{E}_{\mathfrak{L}}$ consist of an additional rank m orbibundle $\mathrm{pr}_i'\colon \mathfrak{E}(i)\longrightarrow\mathcal{Y}(i)$, for every orbifold $\mathcal{Y}(i)$ in $\mathfrak{L}$, such that compatibility conditions similar to Conditions (3), (6), and (7) of Definition 3.3.2 hold.

Given an n-dimensional Kuranishi structure $\mathcal{K}$ on M, and a rank m Kuranishi vector bundle $\mathfrak{E}$ on $\mathcal{K}$, we build a "new" Kuranishi structure $\mathcal{K}^{\mathfrak{E}}$ of dimension $n-m$ on M in the following way. For every $\overline{p}\in M$, we define the Kuranishi chart $\mathcal{U}_p^{\mathfrak{E}}$ of $\mathcal{K}^{\mathfrak{E}}$ centered at $\overline{p}$ to be

$$\mathcal{U}_p^{\mathfrak{E}}=\{\mathrm{pr}_p^{\mathfrak{E}}\equiv\mathrm{pr}_p\oplus\mathrm{pr}_p'\colon U_p^{\mathfrak{E}}\equiv U_p\oplus U_p'\longrightarrow V_p, s_p^{\mathfrak{E}}\equiv s_p\oplus 0,\psi_p\}. \tag{3.97}$$

By definition, $(s_p^{\mathfrak{E}})^{-1}(0)=s_p^{-1}(0)$; therefore, ψ_p still makes sense and has the same footprint. We define the change of coordinate maps similarly. Moreover, by definition, for the resulting coordinate change maps $\Phi^{\mathfrak{E}}_{qp}\colon \mathcal{U}^{\mathfrak{E}}_{q,p} \longrightarrow \mathcal{U}^{\mathfrak{E}}_{p}$, since

$$(U^{\mathfrak{E}}_p/\mathfrak{D}\phi^{\mathfrak{E}}_{qp}(U^{\mathfrak{E}}_{q,p}))|_{\phi^{\mathfrak{E}}_{qp}((s^{\mathfrak{E}})^{-1}_{q,p}(0))} = (U_p/\mathfrak{D}\phi_{qp}(U_{q,p}))|_{\phi_{qp}(s^{-1}_{q,p}(0))},$$

we conclude that if $\mathcal{K}$ has tangent bundle, then $\mathcal{K}^{\mathfrak{E}}$ has tangent bundle as well. Similarly, if $\mathcal{K}$ and $\mathfrak{E}$ are oriented, $\mathcal{K}^{\mathfrak{E}}$ is oriented as well. It is clear from (3.97) and the construction of Section 3.6 that every Kuranishi vector bundle gives rise to an DGB over every associated natural DGS.

We then define the Euler class of $\mathfrak{E}$ to be the VFC of the augmented Kuranishi space $(M, \mathcal{K}^{\mathfrak{E}})$, as defined in Section 4. In Section 6.7, we use vector bundles over Kuranishi structures and their Euler class to define GW invariants involving ψ-classes and other similarly defined insertions.

4. VFC for abstract Kuranishi spaces

The notion of a virtual fundamental class, or simply a VFC, is a generalization of the fundamental class of topological manifolds to more general compact topological spaces with singularities and with components of various dimensions.

For a suitable kind of topological space M, with some extra structure $\mathcal{M}$, a VFC is ideally a homology class

$$[\mathcal{M}]^{\mathrm{vir}} \in H_{\dim_{\mathrm{vir}}(\mathcal{M})}(M),$$

where $\dim_{\mathrm{vir}}(\mathcal{M})$ is the "expected dimension" of $\mathcal{M}$. Depending on the application, the homology theory used above can be the de Rham, singular, or the Čech homology/cohomology. In the case of extracting invariants from moduli spaces, evaluating certain natural cocycles against the VFC class should be thought of as the integration over that moduli space as in (1.8).

For an oriented Kuranishi space $(M, \mathcal{K})$ as in Definition 3.2.9 and 3.2.11, we first build $[\mathcal{K}]^{\mathrm{vir}}$ as a singular homology class inside a thickening $Y(\mathfrak{L})$ as defined in (3.22), where $\mathfrak{L}$ is a dimensionally graded system for $\mathcal{K}$ as in Definition 3.3.2. Therefore, a priori $[\mathcal{K}]^{\mathrm{vir}}$ is not supported in M and depends on the auxiliary data of the thickening. After this step, which is common between various constructions in the literature, there are different methods for getting an VFC independent of the choice of the thickening (and other auxiliary data). We explain two different methods: one by looking at the image cycle under certain natural evaluation maps as in (1.9) that extend to Kuranishi structure, and the other, by taking the inverse limit, due to McDuff-Wehrheim [**39**, Theorem B], which gives us a Čech homology class on M, itself.

4.1. The construction of a VFC in a thickening. Let $(M, \mathcal{K})$ be an oriented n-dimensional Kuranishi space, $\mathfrak{L}$ be an oriented Hausdorff DGS for $\mathcal{K}$ as in (3.16), and $\mathcal{N}$ be a perfect EOB for $\mathfrak{L}$ as in Definition 3.7.9. By Theorem 3.7.13, such an EOB gives rise to a compatible set of sequences of deformations $\mathfrak{s}_a \equiv \{\mathfrak{s}_{i,a}\}_{i\in\mathfrak{B}, a\in\mathbb{N}}$ of Kuranishi maps $s\equiv\{s_i\}_{i\in\mathfrak{B}}$ of $\mathfrak{L}$.

Similar to Section 2.6, for sufficiently large $a\in\mathbb{N}$, after restricting to some shrinking $\mathfrak{L}'$ of $\mathfrak{L}$, we build a singular n-cycle supported in the zero set of $\mathfrak{s}_a$ inside the thickening $Y(\mathfrak{L}')$.

With notation similar to (2.23), let

$$Z_i = Z(s_i) \subset Y(i) \qquad \forall i \in \mathfrak{B}, \qquad M = Z(s) = \bigcup_{i\in\mathfrak{B}} \wp_i(Z_i) \subset Y(\mathfrak{L}),$$

$$Z_{i,a} = Z(\mathfrak{s}_{i,a}) \subset Y(i) \qquad \forall i \in \mathfrak{B},\ a \in \mathbb{N}, \qquad Z(\mathfrak{s}_a) = \bigcup_{i\in\mathfrak{B}} \wp_i(Z_{i,a}) \subset Y(\mathfrak{L}) \quad \forall a \in \mathbb{N}.$$

By Definition 3.7.12.(2),

$$\overline{\phi}_{ij}(Z_{i,a} \cap Y(i,j)) = Z_{j,a} \cap N(i,j) \qquad \forall i,j \in \mathfrak{B},\ i<j,\ a \in \mathbb{N}. \tag{4.1}$$

Let

$$\mathfrak{L}^{(2)} \subset \mathfrak{L}^{(1)} \subset \mathfrak{L}^{(0)} \equiv \mathfrak{L} \tag{4.2}$$

be a sequence of shrinkings of $\mathfrak{L}$. The natural inclusion maps

$$Y(\mathfrak{L}^{(\mathrm{b})}) \subset Y(\mathfrak{L}^{(\mathrm{c})}) \quad \forall\ 0 \le c < b \le 2 \tag{4.3}$$

are relatively compact embeddings, but they are not open[44]. For $b=1,2$, let $s^{(\mathrm{b})}$, $\mathfrak{s}^{(\mathrm{b})}_{i,a}$, etc, be the restriction of corresponding objects to the corresponding sets for $\mathfrak{L}^{(\mathrm{b})}$. By Remark 3.4.5, the induced and the quotient topologies on

$$\mathrm{cl}(Y(\mathfrak{L}^{(\mathrm{b})})) \subset Y(\mathfrak{L}), \quad \mathrm{cl}(Y(\mathfrak{L}^{(\mathrm{b})})) = \bigcup_{i\in\mathfrak{B}} \wp_i\big(\mathrm{cl}\big(Y^{(\mathrm{b})}(i)\big)\big), \quad b=1,2,$$

are the same and are metrizable.

LEMMA 4.1.1. *For sufficiently large $a \in \mathbb{N}$,*

$$Z(\mathfrak{s}^{(2)}_a) = Z(\mathfrak{s}^{(1)}_a);$$

in particular $Z(\mathfrak{s}^{(1)}_a)$ is compact.

PROOF. Since

$$M = Z(s^{(\mathrm{b})}) \subset Y(\mathfrak{L}^{(\mathrm{b})}) \quad \forall b \in \{0,1,2\},$$

for every $p \in M$, there exists a maximal index $i_{b,p} \in \mathfrak{B}$ such that

$$p \in \wp_{i_{b,p}}(\overline{x}_{i_{b,p}}) \quad \text{for some (unique)} \quad \overline{x}_{i_{b,p}} \in Z^{(\mathrm{b})}_{i_{b,p}} \subset Y^{(\mathrm{b})}(i_{b,p}). \tag{4.4}$$

For $b<c$, by (4.2) and (3.21),

$$i_{b,p} \ge i_{c,p}, \quad \overline{x}_{i_{c,p}} \in Y(i_{c,p}, i_{b,p}), \quad \overline{\phi}_{i_{c,p} i_{b,p}}(\overline{x}_{i_{c,p}}) = \overline{x}_{i_{b,p}}. \tag{4.5}$$

For $b=0$, we will write $i_{0,p}$ as i_p. For every $p \in M$, let

$$W_{\overline{x}_{i_p}} \subset Y(i_p)$$

be an open set such that

$$W_{\overline{x}_{i_p}} \cap \mathrm{cl}(Y^{(1)}(i_p, j)) = \emptyset \quad \forall j > i_p, \qquad W_{\overline{x}_{i_p}} \subset \mathcal{N}(j, i_p) \quad \forall j < i_p. \tag{4.6}$$

Although

$$W_p := \wp_{i_p}(W_{\overline{x}_{i_p}}) \subset Y(\mathfrak{L})$$

is not necessarily open, but similarly to the case of (3.31) in the proof of Proposition 3.4.4, for $b=1,2$, the restriction $W_p \cap \mathrm{cl}(Y(\mathfrak{L}^{(\mathrm{b})}))$ is open in $\mathrm{cl}(Y(\mathfrak{L}^{(\mathrm{b})}))$. For a sequence of integers $(a_k)_{k=1}^{\infty}$ with $\lim_{k\to\infty} a_k = \infty$, suppose there exists

$$p_k \in Z(\mathfrak{s}^{(1)}_{a_k}) \quad \text{s.t} \quad p_k \notin Z(\mathfrak{s}^{(2)}_{a_k}) \quad \forall k \in \mathbb{N};$$

[44]Otherwise, the quotient topology would have been the same as induced topology, which by the example before Definition 3.4.1 we know it is not the case.

i.e. $p_k \notin Y(\mathfrak{L}^{(2)})$, for all $k \in \mathbb{N}$. By compactness of $\mathrm{cl}(Y(\mathfrak{L}^{(1)}))$ there exists

$$p \in \mathrm{cl}(Y(\mathfrak{L}^{(1)})) \subset Y(\mathfrak{L})$$

such that $\lim_{k\to\infty} p_k = p$ with respect to a metric for the metrizable topology of $\mathrm{cl}(Y(\mathfrak{L}^{(1)}))$. Since $\mathfrak{s}_{i,a_k}$ is C^1-converging to s_i, for all $i \in \mathfrak{B}$, we conclude that $p \in M = Z(s)$; in particular, $p \in Y(\mathfrak{L}^{(2)})$.

For k sufficiently large, $p_k \in W_p$ and by the left-hand side of (4.6) there exists (a unique) $\overline{x}_k \in W_{\overline{x}_{i_p}}$ such that $\wp_{i_p}(\overline{x}_k) = p_k$. By (4.5) and (4.4),

$$i_{2,p} \leq i_{0,p} \equiv i_p, \quad \overline{x}_{i_{2,p}} \in Y(i_{2,p}, i_p) \cap Y^{(2)}(i_{2,p}), \quad \overline{\phi}_{i_{2,p} i_p}(\overline{x}_{i_{2,p}}) = \overline{x}_{i_p}.$$

By the right-hand side of (4.6),

$$\overline{x}_k \in N(i_{2,p}, i_p);$$

therefore, by (4.1),

$$\overline{x}_k \in \overline{\phi}_{i_{2,p} i_p} Y(i_{2,p}, i_p).$$

Since $\overline{x}_k$ is converging to $\overline{x}_{i_p}$, we conclude that $\overline{x}'_k \equiv \overline{\phi}^{-1}_{i_{2,p} i_p}(\overline{x}_k)$ is converging to $\overline{x}_{i_{2,p}} \in Y^{(2)}(i_{2,p})$. Therefore, $\overline{x}'_k \in Y^{(2)}(i_{2,p})$, for sufficiently large k. This implies that

$$p_k = \wp_{i_{2,p}}(\overline{x}'_k) \in Y(\mathfrak{L}^{(2)})$$

for sufficiently large k, which is a contradiction.

□

In the rest of this section, for every $a \in \mathbb{N}$ such that the conclusion of Lemma 4.1.1 holds, we build a singular n-cycle[45] with $\mathbb{Q}$-coefficients supported in $Z(\mathfrak{s}_a^{(1)})$ whose homology class in $Y(\mathfrak{L})$ is independent of the choice of the perturbation. This gives us a VFC as a singular homology class inside the thickening $Y(\mathfrak{L})$. To this end, we use a generalization of the concept of the resolution from Section 2.5 and construct $\mathrm{VFC}(\mathfrak{L}) \in H_n(Y(\mathfrak{L}), \mathbb{Q})$ similarly to Section 2.6. In the case of pure orbibundle Kuranishi structures, the resulting homology class is simply the Euler class of the orbibundle as in (2.43). The thickening, and hence the resulting VFC homology class depends on the choice of $\mathfrak{L}$. We remove this indeterminacy in Sections 4.2 and 4.3, in two different ways.

Starting with $\mathcal{K}$, $\mathfrak{L}$, and $\mathcal{N}$ as in the beginning of this section, we replace $\mathfrak{L}$ with $\mathfrak{L}^{(1)}$ of Lemma 4.1.1. Similarly, we replace $\mathcal{N}$ with the corresponding shrinking of Lemma 3.7.7. We will also denote $\mathfrak{L}^{(2)}$ by $\mathfrak{L}'$. For the following discussion, let $a \in \mathbb{N}$ be some fixed sufficiently large integer such that the conclusion of Lemma 4.1.1 holds. For simplicity, we write $\mathfrak{s}$ for $\mathfrak{s}_a$. For every $p \in Y(\mathfrak{L})$, let

$$\mathfrak{B}_p = \{i \in \mathfrak{B} : p \in \wp_i(Y(i))\}.$$

For every $p \in Y(\mathfrak{L})$ and $i, j \in \mathfrak{B}$, with $i < j$, if

$$p = \wp_i(\overline{x}_i), \quad \overline{x}_i \in Y(i), \quad \text{and} \quad p = \wp_j(\overline{x}_j), \quad \overline{x}_j \in Y(j),$$

then

$$\overline{x}_i \in Y(i,j), \quad \phi_{ij}(\overline{x}_i) = \overline{x}_j,$$

and by Definition 3.7.12.(3)

$$\mathrm{val}_{\mathfrak{s}_i}(\overline{x}_i) = \mathrm{val}_{\mathfrak{s}_j}(\overline{x}_j), \quad \text{where } \mathfrak{s}_i = \mathfrak{s}|_{\mathcal{Y}(i)} \quad \forall i \in \mathfrak{B}; \tag{4.7}$$

[45] Where n is the dimension of $\mathfrak{L}$.

i.e. the positive integer

$$\mathrm{val}_{\mathfrak{s}}(p) \equiv \mathrm{val}_{\mathfrak{s}_i}(\overline{x}_i), \quad \forall i \in \mathfrak{B}_p, \overline{x}_i \in Y(i) \quad \text{s.t} \quad \wp_i(\overline{x}_i) = p,$$

is well-defined. Similar to Section 2.5, the restricted function $\mathrm{val}_{\mathfrak{s}}\colon Y(\mathfrak{L}') \longrightarrow \mathbb{N}$ is upper semi-continuous. Therefore, $\mathrm{val}_{\mathfrak{s}}$ restricted to the compact set $M = Z(\mathfrak{s}) \subset Y(\mathfrak{L}')$ takes only finitely many values

$$c_1 < \cdots < c_N.$$

Let

$$Z(\mathfrak{s})_\ell = \{p \in Z(\mathfrak{s})\colon \mathrm{val}_{\mathfrak{s}}(p) = c_\ell\} \quad \forall \ell \in [N];$$

in particular, $Z(\mathfrak{s})_N$ is compact.

Let

$$\mathfrak{B} = \{i_1 < \ldots < i_S\}, \quad Z(\mathfrak{s})_{i;\ell} \equiv Z(\mathfrak{s})_\ell \cap \wp_i(\mathrm{cl}(Y'(i))) \subset Y(\mathfrak{L}') \quad \forall i \in \mathfrak{B}, \ell \in [N].$$

By definition, $Z(\mathfrak{s})_{i;\ell} = \wp_i(Z(\mathfrak{s}_i)_\ell)$, where

$$Z(\mathfrak{s}_i)_\ell \equiv \big\{\overline{x} \in Z(\mathfrak{s}_i) \cap \mathrm{cl}(Y'(i))\colon \mathrm{val}_{\mathfrak{s}_i}(\overline{x}) = c_\ell\big\} \subset Y(i) \quad \forall i \in \mathfrak{B}, \ell \in [N]. \tag{4.8}$$

Since $Z(\mathfrak{s})_N$ is compact, $Z(\mathfrak{s}_i)_N \subset Y(i)$, for all $i \in \mathfrak{B}$, is compact.

Starting with maximal elements $i_S \in \mathfrak{B}$ and $N \in [N]$, by Lemma 2.5.4, there exist a neighborhood $W_{i_S;N}$ of

$$K(\mathfrak{s}_{i_S})_N \equiv Z(\mathfrak{s}_{i_S})_N$$

in $Y(i_S)$, and an orbifold $\widetilde{\mathcal{W}}_{i_S;N}$ with a c_N-covering map

$$q_{i_S;N}\colon \widetilde{\mathcal{W}}_{i_S;N} \longrightarrow \mathcal{W}_{i_S;N} = \mathcal{Y}(i_S)|_{W_{i_S;N}},$$

such that the multisection $\mathfrak{s}_{i_S}$, restricted to $W_{i_S;N}$, is equivalent to the weighted push forward, with weight $w_{i_S;N}$, of some orbifold section $s_{i_S;N}$ of

$$\widetilde{\mathcal{E}}_{i_S;N} = q^*_{i_S;N}(\mathcal{E}|_{W_{i_S;N}}).$$

Moving down in $\mathfrak{B}$ to i_{S-1}, let

$$K(i_{S-1})_N := Z(\mathfrak{s}_{i_{S-1}})_N \setminus \overline{\phi}^{-1}_{i_{S-1}i_S}(W_{i_S;N}) \subset Y(i_{S-1}).$$

Once again, by Lemma 2.5.4, there exist a neighborhood $W_{i_{S-1};N}$ of $K(\mathfrak{s}_{i_{S-1}})_N$ in $Y(i_{S-1})$, and an orbifold $\widetilde{\mathcal{W}}_{i_{S-1};N}$ with a c_N-covering map

$$q_{i_{S-1};N}\colon \widetilde{\mathcal{W}}_{i_{S-1};N} \longrightarrow \mathcal{W}_{i_{S-1};N} = \mathcal{Y}(i_{S-1})|_{W_{i_{S-1};N}},$$

such that the multisection $\mathfrak{s}_{i_{S-1}}$, restricted to $W_{i_{S-1};N}$, is equivalent to the weighted push forward, with weight $w_{i_{S-1};N}$, of some orbifold section $s_{i_{S-1};N}$ of

$$\widetilde{\mathcal{E}}_{i_{S-1};N} = q^*_{i_{S-1};N}(\mathcal{E}|_{W_{i_{S-1};N}}).$$

By (4.7) and (4.8),

$$K(i_{S-1})_N \cap \mathrm{cl}(Y'(i_{S-1}, i_S)) = \emptyset;$$

therefore, after possibly replacing $W_{i_{S-1};N}$ with a smaller neighborhood, we may assume that

$$W_{i_{S-1};N} \cap \mathrm{cl}(Y'(i_{S-1}, i_S)) = \emptyset. \tag{4.9}$$

Continuing downward inductively in $\mathfrak{B}$, we obtain a set of c_N-covering maps

$$q_{i;N}\colon \widetilde{\mathcal{W}}_{i;N} \longrightarrow \mathcal{W}_{i;N} = \mathcal{Y}(i)|_{W_{i;N}} \quad \forall i \in \mathfrak{B},$$

such that the multisection $\mathfrak{s}_i$, restricted to $W_{i;N}$, is equivalent to the weighted push forward, with weight $w_{i;N}$, of a natural orbifold section $s_{i;N}$ of

$$\widetilde{\mathcal{E}}_{i;N} = q_{i;N}^{*}(\mathcal{E}|_{W_{i;N}}),$$

and

$$W_{i;N} \cap \mathrm{cl}(Y'(i,j)) = \emptyset \quad \forall i < j.$$

Moreover, by the explicit construction of proof of Lemma 2.5.4, the orbibundle embeddings $\Phi_{ij} = (\mathfrak{D}\phi_{ij}, \phi_{ij})$ in Definition 3.3.2 lift to orbibundle embeddings $\widetilde{\Phi}_{ij;N} = (\mathfrak{D}\widetilde{\phi}_{ij;N}, \widetilde{\phi}_{ij;N})$ between the covering spaces such that the following diagram commutes,

$$\begin{array}{ccc}
\widetilde{\mathcal{E}}_{i;N}|_{\overline{q}_{i;N}^{-1}(\overline{\phi}_{ij}^{-1}(W_{j;N})\cap W_{i;N})} & \xrightarrow{\mathfrak{D}\widetilde{\phi}_{ij;N}} & \widetilde{\mathcal{E}}_{j;N} \\
\uparrow s_{i;N} & & \uparrow s_{j;N} \\
\widetilde{\mathcal{W}}_{i;N}|_{\overline{q}_{i;N}^{-1}(\overline{\phi}_{ij}^{-1}(W_{j;N})\cap W_{i;N})} & \xrightarrow{\widetilde{\phi}_{ij;N}} & \widetilde{\mathcal{W}}_{j;N}\,.
\end{array}$$

We next move down to c_{N-1}; i.e. we move down inductively with respect to the lexicographic order

$$(i_b, c_\ell) < (i_{b'}, c'_\ell) \quad \Leftrightarrow \quad \ell < \ell' \quad \text{or} \quad (\ell = \ell' \quad \text{and} \quad b < b').$$

Let

$$W_N = \bigcup_{i \in \mathfrak{B}} \wp_i(W_{i;N}) \subset Y(\mathfrak{L}); \tag{4.10}$$

this is not necessarily an open set in $Y(\mathfrak{L})$, but similarly to the case of (3.31) in the proof of Proposition 3.4.4, by (4.9), the restriction $W_N \cap Y(\mathfrak{L}')$ is an open set in $Y(\mathfrak{L}')$. Let

$$K_{N-1} = Z(\mathfrak{s})_{N-1} \backslash W_N.$$

Since $Z(\mathfrak{s}) \subset Y(\mathfrak{L}')$, $W_N \cap Y(\mathfrak{L}')$ is open, and $\mathrm{val}_{\mathfrak{s}}$ is upper semi-continuous, K_{N-1} is compact. With $N-1$ instead of N and K_{N-1} instead of $Z(\mathfrak{s})_N$ in the previous argument, we inductively build

$$q_{i;N-1} \colon \widetilde{\mathcal{W}}_{i;N-1} \longrightarrow \mathcal{W}_{i;N-1} = \mathcal{Y}(i)|_{W_{i;N-1}} \quad \forall i \in \mathfrak{B},$$

similarly. Continuing inductively with respect to the lexicographic order on $\mathfrak{B} \times [N]$, we obtain a double-indexed resolution of $\mathfrak{s}$ similar to Definition 2.5.8, described in Definition 4.1.2 below.

DEFINITION 4.1.2. *With $\mathfrak{L}$, $\mathfrak{L}'$, $\mathfrak{s}$, and N as above, a* ***resolution*** *of $\mathfrak{s}$ consists of a set of c_ℓ-covering maps*

$$q_{i;\ell} \colon \widetilde{\mathcal{W}}_{i;\ell} \longrightarrow \mathcal{W}_{i;\ell}, \quad \forall i \in \mathfrak{B}, \ \ell \in [N], \tag{4.11}$$

over open subsets $W_{i;\ell} \subset Y(i)$ such that the following properties hold.

(1) *The image of open sets $W_{i;\ell}$ cover M in the sense that*

$$M = Z(\mathfrak{s}) \subset \bigcup_{i=\ell}^{N} W_\ell, \quad W_\ell = \bigcup_{i \in \mathfrak{B}} \wp_i(W_{i,\ell}).$$

Moreover, for every $\ell \in [N]$, $W_\ell \cap Y(\mathfrak{L}')$ *is open in* $Y(\mathfrak{L}')$,

$$W_\ell \cap \big(Z(\mathfrak{s})_{>\ell} \equiv \{p \in Z(\mathfrak{s})\colon \operatorname{val}_{\mathfrak{s}}(p) > c_\ell\}\big) = \emptyset, \text{ and } \quad Z(\mathfrak{s})_{\geq \ell} \subset \bigcup_{b=\ell}^{N} W_b.$$

(2) *For every* $\ell \in [N]$ *and* $i \in \mathfrak{B}$, *the multisection* $\mathfrak{s}_i$, *restricted to* $W_{i;\ell}$, *is equivalent to the weighted push forward with weight* $w_{i;\ell}$ *of some orbifold section* $s_{i;\ell}$ *of* $\widetilde{\mathcal{E}}_{i;\ell} \equiv q^*_{i;\ell}(\mathcal{E}(i)|_{\mathcal{W}_{i;\ell}})$.

(3) *For every* $i \in \mathfrak{B}$ *and* $\ell > \ell'$, *on the overlap*

$$W_{i;\ell,\ell'} = W_{i;\ell',\ell} = W_{i;\ell'} \cap W_{i;\ell}, \tag{4.12}$$

there exists a component-wise covering map

$$q_{i;\ell,\ell'}\colon \widetilde{\mathcal{W}}_{i;\ell}|_{\overline{q}^{-1}_{i;\ell}(W_{i;\ell,\ell'})} \longrightarrow \widetilde{\mathcal{W}}_{i;\ell'}|_{\overline{q}^{-1}_{i;\ell'}(W_{i;\ell',\ell})},$$

such that the analogue of Definition 2.5.8.(3) *holds.*

(4) *For every* $\ell \in [N]$ *and* $i, j \in \mathfrak{B}$ *with* $i < j$, *on the overlap*

$$W_{i,j;\ell} = \overline{\phi}_{ij}^{-1}(W_{j;\ell}) \cap W_{i;\ell}, \tag{4.13}$$

the orbifold embedding ϕ_{ij} *lifts to an orbifold embedding* $\widetilde{\phi}_{ij;\ell}$

$$\widetilde{\phi}_{ij;\ell}\colon \widetilde{\mathcal{W}}_{i;\ell}|_{\overline{q}^{-1}_{i;\ell}(W_{i,j;\ell})} \longrightarrow \widetilde{\mathcal{W}}_{j;\ell}.$$

(5) *For* $i < j$ *and* $\ell_i < \ell_j$, *there exists a* c_{ℓ_j}*-covering*

$$q_{j;\ell_j|i;\ell_i}\colon \widetilde{\mathcal{W}}_{i;\ell_i|j;\ell_j} \longrightarrow \mathcal{W}_{i;\ell_i|j;\ell_j}, \quad W_{i;\ell_i|j;\ell_j} = \overline{\phi}_{ij}^{-1}(W_{j;\ell_j}) \cap W_{i;\ell_i},$$

such that the orbifold embedding ϕ_{ij} *lifts to an orbifold embedding*

$$\widetilde{\phi}_{i;\ell_i|j;\ell_j}\colon \widetilde{\mathcal{W}}_{i;\ell_i|j;\ell_j} \longrightarrow \widetilde{\mathcal{W}}_{j;\ell_j},$$

commuting with $q_{j;\ell_j|i;\ell_i}$ *on the source and* $q_{j;\ell_j}$ *on the target, and* $q_{j;\ell_j|i;\ell_i}$ *factors through* $q_{i;\ell_i}$:

$$\begin{array}{ccc}
\widetilde{\mathcal{W}}_{i;\ell_i|j;\ell_j} & \xrightarrow{\ \widetilde{\phi}_{i;\ell_i|j;\ell_j}\ } & \widetilde{\mathcal{W}}_{j;\ell_j}|_{\overline{q}^{-1}_{j;\ell_j}(W_{j;\ell_j|i;\ell_i})} \\
\downarrow{\scriptstyle q'_{j;\ell_j|i;\ell_i}} & & \\
\widetilde{\mathcal{W}}_{i;\ell_i}|_{\overline{q}^{-1}_{i;\ell_i}(W_{i;\ell_i|j;\ell_j})} & & \downarrow{\scriptstyle q_{j;\ell_j}} \\
\downarrow{\scriptstyle q_{i;\ell_i}} & & \\
\mathcal{W}_{i;\ell_i|j;\ell_j} & \xrightarrow{\ \phi_{ij}\ } & \mathcal{W}_{j;\ell_j|i;\ell_i}\,.
\end{array}
\quad (\text{curved arrow } q_{j;\ell_j|i;\ell_i}\colon \widetilde{\mathcal{W}}_{i;\ell_i|j;\ell_j} \to \mathcal{W}_{i;\ell_i|j;\ell_j}) \tag{4.14}$$

(6) *For* $i < j$ *and* $\ell_i > \ell_j$, *there exists a* c_{ℓ_j}*-covering*

$$q_{j;\ell_j|i;\ell_i}\colon \widetilde{\mathcal{W}}_{i;\ell_i|j;\ell_j} \longrightarrow \mathcal{W}_{i;\ell_i|j;\ell_j}, \quad W_{i;\ell_i|j;\ell_j} = \overline{\phi}_{ij}^{-1}(W_{j;\ell_j}) \cap W_{i;\ell_i},$$

such that the orbifold embedding ϕ_{ij} *lifts to an orbifold embedding*

$$\widetilde{\phi}_{i;\ell_i|j;\ell_j}\colon \widetilde{\mathcal{W}}_{i;\ell_i|j;\ell_j} \longrightarrow \widetilde{\mathcal{W}}_{j;\ell_j},$$

commuting with $q_{j;\ell_j|i;\ell_i}$ *on the source and* $q_{j;\ell_j}$ *on the target, and* $q_{i;\ell_i}$ *factors through* $q_{j;\ell_j|i;\ell_i}$*:*

$$
\begin{array}{ccc}
\widetilde{\mathcal{W}}_{i;\ell_i}|_{\overline{q}^{-1}_{i;\ell_i}(W_{i;\ell_i|j;\ell_j})} & & \\
\big\downarrow{\scriptstyle q'_{j;\ell_j|i;\ell_i}} & & \\
\widetilde{\mathcal{W}}_{i;\ell_i|j;\ell_j} & \xrightarrow{\ \widetilde{\phi}_{i;\ell_i|j;\ell_j}\ } & \widetilde{\mathcal{W}}_{j;\ell_j}|_{\overline{q}^{-1}_{j;\ell_j}(W_{j;\ell_j|i;\ell_i})} \\
\big\downarrow{\scriptstyle q_{j;\ell_j|i;\ell_i}} & & \big\downarrow{\scriptstyle q_{j;\ell_j}} \\
\mathcal{W}_{i;\ell_i|j;\ell_j} & \xrightarrow{\ \phi_{ij}\ } & \mathcal{W}_{j;\ell_j|i;\ell_i}\,.
\end{array}
\tag{4.15}
$$

(with the curved arrow $q_{i;\ell_i}$ from $\widetilde{\mathcal{W}}_{i;\ell_i}|_{\overline{q}^{-1}_{i;\ell_i}(W_{i;\ell_i|j;\ell_j})}$ to $\mathcal{W}_{i;\ell_i|j;\ell_j}$)

(7) *The overlap maps* $q_{i;\ell,\ell'}$, $q_{i,j;\ell}$, *and* $q_{i;\ell_i|j;\ell_j}$ *are compatible on triple intersections similar*[46] *to* (2.37)*.*
(8) *In* (4)*-*(6)*, the lifted maps commute with the sections* $s_{i;\ell}$ *on the source and the target.*
(9) *The twisted weight-ratio functions* $(\overline{w}_{i;\ell}/\mathfrak{I}_{\mathcal{W}_{i;\ell}})$ *are equal on the overlaps as in* (2.38)*.*

We will not further go into the details of existence of such a resolution and its properties. The argument is similar to that of Section 2.5, but as the rather long statement of this definition indicates, it is more tedious to write down the proof in details.

Finally, similarly to Section 2.6, in order to construct $\mathrm{VFC}(\mathfrak{L})$, we compatibly triangulate the zero set of orbifold sections of a given resolution in the following way. The key point is that at the level of zero sets, the double-indexed data of Definition 4.1.2 simplifies into a set of orbifolds simply indexed by $\ell\in[N]$. For every $\ell\in[N]$, although the orbifolds $\widetilde{\mathcal{W}}_{i,\ell}$ and $\widetilde{\mathcal{W}}_{j,\ell}$, for different $i,j\in\mathfrak{B}$, have different dimensions, the zero sets $\widetilde{Z}(s_{i;\ell})$ and $\widetilde{Z}(s_{j;\ell})$ inherit, possibly non-effective, oriented orbifold structures $\widetilde{\mathcal{Z}}(s_{i;\ell})$ and $\widetilde{\mathcal{Z}}(s_{j;\ell})$ of the same dimension[47] n from $\widetilde{\mathcal{E}}_{i,\ell}\longrightarrow\widetilde{\mathcal{W}}_{i,\ell}$ and $\widetilde{\mathcal{W}}_{j,\ell}\longrightarrow\widetilde{\mathcal{W}}_{j,\ell}$, respectively. Therefore, via the restriction of orbifold embedding maps $\widetilde{\phi}_{ij;\ell}$ in Definition 4.1.2.(4) to $\widetilde{\mathcal{Z}}(s_{i;\ell})$ and $\widetilde{\mathcal{Z}}(s_{j;\ell})$, we can glue these orbifolds and obtain an oriented, and possibly non-effective, orbifold $\widetilde{\mathcal{Z}}_\ell$ as in (2.39).

For $\ell>\ell'$, we can glue the overlap maps $q_{i;\ell,\ell'}$ and $q_{i;\ell|j;\ell'}$, on the intersection of overlaps with zero sets of the orbifold sections, to obtain covering maps

$$q_{\ell,\ell'}\colon \widetilde{\mathcal{Z}}_{\ell,\ell'}\longrightarrow\widetilde{\mathcal{Z}}_{\ell',\ell}$$

similar to the covering maps used in Definition 2.6.3. Therefore, Definition 2.6.3 still applies to the induced resolution of zero sets,

$$\mathfrak{R}_0\equiv\Big(\{\overline{q}_\ell\colon \widetilde{Z}_\ell\longrightarrow Z(\mathfrak{s})_\ell\cap W_\ell\}_{\ell\in[N]},\{q_{\ell,\ell'}\colon\widetilde{\mathcal{Z}}_{\ell,\ell'}\longrightarrow\widetilde{\mathcal{Z}}_{\ell',\ell}\}_{1\le\ell'<\ell\le N}\Big),$$

and gives us an "admissible" triangulation that via Proposition 2.6.4 gives us a singular homology class

$$\mathrm{VFC}(\mathfrak{L})\in H_n(Y(\mathfrak{L}),\mathbb{Q}).$$

---

[46]The precise statement involves some complicated commutative diagrams similar to (4.14) and (4.15).

[47]This is by the transversality assumption, Definition 3.3.2.(1), and the fact that each orbibundle is relatively oriented.

REMARK 4.1.3. We could also define a double-indexed version of admissible triangulation for every resolution in Definition 4.1.2, and generalize Definition 2.6.3, similarly. However, by the trick of previous paragraph, we avoided such a complicated definition by first gluing the zero sets $\widetilde{\mathcal{Z}}(\mathfrak{s}_{i;\ell})$, for every fixed ℓ, into a single orbifold $\widetilde{\mathcal{Z}}_\ell$ to which Definition 2.6.3 and Proposition 2.6.4 apply.

We summarize the results of this section in the following proposition.

PROPOSITION 4.1.4. *Let $\mathfrak{L}$ be an oriented Hausdorff dimensionally graded system. Let $\mathfrak{L}'$ be a sufficiently small shrinking of $\mathfrak{L}$ admitting a perfect EOB $\mathcal{N}$, and $\mathfrak{s}$ be a transversal multisection deformation of the Kuranishi map $s=\{s_i\}_{i\in\mathfrak{B}}$ provided by Theorem* 3.7.13 *with $Z(\mathfrak{s})\subset Y(\mathfrak{L}')$. Then the homology class*

$$\mathrm{VFC}(\mathfrak{L})\in H_n(Y(\mathfrak{L}),\mathbb{Q})$$

constructed above is independent of the choice of $\mathfrak{L}'$, $\mathcal{N}$, $\mathfrak{s}$, a resolution as in Definition 4.1.2, *and the choice of admissible triangulation.*

REMARK 4.1.5. The proof of independence is by considering the trivial deformation equivalence $\mathfrak{L}\times[0,1]$ and extending two given sets of auxiliary data over the boundaries to the entire space. There is however one delicate issue that we need to be careful about. The $\mathrm{val}_{\mathfrak{s}}$ function used in the construction of a resolution does not generally behave well under the restriction to boundary, i.e. if $\mathfrak{s}$ is a multisection of an orbibundle $\mathcal{E}$ over an orbifold $\mathcal{M}$ with boundary $\partial\mathcal{M}$ and $\overline{p}\in\partial M$, then $\mathrm{val}_{\mathfrak{s}}(\overline{p})$ can be larger than $\mathrm{val}_{\mathfrak{s}|_{\partial M}}(\overline{p})$; i.e. a set of branches of $\mathfrak{s}$ which are different in every neighborhood of $\overline{p}$ in M may have equal restrictions over some neighborhood of $\overline{p}$ in ∂M. In this situation, the restriction of a resolution corresponding to a given tuple $(\mathcal{E}\longrightarrow\mathcal{M},\mathfrak{s})$ to the boundary $(\mathcal{E}|_{\partial M}\longrightarrow\partial\mathcal{M},\mathfrak{s}|_{\partial M})$ is not a resolution of the latter. In order to avoid this issue, we may only consider multisections that over a sufficient small bordered chart $V\times[0,a)$ around each point of ∂M are trivial extensions of a multisection on V to the normal direction.

4.2. VFC via evaluation maps. In this section, we consider Kuranishi structures for a pair consisting of a compact metrizable topological space and a map into another space. We use this specialization to define VFC of a given pair as a singular homology class inside the target space. For example, we use this approach to define the GW virtual fundamental class in Theorem 1.4.1 for the pair of moduli space of J-holomorphic maps and evaluation and forgetful maps.

DEFINITION 4.2.1. *Let M be a compact metrizable topological space, W be a topological space, and $\overline{\varrho}\colon M\longrightarrow W$ be a continuous map. A Kuranishi structure $\mathcal{K}$ on M compatible with $\overline{\varrho}$, called a **$\overline{\varrho}$-Kuranishi structure**, is a Kuranishi structure on M as in Definition* 3.2.6 *such the the map $\overline{\varrho}$ consistently lifts to all Kuranishi charts in the following way. For every Kuranishi chart $\mathcal{U}_p=(\mathrm{pr}_p\colon U_p\longrightarrow V_p,G_p,s_p,\psi_p)$, the map*

$$\overline{\varrho}_p=\overline{\varrho}|_{F_p}\colon F_p=\psi_p(s_p^{-1}(0))\longrightarrow W$$

lifts to a continuous map $\overline{\varrho}_p\colon (V_p/G_p)\longrightarrow W$ in a way that

$$\overline{\varrho}_p\circ\overline{\phi}_{qp}=\overline{\varrho}_q,$$

*for every coordinate change map $\Phi_{qp}=(\mathfrak{D}\phi_{qp},\phi_{qp})\colon\mathcal{U}_{q,p}\longrightarrow\mathcal{U}_p$. If W is the underlying topological space of an orbifold $\mathcal{W}$, we say $\mathcal{K}$ is an **$\overline{\varrho}$-smooth Kuranishi***

structure, *if the restrictions $\overline{\varrho}_p$ consistently lift to orbifold smooth maps ϱ_p between $[V_p/G_p]$ and $\mathcal{W}$.*

We define $\overline{\varrho}$-DGS and $\overline{\varrho}$-smooth DGS similarly. It is clear from the proof of Theorem 3.5.3 that if $\mathcal{K}$ is a $\overline{\varrho}$-Kuranishi structure (resp. $\overline{\varrho}$-smooth Kuranishi structure), then every associated natural DGS $\mathfrak{L}$ is canonically an $\overline{\varrho}$-DGS (resp. $\overline{\varrho}$-smooth DGS). It is also clear from the definition of the thickening $Y(\mathfrak{L})$ in (3.22) that if $\mathfrak{L}$ is an $\overline{\varrho}$-DGS, then $\overline{\varrho}$ extends to a continuous map

$$\overline{\varrho}\colon Y(\mathfrak{L}) \longrightarrow W. \tag{4.16}$$

Proposition 4.2.2. *Let M be a compact metrizable topological space, W be a topological space, and $\overline{\varrho}\colon M \longrightarrow W$ be a continuous map. Assume $(M,\mathcal{K})$ is an oriented n-dimensional $\overline{\varrho}$-Kuranishi space and $\mathfrak{L}$ is a natural oriented Hausdorff $\overline{\varrho}$-DGS for $\mathcal{K}$ as in Theorem* 3.5.3. *Let* $\mathrm{VFC}(\mathfrak{L})$ *be the n-dimensional homology class of Proposition* 4.1.4. *Then the push forward*[48] *homology class*

$$\overline{\varrho}_*(\mathrm{VFC}(\mathfrak{L})) \in H_n(W,\mathbb{Q})$$

is independent of the choice of a natural $\overline{\varrho}$-DGS $\mathfrak{L}$ for $\mathcal{K}$ and the cobordism class of $\overline{\varrho}$-Kuranishi structure $\mathcal{K}$; thus, we denote it by $\mathrm{VFC}(\mathcal{K},\overline{\varrho})$.

The invariance part of Proposition 4.2.2 is by extending two given sets of auxiliary data on the boundary of a cobordism between two given Kuranishi structures to the entire cobordism Kuranishi structure. We will not go into the details of such cobordism argument in this article as it is a relative version of what we have done so far.

4.3. Čech homology VFC. The VFC of Proposition 4.1.4 is not a priori supported in the topological space M and depends on the auxiliary choice of $\mathfrak{L}$. Another way of resolving this issue is by taking the inverse limit over an infinite sequence of dimensionally graded systems such that $Y(\mathfrak{L})$ converges to M. This gives us a VFC supported in M, at the expense of replacing singular homology with Čech homology; this is how the VFC is defined in [**38, 39, 44**]. We outline the process here.

Let $(M,\mathcal{K})$ be an oriented n-dimensional Kuranishi space, $\mathfrak{L}$ be an oriented Hausdorff DGS for $\mathcal{K}$, $\mathcal{N}$ be a perfect EOB for $\mathfrak{L}$, and $\mathfrak{s}_a \equiv \{\mathfrak{s}_{i,a}\}_{i\in\mathfrak{B},a\in\mathbb{N}}$ be a compatible set of sequences of deformations of Kuranishi maps $s\equiv\{s_i\}_{i\in\mathfrak{B}}$ of $\mathfrak{L}$ as in the beginning of Section 4.1.

Let

$$\mathfrak{L}\equiv\mathfrak{L}^{(0)}\supset\mathfrak{L}^{(1)}\supset\mathfrak{L}^{(2)}\supset\cdots \tag{4.17}$$

be an infinite sequence of shrinkings of $\mathfrak{L}$ such that

$$\bigcap_{b=1}^{\infty} Y(\mathfrak{L}^{(\mathrm{b})}) = M.$$

By Lemma 4.1.1, for every $b\in\mathbb{N}$, there exists $m_b\in\mathbb{N}$ such that

$$Z(\mathfrak{s}_a^{(\mathrm{b})}) = Z(\mathfrak{s}_a^{(\mathrm{b+1})}) \quad \forall a>m_b.$$

Thus, by the construction of Section 4.1, we obtain a sequence of homology classes

$$\mathrm{VFC}(\mathfrak{L}^{(\mathrm{b})}) \in H_n(Y(\mathfrak{L}^{(\mathrm{b})}),\mathbb{Q}) \quad \forall b\in\mathbb{N},$$

[48]Via (4.16).

such that

$$(\iota_{bc})_*(\mathrm{VFC}(\mathfrak{L}^{(\mathrm{b})})) = \mathrm{VFC}(\mathfrak{L}^{(\mathrm{c})}) \qquad \forall b>c, \tag{4.18}$$

where

$$\iota_{bc}\colon Y(\mathfrak{L}^{(\mathrm{b})}) \longrightarrow Y(\mathfrak{L}^{(\mathrm{c})})$$

is the inclusion map. The system of inclusion maps

$$H_n(M,\mathbb{Q}) \longrightarrow H_n(Y(\mathfrak{L}^{(\mathrm{b})}),\mathbb{Q}) \longrightarrow H_n(Y(\mathfrak{L}^{(\mathrm{c})}),\mathbb{Q}) \quad \forall b>c,$$

gives rise to an inclusion map

$$H_n(M,\mathbb{Q}) \longrightarrow \varprojlim H_n(Y(\mathfrak{L}^{(\mathrm{b})}),\mathbb{Q}), \tag{4.19}$$

which is not necessary an isomorphism. Therefore, with this approach, we do not necessarily obtain a singular homology VFC in $H_n(M,\mathbb{Q})$ by applying the inverse limit to (4.18). For every CW-complex Y, the Čech and singular homology groups $\check{H}_n(Y,\mathbb{Q})$ and $H_n(Y,\mathbb{Q})$, respectively, are isomorphic. For every $b\in\mathbb{N}$, $\mathrm{cl}(Y(\mathfrak{L}^{(\mathrm{b})}))$ has the structure[49] of a CW-complex. Therefore, the analogue of (4.19) with Čech homology groups instead is an isomorphism. Thus, after replacing singular homology with Čech homology and taking inverse limit, we obtain a Čech homology virtual fundamental class

$$\check{\mathrm{VFC}}\big((\mathfrak{L}^{(b)})_{b=0}^{\infty}\big) \in \check{H}_n(M,\mathbb{Q}). \tag{4.20}$$

In conclusion, the following theorem is the Čech homology analogue of Proposition 4.1.4; see [**39**, Remark 8.2.4].

PROPOSITION 4.3.1. *Let $(M,\mathcal{K})$ be an oriented n-dimensional Kuranishi space, $\mathfrak{L}$ be a natural oriented Hausdorff DGS for $\mathcal{K}$ as in Theorem* 3.5.3, *$(\mathfrak{s}_a)_{a=1}^{\infty}$ be a compatible set of sequences of deformations of Kuranishi maps of $\mathfrak{L}$ as in Theorem* 3.7.13, *and $(\mathfrak{L}^{(b)})_{b\in\mathbb{N}}$ be a sequence of shrinkings of $\mathfrak{L}$ as in (*4.17*) . Then the resulting homology class (*4.20*) is an invariant of oriented cobordism class of $\mathcal{K}$.*

5. Moduli spaces of stable maps

Stable compactification (or Gromov compactification) $\overline{\mathcal{M}}_{g,k}(M,A)$ of $\mathcal{M}_{g,k}(X,A)$ includes pseudoholomorphic maps with nodal domains. In order to cover the entire $\overline{\mathcal{M}}_{g,k}(X,A)$ with Kuranishi charts, we need to first understand the local structure of the Deligne-Mumford space at an arbitrary nodal curve.

In Section 5.1, we set up some notation and review the definition of the moduli space of stable maps. In Section 5.2, following the analytic approach of [**45**], we review the holomorphic orbifold structure of $\overline{\mathcal{M}}_{g,k}$. We recall the notions of universal families and of universal curve, and collect some well-known results about these objects. In Section 5.3, we fix some more notation and state the Gromov's Convergence Theorem. In Section 6, we will discuss a different realization of the Gromov topology via the notion of ϵ-close maps.

[49] In [**38**, **39**], McDuff-Wehrheim use Poincare duality over the cycle realizing $\mathrm{VFC}(\mathcal{L})$ which has the structure of a CW-complex and then push forward the resulting Čech (co-)homology class.

5.1. Stable curves and stable maps. Let Γ be a graph, possibly with **half edges** or **legs**[50], i.e. edges that have a vertex at one end and are open at the other end. Let $\mathbb{V}=\mathbb{V}(\Gamma)$, $\mathbb{E}=\mathbb{E}(\Gamma)$, and $\mathbb{L}=\mathbb{L}(\Gamma)$ be the set of vertices, edges, and legs of Γ, respectively. Let $\underrightarrow{\mathbb{E}}$ be the set of edges with an orientation. Given $\underrightarrow{e}\in\underrightarrow{\mathbb{E}}$, let $\underleftarrow{e}$ denote the same edge with the opposite orientation. For each $\underrightarrow{e}\in\underrightarrow{\mathbb{E}}$, let $v_1(\underrightarrow{e})$ and $v_2(\underrightarrow{e})$ in $\mathbb{V}$ denote the starting and ending points of the arrow, respectively. For $v,v'\in\mathbb{V}$, let $\mathbb{E}_{v,v'}$ denote the subset of edges between the two vertex, $\underrightarrow{\mathbb{E}}_{v,v'}$ denote the subset of oriented edges from v to v', and $\underrightarrow{\mathbb{E}}_v\subset\underrightarrow{\mathbb{E}}$ be subset of oriented edges starting at v.

A **genus labeling** of Γ is a function $g\colon\mathbb{V}\longrightarrow\mathbb{N}$. An **ordering of the legs** of Γ is a bijection $o\colon\mathbb{L}\longrightarrow\{1,\dots,|\mathbb{L}|\}$. If a genus-labeled graph Γ is connected, the **arithmetic** genus of Γ is

$$g_\Gamma=\sum_{v\in\mathbb{V}}g_v+\operatorname{rank}\,H_1(\Gamma,\mathbb{Z}) \tag{5.1}$$

where $H_1(\Gamma,\mathbb{Z})$ is the first homology group of the underlying topological space of Γ. Similarly, if Γ is possibly disconnected, we can define its (half) Euler characteristic $\chi(\Gamma)$ by

$$\chi(\Gamma)=|\mathbb{E}|-|\mathbb{V}|-\sum_{v\in\mathbb{V}}g_v; \tag{5.2}$$

in the connected case $\chi(\Gamma)=1-g_\Gamma$. Figure 7-left illustrates a labeled graph with 2 legs.

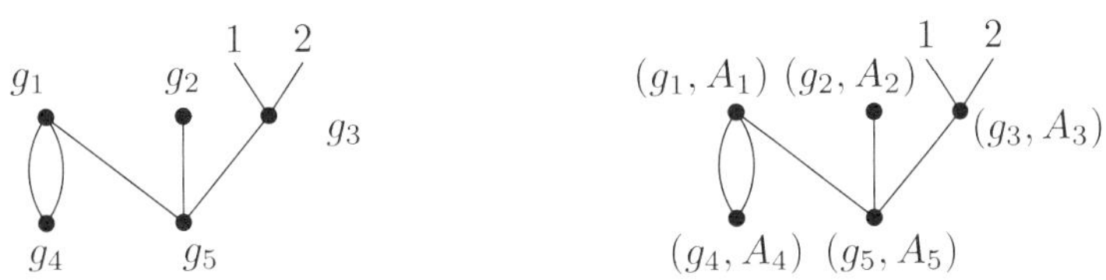

FIGURE 7. Left- a labeled graph Γ representing elements of $\overline{\mathcal{M}}_{g,2}$. Right- a labeled graph Γ representing elements of $\overline{\mathcal{M}}_{g,2}(X,A)$.

Such decorated graphs Γ characterize different topological types of nodal marked surfaces $(\Sigma,\vec{z}=(z^1,\dots,z^k))$ in the following way. Each vertex $v\in\mathbb{V}$ corresponds to a smooth[51] component Σ_v of Σ with genus g_v. Each edge $e\in\mathbb{E}$ corresponds to a node q_e obtained by connecting Σ_v and $\Sigma_{v'}$ at the points $q_{\underrightarrow{e}}\in\Sigma_v$ and $q_{\underleftarrow{e}}\in\Sigma_{v'}$, where $e\in E_{v,v'}$ and $\underrightarrow{e}$ is an orientation on e with $v_1(\underrightarrow{e})=v$. Note that $v_1(\underrightarrow{e})=v$ uniquely determines the orientation $\underrightarrow{e}$ unless e is a loop connecting v to itself. Finally, each leg $l\in\mathbb{L}$ connected to the vertex $v(l)$ corresponds to a marked point $z^{o(l)}\in\Sigma_{v(l)}$ disjoint from the connecting nodes. Then g_Γ is the arithmetic genus of Σ, or equally, $2\chi(\Gamma)$ is the Euler characteristic of Σ. In other words,

$$(\Sigma,\vec{z})=\coprod_{v\in\mathbb{V}}(\Sigma_v,\vec{z}_v,q_v)/\sim,\quad q_{\underrightarrow{e}}\sim q_{\underleftarrow{e}}\quad\forall\ e\in E, \tag{5.3}$$

[50] Also called flags or roots.

[51] We mean a smooth closed oriented surface.

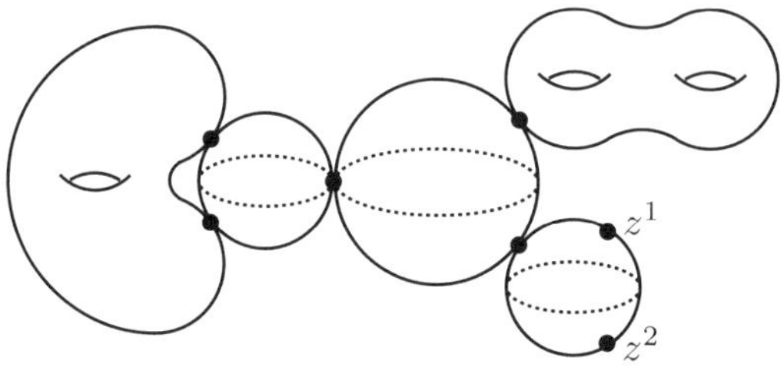

FIGURE 8. A nodal curve in $\overline{\mathcal{M}}_{4,2}$.

where

$$\vec{z}_v = \vec{z} \cap \Sigma_v \quad \text{and} \qquad q_v = \{q_{\underset{\rightarrow}{e}} : \underset{\rightarrow}{e} \in \underset{\rightarrow}{\mathbb{E}}_v\}.$$

We treat q_v as an un-ordered set of marked points on Σ_v. If we fix an ordering on the set q_v, we denote the ordered set by $\vec{q}_v$. For a nodal marked surface as in (5.3), we say Γ is the **dual graph** of $(\Sigma, \vec{z})$.

A complex structure $\mathfrak{j}$ on Σ is a set of complex structures $(\mathfrak{j}_v)_{v\in\mathbb{V}}$ on its components. By a (complex) marked nodal curve, we mean a marked nodal real surface together with a complex structure $(\Sigma, \mathfrak{j}, \vec{z})$. Figure 8 illustrates a nodal curve with $(g_1, g_2, g_3, g_4, g_5) = (0, 2, 0, 1, 0)$ corresponding to Figure 7-left.

Similarly, for nodal marked surfaces mapping into a topological space X, we consider similar decorated graphs where the vertices carry an additional **degree labeling**

$$A : \mathbb{V} \longrightarrow H_2(X, \mathbb{Z}), \quad v \longrightarrow A_v,$$

recording the homology class of the image of the corresponding component. Figure 7-right illustrates a dual graph associated to a marked nodal map over the graph on the left.

For every $v \in \mathbb{V}$, let Γ_v be the decorated graph with $\mathbb{V}(\Gamma_v) = \{v\}$, $\mathbb{E}(\Gamma_v) = \emptyset$, and

$$\mathbb{L}(\Gamma_v) = \{l \in \mathbb{L}(\Gamma) : v(l) = v\} \cup \underset{\rightarrow}{\mathbb{E}}_v(\Gamma). \tag{5.4}$$

An order function on $\mathbb{L}(\Gamma_v)$ depends on an extension of the ordering on the first set on the right-hand side of (5.4) to an ordering of the nodes corresponding to $\underset{\rightarrow}{\mathbb{E}}_v(\Gamma)$. Each sub-graph Γ_v describes the topological type of the component corresponding to v in the normalization of the corresponding nodal marked surface.

With notation as above, for every nodal marked curve $C \equiv (\Sigma, \mathfrak{j}, \vec{z})$ with dual graph Γ, let

$$\begin{aligned} \mathfrak{Aut}(C) = \Big\{ h \equiv \{h_v : \Sigma_v \to \Sigma_{h(v)}\}_{\{v\in\mathbb{V}\}} : h : \Sigma \longrightarrow \Sigma \text{ is continous} \\ h(z^i) = z^i \ \forall z^i \in \vec{z},\ h^*\mathfrak{j}_{h(v)} = \mathfrak{j}_v \ \forall v \in \mathbb{V} \Big\} \end{aligned} \tag{5.5}$$

be the group of biholomorphic automorphisms of C. A marked nodal curve C is called **stable** if $\mathfrak{Aut}(C)$ is finite; this is the case if and only if all $(\Sigma_v, \mathfrak{j}_v, \vec{z}_v \cup \vec{q}_v)$ are stable (for any choice of ordering on q_v).

REMARK 5.1.1. For every $v \in \mathbb{V}$ and $h \in \mathfrak{Aut}(C)$ such that $h(v) = v$, the automorphism h preserves q_v (as a set) but it may permute the points non-trivially; thus, if we fix an ordering $\vec{q}_v$, for all $v \in \mathbb{V}$, $\mathfrak{Aut}(C)$ contains the product

$$\prod_{v\in\mathbb{V}} \mathfrak{Aut}(\Sigma_v, \mathfrak{j}_v, \vec{z}_v \cup \vec{q}_v)$$

but it may have (finitely many) more elements permuting different components or some of q_v.

A nodal map u from $(\Sigma, \mathfrak{j}, \vec{z})$ into a manifold X is a collection of (sufficiently smooth) maps

$$u_v : \Sigma_v \longrightarrow X \quad \forall v \in \mathbb{V} \quad \text{s.t.} \quad u_v(q_{\underrightarrow{e}}) = u_{v'}(q_{\underleftarrow{e}}) \quad \forall\ v, v' \in \mathbb{V},\ \underrightarrow{e} = \underrightarrow{\mathbb{E}}_{v,v'};$$

we denote this common value by $u(q_e)$. If (X, J) is an almost complex manifold, then a nodal map is **J-holomorphic** if every u_v is $(J, \mathfrak{j}_v)$-holomorphic as in (1.2). Similarly, for a marked nodal $(J, \mathfrak{j})$-holomorphic map $f = (u, \Sigma, \mathfrak{j}, \vec{z})$, let

$$\mathfrak{Aut}(f) = \{h \in \mathfrak{Aut}(\Sigma, \mathfrak{j}, \vec{z}) : u \circ h = u\}; \tag{5.6}$$

such an f is called **stable** if $\mathfrak{Aut}(f)$ is finite. We can think of a stable curve as a stable map with the trivial map to a point.

Let (X, ω) be a closed symplectic manifold. Given $g, k \in \mathbb{N}$, $J \in \mathcal{J}(X, \omega)$, and $A \in H_2(X, \mathbb{Z})$, the moduli space $\overline{\mathcal{M}}_{g,k}(X, A, J)$ is the set of equivalence classes of stable J-holomorphic maps from nodal k-marked genus g (not necessarily stable) curves into X. If X is a point pt,

$$\overline{\mathcal{M}}_{g,k} \equiv \overline{\mathcal{M}}_{g,k}(\mathrm{pt}, 0, \mathrm{null})$$

is the Deligne-Mumford space of stable k-marked genus g curves. For simplicity, especially when J is fixed throughout the discussion, we will often write $\overline{\mathcal{M}}_{g,k}(X, A)$ instead of $\overline{\mathcal{M}}_{g,k}(X, A, J)$.

If $2g + k < 3$ (resp. $A = 0$ and $2g + k < 3$), there is no stable curve (resp. map) of type (g, k). In these cases, for our notion to be consistent, we define $\overline{\mathcal{M}}_{g,k}$ (resp. $\overline{\mathcal{M}}_{g,k}(X, 0)$) to be a single point (resp. X). The moduli space $\overline{\mathcal{M}}_{g,k}(X, A)$ can be decomposed into pieces $\mathcal{M}_{g,k}(X, A)_\Gamma$, called **strata**, depending on the dual graph of the map. For $J \in \mathcal{J}(X, \omega)$, the symplectic area (energy) of a J-holomorphic map $u : (\Sigma, \mathfrak{j}) \longrightarrow X$ is given by

$$\omega(u) = \sum_{v\in\mathbb{V}} \int_{\Sigma_v} u_v^*\omega = \frac{1}{2}\sum_{v\in\mathbb{V}} \int_{\Sigma_v} |du_v|_J^2 \ \mathrm{dvol}_{\Sigma_v}. \tag{5.7}$$

Since the symplectic area of non-trivial J-holomorphic maps has a positive lower bound, see [**37**, Proposition 4.1.4], it follows from the Gromov's Compactness Theorem in Section 5.3 that the stratification

$$\overline{\mathcal{M}}_{g,k}(X, A) = \bigcup_{\Gamma} \mathcal{M}_{g,k}(X, A)_\Gamma \tag{5.8}$$

is finite. The forgetful map (1.5) takes each stratum $\mathcal{M}_{g,k}(X, A)_\Gamma$ to the stratum $\mathcal{M}_{\mathrm{st}(\Gamma)} \subset \overline{\mathcal{M}}_{g,k}$, where $\mathrm{st}(\Gamma)$ (in the stable range $2g + k \geq 3$) is the unique dual graph obtained via the stabilization process of shrinking unstable components and may have fewer vertices and edges but equal number of legs. By assumption, if

$2g+k<3$, $\mathrm{st}(\Gamma)$ is the point $\overline{\mathcal{M}}_{g,k}$. Note that each stratum $\mathcal{M}_\Gamma$ (and similarly $\mathcal{M}_{g,k}(X,A)_\Gamma$) is of the form

$$\Big(\prod_{v\in\mathbb{V}}\mathcal{M}_{g_v,k_v}\Big)/\mathfrak{Aut}(\Gamma)$$

where $k_v=|\mathbb{L}(\Gamma_v)|$ (see (5.4)) and $\mathfrak{Aut}(\Gamma)$ is the symmetry group of Γ; see Remark 5.1.1.

5.2. Orbifold structure of the Deligne-Mumford space. In this section, following the construction of Robbin-Salamon [**45**], we review the holomorphic orbifold structure of the Deligne-Mumford space. Except the two cases of $(g,k)=(1,1),(2,0)$, $\overline{\mathcal{M}}_{g,k}$ has the structure of a holomorphic effective orbifold. In those two special cases, the orbifold structure is not effective but it has an effective reduction; see Example 2.1.6. From the algebraic geometry point of view, $\overline{\mathcal{M}}_{g,k}$ has the structure of a complex projective variety representing families of stable marked nodal curves modulo isomorphism; see [**25**].

A **holomorphic family** of complex curves consists of connected open complex manifolds $\mathcal{B}$ and $\mathcal{C}$, with $\dim\mathcal{C}=\dim_{\mathbb{C}}\mathcal{B}+1$, and a proper holomorphic projection map $\pi\colon\mathcal{C}\longrightarrow\mathcal{B}$ with complex one-dimensional fibers. By the holomorphic Implicit Function Theorem, for every regular point $p\in\mathcal{C}$, there are local holomorphic coordinates $w=(w_0,\ldots,w_n)$ around p, with $n=\dim\mathcal{B}$, such that $\pi(w)=(w_1,\ldots,w_n)$. A critical point p of π is called a **nodal point** if there are local holomorphic coordinates $w=(w_0,\ldots,w_n)$ around p such that $\pi(w)=(w_0w_1,w_2,\ldots,w_n)$. A holomorphic family $\mathcal{C}\longrightarrow\mathcal{B}$ is called **nodal** if all the critical points are nodal.

Given $k\in\mathbb{N}$, a ***k*-marked nodal family** of complex curves consists of a nodal family $\pi\colon\mathcal{C}\longrightarrow\mathcal{B}$ and an ordered set of holomorphic sections

$$\vec{\mathfrak{z}}=\{\mathfrak{z}^i\colon\mathcal{B}\longrightarrow\mathcal{C}\quad\forall i\in[k]\}$$

away from the nodal points, such that the graph of every two $\mathfrak{z}^i$ are mutually disjoint. For every $a\in\mathcal{B}$, let $\Sigma_a=\pi^{-1}(a)$, $\mathfrak{j}_a$ be the restriction of complex structure of $\mathcal{C}$ to Σ_a, and

$$\vec{z}_a=(\mathfrak{z}^1(a),\ldots,\mathfrak{z}^k(a))\subset\Sigma_a;$$

then, $C_a=(\Sigma_a,\mathfrak{j}_a,\vec{z}_a)$ is a connected nodal curve of some non-negative genus g with k-marked points. The arithmetic genus of fibers of a nodal family is constant, cf. [**45**, Lemma 4.6]; therefore, the genus of a nodal family is well-defined. A marked family is called **stable** if each of its fibers is stable. The stability is an open condition.

Given two k-marked nodal families

$$\mathcal{V}_i=\big(\pi_i\colon\mathcal{C}_i\longrightarrow\mathcal{B}_i,\quad\vec{\mathfrak{z}}_i\colon\mathcal{B}_i\longrightarrow\mathcal{C}_i\big),\quad\text{with } i=1,2,$$

a **morphism** between $\mathcal{V}_1$ and $\mathcal{V}_2$ is a commutative diagram of holomorphic maps

$$\begin{array}{ccc}\mathcal{C}_1 & \xrightarrow{\varphi^{\mathcal{C}}} & \mathcal{C}_2\\ \pi_1\big\downarrow\big\uparrow(\mathfrak{z}_1^i)_{i\in[k]} & & \pi_2\big\downarrow\big\uparrow(\mathfrak{z}_2^i)_{i\in[k]}\\ \mathcal{B}_1 & \xrightarrow{\varphi^{\mathcal{B}}} & \mathcal{B}_2\end{array}$$

such that the restriction of $\varphi^{\mathcal{C}}$ to each fiber of $\mathcal{C}_1$ is a holomorphic isomorphism.

DEFINITION 5.2.1. *Given a marked nodal curve* $C=(\Sigma,\mathfrak{j},\vec{z})$, *an* ***unfolding*** *of* C *consists of a marked nodal family*

$$(\pi\colon \mathcal{C}\longrightarrow \mathcal{B},\vec{\mathfrak{z}},b\in\mathcal{B}) \tag{5.9}$$

and an identification of C *with the base curve* $C_b=(\Sigma_b,\mathfrak{j}_b,\vec{z}_b)$. *An unfolding is called a* ***universal family*** *around* C *if for every other unfolding of* C,

$$(\pi'\colon \mathcal{C}'\longrightarrow \mathcal{B}',\vec{\mathfrak{z}}',b'\in\mathcal{B}'),$$

and any holomorphic identification

$$\varphi\colon C'_{b'}\longrightarrow C_b, \tag{5.10}$$

there is an open neighborhood $\mathcal{B}''$ *of* $b'\in\mathcal{B}'$ *and a unique base point preserving morphism (a germ of a morphism)*

$$(\varphi^{\mathcal{C}},\varphi^{\mathcal{B}})\colon (\pi'\colon \mathcal{C}'|_{\mathcal{B}''}\longrightarrow \mathcal{B}'',\vec{\mathfrak{z}}'|_{\mathcal{B}''},b\in\mathcal{B}'')\longrightarrow(\pi\colon \mathcal{C}\longrightarrow\mathcal{B},\vec{\mathfrak{z}},b\in\mathcal{B}) \tag{5.11}$$

with $\varphi^{\mathcal{C}}|_{C'_{b'}}=\varphi$.

By [**45**, Theorems 5.5,5.6], every stable marked curve admits a unique (up to isomorphism) germ of universal family. The base $\mathcal{B}$ of such universal families cover $\overline{\mathcal{M}}_{g,k}$ and form an (possibly non-effective) orbifold atlas of the complex dimension $3g-3+k$ for $\overline{\mathcal{M}}_{g,k}$. In fact, for every $h\in\mathfrak{Aut}(C_b)$, with $\mathcal{C}'=\mathcal{C}$, $b'=b$, and $\varphi=h$ in (5.10), let[52]

$$(\varphi^{\mathcal{C}}_h,\varphi^{\mathcal{B}}_h)\colon (\mathcal{C},\mathcal{B})\longrightarrow(\mathcal{C},\mathcal{B}) \tag{5.12}$$

be the resulting (germ of) isomorphism of (5.11). This gives us a holomorphic action of $\mathfrak{Aut}(C_b)$ on the germ of $\mathcal{B}$ (resp. $\mathcal{C}$) such that

$$[\mathcal{B}/\mathfrak{Aut}(C_b)] \tag{5.13}$$

is an orbifold chart centered at $[C_b]\in\overline{\mathcal{M}}_{g,k}$. By [**45**, Theorems 5.3,5.4], if (5.9) is a universal family around C_b, then for every $b'\in\mathcal{B}$ sufficiently close to b, (5.9) is also a universal family around $C_{b'}$. Moreover, the equality

$$\mathfrak{Aut}(C_{b'})=\{\varphi^{\mathcal{C}}_h|_{C_{b'}}:\ \varphi^{\mathcal{B}}_h(b')=b',\ h\in\mathfrak{Aut}(C_b)\} \tag{5.14}$$

identifies $\mathfrak{Aut}(C_{b'})$ with a subgroup of $\mathfrak{Aut}(C_b)$; thus, the size of automorphism group is an upper semi-continuous function.

If C is smooth, there exists a natural identification

$$T_b\mathcal{B}=H^1(\mathcal{T}\Sigma_{\mathfrak{j}}(-\vec{z}))\cong \mathrm{Ext}^1(\Omega^{1,0}_{\Sigma,\mathfrak{j}}(\vec{z}),\mathcal{O}_{\Sigma,\mathfrak{j}}) \tag{5.15}$$

where $\mathcal{T}\Sigma_{\mathfrak{j}}(-\vec{z})$ is the logarithmic[53] holomorphic[54] tangent bundle of $(\Sigma,\mathfrak{j},\vec{z})$, $\Omega^{1,0}_{\Sigma,\mathfrak{j}}(\vec{z})$ is the sheaf of logarithmic holomorphic 1-forms with at most simple poles at $\vec{z}$, $\mathcal{O}_{\Sigma,\mathfrak{j}}$ is the structure sheaf of $(\Sigma,\mathfrak{j})$, and Ext^* are hyper-cohomology functors. We will avoid using these facts as much as possible in our calculations. The second identification has an analogue for nodal curves as well. The conclusion is that $\overline{\mathcal{M}}_{g,k}$ is a holomorphic orbifold of the complex dimension

$$3(g-1)+k=h^1(\mathcal{T}\Sigma_{\mathfrak{j}}(-\vec{z})).$$

[52] By abuse of notation, and after possibly replacing $\mathcal{B}$ with a smaller neighborhood of b, we write $\mathcal{B}$ instead of $\mathcal{B}''$.

[53] $\mathcal{T}\Sigma_{\mathfrak{j}}(-\vec{z})$ is isomorphic to the tensor product of $\mathcal{T}\Sigma_{\mathfrak{j}}$ and the line bundle $\mathcal{O}_{\Sigma,\mathfrak{j}}(-\vec{z})$. It corresponds to the sheaf of holomorphic tangent vector fields vanishing at $\vec{z}$.

[54] With respect to the complex structure $\mathfrak{j}$.

We refer to Definition 6.2 and 6.4, Proposition 6.3, and Theorem 6.5 and 6.6 in [**45**] for further details on the (Lie groupoid) orbifold structure of $\overline{\mathcal{M}}_{g,k}$ obtained via the base of universal families above.

Considering the total space of universal families $\mathcal{C}$ above, instead of just the bases $\mathcal{B}$, and identifying them along their intersections, we obtain a so called **universal curve** $\overline{\mathcal{C}}_{g,k}$ over $\overline{\mathcal{M}}_{g,k}$ which is an effective holomorphic orbifold of one complex dimension higher. The universal curve admits a projection map $\pi\colon \overline{\mathcal{C}}_{g,k} \longrightarrow \overline{\mathcal{M}}_{g,k}$ such that for every marked curve $C_a = (\Sigma_a, \mathfrak{j}_a, \vec{z}_a)$ representing $[C_a]\in\overline{\mathcal{M}}_{g,k}$,

$$\pi^{-1}([C_a]) \cong [\Sigma_a/\mathfrak{Aut}(C_a)].$$

In fact, $\overline{\mathcal{C}}_{g,k} \cong \overline{\mathcal{M}}_{g,k+1}$ with $\pi=\pi_{k+1}$ as in (1.6).

REMARK 5.2.2. Let $C=(\Sigma, \mathfrak{j}, \vec{z})$ be a smooth stable marked curve; i.e. $[C]\in \mathcal{M}_{g,k}$. In this case, there exists a sufficiently small universal family $(\pi\colon \mathcal{C} \longrightarrow \mathcal{B}, \vec{\mathfrak{z}}, b)$ around C such that $\mathcal{C}$ is smoothly trivial, i.e. there exists an $\mathfrak{Aut}(C)$-equivariant[55] diffeomorphism

$$\varphi\colon \mathcal{C} \longrightarrow \Sigma\times\mathcal{B}$$

such that $\pi\circ\varphi^{-1}$ is the projection to the second component,

$$(\varphi\circ\mathfrak{z}^i)(x) \equiv (z^i\times x, x) \quad \forall x\in\mathcal{B},\ i\in[k],$$

(i.e. every section $\varphi_*(\mathfrak{z}^i)$ is constant), $\varphi_b=\mathrm{id}_\Sigma$, and the action of $\mathfrak{Aut}(C)$ on $\Sigma\times\mathcal{B}$ is a semidirect-product action. Here, for every $a\in\mathcal{B}$, φ_a is the restriction of φ to $\Sigma_a=\pi^{-1}(a)\subset\mathcal{C}$. In this sense, we can think of $\mathcal{B}$ as a smooth family of complex structures $\{\mathfrak{j}_{\varphi,a}=(\varphi_a)_*(\mathfrak{j}_a)\}_{a\in\mathcal{B}}$ on the fixed marked surface $(\Sigma,\vec{z})$ with $\mathfrak{j}_{\varphi,b}=\mathfrak{j}_b=\mathfrak{j}$. Note that the $\mathfrak{Aut}(C)$-equivariance of φ implies that for every $a\in\mathcal{B}$ and $h\in\mathfrak{Aut}(C)$, with $a'=\varphi_h^{\mathcal{B}}(a)$ as in (5.12), the map

$$h\colon (\Sigma,\mathfrak{j}_{\varphi,a})\longrightarrow(\Sigma,\mathfrak{j}_{\varphi,a'}) \tag{5.16}$$

is a holomorphic isomorphism.

REMARK 5.2.3. In what follows, we also need to consider marked curves $C=(\Sigma,\mathfrak{j},\vec{z},q)$ where $q=\{q_1,\dots,q_\ell\}$ is an un-ordered set of disjoint marked points away from $\vec{z}$; e.g. they appear in decomposition of nodal curves, in which case q is the pre-image of nodal points in a smooth component. Let

$$\mathfrak{Aut}(\Sigma,\mathfrak{j},\vec{z},q)=\big\{h\in\mathfrak{Aut}(\Sigma,\mathfrak{j},\vec{z}) : h(\{q_1,\dots,q_\ell\})=\{q_1,\dots,q_\ell\}\big\}. \tag{5.17}$$

After fixing an ordering $\vec{q}$ of q, the automorphism group $\mathfrak{Aut}(\Sigma,\mathfrak{j},\vec{z}\cup\vec{q})$ may become a proper subgroup[56] of (5.17); see Remark 5.1.1. Assume $C'=(\Sigma,\mathfrak{j},\vec{z}\cup\vec{q})$ is stable and let

$$(\pi\colon\mathcal{C}\longrightarrow\mathcal{B},\vec{\mathfrak{z}}\cup\vec{\mathfrak{q}},b\in\mathcal{B})$$

be a universal family around C'. After forgetting the ordering on $\vec{\mathfrak{q}}$, the action of $\mathfrak{Aut}(C')$ on $\mathcal{C}$ discussed prior to (5.13) extends to an action of (5.17) on

$$(\pi\colon\mathcal{C}\longrightarrow\mathcal{B},\ \vec{\mathfrak{z}},\mathfrak{q}\colon\mathcal{B}\longrightarrow\mathcal{C},\ b\in\mathcal{B}).$$

[55] The first factor of the map φ could be defined to send every $x\in\mathcal{C}$ to the closest point in Σ with respect to some properly defined metric $\mathcal{G}$ on $\mathcal{C}$. If we choose $\mathcal{G}$ to be $\mathfrak{Aut}(C)$-invariant, then φ becomes equivariant.

[56] This subgroup is independent of choice of ordering.

Therefore, by a Universal family around C we mean a universal family around C', in which we forget the added data of ordering on $\vec{\mathfrak{q}}$ and extend the group action to (5.17).

Let $C=(\Sigma,\mathfrak{j},\vec{z})$ be a stable nodal curve with the dual graph Γ, with the set of vertices $\mathbb{V}$, edges $\mathbb{E}$, and legs $\mathbb{L}$. For every $v\in\mathbb{V}$, let

$$(\pi_v\colon \mathcal{C}_v\to\mathcal{B}_v,\vec{\mathfrak{z}}_v,\mathfrak{q}_v,b_v) \tag{5.18}$$

be a sufficiently small universal family around $C_v\equiv(\Sigma_v,\mathfrak{j}_v,\vec{z}_v,q_v)$ as above that includes no nodal curve, see Remark 5.2.2. Then there exists a **standard** extension of (5.18) to a universal family $(\pi\colon\mathcal{C}\longrightarrow\mathcal{B},\vec{\mathfrak{z}},b)$ around $C=C_b$ with

$$\mathcal{B}=\prod_{v\in\mathbb{V}}\mathcal{B}_v\times\Delta^{\mathbb{E}}\quad\text{and}\quad b=(b_v)_{v\in\mathbb{V}}\times\{0\}^{\mathbb{E}}\in\mathcal{B}, \tag{5.19}$$

where Δ is a disk of sufficiently small radius δ in $\mathbb{C}$ and $\Delta^{\mathbb{E}}=\prod_{e\in\mathbb{E}}\Delta$. More precisely, $\mathcal{C}$ is obtained from $\{\mathcal{C}_v\}_{v\in\mathbb{V}}$ in the following way.

For every $v\in\mathbb{V}$ and $\underrightarrow{e}\in\underrightarrow{\mathbb{E}}_v$, the section

$$\mathcal{D}_{\underrightarrow{e}}=\mathfrak{q}_v^{\underrightarrow{e}}(\mathcal{B}_v)\subset\mathcal{C}_v \tag{5.20}$$

corresponding to the leg[57] (in other words, to the marked point associated to that leg) $\underrightarrow{e}\in\mathbb{L}(\Gamma_v)$ as in (5.4) is a (Cartier) divisor. Therefore, for $\{\mathcal{B}_v\}_{v\in\mathbb{V}}$ sufficiently small, there exist

- $\varepsilon>0$,
- a set of mutually disjoint[58] open neighborhoods $U_{\underrightarrow{e}}$ of $\mathcal{D}_{\underrightarrow{e}}$ in $\mathcal{C}_v$, for all $v\in\mathbb{V}$ and $\underrightarrow{e}\in\underrightarrow{\mathbb{E}}_v$,
- and holomorphic functions

$$w_{\underrightarrow{e}}\colon U_{\underrightarrow{e}}\longrightarrow\mathbb{C}\quad\text{s.t.}\quad\mathcal{D}_{\underrightarrow{e}}=w_{\underrightarrow{e}}^{-1}(0)\quad\text{and}\quad\mathrm{cl}(U_{\underrightarrow{e}}^{\varepsilon})\subset U_{\underrightarrow{e}}, \tag{5.21}$$

where $U_{\underrightarrow{e}}^{\varepsilon}=\{y\in U_{\underrightarrow{e}}\colon|w_{\underrightarrow{e}}(y)|<\varepsilon\}$.

LEMMA 5.2.4. *For every stable nodal curve C with dual graph Γ, there exists a choice of universal families $\{\mathcal{C}_v\}_{v\in\mathbb{V}}$ and holomorphic functions $\{w_{\underrightarrow{e}}\}_{\underrightarrow{e}\in\underrightarrow{\mathbb{E}}}$ as in (5.21) which are compatible with the action of $\mathfrak{Aut}(C)$ in the following sense.*

(1) *For $v\in\mathbb{V}$ and $h\in\mathfrak{Aut}(C)$, we have $\mathcal{C}_v\cong\mathcal{C}_{h(v)}$.*
(2) *For every $h\in\mathfrak{Aut}(C)$*

$$U_{h(\underrightarrow{e})}=\varphi_h^{\mathcal{C}}(U_{\underrightarrow{e}})\qquad\forall\ \underrightarrow{e}\in\underrightarrow{\mathbb{E}},$$

where, for each $v(=v_1(\underrightarrow{e}))\in\mathbb{V}$, $\varphi_h^{\mathcal{C}}\colon\mathcal{C}_v\longrightarrow\mathcal{C}_{h(v)}\cong\mathcal{C}_v$ is the isomorphism corresponding to h in (5.12).
(3) *For every $e\in\mathbb{E}$ and $h\in\mathfrak{Aut}(C)$ of order m_h, there exists an m_h-th root of unity $\mu_{h;e}\in\mathbb{C}^*$ such that*

$$(w_{h(\underrightarrow{e})}\circ\varphi_h^{\mathcal{C}})\cdot(w_{h(\underleftarrow{e})}\circ\varphi_h^{\mathcal{C}})=\mu_{h;e}\ w_{\underrightarrow{e}}\cdot w_{\underleftarrow{e}}. \tag{5.22}$$

PROOF. Only the third condition is somewhat non-trivial. Assuming Conditions (1) and (2) hold, we construct $\{w_{\underrightarrow{e}}\}_{\underrightarrow{e}\in\underrightarrow{\mathbb{E}}}$ in a way that Condition (3) holds as well.

---

[57]i.e. in (5.20), via (5.4), we think of $\underrightarrow{e}$ as a root in $\mathbb{L}(\Gamma_v)$ and $\mathfrak{q}_v^{\underrightarrow{e}}$ is the corresponding section of $\mathcal{C}_v$.

[58]Also disjoint from the sections corresponding to the ordinary marked points.

Fix an oriented edge $\underset{\rightarrow}{e}\in\underset{\rightarrow}{\mathbb{E}}$ and let $\mathfrak{Aut}(C)\cdot\underset{\rightarrow}{e}$ be the orbit space of the action of $\mathfrak{Aut}(C)$ on $\underset{\rightarrow}{e}$. Fix a function $w'_{\underset{\rightarrow}{e}}$ as in (5.21) and let $\mathfrak{Aut}(C)_{\underset{\rightarrow}{e}}\subset\mathfrak{Aut}(C)$ be the subgroup of the authomorphisms fixing $\underset{\rightarrow}{e}$. For every $h\in\mathfrak{Aut}(C)_{\underset{\rightarrow}{e}}$, $\varphi^{\mathcal{C}}_h(\mathcal{D}_{\underset{\rightarrow}{e}})=\mathcal{D}_{\underset{\rightarrow}{e}}$ and $U_{\underset{\rightarrow}{e}}$ is $\varphi^{\mathcal{C}}_h$-invariant by the second assumption. Let

$$\widetilde{w}_{\underset{\rightarrow}{e}}=\prod_{h\in\mathfrak{Aut}(C)_{\underset{\rightarrow}{e}}} w'_{\underset{\rightarrow}{e}}\circ\varphi^{\mathcal{C}}_h. \tag{5.23}$$

If $m=|\mathfrak{Aut}(C)_{\underset{\rightarrow}{e}}|$, choose $w_{\underset{\rightarrow}{e}}$ to be any branch of $\sqrt[m]{\widetilde{w}_{\underset{\rightarrow}{e}}}$. If $\underset{\leftarrow}{e}\in\mathfrak{Aut}(C)\cdot\underset{\rightarrow}{e}$, fix some h such that $\underset{\leftarrow}{e}=h(\underset{\rightarrow}{e})$, set $\mu_{h;e}=1$, and define

$$w_{\underset{\leftarrow}{e}}=w_{\underset{\rightarrow}{e}}\circ(\varphi^{\mathcal{C}}_h)^{-1};$$

otherwise, choose an arbitrary function $w'_{\underset{\leftarrow}{e}}$ as in (5.21) and construct $w_{\underset{\leftarrow}{e}}$ as above. Then, by (5.23), for every $h\in\mathfrak{Aut}(C)_{\underset{\rightarrow}{e}}=\mathfrak{Aut}(C)_{\underset{\leftarrow}{e}}$, there exists an associated m_h-th root of unity $\mu_{h;e}\in\sqrt[m_h]{1}\subset\sqrt[m]{1}$, such that

$$(w_{\underset{\rightarrow}{e}}\circ\varphi^{\mathcal{C}}_h)\cdot(w_{\underset{\leftarrow}{e}}\circ\varphi^{\mathcal{C}}_h)=\mu_{h;e}\,w_{\underset{\rightarrow}{e}}\cdot w_{\underset{\leftarrow}{e}}\,.$$

For every other[59] $\underset{\rightarrow}{e}'\in\mathfrak{Aut}(C)\cdot\underset{\rightarrow}{e}$, fix some $h'\in\mathfrak{Aut}(C)$ such that $\underset{\rightarrow}{e}'=h'(\underset{\rightarrow}{e})$; this also implies $\underset{\leftarrow}{e}'=h'(\underset{\leftarrow}{e})$. Let

$$w_{\underset{\rightarrow}{e}'}=w_{\underset{\rightarrow}{e}}\circ(\varphi^{\mathcal{C}}_{h'})^{-1},\qquad w_{\underset{\leftarrow}{e}'}=w_{\underset{\leftarrow}{e}}\circ(\varphi^{\mathcal{C}}_{h'})^{-1},\quad \mu_{h';e}=1.$$

If h'' is another automorphism satisfying $\underset{\rightarrow}{e}'=h''(\underset{\rightarrow}{e})$, there exists a unique $h\in\mathfrak{Aut}(C)_{\underset{\rightarrow}{e}}$ such that $h''=h'\circ h$. Thus, together with the choices made above, via

$$\mu_{h'';e}=\mu_{h';e}\cdot\mu_{h;e}=\mu_{h;e}$$

we obtain $\mu_{h'';e}$ for all other elements of $\mathfrak{Aut}(C)$ satisfying (5.22); i.e.

$$\begin{aligned}&(w_{h''(\underset{\rightarrow}{e})}\cdot w_{h''(\underset{\leftarrow}{e})})\circ\varphi^{\mathcal{C}}_{h''}=(w_{h'(\underset{\rightarrow}{e})}\cdot w_{h'(\underset{\leftarrow}{e})})\circ\varphi^{\mathcal{C}}_{h''}=\\&\big((w_{h'(\underset{\rightarrow}{e})}\cdot w_{h'(\underset{\leftarrow}{e})})\circ\varphi^{\mathcal{C}}_{h'}\big)\circ\varphi^{\mathcal{C}}_h=(w_{\underset{\rightarrow}{e}}\cdot w_{\underset{\leftarrow}{e}})\circ\varphi^{\mathcal{C}}_h=\\&\mu_{h;e}\,w_{\underset{\rightarrow}{e}}\cdot w_{\underset{\leftarrow}{e}}=\mu_{h'';e}\,w_{\underset{\rightarrow}{e}}\cdot w_{\underset{\leftarrow}{e}}.\end{aligned}$$

□

Going back to the construction of $\mathcal{C}$, for every $e\in\mathbb{E}$, let

$$\mathcal{C}_e=\Big\{(y_{\underset{\rightarrow}{e}},y_{\underset{\leftarrow}{e}},\lambda_e)\in U_{\underset{\rightarrow}{e}}\times U_{\underset{\leftarrow}{e}}\times\Delta^e:\,w_{\underset{\rightarrow}{e}}(y_{\underset{\rightarrow}{e}})\cdot w_{\underset{\leftarrow}{e}}(y_{\underset{\leftarrow}{e}})=\lambda_e\Big\}\times\prod_{\substack{v\in\mathbb{V}\\ v\notin e}}\mathcal{B}_v\times\Delta^{\mathbb{E}\setminus\{e\}},$$

$$\mathcal{C}_{\underset{\rightarrow}{e}}=\mathcal{C}_e\cap\Big(\big(U_{\underset{\rightarrow}{e}}\setminus\mathrm{cl}(U^{\varepsilon}_{\underset{\rightarrow}{e}})\big)\times U_{\underset{\leftarrow}{e}}\times\Delta^e\times\prod_{\substack{v\in\mathbb{V}\\ v\notin e}}\mathcal{B}_v\times\Delta^{\mathbb{E}\setminus\{e\}}\Big).$$

For every $v\in\mathbb{V}$, let

$$\mathcal{C}^{\circ}_v=\big(\mathcal{C}_v\setminus\bigcup_{\underset{\rightarrow}{e}\in\underset{\rightarrow}{\mathbb{E}}_v}\mathrm{cl}(U^{\varepsilon}_{\underset{\rightarrow}{e}})\big)\times\prod_{v'\in\mathbb{V}\setminus\{v\}}\mathcal{B}_{v'}\times\Delta^{\mathbb{E}}. \tag{5.24}$$

Then, for sufficiently small δ,

$$\mathcal{C}=\Big(\coprod_{v\in\mathbb{V}}\mathcal{C}^{\circ}_v\coprod_{e\in\mathbb{E}}\mathcal{C}_e\Big)/\sim\quad\text{and}\quad\vec{\mathfrak{z}}=(\mathfrak{z}^l)_{l\in\mathbb{L}(\Gamma)}, \tag{5.25}$$

[59] i.e. besides $\underset{\rightarrow}{e}$ and $\underset{\leftarrow}{e}$.

with the natural projection map, where

$$\mathfrak{z}^l\big((x_v)_{v\in\mathbb{V}},(\lambda_e)_{e\in\mathbb{E}}\big)=[\mathfrak{z}^l_{v(l)}(x_{v(l)}),(x_{v'})_{v'\in\mathbb{V}\setminus\{v\}},(\lambda_e)_{e\in\mathbb{E}}]\in[\mathcal{C}^\circ_v]\quad\forall\, l\in\mathbb{L}(\Gamma)$$

$$\text{and}\quad \mathcal{C}_{\underrightarrow{e}}\ni\Big(\big(y_{\underrightarrow{e}},y_{\underleftarrow{e}},\lambda_e\big),(x_v)_{v\in V,v\neq v_1,v_2},(\lambda_{e'})_{e'\in\mathbb{E}\setminus\{e\}}\Big)\sim$$

$$\Big(y_{\underrightarrow{e}},\pi_{v_2}(y_{\underleftarrow{e}}),(x_v)_{v\in\mathbb{V},v\neq v_1,v_2},(\lambda_e)_{e\in\mathbb{E}}\Big)\in\mathcal{C}^\circ_{v_1}\quad\forall\, v_1,v_2\in\mathbb{V},\underrightarrow{e}\in\underrightarrow{E}_{v_1,v_2},$$

is a universal family around C. In particular, the curve over $((x_v)_{v\in\mathbb{V}},(\lambda_e)_{e\in\mathbb{E}})$ is obtained from the curves $\{\pi_v^{-1}(x_v)\}_{v\in\mathbb{V}}$ via the gluing identifications

$$w_{\underrightarrow{e}}\cdot w_{\underleftarrow{e}}=\lambda_e\quad\forall e\in\mathbb{E}.$$

If $\{w_{\underrightarrow{e}}\}_{\underrightarrow{e}\in\underrightarrow{\mathbb{E}}}$ is compatible with $\mathfrak{Aut}(C)$ as in Lemma 5.2.4, the action of $\mathfrak{Aut}(C)$ on $\prod_{v\in\mathbb{V}}\mathcal{B}_v$ and $\prod_{v\in\mathbb{V}}\mathcal{C}_v$ described in the argument after Definition 5.2.1 and Remark 5.2.3 naturally extends to (5.19) and (5.25) such that

$$h(\lambda_e)=\mu_{h;e}\lambda_{h(e)}\qquad\forall\, h\in\mathfrak{Aut}(C),\ e\in\mathbb{E}.\tag{5.26}$$

In Section 6, we will use universal families of standard type as in (5.25) to build Kuranishi charts around maps with nodal domain.

5.3. Gromov Topology. The underlying topological space of the orbifold structure, constructed in [**45**], on the Deligne-Mumford space $\overline{\mathcal{M}}_{g,k}$, has a basis for the open sets the collection of all sets

$$(\mathcal{B}/\mathfrak{Aut}(C))\subset\overline{\mathcal{M}}_{g,k}\tag{5.27}$$

where each $\mathcal{B}$ is the base of a universal family as in Definition 5.2.1. With respect to this topology $\overline{\mathcal{M}}_{g,k}$ is compact and Hausdorff; the proof uses the notion of DM-convergence of a sequence of marked nodal curves, see [**45**, Section 14]. In this section, after a quick review of the convergence problem in the Deligne-Mumford space, we recall the more general notion of Gromov convergence for a sequence of marked nodal stable maps. This sequential convergence topology is also compact and metrizable.

DEFINITION 5.3.1. *Given a k-marked genus g (possibly not stable) nodal surface $C\equiv(\Sigma,\vec{z})$ with dual graph Γ, a* ***cutting configuration*** *with dual graph Γ' is a set of disjoint embedded circles*

$$\gamma\equiv\{\gamma_e\}_{e\in\mathbb{E}(\Gamma'/\Gamma)}\subset\Sigma,$$

away from the nodes and marked points, such that the nodal marked surface $(\Sigma',\vec{z}')$ obtained by pinching every γ_e into a node q_e has dual graph Γ'.

Thus, a cutting configuration corresponds to a continuous map

$$\varphi_\gamma\colon C\longrightarrow C',$$

denoted by a γ-**degeneration**[60] in what follows, onto a k-marked genus g nodal surface C' with dual graph Γ' such that $\vec{z}'=\varphi_\gamma(\vec{z})$, the preimage of every node of Σ is either a node in Σ' or a circle in γ, and the restriction

$$\varphi_\gamma\colon\Sigma\backslash\gamma\longrightarrow\Sigma'\backslash(\varphi_\gamma(\gamma)\equiv\{q_e\}_{e\in\mathbb{E}(\Gamma'/\Gamma)})$$

[60]It is called deformation in [**45**].

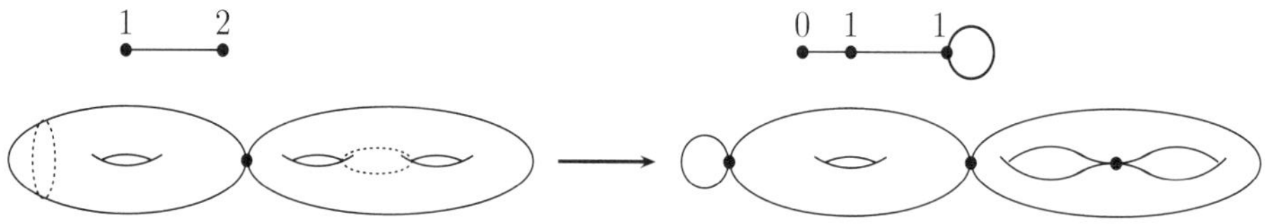

FIGURE 9. Left- a 1-nodal curve of genus 3 and a cutting set made of two circles. Right- the resulting pinched curve.

is a diffeomorphism. Let

$$\gamma^*\colon \Gamma' \longrightarrow \Gamma \tag{5.28}$$

be the map corresponding to φ_γ between the dual graphs. We have

$$\mathbb{E}(\Gamma') \cong \mathbb{E}(\Gamma) \cup \mathbb{E}(\Gamma'/\Gamma), \qquad \mathbb{L}(\Gamma') \cong \mathbb{L}(\Gamma),$$

such that $\gamma^*|_{\mathbb{E}(\Gamma)\subset\mathbb{E}(\Gamma')}$ and $\gamma^*|_{\mathbb{L}(\Gamma')}$ are isomorphisms and

$$\gamma^*\colon \mathbb{E}(\Gamma'/\Gamma) \longrightarrow \mathbb{V}(\Gamma)$$

sends the edge e corresponding to γ_e to v, if $\gamma_e \subset \Sigma_v$. For every $v' \in \mathbb{V}(\Gamma')$ there exists a unique $v \in \mathbb{V}(\Gamma)$ and a connected component $U_{v'}$ of $\Sigma_v \setminus \{\gamma_e\}_{e\in\mathbb{E}(\Gamma'/\Gamma)}$ such that $\Sigma'_{v'} \subset \Sigma'$ is obtained by collapsing the boundaries of $\mathrm{cl}(U_{v'})$. This identification determines the surjective map

$$\gamma^*\colon \mathbb{V}(\Gamma') \longrightarrow \mathbb{V}(\Gamma), \quad v' \longrightarrow v.$$

From another perspective, a cutting configuration corresponds to expanding each vertex $v \in \Gamma$ into a sub-graph $\Gamma'_v \subset \Gamma'$ (some times, this is just adding more loops to the existing graph) with the set of vertices and edges

$$\mathbb{V}(\Gamma'_v) = (\gamma^*)^{-1}(v) \quad \text{and} \quad \mathbb{E}(\Gamma'_v) = (\gamma^*)^{-1}(v) \cap \mathbb{E}(\Gamma'/\Gamma).$$

Moreover, $g_v = g_{\Gamma'_v}$, the ordering of marked points are as before, and

$$A_v = \sum_{v' \in \mathbb{V}(\Gamma'_v)} A_{v'}. \tag{5.29}$$

Figure 9 illustrates a cutting configuration over a 1-nodal curve of genus 3 and the corresponding dual graphs.

A sequence $\{\varphi_{\gamma_a}\colon C_a \longrightarrow C'\}_{a\in\mathbb{N}}$ of degenerations of marked nodal curves is called **monotonic** if $\Gamma(C_a) \cong \Gamma$ for some fixed Γ and the induced maps $\gamma_a^*\colon \Gamma \longrightarrow \Gamma'$ are all the same. In this situation, the underlying marked nodal surfaces are diffeomorphic; i.e.,

$$(C_a, \gamma_a) \cong \big((\Sigma, \mathfrak{j}_a, \vec{z}), \gamma\big) \quad \forall\, a \in \mathbb{N}, \tag{5.30}$$

for some fixed marked surface $(\Sigma, \vec{z})$ with dual graph Γ and cutting configuration γ. In addition, with this identification, the set of nodes

$$\{q_e\}_{e\in\mathbb{E}(\Gamma'/\Gamma)} \subset \Sigma'$$

is independent of $a \in \mathbb{N}$; we denote the complement by Σ'_*.

DEFINITION 5.3.2 ([**45**, Definition 13.3]). *A sequence* $\{C_a \equiv (\Sigma_a, \mathfrak{j}_a, \vec{z}_a)\}_{a\in\mathbb{N}}$ *of genus g k-marked nodal curves* ***monotonically*** *converges to* $C' \equiv (\Sigma', \mathfrak{j}', \vec{z}')$, *if there exist a sequence of cutting configurations* γ_a *on* C_a *of type* Γ' *and a monotonic sequence* $\varphi_{\gamma_a}\colon C_a \longrightarrow C'$ *of* γ_a*-degenerations such that the sequence* $(\varphi_{\gamma_a}|_{\Sigma_a\setminus\gamma_a})_*\mathfrak{j}_a$ *converges to* $\mathfrak{j}'|_{\Sigma'_*}$ *in the* C^∞*-topology*[61].

By [**45**, Section 13], the topology underlying the holomorphic orbifold structure of Section 5.2 on $\overline{\mathcal{M}}_{g,k}$ is equivalent to the sequential DM-convergence topology: a sequence $\{C_a\}_{a\in\mathbb{N}}$ of genus g k-marked stable nodal curves DM-converges to C' if a subsequence of that monotonically converges to C'.

EXAMPLE 5.3.3. For $a \in \mathbb{N}$ sufficiently large, let $(\Sigma_a, \mathfrak{j}_a) = \mathbb{P}^1$ with $\vec{z}_a = (0, \frac{1}{a}, 1, \infty)$,

$$(\Sigma, \mathfrak{j}) = \mathbb{P}^1_0 \; {}_\infty\cup_0 \, \mathbb{P}^1_\infty$$

be a 1-nodal genus 0 curve made of two copies $\mathbb{P}^1_0$ and $\mathbb{P}^1_\infty$ of $\mathbb{P}^1$ and

$$\vec{z} = \big((0,1) \subset (\mathbb{P}^1_0)^2\big) \cup \big((1,\infty) \subset (\mathbb{P}^1_\infty)^2\big),$$

and γ_a be the cutting configuration given by the circle

$$\{z \in \mathbb{P}^1 \cong \mathbb{C} \cup \{\infty\} \colon |z| = \frac{1}{\sqrt{a}}\}.$$

Let $\eta\colon \mathbb{R}^{\geq 0} \longrightarrow \mathbb{R}^{\geq 0}$ be an increasing cutoff function such that $\eta \equiv 0$ on $[0,1]$ and $\eta \equiv 1$ on $[2,\infty)$. Define $\varphi_{\gamma_a}\colon \Sigma_a \longrightarrow \Sigma$ by

$$\varphi_{\gamma_a}(z) = \begin{cases} \eta(\sqrt{a}|z|)z \in \mathbb{P}^1_\infty & \text{if } |z| > \frac{1}{\sqrt{a}}, \\ \eta((\sqrt{a}|z|)^{-1})az \in \mathbb{P}^1_0 & \text{if } |z| < \frac{1}{\sqrt{a}}, \\ \mathbb{P}^1_0 \ni \infty \sim 0 \in \mathbb{P}^1_\infty & \text{if } |z| = \frac{1}{\sqrt{a}}. \end{cases}$$

Then $\varphi_{\gamma_a}(\vec{z}_a) = \vec{z}$ and φ_{γ_a} is holomorphic away from the neck

$$A_a = \{z \colon \frac{1}{2\sqrt{a}} \leq |z| \leq \frac{2}{\sqrt{a}}\}.$$

Since the image of $\mathbb{P}^1 - A_a$ is exhausting, the sequence $(\varphi_{\gamma_a}|_{\Sigma_a\setminus\gamma_a})_*\mathfrak{j}_a$ converges to $\mathfrak{j}|_{(P^1_0-\infty)\cup(\mathbb{P}^1_\infty-0)}$ in the C^∞-topology. Therefore, $\{C_a\}_{a=1}^\infty$ is DM-converging to C.

The following result, known as Gromov's Compactness Theorem [**21**, Theorem 1.5.B], describes a convergence topology on $\overline{\mathcal{M}}_{g,k}(X,A,J)$ which is compact and metrizable; see also [**30**, Theorem 1.2], [**57**, Theorem 0.1], [**2**, Theorem 2.2.1], and [**37**, Theorem 5.3.1] for the genus 0 case. In the special case of Deligne-Mumford space, Gromov convergence is equal to DM-convergence discussed above.

THEOREM 5.3.4. *Let* (X,ω) *be a compact symplectic manifold and* $J \in \mathcal{J}(X,\omega)$. *Let*

$$\{f_a \equiv (u_a, C_a \equiv (\Sigma_a, \mathfrak{j}_a, \vec{z}_a))\}_{a\in\mathbb{N}}$$

be a sequence of k-marked genus g stable J-holomorphic maps of bounded symplectic area into X. After passing to a subsequence, still denoted by $\{f_a\}_{a\in\mathbb{N}}$, *there exists a unique (up to automorphism) k-marked genus g stable J-holomorphic map*

$$f' \equiv (u', C' \equiv (\Sigma', \mathfrak{j}', \vec{z}'))$$

such that $\{C_a\}_{a\in\mathbb{N}}$ *monotonically converges to* C', *and the corresponding monotonic subsequence, still denoted by* $\{f_a\}_{a\in\mathbb{N}}$, *has the following properties.*

[61]Uniform convergence on compact sets with all derivatives.

(1) *We can choose the γ_a-degeneration maps $\varphi_{\gamma_a}\colon C_a \longrightarrow C'$ such that the restriction*

$$u_a|_{\Sigma_a\setminus\gamma_a} \circ \varphi_{\gamma_a}^{-1}|_{\Sigma'_*}$$

uniformly converges to $u|_{\Sigma'_*}$ over compact sets.

(2) *With $\Gamma \cong \Gamma(C_a)$ and $\Gamma' = \Gamma(C')$ as in the definition of monotonic sequences,*

$$\lim_{a\longrightarrow\infty} u_a(\gamma_{a,e}) = u'(q_e), \qquad q_e \equiv \varphi_{\gamma_a}(\gamma_{a,e}) \qquad \forall\ e \in \mathbb{E}(\Gamma'/\Gamma).$$

(3) *Symplectic area of f' coincides with the symplectic area of f_a, for all $a \in \mathbb{N}$.*

It follows from the properties (1) and (3) that for every $v' \in \Gamma'$, with $U_{a,v'} \subset \Sigma_a$ as in the definition of $\gamma_a^*(v')$,

$$\lim_{a\longrightarrow\infty} \int_{\mathrm{cl}(U_{a;v'})} {u_a}^*\omega = \int_{\Sigma'_{v'}} (u')^*\omega.$$

Moreover, the stronger identity (5.29) holds. With respect to the identification of the domains and degeneration maps

$$(\varphi_{\gamma_a}\colon \Sigma_a \longrightarrow \Sigma') \cong (\varphi_\gamma\colon \Sigma \longrightarrow \Sigma')$$

as in (5.30), the second property implies that the sequence $(u_a\colon \Sigma \longrightarrow X)_{a\in\mathbb{N}}$ C^0-converge to $u \circ \varphi_\gamma$. The construction of primary Kuranishi charts in Section 6 gives a rather different description of the Gromov topology, generalizing open sets of the form (5.27); see Remarks 6.2.11 and 6.3.3, and Section 6.4.

6. Kuranishi structure over moduli space of stable maps

In this section, for every $\overline{\mathcal{M}}_{g,k}(X,A,J)$, we sketch the construction of a class of natural Kuranishi structures that together with the results of Section 4, gives us the VFC of Theorem 1.4.1.

To this end, in Sections 6.2, 6.3, and 6.4, we construct a class of Kuranishi charts, called **primary charts**, around maps with smooth stable domain, nodal stable domain, and unstable domain, respectively. The construction of every primary chart depends (at most) on the choice of a so called obstruction space, stabilizing data, slicing divisors, smooth trivializations of the universal families, and holomorphic coordinates at the nodes in (5.21); see the list of all auxiliary data in Page 236.

In Section 6.5, we construct a class of natural Kuranishi structures on $\overline{\mathcal{M}}_{g,k}(X,A,J)$ in the sense that every Kuranishi chart is obtained from the auxiliary data of a finite set of primary charts. We call these **induced charts**. In Section 6.6, we show that the induced charts admit coordinate change maps.

The construction of Kuranishi charts around maps with nodal domain involves a gluing theorem with respect to the gluing parameters of (5.19). In order for the Kuranishi and change of coordinate maps to be smooth, we need a strong gluing theorem and few other analytical results that, due to the lack of space, we will only recall from [**14**] in Sections 6.3 and 6.5.

6.1. Analytics preliminaries. Let (X,ω) be a closed symplectic manifold. Fix $J\in\mathcal{J}(X,\omega)$ and let g_J be the associated metric[62], ∇ be the Levi-Civita connection of g_J, and

$$\widetilde{\nabla}_v w := \nabla_v w - \frac{1}{2}J(\nabla_v J)w \quad \forall v,w\in\Gamma(X,TX)$$

be the associated complex linear **Hermitian** connection. The Hermitian connection $\widetilde{\nabla}$ coincides with ∇ if and only if (X,ω,J) is Kähler, i.e. $\nabla J\equiv 0$. The torsion $\widetilde{T}$ of $\widetilde{\nabla}$ is related to Nijenhueis tensor (1.1) by

$$\widetilde{T}(w,v) = -\frac{1}{4}N_J(w,v) \quad \forall w,v\in TX;$$

see [**37**, Section 2.1].

Fix $\ell\in\mathbb{N}$ and $p>1$ such that

$$\ell p>2; \tag{6.1}$$

for the necessity of this assumptions see [**37**, Section 3.1]. For the proof of gluing theorem in [**14**], the authors consider $p=2$ and ℓ sufficiently large. For a fixed domain Σ, let $W^{\ell,p}(\Sigma,X)$ be the separable Banach manifold of (ℓ,p)-smooth maps[63] from Σ into X. The local charts[64] of the Banach manifold structure $W^{\ell,p}(\Sigma,X)$ are constructed in the following way; see the proof of [**37**, Proposition 3.2.1] for more details.

REMARK 6.1.1. If Σ is the underlying smooth domain of a marked curve C, we will often write $W^{\ell,p}(\Sigma,X)$ as $W^{\ell,p}(C,X)$. The Banach manifold structure does not depend on the choice of the marked complex structure $(\mathfrak{j},\vec{z})$, but later, we will consider certain additional structures and evaluation maps on this space which do depend on the extra data. For instance, the group $\mathfrak{Aut}(C)$ of the automorphisms of C acts on $W^{\ell,p}(C,X)$ by

$$u\longrightarrow u\circ h \qquad \forall\, h\in\mathfrak{Aut}(C),\ u\in W^{\ell,p}(C,X).$$

Given a map $u\colon\Sigma\longrightarrow X$, for every $x\in\Sigma$, let

$$\exp_{u(x)}\colon T_{u(x)}X\longrightarrow X$$

be the exponentiation map of $\widetilde{\nabla}$. For $\epsilon>0$ sufficiently small, depending on the injectivity radius of g_J across Σ, $\exp_{u(x)}$ is a diffeomorphism from the ϵ-ball $B_\epsilon(x)$ around $0\in T_{u(x)}X$ onto its image, for all $x\in\Sigma$. For $\epsilon>0$ sufficiently small, given an (ℓ,p)-smooth vector field

$$\zeta\in\Gamma^{\ell,p}(\Sigma,u^*TX), \quad \text{with} \quad |\zeta(x)|<\epsilon, \tag{6.2}$$

let

$$u_\zeta\equiv\exp_u\zeta \tag{6.3}$$

be the associated close-by (ℓ,p)-smooth map[65] in $W^{\ell,p}(\Sigma,X)$. Therefore, for ϵ sufficiently small, the exponential map (6.3) gives an identification

$$\exp\colon \big(B_\epsilon(0)\subset\Gamma^{\ell,p}(\Sigma,u^*TX)\big)\longrightarrow\big(B_\epsilon(u)\subset W^{\ell,p}(\Sigma,X)\big) \tag{6.4}$$

[62] Not to be confused with the genus.

[63] For (ℓ,p) as in (6.1), with $\ell-\frac{2}{p}>m$, by Sobolev inequality, every such $W^{\ell,p}$-smooth map is at least C^m-smooth; see [**37**, Theorem B.1.11].

[64] With smooth transition maps.

[65] Note that two different maps in $W^{\ell,p}(\Sigma,X)$ can have identical images in X.

of an ϵ-neighborhood of 0 in the Banach space $\Gamma^{\ell,p}(\Sigma, u^*TX)$ with a neighborhood, denoted by $B_\epsilon(u)$, of the given map u in $W^{\ell,p}(\Sigma, X)$.

For every differentiable map $u\colon \Sigma \longrightarrow X$ and complex structure $\mathfrak{j}$ on Σ,

$$du \in \Gamma(\Sigma, \Omega^1_\Sigma \otimes u^*TX)$$

decomposes into complex linear and anti-complex linear parts,

$$\begin{aligned} \partial u &\equiv \partial_{J,\mathfrak{j}} u = \frac{1}{2}(du - J \circ du \circ \mathfrak{j}) \in \Gamma(\Sigma, \Omega^{1,0}_{\Sigma,\mathfrak{j}} \otimes_{\mathbb{C}} u^*TX) \quad \text{and} \\ \bar\partial u &\equiv \bar\partial_{J,\mathfrak{j}} u = \frac{1}{2}(du + J \circ du \circ \mathfrak{j}) \in \Gamma(\Sigma, \Omega^{0,1}_{\Sigma,\mathfrak{j}} \otimes_{\mathbb{C}} u^*TX), \end{aligned} \tag{6.5}$$

respectively, where $\Omega^{1,0}_{\Sigma,\mathfrak{j}}$ and $\Omega^{0,1}_{\Sigma,\mathfrak{j}}$ are the sheaves/vector bundles of $(1,0)$ and $(0,1)$-forms on Σ with respect to $\mathfrak{j}$, respectively; u is $(J,\mathfrak{j})$-holomorphic whenever $\bar\partial u \equiv 0$.

REMARK 6.1.2. If $f = (u, C)$ is a J-holomorphic map, the automorphism group $\mathfrak{Aut}(f)$ of f, defined in (5.6), acts on $\Gamma^{\ell,p}(\Sigma, u^*TX)$ by
(6.6)

$$\zeta \longrightarrow h^*\zeta \in \Gamma^{\ell,p}(\Sigma, h^*u^*TX) = \Gamma^{\ell,p}(\Sigma, (u\circ h)^*TX)) = \Gamma^{\ell,p}(\Sigma, u^*TX) \quad \forall h \in \mathfrak{Aut}(f).$$

The exponential map identification of (6.4) is $\mathfrak{Aut}(f)$-equivariant with respect to (6.6) and the action of $\mathfrak{Aut}(f) \subset \mathfrak{Aut}(C)$ on $W^{\ell,p}(C, X)$ as in Remark 6.1.1, i.e.

$$u_{h^*\zeta} = u_\zeta \circ h \qquad \forall h \in \mathfrak{Aut}(f),\ \zeta \in \Gamma(\Sigma, u^*TX).$$

Motivated by (6.5), for a fixed curve $C = (\Sigma, \mathfrak{j}, \vec z)$, let

$$E^{\ell-1,p}(C, X) \longrightarrow W^{\ell,p}(C, X) \tag{6.7}$$

be the infinite dimensional smooth Banach bundle whose fiber over every map $u\colon C \longrightarrow X$ is the Banach space of $(\ell-1, p)$-smooth u^*TX-valued $(0,1)$-forms

$$\Gamma^{\ell-1,p}(\Sigma, \Omega^{0,1}_{\Sigma,\mathfrak{j}} \otimes_{\mathbb{C}} u^*TX).$$

Similarly, the tangent bundle of $W^{\ell,p}(\Sigma, X)$,

$$TW^{\ell,p}(\Sigma, X) \longrightarrow W^{\ell,p}(\Sigma, X), \tag{6.8}$$

is the infinite dimensional smooth Banach bundle whose fiber over every map $u\colon \Sigma \longrightarrow X$ is the Banach space of (ℓ, p)-smooth vector fields $\Gamma^{\ell,p}(\Sigma, u^*TX)$. Similarly to Remark 6.1.1 and (6.6), the action of $\mathfrak{Aut}(C)$ on C extends to $E^{\ell-1,p}(C, X)$ and $TW^{\ell,p}(C, X)$ such the projection maps to $W^{\ell,p}(C, X)$ are $\mathfrak{Aut}(C)$-equivariant. For instance, the action of $\mathfrak{Aut}(C)$ on $E^{\ell-1,p}(C, X)$ is given by

$$(\eta, u) \longrightarrow (h^*\eta, u \circ h) \qquad \forall h \in \mathfrak{Aut}(C). \tag{6.9}$$

The local defining charts for $E^{\ell-1,p}(C, X)$ (and similarly for $TW^{\ell,p}(\Sigma, X)$) around a fixed J-holomorphic map $f = (u, C)$ are constructed in the following way. For a fixed map $u\colon C \longrightarrow X$, ζ as in (6.2), and u_ζ as in (6.3), let

$$\mathrm{P}_{u,\zeta}\colon u^*TX \longrightarrow u_\zeta^*TX$$

denote the complex line bundle isomorphism given by the parallel transport along the geodesic

$$\big(s \longrightarrow \exp_{u(x)}(s\zeta(x))\big)_{s\in[0,1]} \quad \forall x \in \Sigma$$

with respect to the complex linear connection $\widetilde\nabla$. This induces a similarly denoted complex linear isomorphism

$$\mathrm{P}_{u,\zeta}\colon E^{\ell-1,p}(C, X)|_u \longrightarrow E^{\ell-1,p}(C, X)|_{u_\zeta}. \tag{6.10}$$

These parallel translation maps, together with the local charts (6.4) for $W^{\ell,p}(C,X)$, give us local product charts

$$\begin{aligned}(6.11)\qquad &\Gamma^{\ell-1,p}(\Sigma,\Omega^{0,1}_{\Sigma,\mathfrak{j}}\otimes_{\mathbb{C}}u^*TX)\times B_\epsilon(0)\xrightarrow{P_u}E^{\ell-1,p}(C,X)|_{B_\epsilon(u)},\\ &(\eta,\zeta)\longrightarrow(\eta',u'),\quad u'=u_\zeta,\quad \eta'=\mathrm{P}_{u,\zeta}(\eta),\end{aligned}$$

for the Banach bundle (6.7). Since $\mathrm{P}_{u,h^*\zeta}(h^*\eta)=h^*\mathrm{P}_{u,\zeta}(\eta)$, similarly to Remark 6.1.2, the map P_u is $\mathfrak{Aut}(f)$-equivariant.

We now discuss[66] the case where the marked complex structure $(\mathfrak{j},\vec{z})$ on the domain Σ can vary as well. Assume $C\equiv(\Sigma,\mathfrak{j},\vec{z})$ is a smooth stable marked curve and let

$$(6.12)\qquad (\pi\colon\mathcal{C}\longrightarrow\mathcal{B},\vec{\mathfrak{z}},b\in\mathcal{B})$$

be a sufficiently small universal family around $[C]\in\mathcal{M}_{g,k}$ as in Section 5.2. Let

$$(6.13)\qquad W^{\ell,p}(\mathcal{C},X):=\big\{(u,a)\colon a\in\mathcal{B},\ u\in W^{\ell,p}(\mathcal{C}_a,X)\big\},$$

be the set of pairs (u,a) such that a is a point of $\mathcal{B}$, $\Sigma_a=\pi^{-1}(a)\subset\mathcal{C}$, and u is an (ℓ,p)-smooth map from Σ_a to X. Recall from (5.12) that (for sufficiently small $\mathcal{C}$) the action of $\mathfrak{Aut}(C)$ on C (uniquely) extends to $\mathcal{C}$. This action further extends to $W^{\ell,p}(\mathcal{C},X)$ by

$$(u,a)\longrightarrow(u\circ\varphi_h^{\mathcal{C}},(\varphi_h^{\mathcal{B}})^{-1}(a))\qquad\forall h\in\mathfrak{Aut}(C),$$

where $(\varphi_h^{\mathcal{C}},\varphi_h^{\mathcal{B}})\colon(\mathcal{C},\mathcal{B})\longrightarrow(\mathcal{C},\mathcal{B})$ is the automorphism corresponding to h in (5.12).

By definition, each fiber of the projection map $W^{\ell,p}(\mathcal{C},X)\longrightarrow\mathcal{B}$ has a Banach manifold structure but the total space does not a priori come with any natural smooth structure. In order to define a Banach manifold structure on (6.13), we need to fix a smooth trivialization of $\mathcal{C}$. Such a trivialization gives us a trivialization of $W^{\ell,p}(\mathcal{C},X)$ and thus a product Banach manifold structure on that. More precisely, by Remark 5.2.2, for $\mathcal{B}$ sufficiently small, $\mathcal{C}$ is smoothly trivial, i.e. there exists an $\mathfrak{Aut}(C)$-equivariant diffeomorphism

$$(6.14)\qquad \varphi\colon\mathcal{C}\longrightarrow\Sigma\times\mathcal{B}$$

such that $\pi\circ\varphi^{-1}$ is the projection onto the second factor, each section $\varphi\circ\mathfrak{z}^i$ is constant, and $\varphi_b=\mathrm{id}_\Sigma$. In this sense, we can think of $\mathcal{B}$ as a smooth family of complex structures $\{\mathfrak{j}_{\varphi,a}=(\varphi_a)_*(\mathfrak{j}_a)\}_{a\in\mathcal{B}}$ on the fixed marked surface $(\Sigma,\vec{z})$ with $\mathfrak{j}_{\varphi,b}=\mathfrak{j}_b=\mathfrak{j}$. A choice of such φ induces an $\mathfrak{Aut}(C)$-equivariant trivialization

$$(6.15)\qquad W^{\ell,p}(C,X)\times\mathcal{B}\xrightarrow{T_\varphi}W^{\ell,p}(\mathcal{C},X),\quad (u,a)\longrightarrow(u\circ\varphi_a,a),$$

and thus a Banach manifold structure which we denote by $W^{\ell,p}_\varphi(\mathcal{C},X)$. In particular, for a fixed J-holomorphic map $u\colon C\longrightarrow X$, the exponentiation map of (6.4) and the product structure (6.15) give us $\mathfrak{Aut}(u,C)$-equivariant local charts

$$(6.16)\quad (B_\epsilon(0)\subset\Gamma^{\ell,p}(\Sigma,u^*TX))\times\mathcal{B}\xrightarrow{\exp_\varphi}W^{\ell,p}_\varphi(\mathcal{C},X),\quad(\zeta,a)\longrightarrow(u_\zeta\circ\varphi_a,a).$$

If φ' is another such smooth trivialization, then $\varphi=\rho\circ\varphi'$, where

$$(6.17)\qquad \rho\colon\Sigma\times\mathcal{B}\longrightarrow\Sigma\times\mathcal{B},\quad \rho(x,a)=(\rho_a(x),a),$$

[66]Under the assumption that the domain is stable.

is a fiber-preserving diffeomorphism fixing the constant sections $\vec{z}$. Unless ρ_a is constant in $a\in\mathcal{B}$, the change of trivialization map

$$\begin{array}{ccc} W^{\ell,p}(C,X)\times\mathcal{B} & \xrightarrow{\ T_\varphi\ } & W^{\ell,p}(\mathcal{C},X)\,, \\ \Big\downarrow{\scriptstyle T_\rho} & \nearrow{\scriptstyle T_{\varphi'}} & \\ W^{\ell,p}(C,X)\times\mathcal{B} & & \end{array} \qquad T_\rho(u,a)=(u\circ\rho_a,a), \tag{6.18}$$

is continuous but not smooth. Therefore, the Banach manifold structure (6.15) depends on a choice of a smooth trivialization. Whenever the choice of φ is fixed or irrelevant in a discussion, for simplicity, we may omit φ from $W^{\ell,p}_\varphi(\mathcal{C},X)$ and simply write it as $W^{\ell,p}(\mathcal{C},X)$. For more on this differentiability issue see [**39**, Section 3.1].

Similarly to (6.13), for a universal family $\mathcal{C}\longrightarrow\mathcal{B}$ around C as in (6.12), let

$$\begin{aligned} E^{\ell-1,p}(\mathcal{C},X) &:= \big\{(\eta,u,a)\colon (u,a)\in W^{\ell,p}(\mathcal{C},X),\ \eta\in E^{\ell-1,p}(C_a,X)|_u\big\} \quad \text{and} \\ TW^{\ell,p}(\mathcal{C},X) &:= \big\{(\zeta,u,a)\colon (u,a)\in W^{\ell,p}(\mathcal{C},X),\ \zeta\in TW^{\ell,p}(C_a,X)|_u\big\}. \end{aligned} \tag{6.19}$$

For $\mathcal{B}$ sufficiently small, a choice of smooth trivialization φ as in (6.14) also gives us trivializations[67]

(6.20)

$$\begin{aligned} &E^{\ell-1,p}(C,X)\times\mathcal{B} \xrightarrow{T_\varphi} E^{\ell-1,p}_\varphi(\mathcal{C},X), \quad (\eta,u,a)\longrightarrow\big(\varphi_a^*[\eta]^{0,1}_{\mathfrak{j}_{\varphi,a}},u\circ\varphi_a,a\big), \quad \text{and} \\ &TW^{\ell,p}(C,X)\times\mathcal{B} \xrightarrow{T_\varphi} TW^{\ell,p}_\varphi(\mathcal{C},X), \quad (\zeta,u,a)\longrightarrow\big(\varphi_a^*\zeta,u\circ\varphi_a,a\big), \end{aligned}$$

where

$$[\eta]^{0,1}_{\mathfrak{j}_{\varphi,a}} = \frac{1}{2}\big(\eta + J\circ\eta\circ\mathfrak{j}_{\varphi,a}\big) \tag{6.21}$$

is the $(0,1)$-part of η with respect to the complex structure $\mathfrak{j}_{\varphi,a}$ on Σ. For $\mathfrak{j}_{\varphi,a}$ sufficiently close to $\mathfrak{j}$, the map $\eta\longrightarrow[\eta]^{0,1}_{\mathfrak{j}_{\varphi,a}}$ is an isomorphism between the Banach spaces

$$\Gamma^{\ell-1,p}(\Sigma,\Omega^{0,1}_{\Sigma,\mathfrak{j}}\otimes_{\mathbb{C}}u^*TX) \quad \text{and} \quad \Gamma^{\ell-1,p}(\Sigma,\Omega^{0,1}_{\Sigma,\mathfrak{j}_{\varphi,a}}\otimes_{\mathbb{C}}u^*TX).$$

Also note that $\varphi_a^*[\eta]^{0,1}_{\mathfrak{j}_{\varphi,a}}=[\varphi_a^*\eta]^{0,1}_{\mathfrak{j}_a}$. For every $h\in\mathfrak{Aut}(C)$ and $a\in\mathcal{B}$, with $a=\varphi_h^{\mathcal{B}}(a')$, by (5.16) and the $\mathfrak{Aut}(C)$-equivariance of φ we have

$$[h^*\eta]^{0,1}_{\mathfrak{j}_{\varphi,a'}} = h^*[\eta]^{0,1}_{\mathfrak{j}_{\varphi,a}} \quad \text{and} \quad \varphi_a\circ\varphi_h^{\mathcal{C}} = h\circ\varphi_{a'};$$

i.e. the trivialization maps in (6.20) are $\mathfrak{Aut}(C)$-equivariant. With respect to these smooth structures

$$\begin{aligned} E^{\ell-1,p}_\varphi(\mathcal{C},X) &\longrightarrow W^{\ell,p}_\varphi(\mathcal{C},X), \quad (\eta,u,a)\longrightarrow(u,a), \quad \text{and} \\ TW^{\ell,p}_\varphi(\mathcal{C},X) &\longrightarrow W^{\ell,p}_\varphi(\mathcal{C},X), \quad (\zeta,u,a)\longrightarrow(u,a), \end{aligned} \tag{6.22}$$

are smooth $\mathfrak{Aut}(C)$-equivariant Banach bundles. In particular, for a fixed map $u\colon C\longrightarrow X$, with notation as in (6.16), local trivializations of $E^{\ell-1,p}(C,X)$ as in (6.11) and the product structure (6.20) give us $\mathfrak{Aut}(f)$-equivariant local charts

(6.23)

$$\begin{aligned} &\Gamma^{\ell-1,p}(\Sigma,\Omega^{0,1}_{\Sigma,\mathfrak{j}}\otimes_{\mathbb{C}}u^*TX)\times B_\epsilon(0)\times\mathcal{B} \xrightarrow{P_\varphi} E^{\ell-1,p}_\varphi(\mathcal{C},X)|_{\exp_\varphi(B_\epsilon(0)\times\mathcal{B})}, \\ &(\eta,\zeta,a)\longrightarrow(\eta',u',a), \quad u'=u_\zeta\circ\varphi_a, \quad \eta'=\varphi_a^*P_{u,\zeta,a}(\eta):=\varphi_a^*[P_{u,\zeta}(\eta)]^{0,1}_{\mathfrak{j}_{\varphi,a}}. \end{aligned}$$

[67]For simplicity, we use T_φ for both the trivialization map of $W^{\ell,p}_\varphi(\mathcal{C},X)$ and $E^{\ell-1,p}_\varphi(\mathcal{C},X)$.

REMARK 6.1.3. Similarly to (6.18), for two different trivializations φ' and φ, with the change of parametrization map ρ as in (6.17), the resulting change of trivialization map T_ρ in the commutative diagram below

$$\begin{array}{ccc} E^{\ell-1,p}(C,X)\times\mathcal{B} & \xrightarrow{\;T_\varphi\;} & E^{\ell-1,p}(\mathcal{C},X) \\ \big\downarrow T_\rho & \nearrow_{T_{\varphi'}} & \\ E^{\ell-1,p}(C,X)\times\mathcal{B} & & \end{array} \tag{6.24}$$

is simply the pull-back map $(\eta,u,a)\longrightarrow(\rho_a^*\eta,u\circ\rho_a,a)$. However, it is rather impossible to write down explicit equations for the change of coordinate maps $\exp_{\varphi'}^{-1}\circ\exp_\varphi$ corresponding to (6.16) and $P_{\varphi'}^{-1}\circ P_\varphi$ corresponding to (6.24).

6.2. Case of smooth stable domain. In this section, we outline the construction of primary Kuranishi charts around maps with smooth stable domain. Assume $C\equiv(\Sigma,\mathfrak{j},\vec{z})$ is a smooth stable marked curve and let

$$(\pi\colon\mathcal{C}\longrightarrow\mathcal{B},\vec{\mathfrak{z}},b\in\mathcal{B})$$

be a sufficiently small universal family around $[C]\in\mathcal{M}_{g,k}$ as in (6.12). The non-linear Cauchy-Riemann operator

$$\bar\partial\colon W^{\ell,p}(\mathcal{C},X)\longrightarrow E^{\ell-1,p}(\mathcal{C},X),\quad u\longrightarrow\bar\partial u, \tag{6.25}$$

given by (6.5), is an $\mathfrak{Aut}(C)$-equivariant section of (6.22) whose zero locus is the set of J-holomorphic maps on the fibers of $\mathcal{C}\longrightarrow\mathcal{B}$. For any choice of smooth trivialization φ of (6.12) as in (6.14), the Cauchy-Riemann section (6.25) is smooth with respect to the smooth structures $W_\varphi^{\ell,p}(\mathcal{C},X)$ and $E_\varphi^{\ell-1,p}(\mathcal{C},X)$; see [**37**, Section 3] for more details.

Deformation theory of J-holomorphic maps close to a particular J-holomorphic map u in $W^{\ell,p}(C,X)$ (i.e. if the domain is fixed) is described by the first order $\mathfrak{Aut}(u,C)$-equivariant[68] approximation or linearization of (6.25)[69], which is of the form

$$\begin{gathered} \mathrm{D}_u\bar\partial\colon TW^{\ell,p}(C,X)|_u\longrightarrow E^{\ell-1,p}(C,X)|_u, \\ \mathrm{D}_u\bar\partial(\zeta)=\frac{1}{2}(\widetilde\nabla\zeta+J\widetilde\nabla\zeta\circ\mathfrak{j})+\frac{1}{4}N_J(\zeta,\partial u); \end{gathered} \tag{6.26}$$

see [**37**, Chapter 3.1]. The connection

$$\widetilde\nabla^{0,1}\equiv\frac{1}{2}(\widetilde\nabla+J\widetilde\nabla\circ\mathfrak{j})$$

is the $(0,1)$-part of $\widetilde\nabla$ and defines a $\bar\partial$-operator, thus a holomorphic structure, on u^*TX; see [**37**, Remark C.1.1]. The Nijenhuis term in $\mathrm{D}_u\bar\partial$ is a zeroth order complex anti-linear compact operator. Therefore, $\mathrm{D}_u\bar\partial$ is a deformation of some $\bar\partial$-operator by a compact operator; see [**37**, Definition C.1.5]. It follows that $\mathrm{D}_u\bar\partial$ is Fredholm, cf. [**37**, Theorem A.1.5], i.e. it has a finite dimensional kernel and cokernel.

Define

$$\mathrm{Def}(u)=\ker(\mathrm{D}_u\bar\partial)\quad\text{and}\quad\mathrm{Obs}(u)=\mathrm{coker}(\mathrm{D}_u\bar\partial).$$

---

[68] Recall from Section 6.1 that the finite group $\mathfrak{Aut}(u,C)$ acts on both the domain and target of $\mathrm{D}_u\bar\partial$.

[69] Along the zero set of the section $\bar\partial$, we have a canonical decomposition $TE^{\ell-1,p}(C,X)\cong E^{\ell-1,p}(C,X)\oplus TW^{\ell,p}(C,X)$.

Both $\mathrm{Def}(u)$ and $\mathrm{Obs}(u)$ are finite dimensional $\mathfrak{Aut}(u,C)$-invariant real vector spaces. The first space corresponds to infinitesimal deformations of u (over the fixed domain C) and the second one is the obstruction space for integrating elements of $\mathrm{Def}(u)$ to actual deformations. By Riemann-Roch Theorem [**37**, Theorem C.1.10],

$$\dim_{\mathbb{R}} \mathrm{Def}(u)-\dim_{\mathbb{R}} \mathrm{Obs}(u)=2\big(c_1(u^*TX)+\dim_{\mathbb{C}}X(1-g)\big). \tag{6.27}$$

If $\mathrm{Obs}(u)\equiv 0$, by Implicit Function Theorem [**37**, Theorem A.3.3], in a small neighborhood $B_\epsilon(u)$ of u in $W^{\ell,p}(C,X)$ as in (6.4), the set of $(J,\mathfrak{j})$-holomorphic maps

$$V_u \equiv \bar{\partial}^{-1}(0)\cap B_\epsilon(u) \tag{6.28}$$

is a smooth $\mathfrak{Aut}(u,C)$-invariant manifold of real dimension $\dim_{\mathbb{R}} \mathrm{Def}(u)$, all the elements of $\mathrm{Def}(u)$ are smooth (i.e. independent of choice of ℓ and p), and $T_uV_u \cong \mathrm{Def}(u)$; cf. [**37**, Theorem 3.1.5]. The manifold V_u carries a natural orientation: starting with the $\bar\partial$-part of $\mathrm{D}_u\bar\partial$, both the kernel and cokernel are complex linear and thus naturally oriented. By going from $t=0$ to $t=1$ in the one-parameter family of Fredholm operators

$$\big(\widetilde{\nabla}^{0,1}+t\frac{1}{4}N_J(\cdot,\partial u)\big)_{t\in[0,1]},$$

[**37**, Proposition A.2.4] gives us a natural orientation on $\mathrm{Def}(u)$.

REMARK 6.2.1. Let $f=(u,C)$. More generally, deformation theory of f, i.e. if we allow deformations of both u and C, is given by a long exact sequence of the form

$$\begin{aligned} 0 \longrightarrow \mathrm{Def}(u) \longrightarrow \mathrm{Def}(f) \longrightarrow \mathrm{Def}(C) &\stackrel{\delta}{\longrightarrow} \\ \mathrm{Obs}(u) \longrightarrow \mathrm{Obs}(f) &\longrightarrow 0, \end{aligned} \tag{6.29}$$

where

$$\mathrm{Def}(C)\cong T_b\mathcal{B}\cong H^1(\mathcal{T}\Sigma_{\mathfrak{j}}(-\vec{z}))$$

is as in (5.15). The long exact sequence (6.29) is the hypercohomology of a short exact sequence of complexes of fine sheaves constructed in the following way.

For every holomorphic or complex vector bundle $E\longrightarrow(\Sigma,\mathfrak{j})$, let $\Omega^0(E)$ and $\Omega^{0,1}_{\Sigma,\mathfrak{j}}(E)$ denote the associated fine sheaves of smooth sections of E and of smooth E-valued $(0,1)$-forms, respectively.

The map $du\colon T\Sigma\longrightarrow TX$ gives rise to a similarly denoted map

$$du\colon \Omega^0(\mathcal{T}\Sigma_{\mathfrak{j}}(-\vec{z}))\longrightarrow \Omega^0(u^*TX).$$

Away from the marked points $\vec{z}$, it maps a local generating section $\frac{\partial}{\partial w}$ to $du(\frac{\partial}{\partial w})$[70], and in a holomorphic chart w around a marked point $z^i\equiv(w=0)$, it maps the local generating section $w\frac{\partial}{\partial w}$ to $wdu(\frac{\partial}{\partial w})$. The following commutative diagram has exact

[70] With $\partial w=\frac{1}{2}(\frac{\partial}{\partial x}-\mathrm{i}\frac{\partial}{\partial y})$, $du(\frac{\partial}{\partial w})$ means $\frac{1}{2}(du(\frac{\partial}{\partial x})-Jdu(\frac{\partial}{\partial y}))=du(\frac{\partial}{\partial x})$.

rows:

$$\begin{array}{ccccc}
0 & \longrightarrow & \Omega^0(\mathcal{T}\Sigma_{\mathfrak{j}}(-\vec{z})) & \longrightarrow & \Omega^0(\mathcal{T}\Sigma_{\mathfrak{j}}(-\vec{z})) \\
\big\downarrow & & \big\downarrow{\scriptstyle \mathrm{d}u\oplus\bar{\partial}} & & \big\downarrow{\scriptstyle \bar{\partial}} \\
\Omega^0(u^*TX) & \longrightarrow & \Omega^0(u^*TX)\oplus\Omega^{0,1}_{\Sigma,\mathfrak{j}}(\mathcal{T}\Sigma_{\mathfrak{j}}(-\vec{z})) & \longrightarrow & \Omega^{0,1}_{\Sigma,\mathfrak{j}}(\mathcal{T}\Sigma_{\mathfrak{j}}(-\vec{z})) \\
\big\downarrow{\scriptstyle \mathrm{D}_u\bar{\partial}} & & \big\downarrow{\scriptstyle \mathrm{D}_u\bar{\partial}-\mathrm{d}u} & & \big\downarrow \\
\Omega^{0,1}_{\Sigma,\mathfrak{j}}(u^*TX) & \longrightarrow & \Omega^{0,1}_{\Sigma,\mathfrak{j}}(u^*TX) & \longrightarrow & 0\,;
\end{array}$$

i.e. it is an exact sequence of chain complexes given by the columns. Then (6.29) is the hypercohomology of this diagram. We refer to the article of Siebert-Tian [**3**, Section 3.2] for an explicit and more detailed description of (6.29); also see [**29**, Section 24.4] for more details in the algebraic case.

In the light of Remark 6.2.1, similarly to (6.28), if $\mathrm{Obs}(u)=0$, then

$$\mathrm{Def}(f)\cong\mathrm{Def}(u)\oplus\mathrm{Def}(C) \tag{6.30}$$

and every tangent direction in $\mathrm{Def}(f)$ is integrable; i.e. in a small neighborhood $B_\epsilon(f)$ of f in[71] $W^{\ell,p}(\mathcal{C},X)$, the set of J-holomorphic maps $V_f\equiv\bar{\partial}^{-1}(0)\cap B_\epsilon(f)$ is a smooth $\mathfrak{Aut}(f)$-invariant manifold of real dimension $\dim_{\mathbb{R}}\mathrm{Def}(f)$, all the elements of $\mathrm{Def}(f)$ are smooth (i.e. independent of choice of ℓ and p), and $T_fV_f\cong\mathrm{Def}(f)$. A similar conclusion holds if $\mathrm{Obs}(f)=0$, i.e. if the map δ in (6.29) is surjective; however, we will work with $\mathrm{Obs}(u)$ for simplicity[72].

The main idea of this article is that if $\mathrm{Obs}(u)\neq 0$, we can still reduce the right-hand side of the first equation in (6.26) to a finite dimensional sub-space modulo which $\mathrm{D}_u\bar{\partial}$ is surjective and then we define V_u and V_f similarly. The auxiliary finite dimensional space then appears as an orbibundle over V_f and gives rise to a natural Kuranishi chart on the moduli space in the sense of Definition 3.2.1 .

DEFINITION 6.2.2. *Let $C=(\Sigma,\mathfrak{j},\vec{z})$ be a smooth marked curve[73] and $u\colon C\longrightarrow X$ be a J-holomorphic map. An **obstruction space** E_f for $f=(u,C)$ is a finite dimensional complex linear (with respect to the induced action of J) sub-vector space*

$$E_f\subset\Gamma(\Sigma,\Omega^{0,1}_{\Sigma,\mathfrak{j}}\otimes_{\mathbb{C}}u^*TX)$$

satisfying the following conditions.

(1) *Every $\eta\in E_f$ is smooth;*
(2) *with respect to the projection map*

$$\pi_{E_f}\colon\Gamma^{\ell-1,p}(\Sigma,\Omega^{0,1}_{\Sigma,\mathfrak{j}}\otimes_{\mathbb{C}}u^*TX)\longrightarrow\Gamma^{\ell-1,p}(\Sigma,\Omega^{0,1}_{\Sigma,\mathfrak{j}}\otimes_{\mathbb{C}}u^*TX)/E_f,$$

the composition $\pi_{E_f}\circ\mathrm{D}_u\bar{\partial}$ is surjective;
(3) *and E_f is $\mathfrak{Aut}(u,C)$-invariant.*

[71] For any choice of smooth structure $W^{\ell,p}_{\varphi}(\mathcal{C},X)$ on $W^{\ell,p}(\mathcal{C},X)$.

[72] As Remark 6.2.1 indicates, a generalization of $\mathrm{D}_u\bar{\partial}$ in (6.26) to some Fredholm map $\mathrm{D}_f\bar{\partial}$ that includes deformations of complex structure of the domain is rather difficult to define and work with.

[73] Not necessarily stable.

REMARK 6.2.3. If $\dim X > 0$, every J-holomorphic map admits a plethora of obstruction spaces. In fact, for any invariant open set $U \subset \Sigma$, there exists an obstruction space supported in U. For a proof see [**16**, Lemma 17.11]; it can also be deduced from the proof of [**37**, Proposition 3.2.1].

REMARK 6.2.4. In the light of Remark 6.2.1, it is often more practical for calculations to replace Definition 6.2.2.(2) with the weaker assumption that

$$[E_f] \longrightarrow \mathrm{Obs}(f), \quad [E_f] = E_f/\mathrm{Image}(D_u\bar\partial) \subset \mathrm{Obs}(u),$$

is surjective. For example, it sometimes happens that the moduli space $\overline{\mathcal{M}}_{g,k}(X,A)$ has the structure of an oriented orbifold of larger than expected dimension, $\mathrm{Obs}(f)$ has constant rank throughout $\overline{\mathcal{M}}_{g,k}(X,A)$, and they are fibers of an orbibundle on the moduli space. In these examples, we can often[74] take the obstruction space to be precisely (isomorphic to) $\mathrm{Obs}(f)$[75] and we obtain a pure orbibundle Kuranishi structure on the moduli space (with Kuranishi map $s \equiv 0$) that significantly simplifies the calculations; see the examples in Sections 7.1 and 7.2.

Assume $f = (u, C)$ is a fixed J-holomorphic map with smooth stable domain, E_f is an obstruction space for f, $(\pi\colon \mathcal{C} \longrightarrow \mathcal{B}, b \in \mathcal{B})$ is a universal family around C, and φ is a smooth trivialization for $\mathcal{C}$ as in (6.14). With respect to the local charts (6.16) for $W^{\ell,p}(\mathcal{C},X)$ and (6.20) for $E^{\ell-1,p}(\mathcal{C},X)$, let

$$\begin{aligned} &\widetilde{V}_f = \widetilde{V}_f(\varphi) = \{(\zeta,a) \in B_\epsilon(0)\times\mathcal{B} \colon \mathrm{P}^{-1}_{u,\zeta,a}(\bar\partial_{J,\mathfrak{j}_{\varphi,a}} u_\zeta) \in E_f\} \quad \text{and} \\ &\widetilde{s}_f\colon \widetilde{V}_f \longrightarrow E_f, \quad (\zeta,a) \longrightarrow \mathrm{P}^{-1}_{u,\zeta,a}(\bar\partial_{J,\mathfrak{j}_{\varphi,a}} u_\zeta). \end{aligned} \tag{6.31}$$

In other form, with notation as in (6.23) and

$$B_\epsilon(f) = \exp_\varphi(B_\epsilon(0)\times\mathcal{B}), \tag{6.32}$$

by Definition 6.2.2.(3) and $\mathfrak{Aut}(C)$-equivariance of P_φ, we obtain a finite rank $\mathfrak{Aut}(f)$-equivariant *obstruction bundle* $E_{f,\varphi}$,

$$\begin{array}{ccc} E_{f,\varphi} = P_\varphi(E_f \times B_\epsilon(0)\times\mathcal{B}) & \hookrightarrow & E^{\ell-1,p}_\varphi(\mathcal{C},X)|_{B_\epsilon(f)} \\ & \searrow & \downarrow \\ & & B_\epsilon(f) \subset W^{\ell,p}(\mathcal{C},X). \end{array} \tag{6.33}$$

Let

$$\begin{aligned} &V_f = \{f' \in B_\epsilon(f) \colon \bar\partial f' \in E_{f,\varphi}\}, \qquad U_f = E_{f,\varphi}|_{V_f}, \\ &\text{and} \quad s_f\colon V_f \longrightarrow U_f, \quad f' \longrightarrow \bar\partial f'. \end{aligned} \tag{6.34}$$

Then the tuple[76] $(\widetilde{U}_f = E_f \times \widetilde{V}_f \longrightarrow \widetilde{V}_f, \widetilde{s}_f)$ in (6.31) is the $\mathfrak{Aut}(f)$-equivariant trivialization of $(U_f \longrightarrow V_f, s_f = \bar\partial)$ with respect to the local charts (6.16) and (6.23), i.e.

$$V_f = \exp_\varphi(\widetilde{V}_f), \quad U_f = P_\varphi(E_f \times \widetilde{V}_f), \quad \text{and} \quad s_f = P_\varphi \circ \widetilde{s}_f.$$

For every choice of obstruction space as in Definition 6.2.2, the analogue of Riemann-Roch formula (6.27) asserts that

$$\dim_{\mathbb{R}} \ker(\pi_{E_f} \circ \mathrm{D}_u\bar\partial) = 2\big(c_1(u^*TX) + \dim_{\mathbb{C}} X(1-g) + \dim_{\mathbb{C}} E_f\big). \tag{6.35}$$

[74] But not always.

[75] i.e. we take the minimal possible choice.

[76] By abuse of notation, we use $\widetilde{s}_f$ to also denote for the section $x \in \widetilde{V}_f \longrightarrow (x, \widetilde{s}_f(x)) \in \widetilde{V}_f \times \widetilde{E}_f$.

Then, similarly to the case of $\mathrm{Obs}(u)=0$ before, for sufficiently small ϵ and $\mathcal{B}$, it follows from Definition 6.2.2.(2) and Implicit Function Theorem that V_f is a smooth $\mathfrak{Aut}(f)$-invariant manifold of real dimension

$$2\big(c_1(u^*TX)+(\dim_{\mathbb{C}}X-3)(1-g)+k+\dim_{\mathbb{C}} E_f\big) \tag{6.36}$$

independent of choice (ℓ,p) and

$$T_f V_f \cong T_{0\times\{b\}}\widetilde{V}_f \cong \ker(\pi_{E_f}\circ \mathrm{D}_u\bar\partial)\oplus T_b\mathcal{B}.$$

The zero set of s_f is a set of actual J-holomorphic maps close to f; see Remark 6.2.11. Putting together, by (6.36), we obtain a Kuranishi chart

$$\mathcal{U}_f \equiv (\mathrm{pr}_f\colon U_f \longrightarrow V_f, G_f=\mathfrak{Aut}(f), s_f, \psi_f), \qquad \psi_f(f')=[f'] \quad \forall f'\in s_f^{-1}(0), \tag{6.37}$$

of the expected dimension (1.7) centered at $[f]\in\mathcal{M}_{g,k}(X,A)$. We will denote the trivialization (6.31) of $\mathcal{U}_f$ by

$$\begin{aligned} &\widetilde{\mathcal{U}}_f \equiv (\widetilde{\mathrm{pr}}_f\colon \widetilde{U}_f = E_f\times \widetilde{V}_f \longrightarrow \widetilde{V}_f, G_f=\mathfrak{Aut}(f), \widetilde{s}_f, \widetilde{\psi}_f),\\ &\widetilde{\psi}_f(\zeta,a)=[u_\zeta, C_a] \quad \forall(\zeta,a)\in \widetilde{s}_f^{-1}(0); \end{aligned} \tag{6.38}$$

Depending on the situation, it is some times more convenient to work with $\widetilde{\mathcal{U}}_f$ and vice versa. The manifold V_f admits an $\mathfrak{Aut}(f)$-equivariant smooth fibration

$$\mathrm{st}_f\colon V_f \longrightarrow \mathcal{B}, \quad \mathrm{st}_f((u',C_a))=a,$$

where the fiber over each point is as in (6.28). Therefore, since E_f and $T_b\mathcal{B}$ are complex linear, V_f carries a natural orientation similarly to the case of (6.28).

Remark 6.2.5. Similarly to (5.14), for sufficiently small V_f, it follows from Theorem 5.3.4 and (5.14) that for every J-holomorphic map $f'=(u',C_a)\in(s_f^{-1}(0)\cap V_f)$ we have an identification

$$\mathfrak{Aut}(f')\cong\{h\in\mathfrak{Aut}(f)\subset\mathfrak{Aut}(C)\colon \varphi_h^{\mathcal{B}}(a)=a,\ u'\circ\varphi_h^{\mathcal{C}}|_{C_a}=u'\}$$

between $\mathfrak{Aut}(f')$ and a subgroup of $\mathfrak{Aut}(f)$. Therefore, for a sufficiently small open neighborhood $V_{f;f'}\subset V_f$, the restriction

$$\mathcal{U}_{f;f'}\equiv(\mathrm{pr}_f\colon U_f|_{V_{f;f'}}\longrightarrow V_{f;f'},\mathfrak{Aut}(f'),s_f|_{V_{f;f'}},\psi_f|_{V_{f;f'}})$$

defines a sub-chart of $\mathcal{U}_f$ around $[f']\in\mathcal{M}_{g,k}(X,A)$ in the sense of Definition 3.1, with the correct isotropy group. The same argument holds for the primary Kuranishi charts of the next two sections.

Remark 6.2.6. If φ' is another trivialization of the universal curve, the obstruction bundle $E_{f,\varphi'}$ would be different from $E_{f,\varphi}$ and thus there is no obvious identification of $V_f(\varphi')$ and $V_f(\varphi)$ away from the subset of J-holomorphic maps.

Remark 6.2.7. The evaluation and forgetful maps in (1.9) naturally lift to $W^{\ell,p}(\mathcal{C},X)$ and thus to V_f. The lifts are smooth. Therefore, the resulting Kuranishi charts satisfy the requirements of Definition 4.2.1. Since the domain is stable in this case, the lift of forgetful map is the composition of smooth projection map st_f above and the quotient map $\mathcal{B}\longrightarrow\mathcal{M}_{g,k}$. If $\dim X>0$, for sufficiently large E_f, the lifted evaluation map

$$\mathrm{ev}\colon V_f\longrightarrow X^k, \quad \mathrm{ev}((u',\Sigma_a,\mathrm{j}_a,\vec{z}_a))=u'(\vec{z}_a),$$

becomes a submersion.

REMARK 6.2.8. If $\dim X > 0$, for appropriate choices of E_f, the action of $\mathfrak{Aut}(f)$ on V_f would be effective. This way, we can fortify the possible non-effectiveness of the action of $\mathfrak{Aut}(f) \subset \mathfrak{Aut}(C)$ on $\mathcal{B}$. However, for simplicity, in the calculations of Section 7, we avoid such enlargements and allow non-effective actions in the natural way.

REMARK 6.2.9. For every fixed J-holomorphic map, inclusion defines a partial order on the set of admissible obstruction spaces in Definition 6.2.2. If $E'_f \subset E_f$, then the germ of Kuranishi chart $\mathcal{U}'_f$ corresponding to E'_f as in (6.37) embeds in the Kuranishi chart $\mathcal{U}_f$ corresponding to E_f (with respect to the same fixed φ) as in Definition 3.2.2. This embedding satisfies the tangent bundle condition of Definition 3.2.9. In fact, if $\widetilde{s}_f$ and $\widetilde{s}'_f$ are the Kuranishi maps of $\widetilde{\mathcal{U}}_f$ and $\widetilde{\mathcal{U}}'_f$ as in (6.31), respectively, then the corresponding normal direction derivative map in (3.10),

$$d\widetilde{s}_f/\widetilde{s}'_f : \mathcal{N}_{\widetilde{V}'_f}\widetilde{V}_f|_f \longrightarrow E_f/E'_f,$$

is a restriction of $\mathrm{D}_u\bar{\partial}$. Since $\mathrm{D}_u\bar{\partial}$ is surjective modulo E'_f, $d\widetilde{s}_f/\widetilde{s}'_f$ is surjective and thus an isomorphism (by dimensional reason).

The natural Kuranishi structure constructed in Section 6.5 is made of charts **induced** by the defining equations of the charts V_f constructed in this section and Sections 6.3 and 6.4. For this reason, we call the Kuranishi charts of Sections 6.2, 6.3, and 6.4, **primary charts**. The following observation explains the meaning and necessity of the induced charts.

Assume $f' = (u', C_a)$ is a J-holomorphic map in V_f sufficiently close to f. Then the subspace $E_{f'} = E_{f,\varphi}|_{f'}$ is an obstruction space for f'; see Remark 6.2.5 and [**16**, Proposition 17.22]. By the argument after Definition 5.2.1, for $a \in \mathcal{B}$ sufficiently close to b, $\mathcal{C} \longrightarrow \mathcal{B}$ is also a universal family around C_a. However, unless $\widetilde{\nabla}$ is a flat connection, the manifold

$$V_{f'} \subset W^{\ell,p}(\mathcal{C}, X)$$

constructed via $E_{f'}$ is not naturally isomorphic to a neighborhood of f' in V_f. In this simple case, an "induced" chart at f' is the restriction to some neighborhood $V_{f;f'} \subset V_f$ of $\mathcal{U}_f$; see Remark 6.2.5. In Section 6.5, we will generalize this trick to the cases where f' belongs to several primary charts $\{V_{f_i}\}_{i=1}^N$. The defining equation of an induced chart around f' will be a sum of Cauchy-Riemann equations centered at f_i, with $i \in [N]$.

REMARK 6.2.10. Similarly to Remark 5.2.3, we also need to consider marked maps $(u, C = (\Sigma, \mathfrak{j}, \vec{z}, q))$ where $q = \{q_1, \ldots, q_\ell\}$ is an un-ordered set of disjoint marked points away from $\vec{z}$. In certain cases, we will require the obstruction space to be supported away from q. Other than that, the construction of primary charts above readily extends to this situation. In this case, we use a universal family around $C = (\Sigma, \mathfrak{j}, \vec{z}, q)$ as in Remark 5.2.3 and construct a Kuranishi chart via such a universal family, similarly. The only difference between a chart for $(u, (\Sigma, \mathfrak{j}, \vec{z}, \vec{q}))$ and the resulting chart for $(u, (\Sigma, \mathfrak{j}, \vec{z}, q))$ is that the latter may have a larger group of automorphisms.

REMARK 6.2.11. Although the exponentiation map identification

$$B_\epsilon(f) \overset{\exp_\varphi}{\cong} \{(\zeta, a) \in \Gamma^{\ell,p}(\Sigma, u^*TX) \times \mathcal{B} \colon |\zeta| < \epsilon\} \tag{6.39}$$

of a neighborhood of f in $W^{\ell,p}(\mathcal{C},X)$ with a neighborhood of

$$0\times\{b\}\in\Gamma^{\ell,p}(\Sigma,u^*TX)\times\mathcal{B}$$

depends on the choice of a smooth trivialization φ in (6.14), as we pointed out earlier, the topology defined by the neighborhoods $B_\epsilon(f)$ is independent of the choice of φ. Therefore, the following notion of convergence to f is independent of the choice of φ. We say a sequence $(f_i=(u_i,C_i=(\Sigma_i,\mathfrak{j}_i,\vec{z}_i))_{i=1}^\infty$ of J-holomorphic maps **ϵ-converges** to f (or they are **ϵ-close** to f) if for every $i\gg 1$ there exists $a_i\in\mathcal{B}$ and $\zeta_i\in u^*TX$ such that $C_i\cong C_{a_i}$, $u_i=\exp_u\zeta_i\circ\varphi_{a_i}$, and

$$\lim_{i\longrightarrow\infty}a_i=b,\qquad\lim_{i\longrightarrow\infty}|\zeta_i|_{\ell,p}=0.$$

This notion of convergence coincides with the Gromov's convergence of Theorem 5.3.4; see [**57**]. Therefore, with notation as in (6.37), the topology on $\mathcal{M}_{g,k}(X,A)$ determined by[77] "open sets" $\psi_f(s_f^{-1}(0))$ coincides with the Gromov topology. In Sections 6.3 and 6.4, we will extend this notion of convergence to nodal maps with possibly unstable domain; see Remark 6.3.3. This alternative description of the Gromov convergence is the description of the choice in [**17, 20**] and directly shows that our footprint maps are homeomorphisms onto open sets of $\overline{\mathcal{M}}_{g,k}(X,A)$. The identification (6.39) also shows that the Gromov topology is locally (and thus by Remark 3.4.6 globally) metrizable.

6.3. Case of stable nodal domain. In this section we extend the construction of previous section to primary charts for nodal maps with stable domain. The construction depends on the choice of an obstruction space E_f, smooth trivializations of universal families of components of the nodal domain, and the holomorphic functions in (5.21).

With notation as in Sections 5.1 and 5.2, let $C=(\Sigma,\mathfrak{j},\vec{z})$ be a stable marked nodal curve with dual graph[78] Γ and $u\colon C\longrightarrow X$ be a J-holomorphic map. The analogue of linearization map (6.26) for $f=(u,C)$, still denoted by $\mathrm{D}_u\bar\partial$, is the restriction of $\bigoplus_{v\in\mathbb{V}}\mathrm{D}_{u_v}\bar\partial$ to

$$\mathrm{Domain}(\mathrm{D}_u\bar\partial)=TW^{\ell,p}(C,X)|_u\equiv\Gamma^{\ell,p}(\Sigma,u^*TX):=\Big\{\zeta\equiv(\zeta_v)_{v\in\mathbb{V}}\in$$
$$\bigoplus_{v\in\mathbb{V}}\Gamma^{\ell,p}(\Sigma,u_v^*TX)\colon\zeta_v(q_{\underset{\rightarrow}{e}})=\zeta_{v'}(q_{\underset{\leftarrow}{e}})\in T_{u(q_e)}X\quad\forall\,v,v'\in\mathbb{V},\ \underset{\leftarrow}{e}\in\underset{\rightarrow}{\mathbb{E}}_{v,v'}\Big\}$$

with image in the direct sum

$$\mathrm{Target}(\mathrm{D}_u\bar\partial)=\Gamma^{\ell-1,p}(\Sigma,\Omega^{0,1}_{\Sigma,\mathfrak{j}}\otimes_{\mathbb{C}}u^*TX):=\bigoplus_{v\in\mathbb{V}}\Gamma^{\ell-1,p}(\Sigma_v,\Omega^{0,1}_{\Sigma_v,\mathfrak{j}_v}\otimes_{\mathbb{C}}u_v^*TX).\tag{6.40}$$

Recall from the notation of (5.3) that $q_{\underset{\rightarrow}{e}}$ is the special point on Σ_v corresponding to the node $q_e\in\Sigma$ and the set of points $q_v=\{q_{\underset{\rightarrow}{e}}\}_{\underset{\rightarrow}{e}\in\mathbb{E}_v}$ form a set of un-ordered marked points on Σ_v away from $\vec{z}_v$. Note that by [**37**, Theorem B.1.11] and (6.1), every ζ_v is continuous and thus the evaluation $\zeta_v(q_{\underset{\rightarrow}{e}})$ is well-defined. Similarly to (6.6) and (6.9), the automorphism group $\mathfrak{Aut}(f)$, defined in (5.6), acts on the domain and target of $D_u\bar\partial$ such that $D_u\bar\partial$ is $\mathfrak{Aut}(f)$-equivariant.

[77] If one f in $\mathcal{M}_{g,k}(X,A)$ is domain-stable, then every element of $\mathcal{M}_{g,k}(X,A)$ is domain-stable.

[78] With vertices $\mathbb{V}$, edges $\mathbb{E}$, and legs $\mathbb{L}$.

DEFINITION 6.3.1. *With $f=(u,C)$ as above, an* ***obstruction space*** *for f is a direct sum*

$$E_f=\bigoplus_{v\in\mathbb{V}} E_{f_v} \subset \Gamma^{\ell-1,p}(\Sigma,\Omega^{0,1}_{\Sigma,\mathfrak{j}}\otimes_{\mathbb{C}} u^*TX),$$

satisfying the following conditions.

(1) *For every $v\in\mathbb{V}$,*

$$E_{f_v}\subset\Gamma^{\ell-1,p}(\Sigma_v,\Omega^{0,1}_{\Sigma_v,\mathfrak{j}_v}\otimes_{\mathbb{C}} u_v^*TX)$$

is finite dimensional, made of smooth sections, and is supported away from q_v (see Remark 6.2.10).

(2) *With respect to the projection map*

$$\pi_{E_f}:\Gamma^{\ell-1,p}(\Sigma,\Omega^{0,1}_{\Sigma,\mathfrak{j}}\otimes_{\mathbb{C}} u^*TX)\longrightarrow\Gamma^{\ell-1,p}(\Sigma,\Omega^{0,1}_{\Sigma,\mathfrak{j}}\otimes_{\mathbb{C}} u^*TX)/E_f,$$

the composition $\pi_{E_f}\circ \mathrm{D}_u\bar\partial$ is surjective.

(3) *E_f is $\mathfrak{Aut}(f)$-invariant.*

Fix an obstruction space E_f. With notation as in Section 5.2, let

$$\big(\pi\colon \mathcal{C}\longrightarrow\mathcal{B},\vec{\mathfrak{z}}=(\mathfrak{z}^l)_{l\in\mathbb{L}(\Gamma)},b\big) \tag{6.41}$$

be a sufficiently small standard universal family as in (5.25) around $C_b\equiv C$ such that

$$\mathrm{Support}(E_{f_v})\subset\Sigma_v\setminus\bigcup_{\underset{\rightarrow}{e}\in\underset{\rightarrow}{\mathbb{E}}_v}(\Sigma_v\cap U_{\underset{\rightarrow}{e}})\qquad\forall v\in\mathbb{V}. \tag{6.42}$$

In this equation, $U_{\underset{\rightarrow}{e}}$ are the domains[79] of holomorphic functions $w_{\underset{\rightarrow}{e}}$ in (5.21), which we used to smooth out the nodes. Recall from the construction following (5.18) that $\mathcal{C}$ is explicitly made from a set of universal families

$$(\pi_v\colon\mathcal{C}_v\to\mathcal{B}_v,\vec{\mathfrak{z}}_v,\mathfrak{q}_v,b_v)\qquad\forall v\in\mathbb{V},$$

by gluing them at the nodal points (sections) $\mathfrak{q}_v$, $\mathcal{B}=\prod_{v\in\mathbb{V}}\mathcal{B}_v\times\Delta^{\mathbb{E}}$, and the action of $\mathfrak{Aut}(C)$ on $\prod_{v\in\mathbb{V}}\mathcal{B}_v$ and $\prod_{v\in\mathbb{V}}\mathcal{C}_v$ extends to $(\mathcal{C},\mathcal{B})$ according to (5.26).

For every $v\in\mathbb{V}$, let

$$\mathcal{U}_{f_v}\equiv(\mathrm{pr}_v\colon U_v\longrightarrow V_v,G_v,s_v,\psi_v)$$

be the primary Kuranishi chart[80] of (6.37) centered at $[f_v]$ with respect to the obstruction space E_{f_v}, the universal family $\mathcal{C}_v$ around C_v, and a fixed choice of smooth trivializations φ_v of $\mathcal{C}_v$ as in (6.14). Recall from (6.34) that each V_v is a finite dimensional submanifold of $W^{\ell,p}_{\varphi_v}(\mathcal{C}_v,X)$ and we denote the trivialization of this manifold in $\Gamma^{\ell,p}(\Sigma_v,u_v^*TX)\times\mathcal{B}_v$ under the exponential map $\exp_{\varphi_v}$ by $\widetilde{V}_v$; see (6.31). In the following discussion, depending on the situation and for the sake of convenience, we will switch from $\mathcal{U}_v$ to its trivialization $\widetilde{\mathcal{U}}_v$ as in (6.38) and vice versa.

Let

$$\mathrm{ev}_v\colon V_v\longrightarrow X^{\underset{\rightarrow}{\mathbb{E}}_v}=\prod_{\underset{\rightarrow}{e}\in\underset{\rightarrow}{\mathbb{E}}_v}X$$

[79] Open sets around nodal points.

[80] For simplicity, we write $(\mathrm{pr}_v\colon U_v\longrightarrow V_v,G_v,\psi_v,s_v)$ instead of $(\mathrm{pr}_{f_v}\colon U_{f_v}\longrightarrow V_{f_v},G_{f_v},\psi_{f_v},s_v)$ in (6.37).

(and similarly denoted $\mathrm{ev}_v\colon \widetilde{V}_v \longrightarrow X^{\mathbb{E}_v}$) be the evaluation map at the nodal points $\mathfrak{q}_v$ in $\mathcal{C}_v$, i.e.

$$\mathrm{ev}_{\vec{v}}^{\underline{e}}(f'_v) = u'(\mathfrak{q}_{\underline{e}}(a_v)) \quad \forall f'_v = (u'_v, C_{a_v}) \in V_v,\ v\in\mathbb{V}.$$

Let $\Delta_X \subset X\times X$ be the diagonal submanifold,

$$\mathrm{Diag}_\Gamma(X) = (\Delta_X)^{\mathbb{E}} \subset (X\times X)^{\mathbb{E}},$$

and

(6.43)

$$V_{f;0} = \left\{ (f'_v)_{v\in\mathbb{V}} \in \prod_{v\in\mathbb{V}} V_v : \mathrm{ev}_{\vec{v}_1}^{\underline{e}}(f'_{v_1}) = \mathrm{ev}_{\vec{v}_2}^{\underline{e}}(f'_{v_2}) \in X \quad \forall v_1, v_2 \in \mathbb{V}, \underline{e}\in \vec{\mathbb{E}}_{v_1,v_2} \right\}$$

be the fiber product of $\{V_v\}_{v\in\mathbb{V}}$ over $\mathrm{Diag}_\Gamma(X)$; i.e.

$$V_{f;0} = \Big(\prod_{v\in\mathbb{V}} V_v\Big) \cap \Big(\prod_{\underline{e}\in\vec{\mathbb{E}}_{v_1,v_2}} \mathrm{ev}_{\vec{v}_1}^{\underline{e}} \times \mathrm{ev}_{\vec{v}_2}^{\underline{e}}\Big)^{-1}(\mathrm{Diag}_\Gamma(X)). \tag{6.44}$$

In other form, by (6.31), via the exponential map identification of (6.16) we have

$$V_{f;0} \overset{(\exp_{\varphi_v})_{v\in\mathbb{V}}}{\cong} \widetilde{V}_{f;0} = \Big\{ (\zeta_v, a_v)_{v\in\mathbb{V}} \in \mathrm{Domain}(\mathrm{D}_u\bar\partial) \times \prod_{v\in\mathbb{V}} \mathcal{B}_v : \\ \mathrm{P}^{-1}_{u_v,\zeta_v,a_v}(\bar\partial_{J,\mathfrak{j}_{\varphi_v,a_v}}(u_v)_{\zeta_v}) \in E_{f_v},\ |\zeta_v| < \epsilon,\ \forall v\in\mathbb{V} \Big\}. \tag{6.45}$$

Similarly to (6.31), in what follows, we will write the defining equations on the right-hand side of (6.45) as

$$\mathrm{P}^{-1}_{u,\zeta,a}(\bar\partial_{J,\mathfrak{j}_{\varphi,a}} u_\zeta) \in E_f. \tag{6.46}$$

REMARK 6.3.2. Condition (2) in Definition 6.3.1 is equivalent to the intersection (6.44) to be transverse to the diagonal. In [**37**, Definition 10.1.1] and [**16**, Assumption 10.3], the transversally condition is assumed instead of Definition 6.3.1.(2). That the transversality condition implies Definition 6.3.1.(2) is discussed after [**37**, Proposition 10.5.1]; the other direction follows from a simple dimension counting argument.

By Definition 6.3.1 and (6.44), for sufficiently small $\{\mathcal{B}_v\}_{v\in\mathbb{V}}$ and ϵ, $V_{f;0}$ (isomorphically $\widetilde{V}_{f;0}$) is a smooth $\mathfrak{Aut}(f)$-invariant oriented submanifold of the fiber product

$$B_\epsilon(f;0) = \Big(\prod_{v\in\mathbb{V}} B_\epsilon(f_v)\Big) \cap \Big(\prod_{\underline{e}\in\vec{\mathbb{E}}_{v_1,v_2}} \mathrm{ev}_{\vec{v}_1}^{\underline{e}} \times \mathrm{ev}_{\vec{v}_2}^{\underline{e}}\Big)^{-1}(\mathrm{Diag}_\Gamma(X)). \tag{6.47}$$

(isomorphically $\mathrm{Domain}(\mathrm{D}_u\bar\partial)\times\prod_{v\in\mathbb{V}}\mathcal{B}_v$) with

$$T_f V_{f;0} \cong T_{0\times\{b_v\}_{v\in\mathbb{V}}} \widetilde{V}_{f;0} \cong \ker(\pi_{E_f}\circ \mathrm{D}_u\bar\partial) \oplus \bigoplus_{v\in\mathbb{V}} T_{b_v}\mathcal{B}_v.$$

Let

$$\mathrm{pr}_{f;0}\colon U_{f;0} = \boxtimes_{v\in\mathbb{V}} U_v \longrightarrow V_{f;0} \quad \text{and} \quad s_{f;0} = \oplus_{v\in\mathbb{V}} s_v\colon V_{f;0} \longrightarrow U_{f;0} \tag{6.48}$$

be the box-product[81] Kuranishi bundle and direct sum Kuranishi map, respectively. With respect to the trivialization (6.45) and (6.23), (6.48) is isomorphic to the

[81] With respect to the projection maps $\rho_v\colon V_{f;0}\longrightarrow V_v$, the box-product bundle is defined by $\boxtimes_{v\in\mathbb{V}} U_v = \bigoplus_{v\in\mathbb{V}} \rho_v^* U_v$.

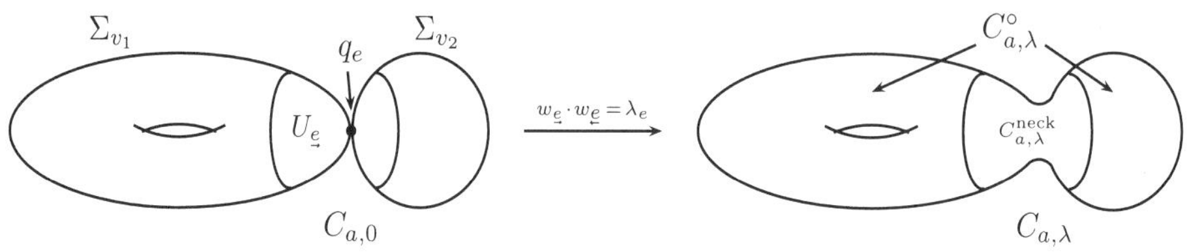

FIGURE 10. Neck-Core decomposition of a smoothing.

product bundle and direct sum Kuranishi map

(6.49) $\quad \widetilde{\mathrm{pr}}_{f;0}\colon \widetilde{U}_{f;0} = \widetilde{V}_{f;0} \times E_f \longrightarrow \widetilde{V}_{f;0} \quad \text{and} \quad \widetilde{s}_{f;0} = \oplus_{v\in\mathbb{V}} \widetilde{s}_v \colon \widetilde{V}_{f;0} \longrightarrow \widetilde{U}_{f;0}.$

The forgetful map

$$\mathrm{st}_{f;0}\colon V_{f;0} \longrightarrow \mathcal{B}$$

has image the complex-codimension $|\mathbb{E}|$ submanifold

$$\prod_{v\in\mathbb{V}} \mathcal{B}_v \times 0^{\mathbb{E}} \subset \mathcal{B}$$

of marked nodal curves of topological type Γ sufficiently close to C. Similarly, the footprint map

(6.50) $$\psi_{f;0} = (\psi_v)_{v\in\mathbb{V}}\colon \big(V_{f;0} \cap s^{-1}_{f;0}(0)\big) \longrightarrow \overline{\mathcal{M}}_{g,k}(X, A)$$

has image the set of J-holomorphic maps in the stratum $\mathcal{M}_{g,k}(X, A)_\Gamma$ (see (5.8)) sufficiently close to f. By Theorem 6.3.5 below, we $\mathfrak{Aut}(f)$-equivariantly extend the Kuranishi chart $\mathcal{U}_{f;0} = (\mathrm{pr}_{f;0}\colon U_{f;0} \longrightarrow V_{f;0}, \mathfrak{Aut}(f), s_{f;0}, \psi_{f;0})$ on $\mathcal{M}_{g,k}(X, A)_\Gamma$ to a Kuranishi chart for $\overline{\mathcal{M}}_{g,k}(X, A)$. In order to state the gluing theorem and describe the extended chart, we need to fix some notation.

In Section 5.2, for every $(a = (a_v), \lambda \equiv (\lambda_e)) \in \mathcal{B}$ the marked curve

$$C_{a,\lambda} = \big(\Sigma_{a,\lambda} = \pi^{-1}(a, \lambda), \vec{z}_{a,\lambda} = \mathfrak{z}(a, \lambda)\big) \subset (\mathcal{C}, \vec{\mathfrak{z}}(\mathcal{B}))$$

is obtained from the curves $\{C_{a_v}\}_{v\in\mathbb{V}}$ via the gluing identifications

(6.51) $$w_{\underset{\leftarrow}{e}} \cdot w_{\underset{\rightarrow}{e}} = \lambda_e \quad \forall\, e \in \mathbb{E}.$$

For sufficiently small δ in the gluing construction of $\mathcal{C}$, the defining equation (5.25) gives us a holomorphic identification between the **core** regions

$$C^\circ_{a,\lambda} \equiv C_{a,\lambda} \cap \bigcup_{v\in\mathbb{V}} \mathcal{C}^\circ_v \quad \text{and} \quad C^\circ_a = C^\circ_{a,0};$$

Therefore, in the following equations we will often write C°_a instead of $C^\circ_{a,\lambda}$. We refer to the complement

$$C^{\text{neck}}_{a,\lambda} = C_{a,\lambda} \setminus C^\circ_{a,\lambda} \subset C_{a,\lambda} \cap \bigcup_{e\in\mathbb{E}} \mathcal{C}_e$$

as the **neck** region. Let

$$C^\circ_a = \bigcup_{v\in\mathbb{V}} C^\circ_{a_v}, \quad C^\circ_{a_v} = C_a \cap \mathcal{C}^\circ_v \quad \forall v \in \mathbb{V},\ a \in \mathcal{B},$$

be the decomposition of the core region $C^\circ_a \cong C^\circ_{a,\lambda}$ into connected components labeled by $v \in \mathbb{V}$; see Figure 10.

On the other hand, by (5.24), for suitable choices of $U_{\underline{e}}$ in (5.21), the smooth trivializations $\{\varphi_v\}_{v\in\mathbb{V}}$ of $\{\mathcal{C}_v\}_{v\in\mathbb{V}}$ give a smooth trivialization

$$\varphi=(\varphi_v)_{v\in\mathbb{V}}\colon \big(\mathcal{C}^\circ \longrightarrow \mathcal{B},\vec{\mathfrak{z}},b\big)\longrightarrow \big(\Sigma^\circ\times\mathcal{B}\longrightarrow\mathcal{B},\vec{z},b\big)$$

of the core region

$$\mathcal{C}^\circ=\bigcup_{v\in\mathbb{V}}\mathcal{C}^\circ_v$$

of the universal family $\mathcal{C}$. Thus

$$(\Sigma^\circ_{a_v},\vec{\mathfrak{z}}_v(a_v))\cong(\Sigma^\circ_v,\vec{z}_v)$$

as marked smooth surfaces; i.e. the smooth structure of $C^\circ_{a,\lambda}$ does not depend on the choice of λ or a and thus we may identify the underlying smooth domains with Σ° and write $(\Sigma^\circ_{a,\lambda},\mathfrak{j}_{a,\lambda},\vec{\mathfrak{z}}(a,\lambda))$ as[82] $(\Sigma^\circ,\mathfrak{j}_{\varphi,a,\lambda},\vec{z})$, with $(a,\lambda)\in\mathcal{B}$. The following construction depends on the choice of smooth trivialization φ.

As a generalization of (6.32), and in order to extend (6.47) over $\mathcal{C}$, define $B_\epsilon(f)$ to be the set of (ℓ,p)-smooth maps $u'\colon C_{a,\lambda}\longrightarrow X$, with $(a,\lambda)\in\mathcal{B}$, such that

(1) $u'|_{C^\circ_{a,\lambda}}=\exp_u(\xi)\circ\varphi$ for some $\xi\in\Gamma^{\ell,p}(\Sigma^\circ,u^*TX)$ with $|\xi|<\epsilon$,

(2) and

$$\operatorname{dist}(u(q_e),u'|_{C^e_{a,\lambda}})<\epsilon, \tag{6.52}$$

where $C^e_{a,\lambda}=C^{\rm neck}_{a,\lambda}\cap\mathcal{C}_e$ and (6.52) means that the image of u' restricted to the neck region $C^e_{a,\lambda}$ lies in a ball of radius ϵ around $u(q_e)$.

Let $B^{\rm hol}_\epsilon(f)\subset B_\epsilon(f)$ be the subset where in addition

(3) $\bar\partial u'\equiv 0$ on $C^{\rm neck}_{a,\lambda}$.

The automorphism group $\mathfrak{Aut}(f)$ acts on both $B_\epsilon(f)$ and $B^{\rm hol}_\epsilon(f)$. Since the obstruction space E_f is supported away from the nodes, $B^{\rm hol}_\epsilon(f)$ includes $B_\epsilon(f;0)$ (see (6.47)).

Remark 6.3.3. For sufficiently small ϵ (etc.[83]), given $(a,\lambda)\in\mathcal{B}$ and $\xi\in\Gamma^{\ell,p}(\Sigma^\circ,u^*TX)$, there is at most one extension $u'\colon C_{a,\lambda}\longrightarrow X$ such that $(u',C_{a,\lambda})\in B^{\rm hol}_\epsilon(f)$ and $u'|_{C^\circ_{a,\lambda}}=\exp_u(\xi)\circ\varphi$. In other words, there is at most one pseudoholomorphic extension of a given map on the core region to the neck region. This is a consequence of Aronszajn's Unique Continuation Theorem [**37**, Theorem 2.3.4] applied to each neck, as every neck is isomorphic to an annulus in $\mathbb{C}$. Therefore, we get an $\mathfrak{Aut}(f)$-equivariant embedding of $B^{\rm hol}_\epsilon(f)$ into the Banach manifold[84]

$$W^{\ell,p}(\Sigma^\circ,X)\times\mathcal{B}\cong W^{\ell,p}_\varphi(\mathcal{C}^\circ,X). \tag{6.53}$$

While this embedding can be very wild, similarly to Remark 6.2.11, it defines a topology on $B^{\rm hol}_\epsilon(f)$ that restricted to the subset of pseudoholomorphic maps gives an open neighborhood of $[f]$ in $\overline{\mathcal{M}}_{g,k}(X,A)$. The resulting topology coincides with Gromov topology. This extends the notion of ϵ-close maps and ϵ-convergence in Remark 6.2.11 to a neighborhood of maps with stable nodal domain in the moduli

[82] By the first holomorphic identification above, restricted to the core region, the complex structure is determined only by a.

[83] There are various other parameters in the construction of $\mathcal{C}$ that we want them to be sufficiently small as well.

[84] Since Σ° is open, by abuse of notation in (6.53), $W^{\ell,p}(\Sigma^\circ,X)$ means the space of maps on the closure $\mathrm{cl}(\Sigma^\circ)$.

space. From this characterization of the Gromov topology we get the last claim before Theorem 6.3.5 below.

Similarly to (6.33), via trivialization (6.45), we obtain an $\mathfrak{Aut}(f)$-equivariant obstruction bundle

$$E_{\varphi,f} \longrightarrow B_\epsilon(f) \tag{6.54}$$

whose fiber over each point is obtained from parallel translations of E_v over the corresponding component of the core region. In other words, for $f'=(u',C_{a,\lambda})\in B_\epsilon(f)$ with

$$u'|_{C^\circ_{a_v}} = \exp_{u_v}(\xi_v)|_{\Sigma^\circ_v} \circ \varphi|_{C^\circ_{a_v}},$$

and $P_\varphi=(P_{\varphi_v})_{v\in\mathbb{V}}$ as in (6.23), let

$$E_{\varphi,f}|_{f'} = P_\varphi(E_f) := \bigoplus_{v\in\mathbb{V}} P_{\varphi_v}(E_v) \subset \Gamma^{\ell-1,p}(\Sigma_{a,\lambda}, \Omega^{0,1}_{\Sigma_{a,\lambda},\mathfrak{j}_{a,\lambda}} \otimes_{\mathbb{C}} (u')^*TX), \tag{6.55}$$

where the right-hand side is defined as in (6.40) and the direct sum extends by zero to the neck region. The latter is well-defined by the first condition in Definition 6.3.1. Restricted to $V_{f;0}\subset B_\epsilon(f)$, $E_{\varphi,f}$ coincides with the obstruction bundle $U_{f;0}$ in (6.48). Similarly to the paragraph before Remark 6.2.10, by [**16**, Proposition 17.22], if $\mathcal{B}$ and ϵ are sufficiently small, for every $f'\in B_\epsilon(f)$ with $\bar\partial u'=0$, the subspace (6.55) is an obstruction space for f' supported away from the neck region. Define

$$\mathcal{U}_f = (\mathrm{pr}_f\colon U_f \longrightarrow V_f, \mathfrak{Aut}(f), s_f, \psi_f) \tag{6.56}$$

as in (6.34), i.e.

$$V_f = \bar\partial^{-1}(E_{\varphi,f}) \subset B^{\mathrm{hol}}_\epsilon(f), \quad U_f = E_{\varphi,f}|_{V_f}, \qquad s_f = \bar\partial|_{V_f}, \tag{6.57}$$

the action of $\mathfrak{Aut}(f)$ is the restriction of the action of $\mathfrak{Aut}(f)$ to the zero set of $\mathfrak{Aut}(f)$-equivariant section $\bar\partial$, and ψ_f sends f' to $[f']$. The restriction of $\mathcal{U}_f$ to $\lambda=0\equiv(0)_{e\in\mathbb{E}}\in\Delta^{\mathbb{E}}$ coincides with the smooth fiber product chart $\mathcal{U}_{f;0}$. The inclusion $\bar\partial^{-1}(E_{\varphi,f})\subset B^{\mathrm{hol}}_\epsilon(f)$ follows from (6.55) and the third assumption in the defintion of $B_\epsilon(f)$ above. In the following, for $f'=(u',C_{a,\lambda})\in B_\epsilon(f)$, we write

$$\bar\partial u' \equiv 0 \text{ modulo } E_f \quad \text{or} \quad \bar\partial u' \overset{E_f}{\equiv} 0, \tag{6.58}$$

whenever $f'\in V_f$.

REMARK 6.3.4. Recall from Remark 6.3.3 that unlike in Section 6.2, the ϵ-neighborhood $B_\epsilon(f)$ defined in Page (2) does not have the structure of a Banach manifold; it just wildly embeds inside a Banach space. Therefore, we can not directly apply the Implicit Function Theorem to conclude that V_f is a smooth manifold of the expected dimension. The following theorem gives us an $\mathfrak{Aut}(f)$-equivariant product (smooth) structure on (6.56).

THEOREM 6.3.5 (Gluing Theorem). *For $\{\mathcal{B}_v\}_{v\in\mathbb{V}}$ and ϵ sufficiently small, and $\varepsilon\ll\epsilon$ and $\delta\ll\varepsilon$*[85] *in the construction of $\mathcal{C}$, there exists a "natural" (class of) $\mathfrak{Aut}(f)$-equivariant continuous one-to-one map*

$$\mathfrak{gl}\colon V_{f;0}\times\Delta^{\mathbb{E}} \longrightarrow V_f \tag{6.59}$$

[85] The construction of $\mathcal{B}$ depends on the parameters ε and δ and the definition of $B_\epsilon(f)$ depends additionally on ϵ.

such that the restriction of $\mathfrak{gl}$ *to* $V_{f;0}\times\{0\}^{\mathbb{E}}\cong V_{f;0}$ *is the inclusion map and the action of* $\mathfrak{Aut}(f)\subset\mathfrak{Aut}(C)$ *on* $\Delta^{\mathbb{E}}$ *is given by* (5.26).

In simple words, Theorem 6.3.5 states that for every $(u',C_a)\in V_{f;0}$ and every deformation $C_{a,\lambda}$ of the domain in the normal direction to the stratum $\mathcal{M}_\Gamma\subset\mathcal{M}_{g,k}$, there exists a systematic way of deforming (u',C_a) to $(u'_\lambda,C_{a,\lambda})$ satisfying

$$\bar{\partial}u'_\lambda \overset{E_f}{\equiv} 0.$$

Here the word "natural" means that u'_λ is constructed from u' via some standard procedure that depends on some choices, but modulo those choices, the map $\mathfrak{gl}$ is uniquely determined. More precisely, the construction of gluing map is by constructing first a pre-gluing map [**14**, Definition 5.1] via a set of appropriate cut-off functions [**14**, (4.1)-(4.6)] and then deforming that by an *alternating method* [**14**, Section 5II] into an actual solution of (6.58). Therefore, the construction of $\mathfrak{gl}$ depends on the choice of holomorphic coordinates (5.21), cut-off functions, etc. Nevertheless, the "weak" smooth structure on V_f given by (6.61) bellow is independent of these choices; see [**14**, Section 8].

By (6.55), (6.44), and Theorem 6.3.5, the corresponding trivialization

$$(\widetilde{\mathrm{pr}}_f\colon \widetilde{U}_f\longrightarrow \widetilde{V}_f,\widetilde{s}_f)$$

of

$$(\mathrm{pr}_f\colon U_f\longrightarrow V_f,s_f)$$

is of the form

$$\begin{aligned}&\widetilde{V}_f=\widetilde{V}_{f;0}\times\Delta^{\mathbb{E}},\quad \widetilde{U}_f=\widetilde{V}_f\times E_f=\widetilde{V}_{f;0}\times\Delta^{\mathbb{E}}\times E_f=\widetilde{U}_{f;0}\times\Delta^{\mathbb{E}},\\ &\text{and}\quad \widetilde{s}_f(\zeta,a,\lambda)=\mathrm{P}^{-1}_{u,\zeta',a}(\bar{\partial}_{J,\mathfrak{j}_{\varphi,a}}u')\in E_f,\end{aligned}\tag{6.60}$$

such that $\mathfrak{gl}(\zeta,a,\lambda)=(u',C_{a,\lambda})$ with

$$u'|_{C^\circ_{a,\lambda}}=u'|_{C^\circ_a}=u_{\zeta'}\quad\text{for some}\quad \zeta'=\zeta'(\zeta,a,\lambda)\in\Gamma^{\ell,p}(\Sigma^\circ,u^*TX)$$

and $\mathrm{P}_{u,\zeta',a}$ is (the restriction to the core region of) the parallel translation map in (6.46). Therefore, Theorem 6.3.5 defines an $\mathfrak{Aut}(f)$-equivariant product smooth structure on $\mathrm{pr}_f\colon U_f\longrightarrow V_f$.

In Definition 3.2.1.(2), we require the Kuranishi map to be smooth. While $s_{f;0}$ in the gluing theorem is smooth, it is not clear whether the extended Kuranishi map s_f remains smooth or not (with respect to the product smooth structure on V_f). For this reason, and also to make coordinate change maps smooth, we will equip V_f and U_f with a weaker product smooth structure in the normal direction.

Assuming $\delta<1$, define

$$T\colon\Delta\longrightarrow\mathbb{C},\quad T(z)=\frac{-1}{\log(r)}e^{i\theta}\quad\forall z=re^{i\theta}\in\mathbb{C}^*,\ T(0)=0.\tag{6.61}$$

We define the **weak** smooth structure Δ_{weak} on Δ to be the smooth structure given by the coordinate chart $T(z)$.

THEOREM 6.3.6 (Smoothness Theorem). *With respect to the "weak" smooth structures* V_f^{weak} *and* U_f^{weak} *given by* Δ_{weak} *instead of* Δ *in* (6.60), *the Kuranishi map of Theorem* 6.3.5 *is smooth.*

We do not go into the proofs of Theorems 6.3.5 and 6.3.6 in this article and refer the interested reader to [**14**] which is dedicated to the proof of the gluing theorem and its properties (including the case of bordered pseudoholomorphic maps). In the rest of this section, we will only recall some of the main steps and unique features of the proofs.

REMARK 6.3.7. In the upcoming sections, any argument involving a smooth structure on the Kuranishi chart (6.56), e.g. transversality of certain evaluation maps in Section 6.4, refers to the weak smooth structure provided by Theorem 6.3.6. Since this is the only way we put a smooth structure on (6.56), we will omit the superscript "weak" from the notation for simplicity.

The proof of Theorem 6.3.5 in [**16**, **20**] and with more details in [**14**] uses weighted Sobolev spaces [**14**, Definition 3.4] with $\ell \gg 1$ and $p=2$. The choice of $p=2$ allows us to define an inner product [**14**, (4.11)] on the source and target of $\mathrm{D}_u\bar{\partial}$. While the overall process is similar to the classical proof of [**37**, Chapter 10], the weighted norms and cut-off functions used in the proof of the gluing theorem in [**14**] are different than those of [**37**, Section 10.3]; see [**14**, Remark 6.17]. They allow for a better control of the gluing process near the nodes which is essential for establishing Theorem 6.3.6. Theorem 6.3.5 in [**14**] is spread among few statements which we have gathered into one theorem here. The case where the gluing parameter λ and the complex structure of the domain is fixed is covered in [**14**, Theorem 3.13]. The injectivity and surjectivity of the gluing map is discussed in [**14**, Section 7]. The argument uses the alternative description of the Gromov topology via the notion of ϵ-close maps or ϵ-convergence, see Remarks 6.2.11 and 6.3.3. The necessary adjustments for the case where the complex structure of the domain and the gluing parameter are allowed to change are discussed in [**14**, Section 8.1]. Theorem 6.3.6 is a consequence of Exponential Decay Theorem [**14**, Theorem 6.4] in simple words; see [**14**, Proposition 8.31].

REMARK 6.3.8. In [**20**] and its follow-ups, the authors make use of cylindrical coordinates

$$(-s_{\underrightarrow{e}}+\mathrm{i}\theta_{\underrightarrow{e}})=\log(w_{\underrightarrow{e}})\in(-\infty,\log(\varepsilon)]\times S^1 \quad \forall\ \underrightarrow{e}\in\underrightarrow{\mathbb{E}}$$

at the nodes. In these coordinates, Equation 6.51 takes the form

$$s_{\underrightarrow{e}}+s_{\underleftarrow{e}}=R_e,\quad \theta_{\underrightarrow{e}}+\theta_{\underleftarrow{e}}=\theta_e,\quad \text{where } \log(\lambda_e)=-R_e+\mathrm{i}\theta_e. \tag{6.62}$$

Geometrically, it corresponds to removing $(-\infty,-R_e)\times S^1$ from $(-\infty,\log(\varepsilon)]\times S^1$ at each node and identifying the remaining finite length cylinders $[-R_e,\log(\varepsilon)]\times S^1$ according to (6.62). It is easier to express the weighted norms, cut-off functions, and the gluing process in terms of the cylindrical coordinates. In cylindrical coordinates, the weak smooth structure is given by $T(\lambda_e)=\mathrm{e}^{\mathrm{i}\theta}/R_e$. Since we do not go into the details of the gluing theorem in this article, we will stay with the usual holomorphic coordinates for simplicity.

REMARK 6.3.9. Similarly to Remark 6.2.9, if $E'_f\subset E_f$, then the (germ of) Kuranishi chart $\mathcal{U}'_f$ in (6.56) corresponding to E'_f embeds in the Kuranishi chart $\mathcal{U}_f$ corresponding to E_f (defined via the same trivialization φ) as in Definition 3.2.2. This embedding also satisfies the tangent bundle condition of Definition 3.2.9.

6.4. Case of un-stable domain. In this section, we extend the construction of Section 6.3 to stable maps with un-stable domain. This is done by adding an invariant set of extra marked points to the domain and intersecting with divisorial slices on the image (at the end) to reduce the dimension back to the expected dimension. In the light of Remarks 6.3.3 and 6.2.11, this stabilization process also allows us to obtain a rather different description of the Gromov topology.

DEFINITION 6.4.1. *Given a stable map $f=(u,\Sigma,\mathfrak{j},\vec{z})$, an un-ordered $\mathfrak{Aut}(f)$-invariant set of marked points $w\subset\Sigma$ disjoint from $\vec{z}$ and the nodes of Σ is called a* ***stabilizing*** *set if*

$$\mathfrak{Aut}(\Sigma,\mathfrak{j},\vec{z},w)=\mathfrak{Aut}(f) \quad \text{and} \quad du(w^i)\neq 0 \qquad \forall w^i\in w, \tag{6.63}$$

where $\mathfrak{Aut}(\Sigma,\mathfrak{j},\vec{z},w)$ is the automorphism group defined in (5.17).

Since the restriction of u to every unstable component is non-trivial and the set of regular values is open dense, every stable map admits plenty of stabilizing sets.

EXAMPLE 6.4.2. Let $(\Sigma,\mathfrak{j})\cong\mathbb{P}^1=\mathbb{C}\cup\{\infty\}$, $\vec{z}=\emptyset$, and

$$u\colon\mathbb{P}^1\longrightarrow X=\mathbb{P}^1, \quad u(z)=z^2;$$

thus, $[f=(u,\mathbb{P}^1)]\in\mathcal{M}_{0,0}(\mathbb{P}^1,[2])$. Then any quadruple of points $w=\{a,b,-a,-b\}$, with $a,b\in\mathbb{C}^*$ and $a\neq\pm b$, is a stabilizing set. These are the minimal stabilizing sets.

Let $f=(u,\Sigma,\mathfrak{j},\vec{z})$ be a stable J-holomorphic map and w be a stabilizing set for f. By the second assumption in (6.63), $\mathfrak{Aut}(f)$ does not fix any of w^i and w decomposes to finitely many $\mathfrak{Aut}(f)$-orbits $[w^i]$, each of which is isomorphic to $\mathfrak{Aut}(f)$; i.e. if $w\neq\emptyset$, then the first assumption in (6.63) follows from the second one. For every orbit $[w^i]\in w/\mathfrak{Aut}(f)$, let $H_{[w^i]}$ be a sufficiently small codimension 2 submanifold of X which J-orthogonally[86] intersects $u(\Sigma)$ only at $u(w^i)$. A set of **slicing divisors** is a collection of disjoint submanifolds

$$H=\{H_{[w^i]}\}_{[w^i]\in w/\mathfrak{Aut}(f)}$$

with this property. In what follows, by a **stabilizing pair** or **stabilizing data** we mean a pair (w,H) consisting of a stabilizing set of points together with a corresponding set of slicing divisors for a fixed J-holomorphic f.

Given a stable J-holomorphic map $f=(u,\Sigma,\mathfrak{j},\vec{z})$ and a stabilizing pair (w,H), we construct a (class of) primary Kuranishi chart centered at $[f]$ in the following way. In the following argument let $H_{w^i}:=H_{[w^i]}$. Choose an obstruction space E_f for f as in Definition 6.3.1. Such E_f can also be thought[87] of as an obstruction space for $f^+=(u,\Sigma,\mathfrak{j},\vec{z},w)$. Take a (standard) universal family $\mathcal{C}_+$ around $C^+=(\Sigma,\mathfrak{j},\vec{z},w)$ and let

$$\mathcal{U}^+=(\mathrm{pr}^+\colon U^+\longrightarrow V^+,\mathfrak{Aut}(f^+)=\mathfrak{Aut}(f),s^+,\psi^+) \tag{6.64}$$

be the resulting primary Kuranishi chart constructed in Section 6.3 around $[f^+]\in\overline{\mathcal{M}}_{g,k+\ell}(X,A)$. Recall that if Σ is nodal, the smooth structure on V^+ considered in the following argument is the weak smooth structure of Theorem 6.3.6; see Remark 6.3.7.

[86]i.e. $T_{u(w^i)}H_{[w^i]}$ is the J-orthogonal complement of $T_{u(w^i)}u(\Sigma)$.

[87]Note that the support of E_f is allowed to have overlap with the stabilizing points. It should only be away from nodal points.

REMARK 6.4.3. The map f^+ comes with an un-ordered set of marked points. Therefore, $[f^+]$ is not actually an element of $\overline{\mathcal{M}}_{g,k+\ell}(X,A)$, unless we let the notation $\overline{\mathcal{M}}_{g,k+\ell}(X,A)$ denote the moduli of genus g maps with k ordered and ℓ un-ordered marked points. Fix an ordering $\vec{w}$ on w. Then the Kuranishi chart $\mathcal{U}^+$ is a Kuranishi chart for $[u,\Sigma,\mathfrak{j},\vec{z}\cup\vec{w}]\in\overline{\mathcal{M}}_{g,k+\ell}(X,A)$ with a possibly larger group action; cf. Remarks 5.2.3 and 6.2.10.

For sufficiently small V^+, the evaluation map at w-points[88],

$$\mathrm{ev}_w\colon V^+\longrightarrow X^w\equiv\prod_{w^i\in w}X, \tag{6.65}$$

remains transverse to $(H_{w^i})_{w^i\in w}\subset X^w$. By abuse of notation, we will also denote the last tuple by H. The choice of weak smooth structure has no effect on this transversality. In order to cancel out the effect of the added points to the dimension of $\mathcal{U}^+$, we replace V^+ with

$$V=\mathrm{ev}_w^{-1}(H).$$

Then V is $\mathfrak{Aut}(f)$-invariant and

$$\mathcal{U}\equiv(\mathrm{pr}=\mathrm{pr}^+|_U\colon U=U^+|_V\longrightarrow V,\mathfrak{Aut}(f),s=s^+|_V,\psi), \tag{6.66}$$

where ψ is the composition of

$$\psi^+|_{(V\cap(s^+)^{-1}(0))}\colon(V\cap(s^+)^{-1}(0))\longrightarrow\overline{\mathcal{M}}_{g,k+\ell}(X,A)$$

with the forgetful map of removing w-points $\overline{\mathcal{M}}_{g,k+\ell}(X,A)\longrightarrow\overline{\mathcal{M}}_{g,k}(X,A)$ as in (1.6), defines a natural primary Kuranishi chart centered at $[f]$.

Since the local slicing divisors $\{H_{w^i}\}$ are intersecting J-orthogonally, the normal bundles

$$N_{H_{w^i}}X=\frac{TX|_{H_{w^i}}}{TH_{w^i}}\cong TH_{w^i}^{\perp\omega}$$

are canonically oriented. We equip TV with the natural orientation given by the natural orientation of V^+ and the exact sequence

$$0\longrightarrow TV\longrightarrow TV^+|_V\xrightarrow{[\mathrm{dev}_w]}\bigoplus_{w^i\in w}N_{H_{w^i}}X\longrightarrow 0\,.$$

Since, by assumption, u is interesting every H_{w^i} positively, the induced linear map

$$[du]\colon T_{w^i}\Sigma\longrightarrow N_{H_{w^i}}X$$

is an orientation preserving isomorphism. The same holds for every other map in V. This implies that our orientation scheme is compatible with the forgetful maps that arise in the following arguments.

REMARK 6.4.4. Let $f=(u,C=(\Sigma,\mathfrak{j},\vec{z}))$ be a J-holomorphic map with smooth domain, and (w_0,H_0) and (w_1,H_1) be two stabilizing sets with $w_0\subsetneq w_1$. Let

$$(\mathcal{C}_0\longrightarrow\mathcal{B}_0,\ \vec{\mathfrak{z}},\mathfrak{w}_0\colon\mathcal{B}_0\longrightarrow\mathcal{C}_0),\quad\text{and}\quad(\mathcal{C}_1\longrightarrow\mathcal{B}_1,\ \vec{\mathfrak{z}},\mathfrak{w}_1\colon\mathcal{B}_1\longrightarrow\mathcal{C}_1)$$

be universal families around $C_0=(\Sigma,\mathfrak{j},\vec{z},w_0)$ and $C_1=(\Sigma,\mathfrak{j},\vec{z},w_1)$, respectively. After possibly restricting to smaller subfamilies, the process of removing the extra

[88]More precisely, we again need to fix an ordering on w but the conclusion is independent of the choice of that.

marked points $\mathfrak{w}_1 \setminus \mathfrak{w}_0$ (in other words, extra sections) gives us a holomorphic $\mathfrak{Aut}(f)$-equivariant forgetful map

$$(\pi^{\mathcal{C}}, \pi^{\mathcal{B}}) \colon (\mathcal{C}_1, \mathcal{B}_1) \longrightarrow (\mathcal{C}_0, \mathcal{B}_0),$$

and thus a projection map

$$\pi_{1,0} \colon W^{\ell,p}(\mathcal{C}_1, X) \longrightarrow W^{\ell,p}(\mathcal{C}_0, X).$$

Let φ_0 and φ_1 be smooth trivializations for $\mathcal{C}_0$ and $\mathcal{C}_1$ as in (6.14), respectively. Then, with notation as in (6.15), the resulting projection map $\pi_{\varphi_1,\varphi_0}$ given by

$$\begin{array}{ccc} W^{\ell,p}(C_1, X) \times \mathcal{B}_1 & \xrightarrow{T_{\varphi_1}} & W^{\ell,p}(\mathcal{C}_1, X) \\ \big\downarrow \pi_{\varphi_1,\varphi_0} & & \big\downarrow \pi_{1,0} \\ W^{\ell,p}(C_0, X) \times \mathcal{B}_0 & \xrightarrow{T_{\varphi_0}} & W^{\ell,p}(\mathcal{C}_0, X) \end{array}$$

can not be identity on the first component. In fact, for each $a \in \mathcal{B}_0$,

$$\pi_{\varphi_1,\varphi_0} \colon W^{\ell,p}(\Sigma, X) \times (\pi^{\mathcal{B}})^{-1}(a) \longrightarrow W^{\ell,p}(\Sigma, X) \times \{a\}$$

corresponds to twisting with the non-constant $(\pi^{\mathcal{B}})^{-1}(a)$-family of diffeomorphisms

$$\rho_{a';a} \colon \varphi_0 \circ \pi^{\mathcal{C}} \circ \varphi_1^{-1}|_{\Sigma \times \{a'\}} \colon \Sigma \times \{a'\} \longrightarrow \Sigma \times \{a\} \quad \forall a' \in (\pi^{\mathcal{B}})^{-1}(a),$$

each of which fixes $\vec{z}$ and $w_0 \subset w_1$ but possibly maps $w_1 \setminus w_0$ to a variation of these points. In other words, $\pi_{\varphi_1,\varphi_0}$ is of the form

$$(u', a') \longrightarrow (u' \circ \rho_{a';a}^{-1}, a = \pi^{\mathcal{B}}(a')) \quad \forall (u', a') \in W^{\ell,p}(\Sigma, X) \times \mathcal{B}_1.$$

Since the map depends on a non-trivial reparametrization of the domain, the forgetful map

$$\pi_{1,0} \colon W^{\ell,p}_{\varphi_1}(\mathcal{C}_1, X) \longrightarrow W^{\ell,p}_{\varphi_0}(\mathcal{C}_0, X)$$

is continuous but not smooth. Similarly to (6.18), there will be a loss of differentiability caused by the derivative of the reparametrization diffeomorphism. However, if $E_0 \subset E^{\ell-1,p}_{\varphi_0}(\mathcal{C}_0, X)$ is a "nice" smooth finite dimensional sub-bundle where each fiber is made of smooth sections (e.g. an obstruction bundle E_{φ_0,f_0} as in (6.33)), the pull back bundle $E_1 = \pi_{1,0}^* E_0$ still has the structure of a smooth finite dimensional sub-bundle of $E^{\ell-1,p}_{\varphi_1}(\mathcal{C}_1, X)$. We will extensively use this argument in the next section to define the induced Kuranishi charts.

REMARK 6.4.5. Proceeding with the notation of Remark 6.4.4, let

$$f_0 = (u, C = (\Sigma, \mathfrak{j}, \vec{z}, w_0)) \quad \text{and} \quad f_1 = (u, C = (\Sigma, \mathfrak{j}, \vec{z}), w_1).$$

Given an obstruction space E_f for f, since E_f can also be thought of as an obstruction space for f_0 and f_1, let

$$E_{f_1,\varphi_1} \longrightarrow B_\epsilon(f_1) \subset W^{\ell,p}_{\varphi_1}(\mathcal{C}_1, X) \quad \text{and} \quad E_{f_0,\varphi_0} \longrightarrow B_\epsilon(f_0) \subset W^{\ell,p}_{\varphi_0}(\mathcal{C}_0, X)$$

be the corresponding obstruction bundles of (6.33). Then in general, because of the reparametrization map ρ in Remark 6.4.4, $\pi_{1,0}^* E_{f_0,\varphi_0}$ is different from E_{f_1,φ_1} and there is no clear relation between the two. Therefore, there exists no natural map between the resulting manifolds $V_{f_1}(\varphi_1)$ and $V_{f_0}(\varphi_0)$ away from the zero set of the Kuranishi maps (i.e. the subset of actual J-holomorphic maps); see also

Remark 6.2.6. Nevertheless, restricted to the subsets $s_{f_1}^{-1}(0)$ and $s_{f_0}^{-1}(0)$ of actual J-holomorphic maps, the projection map

$$\pi_{\varphi_1,\varphi_0}\colon (V_{f_1}(\varphi_1)\cap s_{f_1}^{-1}(0)) \longrightarrow (V_{f_0}(\varphi_1)\cap s_{f_0}^{-1}(0))$$

is a continuous fiber bundle. After interesting with the slicing divisors, the resulting projection map

$$\pi_{\varphi_1,\varphi_0}\colon (V_{f_1}(\varphi_1)\cap s_{f_1}^{-1}(0)\cap \mathrm{ev}_{w_1}^{-1}(H_1)) \longrightarrow (V_{f_0}(\varphi_1)\cap s_{f_0}^{-1}(0)\cap \mathrm{ev}_{w_0}^{-1}(H_0))$$

is a homeomorphism (between sufficiently small neighborhoods of f_1 and f_0). Therefore, with the Kuranishi chart $\mathcal{U}$ as in (6.66), the topology defined via open sets $\psi(s^{-1}(0))$ in a neighborhood of $[f]\in\overline{\mathcal{M}}_{g,k}(X,A)$ is independent of the choice of a stabilizing pair (H,w) used to obtain the Kuranishi chart (6.66). This topology coincides with the Gromov topology and extends the notion of ϵ-closed topology in Remarks 6.2.11 and 6.3.3 to maps with unstable domain.

Let us summarize the outcome of our construction so far in the following statement. For every stable marked J-holomorphic map $f=(u,C=(\Sigma,\mathfrak{j},\vec{z}))$ with dual graph Γ, our construction of a primary Kuranishi chart $\mathcal{U}_f=(\mathrm{pr}_f\colon U_f\longrightarrow V_f,G_f,s_f,\psi_f)$ centered at f depends (at most) on a set of **auxiliary data** $\mathrm{Aux}(f)$ consisting of

(1) an obstruction space E_f,
(2) a set of stabilizing points w,
(3) a set of slicing divisors H,
(4) a set of sufficiently small universal families $\{\mathcal{C}_v^+\longrightarrow\mathcal{B}_v^+\}_{v\in\mathbb{V}}$ for each smooth component $C_v^+=(\Sigma_v,\mathfrak{j}_v,\vec{z}_v,w_v)$ of $C^+=(\Sigma,\mathfrak{j},\vec{z},w)$,
(5) smooth trivializations $\varphi=\{\varphi_v\}_{v\in\mathbb{V}}$ of $\{\mathcal{C}_v^+\}_{v\in\mathbb{V}}$,
(6) a standard extension of the universal families and trivializations in (4) and (5) above to a universal family $\mathcal{C}^+\longrightarrow\mathcal{B}^+$ around C^+ and a trivialization φ of the core region $(\mathcal{C}^+)^\circ$ via holomorphic functions (5.21) and sufficiently small parameters δ and ε,
(7) and a sufficiently small positive number ϵ which via the exponentiation identification and trivialization φ above determines an ϵ-close neighborhood $B_\epsilon(f)$ of f in the corresponding Banach space.

CONDITION 6.4.6. *The families and parameters in* (4)(6)(7) *are taken sufficiently small such that $\mathcal{C}^+\longrightarrow\mathcal{B}^+$ is a universal family around every $a\in\mathcal{B}^+$, the argument of Remark* 6.2.5 *for every $f'\in(V_f\cap s_f^{-1}(0))$ holds, $E_{\varphi,f}$ restricted to every such f' is an obstruction space, and every $f'\in V_f$ intersects H transversely.*

Our construction of primary charts readily extends to the $\mathcal{J}$-family case of Remark 1.4.2. In this situation, the linearization of Cauchy-Riemann equation would also involve deformations of J. We refer to [**37**, Section 3.1] for the necessary adjustments.

6.5. Induced charts. Finally, in this section and the next, we construct a cobordism class of natural oriented Kuranishi structures on the moduli space of pseudoholomorphic maps. Every chart of a natural Kuranishi structure depends on a fixed finite collection of primary charts constructed in Sections 6.2, 6.3, and 6.4. We call them **induced** charts. We will discuss coordinate change maps in the next section.

THEOREM 6.5.1. *Let* (X^{2n},ω) *be a closed symplectic manifold,* $J\in\mathcal{J}(X,\omega)$ *be an arbitrary compatible (or tame) almost complex structure,* $A\in H_2(X,\mathbb{Z})$*, and* $g,k\in\mathbb{N}$*. With the dimension* d *as in (1.7), the construction of this section provides a class of natural oriented* d*-dimensional Kuranishi structures on* $\overline{\mathcal{M}}_{g,k}(X,A,J)$*, every two of which are deformation equivalent as defined after Definition 3.5.1. If* ω_0 *and* ω_1 *are isotopic symplectic structures on* X*,* $J_0\in\mathcal{J}(X,\omega_0)$*, and* $J_1\in\mathcal{J}(X,\omega_1)$*, then every two of such Kuranishi structures on* $\overline{\mathcal{M}}_{g,k}(X,A,J_0)$ *and* $\overline{\mathcal{M}}_{g,k}(X,A,J_1)$ *are cobordant in the sense of Definition 3.5.1.*

For every J-holomorphic map $[f]\in\overline{\mathcal{M}}_{g,k}(X,A)$, after fixing a choice of auxiliary data $\mathrm{Aux}(f)$, as listed at the end of the last section, let

$$\mathcal{U}_f=(\mathrm{pr}_f\colon U_f\longrightarrow V_f,\mathfrak{Aut}(f),\psi_f,s_f)$$

be the resulting primary Kuranishi chart with the footprint $F_f\subset\overline{\mathcal{M}}_{g,k}(X,A)$. For each f, fix a relatively compact connected open sub-set $F'_f\subset F_f$ around $[f]$. Since $\overline{\mathcal{M}}_{g,k}(X,A)$ is compact, there exists a finite set

$$\mathcal{P}=\{f_\alpha\}_{\alpha\in S}\subset\overline{\mathcal{M}}_{g,k}(X,A)$$

such that $\{F'_{f_\alpha}\}_{\alpha\in S}$ covers the moduli space; we call such $\mathcal{P}$ a **primary collection**.

Fix a primary collection $\mathcal{P}$. For every $f_\alpha\in\mathcal{P}$, let Γ_α denote its dual graph and

$$\mathrm{Aux}(f_\alpha)=\big(E_\alpha,w_\alpha,H_\alpha,\varphi_\alpha,\mathcal{C}_\alpha\longrightarrow\mathcal{B}_\alpha,\epsilon_\alpha\big)\tag{6.67}$$

be the auxiliary data of Items (1)-(3) and (5)-(7) in page 236 involved in the construction of primary chart $\mathcal{U}_{f_\alpha}$.

For every arbitrary $[f=(u,C=(\Sigma,\mathfrak{j},\vec{z}))]\in\overline{\mathcal{M}}_{g,k}(X,A)$ define

$$S_f=\{\alpha\in S:[f]\in\mathrm{cl}(F'_{f_\alpha})\}.\tag{6.68}$$

Then

$$S_{f'}\subset S_f,\tag{6.69}$$

whenever $[f']$ is sufficiently close to $[f]$. Corresponding to every $\alpha\in S_f$ we get a unique set of points $w_{\alpha,f}\subset\Sigma$ at which f transversely intersects the slicing divisors H_α. For every $[f]$, we fix an arbitrary stabilizing set (w_0,H_0) as in Section 6.4, disjoint from $(w_{\alpha,f},H_\alpha)$, for all $\alpha\in S_f$.

Before discussing the general case, which involves several components labeled by dual graph, let us consider the simpler case where Σ is smooth. Let

$$\begin{aligned}&f^{+0}=(u,C^{+0}=(\Sigma,\mathfrak{j},\vec{z},w_0)),\quad f^{+\alpha}=(u,C^{+\alpha}=(\Sigma,\mathfrak{j},\vec{z},w_{\alpha,f})),\\&\text{and}\quad f^{+0\alpha}=(u,C^{+0\alpha}=(\Sigma,\mathfrak{j},\vec{z},w_{0\alpha})),\qquad\forall\alpha\in S_f,\end{aligned}\tag{6.70}$$

where $w_{0\alpha}=w_0\cup w_{\alpha,f}$.

By Condition 6.4.6, for every $\alpha\in S_f$, the universal family $\mathcal{C}_\alpha$ is also a universal family around $C^{+\alpha}$. Fix an $\alpha\in S_f$. Let $\mathcal{C}_0$ and $\mathcal{C}_{0\alpha}$ be sufficiently small universal families around C^{+0} and $C^{+0\alpha}$, respectively. The process of removing w_0 and $w_{\alpha,f}$ from $w_{0\alpha}$ gives us forgetful maps

$$\pi_0\colon(\mathcal{C}_{0\alpha},\mathcal{B}_{0\alpha})\longrightarrow(\mathcal{C}'_\alpha,\mathcal{B}'_\alpha)\quad\text{and}\quad\pi_\alpha\colon(\mathcal{C}_{0\alpha},\mathcal{B}_{0\alpha})\longrightarrow(\mathcal{C}_0,\mathcal{B}_0),\tag{6.71}$$

respectively, where $(\mathcal{C}'_\alpha\longrightarrow\mathcal{B}'_\alpha)$ is a sufficiently small sub-family[89] of $(\mathcal{C}_\alpha\longrightarrow\mathcal{B}_\alpha)$. Recall from Remark 6.2.5 that $\mathfrak{Aut}(f)\subset\mathfrak{Aut}(f_\alpha)$ acts on $(\mathcal{C}'_\alpha,\mathcal{B}'_\alpha)$, $(\mathcal{C}_0,\mathcal{B}_0)$, and $(\mathcal{C}_{0\alpha},\mathcal{B}_{0\alpha})$.

[89] In which all the curves have smooth domain.

For a choice of

$$0<\epsilon'_\alpha \ll \epsilon_\alpha, \quad 0<\epsilon_0 \ll \epsilon'_\alpha, \quad \text{and} \quad 0<\epsilon_{0\alpha} \ll \epsilon_0, \tag{6.72}$$

let

$$B_{\epsilon'_\alpha}(f^{+\alpha}), \quad B_{\epsilon_0}(f^{+0}), \quad \text{and} \quad B_{\epsilon_{0\alpha}}(f^{+0\alpha}) \tag{6.73}$$

be the Banach neighborhoods of (ℓ,p)-smooth maps within ϵ'_α, ϵ_0, and $\epsilon_{0\alpha}$ distances from $f^{+\alpha}$, f^{+0}, and $f^{+0\alpha}$ as in (6.39), respectively. Note that the definition of ϵ-neighborhoods in (6.73) as open subsets of

$$W^{\ell,p}(\mathcal{C}'_\alpha, X), \quad W^{\ell,p}(\mathcal{C}_0, X), \quad \text{and} \quad W^{\ell,p}(\mathcal{C}_{0\alpha}, X),$$

depends on the choices of smooth trivializations φ'_α, φ_0, and $\varphi_{0\alpha}$, for $\mathcal{C}'_\alpha$, $\mathcal{C}_0$, and $\mathcal{C}_{0\alpha}$ as in (6.14), respectively. However, the final construction of the *induced* obstruction bundle over a neighborhood of f^{+0} is independent of the choice of these trivializations.

REMARK 6.5.2. Let Σ_α be the domain of f_α. While $\mathcal{C}'_\alpha$ is a subfamily of $\mathcal{C}_\alpha$, the trivializations

$$\varphi'_\alpha\colon \mathcal{C}'_\alpha \longrightarrow \Sigma \times \mathcal{B}'_\alpha \quad \text{and} \quad \varphi_\alpha\colon \mathcal{C}_\alpha \longrightarrow \Sigma_\alpha \times \mathcal{B}_\alpha \ (\text{or}^{90}\ \Sigma^\circ_\alpha \times \mathcal{B}_\alpha)$$

have often no connection to each other. Even the topological type of Σ_α and Σ could be different; Σ_α could be nodal.

By (6.68), $f^{+\alpha}$ is an element of V_{f_α}. In the following, we will use $f^{+0\alpha}$ to relate a neighborhood of $f^{+\alpha}$ in $B_{\epsilon_\alpha}(f_\alpha)$ to some smaller neighborhood $B_{\epsilon'_0}(f^{+0})$ of f^{+0} in $B_{\epsilon_0}(f^{+0})$. This relation will give us a finite dimensional obstruction bundle

$$E_{0;\alpha} \longrightarrow B_{\epsilon'_0}(f^{+0}), \tag{6.74}$$

whose fiber at each point is obtained from parallel translation of E_α with respect to φ_α as in (6.55)[91]. We will add up these obstruction bundles, for all $\alpha \in S_f$, and this gives us the total obstruction bundle E_0. Then the induced chart $(\mathrm{pr}_f\colon U_f \longrightarrow V_f, s_f)$ around f is given by

$$V_f = (\bar\partial^{-1}(E_0) \subset B_{\epsilon'_0}(f^{+0})) \cap \mathrm{ev}_{w_0}^{-1}(H_0), \quad U_f = E_0|_{V_f}, \quad s_f = \bar\partial.$$

By the first assumption in (6.72) and Remark 6.2.11, $B_{\epsilon'_\alpha}(f^{+\alpha})$ continuously embeds in $B_{\epsilon_\alpha}(f_\alpha)$. Let

$$E_{\varphi_\alpha, f^{+\alpha}} = E_{\varphi_\alpha, f_\alpha}|_{B_{\epsilon'_\alpha}(f^{+\alpha})}$$

be the restriction of the obstruction bundle $E_{\varphi_\alpha,f_\alpha}$ defined in (6.54) to $B_{\epsilon'_\alpha}(f^{+\alpha})$. This is an $\mathfrak{Aut}(f)$-equivariant continuous vector bundle; see Remark 6.2.5. The forgetful maps of (6.71) extend to similarly denoted continuous $\mathfrak{Aut}(f)$-equivariant forgetful maps

$$\pi_0\colon B_{\epsilon_{0\alpha}}(f^{+0\alpha}) \longrightarrow B_{\epsilon'_\alpha}(f^{+\alpha}) \quad \text{and} \quad \pi_\alpha\colon B_{\epsilon_{0\alpha}}(f^{+0\alpha}) \longrightarrow B_{\epsilon_0}(f^{+0}); \tag{6.75}$$

see Remark 6.4.4. Let

$$E_{0\alpha} = \pi_0^* E_{\varphi_\alpha, f^{+\alpha}} \subset E^{\ell-1,p}(\mathcal{C}_{0\alpha}, X)|_{B_{\epsilon_{0\alpha}}(f^{+0\alpha})} \tag{6.76}$$

be the resulting $\mathfrak{Aut}(f)$-equivariant pull-back bundle. At this point one can pursue two different paths:

---

[90]If Σ_α is nodal.

[91]So it is independent of the choice of φ'_α, φ_0, and $\varphi_{0\alpha}$.

(1) we may first consider the fiber product space

$$\times_{\alpha \in S_f}\big(B_{\epsilon_{0\alpha}}(f^{+0\alpha}) \longrightarrow B_{\epsilon_0}(f^{+0})\big),$$

build a Kuranishi chart in this fiber product, and then apply all the slicing conditions H_0 and $\{H_\alpha\}_{\alpha \in S_f}$ at the end;

(2) or we may use the slicing conditions H_α to define an $\mathfrak{Aut}(f)$-equivariant continuous section

$$s_\alpha \colon B_{\epsilon_0'}(f^{+0}) \longrightarrow B_{\epsilon_{0\alpha}}(f^{+0\alpha}) \tag{6.77}$$

of π_α, for some $\epsilon_0' \ll \epsilon_0$, pull-back $E_{0;\alpha}$ to an $\mathfrak{Aut}(f)$-equivariant vector bundle $E_{0;\alpha}$ over $B_{\epsilon_0}(f^{+0})$, and add up the pull-back bundles $E_{0;\alpha}$, for all $\alpha \in S_f$, to construct the total $\mathfrak{Aut}(f)$-equivariant obstruction bundle E_0.

We will pursue the second method since $w_{\alpha,f}$ can not be assumed to be disjoint. Let

$$B_{\epsilon_{0\alpha}}(f^{+0\alpha})_{\cap H_\alpha} = \mathrm{ev}_{w_\alpha}^{-1}(H_\alpha) \subset B_{\epsilon_{0\alpha}}(f^{+0\alpha}), \tag{6.78}$$

where

$$\mathrm{ev}_{w_\alpha} \colon B_{\epsilon_{0\alpha}}(f^{+0\alpha}) \longrightarrow X^{w_\alpha}$$

is the evaluation map at the w_α-points as in (6.65). Since ev_{w_α} remains transverse to H_α in a neighborhood of $f^{+0\alpha}$, the left-hand side of (6.78) is the image of a continuous section (6.77) of the projection map π_α over some smaller neighborhood $B_{\epsilon_0'}(f^{+0})$; i.e. the section s_α in (6.77) obtained by adding the intersection points of f^{+0} with H_α. Finally, define

$$E_{0;\alpha} = s_\alpha^* E_{0\alpha} \subset E^{\ell-1,p}(\mathcal{C}_0, X)|_{B_{\epsilon_0'}(f^{+0})}.$$

Lemma 6.5.3. *For any choice of smooth trivialization φ_0 for $\mathcal{C}_0$, $E_{0;\alpha}$ is an $\mathfrak{Aut}(f)$-equivariant smooth sub-bundle of $E_{\varphi_0}^{\ell-1,p}(\mathcal{C}_0, X)|_{B_{\epsilon_0'}(f^{+0})}$.*

The proof of this lemma relies on the fact that the obstruction space E_α is made of smooth sections and thus the change of trivialization map preserves the smoothness of the corresponding obstruction bundles; see Remark 6.4.4. We skip the proof and refer to [**16**, Section 34.4] and [**14**, Appendix A-B]; especially see [**14**, Figure 14].

Let

$$\mathrm{pr}_0 \colon E_0 = \bigoplus_{\alpha \in S_f} E_{0;\alpha} \longrightarrow B_{\epsilon_0'}(f^{+0}) \tag{6.79}$$

be the resulting induced direct sum $\mathfrak{Aut}(f)$-equivariant obstruction bundle. The finite rank bundle E_0 is a domain-dependent version of the obstruction bundle in (6.33), with the difference that at each point, the α-summand of the fiber is obtained from the obstruction space E_α by a parallel-transport procedure from f_α (instead of f) to that point.

Remark 6.5.4. We say a primary collection $\mathcal{P}$ is **in general position** if

$$E_{0;\alpha}|_{f^{+0}} \cap E_{0;\beta}|_{f^{+0}} = \emptyset \qquad \forall [f] \in \overline{\mathcal{M}}_{g,k}(X, A),\ \alpha, \beta \in S_f;$$

i.e. if the formal direct sum in (6.79) is isomorphic to the usual sum. By [**16**, Lemma 18.8], every $\overline{\mathcal{M}}_{g,k}(X, A)$ admits a plethora of primary collections in general position. In fact, it is shown that any given primary collection $\mathcal{P}$ can be slightly perturbed so that the resulting Kuranishi charts form a primary collection in general position. The proof is somewhat deliberate. Assuming $\mathcal{P}$ is in general position, one can simplify the following construction to some extent; see Remark 6.5.5 below. However, for the sake of generality and avoiding the proof of this argument, we continue **without** this assumption.

Finally, we construct the induced chart at f in the following way. Let

$$V^{+0}=\{(f',(\eta_\alpha)_{\alpha\in S_f})\in E_0\colon \bar\partial f'-\sum_{\alpha\in S_f}\eta_\alpha=0\},\quad U^{+0}=(\mathrm{pr}_0^*E_0)|_{V^{+0}}, \tag{6.80}$$

and

$$s^{+0}\colon V^{+0}\longrightarrow U^{+0},\quad s^{+0}(f',(\eta_\alpha)_{\alpha\in S_f})=\bigoplus_{\alpha\in S_f}\eta_\alpha\in E_0|_{f'}. \tag{6.81}$$

In order to clarify the notation in (6.80), by $(f',(\eta_\alpha)_{\alpha\in S_f})\in E_0$ we mean the point corresponding to the vector $\oplus_{\alpha\in S_f}\eta_\alpha\in E_0|_{f'}$ in the total space of E_0. Then U^{+0} is the pull-back of E_0 to the total space of E_0 restricted to V^{+0}. Restricted to the zero set of s^{+0}, we get the obvious footprint map

$$\psi^{+0}\colon (s^{+0})^{-1}(0)\longrightarrow \overline{\mathcal{M}}_{g,k+|w_0|}(X,A),\quad \psi^{+0}(f',(\eta_\alpha)_{\alpha\in S_f})=[f']; \tag{6.82}$$

i.e. $\mathcal{U}^{+0}=(\mathrm{pr}^{+0}\colon U^{+0}\longrightarrow V^{+0},\mathfrak{Aut}(f),s^{+0},\psi^{+0})$ is a Kuranishi chart around $[f^{+0}]$; see Remark 6.4.3 for the ordering of extra marked points. We will intersect with the H_0 slices to obtain a Kuranishi chart around $[f]$.

REMARK 6.5.5. In the case of primary collections in general position, as in Remark 6.5.4, we have $E_0=\sum_{\alpha\in S_f}E_{0;\alpha}$ and (6.80) simplifies to

$$V^{+0}=\{f'\in B_{\epsilon_0'}(f^{+0})\colon \bar\partial f'\in E_0\},\quad U^{+0}=E_0|_{V^{+0}},$$
$$\text{and}\quad s^{+0}\colon V^{+0}\longrightarrow U^{+0},\quad s^{+0}(f')=\bar\partial f'.$$

In this case, the manifold V^{+0} in (6.80) is the graph of s^{+0} above.

Similarly to the case of primary charts in Section 6.2, by Implicit Function Theorem[92] and Condition 6.4.6, for ϵ_0' and $\mathcal{C}_0$ sufficiently small,

- $\mathrm{pr}^{+0}\colon U^{+0}\longrightarrow V^{+0}$ s a smooth $\mathfrak{Aut}(f)$-equivariant vector bundle;
- the evaluation map
$$\mathrm{ev}_{w_0}\colon V^{+0}\longrightarrow X^{w_0}$$
is $\mathfrak{Aut}(f)$-invariant and transverse to H_0;
- and the **induced chart**
$$\mathcal{U}_f\equiv(\mathrm{pr}_f\colon U_f\longrightarrow V_f,\mathfrak{Aut}(f),s_f,\psi_f), \tag{6.83}$$
where
$$V_f=\mathrm{ev}_{w_0}^{-1}(H_0)\cap V^{+0},\quad U_f=U^{+0}|_{V_f},\quad s_f=s^{+0}|_{V_f},$$
and ψ_f is the composition of ψ^{+0} with the forgetful map of removing w_0-points[93]
$$\overline{\mathcal{M}}_{g,k+|w_0|}(X,A)\longrightarrow\overline{\mathcal{M}}_{g,k}(X,A),$$
is a Kuranishi chart of the expected dimension (1.7) centered at $[f]$.

---

[92] With respect to the smooth structure determined by any choice of smooth trivialization φ_0 for $\mathcal{C}_0$ in Lemma 6.5.3.

[93] See Remark 6.4.3.

We now return to the general case where the domain of f can be nodal. Together with the modification above on each smooth component, the construction is similar to that of Section 6.3. Let Γ be the dual graphs of f with the set of vertices $\mathbb{V}$, edges $\mathbb{E}$, and legs $\mathbb{L}$. With notation as in the special case above, dual graphs of f^{+0}, $f^{+\alpha}$, and $f^{+0\alpha}$ have the same set of vertices and edges; they only have different set of legs. Since the following argument does not involve the legs, by abuse of notation, we will think of Γ as the dual graph of all three maps.

For every $v\in\mathbb{V}$, let $\mathcal{C}_{0;v}\longrightarrow\mathcal{B}_{0;v}$ be a sufficiently small universal family around the smooth component $C^{+0}_v=(\Sigma_v,\mathfrak{j}_v,\vec{z}_v,w_{0;v})$ of C^{+0}. In the product family

$$\mathcal{C}_0(\Gamma)=\prod_{v\in\mathbb{V}}\mathcal{C}_{0;v}\longrightarrow\mathcal{B}_0(\Gamma)=\prod_{v\in\mathbb{V}}\mathcal{B}_{0;v},$$

all the curves have the same fixed dual graph Γ. Let

$$\mathcal{C}_0\longrightarrow\mathcal{B}_0=\mathcal{B}_0(\Gamma)\times\Delta^{\mathbb{E}}$$

be a standard universal family around C^{+0} obtained from $\mathcal{C}_0(\Gamma)$ by smoothing the nodes as in (5.25). Similarly to Section 6.3, we first construct a fiber-product Kuranishi chart covering a neighborhood of $[f]$ in $\mathcal{M}_{g,k}(X,A)_\Gamma$, and then, via a gluing construction, extend it to a Kuranishi chart for $[f]$ in $\overline{\mathcal{M}}_{g,k}(X,A)$.

For every $\alpha\in S_f$, let

$$\mathcal{C}_{0\alpha;v}\longrightarrow\mathcal{B}_{0\alpha;v},\quad \mathcal{C}_{0\alpha}(\Gamma)\longrightarrow\mathcal{B}_{0\alpha}(\Gamma),\quad\text{and}\quad \mathcal{C}_{0\alpha}\longrightarrow\mathcal{B}_{0\alpha}=\mathcal{B}_{0\alpha}(\Gamma)\times\Delta^{\mathbb{E}}$$

be similarly defined universal families associated to $C^{+0\alpha}$. Once again, by Condition 6.4.6, the universal family $\mathcal{C}_\alpha$ of (6.67) is also a universal family (of the standard form) around $C^{+\alpha}$, and the process of removing w_0 and $w_{\alpha,f}$ from $w_{0\alpha}$ (as in (6.71)) gives us forgetful maps

$$\pi_0\colon(\mathcal{C}_{0\alpha},\mathcal{B}_{0\alpha})\longrightarrow(\mathcal{C}'_\alpha,\mathcal{B}'_\alpha)\quad\text{and}\quad\pi_\alpha\colon(\mathcal{C}_{0\alpha},\mathcal{B}_{0\alpha})\longrightarrow(\mathcal{C}_0,\mathcal{B}_0),\tag{6.84}$$

respectively, where $(\mathcal{C}'_\alpha\longrightarrow\mathcal{B}'_\alpha)$ is a sufficiently small sub-family of $(\mathcal{C}_\alpha\longrightarrow\mathcal{B}_\alpha)$. For each $\alpha\in S_f$, the dual graph Γ_α of f_α is a degeneration of Γ as in (5.28); i.e. every neck region of $\mathcal{C}'_\alpha$ (and the other two universal families) is also a neck region of the primary family $\mathcal{C}_\alpha$. Let $\mathcal{C}'_\alpha(\Gamma)\subset\mathcal{C}'_\alpha$ be the sub-family of curves with dual graph Γ; the forgetful maps π_0 and π_α take $\mathcal{C}_{0\alpha}(\Gamma)$ to $\mathcal{C}'_\alpha(\Gamma)$ and $\mathcal{C}_0(\Gamma)$, respectively.

For a choice of ϵ'_α, ϵ_0, and $\epsilon_{0\alpha}$ as in (6.72), let

$$B_{\epsilon'_\alpha}(f^{+\alpha}),\quad B_{\epsilon_0}(f^{+0}),\quad\text{and}\quad B_{\epsilon_{0\alpha}}(f^{+0\alpha})\tag{6.85}$$

be the set of maps ϵ-close to $f^{+\alpha}$, f^{+0}, and $f^{+0\alpha}$, as defined in Page 229, respectively. The same push-pull procedure, as in the case of f with smooth domain above, with (6.84) and (6.85) in place of (6.84) and (6.73), respectively, gives us an obstruction bundle

$$\mathrm{pr}_0\colon E_0=\bigoplus_{\alpha\in S_f}E_{0;\alpha}\longrightarrow B_{\epsilon'_0}(f^{+0})\tag{6.86}$$

as in (6.79). Define

$$\mathcal{U}^{+0}=(\mathrm{pr}^{+0}\colon U^{+0}\longrightarrow V^{+0},\mathfrak{Aut}(f),s^{+0},\psi^{+0})\tag{6.87}$$

as in (6.80)-(6.82). Similarly to Remark 6.3.4, we can not immediately apply the Implicit Function Theorem to conclude that V^{+0} is a smooth manifold of the expected dimension. We again need a gluing theorem to define a smooth structure on V^{+0}.

Let $B_{\epsilon_0'}(f^{+0};0)\subset B_{\epsilon_0'}(f^{+0})$ as in (6.47) be the subset of maps with dual graph the same as the dual graph of f^{+0}. This subspace lies in the fiber-product of Banach manifolds

$$W^{\ell,p}(\mathcal{C}_{0;v},X) \qquad \forall v\in\mathbb{V}$$

over the evaluation map at the nodal points as in (6.43). The restriction

$$\mathrm{pr}_{0;\Gamma}\colon E_{0;\Gamma}=E_0|_{B_{\epsilon_0'}(f^{+0};0)} \longrightarrow B_{\epsilon_0'}(f^{+0};0)$$

is a smooth $\mathfrak{Aut}(f)$-equivariant vector bundle of finite rank as in the case of smooth domain. We define

$$V_\Gamma^{+0}=\big\{(f',(\eta_\alpha)_{\alpha\in S_f})\in E_{0;\Gamma}\colon \bar\partial f'-\sum_{\alpha\in S_f}\eta_\alpha=0\big\},\quad U_\Gamma^{+0}=\mathrm{pr}_{0;\Gamma}^*E_{0;\Gamma}|_{V_\Gamma^{+0}}. \tag{6.88}$$

and $s_\Gamma^{+0}\colon V_\Gamma^{+0}\longrightarrow U_\Gamma^{+0}$ as in (6.80) and (6.81). Restricted to the zero set of s_Γ^{+0}, we get the obvious footprint map

$$\psi^{+0}\colon (s^{+0})^{-1}(0)\longrightarrow \mathcal{M}_{g,k+|w_0|}(X,A)_\Gamma,\quad \psi^{+0}(f',(\eta_\alpha)_{\alpha\in S_f})=[f'].$$

By Implicit Function Theorem, for ϵ_0' and $\mathcal{C}_0(\Gamma)$ sufficiently small, the projection map $\mathrm{pr}_\Gamma^{+0}\colon U_\Gamma^{+0}\longrightarrow V_\Gamma^{+0}$ is a smooth $\mathfrak{Aut}(f)$-equivariant vector bundle. Moreover, the evaluation map

$$\mathrm{ev}_{w_0}\colon V_\Gamma^{+0}\longrightarrow X^{w_0}$$

is $\mathfrak{Aut}(f)$-invariant and transverse to H_0.

The fiber-product Kuranishi chart

$$\mathcal{U}_\Gamma^{+0}\equiv(\mathrm{pr}_\Gamma^{+0}\colon U_\Gamma^{+0}\longrightarrow V_\Gamma^{+0},\mathfrak{Aut}(f),s_\Gamma^{+0},\psi_\Gamma^{+0}),$$

with footprint a neighborhood of $[f^{+0}]$ in[94] $\mathcal{M}_{g,k+|w_0|}(X,A)_\Gamma$ is the analogue of pre-gluing Kuranishi chart in (6.49)-(6.50). Next, by a gluing procedure similar to Section 6.3, we extend this chart to the entire $\mathcal{C}_0$. The following gluing theorem is the analogue of Theorem 6.3.5 for the induced charts. The only difference between Theorem 6.5.6 and Theorem 6.3.5 is in how the obstruction bundles are defined; in the former, the obstruction bundle was defined via parallel translate of the obstruction space at the base map f, while in the latter, the obstruction bundle is a direct sum of several bundles, each of which is obtained from parallel translate of the obstruction spaces E_α at the primary maps f_α.

THEOREM 6.5.6 ([**16**, Theorem 19.3]). *For ϵ_0' and $\mathcal{C}_0$ sufficiently small, there exists a "natural" (class of) $\mathfrak{Aut}(f)$-equivariant continuous one-to-one map*

$$\mathfrak{gl}\colon V_\Gamma^{+0}\times\Delta^{\mathbb{E}}\longrightarrow V^{+0}$$

such that the restriction of $\mathfrak{gl}$ to $V_\Gamma^{+0}\times\{0\}^{\mathbb{E}}\cong V_\Gamma^{+0}$ is the inclusion map and the action of $\mathfrak{Aut}(f)\subset\mathfrak{Aut}(C)$ on $\Delta^{\mathbb{E}}$ is given by (5.26).

Similarly to Theorem 6.3.6, we will use the product smooth structure given by the weak smooth structure on Δ in (6.61). Finally, similarly to the case of smooth domain before, the evaluation map

$$\mathrm{ev}_{w_0}\colon V^{+0}\longrightarrow X^{w_0}$$

is $\mathfrak{Aut}(f)$-invariant and transverse to H_0, and the **induced chart**

$$\mathcal{U}_f\equiv(\mathrm{pr}_f\colon U_f\longrightarrow V_f,\mathfrak{Aut}(f),s_f,\psi_f), \tag{6.89}$$

[94] See Remark 6.4.3.

defined as in (6.83) is a Kuranishi chart of the expected dimension (1.7) centered at $[f]$. This finishes our construction of induced Kuranishi charts covering the entire moduli space $\overline{\mathcal{M}}_{g,k}(X,A)$. We will discuss the existence of coordinate change maps (cf. Definition 3.2.2) in the next section.

6.6. Coordinate change maps. Assume $f_0=(u_0,C_0=(\Sigma_0,\mathfrak{j}_0,\vec{z}_0))$ is a J-holomorphic map in $\overline{\mathcal{M}}_{g,k}(X,A)$, (w_0,H_0) is an arbitrary stabilizing pair for that, and $\mathcal{U}_0=\mathcal{U}_{f_0}(w_0,H_0)$ is the resulting induced Kuranishi chart of (6.89). Let $f_1=(u_1,C_1=(\Sigma_1,\mathfrak{j}_1,\vec{z}_1))$ be a J-holomorphic map in the footprint of $\mathcal{U}_0$, (w_1,H_1) be an arbitrary stabilizing pair for f_1, and $\mathcal{U}_1=\mathcal{U}_{f_1}(w_1,H_1)$ be the resulting induced Kuranishi chart of (6.89). In particular, if $f_1=f_0$, we can choose two different stabilizing pairs (w_0,H_0) and (w_1,H_1) which a priori could result in different charts $\mathcal{U}_0$ and $\mathcal{U}_1$ around the same point $[f_0]$. If f_1 is sufficiently close to f_0, by (6.69), we have $S_{f_1}\subset S_{f_0}$. In order to conclude that the induced charts of Section 6.5 form a Kuranishi structure in the sense of Definition 3.2.6, we show that there exists a natural coordinate change map from a sub-chart of $\mathcal{U}_1$ into $\mathcal{U}_0$, as in Definition 3.2.2, independent of any choice except the data of primary charts and stabilizing pairs $\{(w_i,H_i)\}_{i=0,1}$. The cocycle condition property of Definition 3.2.4 would naturally follow from the construction. Confirming that tangent bundle condition holds is similar to Remark 6.2.9. We refer to [**16**, Section 22-24] for more details on this construction and the properties of coordinate change maps.

With notation as above, the points w_0, through the family $\mathcal{C}_0$, give us a unique set of points $w_{1,0}\subset\Sigma_1$ at which f_1 transversely intersects the slicing divisors H_0. The stabilizing pairs $(w_{1,0},H_0)$ and (w_1,H_1) for f_1 could be different. Let $\mathcal{U}_{1,0}=\mathcal{U}_{f_1}(w_{1,0},H_0)$ be an induced Kuranishi chart as in (6.89) corresponding to the inherited stabilization data $(w_{1,0},H_0)$ at f_1. The coordinate change map $\Phi_{10}\colon\mathcal{U}_1'\longrightarrow\mathcal{U}_0$ as we construct below, where $\mathcal{U}_1'$ is a sufficiently small sub-chart of $\mathcal{U}_1$ around f_1 as in Definition 3.2.2, is a composition of

$$\Phi_{1,10}\colon\mathcal{U}_1'\longrightarrow\mathcal{U}_{1,0}'\quad\text{and}\quad\Phi_{10,0}\colon\mathcal{U}_{1,0}'\longrightarrow\mathcal{U}_0,$$

where $\mathcal{U}_{1,0}'$ is a sufficiently small sub-chart of $\mathcal{U}_{1,0}'$ around f_1, $\Phi_{1,10}$ is a natural isomorphism of Kuranishi charts given by Lemma 6.6.3 below, and $\Phi_{10,0}$ is a canonical embedding of Kuranishi charts given by Lemma 6.6.1 below. That these coordinate change maps satisfy the cocycle condition follows from Lemma 6.6.4.

LEMMA 6.6.1. *Assume f_0 is a J-holomorphic map in $\overline{\mathcal{M}}_{g,k}(X,A)$, (w_0,H_0) is an arbitrary stabilizing pair for that, and $\mathcal{U}_0=\mathcal{U}_{f_0}(w_0,H_0)$ is the resulting induced Kuranishi chart of (6.89). Let f_1 be a J-holomorphic map in the footprint of $\mathcal{U}_0$ sufficiently close to f_0. If $(w_1,H_1)=(w_{1,0},H_0)$, with $w_{1,0}$ and $\mathcal{U}_1=\mathcal{U}_{f_1}(w_1,H_1)$ defined as above, there exists a canonical coordinate change map centered at f_1,*

$$\Phi_{10}\colon\mathcal{U}_1'\longrightarrow\mathcal{U}_0,$$

from a sub-chart $\mathcal{U}_1'$ of $\mathcal{U}_1$ into $\mathcal{U}_0$.

PROOF. Since by assumption $w_1=w_{1,0}$, with notation as in Section 6.5, the germ of universal family $\mathcal{C}_1$ corresponding to $(\Sigma_1,\mathfrak{j}_1,\vec{z}_1,w_1)$ is a subfamily of the universal family $\mathcal{C}_0$ corresponding to $(\Sigma_0,\mathfrak{j}_0,\vec{z}_0,w_0)$. Therefore, for $\epsilon_1'\ll\epsilon_0'$, with notation as in Page 229, we obtain an embedding $B_{\epsilon_1'}(f_1^{+1})\subset B_{\epsilon_0'}(f_0^{+0})$, where

$$f_0^{+0}=(u_0,(\Sigma_0,\mathfrak{j}_0,\vec{z}_0,w_0))\quad\text{and}\quad f_1^{+1}=(u_1,(\Sigma_1,\mathfrak{j}_1,\vec{z}_1,w_1));$$

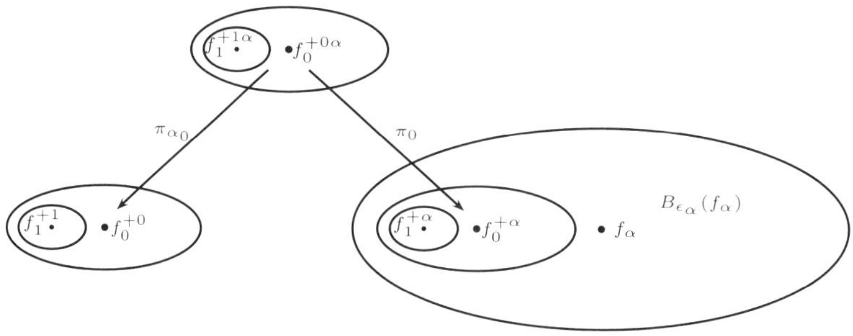

FIGURE 11. The diagram corresponding to the proof Lemma 6.6.1.

this inclusion is illustrated in the left-bottom corner of Figure 11. By assumption $S_{f_1} \subset S_{f_0}$; therefore, similarly, with notation as in (6.75), for every $\alpha \in S_{f_1}$, $\epsilon_{1\alpha} \ll \epsilon_{0\alpha}$ sufficiently small, and $\epsilon_{\alpha_1} \ll \epsilon_{\alpha_0} \ll \epsilon_\alpha$ we have[95]

$$B_{\epsilon_{1\alpha}}(f_1^{+1\alpha}) \subset B_{\epsilon_{0\alpha}}(f_0^{+0\alpha}) \quad \text{and} \quad B_{\epsilon_{\alpha_1}}(f_1^{+\alpha}) \subset B_{\epsilon_{\alpha_0}}(f_0^{+\alpha}) \subset B_{\epsilon_\alpha}(f_\alpha);$$

These inclusions are illustrated in the top and right-bottom corner of Figure 11. The projection maps

$$\pi_0 : B_{\epsilon_{0\alpha}}(f_0^{+0\alpha}) \longrightarrow B_{\epsilon_{\alpha_0}}(f_0^{+\alpha}) \quad \text{and} \quad \pi_1 : B_{\epsilon_{1\alpha}}(f_0^{+1\alpha}) \longrightarrow B_{\epsilon_{\alpha_1}}(f_1^{+\alpha})$$

and

$$\pi_{\alpha_0} : B_{\epsilon_{0\alpha}}(f_0^{+0\alpha}) \longrightarrow B_{\epsilon_0}(f_0^{+0}) \quad \text{and} \quad \pi_{\alpha_1} : B_{\epsilon_{1\alpha}}(f_0^{+1\alpha}) \longrightarrow B_{\epsilon_1}(f_1^{+1})$$

as in (6.75) commute with these emebeddings. Therefore, for every $\alpha \in S_{f_1}$, with

$$E_{0\alpha} \longrightarrow B_{\epsilon_{0\alpha}}(f_0^{+0\alpha}) \quad \text{and} \quad E_{1\alpha} \longrightarrow B_{\epsilon_{1\alpha}}(f_1^{+1\alpha})$$

as in (6.76), we have

$$E_{1\alpha} = E_{0\alpha}|_{B_{\epsilon_{1\alpha}}(f_1^{+1\alpha})}\,.$$

Since we use the same slicing conditions H_α to define the section s_α in (6.77), we have

$$E_{1;\alpha} = E_{0;\alpha}|_{B_{\epsilon'_1}(f_1^{+1})}.$$

With notation as in (6.79), we conclude that $E_1 = \bigoplus_{\alpha \in S_1} E_{1;\alpha}$ embeds in $E_0|_{B_{\epsilon'_1}(f_1^{+1})}$. Therefore, with notation as in (6.87), the germ of[96] pre-slicing Kuranishi chart $\mathcal{U}^{+1}$ corresponding to E_1 embeds in $\mathcal{U}^{+0}$. Since we apply the same slicing condition H_0 to both $\mathcal{U}^{+1}$ and $\mathcal{U}^{+0}$ to obtain $\mathcal{U}_1$ and $\mathcal{U}_0$, we get a canonical embedding of the germ of $\mathcal{U}_1$ around f_1 into $\mathcal{U}_0$, as defined in (6.83). □

LEMMA 6.6.2. *Assume f is a J-holomorphic map in $\overline{\mathcal{M}}_{g,k}(X,A)$, $(w_1,H_1) \subset (w_0,H_0)$ are two stabilizing pairs for f, and $\mathcal{U}_0 = \mathcal{U}_f(w_0,H_0)$ and $\mathcal{U}_1 = \mathcal{U}_f(w_1,H_1)$ are the resulting induced Kuranishi chart of (6.89). Then the forgetful map $\mathcal{U}_0 \longrightarrow \mathcal{U}_1$ is well-defined and is a canonical isomorphism.*

[95] $f_i^{+i\alpha} = (u_i, (\Sigma_i, \mathfrak{j}_i, \vec{z}_i, w_i \cup w_{\alpha,f_i}))$, with $i = 0, 1$.
[96] Because we are possibly restricting to smaller values of ϵ.

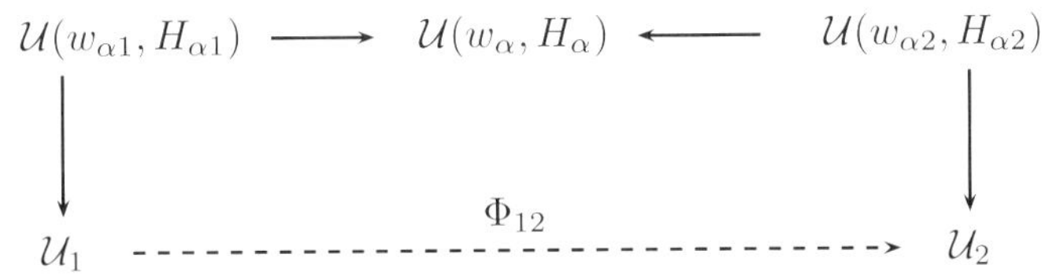

FIGURE 12. Diagram of natural embedding (6.92).

PROOF. The claim follows from the commutativity of the following diagram:

(6.90)

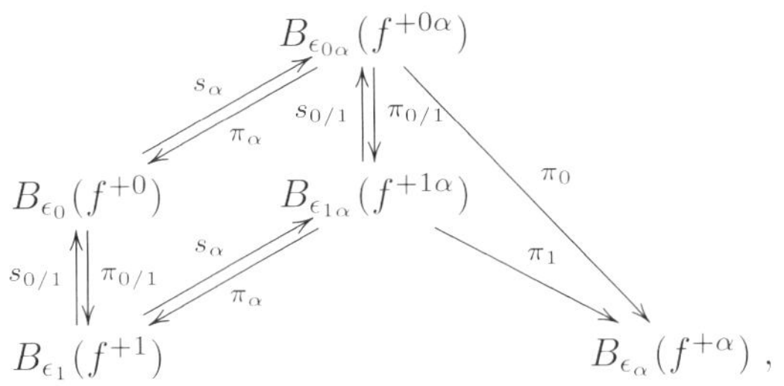

where π_0, π_1, and π_α are the forgetful maps of (6.75), $\pi_{0/1}$ is the forgetful map of removing w_0/w_1 stabilizing points, and s_α and $s_{0/1}$ are the section maps of (6.77). $\square$

For $i = 1, 2$, let (w_i, H_i) be two arbitrary stabilization data for a fixed map f. Let $\{\mathcal{U}_i = \mathcal{U}(w_i, H_i)\}_{i=1,2}$ be the resulting induced Kuranishi charts of (6.89). Another arbitrary stabilizing pair (w_α, H_α) for f disjoint from $\{(w_i, H_i)\}_{i=1,2}$ gives us larger stabilizing pairs

$$(w_{\alpha i}, H_{\alpha i}) = (w_i, H_i) \cup (w_\alpha, H_\alpha) \qquad \forall i = 1, 2,$$

and similarly defined induced Kuranishi charts

$$\{\mathcal{U}(w_{\alpha i}, H_{\alpha i})\}_{i=1,2} \quad \text{and} \quad \{\mathcal{U}(w_\alpha, H_\alpha)\}_{i=1,2}.$$

By Lemma 6.6.2, for $i = 1, 2$, after possibly restricting to sub-charts

$$\mathcal{U}'_i \subset \mathcal{U}_i, \quad \mathcal{U}'(w_{\alpha i}, H_{\alpha i}) \subset \mathcal{U}(w_{\alpha i}, H_{\alpha i}), \quad \text{and} \quad \mathcal{U}'(w_\alpha, H_\alpha) \subset \mathcal{U}(w_\alpha, H_\alpha)$$

around f, the forgetful maps obtained by removing the w_α and w_i points, respectively, give us isomorphisms

$$\Phi_{\alpha i, i}, \Phi_{\alpha i, \alpha} \colon \mathcal{U}'(w_{\alpha i}, H_{\alpha i}) \longrightarrow \mathcal{U}'_i, \mathcal{U}'(w_\alpha, H_\alpha) \tag{6.91}$$

LEMMA 6.6.3. *With notation as in (6.91), the orbibundle isomorphism*

$$\Phi_{12} = \Phi_{\alpha 2, 2} \circ \Phi^{-1}_{\alpha 2, \alpha} \circ \Phi_{\alpha 1, \alpha} \circ \Phi^{-1}_{\alpha 1, 1} \colon \mathcal{U}'_1 \longrightarrow \mathcal{U}'_2, \tag{6.92}$$

is independent of the choice of (w_α, H_α); see Figure 12.

PROOF. The claim follows from the commutativity of a diagram similar to (6.90): for two non-overlapping choices of (w_α, H_α), say $(w_{\alpha_1}, H_{\alpha_1})$ and $(w_{\alpha_2}, H_{\alpha_2})$, we consider the union and the claim follows from Lemma 6.6.2. The general case follows from continuity (or a bigger commutative diagram). $\square$

Consequently, we call Φ_{12} the **natural isomorphism** between the germ of Kuranishi charts $\mathcal{U}_1$ and $\mathcal{U}_2$.

LEMMA 6.6.4. *For $i=1,2,3$, let (w_i,H_i) be arbitrary stabilizing pairs for a fixed map f. Then the natural isomorphisms $\{\Phi_{ij}\}_{1\le i<j\le 3}$ satisfy $\Phi_{13}=\Phi_{23}\circ\Phi_{12}$.*

PROOF. Choose (ω_α,H_α) disjoint from each of $(w_i,H_i)_{i=1,2,3}$. Let

$$\Phi_{ij}=\Phi_{\alpha;ij}\quad \forall 1\le i<j\le 3$$

be the isomorphims of (6.92). Then

$$\Phi_{23}\circ\Phi_{12}=\left(\Phi_{\alpha 3,3}\circ\Phi^{-1}_{\alpha 3,\alpha}\circ\iota_{23}\circ\Phi_{\alpha 2,\alpha}\circ\Phi^{-1}_{\alpha 2,2}\right)\circ\left(\Phi_{\alpha 2,2}\circ\Phi^{-1}_{\alpha 2,\alpha}\circ\iota_{12}\circ\Phi_{\alpha 1,\alpha}\circ\Phi^{-1}_{\alpha 1,1}\right)=$$
$$\Phi_{\alpha 3,3}\circ\Phi^{-1}_{\alpha 3,\alpha}\circ(\iota_{23}\circ\iota_{12})\circ\Phi_{\alpha 1,\alpha}\circ\Phi^{-1}_{\alpha 1,1}=\Phi_{\alpha 3,3}\circ\Phi^{-1}_{\alpha 3,\alpha}\circ\iota_{13}\circ\Phi_{\alpha 1,\alpha}\circ\Phi^{-1}_{\alpha 1,1}=\Phi_{13}.$$

□

REMARK 6.6.5. Recall from Remark 6.4.5 that similar statements do not hold for primary Kuranishi charts; thus, primary charts do not admit coordinate change maps. This is essentially the main reason for considering the more complicated induced charts to put a Kuranishi structure on $\overline{\mathcal{M}}_{g,k}(X,A)$. Since the obstruction bundle at a particular point $[f]\in\overline{\mathcal{M}}_{g,k}(X,A)$ does not depend on the choice of the induced chart covering that point, it enables us to compare different charts corresponding to different stabilization data and other choices.

6.7. GW invariants. Let (X^{2n},ω) be a closed symplectic manifold, $J\in\mathcal{J}(X,\omega)$ be an arbitrary compatible (or tame) almost complex structure, $A\in H_2(X,\mathbb{Z})$, and $g,k\in\mathbb{N}$. With d as in (1.7), the construction of last few sections gives us a class of natural oriented d-dimensional Kuranishi structures on $\overline{\mathcal{M}}_{g,k}(X,A,J)$. Thus it establishes the first statement of Theorem 6.5.1. Let

- ω_0 and ω_1 be isotopic symplectic structures on X,
- $J_0\in\mathcal{J}(X,\omega_0)$ and $J_1\in\mathcal{J}(X,\omega_1)$,
- $\mathcal{P}_0$ and $\mathcal{P}_1$ be arbitrary primary collections for $\overline{\mathcal{M}}_{g,k}(X,A,J_0)$ and $\overline{\mathcal{M}}_{g,k}(X,A,J_1)$,
- and $\mathcal{K}_0$ and $\mathcal{K}_1$ be the resulting Kuranishi structures on $\overline{\mathcal{M}}_{g,k}(X,A,J_0)$ and $\overline{\mathcal{M}}_{g,k}(X,A,J_1)$.

Choose a path $J_{[0,1]}\equiv\{J_t\}_{t\in[0,1]}$ of smooth compatible almost complex structures on (X,ω) connecting J_0 to J_1. For simplicity, we choose a path $J_{[0,1]}$ such that J_t is equal to constant J_0 on $[0,\epsilon]$ and is equal to J_1 on $[1-\epsilon,1]$. Let $\overline{\mathcal{M}}_{g,k}(X,A,J_{[0,1]})$ be the moduli space of all genus g degree A k-marked J_t-holomorphic maps (1.14). When equipped with the Gromov topology, this is again a compact metrizable topology space with boundary

$$\partial\overline{\mathcal{M}}_{g,k}(X,A,J_{[0,1]})=\overline{\mathcal{M}}_{g,k}(X,A,J_0)\cup\overline{\mathcal{M}}_{g,k}(X,A,J_1). \tag{6.93}$$

Every chart $\mathcal{U}_{f_\alpha}$ in $\mathcal{P}_0$ trivially extends to a Kuranishi chart with boundary $\mathcal{U}_{f_\alpha}\times[0,\epsilon)$ covering a neighborhood of $f_\alpha\times\{J_0\}$ in $\overline{\mathcal{M}}_{g,k}(X,A,J_{[0,1]})$. Similarly, $\mathcal{P}_1$ extends to a primary collection covering a neighborhood of

$$\overline{\mathcal{M}}_{g,k}(X,A,J_1)\subset\overline{\mathcal{M}}_{g,k}(X,A,J_{[0,1]}).$$

By adding more primary Kuranishi charts covering a neighborhood of the middle part

$$\overline{\mathcal{M}}_{g,k}(X,A,J_{[\epsilon,1-\epsilon]}),$$

we can extend the primary collections $\mathcal{P}_0$ and $\mathcal{P}_1$ to a primary collection $\mathcal{P}$ of Kuranishi charts which covers the entire $\overline{\mathcal{M}}_{g,k}(X,A,J_{[0,1]})$. The construction of previous two sections then gives a Kuranishi structure with boundary $\mathcal{K}$ on $\overline{\mathcal{M}}_{g,k}(X,A,J_{[0,1]})$

whose restriction to the boundary components in the right-hand side of (6.93) coincide with $\mathcal{K}_0$ and $\mathcal{K}_1$, respectively. This gives us a cobordism between $\mathcal{K}_0$ and $\mathcal{K}_1$ in the sense of Definition 3.5.1 and thus establishes the second part of Theorem 6.5.1.

The evaluation maps ev and forgetful map st in (1.4) and (1.5), respectively, readily lift to continuous maps on every one of the Kuranishi charts constructed in the previous sections. These lifts are compatible with coordinate change maps. We conclude that all the natural Kuranishi structures of Theorem 6.5.1 are indeed (ev×st)-Kuranishi structures in the sense of Definition 4.2.1. Together with Theorem 6.5.1 and Proposition 4.2.2, we obtain a singular homology class

$$[\overline{\mathcal{M}}_{g,k}(X,A,J)]^{\mathrm{VFC}} \in H_d(X^k \times \overline{\mathcal{M}}_{g,k},\mathbb{Q})$$

which only depends on the deformation equivalence class of (X,ω). This establishes Theorem 1.4.1. Gromov-Witten invariants with primary insertions are defined by integration of cohomology classes in X^k and $\overline{\mathcal{M}}_{g,k}$ against this homology class.

In order to define GW invariants involving ψ and Hodge classes, defined in Section 1.4, we start with a Kuranishi structure $\mathcal{K}$ provided by Theorem 6.5.1. The ψ and Hodge classes in the question define a vector bundle $\mathfrak{E}$ on $\mathcal{K}$ in the sense of Definition 3.9.1. This Kuranishi vector bundle gives us an augmented Kuranishi structure $\mathcal{K}^{\mathfrak{E}}$ in the sense of (3.97). The augmented Kuranishi structure $\mathcal{K}^{\mathfrak{E}}$ is still an (ev×st)-Kuranishi structure. Every two of such $\mathcal{K}^{\mathfrak{E}}$ are still cobordant. Let

$$[\mathfrak{E}\to\overline{\mathcal{M}}_{g,k}(X,A,J)]^{\mathrm{VFC}} := [\mathcal{K}^{\mathfrak{E}}]^{\mathrm{VFC}} \in H_d(X^k \times \overline{\mathcal{M}}_{g,k},\mathbb{Q})$$

denote the resulting virtual fundamental class of Section 4.2. Gromov-Witten invariants involving a particular class of type $\mathfrak{E}$ and primary insertions are defined by integration of cohomology classes in X^k and $\overline{\mathcal{M}}_{g,k}$ against $[\mathfrak{E}\to\overline{\mathcal{M}}_{g,k}(X,A,J)]^{\mathrm{VFC}}$. They only depend on the deformation equivalence class of (X,ω) and the type of $\mathfrak{E}$. We will not further go into establishing the functorial properties of these virtual fundamental classes corresponding to tensor product, direct sum, etc. of these bundles.

7. Examples

In this section, we go over some basic examples of moduli spaces of pseudo-holomorphic maps with admit pure orbibundle Kuranishi structure.

7.1. Degree zero maps. Let (X^{2n},ω) be a non-trivial symplectic manifold and $g,k\in\mathbb{N}$, with $2g+k\geq 3$. Every pseudoholomorphic map with trivial homology class $0\in H_2(X,\mathbb{Z})$ is a constant map. Therefore, $\overline{\mathcal{M}}_{g,k}(X,0)$ (which is independent of the choice of J) is homeomorphic to

$$\overline{\mathcal{M}}_{g,k}(X,0)=X\times\overline{\mathcal{M}}_{g,k}. \tag{7.1}$$

In this case, for every $f=(u,C=(\Sigma,\mathfrak{j},\vec{z}))\in\overline{\mathcal{M}}_{g,k}$, we have

$$\mathrm{Im}(u)\equiv p\in X, \quad u^*TX=\Sigma\times T_pX,$$

and

$$\mathrm{D}_u\bar\partial(\zeta)=\bar\partial\zeta \qquad \forall\zeta\colon\Sigma\longrightarrow T_pX\cong\mathbb{C}^n.$$

Therefore

$$\mathrm{Obs}(u)\cong\mathrm{Obs}(f)\cong T_pX\otimes_{\mathbb{C}} H^1(\mathcal{T}\Sigma_{\mathfrak{j}}); \tag{7.2}$$

see Remark 6.2.1. Therefore, if $g=0$, then $\overline{\mathcal{M}}_{g,k}(X,0)$ has the structure of a smooth orbifold given by the right-hand side of (7.1). Let $\mathcal{E}_g$ be the Hodge orbibundle over $\overline{\mathcal{M}}_{g,k}$ defined in (1.13). For $g\geq 1$,

$$\dim_{\mathbb{C}}^{\mathrm{vir}} \overline{\mathcal{M}}_{g,k}(X,0) = \dim_{\mathbb{C}} X(1-g)+\dim_{\mathbb{C}} \overline{\mathcal{M}}_{g,k}$$

and the right-hand side of (7.2) defines an orbibundle

$$TX \boxtimes \mathcal{E}_g \longrightarrow X \times \overline{\mathcal{M}}_{g,k} \tag{7.3}$$

whose fiber at every point, by (7.2), is the minimal possible obstruction space; see Remark 6.2.4. Therefore, (7.3) with Kuranishi map $s\equiv 0$ defines a pure orbibundle Kuranishi structure on $\overline{\mathcal{M}}_{g,k}(X,0)$ whose top Chern class is the VFC of Section 4; see Example 3.2.8.

Note that if $(g,k)=(1,1)$ or $(2,0)$, then the orbifold structure of $\overline{\mathcal{M}}_{g,k}$ is not effective. If $n>0$, we may consider a larger obstruction space to obtain an effective orbifold.

As an example, if $\dim_{\mathbb{C}} X=3$, $k=0$, and $g\geq 2$, then the Gromov-Witten VFC is a zero-dimensional class, thus a rational number, and it is equal to

$$\chi(X)\int_{\overline{\mathcal{M}}_{g,0}} c_{\mathrm{top}}(\mathcal{E}_g) = \chi(X)\frac{B_{2g}}{4g(g-1)},$$

where B_n is the n-th Bernoulli number.

7.2. Elliptic surfaces. Let (X,ω,J) be a Kähler surface which has the structure of an elliptic fiberation $\pi: X\longrightarrow S$ over a smooth curve of genus g. Moreover, assume that π has no singular fiber other than multiple-fibers of order $(m_i)_{i\in[\ell]}$ over $(p_i)_{i\in[\ell]}\subset S$. Let

$$\rho(m)=\sum_{a|m} a \qquad \forall m\in\mathbb{N}.$$

By [4, Theorem 18.2], the complex structure of reduced fibers are all the same; say equal to some fixed $\mathrm{j}\in\mathbb{H}/\mathrm{PSL}(2,\mathbb{Z})$. Let $\mathcal{S}$ denote the orbifold structure on S given by replacing a neighborhood of each p_i with a cyclic m_i-quotient of a disk as in Section 2.1. The moduli space $\overline{\mathcal{M}}_{1,0}(X,dF)$, where F is a generic fiber class and $d\in\mathbb{Z}^+$, has virtual dimension zero. Therefore, the Gromov-Witten virtual fundamental class $[\overline{\mathcal{M}}_{1,0}(X,dF)]^{\mathrm{VFC}}$ is a zero-dimensional cycle in $H_0(X^0\times \mathcal{M}_{1,0},\mathbb{Q})=H_0(\mathrm{pt},\mathbb{Q})\cong\mathbb{Q}$; we denote by N_d to be the degree of this cycle. In the simplest case of $d=1$, the moduli space $\overline{\mathcal{M}}_{1,0}(X,F)$ consists of

- a unique map over every point of $S-\{p_i\}_{i\in[\ell]}$,
- and $\rho(m_i)$ different m_i-covering maps f_{ij} of the reduced fiber over p_i, for every $i\in[\ell]$, only one of which, say f_{i1}, is in the limit of the first group of maps.

Therefore, $\overline{\mathcal{M}}_{1,0}(X,F)$ is homeomorphic to the disjoint union

$$S\cup \bigcup_{\substack{i\in[\ell]\\ 2\leq j\leq \rho(m_i)}} [f_{ij}];$$

this is an example of a moduli space with connected components of various dimension. Over the connected component identified with S, the obstruction space $\mathrm{Obs}(f_s)$ at every map f_s, with $s\in S$, has constant rank and is isomorphic to

$$H^0(\Omega^{0,1}_{\Sigma_{\mathrm{j}}})\otimes T_sS.$$

The automorphism group of every holomorphic map over $S-\{p_i\}_{i\in[\ell]}$ is trivial and f_{ij} has an automorhism group of order m_i. Therefore, over the component identified with S, the moduli space has the structure of a pure orbibundle Kuranishi structure

$$H^0(\Omega^{0,1}_{\Sigma_i})\otimes T\mathcal{S}\cong T\mathcal{S}\longrightarrow\mathcal{S}$$

with the Kuranishi map $s\equiv 0$. On the otherhand, each of the holomorphic maps f_{ij}, with $i\in[\ell]$ and $2\leq j\leq\rho(m_i)$, is regular (i.e. $\mathrm{Obs}(f)=0$). Therefore, the simplest natural Kuranishi structure on every $[f_{ij}]$ is a non-effective orbifold structure on a single point (with trivial obstruction bundle and) with an isotropy group of order m_i. Similarly to before, in order to get an effective orbifold, we may enlarge the obstruction bundle to take care of the non-effective action. Nevertheless, the contribution of every $[f_{ij}]$ to N_1 is $1/m_i$. We conclude that

$$N_1=\chi(\mathcal{S})+\sum_{i=1}^{\ell}\frac{\rho(m_i)-1}{m_i}=(2-2g-\ell)+\sum_{i=1}^{\ell}\frac{\rho(m_i)}{m_i}\in\mathbb{Q}.$$

7.3. Genus zero maps in quintic. In this section, we consider the example of genus 0 maps in a quintic Calabi-Yau 3-fold in $\mathbb{P}^4$.

Let (X,ω,J) be a smooth degree 5 divisor in $\mathbb{P}^4$ (i.e. ω is the restriction of Fubini-Study Kähler form and J is the complex structure). By Lefschetz hyperplane theorem, $H_2(X,\mathbb{Z})\cong H_2(\mathbb{P}^4,\mathbb{Z})\cong\mathbb{Z}$. For every $[d]\in H_2(X,\mathbb{Z})$, the moduli space $\overline{\mathcal{M}}_{0,0}(X,[d])$ has virtual dimension zero. Let

$$N_d=[\overline{\mathcal{M}}_{0,0}(X,[d])]^{\mathrm{VFC}}\in H_0(X^0\times\mathcal{M}_{0,0},\mathbb{Q})\cong H_0(\mathrm{pt},\mathbb{Q})\cong\mathbb{Q}\quad\forall d\in\mathbb{N}$$

be the corresponding GW invariants.

Unlike previous examples, an explicit description of the moduli space and of natural Kuranishi structures is beyond the reach. However, for this (class of) example(s), we can construct an abstract pure orbibundle Kuranishi structure which a priori has no trivial connection to the natural Kuranishi structures of Section 6.5. We first recall the construction of this particular abstract pure orbibundle Kuranishi structure and then discuss the known facts about its relation to the natural Kuranishi structures of Section 6.5.

Suppose X is the zero set of a degree 5 homogenous polynomial $F\in H^0(\mathcal{O}_{\mathbb{P}^4}(5))$, i.e. we think of F as a section of the line bundle $\mathcal{O}_{\mathbb{P}^4}(5)$. Every holomorphic map into X can be thought of as a holomorphic map into $\mathbb{P}^4$ (of the same degree and genus). By [**29**, Exercise 24.3.4(7)], $\overline{\mathcal{M}}_{0,0}(\mathbb{P}^4,[d])$ has the structure of a smooth complex orbifold of complex dimension $5d$. A holomorphic map $f=[u,\Sigma,\mathfrak{j}]\in\overline{\mathcal{M}}_{0,0}(\mathbb{P}^4,[d])$ belongs to $\overline{\mathcal{M}}_{0,0}(X,[d])$ if and only if the pullback section $u^*F\in H^0(u^*\mathcal{O}_{\mathbb{P}^4}(5))$ vanishes. The complex vector space $H^0(u^*\mathcal{O}_{\mathbb{P}^4}(5))$ is of constant rank $5d$. In fact, there exists a rank $5d$ orbibundle

$$\mathcal{E}_{0,d}\longrightarrow\overline{\mathcal{M}}_{0,0}(\mathbb{P}^4,[d]),\tag{7.4}$$

with the fiber $[H^0(u^*\mathcal{O}_{\mathbb{P}^4}(5))/\mathfrak{Aut}(f)]$ over each f, and

$$s_F\colon\overline{\mathcal{M}}_{0,0}(\mathbb{P}^4,[d])\longrightarrow\mathcal{E}_{0,d},\quad s_F(f)=u^*F$$

is a section of this bundle; see [**29**, Section 26.1.3]. The conclusion is that the orbibundle (7.4) together with the orbifold section s_F defines an "abstract" zero-dimensional pure orbibundle Kuranishi structure on $\overline{\mathcal{M}}_{0,0}(X,[d])$. Let N'_d be the orbifold top-Chern number of this orbibundle; see Section 2.6.

Then the main claim is that

$$N_d = N'_d \,, \tag{7.5}$$

i.e. the VFC obtained from the abstract pure orbibundle Kuranishi structure above coincides with that of Section 6.5. The numbers N'_d can be calculated via localization technique; see [**29**, Chapter 29.1]. Therefore, the equality (7.5) provides a way of calculating genus 0 GW invariants of quintic threefold (as well as several other similar examples). This trick does not extend beyond genus 0; see [**59**] for the case of genus 1 invariants and the higher genus case is still a field of active research.

We finish this section and thus this manuscript with few words on the proof of (7.5). In the algebraic case, this is stated as [**29**, Theorem 26.1.1] without proof. Cox-Katz [**8**, Example 7.1.5.1] outlines a scheme of proving that via the algebraic approach of Li-Tian [**34**]. In an argument aimed at the moduli spaces of pseudo-holomorphic disks which also applies to this example, [**43**, Section 5.3] outlines a way of proving (7.5) by extending a Kuranishi structure on $\overline{\mathcal{M}}_{0,0}(X,[d])$ to a Kuranishi structure on $\overline{\mathcal{M}}_{0,0}(\mathbb{P}^4,[d])$ and comparing that with the trivial Kuranishi (orbifold) structure of $\overline{\mathcal{M}}_{0,0}(\mathbb{P}^4,[d])$. However, a complete proof in the analytic setting has yet to be written in details.

References

[1] A. Adem, J. Leida, and Y. Ruan, *Orbifolds and stringy topology*, Cambridge Tracts in Mathematics, vol. 171, Cambridge University Press, Cambridge, 2007. MR2359514

[2] M. Audin and J. Lafontaine (eds.), *Holomorphic curves in symplectic geometry*, Progress in Mathematics, vol. 117, Birkhäuser Verlag, Basel, 1994. MR1274923

[3] D. Auroux, F. Catanese, M. Manetti, P. Seidel, B. Siebert, I. Smith, and G. Tian, *Symplectic 4-manifolds and algebraic surfaces*, Lecture Notes in Mathematics, vol. 1938, Springer-Verlag, Berlin; Fondazione C.I.M.E., Florence, 2008. Lectures from the C.I.M.E. Summer School held in Cetraro, September 2–10, 2003; Edited by Catanese and Tian. MR2463711

[4] W. P. Barth, K. Hulek, C. A. M. Peters, and A. Van de Ven. *Compact complex surfaces*, volume 4 of *Ergebnisse der Mathematik und ihrer Grenzgebiete. 3. Folge. A Series of Modern Surveys in Mathematics [Results in Mathematics and Related Areas. 3rd Series. A Series of Modern Surveys in Mathematics]*. Springer-Verlag, Berlin, second edition, 2004.

[5] P. Candelas, X. C. de la Ossa, P. S. Green, and L. Parkes, *A pair of Calabi-Yau manifolds as an exactly soluble superconformal theory*, Nuclear Phys. B **359** (1991), no. 1, 21–74, DOI 10.1016/0550-3213(91)90292-6. MR1115626

[6] R. Castellano. Genus zero gromov-witten axioms via kuranishi atlases. *arXiv:1601.04048*.

[7] K. Cieliebak and K. Mohnke, *Symplectic hypersurfaces and transversality in Gromov-Witten theory*, J. Symplectic Geom. **5** (2007), no. 3, 281–356. MR2399678

[8] D. A. Cox and S. Katz, *Mirror symmetry and algebraic geometry*, Mathematical Surveys and Monographs, vol. 68, American Mathematical Society, Providence, RI, 1999. MR1677117

[9] P. Deligne and D. Mumford, *The irreducibility of the space of curves of given genus*, Inst. Hautes Études Sci. Publ. Math. **36** (1969), 75–109. MR0262240

[10] S. K. Donaldson, *An application of gauge theory to four-dimensional topology*, J. Differential Geom. **18** (1983), no. 2, 279–315. MR710056

[11] B. Fantechi, *Stacks for everybody*, European Congress of Mathematics, Vol. I (Barcelona, 2000), Progr. Math., vol. 201, Birkhäuser, Basel, 2001, pp. 349–359. MR1905329

[12] A. Floer, *Proof of the Arnol′d conjecture for surfaces and generalizations to certain Kähler manifolds*, Duke Math. J. **53** (1986), no. 1, 1–32, DOI 10.1215/S0012-7094-86-05301-9. MR835793

[13] K. Fukaya. Lectures on kuranishi structures. *http://homepage.divms.uiowa.edu/ mftehrani /Fukaya.html.*

[14] K. Fukaya, Y.-G. Oh, H. Ohta, and K. Ono. Exponential decay estimates and smoothness of the moduli space of pseudoholomorphic curves. *arXiv:1603.07026.*

[15] K. Fukaya, Y.-G. Oh, H. Ohta, and K. Ono. Kuranishi structure, pseudo-holomorphic curve, and virtual fundamental chain: Part 1. *arXiv:1503.07631*.

[16] K. Fukaya, Y.-G. Oh, H. Ohta, and K. Ono. Technical details on kuranishi structure and virtual fundamental chain. *arXiv:1209.4410*.

[17] K. Fukaya, Y.-G. Oh, H. Ohta, and K. Ono, *Lagrangian intersection Floer theory: anomaly and obstruction. Part I*, AMS/IP Studies in Advanced Mathematics, vol. 46, American Mathematical Society, Providence, RI; International Press, Somerville, MA, 2009. MR2553465

[18] K. Fukaya, Y.-G. Oh, H. Ohta, and K. Ono, *Lagrangian Floer theory over integers: spherically positive symplectic manifolds*, Pure Appl. Math. Q. **9** (2013), no. 2, 189–289, DOI 10.4310/PAMQ.2013.v9.n2.a1. MR3138477

[19] K. Fukaya, Y.-G. Oh, H. Ohta, and K. Ono, *Shrinking good coordinate systems associated to Kuranishi structures*, J. Symplectic Geom. **14** (2016), no. 4, 1295–1310, DOI 10.4310/JSG.2016.v14.n4.a10. MR3601890

[20] K. Fukaya and K. Ono, *Arnold conjecture and Gromov-Witten invariant*, Topology **38** (1999), no. 5, 933–1048, DOI 10.1016/S0040-9383(98)00042-1. MR1688434

[21] M. Gromov, *Pseudo holomorphic curves in symplectic manifolds*, Invent. Math. **82** (1985), no. 2, 307–347, DOI 10.1007/BF01388806. MR809718

[22] A. Haefliger, *Homotopy and integrability*, Manifolds–Amsterdam 1970 (Proc. Nuffic Summer School), Lecture Notes in Mathematics, Vol. 197, Springer, Berlin, 1971, pp. 133–163. MR0285027

[23] A. Haefliger, *Groupoïdes d'holonomie et classifiants* (French), Astérisque **116** (1984), 70–97. Transversal structure of foliations (Toulouse, 1982). MR755163

[24] A. Haefliger, *Groupoids and foliations*, Groupoids in analysis, geometry, and physics (Boulder, CO, 1999), Contemp. Math., vol. 282, Amer. Math. Soc., Providence, RI, 2001, pp. 83–100, DOI 10.1090/conm/282/04680. MR1855244

[25] J. Harris and I. Morrison, *Moduli of curves*, Graduate Texts in Mathematics, vol. 187, Springer-Verlag, New York, 1998. MR1631825

[26] M. W. Hirsch, *Differential topology*, Springer-Verlag, New York-Heidelberg, 1976. Graduate Texts in Mathematics, No. 33. MR0448362

[27] H. Hofer and D. A. Salamon, *Floer homology and Novikov rings*, The Floer memorial volume, Progr. Math., vol. 133, Birkhäuser, Basel, 1995, pp. 483–524. MR1362838

[28] H. Hofer, *Polyfolds and Fredholm theory*, Lectures on geometry, Clay Lect. Notes, Oxford Univ. Press, Oxford, 2017, pp. 87–158. MR3676594

[29] K. Hori, S. Katz, A. Klemm, R. Pandharipande, R. Thomas, C. Vafa, R. Vakil, and E. Zaslow, *Mirror symmetry*, Clay Mathematics Monographs, vol. 1, American Mathematical Society, Providence, RI; Clay Mathematics Institute, Cambridge, MA, 2003. With a preface by Vafa. MR2003030

[30] C. Hummel, *Gromov's compactness theorem for pseudo-holomorphic curves*, Progress in Mathematics, vol. 151, Birkhäuser Verlag, Basel, 1997. MR1451624

[31] D. Joyce. D-manifolds, d-orbifolds and derived differential geometry. *arXiv:1208.4948*.

[32] D. Joyce. Kuranishi homology and kuranishi cohomology. *arXiv:0707.3572*.

[33] M. Kuranishi, *On the locally complete families of complex analytic structures*, Ann. of Math. (2) **75** (1962), 536–577, DOI 10.2307/1970211. MR0141139

[34] J. Li and G. Tian, *Virtual moduli cycles and Gromov-Witten invariants of algebraic varieties*, J. Amer. Math. Soc. **11** (1998), no. 1, 119–174, DOI 10.1090/S0894-0347-98-00250-1. MR1467172

[35] J. Li and G. Tian, *Virtual moduli cycles and Gromov-Witten invariants of general symplectic manifolds*, Topics in symplectic 4-manifolds (Irvine, CA, 1996), First Int. Press Lect. Ser., I, Int. Press, Cambridge, MA, 1998, pp. 47–83. MR1635695

[36] D. McDuff, *Groupoids, branched manifolds and multisections*, J. Symplectic Geom. **4** (2006), no. 3, 259–315. MR2314215

[37] D. McDuff and D. Salamon, *J-holomorphic curves and symplectic topology*, American Mathematical Society Colloquium Publications, vol. 52, American Mathematical Society, Providence, RI, 2004. MR2045629

[38] D. McDuff and K. Wehrheim, *Smooth Kuranishi atlases with isotropy*, Geom. Topol. **21** (2017), no. 5, 2725–2809, DOI 10.2140/gt.2017.21.2725. MR3687107

[39] D. McDuff and K. Wehrheim, *The fundamental class of smooth Kuranishi atlases with trivial isotropy*, J. Topol. Anal. **10** (2018), no. 1, 71–243, DOI 10.1142/S1793525318500048. MR3737511

[40] D. Metzler. Topological and smooth stacks. *arXiv:0306176*.

[41] I. Moerdijk, *Orbifolds as groupoids: an introduction*, Orbifolds in mathematics and physics (Madison, WI, 2001), Contemp. Math., vol. 310, Amer. Math. Soc., Providence, RI, 2002, pp. 205–222, DOI 10.1090/conm/310/05405. MR1950948

[42] K. Ono, *On the Arnol′d conjecture for weakly monotone symplectic manifolds*, Invent. Math. **119** (1995), no. 3, 519–537, DOI 10.1007/BF01245191. MR1317649

[43] R. Pandharipande, J. Solomon, and J. Walcher, *Disk enumeration on the quintic 3-fold*, J. Amer. Math. Soc. **21** (2008), no. 4, 1169–1209, DOI 10.1090/S0894-0347-08-00597-3. MR2425184

[44] J. Pardon, *An algebraic approach to virtual fundamental cycles on moduli spaces of pseudo-holomorphic curves*, Geom. Topol. **20** (2016), no. 2, 779–1034, DOI 10.2140/gt.2016.20.779. MR3493097

[45] J. W. Robbin and D. A. Salamon, *A construction of the Deligne-Mumford orbifold*, J. Eur. Math. Soc. (JEMS) **8** (2006), no. 4, 611–699, DOI 10.4171/JEMS/101. MR2262197

[46] Y. Ruan, *Symplectic topology on algebraic 3-folds*, J. Differential Geom. **39** (1994), no. 1, 215–227. MR1258920

[47] Y. Ruan, *Topological sigma model and Donaldson-type invariants in Gromov theory*, Duke Math. J. **83** (1996), no. 2, 461–500, DOI 10.1215/S0012-7094-96-08316-7. MR1390655

[48] Y. Ruan and G. Tian, *A mathematical theory of quantum cohomology*, J. Differential Geom. **42** (1995), no. 2, 259–367. MR1366548

[49] Y. Ruan and G. Tian, *Higher genus symplectic invariants and sigma models coupled with gravity*, Invent. Math. **130** (1997), no. 3, 455–516, DOI 10.1007/s002220050192. MR1483992

[50] I. Satake, *On a generalization of the notion of manifold*, Proc. Nat. Acad. Sci. U.S.A. **42** (1956), 359–363, DOI 10.1073/pnas.42.6.359. MR0079769

[51] B. Siebert, *Symplectic Gromov-Witten invariants*, New trends in algebraic geometry (Warwick, 1996), London Math. Soc. Lecture Note Ser., vol. 264, Cambridge Univ. Press, Cambridge, 1999, pp. 375–424, DOI 10.1017/CBO9780511721540.016. MR1714832

[52] C. H. Taubes, *Self-dual Yang-Mills connections on non-self-dual 4-manifolds*, J. Differential Geom. **17** (1982), no. 1, 139–170. MR658473

[53] W. P. Thurson. The geometry and topology of three-manifolds. *available online: http://library.msri.org/books/gt3m/*.

[54] H. Whitney, *Geometric integration theory*, Princeton University Press, Princeton, N. J., 1957. MR0087148

[55] E. Witten, *Topological sigma models*, Comm. Math. Phys. **118** (1988), no. 3, 411–449. MR958805

[56] S.-T. Yau and E. Zaslow, *BPS states, string duality, and nodal curves on K3*, Nuclear Phys. B **471** (1996), no. 3, 503–512, DOI 10.1016/0550-3213(96)00176-9. MR1398633

[57] R. Ye, *Gromov's compactness theorem for pseudo holomorphic curves*, Trans. Amer. Math. Soc. **342** (1994), no. 2, 671–694, DOI 10.2307/2154647. MR1176088

[58] A. Zinger, *Pseudocycles and integral homology*, Trans. Amer. Math. Soc. **360** (2008), no. 5, 2741–2765, DOI 10.1090/S0002-9947-07-04440-6. MR2373332

[59] A. Zinger, *The reduced genus 1 Gromov-Witten invariants of Calabi-Yau hypersurfaces*, J. Amer. Math. Soc. **22** (2009), no. 3, 691–737, DOI 10.1090/S0894-0347-08-00625-5. MR2505298

Kuranishi Spaces as a 2-category

Dominic Joyce

1. Introduction

Kuranishi spaces were introduced in the work of Fukaya, Oh, Ohta and Ono [**9**–**19**], as the geometric structure on moduli spaces of J-holomorphic curves. We will refer to their most recent definition [**13**, §4] as *FOOO Kuranishi spaces*. They are used to define virtual cycles and virtual chains for such moduli spaces, for applications in symplectic geometry such as Gromov–Witten invariants [**19**] and Lagrangian Floer cohomology [**10**].

In related work, McDuff and Wehrheim [**40**–**43**] define *Kuranishi atlases*, which are quite similar to FOOO 'good coordinate systems' in[**9**–**19**], and Dingyu Yang defines *Kuranishi structures* [**53**–**55**], which are very similar to [**9**–**19**]. An alternative theory, philosophically rather different, but which does essentially the same job, is the *polyfolds* of Hofer, Wysocki and Zehnder [**23**–**30**].

Although FOOO Kuranishi spaces are well defined, and are adequate for the applications in Fukaya et al. [**9**–**19**], they seem not very satisfactory as geometric spaces. For example, there is currently no good notion of morphism of FOOO Kuranishi spaces.

This is a partial survey of the author's paper [**35**] and in-progress book [**36**]. In it we present a new definition of Kuranishi spaces, which form a (weak) 2-category **Kur**. That is, we have objects, the Kuranishi spaces $\boldsymbol{X}, \boldsymbol{Y}, \ldots$, and 1-morphisms $\boldsymbol{f}\colon \boldsymbol{X} \to \boldsymbol{Y}$, and also 2-morphisms $\boldsymbol{\eta}\colon \boldsymbol{f} \Rightarrow \boldsymbol{g}$ between 1-morphisms $\boldsymbol{f}, \boldsymbol{g}\colon \boldsymbol{X} \to \boldsymbol{Y}$.

If $\boldsymbol{X}$ is a FOOO Kuranishi space, we can construct a Kuranishi space $\boldsymbol{X}'$ in our sense with the same topological space, uniquely up to canonical equivalence in **Kur**. The same holds for McDuff and Wehrheim's 'Kuranishi atlases' [**40**–**43**], and Hofer, Wysocki and Zehnder's 'polyfold Fredholm structures' [**23**–**30**].

Our theory of Kuranishi spaces is based on the author's theory of Derived Differential Geometry [**32**–**34**], the study of derived manifolds and orbifolds, where 'derived' is in the sense of the Derived Algebraic Geometry of Lurie [**39**] and Toën–Vezzosi [**52**]. The author's derived manifolds and orbifolds are called 'd-manifolds' and 'd-orbifolds', and form strict 2-categories **dMan**, **dOrb**. For alternative definitions of derived manifolds, see Spivak [**50**] and Borisov–Noel [**6**]. The relations between [**32**–**34**] and [**6**, **50**] are explained by Borisov [**5**].

D-manifolds and d-orbifolds [**32**–**34**], and also the derived manifolds of [**6**, **50**], are defined using C^∞-algebraic geometry, as in [**31**]: they are classes of derived schemes and stacks over C^∞-rings. Their definition looks very different to that of Kuranishi spaces. Nonetheless, there is an equivalence of 2-categories $\mathbf{Kur} \simeq \mathbf{dOrb}$, so Kuranishi spaces and d-orbifolds are essentially the same thing. The author used results on d-orbifolds in [**34**] to design the definition of Kuranishi spaces below, so as to arrange that $\mathbf{Kur} \simeq \mathbf{dOrb}$.

One moral is that *Kuranishi spaces are really derived smooth orbifolds*, which does not seem to be well known amongst symplectic geometers. So we should use ideas from derived geometry to understand Kuranishi spaces. One such lesson is that 'derived' objects generally form higher categories. In fact, even classical orbifolds are best defined to be a 2-category. Hence it should not be surprising that Kuranishi spaces are also a 2-category.

One can turn Kuranishi spaces into an ordinary category by passing to the *homotopy category* $\mathrm{Ho}(\mathbf{Kur})$, with objects Kuranishi spaces $\boldsymbol{X}, \boldsymbol{Y}, \ldots$, and whose morphisms $[\boldsymbol{f}]: \boldsymbol{X} \to \boldsymbol{Y}$ are 2-isomorphism classes of 1-morphisms $\boldsymbol{f}: \boldsymbol{X} \to \boldsymbol{Y}$ in $\mathbf{Kur}$. But the definition of $\mathrm{Ho}(\mathbf{Kur})$ is not much easier than that of $\mathbf{Kur}$, and there are disadvantages to doing this:

- Morphisms $[\boldsymbol{f}]: \boldsymbol{X} \to \boldsymbol{Y}$ in $\mathrm{Ho}(\mathbf{Kur})$ are not local on $\boldsymbol{X}$, that is, they do not form a sheaf on $\boldsymbol{X}$.
- Important constructions such as fibre products $\boldsymbol{X} \times_{\boldsymbol{g},\boldsymbol{Z},\boldsymbol{h}} \boldsymbol{Y}$ in $\mathbf{Kur}$ are characterized by universal properties involving 2-morphisms in $\mathbf{Kur}$. They do not satisfy any universal property in $\mathrm{Ho}(\mathbf{Kur})$.

All the Kuranishi-type structures that we consider are given by an 'atlas of charts' on a topological space X, where the 'charts', called *Kuranishi neighbourhoods*, are quintuples (V, E, Γ, s, ψ) for V a manifold, $E \to V$ a vector bundle, Γ a finite group acting smoothly on V, E, $s: V \to E$ a Γ-equivariant smooth section, and $\psi: s^{-1}(0)/\Gamma \to X$ a continuous map which is a homeomorphism with an open set $\mathrm{Im}\,\psi \subseteq X$.

The important differences between FOOO Kuranishi spaces, and MW Kuranishi atlases, and our Kuranishi spaces, are in the 'coordinate changes' between the charts in the atlas. Since our Kuranishi spaces are a 2-category, our Kuranishi neighbourhoods must be a 2-category too, as one can regard a Kuranishi neighbourhood as a Kuranishi space with only one chart. Thus, we define *1-morphisms* $\Phi_{ij}: (V_i, E_i, \Gamma_i, s_i, \psi_i) \to (V_j, E_j, \Gamma_j, s_j, \psi_j)$ between Kuranishi neighbourhoods on X, which include FOOO and MW coordinate changes as special cases, and also *2-morphisms* $\Lambda_{ij}: \Phi_{ij} \Rightarrow \Phi'_{ij}$ between 1-morphisms. A *coordinate change* is a 1-morphism Φ_{ij} which is invertible up to 2-isomorphism.

Given Kuranishi neighbourhoods $(V_i, E_i, \Gamma_i, s_i, \psi_i), (V_j, E_j, \Gamma_j, s_j, \psi_j)$ on X, our 1- and 2-morphisms between them have the crucial property that they form a *stack*, or *2-sheaf*, on $\mathrm{Im}\,\psi_i \cap \mathrm{Im}\,\psi_j$. This is a 2-categorical version of a sheaf, and means that 1- and 2-morphisms glue nicely on open covers. The stack property is essential in defining compositions $\boldsymbol{g} \circ \boldsymbol{f}: \boldsymbol{X} \to \boldsymbol{Z}$ of 1-morphisms $\boldsymbol{f}: \boldsymbol{X} \to \boldsymbol{Y}$ and $\boldsymbol{g}: \boldsymbol{Y} \to \boldsymbol{Z}$ of Kuranishi spaces. The main reason why FOOO Kuranishi spaces and MW Kuranishi atlases do not have good notions of morphisms is that their coordinate changes lack such a sheaf/stack property.

Our theory of Kuranishi spaces is part of a programme by the author [**36**] to rewrite the foundations of areas of symplectic geometry involving moduli spaces of J-holomorphic curves. For such applications it will be an advantage that our Kuranishi spaces form a well-behaved 2-category. Here are three examples:

(a) Defining a Kuranishi structure on a moduli space $\mathcal{M}$ involves making many arbitrary choices, and it is helpful to know how different choices are related. In our theory, we expect different choices of Kuranishi structure on $\mathcal{M}$ to yield equivalent Kuranishi spaces $\boldsymbol{\mathcal{M}}, \boldsymbol{\mathcal{M}}'$ in the 2-category **Kur**.

Equivalence of Kuranishi spaces should be compared with McDuff and Wehrheim's 'commensurate' Kuranishi atlases, as in §2.3, and Yang's 'R-equivalent' Kuranishi structures, as in §2.5.

(b) Fukaya, Oh, Ohta and Ono use finite group actions on Kuranishi spaces in [**9**], [**17**, §7], and T^n-actions on Kuranishi spaces in [**11, 12**]. In our theory it is easy to define and study actions of Lie groups on Kuranishi spaces, in a more flexible way than in [**9, 11, 12, 17**].

(c) Fukaya [**9**, §3, §5] (see also [**15**, §4.2]) works with a forgetful morphism $\mathfrak{forget}: \boldsymbol{\mathcal{M}}_{l,1}(\beta) \to \boldsymbol{\mathcal{M}}_{l,0}(\beta)$ of J-holomorphic curve moduli spaces. This is a kind of morphism of Fukaya–Oh–Ohta–Ono Kuranishi spaces, but can only be defined when the Kuranishi structures on $\boldsymbol{\mathcal{M}}_{l,1}(\beta), \boldsymbol{\mathcal{M}}_{l,0}(\beta)$ are compatible in a strong sense, as they are in the examples of [**9, 15**]. Our theory defines (1-)morphisms without such compatibility conditions.

In this survey we will for simplicity restrict attention almost entirely to Kuranishi spaces *without boundary*, which we just call Kuranishi spaces. Kuranishi spaces with boundary, and with corners, are studied in [**35, 36**].

Section 2 reviews the definitions of Fukaya–Oh–Ohta–Ono's Kuranishi spaces [**9**–**19**], McDuff–Wehrheim's Kuranishi atlases [**40**–**43**], and Dingyu Yang's Kuranishi structures [**53**–**55**], and we also discuss Hofer–Wysocki–Zehnder's polyfolds [**23**–**30**]. Section 3 explains Kuranishi neighbourhoods and their 1- and 2-morphisms and coordinate changes. Section 4 defines the 2-category of Kuranishi spaces, and explains their relation to FOOO Kuranishi spaces, MW Kuranishi atlases, DY Kuranishi structures, polyfolds, and d-orbifolds. Section 5 discusses the differential geometry of Kuranishi spaces. Finally, Appendix A gives background on 2-categories, and stacks on topological spaces.

Acknowledgements. I would like to thank Lino Amorim, Kenji Fukaya, Helmut Hofer, Dusa McDuff, and Dingyu Yang for helpful conversations. This research was supported by EPSRC grants EP/H035303/1 and EP/J016950/1.

2. Previous definitions of Kuranishi space

We explain *Kuranishi spaces* in the work of Fukaya, Oh, Ohta and Ono [**9**–**19**], and the similar idea of *Kuranishi atlases* in the work of McDuff and Wehrheim [**40**–**43**]. We also discuss Dingyu Yang's *Kuranishi structures* [**53**–**55**], a minor variation on Fukaya–Oh–Ohta–Ono Kuranishi spaces, and the rather different *polyfolds* of Hofer, Wysocki and Zehnder [**23**–**30**]. We have made some small changes in notation compared to our sources, for compatibility with §3–§5. We hope the authors concerned will not mind this.

2.1. Fukaya–Oh–Ohta–Ono's Kuranishi spaces. Kuranishi spaces are used in the work of Fukaya, Oh, Ohta and Ono [**9**–**19**] as the geometric structure on moduli spaces of J-holomorphic curves. Initially introduced by Fukaya and Ono [**19**, §5] in 1999, the definition has changed several times as their work has evolved [**9**–**19**].

This section explains their most recent definition of Kuranishi space, taken from [**13**, §4]. As in the rest of our paper 'Kuranishi neighbourhood', 'coordinate change' and 'Kuranishi space' have a different meaning, we will use the terms 'FOOO

Kuranishi neighbourhood', 'FOOO coordinate change' and 'FOOO Kuranishi space' below to refer to concepts from [**13**].

For the next definitions, let X be a compact, metrizable topological space.

DEFINITION 2.1. A *FOOO Kuranishi neighbourhood* on X is a quintuple (V, E, Γ, s, ψ) such that:

(a) V is a smooth manifold, which may or may not have boundary or corners.
(b) E is a finite-dimensional real vector space.
(c) Γ is a finite group with a smooth, effective action on V, and a linear representation on E.
(d) $s\colon V \to E$ is a Γ-equivariant smooth map.
(e) ψ is a homeomorphism from $s^{-1}(0)/\Gamma$ to an open subset $\operatorname{Im}\psi$ in X, where $\operatorname{Im}\psi = \{\psi(x\Gamma)\colon x \in s^{-1}(0)\}$ is the image of ψ, and is called the *footprint* of (V, E, Γ, s, ψ).

We will write $\bar\psi\colon s^{-1}(0) \to \operatorname{Im}\psi \subseteq X$ for the composition of ψ with the projection $s^{-1}(0) \to s^{-1}(0)/\Gamma$.

Now let $p \in X$. A *FOOO Kuranishi neighbourhood of p in X* is a FOOO Kuranishi neighbourhood $(V_p, E_p, \Gamma_p, s_p, \psi_p)$ with a distinguished point $o_p \in V_p$ such that o_p is fixed by Γ_p, and $s_p(o_p) = 0$, and $\psi_p([o_p]) = p$. Then o_p is unique.

DEFINITION 2.2. Let $(V_i, E_i, \Gamma_i, s_i, \psi_i)$, $(V_j, E_j, \Gamma_j, s_j, \psi_j)$ be FOOO Kuranishi neighbourhoods on X. Suppose $S \subseteq \operatorname{Im}\psi_i \cap \operatorname{Im}\psi_j \subseteq X$ is an open subset of the intersection of the footprints $\operatorname{Im}\psi_i, \operatorname{Im}\psi_j \subseteq X$. We say a quadruple $\Phi_{ij} = (V_{ij}, h_{ij}, \varphi_{ij}, \hat\varphi_{ij})$ is a *FOOO coordinate change from $(V_i, E_i, \Gamma_i, s_i, \psi_i)$ to $(V_j, E_j, \Gamma_j, s_j, \psi_j)$ over S* if:

(a) V_{ij} is a Γ_i-invariant open neighbourhood of $\bar\psi_i^{-1}(S)$ in V_i.
(b) $h_{ij}\colon \Gamma_i \to \Gamma_j$ is an injective group homomorphism.
(c) $\varphi_{ij}\colon V_{ij} \hookrightarrow V_j$ is an h_{ij}-equivariant smooth embedding, such that the induced map $(\varphi_{ij})_*\colon V_{ij}/\Gamma_i \to V_j/\Gamma_j$ is injective.
(d) $\hat\varphi_{ij}\colon V_{ij} \times E_i \hookrightarrow V_j \times E_j$ is an h_{ij}-equivariant embedding of vector bundles over $\varphi_{ij}\colon V_{ij} \hookrightarrow V_j$, viewing $V_{ij} \times E_i \to V_{ij}$, $V_j \times E_j \to V_j$ as trivial vector bundles.
(e) $\hat\varphi_{ij}(s_i|_{V_{ij}}) = \varphi_{ij}^*(s_j)$, in sections of $\varphi_{ij}^*(V_j \times E_j) \to V_{ij}$.
(f) $\psi_i = \psi_j \circ (\varphi_{ij})_*$ on $(s_i^{-1}(0) \cap V_{ij})/\Gamma_i$.
(g) h_{ij} restricts to an isomorphism $\operatorname{Stab}_{\Gamma_i}(v) \to \operatorname{Stab}_{\Gamma_j}(\varphi_{ij}(v))$ for all v in V_{ij}, where $\operatorname{Stab}_{\Gamma_i}(v)$ is the *stabilizer subgroup* $\{\gamma \in \Gamma_i\colon \gamma(v) = v\}$.
(h) For each $v \in s_i^{-1}(0) \cap V_{ij} \subseteq V_{ij} \subseteq V_i$ we have a commutative diagram

$$\begin{array}{ccccccccc} 0 & \longrightarrow & T_vV_i & \xrightarrow{\mathrm{d}\varphi_{ij}|_v} & T_{\varphi_{ij}(v)}V_j & \longrightarrow & N_{ij}|_v & \longrightarrow & 0 \\ & & \big\downarrow{\scriptstyle \mathrm{d}s_i|_v} & & \big\downarrow{\scriptstyle \mathrm{d}s_j|_{\varphi_{ij}(v)}} & & \big\downarrow{\scriptstyle \mathrm{d}_{\mathrm{fibre}}s_j|_v} & & \\ 0 & \longrightarrow & E_i|_v & \xrightarrow{\hat\varphi_{ij}|_v} & E_j|_{\varphi_{ij}(v)} & \longrightarrow & F_{ij}|_v & \longrightarrow & 0 \end{array} \tag{2.1}$$

with exact rows, where $N_{ij} \to V_{ij}$ is the normal bundle of V_{ij} in V_j, and $F_{ij} = \varphi_{ij}^*(E_j)/\hat\varphi_{ij}(E_i|_{V_{ij}})$ the quotient bundle. We require that the induced morphism $\mathrm{d}_{\mathrm{fibre}}s_j|_v$ in (2.1) should be an isomorphism.

Note that $\mathrm{d}_{\mathrm{fibre}}s_j|_v$ an isomorphism in (2.1) is equivalent to the following complex being exact, which is how we write the analogous conditions in §3:

$$0 \longrightarrow T_vV_i \xrightarrow{\mathrm{d}s_i|_v \oplus \mathrm{d}\varphi_{ij}|_v} E_i|_v \oplus T_{\varphi_{ij}(v)}V_j \xrightarrow{\hat\varphi_{ij}|_v \oplus -\mathrm{d}s_j|_{\varphi_{ij}(v)}} E_j|_{\varphi_{ij}(v)} \longrightarrow 0. \tag{2.2}$$

Now let $(V_p, E_p, \Gamma_p, s_p, \psi_p)$, $(V_q, E_q, \Gamma_q, s_q, \psi_q)$ be FOOO Kuranishi neighbourhoods of $p \in X$ and $q \in \operatorname{Im}\psi_p \subseteq X$, respectively. We say a quadruple $\Phi_{qp} = (V_{qp}, h_{qp}, \varphi_{qp}, \hat{\varphi}_{qp})$ is a *FOOO coordinate change* if it is a FOOO coordinate change from $(V_q, E_q, \Gamma_q, s_q, \psi_q)$ to $(V_p, E_p, \Gamma_p, s_p, \psi_p)$ over S_{qp}, where S_{qp} is any open neighbourhood of q in $\operatorname{Im}\psi_q \cap \operatorname{Im}\psi_p$.

REMARK 2.3. Fukaya et al. [**13**] only impose Definition 2.2(h) for Kuranishi spaces 'with a tangent bundle'. As the author knows of no reason for considering Kuranishi spaces 'without tangent bundles', and the notation appears to be merely historical, we will include 'with a tangent bundle' in our definitions of FOOO coordinate changes and FOOO Kuranishi spaces.

DEFINITION 2.4. A *FOOO Kuranishi structure* $\mathcal{K}$ on X of *virtual dimension* n in $\mathbb{Z}$ in the sense of [**13**, §4], including the 'with a tangent bundle' condition, assigns a FOOO Kuranishi neighbourhood $(V_p, E_p, \Gamma_p, s_p, \psi_p)$ for each $p \in X$ and a FOOO coordinate change $\Phi_{qp} = (V_{qp}, h_{qp}, \varphi_{qp}, \hat{\varphi}_{qp})\colon (V_q, E_q, \Gamma_q, s_q, \psi_q) \to (V_p, E_p, \Gamma_p, s_p, \psi_p)$ for each $q \in \operatorname{Im}\psi_p$ such that the following holds:

(a) $\dim V_p - \operatorname{rank} E_p = n$ for all $p \in X$.
(b) If $q \in \operatorname{Im}\psi_p$, $r \in \psi_q((V_{qp} \cap s_q^{-1}(0))/\Gamma_q)$, then for each connected component $(\varphi_{rq}^{-1}(V_{qp}) \cap V_{rp})^\alpha$ of $\varphi_{rq}^{-1}(V_{qp}) \cap V_{rp}$ there exists $\gamma_{rqp}^\alpha \in \Gamma_p$ with

$$h_{qp} \circ h_{rq} = \gamma_{rqp}^\alpha \cdot h_{rp} \cdot (\gamma_{rqp}^\alpha)^{-1}, \qquad \varphi_{qp} \circ \varphi_{rq} = \gamma_{rqp}^\alpha \cdot \varphi_{rp}, \qquad \text{and} \qquad \varphi_{rq}^*(\hat{\varphi}_{qp}) \circ \hat{\varphi}_{rq} = \gamma_{rqp}^\alpha \cdot \hat{\varphi}_{rp}, \tag{2.3}$$

where the second and third equations hold on $(\varphi_{rq}^{-1}(V_{qp}) \cap V_{rp})^\alpha$.

The pair $\boldsymbol{X} = (X, \mathcal{K})$ is called a *FOOO Kuranishi space*, of *virtual dimension* $n \in \mathbb{Z}$, written $\operatorname{vdim} \boldsymbol{X} = n$. If the V_p for all $p \in X$ are manifolds without boundary, or with boundary, or with corners, then we call $\boldsymbol{X}$ a *FOOO Kuranishi space without boundary*, or *with boundary*, or *with corners*, respectively.

FOOO Kuranishi spaces are differential-geometric spaces, generalizations of manifolds or orbifolds. Some of the differential geometry of manifolds extends to FOOO Kuranishi spaces. Here are analogues of the notions of orientations, smooth maps, and transverse fibre products of manifolds.

DEFINITION 2.5. Let $\boldsymbol{X}$ be a FOOO Kuranishi space. Then for each $p \in X$, $q \in \operatorname{Im}\psi_p$ and $v \in s_q^{-1}(0) \cap V_{qp}$, we have an exact sequence (2.2). Taking top exterior powers in (2.2) yields an isomorphism

$$\big(\det T_v V_q\big) \otimes \det\big(E_p|_{\varphi_{qp}(v)}\big) \cong \big(\det E_q|_v\big) \otimes \big(T_{\varphi_{qp}(v)} V_p\big),$$

where $\det W$ means $\Lambda^{\dim W} W$, or equivalently, a canonical isomorphism

$$\big(\det T^* V_p \otimes \det E_p\big)|_{\varphi_{qp}(v)} \cong \big(\det T^* V_q \otimes \det E_q\big)|_v. \tag{2.4}$$

Defining the isomorphism (2.4) requires a suitable sign convention. Sign conventions are discussed in Fukaya et al. [**10**, §8.2]. An *orientation* on $\boldsymbol{X}$ is a choice of orientations on the line bundles

$$\det T^* V_p \otimes \det E_p\big|_{s_p^{-1}(0)} \longrightarrow s_p^{-1}(0)$$

for all $p \in X$, compatible with the isomorphisms (2.4).

Definition 2.6. Let $\boldsymbol{X}$ be a FOOO Kuranishi space, and Y a manifold. A *smooth map* $\boldsymbol{f}\colon \boldsymbol{X} \to Y$ is $\boldsymbol{f} = (f_p\colon p \in X)$ where $f_p\colon V_p \to Y$ is a Γ_p-invariant smooth map for all $p \in X$ (that is, f_p factors via $V_p \to V_p/\Gamma_p \to Y$), and $f_p \circ \varphi_{qp} = f_q|_{V_{qp}}\colon V_{qp} \to Y$ for all $q \in \operatorname{Im}\psi_p$. This induces a unique continuous map $f\colon X \to Y$ with $f_p|_{s_p^{-1}(0)} = f \circ \bar\psi_p$ for all $p \in X$. We call $\boldsymbol{f}$ *weakly submersive* if each f_p is a submersion.

Suppose $\boldsymbol{X}, \boldsymbol{X}'$ are FOOO Kuranishi spaces, Y is a manifold, and $\boldsymbol{f}\colon \boldsymbol{X} \to Y$, $\boldsymbol{f}'\colon \boldsymbol{X}' \to Y$ are weakly submersive. Then as in [**10**, §A1.2] one can define a 'fibre product' Kuranishi space $\boldsymbol{W} = \boldsymbol{X} \times_Y \boldsymbol{X}'$, with topological space $W = \{(p,p') \in X \times X'\colon f(p) = f'(p')\}$, and FOOO Kuranishi neighbourhoods $(V_{p,p'}, E_{p,p'}, \Gamma_{p,p'}, s_{p,p'}, \psi_{p,p'})$ for $(p,p') \in W$, where $V_{p,p'} = V_p \times_{f_p,Y,f'_{p'}} V'_{p'}$, $E_{p,p'} = \pi_{V_p}^*(E_p) \oplus \pi_{V'_{p'}}^*(E'_{p'})$, $\Gamma_{p,p'} = \Gamma_p \times \Gamma'_{p'}$, $s_{p,p'} = \pi_{V_p}^*(s_p) \oplus \pi_{V'_{p'}}^*(s'_{p'})$, and $\psi_{p,p'} = \psi_p \circ (\pi_{V_p})_* \times \psi'_{p'} \circ (\pi_{V'_{p'}})_*$. The weakly submersive condition ensures that $V_{p,p'} = V_p \times_Y V'_{p'}$ is well-defined.

Remark 2.7. **(i)** Fukaya et al. [**9**–**19**] do not define morphisms between general FOOO Kuranishi spaces, so Kuranishi spaces in [**9**–**19**] do not form a category. However, they do work with two special classes of morphisms. Smooth maps $\boldsymbol{f}\colon \boldsymbol{X} \to Y$ from Kuranishi spaces $\boldsymbol{X}$ to manifolds Y are important in their theory. Also Fukaya [**9**, §3, §5] (see also [**15**, §4.2]) works with a forgetful morphism $\mathfrak{forget}\colon \boldsymbol{\mathcal{M}}_{l,1}(\beta) \to \boldsymbol{\mathcal{M}}_{l,0}(\beta)$. This is an example of a kind of morphism $\boldsymbol{f}\colon \boldsymbol{X} \to \boldsymbol{Y}$ of FOOO Kuranishi spaces (though not given a formal definition in [**9**, **15**]), which can only be defined when the Kuranishi structures on $\boldsymbol{X}, \boldsymbol{Y}$ are compatible in a strong sense, as they are by construction in [**9**, **15**].

(ii) The 'fibre product' $\boldsymbol{X} \times_Y \boldsymbol{X}'$ in Definition 2.6 is not a fibre product in the sense of category theory, characterized by a universal property, since Fukaya et al. in [**9**–**19**] do not have a category of FOOO Kuranishi spaces in which to state such a universal property (though we do, see §4 and §5.2). Their 'fibre product' is extra data, an operation defined upon FOOO Kuranishi spaces.

2.2. How FOOO Kuranishi spaces are used. The next two theorems summarize some of the main results of [**10**, **12**–**14**, **16**, **19**].

Theorem 2.8. *The following classes of moduli spaces of J-holomorphic curves in symplectic geometry may be given the structure of compact FOOO Kuranishi spaces of known dimension, after making many choices in the construction:*

- **(a)** *Moduli spaces $\overline{\boldsymbol{\mathcal{M}}}_{g,m}(J,\beta)$ of stable maps $u\colon \Sigma \to S$ from a prestable closed Riemann surface Σ of genus g with m marked points, to a compact symplectic manifold (S,ω) with compatible almost complex structure J, with homology class $u_*([\Sigma]) = \beta \in H_2(S;\mathbb{Z})$.*
- **(b)** *Moduli spaces $\overline{\boldsymbol{\mathcal{M}}}_k(L_1,\ldots,L_l,J,\beta)$ of stable maps $u\colon \Sigma \to S$ from a prestable holomorphic disc Σ with k boundary marked points, to a compact symplectic manifold (S,ω) with compatible almost complex structure J, such that $u(\partial\Sigma)$ lies in the union $L_1 \cup \cdots \cup L_l$ of pairwise transversely intersecting, embedded, compact Lagrangians $L_1,\ldots,L_l$ in S, with relative homology class $u_*([\Sigma]) = \beta \in H_2(S, L_1 \cup \cdots \cup L_l;\mathbb{Z})$.*

Here $\overline{\boldsymbol{\mathcal{M}}}_{g,m}(J,\beta)$ is a FOOO Kuranishi space without boundary used to define Gromov–Witten invariants in [**19**], *and $\overline{\boldsymbol{\mathcal{M}}}_k(L_1,\ldots,L_l,J,\beta)$ a FOOO Kuranishi space with corners used to define Lagrangian Floer cohomology in* [**10**].

The 'evaluation maps' at marked points $\mathbf{ev}_i\colon \bar{\mathcal{M}}_{g,m}(J,\beta)\to S$ *and* $\mathbf{ev}_j\colon \bar{\mathcal{M}}_k(L_1,\dots,L_l,J,\beta)\to L_1\amalg\cdots\amalg L_l$ *may be given the structure of weakly submersive smooth maps.*

After choosing orientations and relative spin structures for $L_1,\dots,L_l$ *in* **(b)**, *there are canonical orientations on* $\bar{\mathcal{M}}_{g,m}(J,\beta)$ *and* $\bar{\mathcal{M}}_k(L_1,\dots,L_l,J,\beta)$.

THEOREM 2.9. *Suppose* $\boldsymbol{X}$ *is a compact, oriented FOOO Kuranishi space of virtual dimension* k, *and* Y *is an oriented manifold of dimension* n, *and* $\boldsymbol{f}\colon \boldsymbol{X}\to Y$ *is a weakly submersive smooth map. Then, after making many choices, one can construct a* ***virtual (co)chain*** $[\boldsymbol{X}]_{\rm virt}$ *for* $\boldsymbol{X}$, *either:*

(i) *in the smooth singular chains* $C_k^{\rm ssi}(Y;\mathbb{Q})$ *of* Y *over* $\mathbb{Q}$; *or*

(ii) *in the compactly-supported de Rham cochains* $C^\infty_{\rm cs}(\Lambda^{n-k}T^*Y)$.

If $\partial\boldsymbol{X}=\emptyset$ *then* $[\boldsymbol{X}]_{\rm virt}$ *is a (co)cycle, and has a (co)homology class* $[[\boldsymbol{X}]_{\rm virt}]$, *the* ***virtual cycle***, *in smooth singular homology* $H_k^{\rm ssi}(Y;\mathbb{Q})$, *or in compactly-supported de Rham cohomology* $H^{n-k}_{\rm dR,cs}(Y;\mathbb{R})$. *This* $[[\boldsymbol{X}]_{\rm virt}]$ *is independent of choices in its construction, and depends on* $(\boldsymbol{X},\boldsymbol{f})$ *only up to oriented bordism.*

Thus, FOOO Kuranishi spaces are used as an intermediate stage in the construction of virtual chains or virtual cycles for moduli spaces of J-holomorphic curves. These virtual chains/cycles must also have other important properties which we have not stated — basically, geometric relationships between moduli spaces should translate to algebraic relationships between their virtual chains/cycles. The virtual chains/cycles are used to define Gromov–Witten theory, Lagrangian Floer cohomology, Fukaya categories, and so on.

The original proofs of Theorems 2.8 and 2.9 in [**10**, **19**] were widely criticized as not wholly correct/complete. Recently, Fukaya et al. [**13**, **14**, **16**, **18**] have been working on providing more careful, detailed, and complete proofs. We should also acknowledge that the work of Fukaya, Oh, Ohta and Ono is very original, even visionary, and full of important ideas which other authors have used since. The work we discuss in §2.3–§2.6 came 5 to 15 years after the first versions of [**10**, **19**], partly as an attempt to resolve these initial problems.

2.3. McDuff–Wehrheim's Kuranishi atlases. Next we discuss an approach to Kuranishi spaces developed by McDuff and Wehrheim [**40**–**43**]. Their main definition is that of a *(weak) Kuranishi atlas* on a topological space X. It is a variation on the notion of *good coordinate system* in the work of Fukaya, Oh, Ohta and Ono, as in [**19**, Def. 6.1], [**10**, Lem. A1.11], [**11**, §15], and [**13**, §5]. Here are [**42**, Def.s 2.2.2 & 2.2.8].

DEFINITION 2.10. An *MW Kuranishi neighbourhood* (V,E,Γ,s,ψ) on a topological space X is the same as a FOOO Kuranishi neighbourhood in Definition 2.1, except that Γ is not required to act effectively on V.

DEFINITION 2.11. Suppose $(V_B,E_B,\Gamma_B,s_B,\psi_B)$, $(V_C,E_C,\Gamma_C,s_C,\psi_C)$ are MW Kuranishi neighbourhoods on a topological space X, and $S\subseteq \operatorname{Im}\psi_B\cap\operatorname{Im}\psi_C\subseteq X$ is open. We say a quadruple $\Phi_{BC}=(\tilde V_{BC},\rho_{BC},\varpi_{BC},\hat\varphi_{BC})$ is an *MW coordinate change from* $(V_B,E_B,\Gamma_B,s_B,\psi_B)$ *to* $(V_C,E_C,\Gamma_C,s_C,\psi_C)$ *over* S if:

(a) $\tilde V_{BC}$ is a Γ_C-invariant embedded submanifold of V_C containing $\bar\psi_C^{-1}(S)$.

(b) $\rho_{BC}\colon\Gamma_C\to\Gamma_B$ is a surjective group morphism, with kernel $\Delta_{BC}\subseteq\Gamma_C$.

There should exist an isomorphism $\Gamma_C\cong\Gamma_B\times\Delta_{BC}$ identifying ρ_{BC} with the projection $\Gamma_B\times\Delta_{BC}\to\Gamma_B$.

(c) $\varpi_{BC}\colon \tilde V_{BC}\to V_B$ is a ρ_{BC}-equivariant étale map, with image $V_{BC}=\varpi_{BC}(\tilde V_{BC})$ a Γ_B-invariant open neighbourhood of $\bar\psi_B^{-1}(S)$ in V_B, such that $\varpi_{BC}\colon \tilde V_{BC}\to V_{BC}$ is a principal Δ_{BC}-bundle.

(d) $\hat\varphi_{BC}\colon E_B\to E_C$ is an injective Γ_C-equivariant linear map, where the Γ_C-action on E_B is induced from the Γ_B-action by ρ_{BC}, so in particular Δ_{BC} acts trivially on E_B.

(e) $\hat\varphi_{BC}\circ s_B\circ\varpi_{BC}=s_C|_{\tilde V_{BC}}\colon \tilde V_{BC}\to E_C$.

(f) $\psi_B\circ(\varpi_{BC})_*=\psi_C$ on $(s_C^{-1}(0)\cap\tilde V_{BC})/\Gamma_C$.

(g) For each $v\in\tilde V_{BC}$ we have a commutative diagram

$$\begin{array}{ccccccccc} 0 & \longrightarrow & T_v\tilde V_{BC} & \xrightarrow{\subset} & T_vV_C & \longrightarrow & N_{BC}|_v & \longrightarrow & 0\\ & & \big\downarrow{\scriptstyle \mathrm{d}(\varpi_{BC}^*(s_B))|_v} & & \big\downarrow{\scriptstyle \mathrm{d}s_C|_v} & & \big\downarrow{\scriptstyle \mathrm{d_{fibre}}s_C|_v} & & \\ 0 & \longrightarrow & E_B & \xrightarrow{\hat\varphi_{BC}} & E_C & \longrightarrow & E_C/\hat\varphi_{BC}(E_B) & \longrightarrow & 0 \end{array} \tag{2.5}$$

with exact rows, where N_{BC} is the normal bundle of $\tilde V_{BC}$ in V_C. We require the induced morphism $\mathrm{d_{fibre}}\,s_C|_v$ in (2.5) to be an isomorphism.

Definition 2.12. Let X be a compact, metrizable topological space. An *MW weak Kuranishi atlas* $\mathcal K=\bigl(A,I,(V_B,E_B,\Gamma_B,s_B,\psi_B)_{B\in I},\Phi_{BC,\,B,C\in I,\,B\subsetneq C}\bigr)$ *on X of virtual dimension* $n\in\mathbb Z$, as in [**42**, Def. 2.3.1], consists of a finite indexing set A, a set I of nonempty subsets of A, MW Kuranishi neighbourhoods $(V_B,E_B,\Gamma_B,s_B,\psi_B)$ on X for all $B\in I$ with $\dim V_B-\operatorname{rank}E_B=n$ and $X=\bigcup_{B\in I}\operatorname{Im}\psi_B$, and MW coordinate changes $\Phi_{BC}=(\tilde V_{BC},\rho_{BC},\varpi_{BC},\hat\varphi_{BC})$ from $(V_B,E_B,\Gamma_B,s_B,\psi_B)$ to $(V_C,E_C,\Gamma_C,s_C,\psi_C)$ on $S=\operatorname{Im}\psi_B\cap\operatorname{Im}\psi_C$ for all $B,C\in I$ with $B\subsetneq C$, satisfying the four conditions:

(a) We have $\{a\}\in I$ for all $a\in A$, and $I=\bigl\{\emptyset\ne B\subseteq A\colon \bigcap_{a\in B}\operatorname{Im}\psi_{\{a\}}\ne\emptyset\bigr\}$. Also $\operatorname{Im}\psi_B=\bigcap_{a\in B}\operatorname{Im}\psi_{\{a\}}$ for all $B\in I$.

(b) We have $\Gamma_B=\prod_{a\in B}\Gamma_{\{a\}}$ for all $B\in I$. If $B,C\in I$ with $B\subsetneq C$ then $\rho_{BC}\colon\Gamma_C\to\Gamma_B$ is the obvious projection $\prod_{a\in C}\Gamma_{\{a\}}\to\prod_{a\in B}\Gamma_{\{a\}}$, with kernel $\Delta_{BC}\cong\prod_{a\in C\setminus B}\Gamma_{\{a\}}$.

(c) We have $E_B=\prod_{a\in B}E_{\{a\}}$ for all $B\in I$, with the obvious representation of $\Gamma_B=\prod_{a\in B}\Gamma_{\{a\}}$. If $B\subsetneq C$ in I then $\hat\varphi_{BC}\colon E_B=\prod_{a\in B}E_{\{a\}}\to E_C=\prod_{a\in C}E_{\{a\}}$ is $\mathrm{id}_{E_{\{a\}}}$ for $a\in B$, and maps to zero in $E_{\{a\}}$ for $a\in C\setminus B$.

(d) If $B,C,D\in I$ with $B\subsetneq C\subsetneq D$ then $\varpi_{BC}\circ\varpi_{CD}=\varpi_{BD}$ on $\tilde V_{BCD}:=\tilde V_{BD}\cap\varpi_{CD}^{-1}(\tilde V_{BC})$. One can show using (b),(c) and Definition 2.11 that $\tilde V_{BD}$ and $\varpi_{CD}^{-1}(\tilde V_{BC})$ are both open subsets in $s_D^{-1}(\hat\varphi_{BD}(E_B))$, which is a submanifold of V_D, so $\tilde V_{BCD}$ is a submanifold of V_D.

We call $\mathcal K=\bigl(A,I,(V_B,E_B,\Gamma_B,s_B,\psi_B)_{B\in I},\Phi_{BC,\,B\subsetneq C}\bigr)$ an *MW Kuranishi atlas on X*, as in [**42**, Def. 2.3.1], if it also satisfies:

(e) If $B,C,D\in I$ with $B\subsetneq C\subsetneq D$ then $\varpi_{CD}^{-1}(\tilde V_{BC})\subseteq\tilde V_{BD}$.

McDuff and Wehrheim also define *orientations* on MW weak Kuranishi atlases, in a very similar way to Definition 2.5.

Two MW weak Kuranishi atlases $\mathcal K,\mathcal K'$ on X are called *directly commensurate* if they are both contained in a third MW weak Kuranishi atlas $\mathcal K''$. They are called *commensurate* if there exist MW weak Kuranishi atlases $\mathcal K=\mathcal K_0,\mathcal K_1,\dots,\mathcal K_m=\mathcal K'$ with $\mathcal K_{i-1},\mathcal K_i$ directly commensurate for $i=1,\dots,m$. This is an equivalence relation on MW weak Kuranishi atlases on X.

McDuff and Wehrheim argue that their concept of MW weak Kuranishi atlas is a more natural, or more basic, idea than a FOOO Kuranishi space, since in analytic moduli problems such as J-holomorphic curve moduli spaces, one has to construct an MW weak Kuranishi atlas (or something close to it) first, and then define the FOOO Kuranishi structure using this.

2.4. How MW Kuranishi atlases are used. McDuff and Wehrheim [**41–43**] prove analogues of Theorems 2.8 and 2.9:

THEOREM 2.13. *Let (S,ω) be a symplectic manifold with tame almost complex structure J, and $\mathcal{M}(\beta,J)$ a compact moduli space of simple J-holomorphic maps $u\colon \mathbb{CP}^1 \to S$ in homology class $\beta \in H_2(S;\mathbb{Z})$ with one marked point, modulo reparametrization. Then one can construct an oriented MW weak Kuranishi atlas $\mathcal{K}$ without boundary on $\mathcal{M}(\beta,J)$.*

The construction depends on many arbitrary choices, but any two such atlases $\mathcal{K},\mathcal{K}'$ resulting from different choices are commensurate.

McDuff and Wehrheim also announce an extension of Theorem 2.13 to moduli spaces $\overline{\mathcal{M}}(\beta,J)$ of genus zero prestable J-holomorphic maps $u\colon \Sigma \to S$, without assuming Σ nonsingular or u simple.

THEOREM 2.14. *Let $\mathcal{K}$ be an oriented MW weak Kuranishi atlas without boundary of dimension n on a compact, metrizable topological space X. Then $\mathcal{K}$ determines a* ***virtual fundamental class*** *$[[X]]_{\rm virt}$ in $\check{H}_n(X;\mathbb{Q})$, where $\check{H}_*(-;\mathbb{Q})$ is Čech homology over $\mathbb{Q}$. Any two commensurate MW weak Kuranishi atlases $\mathcal{K},\mathcal{K}'$ on X yield the same virtual fundamental class.*

One should also expect that many other moduli spaces of J-holomorphic curves admit MW weak Kuranishi atlases, and that if X is compact with an oriented MW weak Kuranishi atlas $\mathcal{K}$ with boundary or corners then $(X,\mathcal{K})$ has (choices of) virtual chains $[X]_{\rm virt}$, but these results are not yet available.

2.5. Dingyu Yang's Kuranishi structures, and Hofer–Wysocki–Zehnder's polyfolds. As part of a project to define a truncation functor from polyfolds to Kuranishi spaces, Dingyu Yang [**53–55**] writes down his own theory of Kuranishi spaces:

DEFINITION 2.15. Let X be a compact, metrizable topological space. A *DY Kuranishi structure* $\mathcal{K}$ on X is a FOOO Kuranishi structure in the sense of Definition 2.4, satisfying two additional conditions [**54**, Def. 1.11] the *maximality condition* and the *topological matching condition*, which are designed to ensure that the V_p/Γ_p in $\mathcal{K}$ for all $p \in X$ can be glued nicely using the h_{qp},φ_{qp} in $\mathcal{K}$ for all $p,q \in X$ to make a Hausdorff topological space. There are a few other small differences — for instance, Yang does not require vector bundles E_p in $(V_p,E_p,\Gamma_p,s_p,\psi_p)$ to be trivial.

The next definition comes from Yang [**53**, §1.6], [**54**, §5], [**55**, §2.4].

DEFINITION 2.16. Let $\mathcal{K},\mathcal{K}'$ be DY Kuranishi structures on a compact topological space X. An *embedding* $\epsilon\colon \mathcal{K} \hookrightarrow \mathcal{K}'$ is a choice of FOOO coordinate change $\epsilon_p\colon (V_p,E_p,\Gamma_p,s_p,\psi_p) \to (V'_p,E'_p,\Gamma'_p,s'_p,\psi'_p)$ with domain V_p for all $p \in X$, commuting with the FOOO coordinate changes Φ_{qp},Φ'_{qp} in $\mathcal{K},\mathcal{K}'$ up to elements of Γ'_p.

An embedding is a *chart refinement* if the ϵ_p come from inclusions of Γ_p-invariant open sets $V_p \hookrightarrow V'_p$.

DY Kuranishi structures $\mathcal{K}, \mathcal{K}'$ on X are called *R-equivalent* (or *equivalent*) if there is a diagram of DY Kuranishi structures on X

$$\mathcal{K} \xleftarrow{\sim} \mathcal{K}_1 \Longrightarrow \mathcal{K}_2 \Longleftarrow \mathcal{K}_3 \xrightarrow{\sim} \mathcal{K}',$$

where arrows $\Longrightarrow$ are embeddings, and $\xrightarrow{\sim}$ are chart refinements. Using facts about existence of good coordinate systems, Yang proves [**53**, Th. 1.6.17], [**54**, §11.2] that R-equivalence is an equivalence relation on DY Kuranishi structures.

Yang emphasizes the idea, which he calls *choice independence*, that when one constructs a (DY) Kuranishi structure $\mathcal{K}$ on a moduli space $\overline{\mathcal{M}}$, it should be independent of choices up to R-equivalence.

One goal of Yang's work is to relate the Kuranishi space theory of Fukaya, Oh, Ohta and Ono [**9**–**19**] to the polyfold theory of Hofer, Wysocki and Zehnder [**23**–**30**]. Here is a very brief introduction to this:

- An *sc-Banach space* $\mathcal{V}$ is a sequence $\mathcal{V} = (\mathcal{V}_0 \supset \mathcal{V}_1 \supset \mathcal{V}_2 \supset \cdots)$, where the $\mathcal{V}_i$ are Banach spaces, the inclusions $\mathcal{V}_{i+1} \hookrightarrow \mathcal{V}_i$ are compact, bounded linear maps, and $\mathcal{V}_\infty = \bigcap_{i \geqslant 0} \mathcal{V}_i$ is dense in every $\mathcal{V}_i$.

  The *tangent space* $T\mathcal{V}$ is $T\mathcal{V} = (\mathcal{V}_1 \oplus \mathcal{V}_0 \supset \mathcal{V}_2 \oplus \mathcal{V}_1 \supset \cdots)$, an sc-Banach space. An *open set* $\mathcal{Q}$ in $\mathcal{V}$ is an open set $\mathcal{Q} \subset \mathcal{V}_0$, and we write $\mathcal{Q}_i = \mathcal{Q} \cap \mathcal{V}_i$ for $i \geqslant 0$. Its *tangent space* is $T\mathcal{Q} = \mathcal{Q}_1 \oplus \mathcal{V}_0$, as an open set in $T\mathcal{V}$.

  An example to bear in mind is if M is a compact manifold, $E \to M$ a smooth vector bundle, $\alpha \in (0,1)$, and $\mathcal{V}_k = C^{k,\alpha}(E)$ for $k = 0, 1, \ldots$.
- Let $\mathcal{V} = (\mathcal{V}_0 \supset \mathcal{V}_1 \supset \cdots)$, $\mathcal{W} = (\mathcal{W}_0 \supset \mathcal{W}_1 \supset \cdots)$ be sc-Banach spaces and $\mathcal{Q} \subseteq \mathcal{V}$, $\mathcal{R} \subseteq \mathcal{W}$ be open. A map $f \colon \mathcal{Q} \to \mathcal{R}$ is called *sc*^{0} if $f(\mathcal{Q}_i) \subseteq \mathcal{R}_i$ and $f|_{\mathcal{Q}_i} \colon \mathcal{Q}_i \to \mathcal{R}_i$ is a continuous map of Banach manifolds for all $i \geqslant 0$.

  An sc^0 map $f \colon \mathcal{Q} \to \mathcal{R}$ is called *sc*^{1} if for each $q \in \mathcal{Q}_1$ there exists a bounded linear map $Df_q \colon \mathcal{V}_0 \to \mathcal{W}_0$, such that $f|_{\mathcal{Q}_1} \colon \mathcal{Q}_1 \to \mathcal{R}_0$ is a C^1 map of Banach manifolds with $\nabla f|_q = Df_q|_{\mathcal{V}_1} \colon \mathcal{V}_1 \to \mathcal{W}_0$ for all $q \in \mathcal{Q}_1$, and $Tf \colon T\mathcal{Q} \to T\mathcal{R}$ mapping $Tf \colon (q, v) \mapsto (f(q), Df_q(v))$ is an sc^0 map.

  By induction on k, we call $f \colon \mathcal{Q} \to \mathcal{R}$ an *sc*^{k} map for $k = 2, 3, \ldots$ if f is sc^1 and $Tf \colon T\mathcal{Q} \to T\mathcal{R}$ is an sc^{k-1} map. We call $f \colon \mathcal{Q} \to \mathcal{R}$ *sc-smooth*, or *sc*$^\infty$, if it is sc^k for all $k = 0, 1, \ldots$. This implies that $f|_{\mathcal{Q}_{i+k}} \colon \mathcal{Q}_{i+k} \to \mathcal{R}_i$ is a C^k-map of Banach manifolds for all $i, k \geqslant 0$.
- Let $\mathcal{V} = (\mathcal{V}_0 \supset \mathcal{V}_1 \supset \cdots)$ be an sc-Banach space and $\mathcal{Q} \subseteq \mathcal{V}$ be open. An *sc*$^\infty$*-retraction* is an sc-smooth map $r \colon \mathcal{Q} \to \mathcal{Q}$ with $r \circ r = r$. Set $\mathcal{O} = \operatorname{Im} r \subset \mathcal{V}$. We call $(\mathcal{O}, \mathcal{V})$ a *local sc-model.*

  If $\mathcal{V}$ is finite-dimensional then $\mathcal{O}$ is just a smooth manifold. But in infinite dimensions, new phenomena occur, and the tangent spaces $T_x\mathcal{O}$ can vary discontinuously with $x \in \mathcal{O}$. This is important for 'gluing'.
- An *M-polyfold chart* $(\mathcal{O}, \mathcal{V}, \psi)$ on a topological space Z is a local sc-model $(\mathcal{O}, \mathcal{V})$ and a homeomorphism $\psi \colon \mathcal{O} \to \operatorname{Im} \psi$ with an open set $\operatorname{Im} \psi \subset Z$.
- M-polyfold charts $(\mathcal{O}, \mathcal{V}, \psi), (\tilde{\mathcal{O}}, \tilde{\mathcal{V}}, \tilde{\psi})$ on Z are *compatible* if $\tilde{\psi}^{-1} \circ \psi \circ r \colon \mathcal{Q} \to \tilde{\mathcal{V}}$ and $\psi^{-1} \circ \tilde{\psi} \circ \tilde{r} \colon \tilde{\mathcal{Q}} \to \mathcal{V}$ are sc-smooth, where $\mathcal{Q} \subset \mathcal{V}$, $\tilde{\mathcal{Q}} \subset \tilde{\mathcal{V}}$ are open and $r \colon \mathcal{Q} \to \mathcal{Q}$, $\tilde{r} \colon \tilde{\mathcal{Q}} \to \tilde{\mathcal{Q}}$ are sc-smooth with $r \circ r = r$, $\tilde{r} \circ \tilde{r} = \tilde{r}$ and $\operatorname{Im} r = \psi^{-1}(\operatorname{Im} \tilde{\psi}) \subseteq \mathcal{O}$, $\operatorname{Im} \tilde{r} = \tilde{\psi}^{-1}(\operatorname{Im} \psi) \subseteq \tilde{\mathcal{O}}$.

- An *M-polyfold* is roughly a metrizable topological space Z with a maximal atlas of pairwise compatible M-polyfold charts.
- *Polyfolds* are the orbifold version of M-polyfolds, proper étale groupoids in M-polyfolds.
- A *polyfold Fredholm structure* $\mathcal{P}$ on a metrizable topological space X writes X as the zeroes of an sc-Fredholm section $\mathfrak{s}\colon \mathfrak{V} \to \mathfrak{E}$ of a strong polyfold vector bundle $\mathfrak{E} \to \mathfrak{V}$ over a polyfold $\mathfrak{V}$.

This is deep mathematics, and all rather complicated. The motivation for local sc-models $(\mathcal{O}, \mathcal{V})$ is that they can be used to describe functional-analytic problems involving 'gluing', 'bubbling', and 'neck-stretching', including moduli spaces of J-holomorphic curves with singularities of various kinds.

Yang proves [**53**, Th. 3.1.7] (see also [**55**, §2.6]):

THEOREM 2.17. *Suppose we are given a 'polyfold Fredholm structure' $\mathcal{P}$ on a compact metrizable topological space X, that is, we write X as the zeroes of an sc-Fredholm section $\mathfrak{s}\colon \mathfrak{V} \to \mathfrak{E}$ of a strong polyfold vector bundle $\mathfrak{E} \to \mathfrak{V}$ over a polyfold $\mathfrak{V}$, where $\mathfrak{s}$ has constant Fredholm index $n \in \mathbb{Z}$. Then we can construct a DY Kuranishi structure $\mathcal{K}$ on X, of virtual dimension n, which is independent of choices up to R-equivalence.*

2.6. How polyfolds are used. Hofer, Wysocki and Zehnder's polyfold programme [**23**–**30**] aims to show that moduli spaces of J-holomorphic curves in symplectic geometry may be given a polyfold Fredholm structure, and that compact spaces with oriented polyfold Fredholm structures have virtual chains and virtual classes. They will then use these virtual chains/classes to define big theories in symplectic geometry, such as Gromov–Witten invariants [**19**, **29**] or Symplectic Field Theory [**8**].

Here is an analogue of Theorems 2.8(a) and 2.13 in polyfold theory, proved by Hofer, Wysocki and Zehnder [**29**].

THEOREM 2.18. *Moduli spaces $\overline{\mathcal{M}}_{g,m}(J, \beta)$ of stable maps $u\colon \Sigma \to S$ from a prestable closed Riemann surface Σ of genus g with m marked points, to a compact symplectic manifold (S, ω) of dimension $2n$ with compatible almost complex structure J, with homology class $u_*([\Sigma]) = \beta \in H_2(S; \mathbb{Z})$, may be given a 'polyfold Fredholm structure'. That is, we may write $\overline{\mathcal{M}}_{g,m}(J, \beta)$ as the zeroes of an sc-Fredholm section $\mathfrak{s}\colon \mathfrak{V} \to \mathfrak{E}$ of a strong polyfold vector bundle $\mathfrak{E} \to \mathfrak{V}$ over a polyfold $\mathfrak{V}$, where $\mathfrak{s}$ has Fredholm index $2\bigl(c_1(S) \cdot \beta + (n-3)(1-g) + m\bigr)$.*

The 'evaluation maps' at marked points $\mathrm{ev}_i\colon \overline{\mathcal{M}}_{g,m}(J, \beta) \to S$ lift to sc-smooth maps of polyfolds $\mathfrak{ev}_i\colon \mathfrak{V} \to S$. The polyfold Fredholm structure $(\mathfrak{V}, \mathfrak{E}, \mathfrak{s})$ has a natural 'orientation'.

Hofer, Wysocki and Zehnder also announce the proofs of existence of polyfold Fredholm structures on other moduli spaces of J-holomorphic curves, in particular those relevant to Symplectic Field Theory [**8**]. Constructing a polyfold Fredholm structure on a moduli space of J-holomorphic curves involves far fewer arbitrary choices than defining a Kuranishi structure.

Combining Theorems 2.17–2.18 gives a DY Kuranishi structure on $\overline{\mathcal{M}}_{g,m}(J, \beta)$, uniquely up to R-equivalence. This is an example of a FOOO Kuranishi structure, giving an alternative proof of Theorem 2.8(a).

For the polyfold analogue of Theorems 2.9 and 2.14, Hofer et al. [**27**] prove:

THEOREM 2.19. *Let $(\mathfrak{V}, \mathfrak{E}, \mathfrak{s})$ be an oriented polyfold Fredholm structure with corners of virtual dimension k on a compact topological space X. Then there exist small perturbations $\tilde{\mathfrak{s}}$ of $\mathfrak{s}$ as an 'sc-smooth multisection', such that $\tilde{X} = \tilde{\mathfrak{s}}^{-1}(0)$ is a compact, oriented, branched, $\mathbb{Q}$-weighted k-orbifold with corners.*

We may triangulate $\tilde{X}$ by $\mathbb{Q}$-weighted k-simplices to define a **virtual class** *$[[X]]_{\rm virt}$ in the relative singular homology $H_k^{\rm si}(\mathfrak{V}, \partial\mathfrak{V}; \mathbb{Q})$. Alternatively, using relative de Rham cohomology we may define $[[X]]^{\rm virt} \in H^k_{\rm dR}(\mathfrak{V}, \partial\mathfrak{V}; \mathbb{R})^*$ mapping $[\omega] \mapsto \int_{\tilde{X}} \omega$. Both of $[[X]]_{\rm virt}, [[X]]^{\rm virt}$ are independent of the choice of $\tilde{\mathfrak{s}}, \tilde{X}$. If $\partial\mathfrak{V} = \emptyset$ then $\partial\tilde{X} = \emptyset$, $[[X]]_{\rm virt} \in H_k^{\rm si}(\mathfrak{V}; \mathbb{Q})$, and $[[X]]^{\rm virt} \in H^k_{\rm dR}(\mathfrak{V}; \mathbb{R})^*$.*

3. Kuranishi neighbourhoods as a 2-category

FOOO Kuranishi spaces and MW Kuranishi atlases in §2 are each built out of Kuranishi neighbourhoods (V, E, Γ, s, ψ) on a topological space X, and 'coordinate changes' $\Phi_{ij} \colon (V_i, E_i, \Gamma_i, s_i, \psi_i) \to (V_j, E_j, \Gamma_j, s_j, \psi_j)$ between them.

Our Kuranishi spaces $\boldsymbol{X}$ will be built out of Kuranishi neighbourhoods (V, E, Γ, s, ψ) on X, and 1-morphisms Φ_{ij} between them (which include FOOO and MW coordinate changes as special cases), and 2-morphisms $\Lambda_{ij} \colon \Phi_{ij} \Rightarrow \Phi'_{ij}$ between 1-morphisms, so that Kuranishi neighbourhoods form a 2-category.

This section explains our notions of 1- and 2-morphisms of Kuranishi neighbourhoods, and their properties. Section 4 will then use these to define the 2-category of Kuranishi spaces. With the exception of §3.6, all of this section, including proofs of quoted results, comes from [**35**, **36**].

3.1. Kuranishi neighbourhoods, 1-morphisms, and 2-morphisms.

We will use following '$O(s)$' and '$O(s^2)$' notation very often:

DEFINITION 3.1. Let V be a manifold, $E \to V$ a vector bundle, and $s \in C^\infty(E)$ a smooth section.

(i) If $F \to V$ is another vector bundle and $t_1, t_2 \in C^\infty(F)$ are smooth sections, we write $t_1 = t_2 + O(s)$ if there exists $\alpha \in C^\infty(E^* \otimes F)$ such that $t_1 = t_2 + \alpha \cdot s$ in $C^\infty(F)$, where the contraction $\alpha \cdot s$ is formed using the natural pairing of vector bundles $(E^* \otimes F) \times E \to F$ over V.

(ii) We write $t_1 = t_2 + O(s^2)$ if there exists $\alpha \in C^\infty(E^* \otimes E^* \otimes F)$ such that $t_1 = t_2 + \alpha \cdot (s \otimes s)$ in $C^\infty(F)$, where $\alpha \cdot (s \otimes s)$ uses the pairing $(E^* \otimes E^* \otimes F) \times (E \otimes E) \to F$.

Now let W be another manifold, and $f, g \colon V \to W$ be smooth maps.

(iii) We write $f = g + O(s)$ if whenever $h \colon W \to \mathbb{R}$ is a smooth map, there exists $\alpha \in C^\infty(E^*)$ such that $h \circ f = h \circ g + \alpha \cdot s$.

(iv) We write $f = g + O(s^2)$ if whenever $h \colon W \to \mathbb{R}$ is a smooth map, there exists $\alpha \in C^\infty(E^* \otimes E^*)$ such that $h \circ f = h \circ g + \alpha \cdot (s \otimes s)$.

(v) If $\Lambda \in C^\infty\big(E^* \otimes f^*(TW)\big)$, we write $f = g + \Lambda \cdot s + O(s^2)$ if whenever $h \colon W \to \mathbb{R}$ is a smooth map, there exists $\alpha \in C^\infty(E^* \otimes E^*)$ such that $h \circ f = h \circ g + \Lambda \cdot (s \otimes f^*({\rm d}h)) + \alpha \cdot (s \otimes s)$. Here $s \otimes f^*({\rm d}h)$ lies in $C^\infty\big(E \otimes f^*(T^*W)\big)$, and so has an obvious pairing with Λ.

Next suppose $f, g \colon V \to W$ with $f = g + O(s)$, and $F \to V$, $G \to W$ are vector bundles, and $t_1 \in C^\infty(F \otimes f^*(G))$, $t_2 \in C^\infty(F \otimes g^*(G))$. We wish to compare t_1, t_2, even though they are sections of different vector bundles.

(vi) We write $t_1 = t_2 + O(s)$ if for all $\beta \in C^\infty(G^*)$ we have $t_1 \cdot f^*(\beta) = t_2 \cdot g^*(\beta) + O(s)$ in sections of $F \to V$, as in (i).

Given any $t_1 \in C^\infty(F \otimes f^*(G))$, there exists $t_2 \in C^\infty(F \otimes g^*(G))$ with $t_1 = t_2 + O(s)$ in the sense of (vi), and if t_2' is an alternative choice then $t_2' = t_2 + O(s)$ in the sense of (i). To prove this, first suppose G is trivial, so that $f^*(G) = g^*(G)$, and $t_2 = t_1$ is a possible choice. In general we can trivialize G locally on Y, and combine the corresponding local choices for t_2 on X by a partition of unity.

The moral is that if $f = g + O(s)$ and we are interested in smooth sections t of $F \otimes f^*(G)$ up to $O(s)$, then we can treat the vector bundles $F \otimes f^*(G)$ and $F \otimes g^*(G)$ as essentially the same.

If instead $f, g\colon V \to W$ with $f = g + O(s^2)$ and F, G, t_1, t_2 are as above

(vii) We write $t_1 = t_2 + O(s^2)$ if for all $\beta \in C^\infty(G^*)$ we have $t_1 \cdot f^*(\beta) = t_2 \cdot g^*(\beta) + O(s^2)$ in sections of $F \to V$, as in (ii).

Given any $t_1 \in C^\infty(F \otimes f^*(G))$, there exists $t_2 \in C^\infty(F \otimes g^*(G))$ with $t_1 = t_2 + O(s^2)$ in the sense of (vii), and if t_2' is an alternative choice then $t_2' = t_2 + O(s^2)$ in the sense of (ii).

Here is our notion of Kuranishi neighbourhood. It is very similar to FOOO and MW Kuranishi neighbourhoods in Definitions 2.1 and 2.10, but it is more general, in that we allow E to be a nontrivial vector bundle.

DEFINITION 3.2. Let X be a topological space. A *Kuranishi neighbourhood* on X is a quintuple (V, E, Γ, s, ψ) such that:

(a) V is a smooth manifold.
(b) $\pi\colon E \to V$ is a real vector bundle over V, called the *obstruction bundle*.
(c) Γ is a finite group with a smooth action on V (not necessarily effective), and a compatible action on E preserving the vector bundle structure.
(d) $s\colon V \to E$ is a Γ-equivariant smooth section of E, called the *Kuranishi section*.
(e) ψ is a homeomorphism from $s^{-1}(0)/\Gamma$ to an open subset $\operatorname{Im}\psi = \{\psi(\Gamma v) : v \in s^{-1}(0)\}$ in X, called the *footprint* of (V, E, Γ, s, ψ).

We will write $\bar\psi\colon s^{-1}(0) \to \operatorname{Im}\psi \subseteq X$ for the composition of ψ with the projection $s^{-1}(0) \to s^{-1}(0)/\Gamma$.

We call (V, E, Γ, s, ψ) a *global* Kuranishi neighbourhood on X if $\operatorname{Im}\psi = X$.

The next two definitions are crucial to our programme.

DEFINITION 3.3. Let X, Y be topological spaces, $f\colon X \to Y$ a continuous map, $(V_i, E_i, \Gamma_i, s_i, \psi_i)$, $(V_j, E_j, \Gamma_j, s_j, \psi_j)$ be Kuranishi neighbourhoods on X, Y respectively, and $S \subseteq \operatorname{Im}\psi_i \cap f^{-1}(\operatorname{Im}\psi_j) \subseteq X$ be an open set. A *1-morphism* $\Phi_{ij} = (P_{ij}, \pi_{ij}, \phi_{ij}, \hat\phi_{ij})\colon (V_i, E_i, \Gamma_i, s_i, \psi_i) \to (V_j, E_j, \Gamma_j, s_j, \psi_j)$ *of Kuranishi neighbourhoods over* (S, f) is a quadruple $(P_{ij}, \pi_{ij}, \phi_{ij}, \hat\phi_{ij})$ satisfying

(a) P_{ij} is a manifold, with commuting smooth actions of Γ_i, Γ_j (that is, with a smooth action of $\Gamma_i \times \Gamma_j$), with the Γ_j-action free.
(b) $\pi_{ij}\colon P_{ij} \to V_i$ is a smooth map which is Γ_i-equivariant, Γ_j-invariant, and étale (a local diffeomorphism). The image $V_{ij} := \pi_{ij}(P_{ij})$ is a Γ_i-invariant open neighbourhood of $\bar\psi_i^{-1}(S)$ in V_i, and the fibres $\pi_{ij}^{-1}(v)$ of π_{ij} for $v \in V_{ij}$ are Γ_j-orbits, so that $\pi_{ij}\colon P_{ij} \to V_{ij}$ is a principal Γ_j-bundle.

We do not require $\bar{\psi}_i^{-1}(S) = V_{ij} \cap s_i^{-1}(0)$, only that $\bar{\psi}_i^{-1}(S) \subseteq V_{ij} \cap s_i^{-1}(0)$.

(c) $\phi_{ij}\colon P_{ij} \to V_j$ is a Γ_i-invariant and Γ_j-equivariant smooth map, that is, $\phi_{ij}(\gamma_i \cdot p) = \phi_{ij}(p)$, $\phi_{ij}(\gamma_j \cdot p) = \gamma_j \cdot \phi_{ij}(p)$ for all $\gamma_i \in \Gamma_i$, $\gamma_j \in \Gamma_j$, $p \in P_{ij}$.

(d) $\hat{\phi}_{ij}\colon \pi_{ij}^*(E_i) \to \phi_{ij}^*(E_j)$ is a Γ_i- and Γ_j-equivariant morphism of vector bundles on P_{ij}, where the Γ_i, Γ_j-actions are induced by the given Γ_i-action and the trivial Γ_j-action on E_i, and vice versa for E_j.

(e) $\hat{\phi}_{ij}(\pi_{ij}^*(s_i)) = \phi_{ij}^*(s_j) + O(\pi_{ij}^*(s_i)^2)$, in the sense of Definition 3.1.

(f) $f \circ \bar{\psi}_i \circ \pi_{ij} = \bar{\psi}_j \circ \phi_{ij}$ on $\pi_{ij}^{-1}(s_i^{-1}(0)) \subseteq P_{ij}$.

If $Y = X$ and $f = \mathrm{id}_X$ then we call Φ_{ij} a 1-*morphism of Kuranishi neighbourhoods over* S, or just a 1-*morphism over* S.

Let $(V_i, E_i, \Gamma_i, s_i, \psi_i)$ be a Kuranishi neighbourhood on X, and $S \subseteq \mathrm{Im}\,\psi_i$ be open. We will define the *identity* 1-*morphism* over S

$$\mathrm{id}_{(V_i,E_i,\Gamma_i,s_i,\psi_i)} = (P_{ii}, \pi_{ii}, \phi_{ii}, \hat{\phi}_{ii})\colon (V_i, E_i, \Gamma_i, s_i, \psi_i) \to (V_i, E_i, \Gamma_i, s_i, \psi_i). \tag{3.1}$$

Since P_{ii} must have two different actions of Γ_i, for clarity we write $\Gamma_i^d = \Gamma_i^t = \Gamma_i$, where Γ_i^d and Γ_i^t mean the copies of Γ_i acting on the domain and target of the 1-morphism in (3.1), respectively.

Define $P_{ii} = V_i \times \Gamma_i$, and let Γ_i^d act on P_{ii} by $\gamma^d\colon (v,\gamma) \mapsto (\gamma^d \cdot v, \gamma(\gamma^d)^{-1})$ and Γ_i^t act on P_{ii} by $\gamma^t\colon (v,\gamma) \mapsto (v, \gamma^t\gamma)$. Define $\pi_{ii}, \phi_{ii}\colon P_{ii} \to V_i$ by $\pi_{ii}\colon (v,\gamma) \mapsto v$ and $\phi_{ii}\colon (v,\gamma) \mapsto \gamma \cdot v$. Then π_{ii} is Γ_i^d-equivariant and Γ_i^t-invariant, and is a Γ_i^t-principal bundle, and ϕ_{ii} is Γ_i^d-invariant and Γ_i^t-equivariant.

At $(v,\gamma) \in P_{ii}$, the morphism $\hat{\phi}_{ii}\colon \pi_{ii}^*(E_i) \to \phi_{ii}^*(E_i)$ must map $E_i|_v \to E_i|_{\gamma \cdot v}$. We have such a map, the lift of the γ-action on V_i to E_i. So we define $\hat{\phi}_{ii}$ on $V_i \times \{\gamma\} \subseteq P_{ii}$ to be the lift to E_i of the γ-action on V_i, for each $\gamma \in \Gamma$.

Remark 3.4. **(i)** We will use 1-morphisms of Kuranishi neighbourhoods for two purposes in §4.1: Kuranishi spaces $\boldsymbol{X}$ involve 'coordinate changes', which are 1-morphisms of Kuranishi neighbourhoods over $\mathrm{id}_X\colon X \to X$ satisfying an invertibility condition. And 1-morphisms of Kuranishi spaces $\boldsymbol{f}\colon \boldsymbol{X} \to \boldsymbol{Y}$ involve 1-morphisms of Kuranishi neighbourhoods over $f\colon X \to Y$.

(ii) As we will explain in §3.4, FOOO coordinate changes in §2.1 and MW coordinate changes in §2.3 give examples of 1-morphisms of Kuranishi neighbourhoods over $\mathrm{id}_X\colon X \to X$.

(iii) Definition 3.3 does not include the analogue of (2.2) being exact (the 'with a tangent bundle' condition in the Fukaya–Oh–Ohta–Ono theory). But as Theorem 3.15 shows, this will be a consequence of the definition of 'coordinate change' in Definition 3.11, which are special 1-morphisms over $\mathrm{id}_X : X \to X$.

(iv) In a 1-morphism $\Phi_{ij} = (P_{ij}, \pi_{ij}, \phi_{ij}, \hat{\phi}_{ij})$, the data $P_{ij}, \pi_{ij}, \phi_{ij}$ are actually a standard way to write 'smooth maps' between orbifolds, using 'bibundles':

$$(P_{ij}, \pi_{ij}, \phi_{ij})\colon [V_{ij}/\Gamma_i] \longrightarrow [V_j/\Gamma_j]$$

is a *Hilsum–Skandalis morphism* between the quotient orbifolds $[V_{ij}/\Gamma_i], [V_j/\Gamma_j]$, as in Lerman [**38**, §3.3] or Henriques and Metzler [**21**].

Thus, we can interpret Φ_{ij} as giving the following data:

- An open neighbourhood $[V_{ij}/\Gamma_i]$ of $\psi_i^{-1}(S)$ in the orbifold $[V_i/\Gamma_i]$;
- A 'smooth map' of orbifolds $\Phi'_{ij} = (P_{ij}, \pi_{ij}, \phi_{ij})\colon [V_{ij}/\Gamma_i] \to [V_j/\Gamma_j]$; and
- A morphism of orbifold vector bundles $\hat{\phi}_{ij}\colon E_i|_{[V_{ij}/\Gamma_i]} \to (\Phi'_{ij})^*(E_j)$.

In our theory, 2-morphisms between 1-morphisms will also be important.

DEFINITION 3.5. Suppose X, Y are topological spaces, $f\colon X \to Y$ is a continuous map, $(V_i, E_i, \Gamma_i, s_i, \psi_i)$, $(V_j, E_j, \Gamma_j, s_j, \psi_j)$ are Kuranishi neighbourhoods on X, Y respectively, $S \subseteq \operatorname{Im}\psi_i \cap f^{-1}(\operatorname{Im}\psi_j) \subseteq X$ is open, and $\Phi_{ij}, \Phi'_{ij}\colon (V_i, E_i, \Gamma_i, s_i, \psi_i) \to (V_j, E_j, \Gamma_j, s_j, \psi_j)$ are two 1-morphisms over (S, f), with $\Phi_{ij} = (P_{ij}, \pi_{ij}, \phi_{ij}, \hat\phi_{ij})$ and $\Phi'_{ij} = (P'_{ij}, \pi'_{ij}, \phi'_{ij}, \hat\phi'_{ij})$.

Consider triples $(\dot P_{ij}, \lambda_{ij}, \hat\lambda_{ij})$ satisfying:

(a) $\dot P_{ij}$ is a Γ_i- and Γ_j-invariant open neighbourhood of $\pi_{ij}^{-1}(\bar\psi_i^{-1}(S))$ in P_{ij}.

(b) $\lambda_{ij}\colon \dot P_{ij} \to P'_{ij}$ is a Γ_i- and Γ_j-equivariant smooth map with $\pi'_{ij} \circ \lambda_{ij} = \pi_{ij}|_{\dot P_{ij}}$. This implies that λ_{ij} is an isomorphism of principal Γ_j-bundles over $\dot V_{ij} := \pi_{ij}(\dot P_{ij})$, so λ_{ij} is a diffeomorphism with a Γ_i- and Γ_j-invariant open set $\lambda_{ij}(\dot P_{ij})$ in P'_{ij}.

(c) $\hat\lambda_{ij}\colon \pi_{ij}^*(E_i)|_{\dot P_{ij}} \to \phi_{ij}^*(TV_j)|_{\dot P_{ij}}$ is a Γ_i- and Γ_j-invariant smooth morphism of vector bundles on $\dot P_{ij}$, satisfying

$$\begin{aligned} &\phi'_{ij} \circ \lambda_{ij} = \phi_{ij}|_{\dot P_{ij}} + \hat\lambda_{ij} \cdot \pi_{ij}^*(s_i) + O\big(\pi_{ij}^*(s_i)^2\big) \quad \text{and} \\ &\lambda_{ij}^*(\hat\phi'_{ij}) = \hat\phi_{ij}|_{\dot P_{ij}} + \hat\lambda_{ij} \cdot \phi_{ij}^*(\mathrm{d}\, s_j) + O\big(\pi_{ij}^*(s_i)\big) \quad \text{on } \dot P_{ij}, \end{aligned} \tag{3.2}$$

in the notation of Definition 3.1. Here $\mathrm{d}\, s_j$ is a shorthand for the derivative ∇s_j in $C^\infty(T^*V_j \otimes E_j)$ with respect to any connection ∇ on E_j, and (3.2) is independent of this choice.

Define a binary relation $\approx$ on such triples by $(\dot P_{ij}, \lambda_{ij}, \hat\lambda_{ij}) \approx (\dot P'_{ij}, \lambda'_{ij}, \hat\lambda'_{ij})$ if there exists an open neighbourhood $\ddot P_{ij}$ of $\pi_{ij}^{-1}(\bar\psi_i^{-1}(S))$ in $\dot P_{ij} \cap \dot P'_{ij}$ with

$$\lambda_{ij}|_{\ddot P_{ij}} = \lambda'_{ij}|_{\ddot P_{ij}} \quad \text{and} \quad \hat\lambda_{ij}|_{\ddot P_{ij}} = \hat\lambda'_{ij}|_{\ddot P_{ij}} + O\big(\pi_{ij}^*(s_i)\big) \quad \text{on } \ddot P_{ij}.$$

Then $\approx$ is an equivalence relation. We also write $\approx_S$ for $\approx$ if we wish to stress the open set S. Write $[\dot P_{ij}, \lambda_{ij}, \hat\lambda_{ij}]$ for the $\approx$-equivalence class of $(\dot P_{ij}, \lambda_{ij}, \hat\lambda_{ij})$. We say that $[\dot P_{ij}, \lambda_{ij}, \hat\lambda_{ij}]\colon \Phi_{ij} \Rightarrow \Phi'_{ij}$ is a *2-morphism of 1-morphisms of Kuranishi neighbourhoods on X over (S, f)*, or just a *2-morphism over (S, f)*. We often write $\Lambda_{ij} = [\dot P_{ij}, \lambda_{ij}, \hat\lambda_{ij}]$.

If $Y = X$ and $f = \mathrm{id}_X$ then we call Λ_{ij} a *2-morphism of Kuranishi neighbourhoods over S*, or just a *2-morphism over S*.

For a 1-morphism $\Phi_{ij} = (P_{ij}, \pi_{ij}, \phi_{ij}, \hat\phi_{ij})$, define the *identity 2-morphism*

$$\mathrm{id}_{\Phi_{ij}} = [P_{ij}, \mathrm{id}_{P_{ij}}, 0]\colon \Phi_{ij} \Longrightarrow \Phi_{ij}.$$

REMARK 3.6. This definition of 2-morphism probably seems rather arbitrary. The best way the author knows to motivate it is via d-orbifolds, as we explain in §3.6. It is also justified by the useful properties of 2-morphisms, in particular Theorems 3.10, 3.14 and 3.15 below.

3.2. Making Kuranishi neighbourhoods into a 2-category. Readers unfamiliar with 2-categories are advised to look at §A.1 at this point. We wish to make Kuranishi neighbourhoods on a topological space X into a weak 2-category. We have already defined the objects (Kuranishi neighbourhoods), 1-morphisms, 2-morphisms, and identity 1- and 2-morphisms, in §3.1. But there remains quite a lot of structure to define, which we do in this section:

- Composition of 1-morphisms;
- Horizontal composition of 2-morphisms;
- Vertical composition of 2-morphisms;
- Coherence 2-isomorphisms $\alpha_{g,f,e}\colon (g\circ f)\circ e\Rightarrow g\circ(f\circ e)$, $\beta_f\colon f\circ\mathrm{id}_X\Rightarrow f$, and $\gamma_f\colon \mathrm{id}_Y\circ f\Rightarrow f$, as in (A.5) and (A.7).

Definition 3.7. Let X,Y,Z be topological spaces, $f\colon X\to Y$, $g\colon Y\to Z$ be continuous maps, $(V_i,E_i,\Gamma_i,s_i,\psi_i),(V_j,E_j,\Gamma_j,s_j,\psi_j),(V_k,E_k,\Gamma_k,s_k,\psi_k)$ be Kuranishi neighbourhoods on X,Y,Z respectively, and $T\subseteq \mathrm{Im}\,\psi_j\cap g^{-1}(\mathrm{Im}\,\psi_k)\subseteq Y$ and $S\subseteq\mathrm{Im}\,\psi_i\cap f^{-1}(T)\subseteq X$ be open. Suppose $\Phi_{ij}=(P_{ij},\pi_{ij},\phi_{ij},\hat\phi_{ij})\colon (V_i,E_i,\Gamma_i,s_i,\psi_i)\to(V_j,E_j,\Gamma_j,s_j,\psi_j)$ is a 1-morphism of Kuranishi neighbourhoods over (S,f), and $\Phi_{jk}=(P_{jk},\pi_{jk},\phi_{jk},\hat\phi_{jk})\colon(V_j,E_j,\Gamma_j,s_j,\psi_j)\to(V_k,E_k,\Gamma_k,s_k,\psi_k)$ is a 1-morphism of Kuranishi neighbourhoods over (T,g).

Consider the diagram of manifolds and smooth maps:

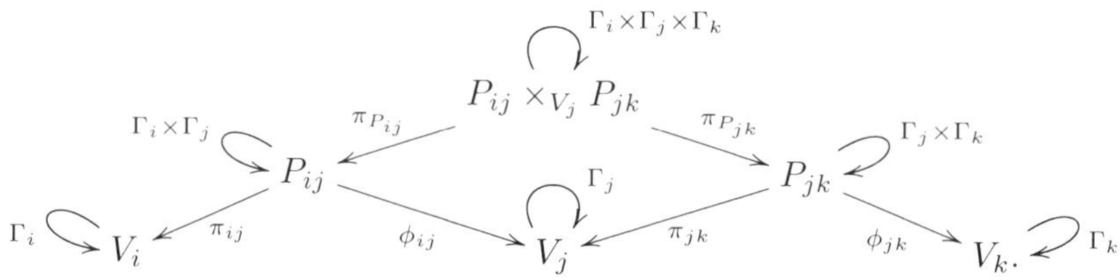

Here the fibre product $P_{ij}\times_{V_j}P_{jk}$ is transverse, and so exists, as π_{jk} is étale. We have shown the actions of various combinations of $\Gamma_i,\Gamma_j,\Gamma_k$ on each space. In fact $\Gamma_i\times\Gamma_j\times\Gamma_k$ acts on the whole diagram, with all maps equivariant, but we have omitted the trivial actions (for instance, Γ_j,Γ_k act trivially on V_i).

As Γ_j acts freely on P_{ij}, it also acts freely on $P_{ij}\times_{V_j}P_{jk}$, so $P_{ik}:=(P_{ij}\times_{V_j}P_{jk})/\Gamma_j$ is a manifold, with projection $\Pi\colon P_{ij}\times_{V_j}P_{jk}\to P_{ik}$. The commuting actions of Γ_i,Γ_k on $P_{ij}\times_{V_j}P_{jk}$ descend to commuting actions of Γ_i,Γ_k on P_{ik} such that Π is Γ_i- and Γ_k-equivariant. As $\pi_{ij}\circ\pi_{P_{ij}}\colon P_{ij}\times_{V_j}P_{jk}\to V_i$ and $\phi_{jk}\circ\pi_{P_{jk}}\colon P_{ij}\times_{V_j}P_{jk}\to V_k$ are Γ_j-invariant, they factor through Π, so there are unique smooth maps $\pi_{ik}\colon P_{ik}\to V_i$ and $\phi_{ik}\colon P_{ik}\to V_k$ such that $\pi_{ij}\circ\pi_{P_{ij}}=\pi_{ik}\circ\Pi$ and $\phi_{jk}\circ\pi_{P_{jk}}=\phi_{ik}\circ\Pi$.

Consider the diagram of vector bundles on $P_{ij}\times_{V_j}P_{jk}$:

$$\begin{array}{ccccccc}
\Pi^*\circ\pi_{ik}^*(E_i) & & \xrightarrow[\Pi^*(\hat\phi_{ik})]{\quad\cdots\cdots\quad} & & & & \Pi^*\circ\phi_{ik}^*(E_k)\\
\| & & & & & & \|\\
\pi_{P_{ij}}^*\circ\pi_{ij}^*(E_i) & \xrightarrow{\pi_{P_{ij}}^*(\hat\phi_{ij})} & \pi_{P_{ij}}^*\circ\phi_{ij}^*(E_j) & = & \pi_{P_{jk}}^*\circ\pi_{jk}^*(E_j) & \xrightarrow{\pi_{P_{jk}}^*(\hat\phi_{jk})} & \pi_{P_{jk}}^*\circ\phi_{jk}^*(E_k).
\end{array}$$

There is a unique morphism on the top line making the diagram commute. As $\hat\phi_{ij},\hat\phi_{jk}$ are Γ_j-equivariant, this is Γ_j-equivariant, so it is the pullback under Π^* of a unique morphism $\hat\phi_{ik}\colon\pi_{ik}^*(E_i)\to\phi_{ik}^*(E_k)$, as shown. It is now easy to check that $(P_{ik},\pi_{ik},\phi_{ik},\hat\phi_{ik})$ satisfies Definition 3.3(a)–(f), and is a 1-morphism $\Phi_{ik}=(P_{ik},\pi_{ik},\phi_{ik},\hat\phi_{ik})\colon(V_i,E_i,\Gamma_i,s_i,\psi_i)\to(V_k,E_k,\Gamma_k,s_k,\psi_k)$ over $(S,g\circ f)$. We write $\Phi_{jk}\circ\Phi_{ij}=\Phi_{ik}$, and call it the *composition of 1-morphisms*.

If we have three such 1-morphisms $\Phi_{ij},\Phi_{jk},\Phi_{kl}$, define

$$\lambda_{ijkl}\colon\bigl[P_{ij}\times_{V_j}\bigl((P_{jk}\times_{V_k}P_{kl})/\Gamma_k\bigr)\bigr]/\Gamma_j\to\bigl[\bigl((P_{ij}\times_{V_j}P_{jk})/\Gamma_j\bigr)\times_{V_k}P_{kl}\bigr]/\Gamma_k \tag{3.3}$$

to be the natural identification. Then we have a 2-isomorphism

$$\alpha_{\Phi_{kl},\Phi_{jk},\Phi_{ij}} := \big[[P_{ij} \times_{V_j} ((P_{jk} \times_{V_k} P_{kl})/\Gamma_k)]/\Gamma_j, \lambda_{ijkl}, 0\big] : (\Phi_{kl} \circ \Phi_{jk}) \circ \Phi_{ij} \Longrightarrow \Phi_{kl} \circ (\Phi_{jk} \circ \Phi_{ij}). \tag{3.4}$$

That is, composition of 1-morphisms is associative up to canonical 2-isomorphism, as for weak 2-categories in §A.1.

For $\Phi_{ij} \colon (V_i, E_i, \Gamma_i, s_i, \psi_i) \to (V_j, E_j, \Gamma_j, s_j, \psi_j)$ as above, define

$$\mu_{ij} \colon ((V_i \times \Gamma_i) \times_{V_i} P_{ij})/\Gamma_i \longrightarrow P_{ij},$$
$$\nu_{ij} \colon (P_{ij} \times_{V_j} (V_j \times \Gamma_j))/\Gamma_j \longrightarrow P_{ij},$$

to be the natural identifications. Then we have 2-isomorphisms

$$\begin{aligned} \beta_{\Phi_{ij}} &:= \big[((V_i \times \Gamma_i) \times_{V_i} P_{ij})/\Gamma_i, \mu_{ij}, 0\big] : \Phi_{ij} \circ \mathrm{id}_{(V_i,E_i,\Gamma_i,s_i,\psi_i)} \Longrightarrow \Phi_{ij}, \\ \gamma_{\Phi_{ij}} &:= \big[(P_{ij} \times_{V_j} (V_j \times \Gamma_j))/\Gamma_j, \nu_{ij}, 0\big] : \mathrm{id}_{(V_j,E_j,\Gamma_j,s_j,\psi_j)} \circ \Phi_{ij} \Longrightarrow \Phi_{ij}, \end{aligned} \tag{3.5}$$

so identity 1-morphisms behave as they should up to canonical 2-isomorphism, as for weak 2-categories in §A.1.

Definition 3.8. Let X, Y be topological spaces, $f \colon X \to Y$ be continuous, $(V_i, E_i, \Gamma_i, s_i, \psi_i)$, $(V_j, E_j, \Gamma_j, s_j, \psi_j)$ be Kuranishi neighbourhoods on X, Y, $S \subseteq \mathrm{Im}\,\psi_i \cap f^{-1}(\mathrm{Im}\,\psi_j) \subseteq X$ be open, and $\Phi_{ij}, \Phi'_{ij}, \Phi''_{ij} \colon (V_i, E_i, \Gamma_i, s_i, \psi_i) \to (V_j, E_j, \Gamma_j, s_j, \psi_j)$ be 1-morphisms over (S, f) with $\Phi_{ij} = (P_{ij}, \pi_{ij}, \phi_{ij}, \hat\phi_{ij})$, $\Phi'_{ij} = (P'_{ij}, \pi'_{ij}, \phi'_{ij}, \hat\phi'_{ij})$, $\Phi''_{ij} = (P''_{ij}, \pi''_{ij}, \phi''_{ij}, \hat\phi''_{ij})$. Suppose $\Lambda_{ij} = [\dot P_{ij}, \lambda_{ij}, \hat\lambda_{ij}] \colon \Phi_{ij} \Rightarrow \Phi'_{ij}$ and $\Lambda'_{ij} = [\dot P'_{ij}, \lambda'_{ij}, \hat\lambda'_{ij}] \colon \Phi'_{ij} \Rightarrow \Phi''_{ij}$ are 2-morphisms over (S, f). We will define the *vertical composition of 2-morphisms over* (S, f), written

$$\Lambda'_{ij} \odot \Lambda_{ij} = [\dot P'_{ij}, \lambda'_{ij}, \hat\lambda'_{ij}] \odot [\dot P_{ij}, \lambda_{ij}, \hat\lambda_{ij}] \colon \Phi_{ij} \Longrightarrow \Phi''_{ij}.$$

Choose representatives $(\dot P_{ij}, \lambda_{ij}, \hat\lambda_{ij}), (\dot P'_{ij}, \lambda'_{ij}, \hat\lambda'_{ij})$ in the $\approx$-equivalence classes $[\dot P_{ij}, \lambda_{ij}, \hat\lambda_{ij}], [\dot P'_{ij}, \lambda'_{ij}, \hat\lambda'_{ij}]$. Define $\dot P''_{ij} = \lambda_{ij}^{-1}(\dot P_{ij}) \subseteq \dot P_{ij} \subseteq P_{ij}$, and $\lambda''_{ij} = \lambda'_{ij} \circ \lambda_{ij}|_{\dot P''_{ij}}$. Consider the morphism of vector bundles

$$\lambda^*_{ij}(\hat\lambda'_{ij}) \colon \pi^*_{ij}(E_i)|_{\dot P''_{ij}} = \lambda^*_{ij} \circ \pi'^*_{ij}(E_i)|_{\dot P''_{ij}} \longrightarrow \lambda^*_{ij} \circ \phi'^*_{ij}(TV_j) = (\phi'_{ij} \circ \lambda_{ij})^*(TV_j)|_{\dot P''_{ij}}.$$

Since $\phi'_{ij} \circ \lambda_{ij}|_{\dot P''_{ij}} = \phi_{ij}|_{\dot P''_{ij}} + O(\pi^*_{ij}(s_i))$ by (3.2), the discussion after Definition 3.1(vi) shows that there exists $\check\lambda'_{ij} \colon \pi^*_{ij}(E_i)|_{\dot P''_{ij}} \to \phi^*_{ij}(TV_j)|_{\dot P''_{ij}}$ with

$$\check\lambda'_{ij} = \lambda_{ij}|^*_{\dot P''_{ij}}(\hat\lambda'_{ij}) + O(\pi^*_{ij}(s_i)), \tag{3.6}$$

as in Definition 3.1(vi), and $\check\lambda'_{ij}$ is unique up to $O(\pi^*_{ij}(s_i))$. By averaging over the $\Gamma_i \times \Gamma_j$-action we can suppose $\check\lambda'_{ij}$ is Γ_i- and Γ_j-equivariant, as $\hat\lambda'_{ij}$ is.

Define $\hat\lambda''_{ij} \colon \pi^*_{ij}(E_i)|_{\dot P''_{ij}} \to \phi^*_{ij}(TV_j)|_{\dot P''_{ij}}$ by $\hat\lambda''_{ij} = \hat\lambda_{ij}|_{\dot P''_{ij}} + \check\lambda'_{ij}$. It is now easy to see that $(\dot P''_{ij}, \lambda''_{ij}, \hat\lambda''_{ij})$ satisfies Definition 3.5(a)–(c) for Φ_{ij}, Φ''_{ij}, using (3.2) for $\hat\lambda_{ij}, \hat\lambda'_{ij}$ and (3.6) to prove (3.2) for $\hat\lambda''_{ij}$. Hence $\Lambda''_{ij} = [\dot P''_{ij}, \lambda''_{ij}, \hat\lambda''_{ij}] \colon \Phi_{ij} \Rightarrow \Phi''_{ij}$ is a 2-morphism over (S, f). It is independent of choices. We define $[\dot P'_{ij}, \lambda'_{ij}, \hat\lambda'_{ij}] \odot [\dot P_{ij}, \lambda_{ij}, \hat\lambda_{ij}] = [\dot P''_{ij}, \lambda''_{ij}, \hat\lambda''_{ij}]$, or $\Lambda'_{ij} \odot \Lambda_{ij} = \Lambda''_{ij}$. All 2-morphisms over (S, f) are invertible under vertical composition, that is, they are 2-isomorphisms.

Write $\mathbf{Hom}_{S,f}\big((V_i,E_i,\Gamma_i,s_i,\psi_i),(V_j,E_j,\Gamma_j,s_j,\psi_j)\big)$ for the groupoid with objects 1-morphisms $\Phi_{ij}\colon (V_i,E_i,\Gamma_i,s_i,\psi_i)\to(V_j,E_j,\Gamma_j,s_j,\psi_j)$ over (S,f), and morphisms 2-morphisms $\Lambda_{ij}\colon \Phi_{ij}\Rightarrow\Phi'_{ij}$ over (S,f). If $X=Y$ and $f=\mathrm{id}_X$, we write this as $\mathbf{Hom}_S\big((V_i,E_i,\Gamma_i,s_i,\psi_i),(V_j,E_j,\Gamma_j,s_j,\psi_j)\big)$.

Definition 3.9. Let X,Y,Z be topological spaces, $f\colon X\to Y$, $g\colon Y\to Z$ be continuous maps, $(V_i,E_i,\Gamma_i,s_i,\psi_i),(V_j,E_j,\Gamma_j,s_j,\psi_j),(V_k,E_k,\Gamma_k,s_k,\psi_k)$ be Kuranishi neighbourhoods on X,Y,Z, and $T\subseteq \operatorname{Im}\psi_j\cap g^{-1}(\operatorname{Im}\psi_k)\subseteq Y$ and $S\subseteq \operatorname{Im}\psi_i\cap f^{-1}(T)\subseteq X$ be open. Suppose $\Phi_{ij},\Phi'_{ij}\colon (V_i,E_i,\Gamma_i,s_i,\psi_i)\to(V_j,E_j,\Gamma_j,s_j,\psi_j)$ are 1-morphisms of Kuranishi neighbourhoods over (S,f), and $\Lambda_{ij}\colon\Phi_{ij}\Rightarrow\Phi'_{ij}$ is a 2-morphism over (S,f), and $\Phi_{jk},\Phi'_{jk}\colon(V_j,E_j,\Gamma_j,s_j,\psi_j)\to(V_k,E_k,\Gamma_k,s_k,\psi_k)$ are 1-morphisms of Kuranishi neighbourhoods over (T,g), and $\Lambda_{jk}\colon\Phi_{jk}\Rightarrow\Phi'_{jk}$ is a 2-morphism over (T,g).

We will define the *horizontal composition of 2-morphisms*, written

$$\Lambda_{jk}*\Lambda_{ij}\colon \Phi_{jk}\circ\Phi_{ij}\Longrightarrow\Phi'_{jk}\circ\Phi'_{ij}\qquad\text{over }(S,g\circ f).\tag{3.7}$$

Use our usual notation for $\Phi_{ij},\ldots,\Lambda_{jk}$, and write $(P_{ik},\pi_{ik},\phi_{ik},\hat\phi_{ik})=\Phi_{jk}\circ\Phi_{ij}$, $(P'_{ik},\pi'_{ik},\phi'_{ik},\hat\phi'_{ik})=\Phi'_{jk}\circ\Phi'_{ij}$, as in Definition 3.7. Choose representatives $(\dot P_{ij},\lambda_{ij},\hat\lambda_{ij})$, $(\dot P_{jk},\lambda_{jk},\hat\lambda_{jk})$ for $\Lambda_{ij}=[\dot P_{ij},\lambda_{ij},\hat\lambda_{ij}]$ and $\Lambda_{jk}=[\dot P_{jk},\lambda_{jk},\hat\lambda_{jk}]$.

Then $P_{ik}=(P_{ij}\times_{V_j}P_{jk})/\Gamma_j$, and $\dot P_{ij}\subseteq P_{ij}$, $\dot P_{jk}\subseteq P_{jk}$ are open and Γ_j-invariant, so $\dot P_{ij}\times_{V_j}\dot P_{jk}$ is open and Γ_j-invariant in $P_{ij}\times_{V_j}P_{jk}$. Define $\dot P_{ik}=(\dot P_{ij}\times_{V_j}\dot P_{jk})/\Gamma_j$, as an open subset of P_{ik}. It is Γ_i- and Γ_k-invariant, as $\dot P_{ij}$, $\dot P_{jk}$ are Γ_i- and Γ_k-invariant, respectively.

The maps $\lambda_{ij}\colon\dot P_{ij}\to P'_{ij}$, $\lambda_{jk}\colon\dot P_{jk}\to P'_{jk}$ satisfy $\phi'_{ij}\circ\lambda_{ij}=\phi_{ij}|_{\dot P_{ij}}\colon\dot P_{ij}\to V_j$ and $\pi'_{jk}\circ\lambda_{jk}=\pi_{jk}|_{\dot P_{jk}}\colon\dot P_{jk}\to V_j$. Hence by properties of fibre products they induce a unique smooth map $\tilde\lambda_{ik}\colon\dot P_{ij}\times_{\phi_{ij},V_j,\pi_{jk}}\dot P_{jk}\to P'_{ij}\times_{\phi'_{ij},V_j,\pi'_{jk}}P'_{jk}$ with $\pi_{P'_{ij}}\circ\tilde\lambda_{ik}=\lambda_{ij}\circ\pi_{\dot P_{ij}}$ and $\pi_{P'_{jk}}\circ\tilde\lambda_{ik}=\lambda_{jk}\circ\pi_{\dot P_{jk}}$. As everything is Γ_j-equivariant, $\tilde\lambda_{ik}$ descends to the quotients by Γ_j. Thus we obtain a unique smooth map

$$\lambda_{ik}\colon\dot P_{ik}=(\dot P_{ij}\times_{V_j}\dot P_{jk})/\Gamma_j\longrightarrow(P'_{ij}\times_{V_j}P'_{jk})/\Gamma_j=P'_{ik}$$

with $\lambda_{ik}\circ\Pi=\Pi'\circ\tilde\lambda_{ik}$, for $\Pi\colon\dot P_{ij}\times_{V_j}\dot P_{jk}\to(\dot P_{ij}\times_{V_j}\dot P_{jk})/\Gamma_j$, $\Pi'\colon P'_{ij}\times_{V_j}P'_{jk}\to(P'_{ij}\times_{V_j}P'_{jk})/\Gamma_j$ the projections.

Define a morphism of vector bundles on $\dot P_{ij}\times_{V_j}\dot P_{jk}$

$$\begin{gathered}\check\lambda_{ik}\colon\Pi^*\circ\pi_{ik}^*(E_i)=(\pi_{ij}\circ\pi_{\dot P_{ij}})^*(E_i)\longrightarrow(\phi_{jk}\circ\pi_{\dot P_{jk}})^*(TV_k)=\Pi^*\circ\phi_{ik}^*(TV_k)\\ \text{by }\ \check\lambda_{ik}=\pi^*_{\dot P_{jk}}(\mathrm{d}\,\phi_{jk}\circ(\mathrm{d}\,\pi_{jk})^{-1})\circ\pi^*_{\dot P_{ij}}(\hat\lambda_{ij})+\pi^*_{\dot P_{jk}}(\hat\lambda_{jk})\circ\pi^*_{\dot P_{ij}}(\hat\phi_{ij}),\end{gathered}$$

where the morphisms are given in the diagram

$$\begin{array}{ccccc}(\pi_{ij}\circ\pi_{\dot P_{ij}})^*(E_i) & \xrightarrow[\pi^*_{\dot P_{ij}}(\hat\phi_{ij})]{} & (\phi_{ij}\circ\pi_{\dot P_{ij}})^*(E_j) & = & (\pi_{jk}\circ\pi_{\dot P_{jk}})^*(E_j)\\ \big\downarrow{\scriptstyle \pi^*_{\dot P_{ij}}(\hat\lambda_{ij})} & & & & \\ (\phi_{ij}\circ\pi_{\dot P_{ij}})^*(TV_j) & & & & \big\downarrow{\scriptstyle \pi^*_{\dot P_{jk}}(\hat\lambda_{jk})}\\ \| & & & & \\ (\pi_{jk}\circ\pi_{\dot P_{jk}})^*(TV_j) & \underset{\pi^*_{\dot P_{jk}}(\mathrm{d}\,\pi_{jk})}{\overset{\pi^*_{\dot P_{jk}}((\mathrm{d}\,\pi_{jk})^{-1})}{\rightleftarrows}} & \pi^*_{\dot P_{jk}}(T\dot P_{jk}) & \xrightarrow{\pi^*_{\dot P_{jk}}(\mathrm{d}\,\phi_{jk})} & (\phi_{jk}\circ\pi_{\dot P_{jk}})^*(TV_k).\end{array}$$

Here $\mathrm{d}\,\pi_{jk}\colon T\dot P_{jk}\to\pi_{jk}^*(TV_j)$ is invertible as π_{jk} is étale. As all the ingredients are $\Gamma_i,\Gamma_j,\Gamma_k$-invariant or equivariant, $\check\lambda_{ik}$ is Γ_j-invariant, and so descends to

$\dot{P}_{ik} = (\dot{P}_{ij} \times_{V_j} \dot{P}_{jk})/\Gamma_j$. That is, there is a unique morphism $\hat{\lambda}_{ik} \colon \pi_{ik}|^*_{\dot{P}_{ik}}(E_i) \to \phi_{ik}|^*_{\dot{P}_{ik}}(TV_k)$ of vector bundles on $\dot{P}_{ik}$ with $\Pi^*(\hat{\lambda}_{ik}) = \check{\lambda}_{ik}$. As $\check{\lambda}_{ik}$ is Γ_i- and Γ_k-equivariant, so is $\hat{\lambda}_{ik}$.

One can now check that $(\dot{P}_{ik}, \lambda_{ik}, \hat{\lambda}_{ik})$ satisfies Definition 3.5(a)–(c), where (3.2) for $\hat{\lambda}_{ik}$ follows from adding the pullbacks to $\dot{P}_{ij} \times_{V_j} \dot{P}_{jk}$ of (3.2) for $\hat{\lambda}_{ij}, \hat{\lambda}_{jk}$, so $\Lambda_{ik} = [\dot{P}_{ik}, \lambda_{ik}, \hat{\lambda}_{ik}]$ is a 2-morphism as in (3.7), which is independent of choices of $(\dot{P}_{ij}, \lambda_{ij}, \hat{\lambda}_{ij})$, $(\dot{P}_{jk}, \lambda_{jk}, \hat{\lambda}_{jk})$. We define $\Lambda_{jk} * \Lambda_{ij} = \Lambda_{ik}$ in (3.7).

We have now defined all the structures of a weak 2-category, except that our 1- and 2-morphisms are defined over an open set $S \subseteq X$, which is not part of the 2-category structure. As in [**35**, §4.1] and [**36**, §6.1], there are two ways to make a genuine 2-category out of this, either (a) working on a fixed topological space X and open $S \subseteq X$, or (b) allowing X to vary but requiring $S = X$ throughout.

THEOREM 3.10. **(a)** *Let X be a topological space and $S \subseteq X$ an open set. The definitions above give a* ***weak*** ***2-category*** $\mathbf{KN}_S(X)$ ***of Kuranishi neighbourhoods over*** *S, with objects Kuranishi neighbourhoods $(V_i, E_i, \Gamma_i, s_i, \psi_i)$ on X with $S \subseteq \operatorname{Im} \psi_i$, and 1- and 2-morphisms 1-morphisms Φ_{ij} of Kuranishi neighbourhoods over S, and 2-morphisms Λ_{ij} over S.*

(b) *The definitions above also give a weak 2-category* **GKN** *of* ***global Kuranishi neighbourhoods****, such that:*

- **(i)** *The objects $\bigl(X, (V, E, \Gamma, s, \psi)\bigr)$ in* **GKN** *are pairs of a topological space X and a Kuranishi neighbourhood (V, E, Γ, s, ψ) on X with $\operatorname{Im} \psi = X$.*
- **(ii)** *The 1-morphisms $(f, \Phi) : \bigl(X, (V, E, \Gamma, s, \psi)\bigr) \to \bigl(Y, (W, F, \Delta, t, \chi)\bigr)$ in* **GKN** *are pairs of a continuous map $f \colon X \to Y$ and a 1-morphism $\Phi \colon (V, E, \Gamma, s, \psi) \to (W, F, \Delta, t, \chi)$ of Kuranishi neighbourhoods over (X, f).*
- **(iii)** *For 1-morphisms $(f, \Phi), (g, \Psi) \colon \bigl(X, (V, E, \Gamma, s, \psi)\bigr) \to \bigl(Y, (W, F, \Delta, t, \chi)\bigr)$, the 2-morphisms $\Lambda \colon (f, \Phi) \Rightarrow (g, \Psi)$ in* **GKN** *exist only if $f = g$, and are 2-morphisms $\Lambda \colon \Phi \Rightarrow \Psi$ of Kuranishi neighbourhoods over (X, f).*

All 2-morphisms in $\mathbf{KN}_S(X)$, **GKN** *are 2-isomorphisms, that is, $\mathbf{KN}_S(X)$,* **GKN** *are $(2,1)$-categories.*

DEFINITION 3.11. Recall from §A.1 that an *equivalence* in a 2-category $\mathcal{C}$ is a 1-morphism $f \colon A \to B$ in $\mathcal{C}$ such that there exist a 1-morphism $g \colon B \to A$ (called a *quasi-inverse*) and 2-isomorphisms $\eta \colon g \circ f \Rightarrow \mathrm{id}_A$ and $\zeta \colon f \circ g \Rightarrow \mathrm{id}_B$. A 1-morphism $f \colon A \to B$ is an equivalence if and only if $[f] \colon A \to B$ is an isomorphism (is invertible) in the homotopy category $\mathrm{Ho}(\mathcal{C})$.

A 1-morphism $\Phi_{ij} \colon (V_i, E_i, \Gamma_i, s_i, \psi_i) \to (V_j, E_j, \Gamma_j, s_j, \psi_j)$ on X over S is a *coordinate change over* S if Φ_{ij} is an equivalence in the 2-category $\mathbf{KN}_S(X)$.

DEFINITION 3.12. Let $T \subseteq S \subseteq X$ be open. Define the *restriction 2-functor* $|_T \colon \mathbf{KN}_S(X) \to \mathbf{KN}_T(X)$ to map objects $(V_i, E_i, \Gamma_i, s_i, \psi_i)$ to exactly the same objects, and 1-morphisms Φ_{ij} to exactly the same 1-morphisms but regarded as 1-morphisms over T, and 2-morphisms $\Lambda_{ij} = [\dot{P}_{ij}, \lambda_{ij}, \hat{\lambda}_{ij}]$ over S to $\Lambda_{ij}|_T = [\dot{P}_{ij}, \lambda_{ij}, \hat{\lambda}_{ij}]|_T$, where $[\dot{P}_{ij}, \lambda_{ij}, \hat{\lambda}_{ij}]|_T$ is the $\approx_T$-equivalence class of any representative $(\dot{P}_{ij}, \lambda_{ij}, \hat{\lambda}_{ij})$ for the $\approx_S$-equivalence class $[\dot{P}_{ij}, \lambda_{ij}, \hat{\lambda}_{ij}]$.

Then $|_T \colon \mathbf{KN}_S(X) \to \mathbf{KN}_T(X)$ commutes with all the structure, so we may we regard $|_T$ as a weak 2-functor as in §A.2 for which the additional 2-isomorphisms $F_{g,f}$ are identities.

If $U \subseteq T \subseteq S \subseteq X$ are open then $|_U \circ |_T = |_U : \mathbf{KN}_S(X) \to \mathbf{KN}_U(X)$. Also $|_S$ gives a functor $|_T : \mathbf{Hom}_S\big((V_i, E_i, \Gamma_i, s_i, \psi_i), (V_j, E_j, \Gamma_j, s_j, \psi_j)\big) \to \mathbf{Hom}_T\big((V_i, E_i, \Gamma_i, s_i, \psi_i), (V_j, E_j, \Gamma_j, s_j, \psi_j)\big)$ when $T \subseteq S \subseteq \operatorname{Im}\psi_i \cap \operatorname{Im}\psi_j$, in the notation of Definition 3.8.

DEFINITION 3.13. So far we have discussed 1- and 2-morphisms of Kuranishi neighbourhoods, and coordinate changes, *over a specified open set* $S \subseteq X$, or over (S, f). We now make the convention that *when we do not specify a domain S for a 1-morphism, 2-morphism, or coordinate change, the domain should be as large as possible.* For example, if we say that $\Phi_{ij} : (V_i, E_i, \Gamma_i, s_i, \psi_i) \to (V_j, E_j, \Gamma_j, s_j, \psi_j)$ is a 1-morphism (or a 1-morphism over $f : X \to Y$) without specifying S, we mean that $S = \operatorname{Im}\psi_i \cap \operatorname{Im}\psi_j$ (or $S = \operatorname{Im}\psi_i \cap f^{-1}(\operatorname{Im}\psi_j)$).

Similarly, if we write a formula involving several 2-morphisms (possibly defined on different domains), without specifying the domain S, we make the convention *that the domain where the formula holds should be as large as possible.* That is, the domain S is taken to be the intersection of the domains of each 2-morphism in the formula, and we implicitly restrict each morphism in the formula to S as in Definition 3.12, to make it make sense.

3.3. Properties of 1- and 2-morphisms. The next result [**35**, Th. 4.14], [**36**, §6.1] is very important in our theory. We will call Theorem 3.14 the *stack property.* We will use it in §4.1 to construct compositions of 1- and 2-morphisms of Kuranishi spaces. *Stacks on topological spaces* are the 2-category analogue of sheaves, and are defined in §A.4.

THEOREM 3.14. **(a)** *Let $(V_i, E_i, \Gamma_i, s_i, \psi_i), (V_j, E_j, \Gamma_j, s_j, \psi_j)$ be Kuranishi neighbourhoods on a topological space X. For each open $S \subseteq \operatorname{Im}\psi_i \cap \operatorname{Im}\psi_j$, set*

$$\begin{aligned}\mathcal{H}om&\big((V_i, E_i, \Gamma_i, s_i, \psi_i), (V_j, E_j, \Gamma_j, s_j, \psi_j)\big)(S)\\ &= \mathbf{Hom}_S\big((V_i, E_i, \Gamma_i, s_i, \psi_i), (V_j, E_j, \Gamma_j, s_j, \psi_j)\big),\end{aligned}$$

as in Definition 3.8, for all open $T \subseteq S \subseteq \operatorname{Im}\psi_i \cap \operatorname{Im}\psi_j$ define a functor

$$\begin{aligned}\rho_{ST} : \ &\mathcal{H}om\big((V_i, E_i, \Gamma_i, s_i, \psi_i), (V_j, E_j, \Gamma_j, s_j, \psi_j)\big)(S) \longrightarrow\\ &\mathcal{H}om\big((V_i, E_i, \Gamma_i, s_i, \psi_i), (V_j, E_j, \Gamma_j, s_j, \psi_j)\big)(T)\end{aligned}$$

by $\rho_{ST} = |_T$, as in Definition 3.12, and for all open $U \subseteq T \subseteq S \subseteq \operatorname{Im}\psi_i \cap \operatorname{Im}\psi_j$ take the obvious isomorphism $\eta_{STU} = \mathrm{id}_{\rho_{SU}} : \rho_{TU} \circ \rho_{ST} \Rightarrow \rho_{SU}$. Then $\mathcal{H}om\big((V_i, E_i, \Gamma_i, s_i, \psi_i), (V_j, E_j, \Gamma_j, s_j, \psi_j)\big)$ is a stack on $\operatorname{Im}\psi_i \cap \operatorname{Im}\psi_j$.

Coordinate changes $(V_i, E_i, \Gamma_i, s_i, \psi_i) \to (V_j, E_j, \Gamma_j, s_j, \psi_j)$ also form a stack $\mathcal{E}qu\big((V_i, E_i, \Gamma_i, s_i, \psi_i), (V_j, E_j, \Gamma_j, s_j, \psi_j)\big)$ on $\operatorname{Im}\psi_i \cap \operatorname{Im}\psi_j$, which is a substack of $\mathcal{H}om\big((V_i, E_i, \Gamma_i, s_i, \psi_i), (V_j, E_j, \Gamma_j, s_j, \psi_j)\big)$.

(b) *Let $f : X \to Y$ be continuous, and $(U_i, D_i, \mathrm{B}_i, r_i, \chi_i), (V_j, E_j, \Gamma_j, s_j, \psi_j)$ be Kuranishi neighbourhoods on X, Y, respectively. Then 1- and 2-morphisms from $(U_i, D_i, \mathrm{B}_i, r_i, \chi_i)$ to $(V_j, E_j, \Gamma_j, s_j, \psi_j)$ over f form a stack $\mathcal{H}om_f\big((U_i, D_i, \mathrm{B}_i, r_i, \chi_i), (V_j, E_j, \Gamma_j, s_j, \psi_j)\big)$ on $\operatorname{Im}\chi_i \cap f^{-1}(\operatorname{Im}\psi_j) \subseteq X$.*

Here [**35**, Th. 4.16], [**36**, §10.5] is a criterion for recognizing when a 1-morphism of Kuranishi neighbourhoods is a coordinate change, that is, is invertible up to 2-isomorphism. Note the similarity of (3.8) to equation (2.2).

THEOREM 3.15. *Let $\Phi_{ij} = (P_{ij}, \pi_{ij}, \phi_{ij}, \hat{\phi}_{ij}) : (V_i, E_i, \Gamma_i, s_i, \psi_i) \to (V_j, E_j, \Gamma_j, s_j, \psi_j)$ be a 1-morphism of Kuranishi neighbourhoods over $S \subseteq X$. Let $p \in$

$\pi_{ij}^{-1}(\bar{\psi}_i^{-1}(S)) \subseteq P_{ij}$, *and set* $v_i = \pi_{ij}(p) \in V_i$ *and* $v_j = \phi_{ij}(p) \in V_j$. *Consider the sequence of real vector spaces:*

$$0 \longrightarrow T_{v_i}V_i \xrightarrow{\mathrm{d}\, s_i|_{v_i} \oplus (\mathrm{d}\,\phi_{ij}|_p \circ \mathrm{d}\,\pi_{ij}|_p^{-1})} E_i|_{v_i} \oplus T_{v_j}V_j \xrightarrow{-\hat{\phi}_{ij}|_p \oplus \mathrm{d}\, s_j|_{v_j}} E_j|_{v_j} \longrightarrow 0. \tag{3.8}$$

Here $\mathrm{d}\,\pi_{ij}|_p \colon T_pP_{ij} \to T_{v_i}V_i$ *is invertible as* π_{ij} *is étale. Definition* 3.3(e) *implies that* (3.8) *is a complex. Also consider the morphism of finite groups*

$$\begin{aligned} &\rho_p \colon \{(\gamma_i, \gamma_j) \in \Gamma_i \times \Gamma_j : (\gamma_i, \gamma_j) \cdot p = p\} \longrightarrow \{\gamma_j \in \Gamma_j : \gamma_j \cdot v_j = v_j\}, \\ &\rho_p \colon (\gamma_i, \gamma_j) \longmapsto \gamma_j. \end{aligned} \tag{3.9}$$

Then Φ_{ij} *is a coordinate change over* S, *in the sense of Definition* 3.11, *if and only if* (3.8) *is exact and* (3.9) *is an isomorphism for all* $p \in \pi_{ij}^{-1}(\bar{\psi}_i^{-1}(S))$.

3.4. Relation to Fukaya–Oh–Ohta–Ono's work. We now relate our definitions in §3.1–§3.3 to FOOO Kuranishi neighbourhoods and FOOO coordinate changes from §2.1.

EXAMPLE 3.16. Compare Definition 2.1 of FOOO Kuranishi neighbourhoods (V, E, Γ, s, ψ), and Definition 3.2 of our Kuranishi neighbourhoods $(\tilde{V}, \tilde{E}, \tilde{\Gamma}, \tilde{s}, \tilde{\psi})$. The differences are that E is a vector space, but $\tilde{E} \to \tilde{V}$ is a vector bundle over $\tilde{V}$. Also Γ must act effectively on V, but $\tilde{\Gamma}$ need not act effectively on $\tilde{V}$.

To make a FOOO Kuranishi neighbourhood (V, E, Γ, s, ψ) into one of our Kuranishi neighbourhoods $(\tilde{V}, \tilde{E}, \tilde{\Gamma}, \tilde{s}, \tilde{\psi})$, take $\tilde{V} = V$, $\tilde{\Gamma} = \Gamma$, $\tilde{\psi} = \psi$, let $\tilde{\pi} \colon \tilde{E} \to \tilde{V}$ be the trivial vector bundle $\pi_V \colon V \times E \to V$ with fibre E, and $\tilde{s} = \mathrm{id} \times s \colon V \to V \times E$. Thus, FOOO Kuranishi neighbourhoods correspond to special examples of our Kuranishi neighbourhoods $(\tilde{V}, \tilde{E}, \tilde{\Gamma}, \tilde{s}, \tilde{\psi})$, in which $\tilde{\pi} \colon \tilde{E} \to \tilde{V}$ is a trivial vector bundle, and $\tilde{\Gamma}$ acts effectively on $\tilde{V}$.

By an abuse of notation, we will sometimes identify FOOO Kuranishi neighbourhoods with the corresponding Kuranishi neighbourhoods in §3.1. That is, we will use E to denote both a vector space, and the corresponding trivial vector bundle over V, and s to denote both a map, and a section of a trivial bundle. Fukaya et al. [**13**, Def. 4.3(4)] also make the same abuse of notation.

EXAMPLE 3.17. Let $\Phi_{ij} = (V_{ij}, h_{ij}, \varphi_{ij}, \hat{\varphi}_{ij}) \colon (V_i, E_i, \Gamma_i, s_i, \psi_i) \to (V_j, E_j, \Gamma_j, s_j, \psi_j)$ be a FOOO coordinate change over S, as in Definition 2.2. As in Example 3.16, regard the FOOO Kuranishi neighbourhoods $(V_i, E_i, \Gamma_i, s_i, \psi_i), (V_j, E_j, \Gamma_j, s_j, \psi_j)$ as examples of Kuranishi neighbourhoods in the sense of §3.1.

Set $P_{ij} = V_{ij} \times \Gamma_j$. Let Γ_i act on P_{ij} by $\gamma_i \colon (v, \gamma) \mapsto (\gamma_i \cdot v, \gamma h_{ij}(\gamma_i)^{-1})$. Let Γ_j act on P_{ij} by $\gamma_j \colon (v, \gamma) \mapsto (v, \gamma_j \gamma)$. Define $\pi_{ij} \colon P_{ij} \to V_i$ and $\phi_{ij} \colon P_{ij} \to V_j$ by $\pi_{ij} \colon (v, \gamma) \mapsto v$ and $\phi_{ij} \colon (v, \gamma) \mapsto \gamma \cdot \varphi_{ij}(v)$. Then π_{ij} is Γ_i-equivariant and Γ_j-invariant. Since φ_{ij} is h_{ij}-equivariant, ϕ_{ij} is Γ_i-invariant, and Γ_j-equivariant.

We will define a vector bundle morphism $\hat{\phi}_{ij} \colon \pi_{ij}^*(E_i) \to \phi_{ij}^*(E_j)$. At $(v, \gamma) \in P_{ij}$, this $\hat{\phi}_{ij}$ must map $E_i|_v \to E_j|_{\gamma \cdot \varphi_{ij}(v)}$. We define $\hat{\phi}_{ij}|_{(v,\gamma)}$ to be the composition of $\hat{\varphi}_{ij}|_v \colon E_i|_v \to E_j|_{\varphi_{ij}(v)}$ with $\gamma \cdot \colon E_j|_{\varphi_{ij}(v)} \to E_j|_{\gamma \cdot \varphi_{ij}(v)}$ from the Γ_j-action on E_j. That is, $\hat{\phi}_{ij}|_{V_{ij} \times \{\gamma\}} = \gamma \cdot \hat{\varphi}_{ij}$ for each $\gamma \in \Gamma_j$.

It is now easy to see that $\tilde{\Phi}_{ij} = (P_{ij}, \pi_{ij}, \phi_{ij}, \hat{\phi}_{ij}) \colon (V_i, E_i, \Gamma_i, s_i, \psi_i) \to (V_j, E_j, \Gamma_j, s_j, \psi_j)$ is a 1-morphism over S, in the sense of §3.1. Theorem 3.15 shows that $\tilde{\Phi}_{ij}$ is a coordinate change over S, in the sense of §3.2.

We will show that the elements $\gamma_{rqp}^{\alpha} \in \Gamma_p$ in Definition 2.4(b) correspond in the setting of §3 to a 2-morphism $\Lambda_{rqp} \colon \tilde\Phi_{qp} \circ \tilde\Phi_{rq} \Rightarrow \tilde\Phi_{rp}$.

EXAMPLE 3.18. **(i)** In the Fukaya–Oh–Ohta–Ono theory [**9–19**], one often relates two FOOO coordinate changes in the following way. Let $\Phi_{ij} = (V_{ij}, h_{ij}, \varphi_{ij}, \hat\varphi_{ij}), \Phi'_{ij} = (V'_{ij}, h'_{ij}, \varphi'_{ij}, \hat\varphi'_{ij}) \colon (V_i, E_i, \Gamma_i, s_i, \psi_i) \to (V_j, E_j, \Gamma_j, s_j, \psi_j)$ be FOOO coordinate changes over S. Suppose there exists $\gamma \in \Gamma_j$ such that

$$h_{ij} = \gamma \cdot h'_{ij} \cdot \gamma^{-1}, \quad \phi_{ij} = \gamma \cdot \phi'_{ij}, \quad \text{and} \quad \hat\phi_{ij} = \gamma \cdot \hat\phi'_{ij}, \tag{3.10}$$

where the second and third equations hold on $\dot V_{ij} := V_{ij} \cap V'_{ij}$.

Let $\tilde\Phi_{ij}, \tilde\Phi'_{ij} \colon (V_i, E_i, \Gamma_i, s_i, \psi_i) \to (V_j, E_j, \Gamma_j, s_j, \psi_j)$ be the 1-morphisms in the sense of §3.1 corresponding to Φ_{ij}, Φ'_{ij} in Example 3.17. Set $\dot P_{ij} = \dot V_{ij} \times \Gamma_j \subseteq P_{ij}$. Define $\lambda_{ij} \colon \dot P_{ij} = \dot V_{ij} \times \Gamma_j \to V'_{ij} \times \Gamma_j = P'_{ij}$ by $\lambda_{ij} \colon (v, \gamma') \mapsto (v, \gamma'\gamma)$, and $\hat\lambda_{ij} = 0$. Then $(\dot P_{ij}, \lambda_{ij}, \hat\lambda_{ij})$ satisfies Definition 3.5(a)–(c), so we have defined a 2-morphism $\Lambda_{ij} = [\dot P_{ij}, \lambda_{ij}, \hat\lambda_{ij}] \colon \tilde\Phi_{ij} \Rightarrow \tilde\Phi'_{ij}$, in the sense of §3.1.

(ii) This enables us to interpret Definition 2.4(b) in terms of a 2-morphism. In the situation of Definition 2.4(b), the composition of the FOOO coordinate changes Φ_{rq}, Φ_{qp} is $\Phi_{qp} \circ \Phi_{qp} = \bigl(\varphi_{rq}^{-1}(V_{qp}), h_{qp} \circ h_{rq}, \varphi_{qp} \circ \varphi_{rq}|_{\varphi_{rq}^{-1}(V_{qp})}, \varphi_{rq}^*(\hat\varphi_{qp}) \circ \hat\varphi_{rq}|_{\varphi_{rq}^{-1}(V_{qp})}\bigr)$. Thus, (2.3) relates $\Phi_{qp} \circ \Phi_{rq}$ to Φ_{rp} in the same way that (3.10) relates Φ_{ij} to Φ'_{ij}, except for allowing γ_{rqp} to vary on different connected components. Hence, if $\tilde\Phi_{rq}, \tilde\Phi_{qp}, \tilde\Phi_{rp}$ are the coordinate changes in the sense of §3.2 associated to $\Phi_{rq}, \Phi_{qp}, \Phi_{rp}$ in Example 3.17, then the method of **(i)** defines a 2-morphism $\Lambda_{pqr} \colon \tilde\Phi_{qp} \circ \tilde\Phi_{rq} \Rightarrow \tilde\Phi_{rp}$, in the sense of §3.1.

(iii) In the situation of Definition 2.4(b), suppose $v \in (\varphi_{rq}^{-1}(V_{qp}) \cap V_{rp})^{\alpha}$ is generic. Then $\mathrm{Stab}_{\Gamma_r}(v) = \{1\}$, as Γ_r acts (locally) effectively on V_r by Definition 2.1(c). Hence $\mathrm{Stab}_{\Gamma_p}(\varphi_{rp}(v)) = \{1\}$ by Definition 2.2(g). Therefore the point $\gamma_{rqp}^{\alpha} \cdot \varphi_{rp}(v) = \varphi_{qp} \circ \varphi_{rq}(v)$ in V_p determines γ_{rqp}^{α} in Γ_p. So the second equation of (2.3) determines $\gamma_{rqp}^{\alpha} \in \Gamma_p$ uniquely, provided it exists. Thus the 2-morphism $\Lambda_{pqr} \colon \tilde\Phi_{qp} \circ \tilde\Phi_{rq} \Rightarrow \tilde\Phi_{rp}$ in **(ii)** is also determined uniquely.

3.5. Relation to McDuff and Wehrheim's work. We can also connect our definitions in §3.1–§3.3 to MW Kuranishi neighbourhoods and MW coordinate changes from §2.3. As in Example 3.16, by an abuse of notation we will regard MW Kuranishi neighbourhoods as examples of our Kuranishi neighbourhoods in §3.1. We relate MW coordinate changes to ours:

EXAMPLE 3.19. Let $\Phi_{BC} = (\tilde V_{BC}, \rho_{BC}, \varpi_{BC}, \hat\varphi_{BC}) \colon (V_B, E_B, \Gamma_B, s_B, \psi_B) \to (V_C, E_C, \Gamma_C, s_C, \psi_C)$ be an MW coordinate change over S, as in Definition 2.11. Regard $(V_B, E_B, \Gamma_B, s_B, \psi_B), (V_C, E_C, \Gamma_C, s_C, \psi_C)$ as Kuranishi neighbourhoods in the sense of §3.1.

Set $P_{BC} = \tilde V_{BC} \times \Gamma_B$. Let Γ_B act on P_{BC} by $\gamma_B \colon (v, \gamma) \mapsto (v, \gamma_B\gamma)$. Let Γ_C act on P_{BC} by $\gamma_C \colon (v, \gamma) \mapsto (\gamma_C \cdot v, \gamma\rho_{BC}(\gamma_C)^{-1})$. Define $\pi_{BC} \colon P_{BC} \to V_B$ and $\phi_{BC} \colon P_{BC} \to V_C$ by $\pi_{BC} \colon (v, \gamma) \mapsto \gamma \cdot \varpi_{BC}(v)$ and $\phi_{BC} \colon (v, \gamma) \mapsto v$. Then π_{BC} is Γ_B-equivariant and Γ_C-invariant, and ϕ_{BC} is Γ_B-invariant and Γ_C-equivariant.

Define $\hat\phi_{BC} : \pi_{BC}^*(V_B \times E_B) \to \phi_{BC}^*(V_C \times E_C)$, as a morphism of trivial vector bundles with fibres E_B, E_C on $P_{BC} = \tilde V_{BC} \times \Gamma_B$, by $\hat\phi_{BC}|_{\tilde V_{BC} \times \{\gamma\}} = \hat\varphi_{BC} \circ (\gamma^{-1} \cdot -)$ for each $\gamma \in \Gamma_B$. It is easy to see that $\tilde\Phi_{BC} = (P_{BC}, \pi_{BC}, \phi_{BC}, \hat\phi_{BC}) \colon (V_B, E_B, \Gamma_B, s_B, \psi_B) \to (V_C, E_C, \Gamma_C, s_C, \psi_C)$ is a 1-morphism over S, in the sense of §3.1. Combining Definition 2.11(g) and Theorem 3.15 shows that $\tilde{\Phi}_{BC}$ is a coordinate change over S, in the sense of §3.2.

We relate Definition 2.12(d) to 2-morphisms in §3.1:

EXAMPLE 3.20. In the situation of Definition 2.12(d), let $\tilde{\Phi}_{BC}, \tilde{\Phi}_{BD}, \tilde{\Phi}_{CD}$ be the coordinate changes in the sense of §3.2 associated to the MW coordinate changes $\Phi_{BC}, \Phi_{BD}, \Phi_{CD}$ in Example 3.19. The composition coordinate change $\tilde{\Phi}_{CD} \circ \tilde{\Phi}_{BC} = (P_{BCD}, \pi_{BCD}, \phi_{BCD}, \hat{\phi}_{BCD})$ from Definition 3.7 has

$$
\begin{aligned}
P_{BCD} &= \big[(\tilde{V}_{BC} \times \Gamma_B) \times_{V_C} (\tilde{V}_{CD} \times \Gamma_C)\big]/\Gamma_C \\
&\cong (\tilde{V}_{BC} \times_{V_C} \tilde{V}_{CD}) \times \Gamma_B \cong \varpi_{CD}^{-1}(\tilde{V}_{BC}) \times \Gamma_B.
\end{aligned}
\tag{3.11}
$$

Define $\dot{P}_{BCD}$ to be the open subset of P_{BCD} identified with $\tilde{V}_{BCD} \times \Gamma_B$ by (3.11), and $\lambda_{BCD} : \dot{P}_{BCD} \to P_{BD} = \tilde{V}_{BD} \times \Gamma_B$ to be the map identified by (3.11) with the inclusion $\tilde{V}_{BCD} \times \Gamma_B \hookrightarrow \tilde{V}_{BD} \times \Gamma_B$, and $\hat{\lambda}_{BCD} = 0$. Then as in Example 3.18(i), we can show that $(\dot{P}_{BCD}, \lambda_{BCD}, \hat{\lambda}_{BCD})$ satisfies Definition 3.5(a)–(c), so we have defined a 2-morphism $\Lambda_{BCD} = [\dot{P}_{BCD}, \lambda_{BCD}, \hat{\lambda}_{BCD}] : \tilde{\Phi}_{CD} \circ \tilde{\Phi}_{BC} \Rightarrow \tilde{\Phi}_{BD}$ on $S_{BCD} = \operatorname{Im}\psi_B \cap \operatorname{Im}\psi_C \cap \operatorname{Im}\psi_D$, in the sense of §3.1.

3.6. Relation to d-orbifolds. The author has developed a theory of Derived Differential Geometry [**32**–**34**], involving 'd-manifolds' and 'd-orbifolds', derived versions of smooth manifolds and orbifolds, where 'derived' is in the sense of Derived Algebraic Geometry. D-manifolds and d-orbifolds form strict 2-categories **dMan**, **dOrb**. For other definitions of derived manifolds, which form ∞-categories, see Spivak [**50**] and Borisov–Noel [**6**]. Borisov [**5**] explains how the d-manifolds of [**32**–**34**] are related to the derived manifolds of Spivak [**50**] and Borisov–Noel [**6**].

D-manifolds are a full 2-subcategory of the strict 2-category of 'd-spaces' **dSpa**, where a d-space $\boldsymbol{X} = (X, \mathcal{O}_X^\bullet)$ is a topological space $\boldsymbol{X}$ with a sheaf $\mathcal{O}_X^\bullet$ of 'square zero dg C^∞-rings', satisfying some conditions. Here a square zero dg C^∞-ring is a 2-categorical, differential-geometric analogue of the commutative differential graded algebras (cdgas) used as basic objects in Derived Algebraic Geometry. D-orbifolds are a Deligne–Mumford stack version of d-manifolds.

The initial definitions of d-manifolds and d-orbifolds broadly follow those of schemes and stacks in algebraic geometry, and look nothing like §3.1–§3.2. But in order to describe d-manifolds and d-orbifolds in classical differential-geometric language, the author [**34**, §3 & §10] proves results on 'standard model' d-manifolds and d-orbifolds, and 'standard model' 1- and 2-morphisms.

In terms of the ideas of §3.1, a 'standard model' d-manifold $\boldsymbol{S}_{V,E,s}$ or d-orbifold $\boldsymbol{S}_{V,E,\Gamma,s}$ is constructed from a Kuranishi neighbourhood (V, E, Γ, s, ψ), with $\Gamma = \{1\}$ in the d-manifold case, and 'standard model' 1- and 2-morphisms are constructed from the data in 1- and 2-morphisms of Kuranishi neighbourhoods. By comparing §3.1–§3.2 with results in [**34**, §3 & §10], we can deduce:

THEOREM 3.21. **(a)** *Suppose (V, E, Γ, s, ψ) is a global Kuranishi neighbourhood on a topological space X, where 'global' means $\operatorname{Im}\psi = X$. Then we can define a 'standard model' d-orbifold $\boldsymbol{S}_{V,E,\Gamma,s}$, with underlying topological space X.*

(b) *Let $(V_i, E_i, \Gamma_i, s_i, \psi_i), (V_j, E_j, \Gamma_j, s_j, \psi_j)$ be global Kuranishi neighbourhoods on X, Y, and $f : X \to Y$ be continuous, and $\Phi_{ij} : (V_i, E_i, \Gamma_i, s_i, \psi_i) \to (V_j, E_j, \Gamma_j, s_j, \psi_j)$ be a 1-morphism of Kuranishi neighbourhoods over (X, f). Then we can define* *a 'standard model' 1-morphism of d-orbifolds in* **dOrb**

$$\boldsymbol{S}_{\Phi_{ij}} : \boldsymbol{S}_{V_i,E_i,\Gamma_i,s_i} \longrightarrow \boldsymbol{S}_{V_j,E_j,\Gamma_j,s_j},$$

with underlying continuous map f.

(c) *Let $\boldsymbol{f}\colon \boldsymbol{S}_{V_i,E_i,\Gamma_i,s_i} \to \boldsymbol{S}_{V_j,E_j,\Gamma_j,s_j}$ be a* 1*-morphism in* **dOrb** *between standard model d-orbifolds. Then $\boldsymbol{f}$ is* 2*-isomorphic in* **dOrb** *to some standard model* 1*-morphism $\boldsymbol{S}_{\Phi_{ij}}$, as in* **(b)**.

(d) *Suppose $\Phi_{ij}, \Phi'_{ij}\colon (V_i, E_i, \Gamma_i, s_i, \psi_i) \to (V_j, E_j, \Gamma_j, s_j, \psi_j)$ are* 1*-morphisms of Kuranishi neighbourhoods over (X, f), with associated* 1*-morphisms of d-orbifolds $\boldsymbol{S}_{\Phi_{ij}}, \boldsymbol{S}_{\Phi'_{ij}}$. Then there is a canonical* 1-1 *correspondence between* 2*-morphisms $\Lambda\colon \Phi_{ij} \Rightarrow \Phi'_{ij}$ of Kuranishi neighbourhoods over (X, f), and* 2*-morphisms $\boldsymbol{\lambda}\colon \boldsymbol{S}_{\Phi_{ij}} \Rightarrow \boldsymbol{S}_{\Phi'_{ij}}$ of d-orbifolds.*

These maps from Kuranishi neighbourhoods, 1*- and* 2*-morphisms to d-orbifolds,* 1*- and* 2*-morphisms are compatible with the rest of the* 2*-category structures.*

This implies that the full strict 2-subcategory **SMod** $\subset$ **dOrb** with objects 'standard model' d-orbifolds is equivalent to the weak 2-category **GKN** of global Kuranishi neighbourhoods in Theorem 3.10.

In fact the author deliberately wrote §3.1–§3.2 by translating facts about d-orbifolds into Kuranishi-style, differential-geometric language, which makes Theorem 3.21 almost a tautology. But the facts about 'standard model' d-manifolds and d-orbifolds on which Theorem 3.21 is based are genuine theorems, not just simple consequences of the definition of d-orbifolds.

4. The weak 2-category of Kuranishi spaces

We now define our 2-category of Kuranishi spaces **Kur**. The material of this section, and the proofs of the theorems, have a different character to those of §3. Section 3 was largely about differential geometry: in Kuranishi neighbourhoods $(V_i, E_i, \Gamma_i, s_i, \psi_i)$ it really mattered that V_i is a manifold, and so on, and the proofs of Theorems 3.10, 3.14 and 3.15 use many properties of manifolds.

This section involves a lot of stack theory from algebraic geometry, but very little differential geometry. When we define Kuranishi structures in §4.1, in Kuranishi neighbourhoods $(V_i, E_i, \Gamma_i, s_i, \psi_i)$ we do not care that V_i is a manifold, it only matters that $(V_i, E_i, \Gamma_i, s_i, \psi_i)$ is an object in a 2-category $\mathbf{KN}_S(X)$ for open $S \subseteq \operatorname{Im}\psi_i \subseteq X$, which has a stack property as in Theorem 3.14.

Because of this, the framework below works whenever we have some class of 'charts' on a topological space satisfying the analogues of Theorems 3.10 and 3.14, so for instance we can easily define Kuranishi spaces with boundary and corners, or Kuranishi spaces associated to other categories of manifolds.

All of this section, including proofs of quoted results, comes from [**35, 36**].

4.1. Kuranishi spaces, 1-morphisms, and 2-morphisms. Here is one of the main definitions of the paper:

DEFINITION 4.1. Let X be a Hausdorff, second countable topological space (not necessarily compact), and $n \in \mathbb{Z}$. A *Kuranishi structure $\mathcal{K}$ on X of virtual dimension n* is data $\mathcal{K} = \bigl(I, (V_i, E_i, \Gamma_i, s_i, \psi_i)_{i\in I}, \Phi_{ij,\ i,j\in I}, \Lambda_{ijk,\ i,j,k\in I}\bigr)$, where:

(a) I is an indexing set (not necessarily finite).

(b) $(V_i, E_i, \Gamma_i, s_i, \psi_i)$ is a Kuranishi neighbourhood on X for each $i \in I$, with $\dim V_i - \operatorname{rank} E_i = n$.

(c) $\Phi_{ij} = (P_{ij}, \pi_{ij}, \phi_{ij}, \hat{\phi}_{ij}) : (V_i, E_i, \Gamma_i, s_i, \psi_i) \to (V_j, E_j, \Gamma_j, s_j, \psi_j)$ is a coordinate change for all $i, j \in I$ (as usual, defined over $S = \operatorname{Im}\psi_i \cap \operatorname{Im}\psi_j$).

(d) $\Lambda_{ijk} = [\dot{P}_{ijk}, \lambda_{ijk}, \hat{\lambda}_{ijk}] : \Phi_{jk} \circ \Phi_{ij} \Rightarrow \Phi_{ik}$ is a 2-morphism for all $i, j, k \in I$ (as usual, defined over $S = \operatorname{Im}\psi_i \cap \operatorname{Im}\psi_j \cap \operatorname{Im}\psi_k$).

(e) $\bigcup_{i \in I} \operatorname{Im}\psi_i = X$.

(f) $\Phi_{ii} = \mathrm{id}_{(V_i, E_i, \Gamma_i, s_i, \psi_i)}$ for all $i \in I$.

(g) $\Lambda_{iij} = \boldsymbol{\beta}_{\Phi_{ij}}$ and $\Lambda_{ijj} = \boldsymbol{\gamma}_{\Phi_{ij}}$ for all $i, j \in I$, for $\boldsymbol{\beta}_{\Phi_{ij}}, \boldsymbol{\gamma}_{\Phi_{ij}}$ as in (3.5).

(h) The following diagram of 2-morphisms over $S = \operatorname{Im}\psi_i \cap \operatorname{Im}\psi_j \cap \operatorname{Im}\psi_k \cap \operatorname{Im}\psi_l$ commutes for all $i, j, k, l \in I$, for $\boldsymbol{\alpha}_{\Phi_{kl}, \Phi_{jk}, \Phi_{ij}}$ as in (3.4):

$$\begin{array}{ccccc}
(\Phi_{kl} \circ \Phi_{jk}) \circ \Phi_{ij} & & \xrightarrow{\quad\Lambda_{jkl} * \mathrm{id}_{\Phi_{ij}}\quad} & & \Phi_{jl} \circ \Phi_{ij} \\
\Big\Downarrow {\scriptstyle \boldsymbol{\alpha}_{\Phi_{kl}, \Phi_{jk}, \Phi_{ij}}} & & & & {\scriptstyle \Lambda_{ijl}} \Big\Downarrow \\
\Phi_{kl} \circ (\Phi_{jk} \circ \Phi_{ij}) & \xrightarrow{\mathrm{id}_{\Phi_{kl}} * \Lambda_{ijk}} & \Phi_{kl} \circ \Phi_{ik} & \xrightarrow{\quad\Lambda_{ikl}\quad} & \Phi_{il}.
\end{array}$$

We call $\boldsymbol{X} = (X, \mathcal{K})$ a *Kuranishi space*, of *virtual dimension* $\operatorname{vdim} \boldsymbol{X} = n$.

When we write $x \in \boldsymbol{X}$, we mean that $x \in X$. We can define *orientations* on Kuranishi spaces in a very similar way to Definition 2.5.

EXAMPLE 4.2. Let V be a manifold, $E \to V$ a vector bundle, Γ a finite group with a smooth action on V and a compatible action on E preserving the vector bundle structure, and $s : V \to E$ a Γ-equivariant smooth section. Set $X = s^{-1}(0)/\Gamma$, with the quotient topology induced from the closed subset $s^{-1}(0) \subseteq V$. Then X is Hausdorff and second countable, as V is and Γ is finite.

Define a Kuranishi structure $\mathcal{K} = \big(\{0\}, (V_0, E_0, \Gamma_0, s_0, \psi_0), \Phi_{00}, \Lambda_{000}\big)$ on X with indexing set $I = \{0\}$, one Kuranishi neighbourhood $(V_0, E_0, \Gamma_0, s_0, \psi_0)$ with $V_0 = V$, $E_0 = E$, $\Gamma_0 = \Gamma$, $s_0 = s$ and $\psi_0 = \mathrm{id}_X$, one coordinate change $\Phi_{00} = \mathrm{id}_{(V_0, E_0, \Gamma_0, s_0, \psi_0)}$, and one 2-morphism $\Lambda_{000} = \mathrm{id}_{\Phi_{00}}$. Then $\boldsymbol{X} = (X, \mathcal{K})$ is a Kuranishi space, with $\operatorname{vdim} \boldsymbol{X} = \dim V - \operatorname{rank} E$. We write $\boldsymbol{S}_{V,E,\Gamma,s} = \boldsymbol{X}$.

We will need notation to distinguish Kuranishi neighbourhoods, coordinate changes, and 2-morphisms on different Kuranishi spaces. We will often use the following notation for Kuranishi spaces $\boldsymbol{W}, \boldsymbol{X}, \boldsymbol{Y}, \boldsymbol{Z}$:

$$\begin{aligned}
&\boldsymbol{W} = (W, \mathcal{H}), \quad \mathcal{H} = \big(H, (T_h, C_h, \mathrm{A}_i, q_h, \varphi_h)_{h \in H}, \Sigma_{hh'} = (O_{hh'}, \pi_{hh'}, \sigma_{hh'}, \\
&\qquad \hat{\sigma}_{hh'})_{h,h' \in H}, \mathrm{I}_{hh'h''} = [\dot{O}_{hh'h''}, \iota_{hh'h''}, \hat{\iota}_{hh'h''}]_{h,h',h'' \in H}\big), \qquad (4.1)\\
&\boldsymbol{X} = (X, \mathcal{I}), \quad \mathcal{I} = \big(I, (U_i, D_i, \mathrm{B}_i, r_i, \chi_i)_{i \in I}, \mathrm{T}_{ii'} = (P_{ii'}, \pi_{ii'}, \tau_{ii'}, \\
&\qquad \hat{\tau}_{ii'})_{i,i' \in I}, \mathrm{K}_{ii'i''} = [\dot{P}_{ii'i''}, \kappa_{ii'i''}, \hat{\kappa}_{ii'i''}]_{i,i',i'' \in I}\big), \qquad (4.2)\\
&\boldsymbol{Y} = (Y, \mathcal{J}), \quad \mathcal{J} = \big(J, (V_j, E_j, \Gamma_j, s_j, \psi_j)_{j \in J}, \Upsilon_{jj'} = (Q_{jj'}, \pi_{jj'}, \upsilon_{jj'}, \\
&\qquad \hat{\upsilon}_{jj'})_{j,j' \in J}, \Lambda_{jj'j''} = [\dot{Q}_{jj'j''}, \lambda_{jj'j''}, \hat{\lambda}_{jj'j''}]_{j,j',j'' \in J}\big), \qquad (4.3)\\
&\boldsymbol{Z} = (Z, \mathcal{K}), \quad \mathcal{K} = \big(K, (W_k, F_k, \Delta_k, t_k, \omega_k)_{k \in K}, \Phi_{kk'} = (R_{kk'}, \pi_{kk'}, \phi_{kk'}, \\
&\qquad \hat{\phi}_{kk'})_{k,k' \in K}, \mathrm{M}_{kk'k''} = [\dot{R}_{kk'k''}, \mu_{kk'k''}, \hat{\mu}_{kk'k''}]_{k,k',k'' \in K}\big). \qquad (4.4)
\end{aligned}$$

Next we define 1- and 2-morphisms of Kuranishi spaces. Note a possible confusion: we will be defining 1-morphisms of Kuranishi spaces $\boldsymbol{f}, \boldsymbol{g} : \boldsymbol{X} \to \boldsymbol{Y}$ and 2-morphisms of Kuranishi spaces $\boldsymbol{\eta} : \boldsymbol{f} \Rightarrow \boldsymbol{g}$, but these will be built out of 1-morphisms of Kuranishi neighbourhoods $\boldsymbol{f}_{ij}, \boldsymbol{g}_{ij} : (U_i, D_i, \mathrm{B}_i, r_i, \chi_i) \to (V_j, E_j, \Gamma_j, s_j, \psi_j)$ and

2-morphisms of Kuranishi neighbourhoods $\boldsymbol{\eta}_{ij}\colon \boldsymbol{f}_{ij}\Rightarrow\boldsymbol{g}_{ij}$ in the sense of §3.1, so '1-morphism' and '2-morphism' can mean two things.

DEFINITION 4.3. Let $\boldsymbol{X}=(X,\mathcal{I})$ and $\boldsymbol{Y}=(Y,\mathcal{J})$ be Kuranishi spaces, with notation (4.2)–(4.3). A *1-morphism of Kuranishi spaces* $\boldsymbol{f}\colon\boldsymbol{X}\to\boldsymbol{Y}$ is data

$$\boldsymbol{f}=\bigl(f,\boldsymbol{f}_{ij,\,i\in I,\,j\in J},\ \boldsymbol{F}^{j,\,j\in J}_{ii',\,i,i'\in I},\ \boldsymbol{F}^{jj',\,j,j'\in J}_{i,\,i\in I}\bigr),$$

satisfying the conditions:

(a) $f\colon X\to Y$ is a continuous map.

(b) $\boldsymbol{f}_{ij}=(P_{ij},\pi_{ij},f_{ij},\hat f_{ij})\colon(U_i,D_i,\mathrm{B}_i,r_i,\chi_i)\to(V_j,E_j,\Gamma_j,s_j,\psi_j)$ is a 1-morphism of Kuranishi neighbourhoods over f for all $i\in I$, $j\in J$ (defined over $S=\operatorname{Im}\chi_i\cap f^{-1}(\operatorname{Im}\psi_j)$, as usual).

(c) $\boldsymbol{F}^j_{ii'}=[\dot P^j_{ii'},F^j_{ii'},\hat F^j_{ii'}]\colon\boldsymbol{f}_{i'j}\circ\mathrm{T}_{ii'}\Rightarrow\boldsymbol{f}_{ij}$ is a 2-morphism over f for all $i,i'\in I$ and $j\in J$ (defined over $S=\operatorname{Im}\chi_i\cap\operatorname{Im}\chi_{i'}\cap f^{-1}(\operatorname{Im}\psi_j)$).

(d) $\boldsymbol{F}^{jj'}_i=[\dot P^{jj'}_i,F^{jj'}_i,\hat F^{jj'}_i]\colon\Upsilon_{jj'}\circ\boldsymbol{f}_{ij}\Rightarrow\boldsymbol{f}_{ij'}$ is a 2-morphism over f for all $i\in I$ and $j,j'\in J$ (defined over $S=\operatorname{Im}\chi_i\cap f^{-1}(\operatorname{Im}\psi_j\cap\operatorname{Im}\psi_{j'})$).

(e) $\boldsymbol{F}^j_{ii}=\boldsymbol{\beta}_{\boldsymbol{f}_{ij}}$ and $\boldsymbol{F}^{jj}_i=\boldsymbol{\gamma}_{\boldsymbol{f}_{ij}}$ for all $i\in I$, $j\in J$, for $\boldsymbol{\beta}_{\boldsymbol{f}_{ij}},\boldsymbol{\gamma}_{\boldsymbol{f}_{ij}}$ as in (3.5).

(f) The following commutes for all $i,i',i''\in I$ and $j\in J$:

$$\begin{array}{ccccc} (\boldsymbol{f}_{i''j}\circ\mathrm{T}_{i'i''})\circ\mathrm{T}_{ii'} & & \xRightarrow{\boldsymbol{F}^j_{i'i''}*\mathrm{id}_{\mathrm{T}_{ii'}}} & & \boldsymbol{f}_{i'j}\circ\mathrm{T}_{ii'} \\ \Downarrow{\scriptstyle\boldsymbol{\alpha}_{\boldsymbol{f}_{i''j},\mathrm{T}_{i'i''},\mathrm{T}_{ii'}}} & & & & {\scriptstyle\boldsymbol{F}^j_{ii'}}\Downarrow \\ \boldsymbol{f}_{i''j}\circ(\mathrm{T}_{i'i''}\circ\mathrm{T}_{ii'}) & \xRightarrow{\mathrm{id}_{\boldsymbol{f}_{i''j}}*\mathrm{K}_{ii'i''}} & \boldsymbol{f}_{i''j}\circ\mathrm{T}_{ii''} & \xRightarrow{\boldsymbol{F}^j_{ii''}} & \boldsymbol{f}_{ij}. \end{array}$$

(g) The following commutes for all $i,i'\in I$ and $j,j'\in J$:

$$\begin{array}{ccccc} (\Upsilon_{jj'}\circ\boldsymbol{f}_{i'j})\circ\mathrm{T}_{ii'} & & \xRightarrow{\boldsymbol{F}^{jj'}_{i'}*\mathrm{id}_{\mathrm{T}_{ii'}}} & & \boldsymbol{f}_{i'j'}\circ\mathrm{T}_{ii'} \\ \Downarrow{\scriptstyle\boldsymbol{\alpha}_{\Upsilon_{jj'},\boldsymbol{f}_{i'j},\mathrm{T}_{ii'}}} & & & & {\scriptstyle\boldsymbol{F}^{j'}_{ii'}}\Downarrow \\ \Upsilon_{jj'}\circ(\boldsymbol{f}_{i'j}\circ\mathrm{T}_{ii'}) & \xRightarrow{\mathrm{id}_{\Upsilon_{jj'}}*\boldsymbol{F}^j_{ii'}} & \Upsilon_{jj'}\circ\boldsymbol{f}_{ij} & \xRightarrow{\boldsymbol{F}^{jj'}_i} & \boldsymbol{f}_{ij'}. \end{array}$$

(h) The following commutes for all $i\in I$ and $j,j',j''\in J$:

$$\begin{array}{ccccc} (\Upsilon_{j'j''}\circ\Upsilon_{jj'})\circ\boldsymbol{f}_{ij} & & \xRightarrow{\Lambda_{jj'j''}*\mathrm{id}_{\boldsymbol{f}_{ij}}} & & \Upsilon_{jj''}\circ\boldsymbol{f}_{ij} \\ \Downarrow{\scriptstyle\boldsymbol{\alpha}_{\Upsilon_{j'j''},\Upsilon_{jj'},\boldsymbol{f}_{ij}}} & & & & {\scriptstyle\boldsymbol{F}^{jj''}_i}\Downarrow \\ \Upsilon_{j'j''}\circ(\Upsilon_{jj'}\circ\boldsymbol{f}_{ij}) & \xRightarrow{\mathrm{id}_{\Upsilon_{j'j''}}*\boldsymbol{F}^{jj'}_i} & \Upsilon_{j'j''}\circ\boldsymbol{f}_{ij'} & \xRightarrow{\boldsymbol{F}^{j'j''}_i} & \boldsymbol{f}_{ij''}. \end{array}$$

If $x\in\boldsymbol{X}$ (i.e. $x\in X$), we will write $\boldsymbol{f}(x)=f(x)\in\boldsymbol{Y}$.

When $\boldsymbol{Y}=\boldsymbol{X}$, define the *identity 1-morphism* $\mathbf{id}_{\boldsymbol{X}}\colon\boldsymbol{X}\to\boldsymbol{X}$ by

$$\mathbf{id}_{\boldsymbol{X}}=\bigl(\mathrm{id}_X,\mathrm{T}_{ij,\,i,j\in I},\ \mathrm{K}^{\,j\in I}_{ii'j,\,i,i'\in I},\ \mathrm{K}^{\,j,j'\in I}_{ijj',\,i\in I}\bigr),$$

Then Definition 4.1(h) implies that (f)–(h) above hold.

DEFINITION 4.4. Let $\boldsymbol{X}=(X,\mathcal{I})$ and $\boldsymbol{Y}=(Y,\mathcal{J})$ be Kuranishi spaces, with notation as in (4.2)–(4.3), and $\boldsymbol{f},\boldsymbol{g}\colon\boldsymbol{X}\to\boldsymbol{Y}$ be 1-morphisms. Suppose the continuous maps $f,g\colon X\to Y$ in $\boldsymbol{f},\boldsymbol{g}$ satisfy $f=g$. A *2-morphism of Kuranishi spaces* $\boldsymbol{\eta}\colon\boldsymbol{f}\Rightarrow\boldsymbol{g}$ is data $\boldsymbol{\eta}=\bigl(\boldsymbol{\eta}_{ij,\,i\in I,\,j\in J}\bigr)$, where $\boldsymbol{\eta}_{ij}=[\dot P_{ij},\eta_{ij},\hat\eta_{ij}]\colon\boldsymbol{f}_{ij}\Rightarrow\boldsymbol{g}_{ij}$ is a 2-morphism of Kuranishi neighbourhoods over $f=g$ (defined over $S=\operatorname{Im}\chi_i\cap f^{-1}(\operatorname{Im}\psi_j)$, as usual), satisfying the conditions:

(a) $\boldsymbol{G}^j_{ii'}\odot(\boldsymbol{\eta}_{i'j}*\mathrm{id}_{\mathrm{T}_{ii'}})=\boldsymbol{\eta}_{ij}\odot\boldsymbol{F}^j_{ii'}\colon\boldsymbol{f}_{i'j}\circ\mathrm{T}_{ii'}\Rightarrow\boldsymbol{g}_{ij}$ for all $i,i'\in I$, $j\in J$.

(b) $\boldsymbol{G}_i^{jj'} \odot (\mathrm{id}_{\Upsilon_{jj'}} * \boldsymbol{\eta}_{ij}) = \boldsymbol{\eta}_{ij'} \odot \boldsymbol{F}_i^{jj'} : \Upsilon_{jj'} \circ \boldsymbol{f}_{ij} \Rightarrow \boldsymbol{g}_{ij'}$ for all $i \in I$, $j, j' \in J$.

Note that by definition, 2-morphisms $\boldsymbol{\eta} \colon \boldsymbol{f} \Rightarrow \boldsymbol{g}$ only exist if $f = g$.

If $\boldsymbol{f} = \boldsymbol{g}$, the *identity 2-morphism* is $\mathbf{id}_{\boldsymbol{f}} = \big(\mathrm{id}_{\boldsymbol{f}_{ij}},\, {}_{i\in I,\, j\in J}\big) \colon \boldsymbol{f} \Rightarrow \boldsymbol{f}$.

4.2. Making Kuranishi spaces into a 2-category. We will make Kuranishi spaces into a weak 2-category. We have already defined the objects, 1-morphisms, 2-morphisms, and identity 1- and 2-morphisms in §4.1. As for §3.2, it remains to explain:

- Composition of 1-morphisms;
- Horizontal composition of 2-morphisms;
- Vertical composition of 2-morphisms; and
- Coherence 2-isomorphisms $\boldsymbol{\alpha}_{\boldsymbol{g},\boldsymbol{f},\boldsymbol{e}} \colon (\boldsymbol{g}\circ\boldsymbol{f})\circ\boldsymbol{e} \Rightarrow \boldsymbol{g}\circ(\boldsymbol{f}\circ\boldsymbol{e})$, $\boldsymbol{\beta}_{\boldsymbol{f}} \colon \boldsymbol{f}\circ\mathbf{id}_{\boldsymbol{X}} \Rightarrow \boldsymbol{f}$, and $\boldsymbol{\gamma}_{\boldsymbol{f}} \colon \mathbf{id}_{\boldsymbol{Y}} \circ \boldsymbol{f} \Rightarrow \boldsymbol{f}$, as in (A.5) and (A.7).

We will define composition of 1-morphisms in Proposition 4.5 and Definition 4.6. The definition involves some interesting issues, so we discuss these first. Suppose $\boldsymbol{f} \colon \boldsymbol{X} \to \boldsymbol{Y}$, $\boldsymbol{g} \colon \boldsymbol{Y} \to \boldsymbol{Z}$ are 1-morphisms of Kuranishi spaces. We want to define the composition $\boldsymbol{g} \circ \boldsymbol{f} \colon \boldsymbol{X} \to \boldsymbol{Z}$. Use notation (4.2)–(4.4) for $\boldsymbol{X}, \boldsymbol{Y}, \boldsymbol{Z}$. Then Kuranishi neighbourhoods on X are parametrized by $i \in I$, where $\{\mathrm{Im}\,\chi_i \colon i \in I\}$ is an open cover of X, and similarly $\{\mathrm{Im}\,\psi_j \colon j \in J\}$ is an open cover of Y, and $\{\mathrm{Im}\,\omega_k \colon k \in K\}$ is an open cover of Z.

To make $\boldsymbol{g} \circ \boldsymbol{f}$, for each $i \in I$ and $k \in K$ we need a 1-morphism of Kuranishi neighbourhoods $(\boldsymbol{g} \circ \boldsymbol{f})_{ik}$ defined on $\mathrm{Im}\,\chi_i \cap (g\circ f)^{-1}(\mathrm{Im}\,\omega_k) \subseteq X$. But we actually have 1-morphisms $\boldsymbol{g}_{jk} \circ \boldsymbol{f}_{ij}$ defined on $\mathrm{Im}\,\chi_i \cap f^{-1}(\psi_j) \cap (g\circ f)^{-1}(\mathrm{Im}\,\omega_k) \subseteq X$ for all $j \in J$. We will construct $(\boldsymbol{g} \circ \boldsymbol{f})_{ik}$ by gluing together the $\boldsymbol{g}_{jk} \circ \boldsymbol{f}_{ij}$ for all $j \in J$ using the stack property of Kuranishi neighbourhoods, Theorem 3.14.

The stack property is crucial, as without it we could not define composition of 1-morphisms of Kuranishi spaces, and could not make Kuranishi spaces into a 2-category. For FOOO Kuranishi spaces in §2.1, or MW weak Kuranishi atlases in §2.3, one could write down notions of morphisms similar to Definition 4.3. However, as FOOO/MW coordinate changes do not form sheaves or stacks, there seems no obvious way to define compositions of such morphisms.

In the next proposition, taken from [**35**, §4.3], [**36**, §6.2], part (a) constructs candidates $\boldsymbol{h}$ for $\boldsymbol{g} \circ \boldsymbol{f}$, part (b) shows such $\boldsymbol{h}$ are unique up to canonical 2-isomorphism, and part (c) that $\boldsymbol{f}, \boldsymbol{g}$ are allowed candidates for $\mathbf{id}_{\boldsymbol{Y}} \circ \boldsymbol{f}, \boldsymbol{g} \circ \mathbf{id}_{\boldsymbol{Y}}$.

Proposition 4.5. **(a)** *Let $\boldsymbol{X} = (X, \mathcal{I}), \boldsymbol{Y} = (Y, \mathcal{J}), \boldsymbol{Z} = (Z, \mathcal{K})$ be Kuranishi spaces with notation* (4.2)–(4.4), *and $\boldsymbol{f} \colon \boldsymbol{X} \to \boldsymbol{Y}$, $\boldsymbol{g} \colon \boldsymbol{Y} \to \boldsymbol{Z}$ be 1-morphisms, with $\boldsymbol{f} = \big(f, \boldsymbol{f}_{ij}, \boldsymbol{F}_{ii'}^{j}, \boldsymbol{F}_i^{jj'}\big)$, $\boldsymbol{g} = \big(g, \boldsymbol{g}_{jk}, \boldsymbol{G}_{jj'}^{k}, \boldsymbol{G}_j^{kk'}\big)$. Then there exists a 1-morphism $\boldsymbol{h} \colon \boldsymbol{X} \to \boldsymbol{Z}$ with $\boldsymbol{h} = \big(h, \boldsymbol{h}_{ik}, \boldsymbol{H}_{ii'}^{k}, \boldsymbol{H}_i^{kk'}\big)$, such that $h = g \circ f \colon X \to Z$, and for all $i \in I$, $j \in J$, $k \in K$ we have 2-morphisms over h*

$$\Theta_{ijk} \colon \boldsymbol{g}_{jk} \circ \boldsymbol{f}_{ij} \Longrightarrow \boldsymbol{h}_{ik}, \tag{4.5}$$

where as usual (4.5) *holds over* $S = \operatorname{Im}\chi_i \cap f^{-1}(\operatorname{Im}\psi_j) \cap h^{-1}(\operatorname{Im}\omega_k)$, *and for all* $i, i' \in I$, $j, j' \in J$, $k, k' \in K$ *the following commute:*

$$\begin{array}{ccccc}
(\boldsymbol{g}_{jk} \circ \boldsymbol{f}_{i'j}) \circ \mathrm{T}_{ii'} & & \xRightarrow{\Theta_{i'jk} * \mathrm{id}_{\mathrm{T}_{ii'}}} & & \boldsymbol{h}_{i'k} \circ \mathrm{T}_{ii'} \\
\Downarrow \boldsymbol{\alpha}_{\boldsymbol{g}_{jk}, \boldsymbol{f}_{i'j}, \mathrm{T}_{ii'}} & & & & \boldsymbol{H}^k_{ii'} \Downarrow \\
\boldsymbol{g}_{jk} \circ (\boldsymbol{f}_{i'j} \circ \mathrm{T}_{ii'}) & \xRightarrow{\mathrm{id}_{\boldsymbol{g}_{jk}} * \boldsymbol{F}^j_{ii'}} & \boldsymbol{g}_{jk} \circ \boldsymbol{f}_{ij} & \xRightarrow{\Theta_{ijk}} & \boldsymbol{h}_{ik},
\end{array}$$

$$\begin{array}{ccccc}
(\boldsymbol{g}_{j'k} \circ \Upsilon_{jj'}) \circ \boldsymbol{f}_{ij} & & \xRightarrow{\boldsymbol{G}^k_{jj'} * \mathrm{id}_{\boldsymbol{f}_{ij}}} & & \boldsymbol{g}_{jk} \circ \boldsymbol{f}_{ij} \\
\Downarrow \boldsymbol{\alpha}_{\boldsymbol{g}_{j'k}, \Upsilon_{jj'}, \boldsymbol{f}_{ij}} & & & & \Theta_{ijk} \Downarrow \\
\boldsymbol{g}_{j'k} \circ (\Upsilon_{jj'} \circ \boldsymbol{f}_{ij}) & \xRightarrow{\mathrm{id}_{\boldsymbol{g}_{j'k}} * \boldsymbol{F}^{jj'}_i} & \boldsymbol{g}_{j'k} \circ \boldsymbol{f}_{ij'} & \xRightarrow{\Theta_{ij'k}} & \boldsymbol{h}_{ik},
\end{array}$$

$$\begin{array}{ccccc}
(\Phi_{kk'} \circ \boldsymbol{g}_{jk}) \circ \boldsymbol{f}_{ij} & & \xRightarrow{\boldsymbol{G}^{kk'}_j * \mathrm{id}_{\boldsymbol{f}_{ij}}} & & \boldsymbol{g}_{jk'} \circ \boldsymbol{f}_{ij} \\
\Downarrow \boldsymbol{\alpha}_{\Phi_{kk'}, \boldsymbol{g}_{jk}, \boldsymbol{f}_{ij}} & & & & \Theta_{ijk'} \Downarrow \\
\Phi_{kk'} \circ (\boldsymbol{g}_{jk} \circ \boldsymbol{f}_{ij}) & \xRightarrow{\mathrm{id}_{\Phi_{kk'}} * \Theta_{ijk}} & \Phi_{kk'} \circ \boldsymbol{h}_{ik} & \xRightarrow{\boldsymbol{H}^{kk'}_i} & \boldsymbol{h}_{ik'}.
\end{array}$$

(b) *If* $\tilde{\boldsymbol{h}} = \bigl(h, \tilde{\boldsymbol{h}}_{ik}, \tilde{\boldsymbol{H}}^k_{ii'}, \tilde{\boldsymbol{H}}^{kk'}_i\bigr), \tilde{\Theta}_{ijk}$ *are alternative choices for* $\boldsymbol{h}, \Theta_{ijk}$ *in* **(a)**, *then there is a unique* 2*-morphism of Kuranishi spaces* $\boldsymbol{\eta} = (\boldsymbol{\eta}_{ik}) \colon \boldsymbol{h} \Rightarrow \tilde{\boldsymbol{h}}$ *satisfying* $\boldsymbol{\eta}_{ik} \odot \Theta_{ijk} = \tilde{\Theta}_{ijk} \colon \boldsymbol{g}_{jk} \circ \boldsymbol{f}_{ij} \Rightarrow \tilde{\boldsymbol{h}}_{ik}$ *for all* $i \in I$, $j \in J$, $k \in K$.

(c) *If* $\boldsymbol{X} = \boldsymbol{Y}$ *and* $\boldsymbol{f} = \mathbf{id}_{\boldsymbol{Y}}$ *in* **(a)**, *so that* $I = J$, *then a possible choice for* $\boldsymbol{h}, \Theta_{ijk}$ *in* **(a)** *is* $\boldsymbol{h} = \boldsymbol{g}$ *and* $\Theta_{ijk} = \boldsymbol{G}^k_{ij}$.

Similarly, if $\boldsymbol{Z} = \boldsymbol{Y}$ *and* $\boldsymbol{g} = \mathbf{id}_{\boldsymbol{Y}}$ *in* **(a)**, *so that* $K = J$, *then a possible choice for* $\boldsymbol{h}, \Theta_{ijk}$ *in* **(a)** *is* $\boldsymbol{h} = \boldsymbol{f}$ *and* $\Theta_{ijk} = \boldsymbol{F}^{jk}_i$.

Proposition 4.5(a) gives possible values $\boldsymbol{h}$ for the composition $\boldsymbol{g} \circ \boldsymbol{f} \colon \boldsymbol{X} \to \boldsymbol{Z}$. Since there is no distinguished choice, we choose $\boldsymbol{g} \circ \boldsymbol{f}$ arbitrarily.

Definition 4.6. For all pairs of 1-morphisms of Kuranishi spaces $\boldsymbol{f} \colon \boldsymbol{X} \to \boldsymbol{Y}$ and $\boldsymbol{g} \colon \boldsymbol{Y} \to \boldsymbol{Z}$, use the Axiom of Global Choice (a strong form of the Axiom of Choice working for classes as well as sets, see Shulman [**49**, §7] or Herrlick and Strecker [**22**, §1.2]) to choose possible values of $\boldsymbol{h} \colon \boldsymbol{X} \to \boldsymbol{Z}$ and Θ_{ijk} in Proposition 4.5(a), and write $\boldsymbol{g} \circ \boldsymbol{f} = \boldsymbol{h}$, and for $i \in I$, $j \in J$, $k \in K$

$$\Theta^{\boldsymbol{g}, \boldsymbol{f}}_{ijk} = \Theta_{ijk} \colon \boldsymbol{g}_{jk} \circ \boldsymbol{f}_{ij} \Longrightarrow (\boldsymbol{g} \circ \boldsymbol{f})_{ik}.$$

We call $\boldsymbol{g} \circ \boldsymbol{f}$ the *composition of* 1*-morphisms of Kuranishi spaces*.

For general $\boldsymbol{f}, \boldsymbol{g}$ we make these choices arbitrarily. However, if $\boldsymbol{X} = \boldsymbol{Y}$ and $\boldsymbol{f} = \mathbf{id}_{\boldsymbol{Y}}$ then we choose $\boldsymbol{g} \circ \mathbf{id}_{\boldsymbol{Y}} = \boldsymbol{g}$ and $\Theta^{\boldsymbol{g}, \mathbf{id}_{\boldsymbol{Y}}}_{jj'k} = \boldsymbol{G}^k_{jj'}$, and if $\boldsymbol{Z} = \boldsymbol{Y}$ and $\boldsymbol{g} = \mathbf{id}_{\boldsymbol{Y}}$ then we choose $\mathbf{id}_{\boldsymbol{Y}} \circ \boldsymbol{f} = \boldsymbol{f}$ and $\Theta^{\mathbf{id}_{\boldsymbol{Y}}, \boldsymbol{f}}_{ijj'} = \boldsymbol{F}^{jj'}_i$. This is allowed by Proposition 4.5(c).

The definition of a weak 2-category in §A.1 includes 2-isomorphisms $\boldsymbol{\beta}_{\boldsymbol{f}} \colon \boldsymbol{f} \circ \mathbf{id}_{\boldsymbol{X}} \Rightarrow \boldsymbol{f}$ and $\boldsymbol{\gamma}_{\boldsymbol{f}} \colon \mathbf{id}_{\boldsymbol{Y}} \circ \boldsymbol{f} \Rightarrow \boldsymbol{f}$ in (A.7), since one does not require $\boldsymbol{f} \circ \mathbf{id}_{\boldsymbol{X}} = \boldsymbol{f}$ and $\mathbf{id}_{\boldsymbol{Y}} \circ \boldsymbol{f} = \boldsymbol{f}$ in a general weak 2-category. We define

$$\boldsymbol{\beta}_{\boldsymbol{f}} = \mathbf{id}_{\boldsymbol{f}} \colon \boldsymbol{f} \circ \mathbf{id}_{\boldsymbol{X}} \Longrightarrow \boldsymbol{f}, \quad \boldsymbol{\gamma}_{\boldsymbol{f}} = \mathbf{id}_{\boldsymbol{f}} \colon \mathbf{id}_{\boldsymbol{Y}} \circ \boldsymbol{f} \Longrightarrow \boldsymbol{f}.$$

Since composition of 1-morphisms $\boldsymbol{g} \circ \boldsymbol{f}$ is natural only up to canonical 2-isomorphism, as in Proposition 4.5(b), composition is associative only up to canonical 2-isomorphism. The next proposition comes from [**35**, §4.3], [**36**, §6.2].

PROPOSITION 4.7. *Let* $e\colon W \to X$, $f\colon X \to Y$, $g\colon Y \to Z$ *be* 1*-morphisms of Kuranishi spaces, and define composition of* 1*-morphisms as in Definition* 4.6. *Then using notation* (4.1)–(4.4), *there is a unique* 2*-morphism*

$$\boldsymbol{\alpha}_{g,f,e}\colon (\boldsymbol{g}\circ\boldsymbol{f})\circ\boldsymbol{e} \Longrightarrow \boldsymbol{g}\circ(\boldsymbol{f}\circ\boldsymbol{e})$$

with the property that for all $h \in H$, $i \in I$, $j \in J$ *and* $k \in K$ *we have*

$$(\boldsymbol{\alpha}_{g,f,e})_{hk} \odot \Theta^{g\circ f,e}_{hik} \odot (\Theta^{g,f}_{ijk} * \mathrm{id}_{e_{hi}}) = \Theta^{g,f\circ e}_{hjk} \odot (\mathrm{id}_{g_{jk}} * \Theta^{f,e}_{hij}) \odot \boldsymbol{\alpha}_{g_{jk},f_{ij},e_{hi}}.$$

We define vertical and horizontal composition of 2-morphisms:

DEFINITION 4.8. Let $\boldsymbol{f},\boldsymbol{g},\boldsymbol{h}\colon \boldsymbol{X} \to \boldsymbol{Y}$ be 1-morphisms of Kuranishi spaces, using notation (4.2)–(4.3), and $\boldsymbol{\eta} = (\boldsymbol{\eta}_{ij})\colon \boldsymbol{f} \Rightarrow \boldsymbol{g}$, $\boldsymbol{\zeta} = (\boldsymbol{\zeta}_{ij})\colon \boldsymbol{g} \Rightarrow \boldsymbol{h}$ be 2-morphisms. Define a 2-morphism of Kuranishi spaces $\boldsymbol{\zeta}\odot\boldsymbol{\eta}\colon \boldsymbol{f} \Rightarrow \boldsymbol{h}$ called the *vertical composition of* 2*-morphisms* by

$$\boldsymbol{\zeta}\odot\boldsymbol{\eta} = \big(\boldsymbol{\zeta}_{ij}\odot\boldsymbol{\eta}_{ij},\ i \in I,\ j \in J\big).$$

Next let $\boldsymbol{e},\boldsymbol{f}\colon \boldsymbol{X} \to \boldsymbol{Y}$ and $\boldsymbol{g},\boldsymbol{h}\colon \boldsymbol{Y} \to \boldsymbol{Z}$ be 1-morphisms of Kuranishi spaces, using notation (4.2)–(4.4), and $\boldsymbol{\eta} = (\boldsymbol{\eta}_{ij})\colon \boldsymbol{e} \Rightarrow \boldsymbol{f}$, $\boldsymbol{\zeta} = (\boldsymbol{\zeta}_{jk})\colon \boldsymbol{g} \Rightarrow \boldsymbol{h}$ be 2-morphisms. As in [**35**, §4.3], [**36**, §6.2], using the stack property Theorem 3.14(b) we can show that there is a unique 2-morphism of Kuranishi spaces $\boldsymbol{\zeta}*\boldsymbol{\eta}\colon \boldsymbol{g}\circ\boldsymbol{e} \Rightarrow \boldsymbol{h}\circ\boldsymbol{f}$ called the *horizontal composition of* 2*-morphisms*, such that for all $i \in I$, $j \in J$ and $k \in K$ we have

$$(\boldsymbol{\zeta}*\boldsymbol{\eta})_{ik} \odot \Theta^{g,e}_{ijk} = \Theta^{h,f}_{ijk} \odot (\boldsymbol{\zeta}_{jk} * \boldsymbol{\eta}_{ij}).$$

We have now defined all the structures of a weak 2-category of Kuranishi spaces **Kur**. As in [**35**, §4.3], [**36**, §6.2], the 2-category axioms are satisfied.

THEOREM 4.9. *The definitions and propositions above define the* ***weak*** **2*-category of Kuranishi spaces*** **Kur**.

4.3. Manifolds, orbifolds, and m-Kuranishi spaces. By imposing conditions on the Kuranishi neighbourhoods $(V_i, E_i, \Gamma_i, s_i, \psi_i)$ on objects $\boldsymbol{X}$ in **Kur**, we can define interesting 2-subcategories of **Kur**:

- There is a full and faithful (2-)functor $F^{\mathbf{Kur}}_{\mathbf{Man}}\colon \mathbf{Man} \to \mathbf{Kur}$ which identifies **Man** with the full 2-subcategory of objects $\boldsymbol{X} = (X, \mathcal{K})$ in **Kur** in which $\mathcal{K}$ has indexing set $I = \{0\}$, and one Kuranishi neighbourhood $(V_0, E_0, \Gamma_0, s_0, \psi_0)$ with E_0 the zero vector bundle and $\Gamma_0 = \{1\}$.
- The full 2-subcategory $\mathbf{Orb}_{\mathrm{Kur}}$ of $\boldsymbol{X} = (X, \mathcal{K})$ in **Kur** for which E_i is the zero vector bundle for all Kuranishi neighbourhoods $(V_i, E_i, \Gamma_i, s_i, \psi_i)$ in $\mathcal{K}$ is equivalent as a 2-category to the 2-category of orbifolds **Orb**.
- Write $\mathbf{mKur} \subset \mathbf{Kur}$ for the full 2-subcategory of $\boldsymbol{X} \in \mathbf{Kur}$ for which $\Gamma_i = \{1\}$ for all Kuranishi neighbourhoods $(V_i, E_i, \Gamma_i, s_i, \psi_i)$ in $\mathcal{K}$. We call such $\boldsymbol{X}$ *m-Kuranishi spaces*, where 'm-' is short for 'manifold'. They are a kind of derived manifold, in the sense of [**32**–**34**], just as Kuranishi spaces are a kind of derived orbifold.
- The full 2-subcategory of $\boldsymbol{X} = (X, \mathcal{K})$ in **Kur** with only one Kuranishi neighbourhood $(V_i, E_i, \Gamma_i, s_i, \psi_i)$ in $\mathcal{K}$, so that $|I| = 1$, is equivalent to the 2-category of global Kuranishi neighbourhoods **GKN** from Theorem 3.10.

We discuss the first three in more detail. Weak 2-functors are defined in §A.2.

Definition 4.10. We will define a weak 2-functor $F_{\mathbf{Man}}^{\mathbf{Kur}}\colon \mathbf{Man} \to \mathbf{Kur}$. If X is a manifold, define a Kuranishi space $F_{\mathbf{Man}}^{\mathbf{Kur}}(X) = \boldsymbol{X} = (X, \mathcal{K})$ with topological space X and Kuranishi structure $\mathcal{K} = \big(\{0\}, (V_0, E_0, \Gamma_0, s_0, \psi_0), \Phi_{00}, \Lambda_{000}\big)$, with indexing set $I = \{0\}$, one Kuranishi neighbourhood $(V_0, E_0, \Gamma_0, s_0, \psi_0)$ with $V_0 = X$, $E_0 \to V_0$ the zero vector bundle, $\Gamma_0 = \{1\}$, $s_0 = 0$, and $\psi_0 = \mathrm{id}_X$, one coordinate change $\Phi_{00} = \mathrm{id}_{(V_0,E_0,\Gamma_0,s_0,\psi_0)}$, and one 2-morphism $\Lambda_{000} = \mathrm{id}_{\Phi_{00}}$.

On (1-)morphisms, if $f\colon X \to Y$ is a smooth map of manifolds and $\boldsymbol{X} = F_{\mathbf{Man}}^{\mathbf{Kur}}(X)$, $\boldsymbol{Y} = F_{\mathbf{Man}}^{\mathbf{Kur}}(Y)$, define a 1-morphism $F_{\mathbf{Man}}^{\mathbf{Kur}}(f) = \boldsymbol{f}\colon \boldsymbol{X} \to \boldsymbol{Y}$ by $\boldsymbol{f} = (f, \boldsymbol{f}_{00}, \boldsymbol{F}_{00}^{0}, \boldsymbol{F}_{0}^{00})$, where $\boldsymbol{f}_{00} = (P_{00}, \pi_{00}, f_{00}, \hat{f}_{00})$ with $P_{00} = X$, $\pi_{00} = \mathrm{id}_X$, $f_{00} = f$, and $\hat{f}_{00}$ is the zero map on zero vector bundles, and $\boldsymbol{F}_{00}^{0} = \boldsymbol{F}_{0}^{00} = \mathrm{id}_{\boldsymbol{f}_{00}}$.

On 2-morphisms, regarding **Man** as a 2-category, the only 2-morphisms are identity morphisms $\mathrm{id}_f\colon f \Rightarrow f$ for (1-)morphisms $f\colon X \to Y$ in **Man**. We define $F_{\mathbf{Man}}^{\mathbf{Kur}}(\mathrm{id}_f) = \mathbf{id}_{F_{\mathbf{Man}}^{\mathbf{Kur}}(f)}$.

If $f\colon X \to Y$, $g\colon Y \to Z$ are (1-)morphisms in **Man**, there is a unique 2-morphism in **Kur**

$$(F_{\mathbf{Man}}^{\mathbf{Kur}})_{g,f}\colon F_{\mathbf{Man}}^{\mathbf{Kur}}(g) \circ F_{\mathbf{Man}}^{\mathbf{Kur}}(f) \Longrightarrow F_{\mathbf{Man}}^{\mathbf{Kur}}(g \circ f).$$

For any object X in **Man**, define

$$(F_{\mathbf{Man}}^{\mathbf{Kur}})_X := \mathbf{id}_{\mathbf{id}_{\boldsymbol{X}}}\colon F_{\mathbf{Man}}^{\mathbf{Kur}}(\mathrm{id}_X) = \mathbf{id}_{\boldsymbol{X}} \Longrightarrow \mathbf{id}_{\boldsymbol{X}} = \mathbf{id}_{F_{\mathbf{Man}}^{\mathbf{Kur}}(X)}.$$

As in [**35**, §4.3], this defines a full and faithful weak 2-functor $F_{\mathbf{Man}}^{\mathbf{Kur}}\colon \mathbf{Man} \to \mathbf{Kur}$. Thus we can regard manifolds X as special examples of Kuranishi spaces.

We say that a Kuranishi space $\boldsymbol{X}$ *is a manifold* if $\boldsymbol{X} \simeq F_{\mathbf{Man}}^{\mathbf{Kur}}(X')$ in **Kur**, for some manifold X'.

The next remark reviews different definitions of orbifolds in the literature.

Remark 4.11. Orbifolds are generalizations of manifolds locally modelled on $\mathbb{R}^n/G$, for G a finite group acting linearly on $\mathbb{R}^n$. They were introduced by Satake [**48**], who called them 'V-manifolds'. Later they were studied by Thurston [**51**, Ch. 13] who gave them the name 'orbifold'.

As for Kuranishi spaces, defining orbifolds $\mathfrak{X}, \mathfrak{Y}$ and smooth maps $\mathfrak{f}\colon \mathfrak{X} \to \mathfrak{Y}$ was initially problematic, and early definitions of ordinary categories of orbifolds [**48**, **51**] had some bad differential-geometric behaviour (e.g. for some definitions, one cannot define pullbacks $\mathfrak{f}^*(\mathfrak{E})$ of orbifold vector bundles $\mathfrak{E} \to \mathfrak{Y}$). It is now generally agreed that it is best to define orbifolds to be a 2-category. See Lerman [**38**] for a good overview of ways to define orbifolds.

There are three main definitions of ordinary categories of orbifolds:

(a) Satake [**48**] and Thurston [**51**] defined an orbifold $\mathfrak{X}$ to be a Hausdorff topological space X with an atlas $\big\{(V_i, \Gamma_i, \psi_i)\colon i \in I\big\}$ of orbifold charts (V_i, Γ_i, ψ_i), where V_i is a manifold, Γ_i a finite group acting smoothly (and locally effectively) on V_i, and $\psi_i\colon V_i/\Gamma_i \to X$ a homeomorphism with an open set in X, and pairs of charts $(V_i, \Gamma_i, \phi_i), (V_j, \Gamma_j, \phi_j)$ satisfy compatibility conditions on their overlaps in X. Smooth maps $\mathfrak{f}\colon \mathfrak{X} \to \mathfrak{Y}$ between orbifolds are continuous maps $f\colon X \to Y$ of the underlying spaces, which lift locally to smooth maps on the charts, giving a category $\mathbf{Orb}_{\mathrm{ST}}$.

(b) Chen and Ruan [**7**, §4] defined orbifolds $\mathfrak{X}$ in a similar way to [**48**, **51**], but using germs of orbifold charts (V_p, Γ_p, ψ_p) for $p \in X$. Their morphisms $\mathfrak{f}\colon \mathfrak{X} \to \mathfrak{Y}$ are called *good maps*, giving a category $\mathbf{Orb}_{\mathrm{CR}}$.

(c) Moerdijk and Pronk [**45, 46**] defined a category of orbifolds $\mathbf{Orb}_{\rm MP}$ as *proper étale Lie groupoids* in **Man**. Their definition of smooth map $\mathfrak{f}\colon \mathfrak{X} \to \mathfrak{Y}$, called *strong maps* [**46**, §5], is complicated: it is an equivalence class of diagrams $\mathfrak{X} \xleftarrow{\phi} \mathfrak{X}' \xrightarrow{\psi} \mathfrak{Y}$, where $\mathfrak{X}'$ is a third orbifold, and ϕ, ψ are morphisms of groupoids with ϕ an equivalence (loosely, a diffeomorphism).

A book on orbifolds in the sense of [**7, 45, 46**] is Adem, Leida and Ruan [**1**].

There are four main definitions of 2-categories of orbifolds:

(i) Pronk [**47**] defines a strict 2-category **LieGpd** of Lie groupoids in **Man** as in (c), with the obvious 1-morphisms of groupoids, and localizes by a class of weak equivalences $\mathcal{W}$ to get a weak 2-category $\mathbf{Orb}_{\rm Pr} = \mathbf{LieGpd}[\mathcal{W}^{-1}]$.

(ii) Lerman [**38**, §3.3] defines a weak 2-category $\mathbf{Orb}_{\rm Le}$ of Lie groupoids in **Man** as in (c), with a non-obvious notion of 1-morphism called 'Hilsum–Skandalis morphisms' involving 'bibundles', and does not need to localize.
Henriques and Metzler [**21**] also use Hilsum–Skandalis morphisms.

(iii) Behrend and Xu [**3**, §2], Lerman [**38**, §4] and Metzler [**44**, §3.5] define a strict 2-category of orbifolds $\mathbf{Orb}_{\rm ManSta}$ as a class of Deligne–Mumford stacks on the site $(\mathbf{Man}, \mathcal{J}_{\mathbf{Man}})$ of manifolds with Grothendieck topology $\mathcal{J}_{\mathbf{Man}}$ coming from open covers.

(iv) The author [**31**] defines a strict 2-category of orbifolds $\mathbf{Orb}_{C^\infty{\rm Sta}}$ as a class of Deligne–Mumford stacks on the site $(\mathbf{C^\infty Sch}, \mathcal{J}_{\mathbf{C^\infty Sch}})$ of C^∞-schemes.

As in Behrend and Xu [**3**, §2.6], Lerman [**38**], Pronk [**47**], and the author [**31**, Th. 9.26], approaches (i)–(iv) give equivalent weak 2-categories $\mathbf{Orb}_{\rm Pr}, \mathbf{Orb}_{\rm Le}, \mathbf{Orb}_{\rm ManSta}, \mathbf{Orb}_{C^\infty{\rm Sta}}$. As they are equivalent, the differences between them are not of mathematical importance, but more a matter of convenience or taste. Properties of localization also imply that $\mathbf{Orb}_{\rm MP} \simeq \mathrm{Ho}(\mathbf{Orb}_{\rm Pr})$. Thus, all of (c) and (i)–(iv) are equivalent at the level of homotopy categories.

We now give a fifth definition of a weak 2-category of orbifolds.

DEFINITION 4.12. Define the weak 2-category $\mathbf{Orb}_{\rm Kur} \subset \mathbf{Kur}$ of *Kuranishi orbifolds*, or just *orbifolds*, to be the full 2-subcategory of $\boldsymbol{X} = (X, \mathcal{K})$ in **Kur** for which E_i is the zero vector bundle for all Kuranishi neighbourhoods $(V_i, E_i, \Gamma_i, s_i, \psi_i)$ in $\mathcal{K}$.

This allows us to simplify the notation a lot. Equations in §3.1 involving error terms $O\big(\pi_{ij}^*(s_i)\big)$ or $O\big(\pi_{ij}^*(s_i)^2\big)$ become exact, as $s_i = 0$. As E_i, s_i are zero we can take 'orbifold charts' to be (V_i, Γ_i, ψ_i). As $\hat\phi_{ij} = 0$ we can take coordinate changes to be $\Phi_{ij} = (P_{ij}, \pi_{ij}, \phi_{ij})$, and we can also take $V_{ij} = \pi_{ij}(P_{ij})$ to be equal to $\bar\psi_i^{-1}(S)$, rather than just an open neighbourhood of $\bar\psi_i^{-1}(S)$ in V_i, since $\bar\psi_i^{-1}(S)$ is open in V_i when $s_i = 0$. For 2-morphisms $\Lambda_{ij} = [\dot P_{ij}, \lambda_{ij}, \hat\lambda_{ij}]\colon \Phi_{ij} \Rightarrow \Phi'_{ij}$ in §3.1, we have $\hat\lambda_{ij} = 0$, and we are forced to take $\dot P_{ij} = P_{ij}$, and the equivalence relation $\approx$ in Definition 3.5 becomes trivial, so we can take 2-morphisms to be just λ_{ij}.

We say that a Kuranishi space $\boldsymbol{X}$ *is an orbifold* if $\boldsymbol{X} \simeq \boldsymbol{X}'$ in **Kur** for some $\boldsymbol{X}' \in \mathbf{Orb}_{\rm Kur} \subset \mathbf{Kur}$.

The next theorem is proved in [**35**, §4.5], [**36**, §6.6], by giving a full and faithful embedding $\mathbf{Orb}_{\rm Kur} \hookrightarrow \mathbf{Orb}_{\rm Le}$, and showing it is surjective on equivalence classes.

THEOREM 4.13. *The weak 2-category of Kuranishi orbifolds* $\mathbf{Orb}_{\rm Kur}$ *above is equivalent to the 2-categories of orbifolds* $\mathbf{Orb}_{\rm Pr}, \mathbf{Orb}_{\rm Le}, \mathbf{Orb}_{\rm ManSta}, \mathbf{Orb}_{C^\infty{\rm Sta}}$ *in*

[**3**, **31**, **38**, **44**, **47**] *described in Remark* 4.11. *Also there is an equivalence of categories* $\mathrm{Ho}(\mathbf{Orb}_{\mathrm{Kur}}) \simeq \mathbf{Orb}_{\mathrm{MP}}$, *for* $\mathbf{Orb}_{\mathrm{MP}}$ *the category of orbifolds from Moerdijk and Pronk* [**45**, **46**].

Fukaya et al. [**13**, §9] and McDuff [**40**] also define (effective) orbifolds as special examples of FOOO Kuranishi spaces and MW Kuranishi atlases.

DEFINITION 4.14. As in [**35**, §4.7], [**36**, Ch. 4 & §6.2], define the weak 2-category $\mathbf{mKur} \subset \mathbf{Kur}$ of *m-Kuranishi spaces* to be the full 2-subcategory of $\boldsymbol{X} = (X, \mathcal{K})$ in **Kur** for which $\Gamma_i = \{1\}$ for all Kuranishi neighbourhoods $(V_i, E_i, \Gamma_i, s_i, \psi_i)$ in $\mathcal{K}$.

As in Definition 4.12, this allows us to simplify the notation a lot. We take 'm-Kuranishi neighbourhoods' to be (V_i, E_i, s_i, ψ_i) rather than $(V_i, E_i, \Gamma_i, s_i, \psi_i)$. In coordinate changes, $\pi_{ij} : P_{ij} \to V_{ij}$ in §3.1 is a diffeomorphism, since it is a principal Γ_j-bundle for $\Gamma_j = \{1\}$, so we can replace P_{ij} by V_{ij} and take coordinate changes to be $(V_{ij}, \phi_{ij}, \hat{\phi}_{ij})$, and 2-morphisms to be $[\dot{V}_{ij}, \hat{\lambda}_{ij}]$ for open $\dot{V}_{ij} \subseteq V_{ij} \subseteq V_i$. The m-Kuranishi space analogues $\mathbf{mKN}_S(X), \mathbf{GmKN}$ of $\mathbf{KN}_S(X), \mathbf{GKN}$ in §3.1 are then *strict* 2-categories, as the canonical identification λ_{ijkl} in (3.3) is replaced by the identity map.

4.4. Relation to FOOO, MW, DY, polyfolds, and d-orbifolds.

The next four theorems are proved in [**35**, §4.8], [**36**, Ch. 7]. Theorem 4.18 is a corollary of Theorem 4.17 and Theorem 2.17 from Yang [**53**, Th. 3.1.7].

THEOREM 4.15. *Suppose* $\boldsymbol{X} = (X, \mathcal{K})$ *is a FOOO Kuranishi space without boundary, as in Definition* 2.4. *Then we can construct a Kuranishi space* $\boldsymbol{X}' = (X, \mathcal{K}')$ *in the sense of* §4.1 *with* $\mathrm{vdim}\, \boldsymbol{X}' = \mathrm{vdim}\, \boldsymbol{X}$, *with the same compact topological space* X, *and* $\boldsymbol{X}'$ *is natural up to equivalence in the* 2*-category* **Kur**.

THEOREM 4.16. *Suppose* X *is a compact, metrizable topological space with an MW weak Kuranishi atlas* $\mathcal{K}$ *without boundary, of virtual dimension* $n \in \mathbb{Z}$, *in the sense of Definition* 2.12. *Then we can make* X *into a Kuranishi space* $\boldsymbol{X}' = (X, \mathcal{K}')$ *in the sense of* §4.1 *with* $\mathrm{vdim}\, \boldsymbol{X}' = n$, *and* $\boldsymbol{X}'$ *is natural up to equivalence in the* 2*-category* **Kur**. *Commensurate MW weak Kuranishi atlases* $\mathcal{K}, \tilde{\mathcal{K}}$ *on* $\boldsymbol{X}$ *yield equivalent Kuranishi spaces* $\boldsymbol{X}', \tilde{\boldsymbol{X}}'$.

THEOREM 4.17. *Suppose* X *is a compact, metrizable topological space with a DY Kuranishi structure* $\mathcal{K}$ *without boundary in the sense of Definition* 2.15. *Then we can construct a Kuranishi space* $\boldsymbol{X}' = (X, \mathcal{K}')$ *in the sense of* §4.1 *with* $\mathrm{vdim}\, \boldsymbol{X}' = \mathrm{vdim}\, \boldsymbol{X}$, *with the same topological space* X, *and* $\boldsymbol{X}'$ *is natural up to equivalence in the* 2*-category* **Kur**. *R-equivalent DY Kuranishi structures* $\mathcal{K}, \tilde{\mathcal{K}}$ *on* $\boldsymbol{X}$ *yield equivalent Kuranishi spaces* $\boldsymbol{X}', \tilde{\boldsymbol{X}}'$.

THEOREM 4.18. *Suppose we are given a 'polyfold Fredholm structure'* $\mathcal{P}$ *on a compact metrizable topological space* X *in the sense of* [**23**–**30**], *that is, we write* X *as the zeroes of an sc-Fredholm section* $\mathfrak{s} : \mathfrak{V} \to \mathfrak{E}$ *of a strong polyfold vector bundle* $\mathfrak{E} \to \mathfrak{V}$ *over a polyfold* $\mathfrak{V}$, *where* $\mathfrak{s}$ *has constant Fredholm index* $n \in \mathbb{Z}$. *Then we can make* X *into a Kuranishi space* $\boldsymbol{X} = (X, \mathcal{K})$ *in the sense of* §4.1 *with* $\mathrm{vdim}\, \boldsymbol{X} = n$, *and* $\boldsymbol{X}$ *is natural up to equivalence in the* 2*-category* **Kur**.

The constructions of Theorems 4.15–4.18 are compatible with other geometric structures. In particular, orientations on FOOO Kuranishi spaces or MW weak

Kuranishi atlases or polyfold Fredholm structures correspond canonically to orientations on the corresponding Kuranishi spaces $\boldsymbol{X}'$. Also smooth maps $\boldsymbol{f}\colon \boldsymbol{X} \to Y$ from a FOOO Kuranishi space $\boldsymbol{X}$ to a manifold Y correspond to 1-morphisms $\boldsymbol{f}'\colon \boldsymbol{X}' \to Y$ in **Kur**, uniquely up to 2-isomorphism. Theorems 4.15–4.18 have applications to symplectic geometry.

The next theorem will be proved in [**36**, Ch. 8].

THEOREM 4.19. *There are equivalences of weak 2-categories*

$$\mathbf{mKur} \simeq \mathbf{dMan} \qquad \textit{and} \qquad \mathbf{Kur} \simeq \mathbf{dOrb},$$

where **dMan**, **dOrb** *are the strict 2-categories of 'd-manifolds' and 'd-orbifolds', as in the author* [**32–34**] *and* §3.6.

Thus, m-Kuranishi spaces and d-manifolds, and Kuranishi spaces and d-orbifolds, are for most purposes interchangeable.

To understand the relationship between Kuranishi spaces and d-orbifolds, consider the two equivalent ways of defining ordinary manifolds:

(a) As a Hausdorff, second countable topological space X equipped with an atlas of charts $\{(V_i, \phi_i)\colon i \in I\}$; or
(b) As a Hausdorff, second countable topological space X equipped with a sheaf of $\mathbb{R}$-algebras $\mathcal{O}_X^{C^\infty}$ which is embedded as a subsheaf $\mathcal{O}_X^{C^\infty} \subset \mathcal{O}_X^{C^0}$ of the sheaf $\mathcal{O}_X^{C^0}$ of continuous functions $f\colon X \to \mathbb{R}$, such that $(X, \mathcal{O}_X^{C^\infty})$ is locally modelled on $(\mathbb{R}^n, \mathcal{O}_{\mathbb{R}^n}^{C^\infty})$, where $\mathcal{O}_{\mathbb{R}^n}^{C^\infty}$ is the sheaf of smooth functions $f\colon \mathbb{R}^n \to \mathbb{R}$.

If we try to define 'derived manifolds' and 'derived orbifolds' by generalizing definition (a), we end up with (m-)Kuranishi spaces (or something similar); if we try to do it by generalizing definition (b), we end up with d-manifolds and d-orbifolds (or something similar).

Kuranishi spaces are more convenient than d-orbifolds if we wish to define 'derived' versions of categories of 'manifolds' which are not ordinary smooth manifolds, for instance, manifolds with corners of various kinds, or some classes of singular manifolds. This is because in the Kuranishi neighbourhoods $(V_i, E_i, \Gamma_i, s_i, \psi_i)$ on a Kuranishi space $\boldsymbol{X}$, it is easy to take V_i to be some different kind of 'manifold', but smooth manifolds are built into the theory of d-manifolds and d-orbifolds at a deep level, in the notion of C^∞-ring.

5. Differential geometry of Kuranishi spaces

5.1. Isotropy groups, and tangent and obstruction spaces. The next definition is taken from [**35**, §4.6], [**36**, §6.5].

DEFINITION 5.1. Let $\boldsymbol{X} = (X, \mathcal{K})$ be a Kuranishi space, with $\mathcal{K} = \big(I, (V_i, E_i, \Gamma_i, s_i, \psi_i)_{i\in I}, \Phi_{ij,\, i,j\in I}, \Lambda_{ijk,\, i,j,k\in I}\big)$, and let $x \in \boldsymbol{X}$. Choose an arbitrary $i \in I$ with $x \in \operatorname{Im}\psi_i$, and choose $v_i \in s_i^{-1}(0) \subseteq V_i$ with $\bar\psi_i(v_i) = x$.

Define a finite group $G_x\boldsymbol{X}$ called the *isotropy group of* $\boldsymbol{X}$ *at* x by

$$G_x\boldsymbol{X} = \big\{\gamma \in \Gamma_i : \gamma \cdot v_i = v_i\big\} = \operatorname{Stab}_{\Gamma_i}(v_i), \tag{5.1}$$

as a subgroup of Γ_i. Define finite-dimensional real vector spaces $T_x\boldsymbol{X}$, the *tangent space of* $\boldsymbol{X}$ *at* x, and $O_x\boldsymbol{X}$, the *obstruction space of* $\boldsymbol{X}$ *at* x, to be the kernel and cokernel of $\mathrm{d}\, s_i|_{v_i}$, so that they fit into an exact sequence

$$0 \longrightarrow T_x\boldsymbol{X} \longrightarrow T_{v_i}V_i \xrightarrow{\mathrm{d}\, s_i|_{v_i}} E_i|_{v_i} \longrightarrow O_x\boldsymbol{X} \longrightarrow 0.$$

The actions of Γ_i on V_i, E_i induce linear actions of $G_x\boldsymbol{X}$ on $T_x\boldsymbol{X}, O_x\boldsymbol{X}$, by the commutative diagram for each $\gamma \in G_x\boldsymbol{X}$:

$$\begin{array}{ccccccccc}
0 \longrightarrow & T_x\boldsymbol{X} & \longrightarrow & T_{v_i}V_i & \xrightarrow{\mathrm{d}\, s_i|_{v_i}} & E_i|_{v_i} & \longrightarrow & O_x\boldsymbol{X} & \longrightarrow 0 \\
& \Big\downarrow{\scriptstyle \gamma\cdot} & & \Big\downarrow{\scriptstyle \mathrm{d}(\gamma\cdot)} & & \Big\downarrow{\scriptstyle \gamma\cdot} & & \Big\downarrow{\scriptstyle \gamma\cdot} & \\
0 \longrightarrow & T_x\boldsymbol{X} & \longrightarrow & T_{v_i}V_i & \xrightarrow{\mathrm{d}\, s_i|_{v_i}} & E_i|_{v_i} & \longrightarrow & O_x\boldsymbol{X} & \longrightarrow 0.
\end{array}$$

This makes $T_x\boldsymbol{X}, O_x\boldsymbol{X}$ into representations of $G_x\boldsymbol{X}$. Definition 4.1(b) yields

$$\dim T_x\boldsymbol{X} - \dim O_x\boldsymbol{X} = \operatorname{vdim} \boldsymbol{X}.$$

The triple $(G_x\boldsymbol{X}, T_x\boldsymbol{X}, O_x\boldsymbol{X})$ depends on the choice of i, v_i. In [**35**, §4.6] and [**36**, §6.5] we show two choices i, v_i and $i', v'_{i'}$ yield triples $(G_x\boldsymbol{X}, T_x\boldsymbol{X}, O_x\boldsymbol{X})$, $(G_x\boldsymbol{X}', T_x\boldsymbol{X}', O_x\boldsymbol{X}')$ which have a finite, nonempty class of natural isomorphisms. Two such isomorphisms $(G_x\boldsymbol{X}, T_x\boldsymbol{X}, O_x\boldsymbol{X}) \to (G'_x\boldsymbol{X}, T'_x\boldsymbol{X}, O'_x\boldsymbol{X})$ differ by conjugation by an element of $G'_x\boldsymbol{X}$, and behave as expected under composition.

We discuss functoriality of the $G_x\boldsymbol{X}, T_x\boldsymbol{X}, O_x\boldsymbol{X}$ under 1- and 2-morphisms:

DEFINITION 5.2. Let $\boldsymbol{f}\colon \boldsymbol{X} \to \boldsymbol{Y}$ be a 1-morphism of Kuranishi spaces, with notation (4.2)–(4.3), and let $x \in \boldsymbol{X}$ with $\boldsymbol{f}(x) = y \in \boldsymbol{Y}$. Then Definition 5.1 gives $G_x\boldsymbol{X}, T_x\boldsymbol{X}, O_x\boldsymbol{X}$, defined using $i \in I$ and $u_i \in U_i$ with $\bar\chi_i(u_i) = x$, and $G_y\boldsymbol{Y}, T_y\boldsymbol{Y}, O_y\boldsymbol{Y}$, defined using $j \in J$ and $v_j \in V_j$ with $\bar\chi_j(v_j) = y$. In $\boldsymbol{f}$ we have a 1-morphism $\boldsymbol{f}_{ij} = (P_{ij}, \pi_{ij}, f_{ij}, \hat f_{ij})$ over f. Define

$$S_{x,\boldsymbol{f}} = \bigl\{p \in P_{ij} \colon \pi_{ij}(p) = u_i,\ \ f_{ij}(p) = v_j\bigr\}.$$

Then $S_{x,\boldsymbol{f}}$ is invariant under the commuting actions of $G_x\boldsymbol{X} = \mathrm{Stab}_{\mathrm{B}_i}(u_i) \subseteq \mathrm{B}_i$ and $G_y\boldsymbol{Y} = \mathrm{Stab}_{\Gamma_j}(v_j) \subseteq \Gamma_j$ on P_{ij} induced by the B_i, Γ_j-actions on P_{ij}, where $G_y\boldsymbol{Y}$ acts freely and transitively on $S_{x,\boldsymbol{f}}$, but the action of $G_x\boldsymbol{X}$ need not be free or transitive. Choose an arbitrary point $p_0 \in S_{x,\boldsymbol{f}}$. Our definitions of $G_x\boldsymbol{f}, T_x\boldsymbol{f}, O_x\boldsymbol{f}$ will depend on this choice.

Define morphisms $G_x\boldsymbol{f}\colon G_x\boldsymbol{X} \to G_y\boldsymbol{Y}$, $T_x\boldsymbol{f}\colon T_x\boldsymbol{X} \to T_y\boldsymbol{Y}$, $O_x\boldsymbol{f}\colon O_x\boldsymbol{X} \to O_y\boldsymbol{Y}$ by $G_x\boldsymbol{f}(\gamma) = \gamma'$ if $\gamma \cdot p_0 = (\gamma')^{-1} \cdot p_0$ in $S_{x,\boldsymbol{f}}$, using the actions of $G_x\boldsymbol{X}, G_y\boldsymbol{Y}$ on $S_{x,\boldsymbol{f}}$ with $G_y\boldsymbol{Y}$ free and transitive, and the commutative diagram

$$\begin{array}{ccccccccc}
0 \longrightarrow & T_x\boldsymbol{X} & \longrightarrow & T_{u_i}U_i & \xrightarrow[\mathrm{d}\, r_i|_{u_i}]{} & D_i|_{u_i} & \longrightarrow & O_x\boldsymbol{X} & \longrightarrow 0 \\
& \Big\downarrow{\scriptstyle T_x\boldsymbol{f}} & & \Big\downarrow{\scriptstyle \mathrm{d} f_{ij}|_{p_0} \circ (\mathrm{d}\, \pi_{ij}|_{p_0})^{-1}} & & \Big\downarrow{\scriptstyle \hat f_{ij}|_{p_0}} & & \Big\downarrow{\scriptstyle O_x\boldsymbol{f}} & \\
0 \longrightarrow & T_y\boldsymbol{Y} & \longrightarrow & T_{v_j}V_j & \xrightarrow{\mathrm{d}\, s_j|_{v_j}} & E_j|_{v_j} & \longrightarrow & O_y\boldsymbol{Y} & \longrightarrow 0.
\end{array}$$

Then $T_x\boldsymbol{f}, O_x\boldsymbol{f}$ are $G_x\boldsymbol{f}$-equivariant linear maps.

If $p'_0 \in S_{x,\boldsymbol{f}}$ is an alternative choice for p_0, yielding $G'_x\boldsymbol{f}, T'_x\boldsymbol{f}, O'_x\boldsymbol{f}$, there is a unique $\delta \in G_y\boldsymbol{Y}$ with $\delta \cdot p_0 = p'_0$, and then $G'_x\boldsymbol{f}(\gamma) = \delta(G_x\boldsymbol{f}(\gamma))\delta^{-1}$, $T'_x\boldsymbol{f}(v) = \delta \cdot T_x\boldsymbol{f}(v)$, $O'_x\boldsymbol{f}(w) = \delta \cdot O_x\boldsymbol{f}(w)$ for all $\gamma \in G_x\boldsymbol{X}$, $v \in T_x\boldsymbol{X}$, and $w \in O_x\boldsymbol{X}$. That is, the triple $(G_x\boldsymbol{f}, T_x\boldsymbol{f}, O_x\boldsymbol{f})$ is canonical up to conjugation by an element of $G_y\boldsymbol{Y}$.

Continuing with the same notation, suppose $\boldsymbol{g}\colon \boldsymbol{X} \to \boldsymbol{Y}$ is another 1-morphism and $\boldsymbol{\eta}\colon \boldsymbol{f} \Rightarrow \boldsymbol{g}$ a 2-morphism in **Kur**. Then above we define $G_x\boldsymbol{g}, T_x\boldsymbol{g}, O_x\boldsymbol{g}$ by choosing an arbitrary point $q_0 \in S_{x,\boldsymbol{g}}$, where

$$S_{x,\boldsymbol{g}} = \bigl\{q \in Q_{ij} \colon \pi_{ij}(q) = u_i,\ \ g_{ij}(q) = v_j\bigr\},$$

with $\boldsymbol{g}_{ij} = (Q_{ij}, \pi_{ij}, g_{ij}, \hat{g}_{ij})$ in $\boldsymbol{g}$. In $\boldsymbol{\eta}$ we have $\boldsymbol{\eta}_{ij} = [\dot{P}_{ij}, \lambda_{ij}, \hat{\lambda}_{ij}]$ represented by $(\dot{P}_{ij}, \lambda_{ij}, \hat{\lambda}_{ij})$, where $\dot{P}_{ij} \subseteq P_{ij}$ and $\lambda_{ij} \colon \dot{P}_{ij} \to Q_{ij}$. From the definitions we find that $S_{x,\boldsymbol{f}} \subseteq \dot{P}_{ij}$, and $\lambda_{ij}|_{S_{x,\boldsymbol{f}}} \colon S_{x,\boldsymbol{f}} \to S_{x,\boldsymbol{g}}$ is a bijection. Since $G_y\boldsymbol{Y}$ acts freely and transitively on $S_{x,\boldsymbol{g}}$, there is a unique element $G_x\boldsymbol{\eta} \in G_y\boldsymbol{Y}$ with $G_x\boldsymbol{\eta} \cdot \lambda_{ij}(p_0) = q_0$. Then we have

$$G_x\boldsymbol{g}(\gamma) = (G_x\boldsymbol{\eta})(G_x\boldsymbol{f}(\gamma))(G_x\boldsymbol{\eta})^{-1}, \quad T_x\boldsymbol{g}(v) = G_x\boldsymbol{\eta} \cdot T_x\boldsymbol{f}(v), \quad \text{and}$$
$$O_x\boldsymbol{g}(w) = G_x\boldsymbol{\eta} \cdot O_x\boldsymbol{f}(w) \quad \text{for all } \gamma \in G_x\boldsymbol{X},\ v \in T_x\boldsymbol{X}, \text{ and } w \in O_x\boldsymbol{X}.$$

That is, $(G_x\boldsymbol{g}, T_x\boldsymbol{g}, O_x\boldsymbol{g})$ is conjugate to $(G_x\boldsymbol{f}, T_x\boldsymbol{f}, O_x\boldsymbol{f})$ under $G_x\boldsymbol{\eta} \in G_y\boldsymbol{Y}$, the same indeterminacy as in the definition of $(G_x\boldsymbol{f}, T_x\boldsymbol{f}, O_x\boldsymbol{f})$.

Suppose instead that $\boldsymbol{g} \colon \boldsymbol{Y} \to \boldsymbol{Z}$ is another 1-morphism of Kuranishi spaces and $\boldsymbol{g}(y) = z \in \boldsymbol{Z}$. Then in a similar way we can show there is a canonical element $G_{x,\boldsymbol{g},\boldsymbol{f}} \in G_z\boldsymbol{Z}$ such that for all $\gamma \in G_x\boldsymbol{X}$, $v \in T_x\boldsymbol{X}$, $w \in O_x\boldsymbol{X}$ we have

$$\begin{aligned} G_x(\boldsymbol{g}\circ\boldsymbol{f})(\gamma) &= (G_{x,\boldsymbol{g},\boldsymbol{f}})((G_y\boldsymbol{g}\circ G_x\boldsymbol{f})(\gamma))(G_{x,\boldsymbol{g},\boldsymbol{f}})^{-1}, \\ T_x(\boldsymbol{g}\circ\boldsymbol{f})(v) &= G_{x,\boldsymbol{g},\boldsymbol{f}} \cdot (T_y\boldsymbol{g}\circ T_x\boldsymbol{f})(v), \\ O_x(\boldsymbol{g}\circ\boldsymbol{f})(w) &= G_{x,\boldsymbol{g},\boldsymbol{f}} \cdot (O_y\boldsymbol{g}\circ O_x\boldsymbol{f})(w). \end{aligned}$$

That is, $(G_x(\boldsymbol{g}\circ\boldsymbol{f}), T_x(\boldsymbol{g}\circ\boldsymbol{f}), O_x(\boldsymbol{g}\circ\boldsymbol{f}))$ is conjugate to $(G_y\boldsymbol{g}, T_y\boldsymbol{g}, O_y\boldsymbol{g}) \circ (G_x\boldsymbol{f}, T_x\boldsymbol{f}, O_x\boldsymbol{f})$ under $G_{x,\boldsymbol{g},\boldsymbol{f}} \in G_z\boldsymbol{Z}$.

Since 2-morphisms $\boldsymbol{\eta} \colon \boldsymbol{f} \Rightarrow \boldsymbol{g}$ relate triples $(G_x\boldsymbol{f}, T_x\boldsymbol{f}, O_x\boldsymbol{f})$ and $(G_x\boldsymbol{g}, T_x\boldsymbol{g}, O_x\boldsymbol{g})$ by isomorphisms, if $\boldsymbol{f} \colon \boldsymbol{X} \to \boldsymbol{Y}$ is an equivalence in **Kur** then $G_x\boldsymbol{f}, T_x\boldsymbol{f}, O_x\boldsymbol{f}$ are isomorphisms for all $x \in \boldsymbol{X}$.

We could use the Axiom of (Global) Choice to choose particular values of $G_x\boldsymbol{X}, T_x\boldsymbol{X}, O_x\boldsymbol{X}, G_x\boldsymbol{f}, T_x\boldsymbol{f}, O_x\boldsymbol{f}, G_x\boldsymbol{\eta}$ in Definitions 5.1–5.2 for all $\boldsymbol{X}, x, \boldsymbol{f}, \boldsymbol{\eta}$, but there seems no need for this. All the definitions we give using $G_x\boldsymbol{X}, \ldots, G_x\boldsymbol{\eta}$ will be independent of the arbitrary choices involved.

The next theorem is proved in [**36**, §10.4 & §10.5].

THEOREM 5.3. **(a)** *A Kuranishi space $\boldsymbol{X}$ is a manifold, in the sense of Definition* 4.10*, if and only if $G_x\boldsymbol{X} = \{1\}$ and $O_x\boldsymbol{X} = 0$ for all $x \in \boldsymbol{X}$.*

(b) *A Kuranishi space $\boldsymbol{X}$ is an orbifold, in the sense of Definition* 4.12*, if and only if $O_x\boldsymbol{X} = 0$ for all $x \in \boldsymbol{X}$.*

(c) *A* 1*-morphism $\boldsymbol{f} \colon \boldsymbol{X} \to \boldsymbol{Y}$ in* **Kur** *is an equivalence in* **Kur** *if and only if $G_x\boldsymbol{f} \colon G_x\boldsymbol{X} \to G_{\boldsymbol{f}(x)}\boldsymbol{Y}$, $T_x\boldsymbol{f} \colon T_x\boldsymbol{X} \to T_{\boldsymbol{f}(x)}\boldsymbol{Y}$ and $O_x\boldsymbol{f} \colon O_x\boldsymbol{X} \to O_{\boldsymbol{f}(x)}\boldsymbol{Y}$ are isomorphisms for all $x \in \boldsymbol{X}$, and $f \colon X \to Y$ is a bijection.*

DEFINITION 5.4. Write $\mathbf{Kur_{trG}} \subset \mathbf{Kur}$ for the full 2-subcategory of Kuranishi spaces $\boldsymbol{X}$ with $G_x\boldsymbol{X} = \{1\}$ for all $x \in \boldsymbol{X}$. Note that the 2-subcategory $\mathbf{mKur} \subset \mathbf{Kur}$ in §4.3 has $\mathbf{mKur} \subset \mathbf{Kur_{trG}}$, since if $\boldsymbol{X} = (X, \mathcal{K}) \in \mathbf{mKur}$ then the Kuranishi neighbourhoods $(V_i, E_i, \Gamma_i, s_i, \psi_i)$ in $\mathcal{K}$ have $\Gamma_i = \{1\}$, which implies that $G_x\boldsymbol{X} = \{1\}$ as $G_x\boldsymbol{X} \subseteq \Gamma_i$ by (5.1).

The next theorem is proved in [**35**, §4.7] and [**36**, §6.5].

THEOREM 5.5. *The inclusion* $\mathbf{mKur} \hookrightarrow \mathbf{Kur_{trG}}$ *is an equivalence of weak* 2*-categories.*

Combined with Theorem 4.19 this gives an equivalence $\mathbf{Kur_{trG}} \simeq \mathbf{dMan}$ of weak 2-categories. That is, $\mathbf{dMan}, \mathbf{mKur}, \mathbf{Kur_{trG}}$ are 2-categories of 'derived manifolds', and the theorems say that they are all essentially the same. The moral is that a derived orbifold (Kuranishi space) $\boldsymbol{X}$ is a derived manifold if and only if the isotropy groups $G_x\boldsymbol{X}$ for $x \in \boldsymbol{X}$ are all trivial.

5.2. W-transverse morphisms and fibre products. Smooth maps of manifolds $g\colon X \to Z$ and $h\colon Y \to Z$ are called *transverse* if $T_xg \oplus T_yh\colon T_xX \oplus T_yY \to T_zZ$ is surjective for all $x \in X$ and $y \in Y$ with $g(x) = h(y) = z \in Z$. It is well known that if g, h are transverse then the fibre product $W = X \times_{g,Z,h} Y$ exists in the category **Man**, in the sense of category theory, with $\dim W = \dim X + \dim Y - \dim Z$.

We now explain the analogue of this for Kuranishi spaces. We define two notions of transverse 1-morphisms in **Kur**, a weak and a strong:

DEFINITION 5.6. Let $\boldsymbol{g}\colon \boldsymbol{X} \to \boldsymbol{Z}$ and $\boldsymbol{h}\colon \boldsymbol{Y} \to \boldsymbol{Z}$ be 1-morphisms of Kuranishi spaces. Call $\boldsymbol{g}, \boldsymbol{h}$ *w-transverse* if for all $x \in \boldsymbol{X}$, $y \in \boldsymbol{Y}$ with $\boldsymbol{g}(x) = \boldsymbol{h}(y) = z$ in $\boldsymbol{Z}$, and all $\gamma \in G_z\boldsymbol{Z}$, then $O_x\boldsymbol{g} \oplus (\gamma \cdot O_y\boldsymbol{h})\colon O_x\boldsymbol{X} \oplus O_y\boldsymbol{Y} \to O_z\boldsymbol{Z}$ is surjective.

We call $\boldsymbol{g}, \boldsymbol{h}$ *transverse* if for all $x \in \boldsymbol{X}$, $y \in \boldsymbol{Y}$ with $\boldsymbol{g}(x) = \boldsymbol{h}(y) = z \in \boldsymbol{Z}$, and all $\gamma \in G_z\boldsymbol{Z}$, then $T_x\boldsymbol{g} \oplus (\gamma \cdot T_y\boldsymbol{h})\colon T_x\boldsymbol{X} \oplus T_y\boldsymbol{Y} \to T_z\boldsymbol{Z}$ is surjective, and $O_x\boldsymbol{g} \oplus (\gamma \cdot O_y\boldsymbol{h})\colon O_x\boldsymbol{X} \oplus O_y\boldsymbol{Y} \to O_z\boldsymbol{Z}$ is an isomorphism.

The next theorem is proved in [**36**, Ch. 11].

THEOREM 5.7. *Suppose $\boldsymbol{g}\colon \boldsymbol{X} \to \boldsymbol{Z}$, $\boldsymbol{h}\colon \boldsymbol{Y} \to \boldsymbol{Z}$ are w-transverse 1-morphisms of Kuranishi spaces. Then the fibre product $\boldsymbol{W} = \boldsymbol{X} \times_{\boldsymbol{g},\boldsymbol{Z},\boldsymbol{h}} \boldsymbol{Y}$ exists in the 2-category* **Kur**, *with* $\operatorname{vdim} \boldsymbol{W} = \operatorname{vdim} \boldsymbol{X} + \operatorname{vdim} \boldsymbol{Y} - \operatorname{vdim} \boldsymbol{Z}$. *It is unique up to canonical equivalence in* **Kur**.

This $\boldsymbol{W}$ is an orbifold if and only if $\boldsymbol{g}, \boldsymbol{h}$ are transverse.

The topological space W of $\boldsymbol{W}$ is given as a set by

$$W = \bigl\{(x, y, C)\colon x \in X,\ \ y \in Y,\ \ \boldsymbol{g}(x) = \boldsymbol{h}(y) = z \in Z,\\ C \in G_x\boldsymbol{g}(G_x\boldsymbol{X})\backslash G_z\boldsymbol{Z}/G_y\boldsymbol{h}(G_y\boldsymbol{Y})\bigr\}.$$

For $(x, y, C) \in W$ with $\gamma \in C \subseteq G_z\boldsymbol{Z}$, there is an exact sequence

$$\begin{array}{ccccccc} 0 \longrightarrow T_{(x,y,C)}\boldsymbol{W} & \longrightarrow & T_x\boldsymbol{X} \oplus T_y\boldsymbol{Y} & \xrightarrow{T_x\boldsymbol{g}\oplus(\gamma\cdot T_y\boldsymbol{h})} & T_z\boldsymbol{Z} \\ & & & & \downarrow \\ 0 \longleftarrow O_z\boldsymbol{Z} & \xleftarrow{O_x\boldsymbol{g}\oplus(\gamma\cdot O_y\boldsymbol{h})} & O_x\boldsymbol{X} \oplus O_y\boldsymbol{Y} & \longleftarrow & O_{(x,y,C)}\boldsymbol{W}. \end{array} \tag{5.2}$$

Here fibre products in 2-categories are explained in §A.3. They are characterized by a universal property involving 2-morphisms. In Theorem 5.7, it is essential that Kuranishi spaces **Kur** are a 2-category. The $\boldsymbol{W}$ described in Theorem 5.7 generally does not satisfy a universal property in the homotopy category Ho(**Kur**), and is not a fibre product in Ho(**Kur**) in the category-theoretic sense. This is an important reason for making Kuranishi spaces into a 2-category rather than an ordinary category.

Note that w-transversality is equivalent to exactness of (5.2) at $O_z\boldsymbol{Z}$, and exactness of the rest of (5.2) determines $T_{(x,y,C)}\boldsymbol{W}, O_{(x,y,C)}\boldsymbol{W}$ up to isomorphism. Transversality is equivalent to (5.2) being exact from $T_z\boldsymbol{Z}$ onwards with $O_{(x,y,C)}\boldsymbol{W} = 0$, so the claim that $\boldsymbol{W}$ is an orbifold if and only if $\boldsymbol{g}, \boldsymbol{h}$ are transverse follows from Theorem 5.3(b).

If $\boldsymbol{g}\colon \boldsymbol{X} \to \boldsymbol{Z}$ and $\boldsymbol{h}\colon \boldsymbol{Y} \to \boldsymbol{Z}$ are 1-morphisms in **Kur** with $\boldsymbol{Z}$ a manifold or orbifold then $O_z\boldsymbol{Z} = 0$ for all $z \in \boldsymbol{Z}$ by Theorem 5.3, so $\boldsymbol{g}, \boldsymbol{h}$ are w-transverse by Definition 5.6, and Theorem 5.7 yields:

Corollary 5.8. *Suppose* $\boldsymbol{g}\colon \boldsymbol{X}\to\boldsymbol{Z}$, $\boldsymbol{h}\colon\boldsymbol{Y}\to\boldsymbol{Z}$ *are* 1*-morphisms of Kuranishi spaces, with* $\boldsymbol{Z}$ *a manifold or orbifold. Then the fibre product* $\boldsymbol{W}=\boldsymbol{X}\times_{\boldsymbol{g},\boldsymbol{Z},\boldsymbol{h}}\boldsymbol{Y}$ *exists in* **Kur**, *with* $\operatorname{vdim}\boldsymbol{W}=\operatorname{vdim}\boldsymbol{X}+\operatorname{vdim}\boldsymbol{Y}-\dim\boldsymbol{Z}$.

In symplectic geometry it is often important to consider fibre products of Kuranishi spaces over manifolds or orbifolds, so Corollary 5.8 is very useful.

Example 5.9. If $\boldsymbol{X},\boldsymbol{Y}$ are Kuranishi spaces, and $\boldsymbol{Z}=*$ is the point in **Kur**, then $\boldsymbol{X}\times_*\boldsymbol{Y}$ is the product $\boldsymbol{X}\times\boldsymbol{Y}$, with $\operatorname{vdim}(\boldsymbol{X}\times\boldsymbol{Y})=\operatorname{vdim}\boldsymbol{X}+\operatorname{vdim}\boldsymbol{Y}$. We can define a model for products of Kuranishi spaces explicitly: if $\boldsymbol{X}=(X,\mathcal{I})$, $\boldsymbol{Y}=(Y,\mathcal{J})$ with notation (4.2)–(4.3), then we may write $\boldsymbol{X}\times\boldsymbol{Y}=(X\times Y,\mathcal{K})$, where $\mathcal{K}$ has indexing set $I\times J$ and Kuranishi neighbourhoods

$$\begin{aligned}&(W_{(i,j)},F_{(i,j)},\Delta_{(i,j)},t_{(i,j)},\omega_{(i,j)})=\\&\bigl(U_i\times V_j,\pi_{U_i}^*(D_i)\oplus\pi_{V_j}^*(E_j),\mathrm{B}_i\times\Gamma_j,\pi_{U_i}^*(r_i)\oplus\pi_{V_j}^*(s_j),\chi_i\times\psi_j\bigr).\end{aligned}$$

The remaining data in $\mathcal{K}$ is also easy to write down explicitly.

5.3. Submersions and w-submersions. A smooth map of manifolds $f\colon X\to Y$ is called a *submersion* if $T_xf\colon T_xX\to T_yY$ is surjective for all $x\in X$ with $f(x)=y\in Y$. If $g\colon X\to Z$ and $h\colon Y\to Z$ are smooth with g a submersion then g,h are transverse, so the fibre product $X\times_{g,Z,h}Y$ exists in **Man**. Here is the analogue of this for Kuranishi spaces. We define two notions of submersion in **Kur**, a weak and a strong.

Definition 5.10. A 1-morphism of Kuranishi spaces $\boldsymbol{f}\colon\boldsymbol{X}\to\boldsymbol{Y}$ is called a *weak submersion*, or *w-submersion*, if $O_x\boldsymbol{f}\colon O_x\boldsymbol{X}\to O_{\boldsymbol{f}(x)}\boldsymbol{Y}$ is surjective for all $x\in\boldsymbol{X}$. It is called a *submersion* if $T_x\boldsymbol{f}\colon T_x\boldsymbol{X}\to T_{\boldsymbol{f}(x)}\boldsymbol{Y}$ is surjective and $O_x\boldsymbol{f}\colon O_x\boldsymbol{X}\to O_{\boldsymbol{f}(x)}\boldsymbol{Y}$ is an isomorphism for all $x\in\boldsymbol{X}$. If $\boldsymbol{Y}$ is a manifold or orbifold then any 1-morphism $\boldsymbol{f}\colon\boldsymbol{X}\to\boldsymbol{Y}$ is a w-submersion.

If $\boldsymbol{g}\colon\boldsymbol{X}\to\boldsymbol{Z}$ and $\boldsymbol{h}\colon\boldsymbol{Y}\to\boldsymbol{Z}$ are 1-morphisms in **Kur** with $\boldsymbol{g}$ a w-submersion then $\boldsymbol{g},\boldsymbol{h}$ are w-transverse. If $\boldsymbol{g}$ is a submersion and $\boldsymbol{Y}$ is an orbifold then $\boldsymbol{g},\boldsymbol{h}$ are transverse. Thus Theorem 5.7 implies:

Corollary 5.11. *Suppose* $\boldsymbol{g}\colon\boldsymbol{X}\to\boldsymbol{Z}$, $\boldsymbol{h}\colon\boldsymbol{Y}\to\boldsymbol{Z}$ *are* 1*-morphisms in* **Kur** *with* $\boldsymbol{g}$ *a w-submersion. Then the fibre product* $\boldsymbol{W}=\boldsymbol{X}\times_{\boldsymbol{g},\boldsymbol{Z},\boldsymbol{h}}\boldsymbol{Y}$ *exists in* **Kur**, *with* $\operatorname{vdim}\boldsymbol{W}=\operatorname{vdim}\boldsymbol{X}+\operatorname{vdim}\boldsymbol{Y}-\operatorname{vdim}\boldsymbol{Z}$.

If $\boldsymbol{g}$ *is a submersion and* $\boldsymbol{Y}$ *is an orbifold, then* $\boldsymbol{W}$ *is an orbifold.*

Appendix A. Background from Category Theory and Algebraic Geometry

Sections A.1–A.3 discuss 2-categories, and §A.4 defines sheaves and stacks on topological spaces.

A.1. Basics of 2-categories. We discuss 2-categories, both strict and weak. Some references are Borceux [**4**, §7], Kelly and Street [**37**], and Behrend et al. [**2**, App. B].

Definition A.1. A *strict* 2*-category* $\boldsymbol{\mathcal{C}}$ consists of a class of *objects* $\operatorname{Obj}(\boldsymbol{\mathcal{C}})$, for all $X,Y\in\operatorname{Obj}(\boldsymbol{\mathcal{C}})$ an essentially small category $\operatorname{Hom}(X,Y)$, for all X,Y,Z in

$\mathrm{Obj}(\mathcal{C})$ a functor $\mu_{X,Y,Z}\colon \mathrm{Hom}(X,Y)\times\mathrm{Hom}(Y,Z)\to\mathrm{Hom}(X,Z)$ called *composition*, and for all X in $\mathrm{Obj}(\mathcal{C})$ an object id_X in $\mathrm{Hom}(X,X)$ called the *identity* 1-*morphism*. These must satisfy the *associativity property*, that

$$\mu_{W,Y,Z}\circ(\mu_{W,X,Y}\times\mathrm{id}_{\mathrm{Hom}(Y,Z)})=\mu_{W,X,Z}\circ(\mathrm{id}_{\mathrm{Hom}(W,X)}\times\mu_{X,Y,Z}) \tag{A.1}$$

as functors $\mathrm{Hom}(W,X)\times\mathrm{Hom}(X,Y)\times\mathrm{Hom}(Y,Z)\to\mathrm{Hom}(W,X)$, and the *identity property*, that

$$\mu_{X,X,Y}(\mathrm{id}_X,-)=\mu_{X,Y,Y}(-,\mathrm{id}_Y)=\mathrm{id}_{\mathrm{Hom}(X,Y)} \tag{A.2}$$

as functors $\mathrm{Hom}(X,Y)\to\mathrm{Hom}(X,Y)$.

Objects f of $\mathrm{Hom}(X,Y)$ are called 1-*morphisms*, written $f\colon X\to Y$. For 1-morphisms $f,g\colon X\to Y$, morphisms $\eta\in\mathrm{Hom}_{\mathrm{Hom}(X,Y)}(f,g)$ are called 2-*morphisms*, written $\eta\colon f\Rightarrow g$. Thus, a 2-category has objects X, and two kinds of morphisms: 1-morphisms $f\colon X\to Y$ between objects, and 2-morphisms $\eta\colon f\Rightarrow g$ between 1-morphisms.

A *weak* 2-*category*, or *bicategory*, is like a strict 2-category, except that the equations of functors (A.1), (A.2) are required to hold only up to specified natural isomorphisms. That is, a weak 2-category $\mathcal{C}$ consists of data $\mathrm{Obj}(\mathcal{C}),\mathrm{Hom}(X,Y)$, $\mu_{X,Y,Z},\mathrm{id}_X$ as above, but in place of (A.1), a natural isomorphism of functors

$$\alpha\colon \mu_{W,Y,Z}\circ(\mu_{W,X,Y}\times\mathrm{id}_{\mathrm{Hom}(Y,Z)})\Longrightarrow\mu_{W,X,Z}\circ(\mathrm{id}_{\mathrm{Hom}(W,X)}\times\mu_{X,Y,Z}), \tag{A.3}$$

and in place of (A.2), natural isomorphisms

$$\beta\colon \mu_{X,X,Y}(\mathrm{id}_X,-)\Longrightarrow\mathrm{id}_{\mathrm{Hom}(X,Y)},\quad \gamma\colon\mu_{X,Y,Y}(-,\mathrm{id}_Y)\Longrightarrow\mathrm{id}_{\mathrm{Hom}(X,Y)}. \tag{A.4}$$

These α,β,γ must satisfy identities which we give below in (A.6) and (A.8).

A strict 2-category $\mathcal{C}$ can be regarded as an example of a weak 2-category, in which the natural isomorphisms α,β,γ in (A.3)–(A.4) are the identities.

We now unpack Definition A.1, making it more explicit.

There are three kinds of composition in a 2-category, satisfying various associativity relations. If $f\colon X\to Y$ and $g\colon Y\to Z$ are 1-morphisms then $\mu_{X,Y,Z}(f,g)$ is the *composition of* 1-*morphisms*, written $g\circ f\colon X\to Z$. If $f,g,h\colon X\to Y$ are 1-morphisms and $\eta\colon f\Rightarrow g$, $\zeta\colon g\Rightarrow h$ are 2-morphisms then composition of η,ζ in $\mathrm{Hom}(X,Y)$ gives the *vertical composition of* 2-*morphisms* of η,ζ, written $\zeta\odot\eta\colon f\Rightarrow h$, as a diagram

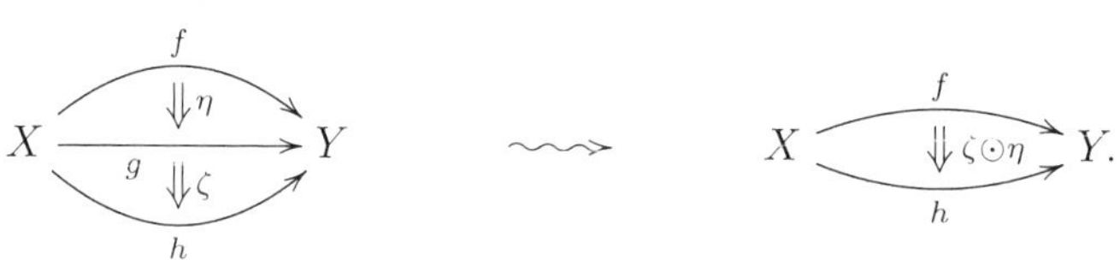

Vertical composition is associative.

If $f,\dot f\colon X\to Y$ and $g,\dot g\colon Y\to Z$ are 1-morphisms and $\eta\colon f\Rightarrow\dot f$, $\zeta\colon g\Rightarrow\dot g$ are 2-morphisms then $\mu_{X,Y,Z}(\eta,\zeta)$ is the *horizontal composition of* 2-*morphisms*, written $\zeta*\eta\colon g\circ f\Rightarrow\dot g\circ\dot f$, as a diagram

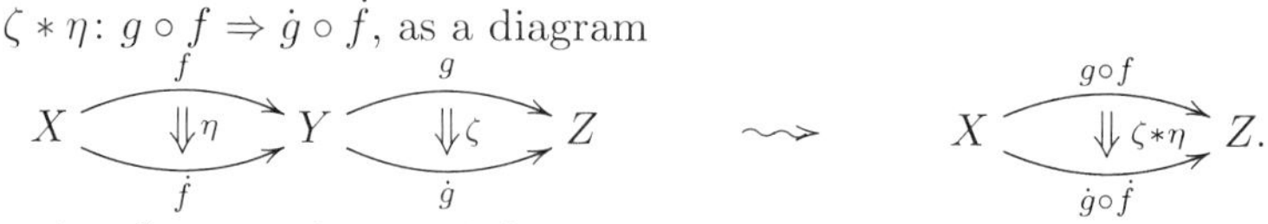

As $\mu_{X,Y,Z}$ is a functor, these satisfy *compatibility of vertical and horizontal composition*: given a diagram of 1- and 2-morphisms

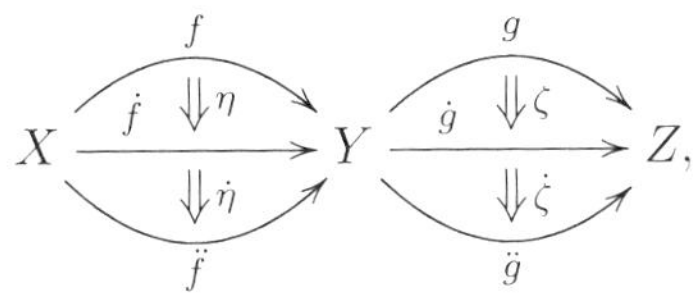

we have

$$(\dot\zeta \odot \zeta) * (\dot\eta \odot \eta) = (\dot\zeta * \dot\eta) \odot (\zeta * \eta) \colon g \circ f \Longrightarrow \ddot g \circ \ddot f.$$

There are also two kinds of identity: *identity 1-morphisms* $\mathrm{id}_X \colon X \to X$ and *identity 2-morphisms* $\mathrm{id}_f \colon f \Rightarrow f$.

In a strict 2-category $\mathcal{C}$, composition of 1-morphisms is strictly associative, $(g \circ f) \circ e = g \circ (f \circ e)$, and horizontal composition of 2-morphisms is strictly associative, $(\zeta * \eta) * \epsilon = \zeta * (\eta * \epsilon)$. In a weak 2-category $\mathcal{C}$, composition of 1-morphisms is associative up to specified 2-isomorphisms. That is, if $e \colon W \to X$, $f \colon X \to Y$, $g \colon Y \to Z$ are 1-morphisms in $\mathcal{C}$ then the natural isomorphism α in (A.3) gives a 2-isomorphism

$$\alpha_{g,f,e} \colon (g \circ f) \circ e \Longrightarrow g \circ (f \circ e). \tag{A.5}$$

As α is a natural isomorphism, given 1-morphisms $e, \dot e \colon W \to X$, $f, \dot f \colon X \to Y$, $g, \dot g \colon Y \to Z$ and 2-morphisms $\epsilon \colon e \Rightarrow \dot e$, $\eta \colon f \Rightarrow \dot f$, $\zeta \colon g \Rightarrow \dot g$ in $\mathcal{C}$, the following diagram of 2-morphisms must commute:

$$\begin{array}{ccc} (g \circ f) \circ e & \xRightarrow{\ \alpha_{g,f,e}\ } & g \circ (f \circ e) \\ \Big\Downarrow {\scriptstyle (\zeta * \eta) * \epsilon} & & {\scriptstyle \zeta * (\eta * \epsilon)} \Big\Downarrow \\ (\dot g \circ \dot f) \circ \dot e & \xRightarrow{\ \alpha_{\dot g,\dot f,\dot e}\ } & \dot g \circ (\dot f \circ \dot e). \end{array}$$

The $\alpha_{g,f,e}$ must satisfy the *associativity coherence axiom*: if $d \colon V \to W$ is another 1-morphism, then the following diagram of 2-morphisms must commute:

$$\begin{array}{ccccc} ((g \circ f) \circ e) \circ d & \xRightarrow{\ \alpha_{g,f,e} * \mathrm{id}_d\ } & (g \circ (f \circ e)) \circ d & \xRightarrow{\ \alpha_{g,f \circ e,d}\ } & g \circ ((f \circ e) \circ d) \\ \Big\Downarrow {\scriptstyle \alpha_{g \circ f,e,d}} & & & & {\scriptstyle \mathrm{id}_g * \alpha_{f,e,d}} \Big\Downarrow \\ (g \circ f) \circ (e \circ d) & & \xRightarrow{\ \alpha_{g,f,d \circ e}\ } & & g \circ (f \circ (e \circ d)). \end{array} \tag{A.6}$$

In a strict 2-category $\mathcal{C}$, given a 1-morphism $f \colon X \to Y$, the identity 1-morphisms $\mathrm{id}_X, \mathrm{id}_Y$ satisfy $f \circ \mathrm{id}_X = \mathrm{id}_Y \circ f = f$. In a weak 2-category $\mathcal{C}$, the natural isomorphisms β, γ in (A.4) give 2-isomorphisms

$$\beta_f \colon f \circ \mathrm{id}_X \Longrightarrow f, \qquad \gamma_f \colon \mathrm{id}_Y \circ f \Longrightarrow f. \tag{A.7}$$

As β, γ are natural isomorphisms, if $\eta \colon f \Rightarrow \dot f$ is a 2-morphism we must have

$$\eta \odot \beta_f = \beta_{\dot f} \odot (\eta * \mathrm{id}_{\mathrm{id}_X}) \colon f \circ \mathrm{id}_X \Rightarrow \dot f,$$

$$\eta \odot \gamma_f = \gamma_{\dot f} \odot (\mathrm{id}_{\mathrm{id}_Y} * \eta) \colon \mathrm{id}_Y \circ f \Rightarrow \dot f.$$

The β_f, γ_f must satisfy the *identity coherence axiom*: if $g \colon Y \to Z$ is another 1-morphism, then the following diagram of 2-morphisms must commute:

$$\begin{array}{ccc} (g \circ \mathrm{id}_Y) \circ f & \xRightarrow{\ \beta_g * \mathrm{id}_f\ } & \\ \Big\Downarrow {\scriptstyle \alpha_{g,\mathrm{id}_Y,f}} & & g \circ f. \\ g \circ (\mathrm{id}_Y \circ f) & \xRightarrow{\ \mathrm{id}_g * \alpha_f\ } & \end{array} \tag{A.8}$$

A basic example of a strict 2-category is the *2-category of categories* $\mathfrak{Cat}$, with objects small categories $\mathcal{C}$, 1-morphisms functors $F\colon \mathcal{C} \to \mathcal{D}$, and 2-morphisms natural transformations $\eta\colon F \Rightarrow G$ for functors $F, G\colon \mathcal{C} \to \mathcal{D}$. Orbifolds naturally form a 2-category (strict or weak, depending on the definition), as in §4.3, and so do stacks in algebraic geometry.

In a 2-category $\boldsymbol{\mathcal{C}}$, there are three notions of when objects X, Y in $\boldsymbol{\mathcal{C}}$ are 'the same': *equality* $X = Y$, and *1-isomorphism*, that is we have 1-morphisms $f\colon X \to Y$, $g\colon Y \to X$ with $g \circ f = \mathrm{id}_X$ and $f \circ g = \mathrm{id}_Y$, and *equivalence*, that is we have 1-morphisms $f\colon X \to Y$, $g\colon Y \to X$ and 2-isomorphisms $\eta\colon g \circ f \Rightarrow \mathrm{id}_X$ and $\zeta\colon f \circ g \Rightarrow \mathrm{id}_Y$. Usually equivalence is the correct notion.

When we say that objects X, Y in a 2-category $\boldsymbol{\mathcal{C}}$ are *canonically equivalent*, we mean that there is a nonempty distinguished class of equivalences $f\colon X \to Y$ in $\boldsymbol{\mathcal{C}}$, and given any two such equivalences $f, g\colon X \to Y$ there is a 2-isomorphism $\eta\colon f \Rightarrow g$. Often there is a distinguished choice of such η.

Commutative diagrams in 2-categories should in general only commute *up to (specified) 2-isomorphisms*, rather than strictly. A simple example of a commutative diagram in a 2-category $\boldsymbol{\mathcal{C}}$ is

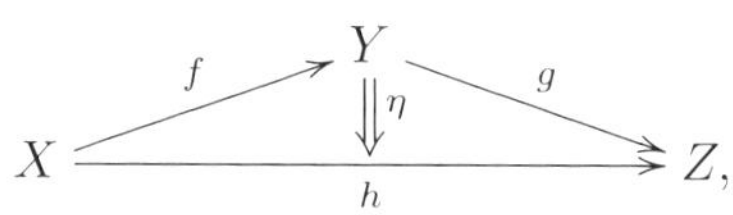

which means that X, Y, Z are objects of $\boldsymbol{\mathcal{C}}$, $f\colon X \to Y$, $g\colon Y \to Z$ and $h\colon X \to Z$ are 1-morphisms in $\boldsymbol{\mathcal{C}}$, and $\eta\colon g \circ f \Rightarrow h$ is a 2-isomorphism.

Let $\boldsymbol{\mathcal{C}}$ be a 2-category. The *homotopy category* $\mathrm{Ho}(\boldsymbol{\mathcal{C}})$ of $\boldsymbol{\mathcal{C}}$ is the category whose objects are objects of $\boldsymbol{\mathcal{C}}$, and whose morphisms $[f]\colon X \to Y$ are 2-isomorphism classes $[f]$ of 1-morphisms $f\colon X \to Y$ in $\boldsymbol{\mathcal{C}}$. Then equivalences in $\boldsymbol{\mathcal{C}}$ become isomorphisms in $\mathrm{Ho}(\boldsymbol{\mathcal{C}})$, 2-commutative diagrams in $\boldsymbol{\mathcal{C}}$ become commutative diagrams in $\mathrm{Ho}(\boldsymbol{\mathcal{C}})$, and so on.

A.2. 2-functors between 2-categories. Next we discuss 2-functors between 2-categories, following Borceux [**4**, §7.2, §7.5] and Behrend et al. [**2**, §B.4].

DEFINITION A.2. Let $\boldsymbol{\mathcal{C}}, \boldsymbol{\mathcal{D}}$ be strict 2-categories. A *strict 2-functor* $F\colon \boldsymbol{\mathcal{C}} \to \boldsymbol{\mathcal{D}}$ assigns an object $F(X)$ in $\boldsymbol{\mathcal{D}}$ for each object X in $\boldsymbol{\mathcal{C}}$, a 1-morphism $F(f) : F(X) \to F(Y)$ in $\boldsymbol{\mathcal{D}}$ for each 1-morphism $f\colon X \to Y$ in $\boldsymbol{\mathcal{C}}$, and a 2-morphism $F(\eta)\colon F(f) \Rightarrow F(g)$ in $\boldsymbol{\mathcal{D}}$ for each 2-morphism $\eta\colon f \Rightarrow g$ in $\boldsymbol{\mathcal{C}}$, such that F preserves all the structures on $\boldsymbol{\mathcal{C}}, \boldsymbol{\mathcal{D}}$, that is,

$$F(g \circ f) = F(g) \circ F(f), \quad F(\mathrm{id}_X) = \mathrm{id}_{F(X)}, \quad F(\zeta * \eta) = F(\zeta) * F(\eta), \tag{A.9}$$

$$F(\zeta \odot \eta) = F(\zeta) \odot F(\eta), \quad F(\mathrm{id}_f) = \mathrm{id}_{F(f)}. \tag{A.10}$$

Now let $\boldsymbol{\mathcal{C}}, \boldsymbol{\mathcal{D}}$ be weak 2-categories. Then strict 2-functors $F\colon \boldsymbol{\mathcal{C}} \to \boldsymbol{\mathcal{D}}$ are not well-behaved. To fix this, we need to relax (A.9) to hold only up to specified 2-isomorphisms. A *weak 2-functor* (or *pseudofunctor*) $F\colon \boldsymbol{\mathcal{C}} \to \boldsymbol{\mathcal{D}}$ assigns an object $F(X)$ in $\boldsymbol{\mathcal{D}}$ for each object X in $\boldsymbol{\mathcal{C}}$, a 1-morphism $F(f)\colon F(X) \to F(Y)$ in $\boldsymbol{\mathcal{D}}$ for each 1-morphism $f\colon X \to Y$ in $\boldsymbol{\mathcal{C}}$, a 2-morphism $F(\eta)\colon F(f) \Rightarrow F(g)$ in $\boldsymbol{\mathcal{D}}$ for each 2-morphism $\eta\colon f \Rightarrow g$ in $\boldsymbol{\mathcal{C}}$, a 2-isomorphism $F_{g,f}\colon F(g) \circ F(f) \Rightarrow F(g \circ f)$ in $\boldsymbol{\mathcal{D}}$ for all 1-morphisms $f\colon X \to Y$, $g\colon Y \to Z$ in $\boldsymbol{\mathcal{C}}$, and a 2-isomorphism $F_X : F(\mathrm{id}_X) \Rightarrow \mathrm{id}_{F(X)}$ in $\boldsymbol{\mathcal{D}}$ for all objects X in $\boldsymbol{\mathcal{C}}$ such that (A.10) holds, and for all $e\colon W \to X$,

$f\colon X \to Y$, $g\colon Y \to Z$ in $\mathcal{C}$ the following diagram of 2-isomorphisms commutes in $\mathcal{D}$:

$$\begin{array}{ccccc} (F(g)\circ F(f))\circ F(e) & \xRightarrow{F_{g,f}*\mathrm{id}_{F(e)}} & F(g\circ f)\circ F(e) & \xRightarrow{F_{g\circ f,e}} & F((g\circ f)\circ e) \\ \Big\Downarrow \alpha_{F(g),F(f),F(e)} & & & & F(\alpha_{g,f,e})\Big\Downarrow \\ F(g)\circ(F(f)\circ F(e)) & \xRightarrow{\mathrm{id}_{F(g)}*F_{f,e}} & F(g)\circ F(f\circ e) & \xRightarrow{F_{g,f\circ e}} & F(g\circ(f\circ e)), \end{array}$$

and for all 1-morphisms $f\colon X \to Y$ in $\mathcal{C}$, the following commute in $\mathcal{D}$:

$$\begin{array}{ccc} F(f)\circ F(\mathrm{id}_X) & \xRightarrow{F_{f,\mathrm{id}_X}} & F(f\circ \mathrm{id}_X) \\ \Big\Downarrow \mathrm{id}_{F(f)}*F_X & & F(\beta_f)\Big\Downarrow \\ F(f)\circ \mathrm{id}_{F(X)} & \xRightarrow{\beta_{F(f)}} & F(f), \end{array} \qquad \begin{array}{ccc} F(\mathrm{id}_Y)\circ F(f) & \xRightarrow{F_{\mathrm{id}_Y,f}} & F(\mathrm{id}_Y\circ f) \\ \Big\Downarrow F_Y*\mathrm{id}_{F(f)} & & F(\gamma_f)\Big\Downarrow \\ \mathrm{id}_{F(Y)}\circ F(f) & \xRightarrow{\gamma_{F(f)}} & F(f), \end{array}$$

and if $f,\dot f\colon X\to Y$ and $g,\dot g\colon Y\to Z$ are 1-morphisms and $\eta\colon f\Rightarrow \dot f$, $\zeta\colon g\Rightarrow \dot g$ are 2-morphisms in $\mathcal{C}$ then the following commutes in $\mathcal{D}$:

$$\begin{array}{ccc} F(g)\circ F(f) & \xRightarrow{F_{g,f}} & F(g\circ f) \\ \Big\Downarrow F(\zeta)*F(\eta) & & F(\zeta*\eta)\Big\Downarrow \\ F(\dot g)\circ F(\dot f) & \xRightarrow{F_{\dot g,\dot f}} & F(\dot g\circ \dot f). \end{array}$$

There are obvious notions of *composition* $G\circ F$ of strict and weak 2-functors $F\colon \mathcal{C}\to\mathcal{D}$, $G\colon \mathcal{D}\to\mathcal{E}$, *identity 2-functors* $\mathrm{id}_{\mathcal{C}}$, and so on.

If $\mathcal{C},\mathcal{D}$ are strict 2-categories, then a strict 2-functor $F\colon \mathcal{C}\to\mathcal{D}$ can be made into a weak 2-functor by taking all $F_{g,f}, F_X$ to be identity 2-morphisms.

We define a weak 2-functor $F\colon \mathcal{C}\to\mathcal{D}$ to be an *equivalence of weak 2-categories*, if for all objects X,Y in $\mathcal{C}$, the functor $F_{X,Y}\colon \mathrm{Hom}_{\mathcal{C}}(X,Y)\to \mathrm{Hom}_{\mathcal{D}}(F(X),F(Y))$ is an equivalence of categories, and the map induced by F from equivalence classes of objects in $\mathcal{C}$ to equivalence classes of objects in $\mathcal{D}$ is surjective (and hence a bijection). There is an alternative definition using 2-natural transformations of 2-functors, which we will not explain.

A.3. Fibre products in 2-categories. We define fibre products in 2-categories, as in Behrend et al. [**2**, Def. B.13].

Definition A.3. Let $\mathcal{C}$ be a strict 2-category and $g\colon X\to Z$, $h\colon Y\to Z$ be 1-morphisms in $\mathcal{C}$. A *fibre product* $X\times_Z Y$ in $\mathcal{C}$ consists of an object W, 1-morphisms $\pi_X\colon W\to X$ and $\pi_Y\colon W\to Y$ and a 2-isomorphism $\eta\colon g\circ\pi_X\Rightarrow h\circ\pi_Y$ in $\mathcal{C}$ with the following universal property: suppose $\pi'_X\colon W'\to X$ and $\pi'_Y\colon W'\to Y$ are 1-morphisms and $\eta'\colon g\circ\pi'_X\Rightarrow h\circ\pi'_Y$ is a 2-isomorphism in $\mathcal{C}$. Then there should exist a 1-morphism $b\colon W'\to W$ and 2-isomorphisms $\zeta_X\colon \pi_X\circ b\Rightarrow \pi'_X$, $\zeta_Y\colon \pi_Y\circ b\Rightarrow\pi'_Y$ such that the following diagram of 2-isomorphisms commutes:

$$\begin{array}{ccc} g\circ\pi_X\circ b & \xRightarrow{\eta*\mathrm{id}_b} & h\circ\pi_Y\circ b \\ \Big\Downarrow \mathrm{id}_g*\zeta_X & & \mathrm{id}_h*\zeta_Y\Big\Downarrow \\ g\circ\pi'_X & \xRightarrow{\eta'} & h\circ\pi'_Y. \end{array} \tag{A.11}$$

Furthermore, if $\tilde b,\tilde\zeta_X,\tilde\zeta_Y$ are alternative choices of b,ζ_X,ζ_Y then there should exist a unique 2-isomorphism $\theta\colon \tilde b\Rightarrow b$ with

$$\tilde\zeta_X=\zeta_X\odot(\mathrm{id}_{\pi_X}*\theta) \quad\text{and}\quad \tilde\zeta_Y=\zeta_Y\odot(\mathrm{id}_{\pi_Y}*\theta).$$

If a fibre product $X \times_Z Y$ in $\mathcal{C}$ exists then it is unique up to canonical equivalence. That is, if W, W' are two choices for $X \times_Z Y$ then there is an equivalence $b\colon W \to W'$ in $\mathcal{C}$, which is canonical up to 2-isomorphism.

If instead $\mathcal{C}$ is a weak 2-category, we must replace (A.11) by

$$\begin{array}{ccccc} (g \circ \pi_X) \circ b & \xrightarrow[\eta * \mathrm{id}_b]{} & (h \circ \pi_Y) \circ b & \xrightarrow[\alpha_{h,\pi_Y,b}]{} & h \circ (\pi_Y \circ b) \\ \Downarrow \alpha_{g,\pi_X,b} & & & & \mathrm{id}_h * \zeta_Y \Downarrow \\ g \circ (\pi_X \circ b) & \xrightarrow{\mathrm{id}_g * \zeta_X} & g \circ \pi'_X & \xrightarrow{\eta'} & h \circ \pi'_Y. \end{array}$$

Orbifolds, and stacks in algebraic geometry, form 2-categories, and Definition A.3 is the right way to define fibre products of orbifolds or stacks.

A.4. Sheaves and stacks on topological spaces. The next definition of sheaves on a topological space, as in Hartshorne [**20**, §II.1], will not actually be used in this paper, but we include it as motivation for the following definition of stacks on a topological space.

DEFINITION A.4. Let X be a topological space. A *presheaf of sets* $\mathcal{E}$ on X consists of the data of a set $\mathcal{E}(S)$ for every open set $S \subseteq X$, and a map $\rho_{ST}\colon \mathcal{E}(S) \to \mathcal{E}(T)$ called the *restriction map* for every inclusion $T \subseteq S \subseteq X$ of open sets, satisfying the conditions that

(i) $\rho_{SS} = \mathrm{id}_{\mathcal{E}(S)}\colon \mathcal{E}(S) \to \mathcal{E}(S)$ for all open $S \subseteq X$; and
(ii) $\rho_{SU} = \rho_{TU} \circ \rho_{ST}\colon \mathcal{E}(S) \to \mathcal{E}(U)$ for all open $U \subseteq T \subseteq S \subseteq X$.

A presheaf of sets $\mathcal{E}$ on X is called a *sheaf* if it also satisfies

(iii) If $S \subseteq X$ is open, $\{T_i : i \in I\}$ is an open cover of S, and $s, t \in \mathcal{E}(S)$ have $\rho_{ST_i}(s) = \rho_{ST_i}(t)$ in $\mathcal{E}(T_i)$ for all $i \in I$, then $s = t$ in $\mathcal{E}(S)$; and
(iv) If $S \subseteq X$ is open, $\{T_i : i \in I\}$ is an open cover of S, and we are given elements $s_i \in \mathcal{E}(T_i)$ for all $i \in I$ such that $\rho_{T_i(T_i \cap T_j)}(s_i) = \rho_{T_j(T_i \cap T_j)}(s_j)$ in $\mathcal{E}(T_i \cap T_j)$ for all $i, j \in I$, then there exists $s \in \mathcal{E}(S)$ with $\rho_{ST_i}(s) = s_i$ for all $i \in I$. This s is unique by (iii).

Suppose $\mathcal{E}, \mathcal{F}$ are presheaves or sheaves of sets on X. A *morphism* $\phi\colon \mathcal{E} \to \mathcal{F}$ consists of a map $\phi(S)\colon \mathcal{E}(S) \to \mathcal{F}(S)$ for all open $S \subseteq X$, such that the following diagram commutes for all open $T \subseteq S \subseteq X$

$$\begin{array}{ccc} \mathcal{E}(S) & \xrightarrow[\phi(S)]{} & \mathcal{F}(S) \\ \downarrow \rho_{ST} & & \rho'_{ST} \downarrow \\ \mathcal{E}(T) & \xrightarrow{\phi(T)} & \mathcal{F}(T), \end{array}$$

where ρ_{ST} is the restriction map for $\mathcal{E}$, and ρ'_{ST} the restriction map for $\mathcal{F}$.

Sheaves are basic objects in algebraic geometry. Informally, something forms a sheaf if it can be defined locally and glued. For example, if X, Y are manifolds then smooth maps $f\colon X \to Y$ form a sheaf on X, because an arbitrary map $f\colon X \to Y$ is smooth if and only if it is a smooth on an open neighbourhood of each point in X. One uses the sheaf property of smooth maps all the time in differential geometry without noticing.

Finally we define stacks on topological spaces. Stacks are the 2-category version of sheaves, roughly, they are sheaves of groupoids rather than sheaves of sets. Informally, something forms a stack if it can be defined locally and glued up to 2-isomorphisms. For example, if X, Y are orbifolds then smooth maps $f\colon X \to Y$

(i.e. 1-morphisms of orbifolds) and their 2-morphisms form a stack on X, but in general they do not form a sheaf on X.

DEFINITION A.5. Let X be a topological space. A *prestack* (or *prestack in groupoids*, or *2-presheaf*) $\mathcal{E}$ on X, consists of the data of a groupoid $\mathcal{E}(S)$ for every open set $S \subseteq X$, and a functor $\rho_{ST}\colon \mathcal{E}(S) \to \mathcal{E}(T)$ called the *restriction map* for every inclusion $T \subseteq S \subseteq X$ of open sets, and a natural isomorphism of functors $\eta_{STU}\colon \rho_{TU} \circ \rho_{ST} \Rightarrow \rho_{SU}$ for all inclusions $U \subseteq T \subseteq S \subseteq X$ of open sets, satisfying the conditions that:

(i) $\rho_{SS} = \mathrm{id}_{\mathcal{E}(S)}\colon \mathcal{E}(S) \to \mathcal{E}(S)$ for all open $S \subseteq X$, and $\eta_{SST} = \eta_{STT} = \mathrm{id}_{\rho_{ST}}$ for all open $T \subseteq S \subseteq X$; and

(ii) $\eta_{SUV} \odot (\mathrm{id}_{\rho_{UV}} * \eta_{STU}) = \eta_{STV} \odot (\eta_{TUV} * \mathrm{id}_{\rho_{ST}})\colon \rho_{UV} \circ \rho_{TU} \circ \rho_{ST} \Longrightarrow \rho_{SV}$ for all open $V \subseteq U \subseteq T \subseteq S \subseteq X$.

A prestack $\mathcal{E}$ on X is called a *stack* (or *stack in groupoids*, or *2-sheaf*) on X if whenever $S \subseteq X$ is open and $\{T_i\colon i \in I\}$ is an open cover of S, then:

(iii) If $\alpha, \beta\colon A \to B$ are morphisms in $\mathcal{E}(S)$ and $\rho_{ST_i}(\alpha) = \rho_{ST_i}(\beta)\colon \rho_{ST_i}(A) \to \rho_{ST_i}(B)$ in $\mathcal{E}(T_i)$ for all $i \in I$, then $\alpha = \beta$.

(iv) If A, B are objects of $\mathcal{E}(S)$ and $\alpha_i\colon \rho_{ST_i}(A) \to \rho_{ST_i}(B)$ are morphisms in $\mathcal{E}(T_i)$ for all $i \in I$ with

$$\begin{aligned}&\eta_{ST_i(T_i\cap T_j)}(B) \circ \rho_{T_i(T_i\cap T_j)}(\alpha_i) \circ \eta_{ST_i(T_i\cap T_j)}(A)^{-1}\\ &\quad = \eta_{ST_j(T_i\cap T_j)}(B) \circ \rho_{T_j(T_i\cap T_j)}(\alpha_j) \circ \eta_{ST_j(T_i\cap T_j)}(A)^{-1}\end{aligned}$$

in $\mathcal{E}(T_i \cap T_j)$ for all $i, j \in I$, then there exists $\alpha\colon A \to B$ in $\mathcal{E}(S)$ (necessarily unique by (iii)) with $\rho_{ST_i}(\alpha) = \alpha_i$ for all $i \in I$.

(v) If $A_i \in \mathcal{E}(T_i)$ for $i \in I$ and $\alpha_{ij}\colon \rho_{T_i(T_i\cap T_j)}(A_i) \to \rho_{T_j(T_i\cap T_j)}(A_j)$ are morphisms in $\mathcal{E}(T_i \cap T_j)$ for all $i, j \in I$ satisfying

$$\begin{aligned}&\eta_{T_k(T_j\cap T_k)(T_i\cap T_j\cap T_k)}(A_k) \circ \rho_{(T_j\cap T_k)(T_i\cap T_j\cap T_k)}(\alpha_{jk}) \circ \eta_{T_j(T_j\cap T_k)(T_i\cap T_j\cap T_k)}(A_j)^{-1}\\ &\circ \eta_{T_j(T_i\cap T_j)(T_i\cap T_j\cap T_k)}(A_j) \circ \rho_{(T_i\cap T_j)(T_i\cap T_j\cap T_k)}(\alpha_{ij}) \circ \eta_{T_i(T_i\cap T_j)(T_i\cap T_j\cap T_k)}(A_i)^{-1}\\ &= \eta_{T_k(T_i\cap T_k)(T_i\cap T_j\cap T_k)}(A_k) \circ \rho_{(T_i\cap T_k)(T_i\cap T_j\cap T_k)}(\alpha_{ik}) \circ \eta_{T_i(T_i\cap T_k)(T_i\cap T_j\cap T_k)}(A_i)^{-1}\end{aligned}$$

for all $i, j, k \in I$, then there exist an object A in $\mathcal{E}(S)$ and morphisms $\beta_i\colon A_i \to \rho_{ST_i}(A)$ for $i \in I$ such that for all $i, j \in I$ we have

$$\eta_{ST_i(T_i\cap T_j)}(A) \circ \rho_{T_i(T_i\cap T_j)}(\beta_i) = \eta_{ST_j(T_i\cap T_j)}(A) \circ \rho_{T_j(T_i\cap T_j)}(\beta_j) \circ \alpha_{ij}.$$

If $\tilde{A}, \tilde{\beta}_i$ are alternative choices then (iii),(iv) imply there is a unique isomorphism $\gamma\colon A \to \tilde{A}$ in $\mathcal{E}(S)$ with $\rho_{ST_i}(\gamma) = \tilde{\beta}_i \circ \beta_i^{-1}$ for all $i \in I$.

In the examples of stacks on topological spaces that will be important to us, we will have $\rho_{TU} \circ \rho_{ST} = \rho_{SU}$ and $\eta_{STU} = \mathrm{id}_{\rho_{SU}}$ for all open $U \subseteq T \subseteq S \subseteq X$. So (ii) is automatic, and all the $\eta_{\cdots}(\cdots)$ terms in (iv),(v) can be omitted.

References

[1] A. Adem, J. Leida, and Y. Ruan, *Orbifolds and stringy topology*, Cambridge Tracts in Mathematics, vol. 171, Cambridge University Press, Cambridge, 2007. MR2359514

[2] K. Behrend, D. Edidin, B. Fantechi, W. Fulton, L. Göttsche and A. Kresch, *Introduction to stacks*, book in preparation, 2010.

[3] K. Behrend and P. Xu, *Differentiable stacks and gerbes*, J. Symplectic Geom. **9** (2011), no. 3, 285–341. MR2817778

[4] F. Borceux, *Handbook of categorical algebra. 1*, Encyclopedia of Mathematics and its Applications, vol. 50, Cambridge University Press, Cambridge, 1994. Basic category theory. MR1291599

[5] D. Borisov, *Derived manifolds and Kuranishi models*, arXiv:1212.1153, 2012.

[6] D. Borisov and J. Noel, *Simplicial approach to derived differential manifolds*, arXiv:1112.0033, 2011.

[7] W. Chen and Y. Ruan, *Orbifold Gromov-Witten theory*, Orbifolds in mathematics and physics (Madison, WI, 2001), Contemp. Math., vol. 310, Amer. Math. Soc., Providence, RI, 2002, pp. 25–85, DOI 10.1090/conm/310/05398. MR1950941

[8] Y. Eliashberg, A. Givental, and H. Hofer, *Introduction to symplectic field theory*, Geom. Funct. Anal. **Special Volume** (2000), 560–673, DOI 10.1007/978-3-0346-0425-3_4. GAFA 2000 (Tel Aviv, 1999). MR1826267

[9] K. Fukaya, *Cyclic symmetry and adic convergence in Lagrangian Floer theory*, Kyoto J. Math. **50** (2010), no. 3, 521–590, DOI 10.1215/0023608X-2010-004. MR2723862

[10] K. Fukaya, Y.-G. Oh, H. Ohta, and K. Ono, *Lagrangian intersection Floer theory: anomaly and obstruction. Part I*, AMS/IP Studies in Advanced Mathematics, vol. 46, American Mathematical Society, Providence, RI; International Press, Somerville, MA, 2009. MR2553465

[11] K. Fukaya, Y.-G. Oh, H. Ohta, and K. Ono, *Lagrangian Floer theory on compact toric manifolds. I*, Duke Math. J. **151** (2010), no. 1, 23–174, DOI 10.1215/00127094-2009-062. MR2573826

[12] K. Fukaya, Y.-G. Oh, H. Ohta, and K. Ono, *Lagrangian Floer theory on compact toric manifolds II: bulk deformations*, Selecta Math. (N.S.) **17** (2011), no. 3, 609–711, DOI 10.1007/s00029-011-0057-z. MR2827178

[13] K. Fukaya, Y.-G. Oh, H. Ohta and K. Ono, *Technical details on Kuranishi structure and virtual fundamental chain*, arXiv:1209.4410, 2012.

[14] K. Fukaya, Y.-G. Oh, H. Ohta and K. Ono, *Kuranishi structure, Pseudo-holomorphic curve, and Virtual fundamental chain: Part 1*, arXiv:1503.07631, 2015.

[15] K. Fukaya, Y.-G. Oh, H. Ohta, and K. Ono, *Lagrangian Floer theory and mirror symmetry on compact toric manifolds* (English, with English and French summaries), Astérisque **376** (2016), vi+340. MR3460884

[16] K. Fukaya, Y.-G. Oh, H. Ohta, and K. Ono, *Shrinking good coordinate systems associated to Kuranishi structures*, J. Symplectic Geom. **14** (2016), no. 4, 1295–1310, DOI 10.4310/JSG.2016.v14.n4.a10. MR3601890

[17] K. Fukaya, Y.-G. Oh, H. Ohta, and K. Ono, *Antisymplectic involution and Floer cohomology*, Geom. Topol. **21** (2017), no. 1, 1–106, DOI 10.2140/gt.2017.21.1. MR3608709

[18] K. Fukaya, Y.-G. Oh, H. Ohta, and K. Ono, *Kuranishi structure, Pseudo-holomorphic curve, and virtual fundamental chain: Part 2*, arXiv:1704.01848, 2017.

[19] K. Fukaya and K. Ono, *Arnold conjecture and Gromov-Witten invariant*, Topology **38** (1999), no. 5, 933–1048, DOI 10.1016/S0040-9383(98)00042-1. MR1688434

[20] R. Hartshorne, *Algebraic geometry*, Springer-Verlag, New York-Heidelberg, 1977. Graduate Texts in Mathematics, No. 52. MR0463157

[21] A. Henriques and D. S. Metzler, *Presentations of noneffective orbifolds*, Trans. Amer. Math. Soc. **356** (2004), no. 6, 2481–2499, DOI 10.1090/S0002-9947-04-03379-3. MR2048526

[22] H. Herrlich and G. E. Strecker, *Category theory: an introduction*, Allyn and Bacon Inc., Boston, Mass., 1973. Allyn and Bacon Series in Advanced Mathematics. MR0349791

[23] H. Hofer, *Polyfolds and Fredholm theory*, Lectures on geometry, Clay Lect. Notes, Oxford Univ. Press, Oxford, 2017, pp. 87–158. MR3676594

[24] H. Hofer, K. Wysocki, and E. Zehnder, *A general Fredholm theory. I. A splicing-based differential geometry*, J. Eur. Math. Soc. (JEMS) **9** (2007), no. 4, 841–876, DOI 10.4171/JEMS/99. MR2341834

[25] H. Hofer, K. Wysocki, and E. Zehnder, *A general Fredholm theory. II. Implicit function theorems*, Geom. Funct. Anal. **19** (2009), no. 1, 206–293, DOI 10.1007/s00039-009-0715-x. MR2507223

[26] H. Hofer, K. Wysocki, and E. Zehnder, *A general Fredholm theory. III. Fredholm functors and polyfolds*, Geom. Topol. **13** (2009), no. 4, 2279–2387, DOI 10.2140/gt.2009.13.2279. MR2515707

[27] H. Hofer, K. Wysocki, and E. Zehnder, *Integration theory on the zero sets of polyfold Fredholm sections*, Math. Ann. **346** (2010), no. 1, 139–198, DOI 10.1007/s00208-009-0393-x. MR2558891

[28] H. Hofer, K. Wysocki, and E. Zehnder, *sc-smoothness, retractions and new models for smooth spaces*, Discrete Contin. Dyn. Syst. **28** (2010), no. 2, 665–788, DOI 10.3934/dcds.2010.28.665. MR2644764

[29] H. Hofer, K. Wysocki, and E. Zehnder, *Applications of polyfold theory I: The polyfolds of Gromov-Witten theory*, Mem. Amer. Math. Soc. **248** (2017), no. 1179, v+218, DOI 10.1090/memo/1179. MR3683060

[30] H. Hofer, K. Wysocki and E. Zehnder, *Polyfold and Fredholm theory*, arXiv:1707.08941, 2017.

[31] D. Joyce, *Algebraic Geometry over C^∞-rings*, to appear in Memoirs of the A.M.S., 2018. arXiv:1001.0023.

[32] L. Brambila-Paz, O. García-Prada, P. Newstead, and R. P. Thomas (eds.), *Moduli spaces*, London Mathematical Society Lecture Note Series, vol. 411, Cambridge University Press, Cambridge, 2014. Papers from the programme held in Cambridge, January 4–July 1, 2011. MR3236896

[33] D. Joyce, *An introduction to d-manifolds and derived differential geometry*, Moduli spaces, London Math. Soc. Lecture Note Ser., vol. 411, Cambridge Univ. Press, Cambridge, 2014, pp. 230–281. MR3221297

[34] D. Joyce, *An introduction to d-manifolds and derived differential geometry*, Moduli spaces, London Math. Soc. Lecture Note Ser., vol. 411, Cambridge Univ. Press, Cambridge, 2014, pp. 230–281. MR3221297

[35] D. Joyce, *A new definition of Kuranishi space*, arXiv:1409.6908, 2014.

[36] D. Joyce, *Kuranishi spaces and Symplectic Geometry*, multiple volume book in progress, 2017-2027. Preliminary versions of volumes I, II available at `http://people.maths.ox.ac.uk/~joyce/Kuranishi.html`.

[37] G. M. Kelly and R. Street, *Review of the elements of* 2*-categories*, Category Seminar (Proc. Sem., Sydney, 1972/1973), Springer, Berlin, 1974, pp. 75–103. Lecture Notes in Math., Vol. 420. MR0357542

[38] E. Lerman, *Orbifolds as stacks?*, Enseign. Math. (2) **56** (2010), no. 3-4, 315–363, DOI 10.4171/LEM/56-3-4. MR2778793

[39] J. Lurie, *Derived algebraic geometry*, ProQuest LLC, Ann Arbor, MI, 2004. Thesis (Ph.D.)–Massachusetts Institute of Technology. MR2717174

[40] D. McDuff, *Strict orbifold atlases and weighted branched manifolds*, J. Symplectic Geom. **15** (2017), no. 2, 507–540, DOI 10.4310/JSG.2017.v15.n2.a3. MR3683790

[41] D. McDuff and K. Wehrheim, *The topology of Kuranishi atlases*, Proc. Lond. Math. Soc. (3) **115** (2017), no. 2, 221–292, DOI 10.1112/plms.12032. MR3684105

[42] D. McDuff and K. Wehrheim, *Smooth Kuranishi atlases with isotropy*, Geom. Topol. **21** (2017), no. 5, 2725–2809, DOI 10.2140/gt.2017.21.2725. MR3687107

[43] D. McDuff and K. Wehrheim, *The fundamental class of smooth Kuranishi atlases with trivial isotropy*, J. Topol. Anal. **10** (2018), no. 1, 71–243, DOI 10.1142/S1793525318500048. MR3737511

[44] D.S. Metzler, *Topological and smooth stacks*, math.DG/0306176, 2003.

[45] I. Moerdijk, *Orbifolds as groupoids: an introduction*, Orbifolds in mathematics and physics (Madison, WI, 2001), Contemp. Math., vol. 310, Amer. Math. Soc., Providence, RI, 2002, pp. 205–222, DOI 10.1090/conm/310/05405. MR1950948

[46] I. Moerdijk and D. A. Pronk, *Orbifolds, sheaves and groupoids*, K-Theory **12** (1997), no. 1, 3–21, DOI 10.1023/A:1007767628271. MR1466622

[47] D. A. Pronk, *Etendues and stacks as bicategories of fractions*, Compositio Math. **102** (1996), no. 3, 243–303. MR1401424

[48] I. Satake, *On a generalization of the notion of manifold*, Proc. Nat. Acad. Sci. U.S.A. **42** (1956), 359–363, DOI 10.1073/pnas.42.6.359. MR0079769

[49] M.A. Shulman, *Set Theory for Category Theory*, arXiv:0810.1279, 2008.

[50] D. I. Spivak, *Derived smooth manifolds*, Duke Math. J. **153** (2010), no. 1, 55–128, DOI 10.1215/00127094-2010-021. MR2641940

[51] W. Thurston, *The geometry and topology of three-manifolds*, Princeton lecture notes, Princeton, 1980. Available at `http://library.msri.org/books/gt3m`.

[52] B. Toën, *Derived algebraic geometry*, EMS Surv. Math. Sci. **1** (2014), no. 2, 153–240, DOI 10.4171/EMSS/4. MR3285853

[53] D. Yang, *A Choice-independent Theory of Kuranishi Structures and the Polyfold–Kuranishi Correspondence*, ProQuest LLC, Ann Arbor, MI, 2014. Thesis (Ph.D.)–New York University. MR3218117

[54] D. Yang, *A Choice-independent Theory of Kuranishi Structures and the Polyfold–Kuranishi Correspondence*, ProQuest LLC, Ann Arbor, MI, 2014. Thesis (Ph.D.)–New York University. MR3218117

[55] D. Yang, *Virtual harmony*, arXiv:1510.06849, 2015.

The Mathematical Institute, Radcliffe Observatory Quarter, Woodstock Road, Oxford, OX2 6GG, U.K.

Email address: joyce@maths.ox.ac.uk.

Selected Published Titles in This Series

237 **Dusa McDuff, Mohammad Tehrani, Kenji Fukaya, and Dominic Joyce,** Virtual Fundamental Cycles in Symplectic Topology, 2019

236 **Bernard Host and Bryna Kra,** Nilpotent Structures in Ergodic Theory, 2018

235 **Habib Ammari, Brian Fitzpatrick, Hyeonbae Kang, Matias Ruiz, Sanghyeon Yu, and Hai Zhang,** Mathematical and Computational Methods in Photonics and Phononics, 2018

234 **Vladimir I. Bogachev,** Weak Convergence of Measures, 2018

233 **N. V. Krylov,** Sobolev and Viscosity Solutions for Fully Nonlinear Elliptic and Parabolic Equations, 2018

232 **Dmitry Khavinson and Erik Lundberg,** Linear Holomorphic Partial Differential Equations and Classical Potential Theory, 2018

231 **Eberhard Kaniuth and Anthony To-Ming Lau,** Fourier and Fourier-Stieltjes Algebras on Locally Compact Groups, 2018

230 **Stephen D. Smith,** Applying the Classification of Finite Simple Groups, 2018

229 **Alexander Molev,** Sugawara Operators for Classical Lie Algebras, 2018

228 **Zhenbo Qin,** Hilbert Schemes of Points and Infinite Dimensional Lie Algebras, 2018

227 **Roberto Frigerio,** Bounded Cohomology of Discrete Groups, 2017

226 **Marcelo Aguiar and Swapneel Mahajan,** Topics in Hyperplane Arrangements, 2017

225 **Mario Bonk and Daniel Meyer,** Expanding Thurston Maps, 2017

224 **Ruy Exel,** Partial Dynamical Systems, Fell Bundles and Applications, 2017

223 **Guillaume Aubrun and Stanisław J. Szarek,** Alice and Bob Meet Banach, 2017

222 **Alexandru Buium,** Foundations of Arithmetic Differential Geometry, 2017

221 **Dennis Gaitsgory and Nick Rozenblyum,** A Study in Derived Algebraic Geometry, 2017

220 **A. Shen, V. A. Uspensky, and N. Vereshchagin,** Kolmogorov Complexity and Algorithmic Randomness, 2017

219 **Richard Evan Schwartz,** The Projective Heat Map, 2017

218 **Tushar Das, David Simmons, and Mariusz Urbański,** Geometry and Dynamics in Gromov Hyperbolic Metric Spaces, 2017

217 **Benoit Fresse,** Homotopy of Operads and Grothendieck–Teichmüller Groups, 2017

216 **Frederick W. Gehring, Gaven J. Martin, and Bruce P. Palka,** An Introduction to the Theory of Higher-Dimensional Quasiconformal Mappings, 2017

215 **Robert Bieri and Ralph Strebel,** On Groups of PL-homeomorphisms of the Real Line, 2016

214 **Jared Speck,** Shock Formation in Small-Data Solutions to 3D Quasilinear Wave Equations, 2016

213 **Harold G. Diamond and Wen-Bin Zhang (Cheung Man Ping),** Beurling Generalized Numbers, 2016

212 **Pandelis Dodos and Vassilis Kanellopoulos,** Ramsey Theory for Product Spaces, 2016

211 **Charlotte Hardouin, Jacques Sauloy, and Michael F. Singer,** Galois Theories of Linear Difference Equations: An Introduction, 2016

210 **Jason P. Bell, Dragos Ghioca, and Thomas J. Tucker,** The Dynamical Mordell–Lang Conjecture, 2016

209 **Steve Y. Oudot,** Persistence Theory: From Quiver Representations to Data Analysis, 2015

For a complete list of titles in this series, visit the
AMS Bookstore at **www.ams.org/bookstore/survseries/**.